T0134084

Discrete Mathematics

Discrete Mathematics

Proofs, Structures, and Applications

Third Edition

Rowan Garnier
John Taylor

CRC Press is an imprint of the
Taylor & Francis Group, an **informa** business
A TAYLOR & FRANCIS BOOK

Taylor & Francis
6000 Broken Sound Parkway NW, Suite 300
Boca Raton, FL 33487-2742

Printed in the United States of America on acid-free paper
10 9 8 7 6 5 4 3 2 1

International Standard Book Number: 978-1-4398-1280-8 (Hardback)

Library of Congress Cataloging-in-Publication Data

Garnier, Rowan.
Discrete mathematics: proofs, structures and applications/ Rowan Garnier, John Taylor. -- 3rd ed.
p. cm.
Includes bibliographical references and index.
ISBN 978-1-4398-1280-8 (hardcover : alk. paper)
1. Computer science--Mathematics. I. Taylor, John, 1957- II. Title.

QA76.9.M35G38 2010
004.01'51--dc22 2009040057

Visit the Taylor & Francis Web site at
http://www.taylorandfrancis.com

and the CRC Press Web site at
http://www.crcpress.com

Contents

Preface to the Third Edition

The most obvious change from the first two editions of this text is its title. We believe that the new title, *Discrete Mathematics: Proofs, Structures and Applications*, provides a better description of the book. This book was originally published under the title *Discrete Mathematics for New Technology*, a title we were never entirely comfortable with for two reasons. Firstly, it was not really clear which 'new technology' was being referred to and, furthermore, one decade's new technology is frequently the next decade's obsolete technology. Secondly, although we had originally conceptualised the text as providing the discrete mathematical background for undergraduate computer science students, it was apparent that the book had a much wider readership including mathematics undergraduates, education students, practising scientists and others.

Our philosophy had always been to provide, so far as we were able, a rigorous and accessible exposition of the mathematics and not to tie the text too closely to any application domain or community. Perhaps this is part of the reason for the wide readership that the book has enjoyed. We have maintained this approach in the current edition. So, whilst we believe that the book continues to provide much of the core mathematical underpinning for computer science, we hope others will continue to find the text accessible, informative and enjoyable.

In the eight years since the publication of the second edition, we have continued to received feedback on the text from users. The feedback has remained complimentary about the clarity of our exposition and, since correcting known errors for the second edition, there have been few comments pointing out errors or suggesting ways in which the text could be improved. Nevertheless, through our own use of the text, we have continued to log errors of substance or style and we have corrected these in the current edition.

The principal changes in this new edition are an expanded chapter 1 and a new chapter 9 on number theory. The revised chapter 1 includes a new section on the formal proof of the validity of arguments in propositional logic. This means that we now consider formal proofs first in the context of propositional logic before moving on to predicate logic. The new chapter 9 covers elementary number theory and congruences. This allows us to explore in a little more depth some of the groups that arise in modular arithmetic. The significant application that we explore is the so-called public key encryption scheme, called RSA encryption, that underpins much of the secure transmission of data on the internet. Although the mathematics behind the RSA system is reasonably straightforward, it does provide a practical, secure and widely used means of encrypting data. As one of our reviewers noted, this encryption scheme represents a premier 'new technology'.

We wish to acknowledge, with thanks, colleagues who have commented on previous versions of the text. John Taylor is also grateful to the University of Brighton for a sabbatical period which was devoted in part to writing the new chapter 9 and the accompanying solutions manual. Nevertheless, as in the previous editions, any remaining shortcomings are ours and we have no one to blame for them but each other.

RG and **JT**
June 2009

Preface to the Second Edition

In the nine years since the publication of the first edition, we have received feedback on the text from a number of users, both teachers and students. Most have been complimentary about the clarity of our exposition, some have pointed out errors of detail or historical accuracy and others have suggested ways in which the text could be improved. In this edition we have attempted to retain the style of exposition, correct the (known) errors and implement various improvements suggested by users.

When writing the first edition, we took a conscious decision not to root the mathematical development in a particular method or language that was current within the formal methods community. Our priority was to give a thorough treatment of the mathematics as we felt this was likely to be more stable over time than particular methodologies. In a discipline like computing which evolves rapidly and where the future direction is uncertain, a secure grounding in theory is important. We have continued with this philosophy in the second edition. Thus, for example, Z made no appearance in the first edition, and the object constraint language (OCL) or the B method make no appearance in this edition. Although the discipline of computing has indeed changed considerably since the publication of the first edition, the core mathematical requirements of the undergraduate curricula have remained surprisingly constant. For example, in the UK, the computing benchmark for undergraduate courses, published by the Quality Assurance Agency for Higher Education (QAA) in April 2000, requires undergraduate programmes to present 'coherent underpinning theory'. In the USA, the joint ACM/IEEE Computer Society Curriculum 2001 project lists 'Discrete Structures' (sets, functions, relations, logic, proof, counting, graphs and trees) as one of the 14 knowledge areas in the computing curriculum 'to emphasize the dependency of computing on discrete mathematics'.

In this edition we have included a new section on typed set theory and subsequently we show how relations and functions fit into the typed world. We have also introduced a specification approach to mathematical operations, via signatures, preconditions and postconditions. Computing undergraduates will be familiar with types from the software design and implementation parts of their course and we hope our use of types will help tie together the mathematical underpinnings more closely with software development practice. For the mathematicians using the text, this work has a payoff in providing a framework in which Russell's paradox can be avoided, for example.

The principal shortcoming reported by users of the first edition was the inclusion of relatively few exercises at a routine level to develop and reinforce the mathematical concepts introduced in the text. In the second edition, we have added many new exercises (and solutions) which we hope will enhance the usefulness of the text to teachers and students alike. Also included are a number of new examples designed to reinforce the concepts introduced.

We wish to acknowledge, with thanks, our colleagues who have commented on and thus improved various drafts of additional material included in the second edition. In particular, we thank Paul Courtney, Gerald Gallacher, John Howse, Brian Spencer and our reviewers for their knowledgeable and thoughtful comments. We would also like to thank those—most notably Peter Kirkegaard—who spotted errors in the first edition or made suggestions for improving the text. Nevertheless, any remaining shortcomings are ours and we have no one to blame for them but each other.

RG and **JT**
April 2001

Preface to the First Edition

This book aims to present in an accessible yet rigorous way the core mathematics requirement for undergraduate computer science students at British universities and polytechnics. Selections from the material could also form a one- or two-semester course at freshman–sophomore level at American colleges. The formal mathematical prerequisites are covered by the GCSE in the UK and by high-school algebra in the USA. However, the latter part of the text requires a certain level of mathematical sophistication which, we hope, will be developed during the reading of the book.

Over 30 years ago the discipline of computer science hardly existed, except as a subdiscipline of mathematics. Computers were seen, to a large extent, as the mathematician's tool. As a result, the machines spent a large proportion of their time cranking through approximate numerical solutions to algebraic and differential equations and the mathematics 'appropriate' for the computer scientist was the theory of equations, calculus, numerical analysis and the like.

Since that time computer science has become a discipline in its own right and has spawned its own subdisciplines. The nature and sophistication of both hardware and software have changed dramatically over the same time period. Perhaps less public, but no less dramatic, has been the parallel development of undergraduate computer science curricula and the mathematics which underpins it. Indeed, the whole relationship between mathematics and computer science has changed so that mathematics is now seen more as the servant of computer science than vice versa as was the case formerly.

Various communities and study groups on both sides of the Atlantic have studied and reported upon the core mathematics requirements for computer scientists educated and trained at various levels. The early emphasis on continuous mathematics in general, and numerical methods in particular, has disappeared. There is now wide agreement that the essential mathematics required for computer scientists comes from the area of 'discrete mathematics'. There is, however, less agreement concerning the detailed content and emphasis of a core mathematics course.

Discrete mathematics encompasses a very wide range of mathematical topics and we have necessarily been selective in our choice of material. Our starting point was a report of the M2 Study Group of the 1986 Undergraduate Mathematics Teaching Conference held at the University of Nottingham. Their report, published in 1987, suggested an outline syllabus for a first-year mathematics course for computer science undergraduates. All the topic areas (with the exception of probability theory) suggested in the outline are covered in this text. We have also been influenced in our selection of material by various courses at the freshman–sophomore level offered by institutions in the USA.

Ultimately the selection, presentation and emphasis of the material in this book were based on our own judgements. We have attempted to include the essential mathematical material required by undergraduate computer scientists in a first course. However, one of our key aims is to develop in students the rigorous logical thinking which, we believe, is essential if computer science graduates are to adapt to the demands of their rapidly developing discipline. Our approach is informal. We have attempted to keep prerequisites to an absolute minimum and to maintain a level of discussion within the reach of the student. In the process, we have not sacrificed the mathematical rigour which we believe to be important if mathematics is to be used in a meaningful way.

Our priority has been to give a sound and thorough treatment of the mathematics. We also felt that it was important to place the theory in context by including a selection of the more salient applications. It is our belief that mathematical applications can be readily assimilated only when a firm mathematical foundation has been laid. Too frequently, students are exposed to concepts requiring mathematical background before the background has been adequately provided. We hope this text will provide such a foundation.

In order to keep the book within manageable proportions and still provide some applications, we have been forced to omit certain topics such as finite state machines and formal languages. Although such topics are relevant to computer scientists and others, we felt that they were not central to the mathematical core of the text. We believe that the book will provide a sound background for readers who wish to explore these and other areas.

As our writing of the text progressed and its content was discussed with colleagues, we became increasingly conscious that we were presenting material which lies at the very foundation of mathematics itself. It seems likely that discrete mathematics will become an increasingly important part of mathematics curricula at all levels in the coming years. Given our emphasis on a sound and thorough development of mathematical concepts, this text would be appropriate for undergraduate mathematicians following a course in discrete mathematics. The first half of the book could also be recommended reading for the aspiring mathematics undergraduate in the summer before he or she enters university.

The approximate interdependence of the various parts of the text is shown in the diagram below. There are various sections which are concerned largely with applications (or further development) of the theory and which may be omitted without jeopardizing the understanding of later material. The most notable of these are §§4.7, 5.5, 5.6 and 8.7.

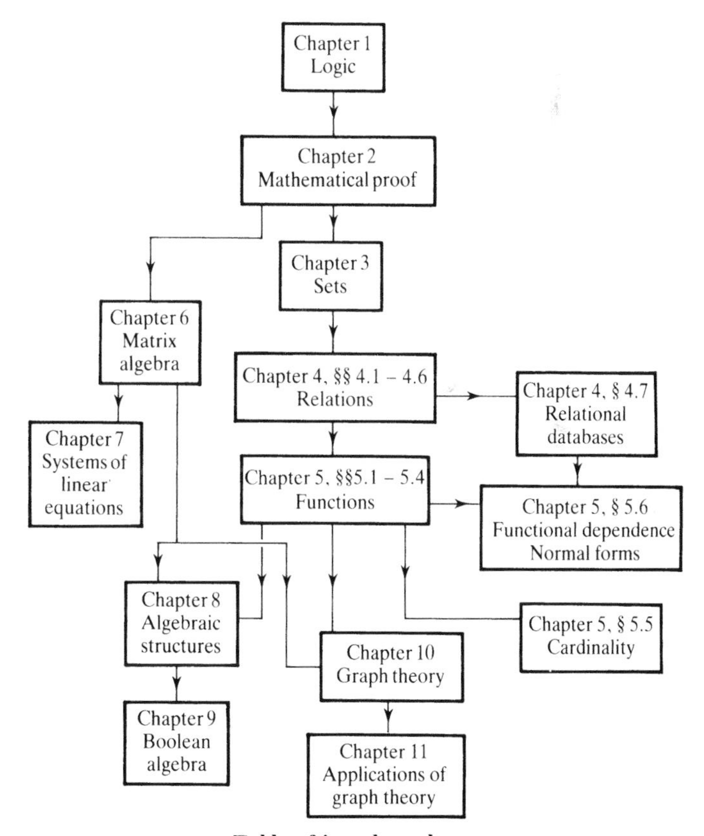

Table of interdependence

We wish to acknowledge with thanks our families, friends and colleagues for their encouragement. In particular we would like to thank Dr Paul Milican, Paul Douglas and Alice Tomič for their advice and comments on various parts of the manuscript. Our reviewers provided many helpful comments and suggestions for which we are grateful. If the text contains any errors or stylistic misjudgements, we can only blame each other. The technical services staff at Richmond College and Jim Revill and Al Troyano at IOP Publishing also deserve our thanks for their patience with us during the development of this text. Last, but not least, we wish to thank Pam Taylor for providing (at short notice) the ideas and sketches for the cartoons.

RG and **JT**
July 1990

List of Symbols

The following is a list of symbols introduced in this book together with their interpretations and the section where each is defined.

Symbol	**Interpretation**	**Section**
$\bar{p}$	negation of the proposition p	1.2
$p \wedge q$	conjunction of the propositions p and q	1.2
$p \vee q$	inclusive disjunction of the propositions p and q	1.2
$p \veebar q$	exclusive disjunction of the propositions p and q	1.2
$p \to q$	conditional proposition 'if p then q'	1.2
$p \leftrightarrow q$	biconditional proposition 'p if and only if q'	1.2
t	tautology	1.3
f	contradiction	1.3
$P \equiv Q$	logical equivalence of P and Q	1.4
$P \vdash Q$	the proposition P logically implies the proposition Q	1.4
$P(x)$	propositional function with variable x	1.4
$\forall$	the universal quantifier	1.8
$\exists$	the existential quantifier	1.8
$\neg$	negation of a propositional function or of a quantified propositional function	1.8
$P \Rightarrow Q$	Q is logically implied by P in conjunction with axioms and theorems which apply to the system	2.3
$P \Leftrightarrow Q$	$P \Rightarrow Q$ and $Q \Rightarrow P$	2.3
$a \in A$	the element a belongs to the set A	3.1
$a \notin A$	the element a does not belong to the set A	3.1
$\varnothing$	the empty set	3.1
$\|A\|$	the cardinality of the set A	3.1

$A \subseteq B$	the set A is a subset of the set B	3.2
$A \subset B$	the set A is a proper subset of the set B	3.2
$A \not\subseteq B$	the set A is not a subset of the set B	3.2
$A \supseteq B$	the set A is a superset of the set B	3.2
$\mathscr{U}$	the universal set	3.2
$\mathbb{N}$	the set of natural numbers	3.2
$\mathbb{Z}$	the set of integers	3.2
$\mathbb{Q}$	the set of rational numbers	3.2
$\mathbb{R}$	the set of real numbers	3.2
$\mathbb{C}$	the set of complex numbers	3.2
$\mathbb{E}$	the set of even numbers	3.2
$\mathbb{O}$	the set of odd numbers	3.2
$\mathbb{Z}^+$	the set of positive integers	3.2
$\mathbb{Z}^-$	the set of negative integers	3.6
$\mathbb{Q}^+$	the set of positive rational numbers	3.2
$\mathbb{R}^+$	the set of positive real numbers	3.2
$\mathbb{E}^+$	the set of positive even numbers	5.4
$\mathbb{O}^+$	the set of positive odd numbers	5.5
$A \cap B$	the intersection of the sets A and B	3.3
$A \cup B$	the union of the sets A and B	3.3
$\bar{A}$	the complement of the set A	3.3
$A - B$	the difference of the sets A and B	3.3
$\bigcap_{r=1}^{n} A_r$	the intersection of the sets $A_1, A_2, \ldots, A_n$	3.3
$\bigcup_{r=1}^{n} A_r$	the union of the sets $A_1, A_2, \ldots, A_n$	3.3
$A * B$	the symmetric difference of the sets A and B	3.5
$\mathscr{P}(A)$	the power set of the set A	3.6
$\bigcap_{i \in I} A_i$	the intersection of the family of sets $\{A_i : i \in I\}$	3.6
$\bigcup_{i \in I} A_i$	the union of the family of sets $\{A_i : i \in I\}$	3.6
$A \times B$	the Cartesian product of the sets A and B	3.7
X^2	the Cartesian product $X \times X$	3.7
$a \mathsf{R} b$	the element a is related to the element b	4.1
$a \not\mathsf{R} b$	the element a is not related to the element b	4.1
I_A	the identity relation on the set A	4.1

U_A	the universal relation on the set A	4.1
R^{-1}	the inverse relation of the relation R	4.1
$\mathsf{S} \circ \mathsf{R}$	the composite of the relations R and S	4.3
$[x]$	the equivalence class of the element x	4.4
$a \equiv_n b$	a is congruent modulo n to b, i.e. $a - b = kn$ for some integer k (see also section 9.3)	4.4
$+_n$	addition modulo n	4.4
$\times_n$	multiplication modulo n	4.4
$\mathbb{Z}_n$	the set of equivalence classes under congruence modulo n, i.e. $\{[0], [1], \ldots, [n-1]\}$	4.4
$\lfloor x \rfloor$	the integer part of the real number x, i.e. the largest integer less than or equal to x	4.4
$[a, b)$	the half-open interval $\{x \in \mathbb{R} : a \leqslant x < b\}$	4.4
$(a, b]$	the half-open interval $\{x \in \mathbb{R} : a < x \leqslant b\}$	4.4
$[a, b]$	the closed interval $\{x \in \mathbb{R} : a \leqslant x \leqslant b\}$	4.5
(a, b)	the open interval $\{x \in \mathbb{R} : a < x < b\}$	4.5
$n \mid m$	n divides m	4.5
$f : A \to B$	a function f from the set A to the set B, i.e. a function with domain A and codomain B	5.1
$f(a)$	the image of the element a under the function f	5.1
$f : a \mapsto b$	for the function f the image of the element a is b	5.1
id_A	the identity function with domain and codomain A	5.1
$im(f)$	the image set of the function f, i.e. the subset of the codomain of f which contains the images of all elements in the domain	5.1
$f(C)$	the image of the set C under the function f	5.1
$f^{-1}(D)$	the inverse image of the set D under the function f	5.1
$f \circ g$	the composite of the functions f and g, where $f \circ g(x) = f[g(x)]$	5.2
i_C	inclusion function of a subset C in a set A	5.2
$f\|_C$	restriction of the function f to a subset C of its domain	5.2
$\mathbb{P}$	the set of prime numbers	5.5
$\aleph_0$	the cardinality of $\mathbb{Z}^+$	5.5
c	the cardinality of $\mathbb{R}$	5.5
a_{ij}	the element in the matrix A occupying the ith row and jth column	6.1
$[a_{ij}]$	the matrix with (i, j)-entry a_{ij}	6.1
$O_{m \times n}$	the $m \times n$ zero matrix	6.2
I_n	the $n \times n$ identity matrix	6.2
A^{T}	the transpose of the matrix A	6.2
$A \sim B$	the matrix A is row-equivalent to the matrix B	6.4

A^{-1}	the multiplicative inverse of the matrix A	6.5
$(A\ B)$	the partitioned matrix with submatrices A and B	6.5
$(A\ \boldsymbol{b})$	the augmented matrix of a system of linear equations with matrix of coefficients A	7.3
e	the identity with respect to a binary operation	8.1
$(S, *)$	the algebraic structure with underlying set S and binary operation $*$	8.2
A^*	the set of all strings over the alphabet A	8.2
λ	the empty string	8.2
$(G, *)$	the group with underlying set G and binary operation $*$	8.3
D_n	the dihedral group of degree n	8.4
S_n	the symmetric group of degree n	8.4
$(G_1, *) \leqslant (G_2, \circ)$	the group $(G_1, *)$ is a subgroup of the group $(G_2, \circ)$	8.5
C_n	the group of rotations of a regular n-sided polygon	8.5
$\|g\|$	the order of an element $g \in G$ of a group $(G, *)$	8.5
$(G_1, *) \cong (G_2, \circ)$	the groups $(G_1, *)$ and $(G_2, \circ)$ are isomorphic	8.6
ker f	the kernel of a morphism $f : G_1 \to G_2$ where $(G_1, *)$ and $(G_2, \circ)$ are groups	8.6
$d(\boldsymbol{x}, \boldsymbol{y})$	the distance between the binary words $\boldsymbol{x}$ and $\boldsymbol{y}$	8.7
B^n	the set of binary words of length n	8.7
$w(\boldsymbol{x})$	the weight of the binary word $\boldsymbol{x}$	8.7
$\boldsymbol{x} \oplus \boldsymbol{y}$	the n bit word whose ith bit is the sum modulo 2 of the ith bits of the n bit words $\boldsymbol{x}$ and $\boldsymbol{y}$	8.7
$b\|a$	the integer b divides the integer a	9.1
$\gcd(a, b)$	the greatest common divisor of a and b	9.1
$a \equiv b \mod n$	a is congruent to b modulo n, i.e. $a - b = kn$ for some integer k (see also section 4.4)	9.3
$\phi(n)$	the number of integers a where $1 \leqslant a \leqslant n$ which are coprime to n	9.5
$(B, \oplus, *, ^{-}, \mathbf{0}, \mathbf{1})$	the Boolean algebra with underlying set B, binary operations $\oplus$ and $*$, complement operation $^{-}$, and identities **0** and **1** under $\oplus$ and $*$ respectively	10.1
$\bar{b}$	the complement of the element $b \in B$, the underlying set of a Boolean algebra	10.1
$m_{e_1 e_2 \ldots e_n}$	the minterm $x_1{}^{e_1} x_2{}^{e_2} \ldots x_n{}^{e_n}$ where $e_1 = 0$ or 1 ($i = 1, 2, \ldots, n$) and $x_i{}^{e_i} = \begin{cases} \bar{x}_i & \text{if } e_i = 0 \\ x_i & \text{if } e_1 = 1 \end{cases}$	10.3

Symbol	Meaning	Section
$M_{e_1 e_2 \dots e_n}$	the maxterm $x_1^{e_1} \oplus x_2^{e_2} \oplus \cdots \oplus x_n^{e_n}$ where $e_1 = 0$ or 1 $(i = 1, 2, \dots, n)$ and $x_i^{e_i} = \begin{cases} \bar{x}_i & \text{if } e_i = 0 \\ x_i & \text{if } e_1 = 1 \end{cases}$	10.3
A (switch)	the switch denoted by A	10.4
$\bar{\mathrm{S}}$	a switch which is always in the opposite state to another switch S	10.4
	AND-gate	10.5
	OR-gate	10.5
	NOT-gate	10.5
	NAND-gate	10.5
	NOR-gate	10.5
$\delta(e)$	the set of vertices incident to the edge e of a graph	11.1
C_n	the cycle graph with n vertices	11.1
W_n	the wheel graph with n vertices	11.1
$deg(v)$	the degree of the vertex v of a graph	11.1
K_n	the complete graph with n vertices	11.1
$K_{n,m}$	the complete bipartite graph on n and m vertices	11.1
$A(\Gamma)$	the adjacency matrix for the graph Γ	11.1
$\Gamma \leqslant \Sigma$	the graph Γ is a subgraph of the graph Σ	11.1
$\Gamma + \Sigma$	the sum of the graphs Γ and Σ	11.1
$\Gamma \cup \Sigma$	the union of the graphs Γ and Σ	11.1
$\Gamma \cong \Sigma$	the graphs Γ and Σ are isomorphic	11.3
$E(v, w)$	the set of edges joining the vertices v and w of a graph	11.3
$\delta(e)$	the ordered pair of initial and final vertices of the (directed) edge e of a directed graph	11.6
(T, v^*)	the rooted tree with root v^*	12.2
$(L, \{v\}, R)$	the binary tree with root v, left subtree L and right subtree R	12.2
$a \leqslant b$	$a \mathrel{\mathsf{R}} b$ where $a, b \in A$ and A is a totally ordered set under the order relation R	12.3
$w(e)$	the weight of the edge e of a weighted graph	12.5
$w(\Gamma')$	the weight of the subgraph Γ' of a weighted graph Γ	12.5
$w(v_1, v_2)$	the weight of the unique edge joining vertices v_1 and v_2 of a complete weighted graph	12.6

Chapter 1

Logic

Logic is used to establish the validity of arguments. It is not so much concerned with what the argument is about but more with providing rules so that the general form of the argument can be judged as sound or unsound. The rules which logic provides allow us to assess whether the conclusion drawn from stated premises is consistent with those premises or whether there is some faulty step in the deductive process which claims to support the validity of the conclusion.

1.1 Propositions and Truth Values

A **proposition** is a declarative statement which is either true or false, but not both simultaneously. (Propositions are sometimes called 'statements'.) Examples of propositions are:

1. This rose is white.
2. Triangles have four vertices.
3. $3 + 2 = 4$.
4. $6 < 24$.
5. Tomorrow is my birthday.

Note that the same proposition may sometimes be true and sometimes false depending on where and when it was stated and by whom. Whilst proposition 5 is true when stated by anyone whose birthday is tomorrow, it is false when stated by anyone else. Further, if anyone for whom it is a true statement today states it on any other day, it will then be false. Similarly, the truth or falsity of proposition 1 depends on the context in which the proposition was stated.

Exclamations, questions and demands are not propositions since they cannot be declared true or false. Thus the following are not propositions:

6. Keep off the grass.
7. Long live the Queen!
8. Did you go to Jane's party?
9. Don't say that.

The truth (T) or falsity (F) of a proposition is called **truth value**. Proposition 4 has a truth value of true (T) and propositions 2 and 3 have truth values of false (F). The truth values of propositions 1 and 5 depend on the circumstances in which the statement was uttered. Sentences 6–9 are not propositions and therefore cannot be assigned truth values.

Propositions are conventionally symbolized using the letters $p, q, r, \ldots$. Any of these may be used to symbolize specific propositions, e.g. p: Manchester is in Scotland, q: Mammoths are extinct. We also use these letters to stand for arbitrary propositions, i.e. as variables for which any particular proposition may be substituted.

1.2 Logical Connectives and Truth Tables

The propositions 1–5 considered in §1.1 are **simple propositions** since they make only a single statement. In this section we look at how simple propositions can be combined to form more complicated propositions called **compound propositions**. The devices which are used to link pairs of propositions are called **logical connectives** and the truth value of any compound proposition is completely determined by (a) the truth values of its component simple propositions, and (b) the particular connective, or connectives, used to link them.

Before we look at the most commonly used connectives we first look at an operation which can be performed on a single proposition. This operation is called **negation** and it has the effect of reversing the truth value of the proposition. We state the negation of a proposition by prefixing it by 'It is not the case that...'. This is not the only way of negating a proposition but what is important is that the negation is false in all circumstances that the proposition is true, and true in all circumstances that the proposition is false.

We can summarize this in a table. If p symbolizes a proposition $\bar{p}$ (or $\neg p$ or $-p$ or $\sim p$) symbolizes the negation of p. The following table shows the relationship

SIMPLE PROPOSITION

COMPOUND PROPOSITION

between the truth values of p and those of $\bar{p}$.

p	$\bar{p}$
T	F
F	T

The left-hand column gives all possible truth values for p and the right-hand column gives the corresponding truth values of $\bar{p}$, the negation of p. A table which summarizes truth values of propositions in this way is called a **truth table**.

There are several alternative ways of stating the negation of a proposition. If we consider the proposition 'All dogs are fierce', some examples of its negation are:

It is not the case that all dogs are fierce.
Not all dogs are fierce.
Some dogs are not fierce.

Note that the proposition 'No dogs are fierce' is not the negation of 'All dogs are fierce'. Remember that to be the negation, the second statement must be false in *all* circumstances that the first is true and vice versa. This is clearly not the case since 'All dogs are fierce' is false if just one dog is not fierce. However, 'No dogs are fierce' is not true in this case. (See §1.8.)

Whilst negation is an operation which involves only a single proposition, logical connectives are used to link pairs of propositions. We shall consider five commonly used logical connectives: conjunction, inclusive disjunction, exclusive disjunction, the conditional and biconditional.

Conjunction

Two simple propositions can be combined by using the word 'and' between them. The resulting compound proposition is called the **conjunction** of its two component simple propositions. If p and q are two propositions $p \wedge q$ (or $p.q$) symbolizes the conjunction of p and q. For example:

$$\begin{aligned} p &: \text{The sun is shining.} \\ q &: \text{Pigs eat turnips.} \\ p \wedge q &: \text{The sun is shining and pigs eat turnips.} \end{aligned}$$

The following truth table gives the truth values of $p \wedge q$ (read as 'p and q') for each possible pair of truth values of p and q.

p	q	$p \wedge q$
T	T	T
T	F	F
F	T	F
F	F	F

From the table it can be seen that the conjunction $p \wedge q$ is true only when both p and q are true. Otherwise the conjunction is false.

Linking two propositions using 'and' is not the only way of forming a conjunction. The following are also conjunctions of p and q even though they have nuances which are slightly different from when the two propositions are joined using 'and'.

The sun shines but pigs eat turnips.
Although the sun shines, pigs eat turnips.
The sun shines whereas pigs eat turnips.

All give the sense that they are true only when each simple component is true. Otherwise they would be judged as false.

Disjunction

The word 'or' can be used to link two simple propositions. The compound proposition so formed is called the **disjunction** of its two component simple propositions. In logic we distinguish two different types of disjunction, the inclusive and exclusive forms. The word 'or' in natural language is ambiguous in conveying which type of disjunction we mean. We return to this point after we have considered the two forms.

Given the two propositions p and q, $p \vee q$ symbolizes the **inclusive disjunction** of p and q. This compound proposition is true when either or both of its components are true and is false otherwise. Thus the truth table for $p \vee q$ is given by:

p	q	$p \vee q$
T	T	T
T	F	T
F	T	T
F	F	F

The **exclusive disjunction** of p and q is symbolized by $p \veebar q$. This compound proposition is true when exactly one (i.e. one or other, but not both) of its components is true. The truth table for $p \veebar q$ is given by:

p	q	$p \veebar q$
T	T	F
T	F	T
F	T	T
F	F	F

When two simple propositions are combined using ‘or’, context will often provide the clue as to whether the inclusive or exclusive sense is intended. For instance, ‘Tomorrow I will go swimming or play golf’ seems to suggest that I will not do both and therefore points to an exclusive interpretation. On the other hand, ‘Applicants for this post must be over 25 or have at least 3 years relevant experience’ suggests that applicants who satisfy both criteria will be considered, and that ‘or’ should therefore be interpreted inclusively.

Where context does not resolve the ambiguity surrounding the word ‘or’, the intended sense can be made clear by affixing ‘or both’ to indicate an inclusive reading, or by affixing ‘but not both’ to make clear the exclusive sense. Where there is no clue as to which interpretation is intended and context does not make this clear, then ‘or’ is conventionally taken in its inclusive sense.

Conditional Propositions

The **conditional** connective (sometimes called **implication**) is symbolized by $\rightarrow$ (or by $\supset$). The linguistic expression of a conditional proposition is normally accepted as utilizing ‘if ... then ...’ as in the following example:

$$\begin{aligned} p &: \text{I eat breakfast.} \\ q &: \text{I don't eat lunch.} \\ p \rightarrow q &: \text{If I eat breakfast then I don't eat lunch.} \end{aligned}$$

Alternative expressions for $p \rightarrow q$ in this example are:

I eat breakfast only if I don’t eat lunch.
Whenever I eat breakfast, I don’t eat lunch.
That I eat breakfast implies that I don’t eat lunch.

The following is the truth table for $p \rightarrow q$:

p	q	$p \rightarrow q$
T	T	T
T	F	F
F	T	T
F	F	T

Notice that the proposition 'if p then q' is false only when p is true and q is false, i.e. a true statement cannot imply a false one. If p is false, the compound proposition is true no matter what the truth value of q. To clarify this, consider the proposition: 'If I pass my exams then I will get drunk'. This statement says nothing about what I will do if I *don't* pass my exams. I may get drunk or I may not, but in either case you could not accuse me of having made a false statement. The only circumstances in which I could be accused of uttering a falsehood is if I pass my exams and don't get drunk.

In the conditional proposition $p \rightarrow q$, the proposition p is sometimes called the **antecedent** and q the **consequent**. The proposition p is said to be a **sufficient condition** for q and q a **necessary condition** for p.

Biconditional Propositions

The **biconditional** connective is symbolized by $\leftrightarrow$, and expressed by 'if and only if . . . then . . . '. Using the previous example:

p : I eat breakfast.

q : I don't eat lunch.

$p \leftrightarrow q$: I eat breakfast if and only if I don't eat lunch (or alternatively, 'If and only if I eat breakfast, then I don't eat lunch').

The truth table for $p \leftrightarrow q$ is given by:

p	q	$p \leftrightarrow q$
T	T	T
T	F	F
F	T	F
F	F	T

Note that for $p \leftrightarrow q$ to be true, p and q must both have the same truth values, i.e. both must be true or both must be false.

Examples 1.1

1. Consider the following propositions:

$$p : \text{Mathematicians are generous.}$$
$$q : \text{Spiders hate algebra.}$$

Write the compound propositions symbolized by:

(i) $p \vee \bar{q}$
(ii) $\overline{(q \wedge p)}$
(iii) $\bar{p} \rightarrow q$
(iv) $\bar{p} \leftrightarrow \bar{q}$.

Solution

(i) Mathematicians are generous or spiders don't hate algebra (or both).
(ii) It is not the case that spiders hate algebra and mathematicians are generous.
(iii) If mathematicians are not generous then spiders hate algebra.
(iv) Mathematicians are not generous if and only if spiders don't hate algebra.

(As we have seen, these are not unique solutions and there are acceptable alternatives.)

2. Let p be the proposition 'Today is Monday' and q be 'I'll go to London'. Write the following propositions symbolically.

(i) If today is Monday then I won't go to London.
(ii) Today is Monday or I'll go to London, but not both.
(iii) I'll go to London and today is not Monday.
(iv) If and only if today is not Monday then I'll go to London.

Solution

(i) $p \rightarrow \bar{q}$
(ii) $p \veebar q$
(iii) $q \wedge \bar{p}$
(iv) $\bar{p} \leftrightarrow q$.

3. Construct truth tables for the following compound propositions.

(i) $\bar{p} \vee q$

(ii) $\bar{p} \wedge \bar{q}$
(iii) $\bar{q} \rightarrow p$
(iv) $\bar{p} \leftrightarrow \bar{q}$.

Solution

(i)

p	q	$\bar{p}$	$\bar{p} \vee q$
T	T	F	T
T	F	F	F
F	T	T	T
F	F	T	T

Note that the truth table is built up in stages. The first two columns give the usual combinations of possible truth values of p and q. The third column gives, for each truth value of p, the truth value of $\bar{p}$. When p is true, $\bar{p}$ is false and vice versa. The last column combines the truth values in columns 3 and 2 using the inclusive disjunction connective. The compound proposition $\bar{p} \vee q$ is true when at least one of its two components is true. This is the case in row 1 (where q is true), row 3 ($\bar{p}$ and q are both true) and row 4 ($\bar{p}$ is true). In the second row, $\bar{p}$ and q are both false and hence $\bar{p} \vee q$ is false.

(ii)

p	q	$\bar{p}$	$\bar{q}$	$\bar{p} \wedge \bar{q}$
T	T	F	F	F
T	F	F	T	F
F	T	T	F	F
F	F	T	T	T

Here we first obtain truth values for $\bar{p}$ and $\bar{q}$ by reversing the corresponding truth values of p and q respectively. Now $\bar{p} \wedge \bar{q}$ is only true when both $\bar{p}$ and $\bar{q}$ are true, i.e. in row 4. In all other cases $\bar{p} \wedge \bar{q}$ is false.

(iii)

p	q	$\bar{q}$	$\bar{q} \rightarrow p$
T	T	F	T
T	F	T	T
F	T	F	T
F	F	T	F

(iv)

p	q	$\bar{p}$	$\bar{q}$	$\bar{p} \leftrightarrow \bar{q}$
T	T	F	F	T
T	F	F	T	F
F	T	T	F	F
F	F	T	T	T

We can construct truth tables for compound propositions involving more than two simple propositions as in the following example.

4. Construct truth tables for:

(i) $p \rightarrow (q \wedge r)$
(ii) $(\bar{p} \vee q) \leftrightarrow \bar{r}$.

Solution

(i)

p	q	r	$q \wedge r$	$p \rightarrow (q \wedge r)$
T	T	T	T	T
T	T	F	F	F
T	F	T	F	F
T	F	F	F	F
F	T	T	T	T
F	T	F	F	T
F	F	T	F	T
F	F	F	F	T

The first three columns list all possible combinations of truth values for p, q and r. Since each proposition can take two truth values there are $2^3 = 8$ possible combinations of truth values for the three propositions. Column 4 gives truth values of $q \wedge r$ by comparing the truth values of q and r individually in columns 2 and 3. Considering the pairs of truth values in columns 1 and 4 gives the truth values for $p \rightarrow (q \wedge r)$. Remember that this compound proposition is false only when p is true and $q \wedge r$ is false, i.e. in rows 2, 3 and 4.

(ii) Again we build up the truth table column by column to obtain the following:

p	q	r	$\bar{p}$	$\bar{r}$	$\bar{p} \vee q$	$(\bar{p} \vee q) \leftrightarrow \bar{r}$
T	T	T	F	F	T	F
T	T	F	F	T	T	T
T	F	T	F	F	F	T
T	F	F	F	T	F	F
F	T	T	T	F	T	F
F	T	F	T	T	T	T
F	F	T	T	F	T	F
F	F	F	T	T	T	T

Exercises 1.1

1. Consider the propositions:

p : Max is sulking.

q : Today is my birthday.

Write in words the compound propositions given by:

(i) $\bar{p} \wedge q$
(ii) $p \vee q$
(iii) $\bar{p} \rightarrow q$
(iv) $q \leftrightarrow p$.

2. Consider the propositions:

p : Mary laughs.

q : Sally cries.

r : Jo shouts.

Write in words the following compound propositions:

(i) $p \rightarrow (q \veebar r)$
(ii) $(r \wedge q) \leftrightarrow p$
(iii) $(p \rightarrow \bar{q}) \wedge (r \rightarrow q)$
(iv) $p \vee (\bar{q} \vee \bar{r})$
(v) $(p \vee r) \leftrightarrow \bar{q}$.

3. Let p, q and r denote the following propositions:

$$p : \text{Bats are blind.}$$
$$q : \text{Gnats eat grass.}$$
$$r : \text{Ants have long teeth.}$$

Express the following compound propositions symbolically.

(i) If bats are blind then gnats don't eat grass.
(ii) If and only if bats are blind or gnats eat grass then ants don't have long teeth.
(iii) Ants don't have long teeth and, if bats are blind, then gnats don't eat grass.
(iv) Bats are blind or gnats eat grass and, if gnats don't eat grass, then ants don't have long teeth.

4. Draw a truth table and determine for what truth values of p and q the proposition $\bar{p} \vee q$ is false.

5. Draw the truth table for the propositions:

(i) $\bar{p} \rightarrow q$
(ii) $\bar{q} \wedge p$
(iii) $(p \vee q) \rightarrow (p \wedge q)$
(iv) $(p \rightarrow q) \veebar \bar{q}$
(v) $\bar{p} \leftrightarrow (p \wedge q)$
(vi) $(\bar{p} \wedge q) \veebar (p \vee \bar{q})$.

6. Consider the two propositions:

$$p : \text{John is rich.}$$
$$q : \text{John is dishonest.}$$

Under what circumstances is the compound proposition 'If John is honest then he is not rich' false?

7. Given the three propositions p, q and r, construct truth tables for:

(i) $(p \wedge q) \rightarrow \bar{r}$
(ii) $(p \veebar r) \wedge \bar{q}$
(iii) $p \wedge (\bar{q} \vee r)$
(iv) $p \rightarrow (\bar{q} \vee \bar{r})$
(v) $(\overline{p \vee q}) \leftrightarrow (r \vee p)$.

1.3 Tautologies and Contradictions

There are certain compound propositions which have the surprising property that they are always true no matter what the truth value of their simple components. Similarly, there are others which are always false regardless of the truth values of their components. In both cases, this property is a consequence of the structure of the compound proposition.

Definition 1.1

A **tautology** is a compound proposition which is true no matter what the truth values of its simple components.

A **contradiction** is a compound proposition which is false no matter what the truth values of its simple components.

We shall denote a tautology by t and a contradiction by f.

Examples 1.2

1. Show that $p \vee \bar{p}$ is a tautology.

   Solution

   Constructing the truth table for $p \vee \bar{p}$, we have:

p	$\bar{p}$	$p \vee \bar{p}$
T	F	T
F	T	T

   Note that $p \vee \bar{p}$ is always true (no matter what proposition is substituted for p) and is therefore a tautology.

2. Show that $(p \wedge q) \vee (\overline{p \wedge q})$ is a tautology.

Solution

The truth table for $(p \wedge q) \vee (\overline{p \wedge q})$ is given below.

p	q	$p \wedge q$	$\overline{p \wedge q}$	$(p \wedge q) \vee (\overline{p \wedge q})$
T	T	T	F	T
T	F	F	T	T
F	T	F	T	T
F	F	F	T	T

The last column of the truth table contains only the truth value T and hence we can deduce that $(p \wedge q) \vee (\overline{p \wedge q})$ is a tautology.

Note that, in the last example, we could have appealed to the result obtained in the first one where we showed that the inclusive disjunction of any proposition and its negation is a tautology. In example 1.2.2 we have a proposition $p \wedge q$ and its negation $(\overline{p \wedge q})$. Hence, by the previous result, the inclusive disjunction $(p \wedge q) \vee (\overline{p \wedge q})$ is a tautology.

The proposition $(p \wedge q) \vee (\overline{p \wedge q})$ is said to be a **substitution instance** of the proposition $p \vee \bar{p}$. The former proposition is obtained from the latter simply by substituting $p \wedge q$ for p throughout. Clearly any substitution instance of a tautology is itself a tautology so that one way of establishing that a proposition is a tautology is to show that it is a substitution instance of another proposition which is known to be a tautology.

Example 1.3

Show that $(p \wedge \bar{q}) \wedge (\bar{p} \vee q)$ is a contradiction.

Solution

p	q	$\bar{q}$	$p \wedge \bar{q}$	$\bar{p}$	$\bar{p} \vee q$	$(p \wedge \bar{q}) \wedge (\bar{p} \vee q)$
T	T	F	F	F	T	F
T	F	T	T	F	F	F
F	T	F	F	T	T	F
F	F	T	F	T	T	F

The last column shows that $(p \wedge \bar{q}) \wedge (\bar{p} \vee q)$ is always false, no matter what the truth values of p and q. Hence $(p \wedge \bar{q}) \wedge (\bar{p} \vee q)$ is a contradiction.

Just as any substitution instance of a tautology is also a tautology, so any substitution instance of a contradiction is also a contradiction. For instance, using a truth table, we can show that $p \wedge \bar{p}$ is a contradiction. Since $(p \to q) \wedge (\overline{p \to q})$ is a substitution instance of $p \wedge \bar{p}$, we can deduce that this compound proposition is also a contradiction.

Exercises 1.2

Determine whether each of the following is a tautology, a contradiction or neither:

1. $p \to (p \vee q)$
2. $(p \to q) \wedge (\bar{p} \vee q)$
3. $(p \vee q) \leftrightarrow (q \vee p)$
4. $(p \wedge q) \to p$
5. $(p \wedge q) \wedge (\overline{p \vee q})$
6. $(p \to q) \to (p \wedge q)$
7. $(\bar{p} \wedge q) \wedge (p \vee \bar{q})$
8. $(p \to \bar{q}) \vee (\bar{r} \to p)$
9. $[p \to (q \wedge r)] \leftrightarrow [(p \to q) \wedge (p \to r)]$
10. $[(p \vee q) \to \bar{r}] \veebar (\bar{p} \vee \bar{q})$.

1.4 Logical Equivalence and Logical Implication

Two propositions are said to be **logically equivalent** if they have identical truth values for every set of truth values of their components. Using P and Q to denote (possibly) compound propositions, we write $P \equiv Q$ if P and Q are logically equivalent. As with tautologies and contradictions, logical equivalence is a consequence of the structures of P and Q.

Example 1.4

Show that $\bar{p} \vee \bar{q}$ and $\overline{p \wedge q}$ are logically equivalent, i.e. that $(\bar{p} \vee \bar{q}) \equiv (\overline{p \wedge q})$.

Solution

We draw up the truth table for $\bar{p} \vee \bar{q}$ and also for $\overline{p \wedge q}$.

p	q	$\bar{p}$	$\bar{q}$	$\bar{p} \vee \bar{q}$	$p \wedge q$	$\overline{p \wedge q}$
T	T	F	F	F	T	F
T	F	F	T	T	F	T
F	T	T	F	T	F	T
F	F	T	T	T	F	T

Comparing the columns for $\bar{p} \vee \bar{q}$ and for $\overline{p \wedge q}$ we note that the truth values are the same. Each is true except in the case where p and q are both true. Hence $\bar{p} \vee \bar{q}$ and $\overline{p \wedge q}$ are logically equivalent propositions.

Note that if two compound propositions are logically equivalent, then the compound proposition formed by joining them using the biconditional connective must be a tautology, i.e. if $P \equiv Q$ then $P \leftrightarrow Q$ is a tautology. This is so because two logically equivalent propositions are either both true or both false. In either of these cases the biconditional is true.

The converse is also the case, i.e. if $P \leftrightarrow Q$ is a tautology, then $P \equiv Q$. This follows from the fact that the biconditional $P \leftrightarrow Q$ is only true when P and Q both have the same truth values.

In example 1.4, we showed that $\bar{p} \vee \bar{q}$ and $\overline{p \wedge q}$ are logically equivalent by constructing their truth tables and comparing truth values. An alternative method would have been to show that $(\bar{p} \vee \bar{q}) \leftrightarrow (\overline{p \wedge q})$ is a tautology and to deduce from this the logical equivalence of $\bar{p} \vee \bar{q}$ and $\overline{p \wedge q}$.

Example 1.5

Show that the following two propositions are logically equivalent.

(i) If it rains tomorrow then, if I get paid, I'll go to Paris.
(ii) If it rains tomorrow and I get paid then I'll go to Paris.

Solution

Define the following simple propositions:

$$p: \text{It rains tomorrow.}$$
$$q: \text{I get paid.}$$
$$r: \text{I'll go to Paris.}$$

We are required to show the logical equivalence of $p \to (q \to r)$ and $(p \wedge q) \to r$. We can do this in one of two ways:

(a) establish that $p \to (q \to r)$ and $(p \wedge q) \to r$ have the same truth values,
or
(b) establish that $[p \to (q \to r)] \leftrightarrow [(p \wedge q) \to r]$ is a tautology.

Using the first method we complete the truth table for $p \to (q \to r)$ and $(p \wedge q) \to r$.

p	q	r	$q \to r$	$p \to (q \to r)$	$p \wedge q$	$(p \wedge q) \to r$
T	T	T	T	T	T	T
T	T	F	F	F	T	F
T	F	T	T	T	F	T
T	F	F	T	T	F	T
F	T	T	T	T	F	T
F	T	F	F	T	F	T
F	F	T	T	T	F	T
F	F	F	T	T	F	T

Since the truth values of $p \to (q \to r)$ and $(p \wedge q) \to r$ are the same for each set of truth values of p, q and r, we can deduce the logical equivalences of these compound propositions. Completing one further column of the truth table for $[p \to (q \to r)] \leftrightarrow [(p \wedge q) \to r]$ would show this to be a tautology and would establish the logical equivalence of the two propositions by the second method.

Another structure-dependent relation which may exist between two propositions is that of logical implication. A proposition P is said to **logically imply** a proposition Q if, whenever P is true, then Q is also true.

Note that the converse does not apply, i.e. Q may also be true when P is false. For logical implication all we insist on is that Q is never false when P is true. We shall symbolize logical implication by $\vdash$ so that 'P logically implies Q' is written $P \vdash Q$.

Example 1.6

Show that $q \vdash (p \vee q)$.

Solution

We must show that, whenever q is true, then $p \vee q$ is true. Constructing the truth table gives:

p	q	$p \vee q$
T	T	T
T	F	T
F	T	T
F	F	F

From a comparison of the second and third columns we note that, whenever q is true (first and third rows), $p \vee q$ is also true. Note that $p \vee q$ is also true when q is false (second row) but this has no relevance in establishing that q logically implies $p \vee q$.

We showed that '$P \equiv Q$' and '$P \leftrightarrow Q$ is a tautology' mean exactly the same. A similar line of argument can be used to establish that '$P \vdash Q$' and '$P \rightarrow Q$ is a tautology' are identical statements. If we have $P \vdash Q$ then Q is never false when P is true. Since this is the only situation where $P \rightarrow Q$ would be false then we must have $P \rightarrow Q$ is a tautology. Conversely, if $P \rightarrow Q$ is a tautology then the truth of P guarantees the truth of Q and hence we have $P \vdash Q$.

Example 1.7

Show that $(p \leftrightarrow q) \wedge q$ logically implies p.

Solution

As with example 1.5 we can show that $[(p \leftrightarrow q) \wedge q] \vdash p$ in one of two ways. We can either show that p is always true when $(p \leftrightarrow q) \wedge q$ is true or we can show that $[(p \leftrightarrow q) \wedge q] \rightarrow p$ is a tautology.

The truth table for $(p \leftrightarrow q) \wedge q$ is given by:

p	q	$p \leftrightarrow q$	$(p \leftrightarrow q) \wedge q$
T	T	T	T
T	F	F	F
F	T	F	F
F	F	T	F

Comparing the fourth column with the first, we see that p is true whenever $(p \leftrightarrow q) \wedge q$ is true (first row only). Therefore $[(p \leftrightarrow q) \wedge q] \vdash p$.

Alternatively, we could complete a further column of of the truth table for $[(p \leftrightarrow q) \wedge q] \rightarrow p$ and show this to be a tautology.

More about conditionals

Given the conditional proposition $p \rightarrow q$, we define the following:

(a) the **converse** of $p \rightarrow q$: $q \rightarrow p$
(b) the **inverse** of $p \rightarrow q$: $\bar{p} \rightarrow \bar{q}$
(c) the **contrapositive** of $p \rightarrow q$: $\bar{q} \rightarrow \bar{p}$.

The following truth table gives values of the conditional together with those for its converse, inverse and contrapositive.

p	q	$p \rightarrow q$	$q \rightarrow p$	$\bar{p} \rightarrow \bar{q}$	$\bar{q} \rightarrow \bar{p}$
T	T	T	T	T	T
T	F	F	T	T	F
F	T	T	F	F	T
F	F	T	T	T	T

From the table we note the following useful result: a conditional proposition $p \rightarrow q$ and its contrapositive $\bar{q} \rightarrow \bar{p}$ are logically equivalent, i.e. $(p \rightarrow q) \equiv (\bar{q} \rightarrow \bar{p})$.

Note that a conditional proposition is *not* logically equivalent to either its converse or inverse. However, the converse and inverse of a proposition are logically equivalent to each other.

Example 1.8

State the converse, inverse and contrapositive of the proposition 'If Jack plays his guitar then Sara will sing'.

Solution

We define: p: Jack plays his guitar
q: Sara will sing
so that: $p \to q$: If Jack plays his guitar then Sara will sing.

Converse: $q \to p$: If Sara will sing then Jack plays his guitar.
Inverse: $\bar{p} \to \bar{q}$: If Jack doesn't play his guitar then Sara won't sing.
Contrapositive: $\bar{q} \to \bar{p}$: If Sara won't sing then Jack doesn't play his guitar.

As we have shown, 'If Jack plays his guitar then Sara will sing' and 'If Sara won't sing then Jack doesn't play his guitar' are equivalent propositions.

Exercises 1.3

1. Prove that $(p \to q) \equiv (\bar{p} \vee q)$.
2. Prove that $(p \wedge q)$ and $(\overline{p \to \bar{q}})$ are logically equivalent propositions.
3. Prove that $(\overline{p \veebar q}) \equiv (p \veebar \bar{q})$.
4. Prove that p logically implies $(q \to p)$.
5. Prove that $(\bar{q} \to \bar{p}) \vdash (p \to q)$.
6. Prove the following logical implications:

   (i) $(p \wedge q) \vdash q$
   (ii) $(p \wedge q) \vdash p$
   (iii) $[(p \to q) \wedge p] \vdash q$
   (iv) $[(p \to q) \wedge (p \vee r)] \vdash (q \vee r)$
   (v) $p \vdash (q \to p)$
   (vi) $[(p \vee q) \wedge \bar{q}] \vdash p$.

7. Prove that the exclusive disjunction of p and q is logically equivalent to the negation of the biconditional proposition $p \leftrightarrow q$.

8. Show that the biconditional proposition $p \leftrightarrow q$ is logically equivalent to the conjunction of the two conditional propositions $p \rightarrow q$ and $q \rightarrow p$. (Thus, in the biconditional $p \leftrightarrow q$, proposition p is a necessary and sufficient condition for q and q is a necessary and sufficient condition for p.)

9. Establish the following logical equivalences:

   (i) $(p \rightarrow q) \equiv (\overline{p \wedge \bar{q}})$

   (ii) $(p \leftrightarrow q) \equiv (\overline{p \wedge \bar{q}}) \wedge (\overline{q \wedge \bar{p}})$

   (iii) $(p \vee q) \equiv (\overline{\bar{p} \wedge \bar{q}})$

   (iv) $(p \veebar q) \equiv \overline{(\overline{p \wedge \bar{q}}) \wedge (\overline{q \wedge \bar{p}})}$.

   (These results show that any compound proposition involving the disjunctive (either form), conditional or biconditional connectives can be written in a logically equivalent form involving only negation and conjunction.)

10. Consider a new connective, denoted by $|$, where $p|q$ is defined by the following truth table:

p	q	$p\|q$
T	T	F
T	F	T
F	T	T
F	F	T

   Show that:

   (i) $\bar{p} \equiv (p|p)$

   (ii) $(p \wedge q) \equiv (p|q)|(p|q)$.

   Use the results for exercise 1.3.9 above to deduce that a proposition involving any of the five familiar connectives can be written in a logically equivalent form which uses only the connective denoted by $|$.

11. State the converse, inverse and contrapositive of the proposition: 'If it's not Sunday then the supermarket is open until midnight'.

1.5 The Algebra of Propositions

The following is a list of some important logical equivalences, all of which can be verified using one of the techniques described in §1.4. (In fact we demonstrated two of these laws in §1.4.) These are often referred to as 'replacement laws' because, as we shall see later, there are situations where it is useful to substitute one proposition for another logically equivalent form. These laws hold for any simple propositions p, q and r and also for any substitution instance of them. We give the name of each law and also, because we shall need to refer to them later, the accepted abbreviation for identifying each one. Recall that we use t to denote a tautology and f to denote a contradiction.

Replacement Laws

Idempotent laws (Idem)

$$p \wedge p \equiv p$$
$$p \vee p \equiv p.$$

Commutative laws (Comm)

$$p \wedge q \equiv q \wedge p$$
$$p \vee q \equiv q \vee p$$
$$p \veebar q \equiv q \veebar p$$
$$p \leftrightarrow q \equiv q \leftrightarrow p.$$

Associative laws (Assoc)

$$(p \wedge q) \wedge r \equiv p \wedge (q \wedge r)$$
$$(p \vee q) \vee r \equiv p \vee (q \vee r)$$
$$(p \veebar q) \veebar r \equiv p \veebar (q \veebar r)$$
$$(p \leftrightarrow q) \leftrightarrow r \equiv p \leftrightarrow (q \leftrightarrow r).$$

Distributive laws (Dist)

$$p \wedge (q \vee r) \equiv (p \wedge q) \vee (p \wedge r)$$
$$p \vee (q \wedge r) \equiv (p \vee q) \wedge (p \vee r).$$

Involution law (Invol)

$$\bar{\bar{p}} \equiv p.$$

De Morgan's† laws (De M)

$$\overline{p \vee q} \equiv \bar{p} \wedge \bar{q}$$
$$\overline{p \wedge q} \equiv \bar{p} \vee \bar{q}.$$

Identity laws (Ident)

$$p \vee f \equiv p$$
$$p \wedge t \equiv p$$
$$p \vee t \equiv t$$
$$p \wedge f \equiv f.$$

Complement laws (Comp)

$$p \vee \bar{p} \equiv t$$
$$p \wedge \bar{p} \equiv f$$
$$\bar{f} \equiv t$$
$$\bar{t} \equiv f.$$

Transposition law (Trans)

$$p \rightarrow q \equiv \bar{q} \rightarrow \bar{p}$$

Material Implication law (Impl)

$$p \rightarrow q \equiv \bar{p} \vee q$$

Material Equivalence laws (Equiv)

$$p \leftrightarrow q \equiv (p \rightarrow q) \wedge (q \rightarrow p)$$
$$p \leftrightarrow q \equiv (p \wedge q) \vee (\bar{p} \wedge \bar{q})$$

† Named after the British mathematician Augustus de Morgan (1806–71) who became the first professor of the new University of London in 1828 and the first president of the London Mathematical Society in 1865.

Exportation law (Exp)

$$(p \wedge q) \rightarrow r \equiv p \rightarrow (q \rightarrow r)$$

The Duality Principle

Given any compound proposition P involving only the connectives denoted by $\wedge$ and $\vee$, the **dual** of that proposition is obtained by replacing $\wedge$ by $\vee$, $\vee$ by $\wedge$, t by f and f by t. For example, the dual of $(p \wedge q) \vee \bar{p}$ is $(p \vee q) \wedge \bar{p}$. The dual of $(p \vee f) \wedge q$ is $(p \wedge t) \vee q$.

Notice that we have not stated how to obtain the dual of a compound proposition containing connectives other than conjunction and inclusive disjunction. This does not matter since we have shown that propositions containing the other connectives can all be written in a logically equivalent form involving only negation and conjunction (see exercise 1.3.9).

DUALITY PRINCIPLE

The **duality principle** states that, if two propositions are logically equivalent, then so are their duals. The principle is evident in several of the laws of the algebra of propositions stated above. In many cases the logical equivalences are stated in pairs where one member of the pair is the dual of the other.

Substitution Rule

Suppose that we have two logically equivalent propositions P_1 and P_2, so that $P_1 \equiv P_2$. Suppose also that we have a compound proposition Q in which P_1 appears. The **substitution rule** says that we may replace P_1 by P_2 and the resulting proposition is logically equivalent to Q. Thus substituting a logically equivalent proposition for another in a compound proposition does not alter the truth value of that proposition.

Although we have not formally proved the substitution rule, it is clearly reasonable if we consider the truth table. Substituting truth values of P_2 for P_1 makes no difference to the truth table since, if P_1 and P_2 are logically equivalent, they have the same truth values for each set of truth values of their components.

The substitution rule and the replacement laws give us a means of establishing logical equivalences between propositions without drawing up a truth table. We demonstrate this in the following example.

Example 1.9

Prove that $(\bar{p} \wedge q) \vee (\overline{p \vee q}) \equiv \bar{p}$.

Solution

$$\begin{aligned} (\bar{p} \wedge q) \vee (\overline{p \vee q}) &\equiv (\bar{p} \wedge q) \vee (\bar{p} \wedge \bar{q}) && \text{(De M)} \\ &\equiv \bar{p} \wedge (q \vee \bar{q}) && \text{(Dist)} \\ &\equiv \bar{p} \wedge t && \text{(Comp)} \\ &\equiv \bar{p}. && \text{(Ident)} \end{aligned}$$

Exercise 1.4

1. Prove each of the following logical equivalences using the method of example 1.9.

   (i) $(p \wedge p) \vee (\bar{p} \vee \bar{p}) \equiv t$.
   (ii) $(p \wedge q) \wedge q \equiv p \wedge q$.
   (iii) $p \rightarrow q \equiv \overline{p \wedge \bar{q}}$.
   (iv) $(p \wedge q) \rightarrow r \equiv (\bar{p} \vee \bar{q}) \vee r$.
   (v) $q \wedge [(p \vee q) \wedge (\overline{\bar{q} \wedge \bar{p}})] \equiv q \wedge (q \vee p)$.

2. Use the method of example 1.9 to show that $p \wedge (q \vee \bar{p})$ is logically equivalent to $p \wedge q$. State the dual of each of these two propositions and show that the two dual propositions are also logically equivalent.

1.6 Arguments

An **argument** consists of a set of propositions called **premises** together with another proposition, purported to follow from the premises, called the **conclusion**. We say that the argument is **valid** if the conjunction of the premises logically implies the conclusion. Otherwise the argument is said to be **invalid**. Thus if we have premises $P_1, P_2, \ldots, P_n$ and a conclusion Q, then the argument is valid if $(P_1 \wedge P_2 \wedge \cdots \wedge P_n) \vdash Q$, i.e. if $(P_1 \wedge P_2 \wedge \cdots \wedge P_n) \rightarrow Q$ is a tautology. What this means (see §1.4) is that whenever $P_1, P_2, \ldots, P_n$ are all true, then Q must be true. This makes sense since it ensures that, in a valid argument, a set of premises all of which are true cannot lead to a false conclusion.

VALID ARGUMENT

Examples 1.10

1. Test the validity of the following argument: 'If you insulted Bob then I'll never speak to you again. You insulted Bob so I'll never speak to you again.'

Solution

We define: p: You insulted Bob.
q: I'll never speak to you again.

The premises in this argument are: $p \rightarrow q$ and p.
The conclusion is: q.

We must therefore investigate the truth table for $[(p \to q) \wedge p] \to q$. If this compound proposition is a tautology, then the argument is valid. Otherwise it is not.

p	q	$p \to q$	$(p \to q) \wedge p$	$[(p \to q) \wedge p] \to q$
T	T	T	T	T
T	F	F	F	T
F	T	T	F	T
F	F	T	F	T

This shows that the argument is valid.

2. Test the validity of the following argument: 'If you are a mathematician then you are clever. You are clever and rich. Therefore if you are rich then you are a mathematician.'

Solution

Define: p: You are a mathematician.
q: You are clever.
r: You are rich.

The premises are: $p \to q$ and $q \wedge r$.
The conclusion is: $r \to p$.

We must test whether or not $[(p \to q) \wedge (q \wedge r)] \to (r \to p)$ is a tautology.

p	q	r	$p \to q$	$q \wedge r$	$(p \to q) \wedge (q \wedge r)$	$r \to p$	$[(p \to q) \wedge (q \wedge r)] \to (r \to p)$
T	T	T	T	T	T	T	T
T	T	F	T	F	F	T	T
T	F	T	F	F	F	T	T
T	F	F	F	F	F	T	T
F	T	T	T	T	T	F	F
F	T	F	T	F	F	T	T
F	F	T	T	F	F	F	T
F	F	F	T	F	F	T	T

From the last column we see that $[(p \to q) \wedge (q \wedge r)] \to (r \to p)$ is not a tautology and hence the argument is not valid.

Exercise 1.5

Test the validity of the following arguments.

1. If you gamble you're stupid. You're not stupid therefore you don't gamble.

2. If I leave college then I'll get a job in a bank. I'm not leaving college so I won't get a job in a bank.

3. James is either a policeman or a footballer. If he's a policeman then he has big feet. James hasn't got big feet so he's a footballer.

4. If I could swim I'd come sailing with you. I can't swim so I'm not coming sailing with you.

5. If you find this difficult then you're stupid or you haven't done your homework. You've done your homework and you're not stupid therefore you won't find this difficult.

6. You can go out if and only if you do the washing up. If you go out then you won't watch television. Therefore you either watch television or wash up but not both.

7. If I graduate in June then I'll go on holiday in the summer. In the summer I'll get a job or I'll go on holiday. I won't go on holiday in the summer so I won't graduate in June.

8. If there are clouds in the sky then the sun doesn't shine and if the sun doesn't shine then the temperature falls. The temperature isn't falling so there are no clouds in the sky.

9. I shall be a lawyer or a banker (but not both). If I become a lawyer then I shall never be rich. Therefore I shall be rich only if I become a banker.

10. If you are eligible for admission then you must be under 25 and if you are not under 25 then you do not qualify for a scholarship. Therefore if you qualify for a scholarship, you are eligible for admission.

1.7 Formal Proof of the Validity of Arguments

Whilst a truth table will always establish whether or not an argument is valid, the method can become tediously lengthy. If there are more than a small number of premises, the number of columns in the table becomes large. Furthermore, the number of rows increases exponentially with the number of simple propositions defined. An argument involving n simple propositions will require 2^n rows in its truth table.

An alternative method of validating an argument consists of constructing a sequence of propositions starting with the premises. Propositions may then be added to the sequence, but only if their truth is guaranteed by the truth of propositions already included in the list. The object in constructing the sequence is to add propositions which will finally justify adding the conclusion of the argument. When this is achieved, the formal proof of validity is complete. At this point we have a list of propositions all of which are known to be true given the truth of the premises. In particular, the truth of the argument's conclusion is guaranteed by the truth of the premises and this is, of course, the condition for the argument to be valid.

How do we recognise which propositions may be added to the list? Clearly, if a proposition is already included, we may add any other proposition to which it is logically equivalent. Therefore the list of logical equivalences given in section 1.5 will be useful. We may also use the substitution rule given in §1.5. If a compound proposition is already in the list, we may add one where part of that proposition is replaced by a logically equivalent form since this substitution does not alter the truth value.

A battery of logical equivalences however, will not be sufficient in itself for us to establish the truth of the argument's conclusion assuming the truth of the premises. Remember that, if a proposition P logically implies a proposition Q, then Q is true whenever P is true. This means that we can add to the list any proposition which is logically implied by an earlier proposition in the sequence. We can extend this principle. Recall that the conjunction of propositions $P_1 \wedge P_2 \wedge \ldots \wedge P_n$ is true only when each of the conjuncts $P_1, P_2, \ldots P_n$ is true. Now if $(P_1 \wedge P_2 \wedge \ldots \wedge P_n) \vdash Q$, then Q is true whenever $P_1 \wedge P_2 \wedge \ldots \wedge P_n$ is true, i.e. when $P_1, P_2, \ldots$ and P_n are all true. This means that we can add to our list any proposition which is logically implied by the conjunction of a set of earlier propositions in the sequence.

The relationship $(P_1 \wedge P_2 \wedge \ldots \wedge P_n) \vdash Q$ means that we can regard Q as the conclusion of a valid argument with premises $P_1, P_2, \ldots P_n$. (This was exactly our definition of a valid argument.) Hence a justification for adding a proposition to the sequence is that it is the conclusion of a valid argument whose premises are already included in the list. It will therefore be useful for us to have a list of valid 'mini-arguments'. Then, when we spot any substitution instance of the premises in our list, we know that we can add the corresponding conclusion if we so wish.

The following is a list of useful valid arguments to which we can appeal when justifying the addition of propositions to the list constituting our formal proof. These are often referred to as 'rules of inference' and there are nine which will prove sufficient for our needs. We list these rules in the table below.

Rules of Inference

Name of rule	Premises	Conclusion
Simplification (Simp)	$P \wedge Q$	P
Addition (Add)	P	$P \vee Q$
Conjunction (Conj)	$P,\ Q$	$P \wedge Q$
Disjunctive syllogism (DS)	$P \vee Q,\ \neg P$	Q
Modus ponens (MP)	$P,\ P \to Q$	Q
Modus tollens (MT)	$P \to Q,\ \bar{Q}$	$\bar{P}$
Hypothetical syllogism (HS)	$P \to Q,\ Q \to R$	$P \to R$
Absorption (Abs)	$P \to Q$	$P \to (P \wedge Q)$
Constructive dilemma (CD)	$P \to Q,\ R \to S,\ P \vee R$	$Q \vee S$

These arguments are valid no matter what propositions are substituted for P, Q and R. For instance an argument with premises $(p \wedge \bar{q}) \to (q \vee r)$ and $p \wedge \bar{q}$ and with conclusion $q \vee r$ is valid because it is a substitution instance of the modus ponens rule. In that rule we simply substitute $p \wedge \bar{q}$ for P and $q \vee r$ for Q. What these rules give us is, in a sense, 'patterns' for certain valid arguments.

There follow some examples to illustrate the construction of a formal proof.

Examples 1.11

1. Construct a formal proof of the validity of the following argument.

   Premises : $p \to q,\ p \wedge r$
   Conclusion : q

We commence the sequence, as always, with the premises:

1.	$p \rightarrow q$	(premise)
2.	$p \wedge r$	(premise)

Our aim is to be able to justify (in terms of the 'truth guarantee' criterion) the addition of the conclusion to the list. This can be done in two steps. The *Simplification* rule applied to the second premise allows us to add p. Now that we have p, the *Modus Ponens* rule applied to this together with the first premise justifies the addition of q, the conclusion of the argument. This completes the proof.

We summarise these two steps below.

3.	p	(2. Simp)
4.	q	$(1, 3.$ MP)

Note that the propositions in the sequence are numbered so that they can be referred to later and also that a justification must be provided for the addition of each proposition to the list.

Putting the steps together, the complete formal proof is the following.

1.	$p \rightarrow q$	(premise)
2.	$p \wedge r$	(premise)
3.	p	(2. Simp)
4.	q	$(1, 3.$ MP)

2. Provide a formal proof of the validity of the following argument:

Premises : $p \rightarrow \bar{q},\ q \vee r$
Conclusion : $p \rightarrow r$

Note that, if we could add $\bar{q} \rightarrow r$ to our sequence, we could apply the *Hypothetical Syllogism* rule to this and the first premise and thereby justify the addition of the conclusion. But how can we sanction the addition of $\bar{q} \rightarrow r$? The second premise provides the clue. If we replace q by $\bar{\bar{q}}$ (justified by the *Involution law*), the second premise becomes $\bar{\bar{q}} \vee r$. Now we can use the *Material Implication* law to justify the logical equivalence of $\bar{\bar{q}} \vee r$ and $\bar{q} \rightarrow r$. Then the conclusion $p \rightarrow r$ follows from $p \rightarrow \bar{q}$ and $\bar{q} \rightarrow r$ by *Hypothetical Syllogism*. This completes the proof.

The formal proof is given below.

1. $p \rightarrow \bar{q}$ (premise)
2. $q \vee r$ (premise)
3. $\bar{\bar{q}} \vee r$ (2. Invol)
4. $\bar{q} \rightarrow r$ (3. Impl)
5. $\bar{q} \rightarrow r$ (2, 3. HS)

3. Construct a formal proof of the following argument.

$$\begin{array}{rc} \textbf{Premises}: & p \rightarrow q,\ r \rightarrow s,\ \bar{q},\ r \\ \textbf{Conclusion}: & \bar{p} \wedge s \end{array}$$

As with the previous example, we shall often find it useful to work backwards by asking ourselves what needs to be added to the list to justify adding the conclusion. Here the conclusion is a conjunction. If the list were to contain each of the conjuncts $\bar{p}$ and s, then we have a rule of inference (*Conjunction*) which will allow us to add $\bar{p} \wedge s$. In this example, each of the conjuncts follows directly from a rule of inference applied to a pair of premises. The proof is given below.

1. $p \rightarrow q$ (premise)
2. $r \rightarrow s$ (premise)
3. $\bar{q}$ (premise)
4. r (premise)
5. $\bar{p}$ (1, 3. MT)
6. s (2, 4. MP)
7. $\bar{p} \wedge s$ (5, 6. Conj)

4. Give a formal proof for the following argument:

$$\begin{array}{rc} \textbf{Premises}: & p \wedge (q \vee r),\ \bar{p} \vee \bar{q} \\ \textbf{Conclusion}: & r \vee q \end{array}$$

The proof here is a little longer and, again, it helps to work backwards. The conclusion here is the inclusive disjunction of r and q. Note that the *Addition* rule would allow us to add the proposition $r \vee q$ if our list contained simply the proposition r. So let's concentrate on how we might justify the inclusion of r.

Now the only source of the proposition r is the disjunction $q \vee r$ in the first premise, so need to extract this out. This can be achieved by applying the *Commutative* law (so that $q \vee r$ becomes the first of the two conjuncts) followed by the *Simplification* rule.

Our proof so far is as follows:

1. $p \wedge (q \vee r)$ (premise)
2. $\bar{p} \vee \bar{q}$ (premise)
3. $(q \vee r) \wedge p$ (1. Comm)
4. $q \vee r$ (3. Simp)

Now, how can we 'extract' r from $q \vee r$? The *Disjunctive Syllogism* rule provides the clue—we need $\bar{q}$ and we can then use this rule to justify the addition of r. To justify the addition of $\bar{q}$, we could use premise 2 along with *Disjunctive Syllogism* again. But for this we need $\bar{\bar{p}}$ or (justified by the *Involution* law) its logically equivalent form p. This can be obtained from premise 1 using the *Simplification* law.

We now have the rest of the steps necessary to complete the proof. The full proof is given below.

1. $p \wedge (q \vee r)$ (premise)
2. $\bar{p} \vee \bar{q}$ (premise)
3. $(q \vee r) \wedge p$ (1. Comm)
4. $q \vee r$ (3. Simp)
5. p (1. Simp)
6. $\bar{\bar{p}}$ (5. Inv)
7. $\bar{q}$ (7, 2. DS)
8. r (7, 4. DS)
9. $r \vee s$ (8. Add)

There are often several alternative formal proofs of a valid argument. For instance, the following is an alternative to the formal proof above.

1. $p \wedge (q \vee r)q$ (premise)
2. $\bar{p} \vee \bar{q}$ (premise)
3. $(p \wedge q) \vee (p \wedge r)$ (1. Dist)
4. $\overline{p \wedge q}$ (2. De M)
5. $p \wedge r$ (3. Simp)
6. $r \wedge p$ (5. Comm)
7. r (6. Simp)
8. $r \vee s$ (8. Add)

Remember that, using a truth table, we were able to establish the validity or invalidity of an argument. Our method of formal proof is really only useful for showing that an argument is valid. If we are unable to find a formal proof of the validity of an argument we cannot be sure whether this is because no such proof exists (because the argument is not valid) or because we have simply failed to find a proof for an argument which is, in fact, valid.

Exercise 1.6

1. Provide a formal proof for each of the following arguments.

(i) Premises: $(p \wedge q) \rightarrow (r \wedge s), p, q$
Conclusion: s

(ii) Premises: $p \rightarrow \bar{q}, q \vee r$
Conclusion: $p \rightarrow r$

(iii) Premises: $p \rightarrow \bar{q}, p \wedge r, q \vee r$
Conclusion: r

(iv) Premises: $p \vee q, r \rightarrow \bar{q}, \bar{p}, (\bar{r} \wedge q) \rightarrow s$
Conclusion: s

(v) Premises: $(p \rightarrow q) \wedge (\bar{p} \rightarrow \bar{r}), s \wedge p$
Conclusion: q

(vi) Premises: $(p \vee q) \rightarrow (r \wedge s), r \rightarrow t, \bar{t}$
Conclusion: $\bar{p}$

(vii) Premises: $(p \vee q) \wedge (q \vee r), \bar{q}$
Conclusion: $p \wedge r$

2. Symbolise the following arguments and provide a formal proof of the validity of each one.

(i) The murder was committed by A or by both B and C. Therefore the murder was committed by A or B.

(ii) I won't play golf only if I go swimming. I won't go swimming. Therefore I'll play golf.

(iii) If Ed goes to the party then I won't go. Either I'll go to the party or I'll stay home and watch TV. Therefore if Ed goes to the party then I'll stay home and watch TV.

(iv) If the summer is hot or wet then my garden flourishes. The summer is hot and windy. Therefore my garden flourishes.

(v) People are happy if and only if they are charitable. Nobody is both happy and charitable. Hence people are unhappy and uncharitable.

(vi) If you go to college or get a good job then you will be successful and respected. You go to college. Therefore you will be respected.

(vii) If I eat cheese I get sick and if I drink wine I get sick. If I go to Ira's I eat cheese or I drink wine. I'm going to Ira's. Therefore I shall get sick.

1.8 Predicate Logic

Consider the following argument: 'Everyone who has green eyes is not to be trusted. Bill has green eyes. Therefore Bill is not to be trusted.' Expressing this in our propositional notation would give us an argument with premises p and q and a conclusion r. Our notation gives us no means of showing that different propositions are making statements about the same thing. Two propositions as similar as 'Bill has green eyes' and 'Jeff has green eyes' would have to be symbolized by p and q. We have as yet no means of expressing the fact that both propositions are about 'green eyes'.

A **predicate** describes a property of one or several objects or individuals. Examples of predicates might be:

(a) ... is red.
(b) ... has long teeth.
(c) ... enjoys standing on his head.
(d) ... has spiky leaves.
(e) ... cannot be tolerated under any circumstances.

The space in front of these predicates can be filled in with the names of objects or individuals where appropriate to form a proposition which may be true or false in the usual way. For instance (a) could be prefixed by 'that door', 'this flower', 'your nose' or any other object. Propositions of this kind consist of a subject together with a predicate describing whatever property the subject is said to possess.

We shall symbolize these propositions in a different way from before so as to distinguish their two component parts. We shall use capital letters to refer to predicates, so that we might define:

R : is red.

T : has long teeth.

$$H : \text{enjoys standing on his head.}$$

Lower-case letters will be used to denote particular objects or individuals. For instance:

$$r : \text{this rose.}$$
$$j : \text{James.}$$

We can then form simple propositions as follows:

$$R(r) : \text{This rose is red.}$$
$$R(j) : \text{James is red.}$$
$$H(j) : \text{James enjoys standing on his head.}$$

Notice that the attribute symbol is written to the left of the symbol representing the particular object or individual. If R is the predicate 'is red', we can write $R(x)$ to denote 'x is red' where x can be replaced by any object or individual. Note that $R(x)$ is not itself a proposition since it cannot be declared true or false. However, it becomes a proposition once x is replaced by a particular object or individual. The letter x is a **variable** which serves as a place marker to indicate where we may substitute the names of objects or individuals in order to form propositions. For this reason, $R(x)$ is called a **propositional function**.

We can negate propositional functions. If $R(x)$ denotes 'x is red' then the negation of $R(x)$, denoted by $\neg R(x)$ (or $\overline{R(x)}$), is the propositional function interpreted as 'x is not red'.

Substituting a particular 'value' for x in a propositional function is not the only way of converting it to a proposition. This can also be achieved through the use of quantifiers.

The Universal Quantifier

Consider the proposition 'All rats are grey'. One way in which we could paraphrase this proposition is: 'For every x, if x is a rat, then x is grey'. This gives us a way of symbolizing the proposition using the predicate symbols described earlier. Suppose we define:

$$R(x) : x \text{ is a rat.}$$
$$G(x) : x \text{ is grey.}$$

We denote 'for every x' by $\forall x$ and we can then write 'All rats are grey' as:

$$\forall x[R(x) \rightarrow G(x)].$$

The symbol $\forall$ is called the **universal quantifier**. The quantified variable $\forall x$ is read as 'for all x' or 'for every x'.

Example 1.12

Symbolize the proposition 'Every day I go jogging'.

Solution

Define the following:

$$D(x) : x \text{ is a day.}$$
$$J(x) : x \text{ is when I go jogging.}$$

Then 'Every day I go jogging' can be paraphrased 'For every x, if x is a day, then x is when I go jogging'. We can express this proposition symbolically by:

$$\forall x[D(x) \rightarrow J(x)].$$

The Existential Quantifier

Consider the proposition 'Some rats are grey'. Here we assert that there is at least one rat which is grey. We could paraphrase this proposition as 'There exists at least one x such that x is a rat and x is grey'. Thus if we define:

$$R(x) : x \text{ is a rat}$$
$$G(x) : x \text{ is grey}$$

and denote 'there exists at least one x' by $\exists x$, then 'Some rats are grey' can be written:

$$\exists x[R(x) \wedge G(x)].$$

The symbol $\exists$ is called the **existential quantifier** and $\exists x$ is read as 'there exists at least one x' or 'for some x'.

Example 1.13

1. Symbolize 'Some people think of no one but themselves'.

Solution

Define: $P(x)$: x is a person
$N(x)$: x thinks of no one but himself.

Then 'Some people think of no one but themselves' can be written:

$$\exists x[P(x) \wedge N(x)].$$

2. Symbolize 'Some of the children didn't apologize'.

Solution

Define: $C(x)$: x is a child
$A(x)$: x apologized.

Then 'Some of the children didn't apologize' can be written using the negation of $A(x)$ thus:

$$\exists x[C(x) \wedge \neg A(x)].$$

3. Symbolize the proposition 'Nobody likes cheats'.

Solution

Define: $P(x)$: x is a person
$C(x)$: x likes cheats.

What we want to say here is that there does *not* exist an x where x is a person and x likes cheats. We can symbolize this by negating the existential quantifier thus:

$$\neg \exists x[P(x) \wedge C(x)].$$

Note that we can use the connectives $\wedge, \vee, \rightarrow$, etc between the propositional functions $P(x)$, $C(x)$ even though these are not propositions. Thus if we define:

$$P(x) : x \text{ is a person}$$
$$C(x) : x \text{ cheats}$$
$$T(x) : x \text{ talks loudly}$$

then the expression

$$\forall x[P(x) \rightarrow \{T(x) \wedge C(x)\}]$$

symbolizes 'Everybody cheats and talks loudly'.

In example 1.13.2 we symbolized the proposition 'Some of the children didn't apologize' as $\exists x[C(x) \wedge \neg A(x)]$. There is a sense in which this proposition seems to refer to some particular group of children which the speaker has in mind rather than children in general. The predicate 'is a child' in this example seems to mean 'is a member of a particular group of children'. This particular group of children is called the **universe of discourse** and we can consider the variable x to be restricted to members of this set.

If we define the universe of discourse carefully, we can shorten the proposition $\exists x[C(x) \wedge \neg A(x)]$ to the simple form $\exists x[\neg A(x)]$ where it is understood that x belongs to the particular group of children that the speaker has in mind. The expression $\exists x[\neg A(x)]$ then states that, within this universe of discourse, at least one x exists who didn't apologize.

When the universe of discourse is not specified it is assumed to be the complete universe of objects or individuals referred to in the proposition. 'All rats are grey' is assumed to be a statement about all rats in the universe unless the context makes it clear that some subset of these is intended.

In determining the truth value of quantified propositions, it is important that we are clear about the universe of discourse. For instance, the proposition 'Some of the children didn't apologize' may be true in one universe of discourse but false in another.

Example 1.14

Define the following:

$$F(x) : x \text{ is greater than five}$$
$$E(x) : x \text{ is an even number}$$
$$N(x) : x \text{ is negative.}$$

Consider the following universes of discourse:

(i) integers (i.e. whole numbers)
(ii) real numbers
(iii) negative integers.

Determine the truth values of each of the following propositions in each universe of discourse.

(a) $\exists x F(x)$
(b) $\forall x N(x)$
(c) $\forall x[F(x) \wedge E(x)]$
(d) $\exists x[\neg N(x)]$.

Solution

(a) This proposition states that there exists an x which is greater than five. This is true for the universe of integers and for the universe of real numbers. It is false if x is restricted to negative integers.

(b) The proposition here is 'For every x, x is negative'. This is false for integers and for real numbers but it is true for the universe of negative integers.

(c) Here we have 'For every x, x is greater than five and even'. This is false in all three universes.

(d) This proposition is 'There exists an x which is not negative'. This is true for integers and for real numbers but is false for negative integers.

Two-Place Predicates

Consider the predicate 'is heavier than'. In order to convert this predicate into a proposition, the names of two objects or individuals are necessary. For instance, using this predicate, we may form the proposition 'A brick is heavier than a hamster'. The predicate 'is heavier than' is an example of a **two-place predicate**. If H denotes this predicate, then $H(x, y)$ denotes the propositional function 'x is heavier than y'.

Two-place predicates can be quantified using the universal and existential quantifiers. However, two quantifiers are necessary to produce a proposition from a two-variable propositional function. The quantified expressions $\forall x F(x, y)$ and $\exists x F(x, y)$ are not propositions but propositional functions of the single variable y.

Suppose we have:

$$P(x, y) : x + y = 7$$

where the universe of discourse for each variable is the real numbers. The following propositions are possible:

1.	$\forall x\, \exists y P(x, y)$	2.	$\exists y\, \forall x P(x, y)$
3.	$\forall y\, \exists x P(x, y)$	4.	$\exists x\, \forall y P(x, y)$
5.	$\forall y\, \forall x P(x, y)$	6.	$\forall x\, \forall y P(x, y)$
7.	$\exists y\, \exists x P(x, y)$	8.	$\exists x\, \exists y P(x, y)$.

Note that the propositions are read from left to right and that the order of quantified variables is important. Consider for instance propositions 1 and 2. The first states that, for every x, there exists at least one y such that, $x + y = 7$. This is clearly true. On the other hand, proposition 2 states that there exists at least one y such that, for every x, $x + y = 7$. This is not true since a single y value cannot be found for every x. Each value of x needs a different value of y to balance the equation $x + y = 7$. Thus the propositions $\forall x\, \exists y P(x, y)$ and $\exists y\, \forall x P(x, y)$ are not equivalent statements. For similar reasons, propositions 3 and 4 are also not equivalent.

The propositions $\forall x\, \forall y F(x, y)$ and $\forall y\, \forall x F(x, y)$ are equivalent for any propositional function $F(x, y)$, i.e. they have identical truth values. In the example above, 5 and 6 are equivalent (false) propositions. Similarly $\exists x\, \exists y F(x, y)$ and $\exists y\, \exists x F(x, y)$ are equivalent propositions for any propositional function $F(x, y)$. Hence 7 and 8 are equivalent (true) propositions.

Negation of Quantified Propositional Functions

The proposition $\forall x F(x)$ states that, for all x in the universe of discourse, x has the property defined by the predicate F. The negation of this proposition, $\neg \forall x F(x)$, states that 'It is not the case that all x have the property defined by F', i.e. there is at least one x that does not have the property F. This is symbolized by $\exists x[\neg F(x)]$. So, for any propositional function $F(x)$, the propositions $\neg \forall x F(x)$ and $\exists x[\neg F(x)]$ have the same truth values and are therefore equivalent, i.e.

$$\neg \forall x F(x) \equiv \exists x[\neg F(x)].$$

Similarly, the negation of $\exists x F(x)$, symbolized by $\neg \exists x F(x)$, states that there does not exist an x within the universe of discourse that has the property defined by F. This is the same as saying that, for all x, x does not have the property F, i.e. $\forall x[\neg F(x)]$. Thus we have

$$\neg \exists x F(x) \equiv \forall x[\neg F(x)]$$

for all propositional functions $F(x)$.

These two equivalences are known as the Rules of Quantification Denial (abbreviated as QD). The rules are summarised below.

Rules of Quantification Denial

Where a universe of discourse is defined for the variable x, then for any propositional function $F(x)$:

$$\neg\forall x F(x) \equiv \exists x[\neg F(x)]$$

and

$$\neg\exists x F(x) \equiv \forall x[\neg F(x)].$$

We can also show that

$$\neg\exists x[\neg F(x)] \equiv \forall x F(x)$$

since

$$\begin{aligned} \neg\exists x[\neg F(x)] &\equiv \forall x[\neg\neg F(x)] && \text{(by the second result above)} \\ &\equiv \forall x F(x) && \text{(by the involution law).} \end{aligned}$$

Similarly we can show that

$$\neg\forall x[\neg F(x)] \equiv \exists x F(x).$$

For doubly quantified propositional functions, equivalences can be established by repeated applications of the rules above. For instance:

$$\begin{aligned} \neg\exists y\, \forall x P(x, y) &\equiv \forall y[\neg\forall x P(x, y)] \\ &\equiv \forall y\, \exists x[\neg P(x, y)]. \end{aligned}$$

The negation of other similar propositions can be obtained in the same way.

Example 1.15

We define the following on the universe of men.

$$M(x) : x \text{ is mortal.}$$
$$C(x) : x \text{ lives in the city.}$$

Symbolize the negations of the following propositions changing the quantifier.

(i) All men are immortal.
(ii) Some men live in the city.

Solution

(i) The proposition given can be symbolized by $\forall x[\neg M(x)]$. The negation of this proposition is given by

$$\neg\forall x[\neg M(x)] \equiv \exists x M(x).$$

The resulting proposition is 'Some men are mortal'.

(ii) 'Some men live in the city' is symbolized by $\exists x C(x)$. Its negation is

$$\neg\exists x C(x) \equiv \forall x[\neg C(x)].$$

That is, 'All men live out of the city'.

Exercises 1.7

1. Suppose the following predicates and individuals are defined:

$$\begin{aligned} m &: \text{Maria} \\ s &: \text{Maria's son} \\ C &: \text{works in the city} \\ B &: \text{rides a bicycle} \\ F &: \text{is a chicken farmer.} \end{aligned}$$

Symbolize the following:

(i) Maria works in the city and her son is a chicken farmer.
(ii) If Maria rides a bicycle then her son works in the city.
(iii) Maria works in the city and rides a bicycle but her son is not a chicken farmer.
(iv) Everyone who works in the city is a chicken farmer.
(v) Everyone who works in the city and doesn't ride a bicycle is a chicken farmer.
(vi) Some people who work in the city and ride a bicycle are not chicken farmers.
(vii) If no-one working in the city rides a bicycle then Maria doesn't work in the city and her son is not a chicken farmer.
(viii) No chicken farmers work in the city and ride a bicycle.

2. Translate the following into symbolic form using one-place predicates. Define predicates used and, where necessary, define the universe of discourse.

(i) All babies cry a lot.
(ii) Nobody can ignore him.
(iii) Some students can't write a good essay.
(iv) Not everybody approves of capital punishment.
(v) There are people who have had a university education and live in poverty.
(vi) Every time it rains I forget my umbrella.
(vii) All of my friends believe in nuclear disarmament.
(viii) All Fred's children are rude or stupid.
(ix) Somebody set off the fire alarm and everybody left the building.
(x) Not all rats are dirty and carry disease.
(xi) Everybody who doesn't like snails has no taste.
(xii) Some toys are dangerous and no child should be given them.

3. Translate the following into symbolic form using two-place predicates.

(i) Everybody loves somebody.
(ii) Somebody loves everybody.
(iii) Everyone is taller than Sam.
(iv) All elephants love buns.

4. Negate each of the following propositions and use the Rules of Quantification Denial to change the quantifier. Express the result as a reasonable English sentence.

(i) Everybody likes strawberry jam.
(ii) There are birds that cannot fly.
(iii) Sometimes I think you are lazy.
(iv) Nobody leaves without my permission.

5. Consider the following predicates:

$$\begin{aligned} P(x,y) &: x > y \\ Q(x,y) &: x \leqslant y \\ R(x) &: x - 7 = 2 \\ S(x) &: x > 9. \end{aligned}$$

If the universe of discourse is the real numbers, give the truth value of each of the following propositions:

(i) $\exists x R(x)$
(ii) $\forall y [\neg S(y)]$
(iii) $\forall x \, \exists y P(x,y)$
(iv) $\exists y \, \forall x Q(x,y)$

(v) $\forall x \forall y[P(x,y) \vee Q(x,y)]$
(vi) $\exists x S(x) \wedge \neg\forall x R(x)$
(vii) $\exists y \forall x[S(y) \wedge Q(x,y)]$
(viii) $\forall x \forall y[\{R(x) \wedge S(y)\} \rightarrow Q(x,y)]$.

1.9 Arguments in Predicate Logic

We return to the argument at the beginning of §1.8: 'Everyone who has green eyes is not to be trusted. Bill has green eyes. Therefore Bill is not to be trusted.'

If we define the following on the universe of all human beings:

$$\begin{aligned} G(x) &: x \text{ has green eyes} \\ T(x) &: x \text{ can be trusted} \\ b &: \text{Bill} \end{aligned}$$

then the premises of this argument are:

$$\forall x[G(x) \rightarrow \neg T(x)] \quad \text{and} \quad G(b)$$

and the conclusion is:

$$\neg T(b).$$

Remember that to establish the validity of an argument, we must show that, whenever all the premises are true, then the conclusion must be true. As before, we shall do this in steps. Assuming the premises to be true will allow us to deduce other true propositions which in turn allow us to guarantee the truth of the conclusion. We shall once again use our Replacement Laws and Rules of Inference (see §1.5 and §1.7). However, with arguments expressed in predicate logic, we shall need to use the truth of quantified propositions to draw conclusions about the truth of propositions relating to specific members of the universe of discourse, and vice versa. We therefore need the following four rules which will allow us to do this.

1. Universal Specification (*US*)

This rule states that if the proposition $\forall x F(x)$ is true, then we can deduce that the proposition $F(a)$ is true for every a in the universe of discourse.

2. Universal Generalization (UG)

If the proposition $F(a)$ is true for every a in the universe of discourse, then we can conclude that $\forall x F(x)$ is true.

3. Existential Specification (ES)

If $\exists x F(x)$ is true, then there is an element a in the universe of discourse such that $F(a)$ is true.

We must be very careful in interpreting this rule. The element a is not arbitrary. It is one of the elements in the universe which has the property F. That at least one such element exists is guaranteed by the truth of $\exists x F(x)$.

4. Existential Generalization (EG)

If $F(a)$ is true for some element a belonging to the universe of discourse then $\exists x F(x)$ is true.

Examples 1.16

1. Show that the following is a valid argument: 'Everyone who has green eyes is not to be trusted. Bill has green eyes. Therefore Bill is not to be trusted.'

Solution

With a universe of discourse of 'people', we have established that, if b denotes 'Bill', the premises are:

$$\forall x[G(x) \to \neg T(x)] \quad \text{and} \quad G(b)$$

and the conclusion is:

$$\neg T(b).$$

Assuming the truth of the premises, we must establish the truth of the conclusion. We do this as follows:

1.	$\forall x[G(x) \to \neg T(x)]$	(premise)
2.	$G(b)$	(premise)
3.	$G(b) \to \neg T(b)$	(1. US)
4.	$\neg T(b)$	(2, 3. MP)

The truth of each of the propositions 1–4 is guaranteed for the reason given. We have shown that the truth of the premises guarantees the truth of the conclusion and hence that the argument is valid.

2. Show that the following is a valid argument: 'All students go to parties. Some students drink too much. Therefore some people who drink too much go to parties.'

Solution

Once again, we take our universe of discourse as 'people'.

Define: $S(x)$: x is a student
$D(x)$: x drinks too much
$P(x)$: x goes to parties.

The premises are:

$$\forall x[S(x) \rightarrow P(x)] \quad \text{and} \quad \exists x[S(x) \wedge D(x)]$$

and the conclusion is:

$$\exists x[D(x) \wedge P(x)].$$

We proceed as follows:

1.	$\exists x[S(x) \wedge D(x)]$	(premise)
2.	$\forall x[S(x) \rightarrow P(x)]$	(premise)
3.	$S(a) \wedge D(a)$	(1. ES)
4.	$S(a) \rightarrow P(a)$	(2. US)
5.	$S(a)$	(3. Simp)
6.	$P(a)$	(4, 5. MP)
7.	$D(a) \wedge S(a)$	(3. Comm)
8.	$D(a)$	(7. Simp)
9.	$D(a) \wedge P(a)$	(6, 8. Conj)
10.	$\exists x[D(x) \wedge P(x)]$	(9. EG)

Note that the a in line 2 is not arbitrary but is an element in the universe which has the properties defined by S and D. The a in line 3 is the same individual for whom we can state $S(a) \rightarrow P(a)$ because we have $\forall x[S(x) \rightarrow P(x)]$, where $S(x) \rightarrow P(x)$ holds for all x in the universe of discourse and hence for a.

3. Show that the following is a valid argument: 'Everyone shouts or cries. Not everyone cries. So some people shout and don't cry.'

Solution

With our universe of discourse as 'people', we define the following:

$$S(x) : x \text{ shouts}$$
$$C(x) : x \text{ cries.}$$

The premises of the argument are: $\forall x[S(x) \vee C(x)]$ and $\neg \forall x C(x)$.

Note that we cannot apply either rule of specification to the second premise in its current negated form. We therefore use the Rule of Quantification Denial to write it in the logically equivalent form: $\exists x \neg C(x)$.

We can now validate the argument as follows.

1.	$\forall x[S(x) \vee C(x)]$	(premise)
2.	$\neg \forall x C(x)$	(premise)
3.	$\exists x \neg C(x)$	(2. QD)
4.	$\neg C(a)$	(3. ES)
5.	$S(a) \vee C(a)$	(1. US)
6.	$C(a) \vee S(a)$	(5. Comm)
7.	$S(a)$	(4, 5. DS)
8.	$S(a) \wedge \neg C(a)$	(7, 4. Conj)
9.	$\exists x[S(x) \wedge \neg C(x)]$	(8. EG)

Note that it was necessary to use the rule of existential specification on the second premise before using universal specification on the first. This is because if we first state $S(a) \vee C(a)$, a is an arbitrary member of the universe. But the property $\neg C(x)$ applies only to certain individuals in the universe, so we cannot assume that it applies to an arbitrary individual a. In other words, an a for which $\neg C(a)$ is true (and the premise asserts that there is at least one such a) must also be one for which $S(a) \vee C(a)$ is true, since the latter property is true for any individual in the universe.

Exercises 1.8

Establish the validity of the following arguments.

1. Some monkeys eat bananas. All monkeys are primates. Therefore some primates eat bananas.

2. All cars are dangerous weapons. No dangerous weapons should be given to children. Therefore cars should not be given to children.

3. No reasonable man approves of wars. Jack approves of wars. Therefore Jack is not a reasonable man.

4. All gamblers are bound for ruin. No one bound for ruin is happy. Therefore no gamblers are happy.

5. All computer scientists are clever or wealthy. No computer scientist is wealthy. Therefore all computer scientists are clever or witty.

6. All those who eat apples have strong teeth. All those who don't eat apples are unhealthy. Betty hasn't strong teeth. Therefore Betty is unhealthy.

7. Some alligators are friendly and sociable. All alligators which are friendly live in a zoo. Therefore some alligators which live in a zoo are sociable.

8. All problems are difficult and frustrating. Some problems are challenging. Hence some problems are frustrating and challenging.

9. All animals with scales are dragons. Some animals which are not dragons have sharp claws. So there are animals without scales which have sharp claws.

10. Everyone who is forty is fat or foolish. No one is foolish and no one is fat. So no one is forty.

Chapter 2

Mathematical Proof

2.1 The Nature of Proof

The discipline of mathematics is characterized by the concept of proof. In this chapter we consider the nature of mathematical proof, some of the different techniques of proof and how a proof should be constructed and written down.

What do mathematicians mean by 'proof'? The popular view of a mathematical proof is probably that of a sequence of steps, almost certainly written mainly in symbols, where each step follows logically from an earlier part of the proof and where the last line is the statement being proved. Associated with this image is probably the notion that a proof is the absolute and rigorous test of mathematical truth. Surprisingly perhaps, this is not quite the view of many mathematicians, although there is by no means unanimity of opinion amongst the mathematical professionals themselves. Many hold a more sociological view of the role of a proof. They see it as essentially an explanation and communication of ideas; a line of argument sufficient to convince a fellow mathematician of the validity of the particular result. As the great English mathematician Godfrey Hardy wrote: 'Strictly speaking there is no such thing as mathematical proof; ... [they are] rhetorical flourishes designed to affect psychology, ...devices to stimulate the imagination of students.'

Which, then, is the 'correct' view of the nature and significance of the proof of a mathematical theorem? Probably the best answer is: both! The word 'proof' is used to cover a wide spectrum of styles. At one extreme we have very formal proofs which are rather like the logical arguments considered in chapter 1. Each step follows from the premises or a previous step by the laws of logic. Indeed, it is possible to write out such a proof using only symbols and no words but, needless to say, this is likely to be very difficult to follow. Away from the formal end of the spectrum are proofs which are more 'reader-friendly'. A less formal proof may use a mixture of words, symbols and diagrams of one kind or another. Most proofs found in mathematical textbooks (and research papers, for that matter) are not formal. They aim to communicate the essential reasons why a particular result holds rather than dwelling on rigorous step-by-step detail.

Any proof exposes certain lines of reasoning to scrutiny by others. As such the mathematical community sets certain standards concerning what should be regarded as an acceptable proof and what should not. Vague descriptions are not allowed. Arguments which are clear and coherent, although somewhat informal, are acceptable even if they gloss over some minor details. It goes without saying that any proof must be 'correct' in that it must not contain any logical errors.

Something which is *not* sanctioned in mathematics is the drawing of conclusions based on large numbers of observations. However many times we square an even number and discover that the result is even, this does not constitute a *proof* that the square of an even number is even. It may, however, strengthen our belief that this is so and encourage us to search for a valid proof. Making judgements about facts on the basis of observation is known as **inductive reasoning**. The type of reasoning where a conclusion is drawn by logical inference is called **deductive reasoning**. For mathematicians, the latter is the only form of reasoning which is acceptable in a proof.

2.2 Axioms and Axiom Systems

To understand more fully what is meant by a proof, formal or informal, we need to look briefly at the nature of modern mathematics. Most mathematicians would agree that their subject has as its mode of operation what is known as the 'axiomatic method'. The use of the axiomatic method was introduced by Euclid in about 300 BC (although the modern view of the nature of axioms differs in important ways from Euclid's).

A mathematical theory, such as set theory, number theory, group theory or whatever, consists of various components of which the most important are the following:

1. Undefined terms.
2. Axioms.
3. Definitions.
4. Theorems.
5. Proofs.

Of these, you probably have a reasonable idea of what we mean by 3, 4 and 5. That we need to have *undefined* terms in mathematics may come as a surprise, but a little reflection should indicate why these are necessary.

Suppose we wish to write the definitive work on, say, set theory. Where do we begin? The obvious starting point is to say precisely what a set is, so we begin: *Definition 1: A set is ...*—what? The problem is that, if we attempt to define 'set', we need to do so in terms of something else (a collection, perhaps?), but now the 'something else' is undefined. If we try to define the something else, we have to do so in terms of something else again, but then the 'something else again' is undefined, and so on. Clearly, we want to avoid an infinite string of definitions (otherwise we could never begin the theory proper) or circularity in our definitions ('a set is a collection; a collection is a set'). This forces us to have some terms which are left undefined. Of course, we can still explain in an intuitive way what we have in mind when using the undefined terms, but this intuitive explanation is not strictly part of the theory itself.

Item 2 in our list above—axioms—also needs some clarification. Just as we cannot define every term which is to be used in a mathematical theory, so we cannot prove every statement about the theory for much the same reason. In order to have somewhere to begin, we need to make some statements which will not be proved. These statements are called **axioms**. They represent, in a sense, the basic properties of the theory, its 'building blocks'.

Note that the truth or falsity of the axioms is not considered; they are merely statements about the undefined terms which serve to 'get the theory going'†. However, they must be consistent amongst themselves in the sense that it must be possible for them all to be true simultaneously. Axioms which contradict each other are not acceptable. When it comes to *applying* a mathematical theory, the undefined terms are given interpretations and the axioms then become propositions which are either true or false. Of course, a mathematical theory can only sensibly be applied if the interpretations of the axioms to the situation under consideration are true propositions.

An axiomatic theory develops by making definitions and proving theorems. Definitions are introduced for the convenience of not having to refer everything back to undefined terms. A theorem is a statement about various terms of the system which follows from the axioms using the kind of logical reasoning introduced in chapter 1. The first theorems are proved directly from the axioms;

† As assumptions about undefined terms, the axioms have neither meaning nor truth. Because mathematics is built from these foundations, it too has no meaning! It was this consequence of the axiomatic method that Bertrand Russell had in mind when he wrote: 'Mathematics may be defined as the subject in which we never know what we are talking about nor whether what we are saying is true.'

more theorems are then proved using these and so on. The theory spreads out further and further 'away from' the original axioms, but ultimately rests solely on them. Theorems and their proofs form the heart of (pure) mathematics.

The axiomatic description of mathematics does seem to imply that the subject is somewhat mechanical. For instance, it should be possible to program a computer with a system of axioms and the rules of logic and then set it off proving theorems. Why, then, has this not been successfully achieved? The missing ingredient (in both the computer and axiomatic description of mathematics) is human intuition. Usually a theorem originates in a **conjecture**—a belief that a certain result holds. Such a belief may arise from the observation of many situations where this was so and none where it was not. On the other hand, many important mathematical conjectures were just 'hunches'—intuitive beliefs that such-and-such must be the case. However it arose, for a conjecture to be promoted to a theorem, a proof must be supplied in which a justification for the conjecture is given. Here again intuition plays an important part in indicating which line of reasoning might lead to a proof. So although the axiomatic method gives a coherent explanation of what mathematics is on a formal level, it does not describe or explain at all the process of *doing* mathematics. Perhaps only psychologists can hope to do that!

To illustrate these general ideas, consider the following example of an axiom system. The example is not one which would be of very much interest for two reasons. Firstly, the axioms are not sufficiently 'rich' for us to be able to prove anything very interesting about the system, and secondly, the system does not have many worthwhile applications. In other words, the example is neither particularly interesting in its own right, nor in terms of its applicability. However, we hope that it will serve to clarify the preceding remarks.

Example 2.1

Undefined terms: 'blub', 'glug' and 'to lie on'.

Axioms: A1. Every blub lies on at least one glug.
A2. For every glug, there are exactly two blubs which lie on it.
A3. There are exactly five blubs.

Figure 2.1 gives an interpretation of the axiom system with blubs represented as points and glugs as lines, with the obvious interpretation of 'to lie on'. Note that, in this interpretation, each of the axioms is a true proposition. A specific interpretation of the undefined terms such that the axioms are true propositions is called a **model** of the axiom system.

In this interpretation there are five glugs, $G_1, G_2, \ldots, G_5$. Is this always the case?

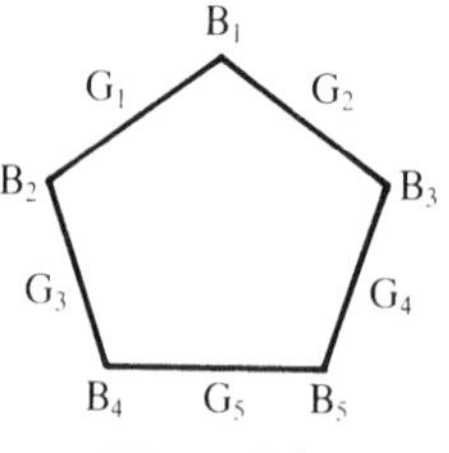

Figure 2.1

In other words, given any interpretation of the axioms, are there always five glugs? If we prove, from the axioms, that there are exactly five glugs then the answer must be 'yes'. However, the answer is in fact 'no'—figure 2.2 gives an alternative model which has 10 glugs.

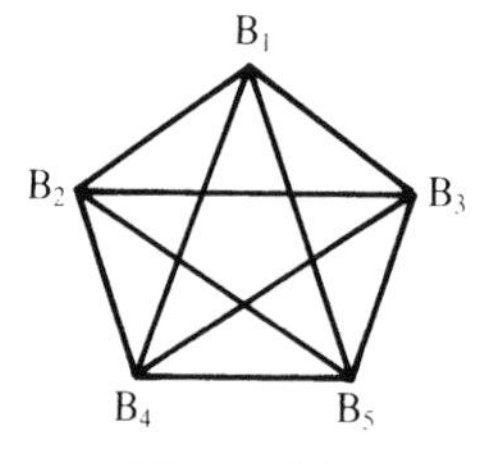

Figure 2.2

We can, however, prove from the axioms that there are at least three glugs. This means that any model of the system must have at least three glugs.

> **Theorem**
>
> There exist at least three glugs.

Proof

Let B_1 be a blub. (Axiom A3 guarantees the existence of a blub.) By axiom A1, B_1 lies on some glug, G_1 say, and, by A2, there is another blub, B_2 say, which also lies on G_1.

There is another blub B_3 which is different from B_1 and B_2 (A3) and B_3 lies on some glug G_2 (A1). G_2 must be different from G_1 because G_1 cannot have three blubs lying on it (A2). Axiom A2 tells us that there is another blub lying on

G_2. There are two possibilities: the other blub of G_2 is either B_1 or B_2, or it is different from B_1 and B_2.

In the first case, there are still two blubs not lying on a glug. In the second case, there is another blub B_4 lying on G_2 which still leaves one blub 'glugless'. In either case there is at least one blub B_5 which does not lie on either G_1 or G_2. Axiom A1 tells us that there must be a third glug on which B_5 lies. Furthermore this glug must be different from G_1 and G_2 by axiom A2.

Therefore there are at least three glugs. □

Having proved our first theorem about blubs and glugs, we could go on and use it to prove further theorems—see exercise 2.1.1. Since the 'blub–glug axiom system' is too restrictive to be of much interest, we shall not dwell on it further here. However, blubs and glugs will reappear in chapter 10 'disguised' as the vertices and edges of graphs. (See exercise 10.1.18 for another model of this axiom system.)

Exercises 2.1

These questions refer to the 'blub–glug axiom system' described in example 2.1.

1. Prove that there exists a blub which lies on (at least) two different glugs. (Hint: the theorem of example 2.1 may be of use here.)

2. Give a model of the axiom system which has more than 10 glugs. Introduce a new axiom to the system which rules out your model. Can you prove from the axioms of the new system that there are at most 10 glugs? (Avoid, if you can, a new axiom which simply says that there are at most 10 glugs.)

2.3 Methods of Proof

As we have seen, formal mathematics is based on the axiomatic method. Beginning with undefined terms and axioms, it develops by proving theorems using the rules of logic. In this section we consider the essential features of a proof and we outline some methods of proof.

To set the scene more precisely, suppose that we are given a system of axioms, $A_1, A_2, \ldots, A_n$. A **theorem** is a statement about the terms of the system which is logically implied by the conjunction of the axioms. We can therefore define a theorem in the system formally as a proposition T such that

$$(A_1 \wedge A_2 \wedge \cdots \wedge A_n) \vdash T.$$

Recall that $P \vdash Q$ if Q is true whenever P is true. In any model of the axiom system, the axioms have interpretations which are true propositions so that every theorem has an interpretation which is a true proposition. Thus theorems are propositions which are true in *every* model of the axiom system.

What then constitutes a proof of a theorem? Informally, a proof is a valid argument in which the theorem is the conclusion. The premises may be axioms or other theorems which have already been proved. Although it must be possible to prove any theorem with only the axioms as premises, this is clearly uneconomical. Once a theorem has been proved, it can be used in conjunction with the axioms to prove other theorems. Hence to prove theorem T we must show that

$$(A_1 \wedge A_2 \wedge \cdots \wedge A_n \wedge T_1 \wedge T_2 \wedge \cdots \wedge T_m) \vdash T$$

where the A_i ($i = 1, 2, \ldots, n$) are axioms and the T_j ($j = 1, 2, \ldots, m$) are theorems which have already been proved. We do this by assuming the truth of the axioms (and hence of the theorems) and showing that this guarantees the truth of T.

Many theorems are, strictly speaking, quantified propositional functions of the form $\forall x T(x)$, where x is a member of a specified universe of discourse. To prove such a theorem, we in fact prove the proposition which is the universal specification of $\forall x T(x)$, i.e. $T(a)$ for every a in the universe of discourse. Having shown that the truth of $T(a)$ follows from the axioms and theorems for any arbitrary a in the universe of discourse, we can then apply universal generalization and conclude that $\forall x T(x)$ is also true (see §1.9).

Before outlining some techniques of proof, there is a piece of notation which needs clarifying. The symbol $\Rightarrow$, read as 'implies that', is used between two propositions where the second 'follows logically' from the first. (If $P \Rightarrow Q$ and also $Q \Rightarrow P$, we write $P \Leftrightarrow Q$.) What we mean by $P \Rightarrow Q$ is that Q is logically implied by the conjunction of P and other statements about the terms of the system, such as axioms and theorems. Hence $P \Rightarrow Q$ is just shorthand for $(A_1 \wedge A_2 \wedge \cdots \wedge A_n \wedge T_1 \wedge T_2 \wedge \cdots \wedge T_m \wedge P) \vdash Q$, where the A_i and T_j are axioms and proved theorems respectively. In a proof, these axioms and theorems may not be referred to explicitly when it can be assumed that those to whom the proof is addressed have some background knowledge of the system. For example, we can write: if x is an arbitrary real number, $x^2 - 2 < 2 \Rightarrow -2 < x < 2$. Note that the truth of the second proposition is not a direct logical consequence of the truth of $x^2 - 2 < 2$ alone. It is also dependent on certain axioms and theorems of the real numbers, such as: for all real numbers a, b, c, if $a < b$, then $a+c < b+c$. Where the real numbers are concerned, many properties reflected in axioms and theorems are so familiar that we apply them without thought. In a proof, the statement $P \Rightarrow Q$ conveys to the reader that the truth of Q follows from the truth of P and the conjunction of other true propositions with which it is assumed he or she is familiar.

Deciding what to justify explicitly and what to assume as background knowledge is part of the art of proof writing. If too much detail is included, the reader will experience a 'can't see the wood for the trees' feeling and the overall structure will be hard to discern. Similarly, too heavy an emphasis on symbols may cause difficulty in understanding the proof. Instead of concentrating on minute levels of detail, it is more useful to explain the important steps, employing a judicious blend of natural language and symbols. Of course, sufficient detail needs to be given to enable the reader to follow the argument and to verify its validity. Exactly how much detail needs to be supplied will depend on such factors as the mathematical sophistication of the intended audience and how novel the approach is.

We now give some examples of methods by which a mathematical statement may be proved. The list is by no means exhaustive but it does give some of the more common techniques. We shall come across plenty of other examples of proofs in later chapters.

Direct Proof of a Conditional Proposition

Many mathematical conjectures can be expressed in the form $P \rightarrow Q$, i.e. as a conditional proposition. Their proof therefore consists of showing that

$$(A_1 \wedge A_2 \wedge \cdots \wedge A_n \wedge T_1 \wedge T_2 \wedge \cdots \wedge T_m) \vdash (P \rightarrow Q)$$

where the A_i and T_j are axioms and theorems as before. This is equivalent to showing that

$$(A_1 \wedge A_2 \wedge \cdots \wedge A_n \wedge T_1 \wedge T_2 \wedge \cdots \wedge T_m) \rightarrow (P \rightarrow Q)$$

is a tautology and, by the logical equivalence of $R \rightarrow (P \rightarrow Q)$ and $(R \wedge P) \rightarrow Q$, that

$$(A_1 \wedge A_2 \wedge \cdots \wedge A_n \wedge T_1 \wedge T_2 \wedge \cdots \wedge T_m \wedge P) \rightarrow Q$$

is a tautology or that

$$(A_1 \wedge A_2 \wedge \cdots \wedge A_n \wedge T_1 \wedge T_2 \wedge \cdots \wedge T_m \wedge P) \vdash Q$$

i.e. that $P \Rightarrow Q$. So, for a direct proof of a theorem of the form $P \rightarrow Q$, we assume the truth of the axioms and hence of any proved theorems. We also assume the truth of P and show that the truth of Q necessarily follows.

Examples 2.2

1. Prove that, for every integer n, if n is even, then n^2 is even. (The integers are the 'whole' numbers.)

Proof

Let n be an even integer. Then 2 is a factor of n, so n can be expressed as $n = 2m$ for some integer m. It follows that $n^2 = (2m)^2 = 4m^2$. Now $4m^2$ can be written as $2(2m^2)$ where $2m^2$ is also an integer. Therefore n^2 is even. This concludes the proof. □

Note that we have omitted reasons for certain steps. For instance no specific reason was given for the fact that $(2m)^2 = 4m^2$. This is because it is assumed that this step is obvious. However, in a more formal proof, missing details would have to be supplied.

The proof can be written using more mathematical notation. This gives the following more concise, but still acceptable, version.

Proof

Let n be an integer.

Then

$$\begin{aligned} & n = 2m \quad \text{for some integer } m \\ \Rightarrow \quad & n^2 = (2m)^2 \\ & \quad = 4m^2 \\ & \quad = 2(2m)^2 \\ \Rightarrow \quad & n^2 \text{ is an even integer.} \end{aligned}$$

□

Strictly speaking, what we are asked to prove here is the proposition: $\forall x[P(x) \rightarrow Q(x)]$, where $P(x)$ is 'x is even', $Q(x)$ is 'x^2 is even' and the universe of discourse is the integers. What we have in fact proved is the proposition which is the universal specification of this quantified proposition: $P(n) \rightarrow Q(n)$ for any n in the universe of discourse. The assumption $P(n)$ is true is that n is an arbitrary even integer, i.e. a 'typical' even integer. The proof shows that the truth of $Q(n)$ follows from this assumption and therefore that $P(n) \rightarrow Q(n)$ for any n in the universe of discourse. Universal generalization allows us to conclude that $\forall x[P(x) \rightarrow Q(x)]$.

2. Prove that, if n and m are integers and 3 is a factor of both n and m, then 3 is a factor of any number of the form $nx + my$ where x and y are integers.

Proof

We are required to prove the conditional proposition

$$[R(n) \wedge R(m)] \rightarrow Q(n, m)$$

where $R(n)$ is '3 is a factor of n' and $R(m)$ is '3 is a factor of m' and m and n are arbitrary integers. The proposition $Q(n, m)$ is given by '3 is a factor of any number of the form $nx+my$, where x and y are integers'. (We are using universal specification here as in the last example.)

We make the assumption that $R(n) \wedge R(m)$ is a true proposition, i.e. that n and m are arbitrary integers such that 3 is a factor of each.

From the truth of '3 is a factor of n' we can deduce that $n = 3p$ for some integer p. Similarly, from '3 is a factor of m' we can write $m = 3q$ for some integer q. Hence $nx + my = 3px + 3qy = 3(px + qy)$. Since $px + qy$ is an integer, we conclude that $nx + my$ can be written as three times an integer and hence is divisible by three. □

This argument may be summarized more symbolically as follows.

Proof

Let m and n be integers both divisible by 3.

Then

$$3 \text{ is a factor of } n \Rightarrow n = 3p, \text{ where } p \text{ is an integer}$$

and

$$3 \text{ is a factor of } m \Rightarrow m = 3q, \text{ where } q \text{ is an integer.}$$

Hence

$$\begin{aligned} nx + my &= 3px + 3qy \\ &= 3(px + qy) \end{aligned}$$

$\Rightarrow$ $nx + my$ is divisible by three. □

3. What is wrong with the following 'proof' that $1 = 2$?

'Proof'

We shall 'prove' the conditional proposition: 'For x a real number, if $x = 2$, then $x = 1$'.

$$
\begin{aligned}
& & x &= 2 \\
\Rightarrow & & x - 1 &= 1 \\
\Rightarrow & & (x-1)^2 &= 1 \\
& & &= x - 1 \\
\Rightarrow & & x^2 - 2x + 1 &= x - 1 \\
\Rightarrow & & x^2 - 2x &= x - 2 \\
\Rightarrow & & x(x-2) &= x - 2 \\
\Rightarrow & & \frac{x(x-2)}{x-2} &= \frac{x-2}{x-2} \\
\Rightarrow & & x &= 1.
\end{aligned}
$$

This example shows that great care needs to be taken when constructing proofs. Each step seems to follow logically from the previous ones, yet there is clearly a flaw in the argument somewhere because it is claiming to prove an absurdity.

The error, in fact, comes right at the end of the proof when both sides of the equation are divided by $x - 2$. This division is not allowed because $x = 2$ and division by zero is not a valid operation. The correct conclusion to draw from the equation $x(x-2) = x - 2$ is: either $x = 2$ or $x = 1$.

Proof of a Conditional Proposition using the Contrapositive

Recall that the contrapositive $\bar{Q} \rightarrow \bar{P}$ is logically equivalent to the conditional proposition $P \rightarrow Q$. Hence, if we can establish the truth of the contrapositive, we can deduce that the conditional is also true. This constitutes an indirect proof of $P \rightarrow Q$ although we may use a direct proof of $\bar{Q} \rightarrow \bar{P}$ since this is itself a conditional proposition. We assume the truth of $\bar{Q}$ (together with the relevant axioms and theorems) and we establish the truth of $\bar{P}$.

Examples 2.3

1. By proving the contrapositive, prove that, for every integer n, if n^2 is even, then n is even.

Proof

The proposition to be proved is $P(n) \rightarrow Q(n)$, where $P(n)$ is 'n^2 is even', $Q(n)$ is 'n is even' and n is an arbitrary integer. The contrapositive is $\neg Q(n) \rightarrow \neg P(n)$: if n is odd then n^2 is odd. We prove this directly by assuming the truth of 'n is odd' and showing that the truth of 'n^2 is odd' follows.

Let n be an odd integer.

Then

$$
\begin{aligned}
& n = 2m + 1 \quad \text{where } m \text{ is an integer} \\
\Rightarrow \quad & n^2 = (2m+1)^2 \\
& \quad = 4m^2 + 4m + 1 \\
& \quad = 2(2m^2 + 2m) + 1 \quad \text{where } 2m^2 + 2m \text{ is an integer} \\
\Rightarrow \quad & n^2 \text{ is odd.}
\end{aligned}
$$

□

2. Prove that, if m and n are positive integers such that $mn = 100$, then either $m \leqslant 10$ or $n \leqslant 10$.

Proof

We shall again prove the contrapositive but we must be a little careful. The proof required is that of $P(m,n) \rightarrow Q(m,n)$ where $P(m,n)$ is 'm and n are arbitrary positive integers such that $mn = 100$' and $Q(m,n)$ is the inclusive disjunction of the two propositions '$m \leqslant 10$' and '$n \leqslant 10$'. By De Morgan's laws $(\overline{p \vee q}) = \bar{p} \wedge \bar{q}$ so that the negation of $Q(m,n)$ is '$m > 10$ and $n > 10$'. The contrapositive $\neg Q(m,n) \rightarrow \neg P(m,n)$ is therefore 'If m and n are arbitrary integers such that $m > 10$ and $n > 10$, then $mn \neq 100$'.

Let m and n be positive integers.

Then

$$
\begin{aligned}
& m > 10 \quad \text{and} \quad n > 10 \\
\Rightarrow \quad & mn > 100
\end{aligned}
$$

$$\Rightarrow \qquad mn \neq 100$$

and the theorem is proved. □

3. The following is given as an example of a common *incorrect* attempt at a proof. The result to be proved is: If x and y are real numbers, $x^2 + y^2 \geqslant 2xy$.

   Suppose that x and y are arbitrary real numbers such that

   $$x^2 + y^2 \geqslant 2xy.$$

   Then

   $$\begin{aligned} & x^2 - 2xy + y^2 \geqslant 0 \\ \Rightarrow \qquad & (x - y)^2 \geqslant 0. \end{aligned}$$

   Since this is clearly always true, we can conclude that $x^2 + y^2 \geqslant 2xy$.

   What has been proved here is $P \to t$, where P is '$x^2 + y^2 \geqslant 2xy$ for arbitrary real numbers x and y' and t is a tautology. But $P \to t$ is not logically equivalent to P and so this does not constitute a valid proof of P.

Proof by Contradiction

Using a truth table we can readily establish the logical equivalence of P and $\bar{P} \to f$, where f is a contradiction (a proposition which is always false). Hence to prove a theorem T we can instead prove the conditional proposition $\bar{T} \to f$. This can be achieved using a direct proof which assumes the truth of axioms and theorems as usual and also assumes the truth of $\bar{T}$ (i.e. the falsity of T). We then show that this implies a proposition which is patently false, i.e. a contradiction. Usually, the contradiction takes the form of the conjunction of a proposition and its negation, $Q \wedge \bar{Q}$. (Recall that $Q \wedge \bar{Q} \equiv f$.) We can then deduce that $\bar{T} \to f$ is true and hence that the theorem T is true.

This method of indirect proof is frequently referred to as 'proof by contradiction' or as *reductio ad absurdum*.

Examples 2.4

1. Prove that $\sqrt{2}$ is not rational. (A rational number is one which can be written in the form p/q where $q \neq 0$ and p and q are integers.)

Proof

The proof of this theorem is a well known example of proof by contradiction. We assume that $\sqrt{2}$ *is* rational and show that this leads to a contradiction.

Suppose that $\sqrt{2}$ is rational, i.e. $\sqrt{2} = m/n$ where m and n are integers and $n \neq 0$. We may assume that the fraction m/n is in its 'lowest terms', i.e. that m and n have no common factors. If they do have common factors we simply cancel them.

Now

$$
\begin{aligned}
& \sqrt{2} = m/n \\
\Rightarrow\quad & 2 = m^2/n^2 \\
\Rightarrow\quad & 2n^2 = m^2 \\
\Rightarrow\quad & m^2 \text{ is even} \\
\Rightarrow\quad & m \text{ is even} \quad \text{(see example 2.3.1)} \\
\Rightarrow\quad & m = 2p \quad \text{for some integer } p \\
\Rightarrow\quad & m^2 = 4p^2.
\end{aligned}
$$

Substituting this result into the equation $2n^2 = m^2$ gives

$$
\begin{aligned}
& 2n^2 = 4p^2 \\
\Rightarrow\quad & n^2 = 2p^2 \\
\Rightarrow\quad & n^2 \text{ is even} \\
\Rightarrow\quad & n \text{ is even.}
\end{aligned}
$$

We have now shown that both m and n are even, i.e. that they have a common factor 2. But m and n have no common factors because any such factors were cancelled at the beginning. Hence we have deduced the conjunction of a proposition and its negation, i.e. a contradiction, and this proves the theorem. □

2. Prove that there are infinitely many prime numbers. (A **prime number** is a positive integer greater than 1 which has no factors other than 1 and itself. We explore the properties of prime numbers in Chapter 9.2.)

   The following is Euclid's proof of the theorem. It is generally regarded as a classic example of proof by contradiction.

Proof

Suppose that there are only a finite number, n say, of prime numbers. This means that we can list all the prime numbers as follows: $P_1, P_2, \ldots, P_n$.

Consider the product of this complete list of prime numbers: $Q = P_1P_2 \ldots P_n$.

Now $Q + 1 = P_1P_2 \ldots P_n + 1$.

The integer $Q+1$ is not prime since it is different from $P_1, P_2, \ldots, P_n$. Therefore $Q + 1$ must be divisible by some prime number, say P, where P is one of the P_i, $i = 1, 2, \ldots, n$. But Q is divisible by P and so clearly P cannot be a factor of $Q + 1$ and here is our contradiction. We conclude that our assumption that the number of prime numbers is finite is false and deduce that there are infinitely many primes. □

Proof of a Biconditional Proposition

To prove a biconditional proposition $P \leftrightarrow Q$, we usually appeal to the logical equivalence of $P \leftrightarrow Q$ and $[(P \rightarrow Q) \wedge (Q \rightarrow P)]$. Commonly, therefore, the proof of a biconditional involves two distinct parts, one proving the result $P \rightarrow Q$ and the other proving $Q \rightarrow P$. It is fairly commonly the case that one of the conditionals will be relatively more straightforward to prove than the other.

Examples 2.5

1. Prove that, for any integers x and y, the product xy is even if and only if x is even or y is even.

Proof

We first prove that, if x is even or y is even then xy is even, using a direct proof.

Suppose x is even, i.e. $x = 2n$, for some integer n. Then $xy = 2ny$ so xy is even.

If y is even, an identical argument shows that xy is even. The word 'similarly' is used to indicate this and to save us having to repeat the argument. We write: similarly, if y is even, then xy is even.

We now prove the converse: if xy is even then x is even or y is even. We shall use a direct proof of the contrapositive: if x and y are odd, then xy is odd.

Now x is odd and y is odd

$$\Rightarrow \qquad x = 2n+1,\ y = 2m+1 \quad \text{for some integers } m \text{ and } n.$$

Then

$$\begin{aligned} xy &= (2n+1)(2m+1) \\ &= 4mn + 2n + 2m + 1 \\ &= 2(2mn + n + m) + 1 \end{aligned}$$

$\Rightarrow$ xy is odd.

This completes the proof. □

2. Prove that $3x^2 - 7x + 4 = x^2 + 3x - 8$ if and only if $x = 2$ or $x = 3$.

Proof

$$\begin{aligned} & 3x^2 - 7x + 4 = x^2 + 3x - 8 \\ \Rightarrow \quad & 2x^2 - 10x + 12 = 0 \\ \Rightarrow \quad & x^2 - 5x + 6 = 0 \\ \Rightarrow \quad & (x-2)(x-3) = 0 \\ \Rightarrow \quad & x = 2 \quad \text{or} \quad x = 3. \end{aligned}$$

To prove the converse we can write:

$$\begin{aligned} & x = 2 \quad \text{or} \quad x = 3 \\ \Rightarrow \quad & (x-2)(x-3) = 0 \\ \Rightarrow \quad & x^2 - 5x + 6 = 0 \\ \Rightarrow \quad & 2x^2 - 10x + 12 = 0 \\ \Rightarrow \quad & 3x^2 - 7x + 4 = x^2 + 3x - 8. \end{aligned}$$

Note that the steps in the second part of the proof are exactly the same as those in the first in reverse. We can therefore summarize both parts of the proof as follows:

$$\begin{aligned} & 3x^2 - 7x + 4 = x^2 + 3x - 8 \\ \Leftrightarrow \quad & 2x^2 - 10x + 12 = 0 \end{aligned}$$

$$\begin{aligned} \Leftrightarrow \quad & x^2 - 5x + 6 = 0 \\ \Leftrightarrow \quad & (x-2)(x-3) = 0 \\ \Leftrightarrow \quad & x = 2 \quad \text{or} \quad x = 3. \end{aligned}$$ □

The methods of proof which we have considered so far all have a similar structure. In each case we start by assuming the truth of one particular proposition. We then show that the truth of another proposition follows given certain background knowledge, i.e. axioms and theorems already proved. We summarize each of these methods of proof in the table below.

Method of proof	Assume	Deduce
Direct proof of $P \to Q$	P; background knowledge	Q
Proof of $P \to Q$ using the contrapositive	$\bar{Q}$; background knowledge	$\bar{P}$
Proof of P by contradiction	$\bar{P}$; background knowledge	A contradiction, f
Proof of the biconditional $P \leftrightarrow Q$	(a) P; background knowledge	Q
	and	
	(b) Q; background knowledge	P

Use of Counter-Examples

Many mathematical conjectures take the form 'all As are Bs' or 'all objects with property A have property B'. This could be rewritten as the universally quantified conditional propositional function $\forall x[A(x) \to B(x)]$, where $A(x)$ is 'x is an (or has the property) A' and $B(x)$ is 'x is a (or has the property) B'. The proof of the conjecture could then take one of the forms described above.

The proposition could also be written $\forall x B(x)$ where x is restricted to the universe of discourse of As (or objects having the property A). As we have already remarked, the inability to find an x which has not the property B does not constitute a proof of the theorem. However many x we find which have the property B, this is no guarantee that we have failed to find an elusive x which does not have this property. However, if the universe of discourse is finite, then (given time if it is large) we can examine every element to check that it has the property in question. If no element fails the test then the theorem is proved. This is called **proof by exhaustion** because it exhausts all the possibilities for x.

PROOF BY EXHAUSTION

On the other hand, to disprove a conjecture of the form $\forall x B(x)$, we need find only one member of the universe of discourse which does not have the property B. We can justify this logically. To disprove $\forall x B(x)$ we must prove the negation $\neg\forall x B(x)$. As we have seen (§1.8), this is equivalent to $\exists x \neg B(x)$, i.e. there is at least one member of the universe which does not have the property B. To prove this, all we need to do is to demonstrate that such an individual exists. This is the essence of what is sometimes called 'proof by counter-example' (although a more accurate title would be 'disproof by counter-example').

Example 2.6

Prove or disprove the proposition: for all positive integers n,

$$f(n) = n^2 - n + 17$$

is prime.

Solution

We begin by trying a few positive integer values: $f(1) = 17$, $f(2) = 19$, $f(3) = 23$, $f(4) = 29$, $f(5) = 37$.

In each of these $f(n)$ is prime, so we might be tempted to suspect that $f(n)$ is always prime and to wonder how this conjecture might be proved. A few more examples might show some pattern developing and give us some insight into a possible method of proof: $f(6) = 47$, $f(7) = 59$, $f(8) = 73$, $f(9) = 89$, $f(10) = 107$.

Since all of these are prime, our conjecture seems well founded and we may feel sufficiently confident to commence the attempt to find a valid proof. However, a little thought together with a degree of mathematical insight will save us wasting our time. It is not too difficult to see that $f(n)$ could not be prime for every positive integer n. An obvious counter-example is:

$$\begin{aligned} f(17) &= 17^2 - 17 + 17 \\ &= 17 \times 17. \end{aligned}$$

(For centuries mathematicians have attempted to find a formula which will generate only prime numbers. Pierre de Fermat (1601–65) thought that he had cracked the problem with the formula $2^{2^n} + 1$ where n is any integer. For $n = 0, 1, 2, 3, 4$ the formula generates the integers 3, 5, 17, 257 and 65 537 all of which are prime. However $n = 5$ gives 4 294 967 297 which has a factor 641. Fermat's conjecture was therefore disproved, although not until nearly 100 years after his death when Euler discovered this counter-example.)

Exercises 2.2

1. Prove that the sum of two consecutive integers is odd.

2. Prove that, if n is an integer, n^2 is odd if and only if n is odd.

3. Prove directly that the product of two consecutive integers is even. Use this result to prove that, if the quadratic equation $x^2 + ax + b = 0$ has solutions which are consecutive integers, then a is odd and b is even.

4. Prove that, if both solutions of $x^2 + ax + b = 0$ are even integers, then a and b are both even integers.

5. Prove that, if m and n and positive integers such that m is a factor of n and n is a factor of m, then $m = n$.

6. By proving the contrapositive, prove that, if n^2 is not divisible by 5, then n is not divisible by 5.

7. Use proof by contradiction to prove that $1 + \sqrt{2}$ is not rational.

8. Prove or disprove that, if a, b and c are integers such that a is a factor of $b + c$, then a is a factor of b or a is a factor of c.

9. Prove that, for any integer n, if $n - 2$ is divisible by four, then $n^2 - 4$ is divisible by 16.

10. Prove that the smallest factor greater than 1 of any integer is prime.

2.4 Mathematical Induction

Despite its title, the method of proof known as 'mathematical induction' is not an inductive proof! It could not be so because, as we have already pointed out, the only acceptable mathematical proofs employ deductive reasoning. Induction has a role in providing us with information as to what is likely to be true and hence what is a reasonable conjecture. The problem with any proof is that we need to know the result before we can commence proving it.

Many mathematical conjectures concern properties of the positive integers. Consider, for example, the following problem: find a formula for the sum of the first n odd integers. A useful starting point might be to write down the sums for some small values of n and see if this gives us any idea as to what might be a possible conjecture.

For $n = 1$, the sum is 1.

For $n = 2$, the sum is $1 + 3 = 4$.

For $n = 3$, the sum is $1 + 3 + 5 = 9$.

For $n = 4$, the sum is $1 + 3 + 5 + 7 = 16$.

At this stage we notice that, so far, for each value of n, the sum is n^2. We try a few more to see if our conjecture is well founded.

For $n = 5$, the sum is $16 + 9 = 25$.

For $n = 6$, the sum is $25 + 11 = 36$.

Inductive reasoning leads us to the conjecture that the sum of the first n odd positive integers is n^2. We must now find a proof, based on deduction, that this is true for all positive integers n.

Mathematical induction is appropriate for proving that a result holds for all positive integers. It consists of the following steps:

(a) Prove that the conjecture holds for $n = 1$.

(b) Prove that, for all $k \geqslant 1$, *if* the result holds for $n = k$, then it must also hold for $n = k + 1$. This is known as the **inductive step**.

To prove the conditional proposition in (b), we call upon the techniques outlined in the previous section. However, the inductive step is most commonly established using a direct proof. We assume that the result holds for $n = k$. (This assumption is sometimes known as the **inductive hypothesis**.) From this we deduce that it also holds for $n = k + 1$. Because it holds for $n = 1$, the inductive step allows us to deduce that it holds for $n = 2$, $n = 3$, etc. The 'principle of mathematical induction' allows us to conclude that the result therefore holds for all positive integers n. (This principle is usually taken as an axiom of the positive integers.)

Principle of Mathematical Induction

Let $S(n)$ be a proposition concerning a positive integer n. If

(a) $S(1)$ is true, and

(b) for every $k \geqslant 1$, the truth of $S(k)$ implies the truth of $S(k + 1)$,

then $S(n)$ is true for all positive integers n.

An analogy to the process of mathematical induction is an infinite line of fireworks connected together so that each is set off by the previous one in the line. Although it has been arranged that the kth firework will ignite the $(k+1)$st, nothing happens until we light the first firework in the line. This sets off the second, which sets off the third and so on to the end of the (infinite) line.

Let us now subject our conjecture, that the sum of the first n odd positive integers is n^2, to proof by mathematical induction.

Examples 2.7

1. Prove that the sum of the first n odd positive integers is n^2.

Proof

We want to prove that

$$\underbrace{1 + 3 + 5 + \cdots}_{n \text{ terms}} = n^2.$$

Note that the last term in the sequence is $2n - 1$ so that we may write our conjecture as

$$1 + 3 + 5 + \cdots + (2n - 1) = n^2.$$

We follow the steps:

(a) Prove that the conjecture is true for $n = 1$.

The sum of the first one odd integer is 1 and, for $n = 1$, $1 = n^2$. So the conjecture holds for $n = 1$.

(b) Assume that the conjecture is true for $n = k$ where $k \geqslant 1$ and show that this implies the truth of the conjecture for $n = k + 1$.

Suppose that $1+3+5+\cdots+(2k-1) = k^2$. Adding the next odd integer, $2k+1$, to each side of the equation, we have

$$\begin{aligned} 1 + 3 + 5 + \cdots + (2k - 1) + (2k + 1) &= k^2 + (2k + 1) \\ &= (k + 1)^2. \end{aligned}$$

The left-hand side of this equation is the sum of the first $k + 1$ odd numbers and we have shown, using the inductive hypothesis, that this sum is $(k + 1)^2$. Hence we have shown that, if the conjecture holds for $n = k$, then it also holds for $n = k + 1$. But we have shown that it holds for $n = 1$, and, by the principle of mathematical induction, it therefore holds for all positive integers n. □

2. Prove that, for every positive integer n, the expression $2^{n+2} + 3^{2n+1}$ is divisible by 7.

Proof

Let $f(n) = 2^{n+2} + 3^{2n+1}$.

For $n = 1$, we have $f(1) = 2^3 + 3^3 = 8 + 27 = 35$ which is divisible by 7. Hence the result holds for $n = 1$.

Assume that, for some integer $k \geqslant 1$,

$$f(k) = 2^{k+2} + 3^{2k+1} = 7a$$

where a is an integer. (This is the inductive hypothesis.)

Now

$$f(k + 1) = 2^{(k+1)+2} + 3^{2(k+1)+1}$$

$$
\begin{aligned}
&= 2^{k+3} + 3^{2k+3} \\
&= 2 \times 2^{k+2} + 3^2 \times 3^{2k+1} \\
&= 2 \times 2^{k+2} + 9 \times 3^{2k+1}.
\end{aligned}
$$

At this point we need to use the inductive hypothesis, $2^{k+2} + 3^{2k+1} = 7a$, to substitute for either 2^{k+2} or 3^{2k+1} (it doesn't matter which).

So, substituting $3^{2k+1} = 7a - 2^{k+2}$ gives

$$
\begin{aligned}
f(k+1) &= 2 \times 2^{k+2} + 9(7a - 2^{k+2}) \\
&= 9 \times 7a + 2 \times 2^{k+2} - 9 \times 2^{k+2} \\
&= 7(9a - 2^{k+2}) \\
&= 7b \quad \text{where } b = 9a - 2^{k+2}.
\end{aligned}
$$

Since b is an integer, we can conclude that $f(k+1)$ is divisible by 7. This completes the inductive step.

Applying the principle of mathematical induction we deduce that $2^{n+2} + 3^{2n+1}$ is divisible by 7 for all positive integers n. □

3. What is wrong with the following 'proof' by induction?

   Conjecture: All computers are the same price.

   'Proof': Let $S(n)$ denote the proposition 'any group of n computers are the same price'.

   Clearly $S(1)$ is true.

   Assume the truth of $S(k)$, i.e. that any group of k computers are the same price, and consider any collection of $k+1$ (distinct) computers denoted by $C_1, C_2, \ldots, C_k, C_{k+1}$. By the inductive hypothesis, all of $C_1, C_2, \ldots, C_k$ are the same price and also $C_2, \ldots, C_k, C_{k+1}$ are the same price. Therefore all of $C_1, C_2, \ldots, C_k, C_{k+1}$ are the same price.

   Since $C_1, C_2, \ldots, C_k, C_{k+1}$ was any collection of $k+1$ computers, we have established the inductive step. Hence all computers are the same price by mathematical induction.

Solution

Empirical evidence shows that the 'proved' statement is false, so the proof contains some error which must be in the inductive step.

The inductive step relies implicitly on the two groups of computers consisting of $C_1, C_2, \ldots, C_k$ and $C_2, \ldots, C_k, C_{k+1}$ having members in common so that the 'same price' property can be transferred from the first group to the second. If $k \geqslant 2$ this is indeed the case, so the inductive step is valid for $k \geqslant 2$. The problem is that the implication 'if $S(1)$ is true then $S(2)$ is true' does not hold as the two groups in question, C_1 and C_2, do not have members in common.

The 'proof' is not valid because we have not established the inductive step for *every* $k \geqslant 1$.

Variations on the Principle of Mathematical Induction

There are various modifications which we can make to the inductive principle. Suppose, for example, that we wish to prove that a proposition $S(n)$ is true for all integers greater than or equal to some fixed integer N. The following simple modification to the principle of induction would achieve this.

(a) Prove that $S(N)$ is true.
(b) Prove that, for every integer $k \geqslant N$, if $S(k)$ is true, then $S(k+1)$ is true.

This is just the standard method of proof by induction except that we 'begin' at N instead of 1.

Note that, even when required to prove $S(n)$ for all positive integers, sometimes it can be simpler to begin the induction at $n = 0$ rather than $n = 1$. If we do begin with $n = 0$, we have in fact proved slightly more than was required, but no one would quibble with that! In example 2.7.2 for instance, $f(0) = 2^2 + 3 = 7$, which is clearly divisible by seven. Continuing with the inductive hypothesis and inductive step as in the example would have shown that $2^{n+2} + 3^{2n+1}$ is divisible by 7 for all positive integers n and also for $n = 0$.

A more substantial modification of the inductive method is provided by the so-called 'second principle of induction'. The essence of this is that, when we come to the inductive step, we assume that $S(r)$ is true for all positive integers r less than or equal to k, rather than just for k itself.

Second Principle of Induction

Let $S(n)$ be a proposition concerning a positive integer n. If

(a) $S(1)$ is true, and
(b) for every $k \geqslant 1$, the truth of $S(r)$ for all $r \leqslant k$ implies the truth of $S(k+1)$,

then $S(n)$ is true for all positive integers.

This second principle of induction may at first appear to be more general than the first because we are allowed to assume rather more in order to deduce the truth of $S(k+1)$. However, if we let $T(n)$ be the proposition '$S(r)$ is true for all positive integers $r \leqslant n$' then the two parts of the second principle are:

(a) $T(1)$ is true, and
(b) for every $k \geqslant 1$, the truth of $T(k)$ implies the truth of $T(k+1)$.

This is just the (first) principle of induction for the proposition $T(n)$. Thus the second principle is no more general than the first although it may be simpler to use in the proofs of certain results.

We summarize the proof of $S(n)$ (where n is a positive integer) using each of the two principles of induction in the table below.

	Assume	Deduce
Proof of $S(n)$ using the (first) principle of induction	(a) Background knowledge	$S(1)$
	and	
	(b) $S(k)$; background knowledge	$S(k+1)$
Proof of $S(n)$ using the second principle of induction	(a) Background knowledge	$S(1)$
	and	
	(b) $S(1), S(2), \ldots, S(k)$; background knowledge	$S(k+1)$

Example 2.8

Prove that every positive integer greater than 1 is either prime or can be expressed as a product of prime numbers.

(This is part of a result which has the grand name 'the fundamental theorem of arithmetic'. The complete statement of this theorem goes on to say that, for any given positive integer, its expression as the product of primes is unique apart from the order in which the prime factors are written.)

Proof

Since the proposition involves integers greater than 1, we begin the induction with $n = 2$. The proposition clearly holds for $n = 2$ since 2 is itself a prime number.

Now suppose that every integer greater than 1 and less than or equal to k is either prime or can be expressed as the product of prime numbers. Consider the integer $k + 1$. There are two possibilities: either it is prime or it is composite (not prime). If it is prime then there is nothing to prove.

If, on the other hand, $k + 1$ is composite, then it can be written as $k + 1 = rs$ where $2 \leqslant r \leqslant k$ and $2 \leqslant s \leqslant k$. Now, by our inductive hypothesis, r and s are prime or can be written as products of prime numbers: $r = p_1 p_2 \ldots p_t$ and $s = q_1 q_2 \ldots q_u$ where the p_i $(i = 1, 2, \ldots, t)$ and q_j $(j = 1, 2, \ldots, u)$ are prime numbers.

Hence

$$\begin{aligned} k + 1 &= rs \\ &= p_1 p_2 \ldots p_t q_1 q_2 \ldots q_u \end{aligned}$$

so that $k+1$ can be expressed as the product of prime numbers. The result follows by the second principle of induction. □

Inductive Definitions

The use of the inductive principles is not confined entirely to proofs of propositions about the positive integers; they can also be used to define mathematical objects or properties which depend upon the positive integers.

Consider the following sequence of 'Fibonacci numbers'†:

$$1, 1, 2, 3, 5, 8, 13, 21, \ldots.$$

† Named after the Italian, Leonardo of Pisa (born *c.* 1170 and know as Fibonacci), who was reputed to have used the sequence to model the increase in a population of rabbits over time. Unfortunately, the model proved to be inaccurate; unrestrained rabbit populations increase more rapidly than the Fibonacci numbers! However, over the years many 'occurrences' of this sequence have been noticed in nature, art and architecture.

Each number in the sequence after the first two is the sum of the two preceding numbers. Denoting the nth Fibonacci number by f_n, we can define the sequence as follows:

$$f_1 = 1, \quad f_2 = 1 \quad \text{and, for } n \geqslant 3, \quad f_n = f_{n-1} + f_{n-2}.$$

This is an example of an inductive definition; we can think of it as a means of making precise the '...' in the sequence of Fibonacci numbers defined above. The astute reader will have noticed that the inductive definition does not quite conform to the principles of induction stated above. To begin the inductive definition, we need to define the first *two* Fibonacci numbers, rather than only the first one. The following describes the general form of an inductive definition of some mathematical object or property A_n which depends on a positive integer n.

Inductive Definition

To define A_n for all positive integers:

(a) Define explicitly A_k for $k = 1, 2, \ldots, r$.
(b) For $k > r$, define A_k in terms of $A_1, \ldots, A_{k-1}$.

To prove propositions about some object or involving some property which has been defined inductively, it is natural to use mathematical induction. We end this chapter with an inductive proof (using the second principle) of a property of Fibonacci numbers. Note that we need to begin the proof with an explicit verification of the result for $n = 1$ *and* $n = 2$. (Why is this?)

Example 2.9

Let f_n denote the nth Fibonacci number defined inductively above. Prove that $f_n < 2^n$.

Proof

First note that $f_1 = 1 < 2 = 2^1$ and $f_2 = 1 < 4 = 2^2$, so the proposition is true for $n = 1$ and $n = 2$.

Now suppose that $f_r < 2^r$ for all positive integers $r \leqslant k$. Then for $k \geqslant 2$,

$$\begin{aligned} f_{k+1} &= f_k + f_{k-1} && \text{(the inductive definition of } f_n\text{)} \\ &< 2^k + 2^{k-1} && \text{(by the inductive hypothesis)} \\ &< 2^k + 2^k && \text{(since } 2^{k-1} < 2^k\text{)} \\ &= 2 \times 2^k \\ &= 2^{k+1}. \end{aligned}$$

This completes the inductive step. We conclude that $f_n < 2^n$ for all positive integers n. □

Exercises 2.3

1. Prove that, for all positive integers n,

$$1 + 2 + \cdots + n = \tfrac{1}{2}n(n+1).$$

2. Prove that $2^n > n$ for all positive integers n.

3. Prove that, for all positive integers n, $5^n - 1$ is divisible by 4.

4. Prove that, for all non-negative integers n,

$$1 + x + x^2 + x^3 + \cdots + x^n = \frac{x^{n+1} - 1}{x - 1}$$

where x is a real number, $x \neq 1$.

5. Prove that the sum of the squares of the first n positive integers is

$$\frac{n(n+1)(2n+1)}{6}.$$

6. Prove that the sum of the cubes of the first n positive integers is

$$\left[\frac{n(n+1)}{2}\right]^2.$$

7. For all positive integers n, A_n is defined inductively as follows:

$$\begin{aligned} A_1 &= 3 \\ A_n &= A_{n-1} + 3 \quad \text{for } n \geqslant 2. \end{aligned}$$

Prove that $A_n = 3n$.

8. Prove that, for all integers $n \geqslant 4$, $n! > 2^n$. ($n! = n(n-1)(n-2)\ldots 1$.)

9. Prove that the sum of the first n even integers is $n(n+1)$.

10. Prove that, for all positive integers n, $4^{2n+1} + 3^{n+2}$ is divisible by 13.

11. A_n is defined inductively as follows:

$$\begin{aligned} A_1 &= 6 \\ A_2 &= 11 \\ A_n &= 3A_{n-1} - 2A_{n-2} \quad \text{for } n \geqslant 3. \end{aligned}$$

Prove that, for $n \geqslant 1$, $A_n = 5 \times 2^{n-1} + 1$.

12. Show that, for the Fibonacci sequence,

$$f_{n+2}^2 - f_{n+1}^2 = f_n f_{n+3} \quad n = 1, 2, \ldots.$$

Chapter 3

Sets

3.1 Sets and Membership

The notion of a 'set' is one of the basic concepts of mathematics—some would say *the* basic concept. Those who have encountered sets in their previous study of mathematics may be tempted to skip this chapter, regarding sets as rather trivial objects. Our advice is: don't! Set theory *is* non-trivial and we shall be using set-theoretic terminology and concepts throughout the book.

We shall make no attempt to give a precise definition of a set†. However, we can describe what we mean by the term: a **set** is to be thought of as any collection of objects whatsoever. The objects can also be anything and they are called **elements** of the set. The elements contained in a given set need not have anything in common (other than the obvious common attribute that they all belong to the given set). Equally, there is no restriction on the number of elements allowed in a set; there may be an infinite number, a finite number or even no elements at all. There is, however, one restriction we insist upon: given a set and an object, we should be able to decide (in principle at least—it may be difficult in practice) whether or not the object belongs to the set. Clearly a concept as general as this has many familiar examples as well as many frivolous ones.

† In §2.2 we explained why undefined terms are necessary in mathematics. In an axiomatic treatment of set theory, it is usual for 'set' to be undefined.

Examples 3.1

1. A set could be defined to contain Picasso, the Eiffel Tower and the number π. This is a (rather strange) finite set.

2. The set containing all the positive, even integers is clearly an infinite set.

3. Consider the 'set' containing the 10 best songs of all time. This is *not* allowed unless we give a precise definition of 'best'. Your best? Mine? Without being more precise this fails the condition that we should be able to decide whether an element belongs to the set.

Notation

We shall generally use upper-case letters to denote sets and lower-case letters to denote elements. (This convention will sometimes be violated, for example when the elements of a particular set are themselves sets.) The symbol $\in$ denotes 'belongs to' or 'is an element of'. Thus

$$a \in A \text{ means (the element) } a \text{ belongs to (the set) } A$$

and

$$a \notin A \text{ means } \neg(a \in A) \text{ or } a \text{ does not belong to } A.$$

Defining Sets

Sets can be defined in various ways. The simplest is by listing the elements enclosed between curly brackets or 'braces' { }. The two (well defined) sets in examples 3.1 could be written:

$$A = \{\text{Picasso, Eiffel Tower}, \pi\}$$
$$B = \{2, 4, 6, 8, \ldots\}.$$

In the second of these we clearly cannot list *all* the elements. Instead we list enough elements to establish a pattern and use '...' to indicate that the list continues indefinitely. Other examples are the following.

For a fixed positive integer n, $C_n = \{1, 2, \ldots, n\}$, the set of the first n positive integers. Again we use '...' to indicate that there are elements in the list which we have omitted to write, although in this case only finitely many are missing.

$D = \{\,\}$, the **empty set** (or **null set**), which contains no elements. This set is usually denoted $\varnothing$.

Listing the elements of a set is impractical except for small sets or sets where there is a pattern to the elements such as B and C_n above. An alternative is to define the elements of a set by a property or predicate (see chapter 1). More precisely, if $P(x)$ is a single-variable propositional function, we can form the set whose elements are all those objects a (and only those) for which $P(a)$ is a true proposition. A set defined in this way is denoted

$$A = \{x : P(x)\}.$$

(This is read: the set of all x such that $P(x)$ (is true).)

Note that 'within A'—that is, if we temporarily regard A as the universe of discourse—the quantified propositional function $\forall x P(x)$ is a true statement.

Examples 3.2

1. The set B above could be defined as $B = \{n : n \text{ is an even, positive integer}\}$, or $B = \{n : n = 2m, \text{ where } m > 0 \text{ and } m \text{ is an integer}\}$, or, with a slight change of notation, $B = \{2m : m > 0 \text{ and } m \text{ is an integer}\}$.

   Note that although the propositional functions used are different, the same elements are generated in each case.

2. The set C_n above could be defined as $C_n = \{p : p \text{ is an integer and } 1 \leqslant p \leqslant n\}$.

3. The set $\{1, 2\}$ could alternatively be defined as $\{x : x^2 - 3x + 2 = 0\}$. We say that $\{1, 2\}$ is the **solution set** of the equation $x^2 - 3x + 2 = 0$.

4. The empty set $\varnothing$ can be defined in this way using any propositional function $P(x)$ which is true for no objects x. For instance, rather frivolously,

$$\varnothing = \{x : x \text{ is a green rabbit with long purple ears}\}.$$

5. $X = \{x : x \text{ is an honest politician}\}$ is *not* a set unless we define 'honest' more precisely.

Equality of Sets

Two sets are defined to be **equal** if and only if they contain the same elements; that is, $A = B$ if $\forall x[x \in A \leftrightarrow x \in B]$ is a true proposition, and conversely. The order in which elements are listed is immaterial. Also, it is the standard convention to disregard repeats of elements in a listing. Thus the following all define the same set:

$$\{1, -\tfrac{1}{2}, 1066, \pi\}$$
$$\{-\tfrac{1}{2}, \pi, 1066, 1\}$$
$$\{1, -\tfrac{1}{2}, -\tfrac{1}{2}, \pi, 1066, -\tfrac{1}{2}, 1\}.$$

We should perhaps note here that there is only one empty set; or, put another way, all empty sets are equal. This is because any two empty sets contain precisely the same elements: none!

Also, if $P(x)$ and $Q(x)$ are propositional functions which are true for precisely the same objects x, then the sets they define are equal, i.e.

$$\{x : P(x)\} = \{x : Q(x)\}.$$

For example, $\{x : (x-1)^2 = 4\} = \{x : (x+1)(x-3) = 0\}$, since the two propositional functions $P(x) : (x-1)^2 = 4$ and $Q(x) : (x+1)(x-3) = 0$ are true for precisely the same values of x, namely -1 and 3.

Definition 3.1

If A is a finite set its **cardinality**, $|A|$, is the number of (distinct) elements which it contains.

If A has an infinite number of elements, we say it has **infinite cardinality**†, and write $|A| = \infty$.

Other notations commonly used for the cardinality of A are $n(A)$, $\#(A)$ and $\bar{\bar{A}}$.

† There is a more sophisticated approach to cardinality of infinite sets which allows different infinite sets to have different cardinality. Thus 'different sizes' of infinite sets can be distinguished! In this theory the set of integers has different cardinality from the set of numbers, for example. See §5.5 for more details of how this distinction can be made.

Examples 3.3

1. $|\varnothing| = 0$ since $\varnothing$ contains no elements.

2. $|\{\pi, 2, \text{Attila the Hun}\}| = 3$.

3. If $X = \{0, 1, \ldots, n\}$ then $|X| = n + 1$.

4. $|\{2, 4, 6, 8, \ldots\}| = \infty$.

Although cardinality appears to be a simple enough concept, determining the cardinality of a given set may be difficult in practice. This is particularly the case when some or all of the elements of the given set are themselves sets. This is a perfectly valid construction: the elements of a set can be anything, so certainly they can be sets.

For example, let $X = \{\{1, 2\}\}$. Then X contains only a single element, namely the set $\{1, 2\}$, so $|X| = 1$. It is clearly important to distinguish between the set $\{1, 2\}$ (which has cardinality 2) and X, the set which has $\{1, 2\}$ as its only element. Similarly, the sets $\varnothing$ and $\{\varnothing\}$ are different. The latter is non-empty since it contains a single element—namely $\varnothing$. Thus $|\{\varnothing\}| = 1$.

Examples 3.4

1. Let $A = \{1, \{1, 2\}\}$. Note that A has two elements, the number 1 and the set $\{1, 2\}$. Therefore, $|A| = 2$.

2. Similarly,

$$\begin{aligned}
|\{1, 2, \{1, 2\}\}| &= 3, \\
|\{\varnothing, \{1, 2\}\}| &= 2, \\
|\{\varnothing, \{\varnothing\}\}| &= 2, \\
|\{\varnothing, \{\varnothing\}, \{1, 2\}\}| &= 3, \\
|\{\varnothing, \{\varnothing, \{\varnothing\}\}\}| &= 2, \text{ etc.}
\end{aligned}$$

Exercises 3.1

1. List the elements of each of the following sets, using the '...' notation where necessary:

   (i) $\{x : x \text{ is an integer and } -3 < x < 4\}$
   (ii) $\{x : x \text{ is a positive (integer) multiple of three}\}$
   (iii) $\{x : x = y^2 \text{ and } y \text{ is an integer}\}$
   (iv) $\{x : (3x-1)(x+2) = 0\}$
   (v) $\{x : x \geqslant 0 \text{ and } (3x-1)(x+2) = 0\}$
   (vi $\{x : x \text{ is an integer and } (3x-1)(x+2) = 0\}$
   (vii) $\{x : x \text{ is a positive integer and } (3x-1)(x+2) = 0\}$
   (viii) $\{x : 2x \text{ is a positive integer}\}$.

2. Let $X = \{0, 1, 2\}$. List the elements of each of the following sets:

   (i) $\{z : z = 2x \text{ and } x \in X\}$
   (ii) $\{z : z = x + y \text{ where } x \in X \text{ and } y \in X\}$
   (iii) $\{z : x = z + y \text{ where } x \in X \text{ and } y \in X\}$
   (iv) $\{z : z \in X \text{ or } -z \in X\}$
   (v) $\{z : z^2 \in X\}$
   (vi) $\{z : z \text{ is an integer and } z^2 \in X\}$.

3. Determine the cardinality of each of the following sets:

   (i) $\{x : x \text{ is an integer and } 1/8 < x < 17/2\}$
   (ii) $\{x : \sqrt{x} \text{ is an integer}\}$
   (iii) $\{x : x^2 = 1 \text{ or } 2x^2 = 1\}$
   (iv) $\{a, b, c, \{a, b, c\}\}$
   (v) $\{a, \{b, c\}, \{a, b, c\}\}$
   (vi) $\{\{a, b, c\}, \{a, b, c\}\}$
   (vii) $\{a, \{a\}, \{\{a\}\}, \{\{\{a\}\}\}\}$
   (viii) $\{\varnothing, \{\varnothing\}, \{\{\varnothing\}\}\}$.

4. Use the notation $\{x : P(x)\}$, where $P(x)$ is a propositional function, to describe each of the following sets.

   (i) $\{1, 2, 3, 4, 5\}$.
   (ii) $\{3, 6, 9, 12, 15, \ldots, 27, 30\}$.
   (iii) $\{1, 3, 5, 7, 9, 11, \ldots\}$.
   (iv) $\{2, 3, 5, 7, 11, 13, 17, 19, 23, \ldots\}$.
   (v) $\{\mathrm{a, e, i, o, u}\}$.

(vi) The set of integers which can be written as the sum of the squares of two integers.
(vii) The set of all integers less than 1000 which are perfect squares.
(viii) The set of all numbers that are an integer multiple of 13.
(ix) {Afghanistan, Albania, Algeria, ..., Zambia, Zimbabwe}.
(x) {*Love's Labour's Lost*, *The Comedy of Errors*, *The Two Gentlemen of Verona*, ..., *The Tempest*, *The Winter's Tale*, *The Famous History of the Life of King Henry VIII*}.

3.2 Subsets

Definition 3.2

The set A is a **subset** of the set B, denoted $A \subseteq B$, if every element of A is also an element of B. Symbolically, $A \subseteq B$ if $\forall x[x \in A \rightarrow x \in B]$ is true, and conversely.

If A is a subset of B, we say that B is a **superset** of A, and write $B \supseteq A$.

Clearly every set B is a subset of itself, $B \subseteq B$. (This is because, for any given x, $x \in B \rightarrow x \in B$ is 'automatically' true.) Any other subset of B is called a **proper subset** of B. The notation $A \subset B$ is used to denote 'A is a proper subset of B'. Thus $A \subset B$ if and only if $A \subseteq B$ and $A \neq B$.

It should also be noted that $\varnothing \subseteq A$ for every set A. In this case definition 3.2 is satisfied in a vacuous way—the empty set has no elements, so certainly each of them belongs to A. Alternatively, for any object x, the proposition $x \in \varnothing$ is false so the conditional $(x \in \varnothing) \rightarrow (x \in A)$ is true.

Examples 3.5

1. $\{2, 4, 6, \ldots\} \subseteq \{1, 2, 3, \ldots\} \subseteq \{0, 1, 2, \ldots\}$. Of course, we could have used the proper subset symbol $\subset$ to link these three sets instead.

2. Similarly: {women} $\subseteq$ {people} $\subseteq$ {mammals} $\subseteq$ {creatures};
{*War and Peace*} $\subseteq$ {novels} $\subseteq$ {works of fiction};
{*Mona Lisa*} $\subseteq$ {paintings} $\subseteq$ {works of art}; etc.

Again, in each of these we could have used $\subset$ instead.

3. Let $X = \{1, \{2, 3\}\}$. Then $\{1\} \subseteq X$ but $\{2, 3\}$ is not a subset of X, which we can denote by $\{2, 3\} \not\subseteq X$. However, $\{2, 3\}$ is an *element* of X, so $\{\{2, 3\}\} \subseteq X$. Care clearly needs to be taken to distinguish between set membership and subset, particularly when a set has elements which are themselves sets.

To prove that two sets are equal, $A = B$, it is sufficient (and frequently very convenient) to show that each is a subset of the other, $A \subseteq B$ and $B \subseteq A$. Essentially, this follows from the following logical equivalence of compound propositions:

$$(P \leftrightarrow Q) \equiv [(P \rightarrow Q) \wedge (Q \rightarrow P)].$$

We know that $A = B$ if $\forall x(x \in A \leftrightarrow x \in B)$ is a true proposition. In chapter 2 we noted that to prove that a biconditional $P \leftrightarrow Q$ is true, it is sufficient to prove both conditionals $P \rightarrow Q$ and $Q \rightarrow P$ are true. It follows that to prove $\forall x(x \in A \leftrightarrow x \in B)$ it is sufficient to prove both $\forall x(x \in A \rightarrow x \in B)$ and $\forall x(x \in B \rightarrow x \in A)$. But $\forall x(x \in A \rightarrow x \in B)$ is true precisely when $A \subseteq B$ and similarly $\forall x(x \in B \rightarrow x \in A)$ is true precisely when $B \subseteq A$. In summary:

Theorem 3.1

Two sets A and B are equal if and only if $A \subseteq B$ and $B \subseteq A$.

Examples 3.6

1. Show that $\{x : 2x^2 + 5x - 3 = 0\} \subseteq \{x : 2x^2 + 7x + 2 = 3/x\}$.

Solution

Let $A = \{x : 2x^2 + 5x - 3 = 0\}$ and $B = \{x : 2x^2 + 7x + 2 = 3/x\}$.

We need to show that every element of A is an element of B. The equation $2x^2 + 5x - 3 = 0$ has solutions $x = \frac{1}{2}$ and $x = -3$, so $A = \{\frac{1}{2}, -3\}$.

When $x = \frac{1}{2}$, $2x^2 + 7x + 2 = \frac{1}{2} + \frac{7}{2} + 2 = 6 = 3/x$, so $\frac{1}{2} \in B$.

When $x = -3$, $2x^2 + 7x + 2 = 18 - 21 + 2 = -1 = 3/x$, so $-3 \in B$.

Therefore every element of A is an element of B, so $A \subseteq B$.

2. Let $A = \{\{1\}, \{2\}, \{1, 2\}\}$ and let B be the set of all non-empty subsets of $\{1, 2\}$. Show that $A = B$.

Solution

$A \subseteq B$ since each of the three elements of A is a non-empty subset of $\{1, 2\}$ and therefore an *element* of B.

$B \subseteq A$ since every non-empty subset of $\{1, 2\}$ (i.e. every element of B) is contained in A.

Using theorem 3.1, we conclude that $A = B$.

3. Prove that if $A \subseteq B$ and $C = \{x : x \in A \vee x \in B\}$, then $C = B$.

Solution

Let $A \subseteq B$. We will show that $B \subseteq C$ and $C \subseteq B$.

Let $x \in B$. Then $x \in A \vee x \in B$ is true, so $x \in C$. Thus every element of B also belongs to C, so $B \subseteq C$.

Now let $x \in C$. Then either $x \in A$ or $x \in B$ (or both). However, if $x \in A$ then it follows that $x \in B$ also, since $A \subseteq B$. Therefore in either case we can conclude $x \in B$. This shows that every element of C also belongs to B, so $C \subseteq B$.

We have now shown $B \subseteq C$ and $C \subseteq B$, so theorem 3.1 allows us to conclude that $B = C$.

Since the concept of a set is such a broad one, it is usual to restrict attention to only those sets which are relevant in a particular context. For example, we would surely wish to discount sets such as {Genghis Khan, Queen Boadicea, Attila the Hun} in a study of expert systems! It is convenient to define some **universal set** which contains as subsets all sets relevant to the current task or study. Anything outside the universal set is simply not considered. The universal set is not something fixed for all time—we can change it to suit different contexts. The universal set is frequently denoted $\mathscr{U}$. The universal set is, of course, essentially the universe of discourse introduced in chapter 1.

Some special sets of numbers which are frequently used as universal sets are the following.

$\mathbb{N} = \{0, 1, 2, 3, \ldots\}$ the set of **natural numbers**.
$\mathbb{Z} = \{\ldots, -2, -1, 0, 1, 2, \ldots\}$ the set of **integers**†.
$\mathbb{Q} = \{p/q : p, q \in \mathbb{Z} \text{ and } q \neq 0\}$ the set of fractions or **rational numbers**.
$\mathbb{R} =$ the set of **real numbers**; real numbers can be thought of as corresponding to points on a number line or as numbers written as (possibly infinite) decimals.
$\mathbb{C} = \{x + iy : x, y \in \mathbb{R} \text{ and } i^2 = -1\}$ the set of **complex numbers**.

IRRATIONAL NUMBERS

Clearly the following subset relations hold amongst these sets:

$$\mathbb{N} \subseteq \mathbb{Z} \subseteq \mathbb{Q} \subseteq \mathbb{R} \subseteq \mathbb{C}.$$

Also frequently used are $\mathbb{Z}^+$, $\mathbb{Q}^+$ and $\mathbb{R}^+$, the sets of *positive* integers, rational numbers and real numbers respectively. Note that $\mathbb{N}$ is *not* equal to $\mathbb{Z}^+$ since 0 belongs to the former but not the latter. In addition, we shall sometimes use $\mathbb{E}$ and $\mathbb{O}$ to denote the sets of even and odd integers respectively:

$$\mathbb{E} = \{2n : n \in \mathbb{Z}\} = \{\ldots, -4, -2, 0, 2, 4, \ldots\}$$

† The notation $\mathbb{Z}$ comes from the German word for numbers: *Zahlen*.

$$\mathbb{O} = \{2n+1 : n \in \mathbb{Z}\} = \{\ldots, -3, -1, 1, 3, 5, \ldots\}.$$

If a universal set has been defined the notation $\{x : P(x)\}$ means the set of all x *in the universal set* satisfying the property $P(x)$. Therefore if our current universal set is $\mathbb{Z}$ then $X = \{x : 2x^2 + 3x - 2 = 0\}$ is the set $\{-2\}$, but if $\mathscr{U}$ is $\mathbb{Q}$ or $\mathbb{R}$ then $X = \{-2, \frac{1}{2}\}$. In the former case we would probably make the restriction more explicit and write

$$X = \{x : x \in \mathbb{Z} \text{ and } 2x^2 + 3x - 2 = 0\}$$

or, using a slight but useful abuse of the notation,

$$X = \{x \in \mathbb{Z} : 2x^2 + 3x - 2 = 0\}.$$

Exercises 3.2

1. State whether each of the following statements is *true* or *false*.

   (i) $2 \in \{1,2,3,4,5\}$
   (ii) $\{2\} \in \{1,2,3,4,5\}$
   (iii) $2 \subseteq \{1,2,3,4,5\}$
   (iv) $\{2\} \subseteq \{1,2,3,4,5\}$
   (v) $\varnothing \subseteq \{\varnothing, \{\varnothing\}\}$
   (vi) $\{\varnothing\} \subseteq \{\varnothing, \{\varnothing\}\}$
   (vii) $0 \in \varnothing$
   (viii) $\{1,2,3,4,5\} = \{5,4,3,2,1\}$.

2. (i) List all the subsets of:

   (a) $\{a, b\}$
   (b) $\{a, b, c\}$
   (c) $\{a\}$.

   Can you conjecture how many subsets a set with n elements will have?

   (ii) Does the empty set have any subsets? Explain your answer. Is your answer consistent with your conjecture from part (i)?

3. In each of the following cases state whether $x \in A$, $x \subseteq A$, both or neither:

   (i) $x = \{1\}$; $A = \{1,2,3\}$
   (ii) $x = \{1\}$; $A = \{\{1\},\{2\},\{3\}\}$
   (iii) $x = \{1\}$; $A = \{1,2,\{1,2\}\}$
   (iv) $x = \{1,2\}$; $A = \{1,2,\{1,2\}\}$
   (v) $x = \{1\}$; $A = \{\{1,2,3\}\}$
   (vi) $x = 1$; $A = \{\{1\},\{2\},\{3\}\}$.

4. Given that $X = \{1, 2, 3, 4\}$, list the elements of each of the following sets:

(i) $\{A : A \subseteq X \text{ and } |A| = 2\}$
(ii) $\{A : A \subseteq X \text{ and } |A| = 1\}$
(iii) $\{A : A \text{ is a proper subset of } X\}$
(iv) $\{A : A \subseteq X \text{ and } 1 \in A\}$.

5. Let $\mathscr{U} = \{x : x \text{ is an integer and } 2 \leqslant x \leqslant 10\}$. In each of the following cases, determine whether $A \subseteq B$, $B \subseteq A$, both or neither:

(i) $A = \{x : x \text{ is odd}\}$ $B = \{x : x \text{ is a multiple of } 3\}$
(ii) $A = \{x : x \text{ is even}\}$ $B = \{x : x^2 \text{ is even}\}$
(iii) $A = \{x : x \text{ is even}\}$ $B = \{x : x \text{ is a power of } 2\}$
(iv) $A = \{x : 2x + 1 > 7\}$ $B = \{x : x^2 > 20\}$
(v) $A = \{x : \sqrt{x} \in \mathbb{Z}\}$ $B = \{x : x \text{ is a power of 2 or } 3\}$
(vi) $A = \{x : \sqrt{x} \leqslant 2\}$ $B = \{x : x \text{ is a perfect square}\}$
(vii) $A = \{x : x^2 - 3x + 2 = 0\}$ $B = \{x : x + 7 \text{ is a perfect square}\}$.

6. In each of the following cases, prove that $A \subseteq B$:

(i) $A = \{x : 2x^2 + 5x = 3\}$
$B = \{x : 2x^2 + 17x + 27 = 18/x\}$

(ii) $A = \{x : x \text{ is a positive integer and } x \text{ is even}\}$
$B = \{x : x \text{ is a positive integer and } x^2 \text{ is even}\}$

(iii) $A = \{x : x \text{ is an integer and } x \text{ is a multiple of } 6\}$
$B = \{x : x \text{ is an integer and } x \text{ is a multiple of } 3\}$.

7. Let A be any set and $P(x)$ be *any* propositional function.

(i) Prove that $B = \{x : x \in A \text{ and } P(x)\}$ is a subset of A.
If $B \subset A$ what can you deduce about $P(x)$?
If $A = B$ what can you deduce about $P(x)$?

(ii) Prove that A is a subset of $C = \{x : x \in A \text{ or } P(x)\}$.
If $A \subset C$ what can you deduce about $P(x)$?
If $A = C$ what can you deduce about $P(x)$?

8. Prove that, if $A \subseteq B$ and $C = \{x : x \in A \wedge x \in B\}$, then $C = A$.

9. Prove that, if A and B have no elements in common and $C = \{x : x \in A \wedge x \in B\}$, then $C = \varnothing$.

10. (i) Prove that, if $A \subseteq B$ and $B \subseteq C$, then $A \subseteq C$.
(ii) Deduce that, if $A \subseteq B$, $B \subseteq C$ and $C \subseteq A$, then $A = B = C$.

11. Given that $A = \{1, 2, 3, 4\}$, determine the cardinality of each of the following sets:

    (i) $\{B : B \subseteq A \text{ and } |B| = 2\}$
    (ii) $\{B : B \subseteq A \text{ and } 1 \in B\}$
    (iii) $\{B : B \subseteq A \text{ and } \{1, 2\} \subseteq B\}$
    (iv) $\{B : B \subseteq A \text{ and } \{1, 2\} \subset B\}$.

12. (Russell's paradox†.) Consider the 'set' R of all sets which are not elements of themselves. That is,

$$R = \{A : A \text{ is a set and } A \notin A\}.$$

    Find a set which is an element of R. Can you find a set which is not an element of R?

    Explain why R is *not* a well defined set. (Hint: is R itself an element of R?)

3.3 Operations on Sets

The **Venn-Euler diagram**‡ is a useful visual representation of sets. In such a diagram sets are represented as regions in the plane and elements which belong to a given set are placed inside the region representing it. Frequently all the sets in the diagram are placed inside a box which represents the universal set. If an element belongs to more than one set in the diagram, the two regions representing the sets concerned must overlap and the element is placed in the overlapping region. In this way the picture represents the relationships between the sets concerned.

For example, if $A \subseteq B$ the region representing A may be enclosed inside the region representing B to ensure that every element in the region representing A is also inside that representing B; see figure 3.1.

† Bertrand Russell, celebrated mathematician, logician, philosopher, politician, peace campaigner, Nobel laureate, etc!, communicated this paradox to his fellow mathematician Frege in 1902 just as Frege had completed a major work in set theory. This and other paradoxes which struck at the heart of set theory created turmoil in the foundations of mathematics at the time.

‡ These diagrams are more commonly called just 'Venn diagrams' after John Venn, the nineteenth-century English mathematician. In fact, diagrams such as figure 3.1 are more properly called 'Euler diagrams' after Leonhard Euler who first introduced them in 1761. Although both Venn and Euler had precise rules for constructing their diagrams, today the term 'Venn diagram' is used informally to denote any diagram that represents sets by regions in the plane.

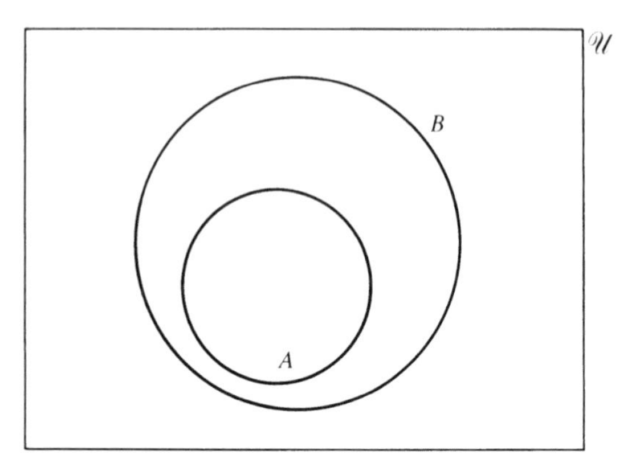

Figure 3.1

Example 3.7

The sets $A = \{$Ann, Alan, Fred, Jack, Mark, Mary, Ruth$\}$
$B = \{$Ann, Janet, Margaret, Mary, Ruth$\}$
$C = \{$Margaret, Mark, Mary, Matthew, Molly$\}$
can be represented by the Venn-Euler diagram shown in figure 3.2.

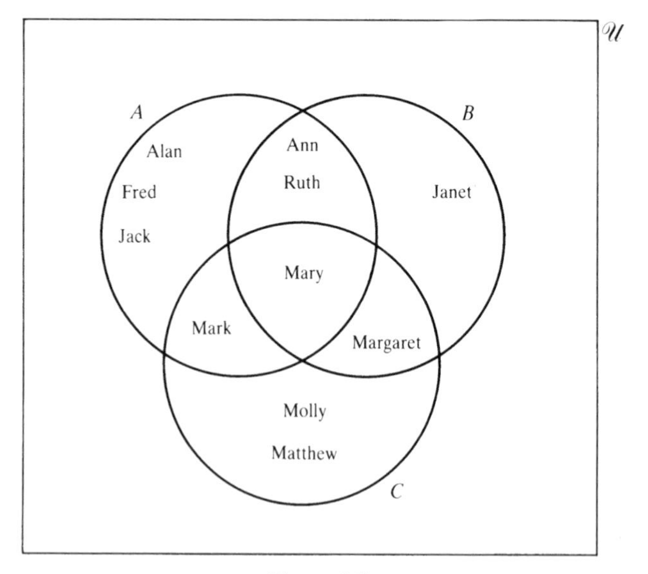

Figure 3.2

Given sets A and B we can define two new sets as follows.

The **intersection** of A and B is the set of all elements which belong both to A and B—it is denoted $A \cap B$.

The **union** of A and B is the set of all elements which belong to A or to B or to both—it is denoted $A \cup B$.

Symbolically:

$$A \cap B = \{x : x \in A \text{ and } x \in B\}$$
$$A \cup B = \{x : x \in A \text{ or } x \in B \text{ or both}\}.$$

There are obvious connections between intersection of sets and conjunction of propositions, and between union of sets and (inclusive) disjunction of propositions. If A and B are defined by propositional functions $P(x)$ and $Q(x)$ respectively, then

$$A \cap B = \{x : P(x) \wedge Q(x)\}$$

and

$$A \cup B = \{x : P(x) \vee Q(x)\}.$$

These sets can best be visualized by the following Venn-Euler diagrams (figures 3.3 and 3.4 respectively) where the regions representing intersection and union are shaded.

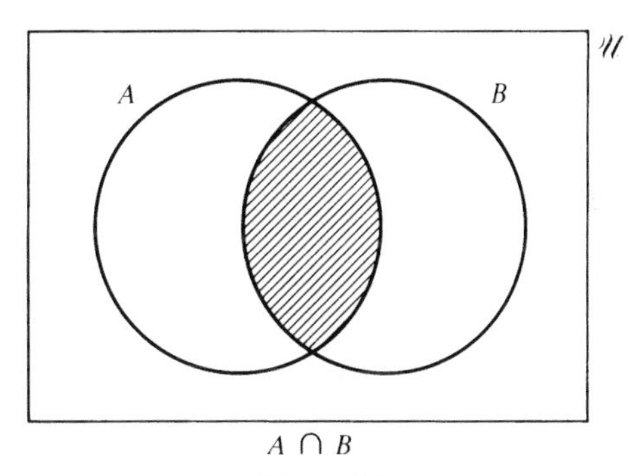

A ∩ B

Figure 3.3

Clearly we can extend the definitions of intersection and union to more than two sets. Let $A_1, A_2, \ldots, A_n$ be sets.

Their intersection is:

$$\begin{aligned}\bigcap_{r=1}^{n} A_r &= A_1 \cap A_2 \cap \cdots \cap A_n \\ &= \{x : x \in A_1 \text{ and } x \in A_2 \text{ and } \ldots \text{ and } x \in A_n\} \\ &= \{x : x \text{ belongs to } \textit{each} \text{ set } A_r, \text{ for } r = 1, 2, \ldots, n\}.\end{aligned}$$

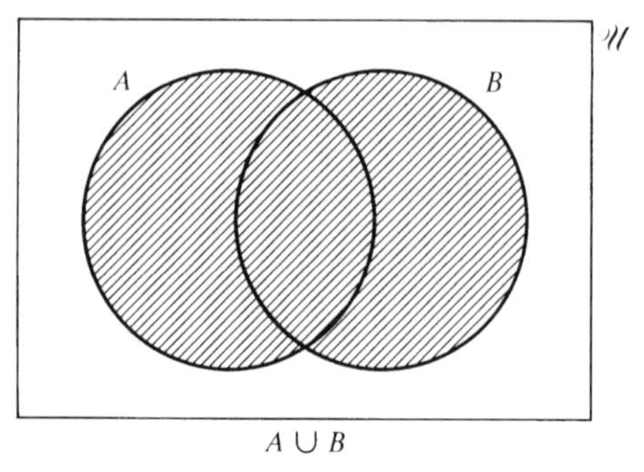

Figure 3.4

Their union is:

$$\begin{aligned}\bigcup_{r=1}^{n} A_r &= A_1 \cup A_2 \cup \cdots \cup A_n \\ &= \{x : x \in A_1 \text{ or } x \in A_2 \text{ or} \ldots \text{ or } x \in A_n\} \\ &= \{x : x \text{ belongs to } \textit{at least one} \text{ set } A_r, r = 1, \ldots, n\}.\end{aligned}$$

Sets A and B are said to be **disjoint** if they have no elements in common; that is, if $A \cap B = \varnothing$. In a Venn-Euler diagram this may be represented by drawing the regions representing the two sets to be non-overlapping, as in figure 3.5.

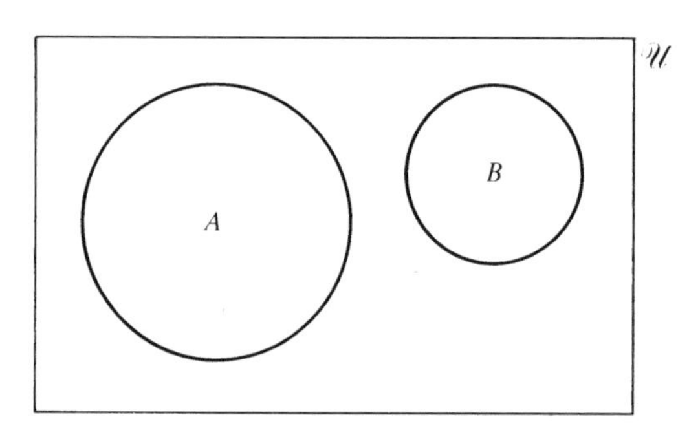

Figure 3.5

Given a set A, another set we can define is its **complement** which consists of all those elements in $\mathscr{U}$ which do *not* belong to A. The complement of A is denoted $\bar{A}$ (or A' or A^{c}). Of course, it is important that a universal set has already been defined; otherwise the complement will not be a well defined set. There is an obvious connection between complement and negation; namely, if $A = \{x : P(x)\}$ then $\bar{A} = \{x : \neg P(x)\}$. The Venn-Euler diagram shown in figure 3.6 illustrates the complement.

Related to the complement of a set is the **difference** or **relative complement** of

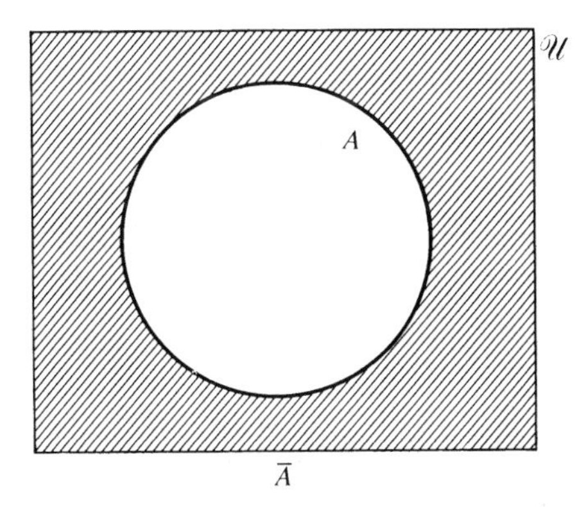

Figure 3.6

two sets A and B, denoted $A - B$ or $A \setminus B$. This set contains all the elements of A which do not belong to B:

$$A - B = \{x : x \in A \text{ and } x \notin B\} = A \cap \bar{B}.$$

Note that the complement of A is given by $\bar{A} = \mathscr{U} - A$. The difference $A - B$ is illustrated in figure 3.7.

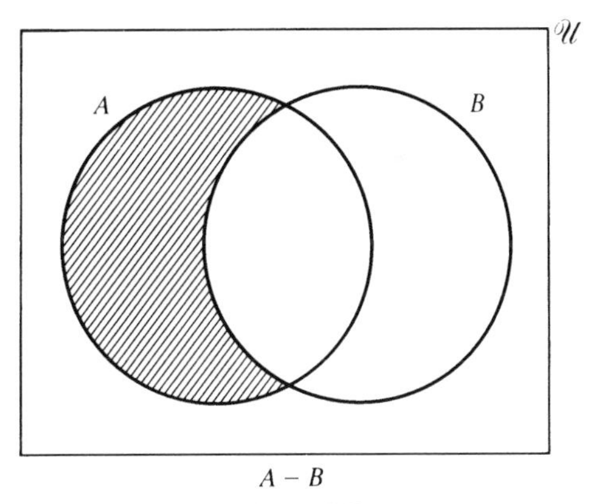

Figure 3.7

Examples 3.8

1. Let $\mathscr{U} = \{1, 2, 3, \ldots, 10\} = \{n : n \in \mathbb{Z}^+ \text{ and } n \leqslant 10\}$,
   $A = \{n \in \mathscr{U} : 1 \leqslant n < 7\}$,
   $B = \{n \in \mathscr{U} : n \text{ is a multiple of } 3\}$.

   Then $A = \{1, 2, 3, 4, 5, 6\}$ and $B = \{3, 6, 9\}$.

Therefore:

$$\begin{aligned}
A \cap B &= \{3, 6\} \\
A \cup B &= \{1, 2, 3, 4, 5, 6, 9\} \\
A - B &= \{1, 2, 4, 5\} \\
B - A &= \{9\} \\
\bar{A} &= \{7, 8, 9, 10\} \\
\bar{B} &= \{1, 2, 4, 5, 7, 8, 10\} \\
\overline{A \cup B} &= \{7, 8, 10\} = \bar{A} \cap \bar{B} \\
\overline{A \cap B} &= \{1, 2, 4, 5, 7, 8, 9, 10\} = \bar{A} \cup \bar{B} \\
\overline{A - B} &= \{3, 6, 7, 8, 9, 10\} = \bar{A} \cup B.
\end{aligned}$$

2. (i) For each of the following, draw a Venn-Euler diagram and shade the region corresponding to the indicated set.

   (a) $A - (B \cap C)$ (b) $(A - B) \cup (A - C)$.

   (ii) Show that $A - (B \cap C) = (A - B) \cup (A - C)$ for all sets A, B and C.

Solution

(i) (a) The region representing $A - (B \cap C)$ is that part of A that lies outside $B \cap C$. This is represented by the following diagram.

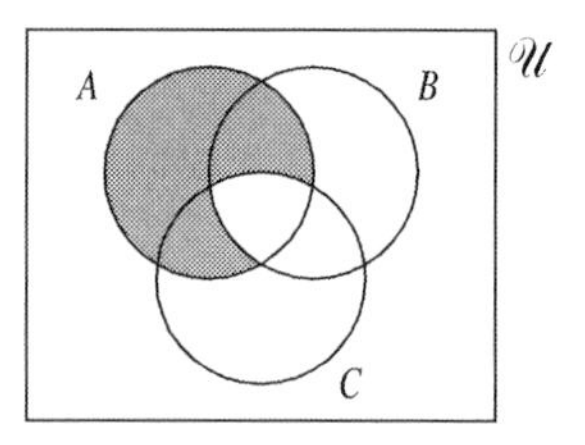

(b) In the following diagram, the regions representing $A - B$ and $A - C$ are shaded differently. Then $(A - B) \cup (A - C)$ is the region which has either shading.

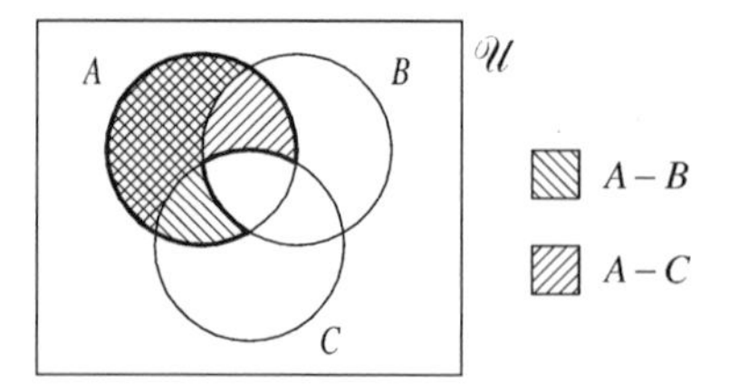

(ii) In the diagrams above, the region representing $A - (B \cap C)$ in part (a) is the same as that representing $(A - B) \cup (A - C)$ in part (b). This suggests that the two sets are equal. However, a pair of diagrams does not constitute a proof, so we now prove this using the technique suggested by theorem 3.1.

Let A, B and C be sets.

First we show $A - (B \cap C) \subseteq (A - B) \cup (A - C)$.

Let $x \in A - (B \cap C)$. Then $x \in A$ and $x \notin B \cap C$. Hence $x \in A$ and either $x \notin B$ or $x \notin C$ (or both). Therefore either $x \in A$ and $x \notin B$ or $x \in A$ and $x \notin C$ (or both). It follows that $x \in A - B$ or $x \in A - C$ (or both). Hence $x \in (A - B) \cup (A - C)$. We have shown that if $x \in A - (B \cap C)$ then $x \in (A - B) \cup (A - C)$. Therefore $A - (B \cap C) \subseteq (A - B) \cup (A - C)$.

Secondly we must show that $(A - B) \cup (A - C) \subseteq A - (B \cap C)$.

Let $x \in (A - B) \cup (A - C)$. Then $x \in A - B$ or $x \in A - C$ (or both) so $x \in A$ and $x \notin B$ or $x \in A$ and $x \notin C$ (or both). Hence $x \in A$ and either $x \notin B$ or $x \notin C$ (or both) which implies $x \in A$ and $x \notin B \cap C$. Therefore $x \in A - (B \cap C)$. We have shown that if $x \in (A - B) \cup (A - C)$ then $x \in A - (B \cap C)$. Therefore $(A - B) \cup (A - C) \subseteq A - (B \cap C)$.

Finally, since we have shown that each set is a subset of the other, we may conclude $(A - B) \cup (A - C) = A - (B \cap C)$.

Exercises 3.3

1. Draw Venn-Euler diagrams and shade the regions representing each of the following sets:

   (i) $\bar{A} \cap B$
   (ii) $\bar{A} \cup B$
   (iii) $(A \cap B) \cup (\overline{A \cup B})$
   (iv) $A \cap (B \cup C)$
   (v) $A \cup (B \cap C)$
   (vi) $(A \cap B) - C$
   (vii) $A - (B \cap C)$
   (viii) $(A \cup B) - C$
   (ix) $A - (B \cup C)$

(x) $(A - B) \cap (A - C)$.

2. Let $\mathscr{U} = \{n : n \in \mathbb{N} \wedge n < 10\}$, $A = \{2, 4, 6, 8\}$, $B = \{2, 3, 5, 7\}$, $C = \{1, 4, 9\}$. Define (for example, by listing elements) each of the following sets.

(i)	$A \cap B$	(vi)	$A \cap (B \cup C)$
(ii)	$A \cup B$	(vii)	$\bar{B} \cup B$
(iii)	$A - B$	(viii)	$\bar{B} \cap B$
(iv)	$B \cap C$	(ix)	$\overline{A \cup C}$
(v)	$\bar{A} \cap B$	(x)	$(A - C) - B$.

3. Consider the sets A, B, C, D and E represented by the following Venn-Euler diagram. (The sets C and E are represented by shaded regions.) For each of the following pairs of sets X and Y, state whether $X \subseteq Y$, $Y \subseteq X$, $X \cap Y = \varnothing$ or none of these.

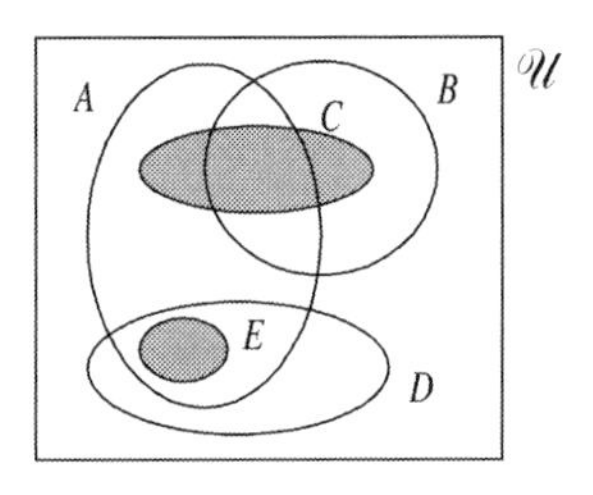

(i)	$X = A \cup B$	$Y = C$
(ii)	$X = A \cap B$	$Y = D$
(iii)	$X = A \cap B$	$Y = C$
(iv)	$X = E$	$Y = A \cap D$
(v)	$X = B \cap C$	$Y = C \cup D$
(vi)	$X = A \cap E$	$Y = D \cup E$
(vii)	$X = C \cup E$	$Y = A \cup D$
(viii)	$X = C - B$	$Y = D \cup E$
(ix)	$X = A \cup D$	$Y = B \cap E$
(x)	$X = A - E$	$Y = A - D$.

4. Let $\mathscr{U} = \{1, 2, 3, \ldots, 9, 10\}$ and define sets A, B, C and D as follows.

$$A = \{2, 4, 6, 8, 10\} \qquad B = \{3, 4, 5, 6\}$$
$$C = \{7, 8, 9, 10\} \qquad D = \{1, 3, 5, 7, 9\}.$$

List the elements of each of the following sets.

(i)	$A \cup B$	(vii)	$\bar{B} \cap \bar{C}$
(ii)	$A \cap D$	(viii)	$A - (B \cap \bar{C})$
(iii)	$\overline{B \cup C}$	(ix)	$(A - B) \cup (D - C)$
(iv)	$A \cap (B \cup D)$	(x)	$\overline{D - C}$
(v)	$B \cup (\bar{A} \cap \bar{D})$	(xi)	$(\bar{A} \cup \bar{B}) - (\overline{A \cup B})$
(vi)	$(\overline{C \cap D}) \cup B$	(xii)	$(\bar{C} - A) \cap (A - \bar{C})$.

5. Let $\mathscr{U} = \{n \in \mathbb{Z} : 1 \leqslant n \leqslant 12\}$, $A = \{n : n \text{ is a divisor of } 12\}$, $B = \{n : n \text{ is a prime number}\}$ and $C = \{n : n \text{ is odd}\}$. (Recall that 1 is not a prime number.)

   (i) Describe in words each of the following sets:
   (a) $A \cap B$
   (b) $A \cap B \cap C$
   (c) $B \cap \bar{C}$
   (d) $A - C$.

   (ii) List the elements of each of the following sets:
   (a) $A \cup B$
   (b) $B \cap C$
   (c) $\overline{A \cup C}$
   (d) $C - A$
   (e) $\overline{A \cap B}$.

6. Given that
$$A = \{x : P(x)\}$$
$$B = \{x : Q(x)\}$$
$$C = \{x : R(x)\}$$

   define each of the following sets in terms of $P(x)$, $Q(x)$ and $R(x)$ (and logical connectives):

   (i) $A \cap \bar{B}$
   (ii) $\overline{A \cup B}$
   (iii) $A \cap (B \cup \bar{C})$
   (iv) $\overline{A - B}$
   (v) $A - (B \cup C)$
   (vi) $\bar{A} - \bar{B}$.

7. For each of the following, draw two Venn-Euler diagrams. On one diagram shade the region represented by the set on the left-hand side of the equality and on the other diagram shade the region represented by the set on the right-hand side of the equality. Then prove that the identity for all sets A, B and C.

   (i) $A - B = A - (A \cap B)$

(ii) $A \cap (B - C) = (A \cap B) - C$
(iii) $(A \cup B) - C = (A - C) \cup (B - C)$
(iv) $A \cup (B - C) = (A \cup B) - (\bar{A} \cap C)$
(v) $(A - B) - C = A - (B \cup C)$.

3.4 Counting Techniques

Some quite complex mathematical results rely for their proofs on counting arguments: counting the numbers of elements of various sets, the number of ways in which a certain outcome can be achieved, etc. Although counting may appear to be a rather elementary exercise, in practice it can be extremely complex and rather subtle. Mathematicians have devised a number of techniques and results to deal with counting problems in a branch of the subject called enumeration theory.

One of the simplest counting results is the following, which says that to count the total number of elements of two *disjoint* sets A and B, we simply count the elements of A, count the elements of B and add them.

Counting Principle 1

If A and B are disjoint sets, then

$$|A \cup B| = |A| + |B|.$$

In many applications, of course, more than two sets are involved. The above principle easily generalizes to the following, which can be proved formally using mathematical induction (see chapter 2).

Counting Principle 2

If $A_1, A_2, \ldots, A_n$ are sets, no pair of which have elements in common, then

$$|A_1 \cup A_2 \cup \cdots \cup A_n| = |A_1| + |A_2| + \cdots + |A_n|.$$

Frequently, the sets whose elements are to be counted will not satisfy the rather stringent condition of the counting principles above—that any pair of them be disjoint. However, in these situations it is often possible to divide the set under consideration into subsets which do satisfy the conditions of the counting principles. One of the simplest results which can be proved in this way is the following.

> **Theorem 3.2 (The inclusion–exclusion principle)**
>
> If A and B are finite sets then
>
> $$|A \cup B| = |A| + |B| - |A \cap B|.$$

Proof

We can divide $A \cup B$ into its subsets $A - B$, $A \cap B$ and $B - A$ which satisfy the condition of counting principle 2; see figure 3.8.

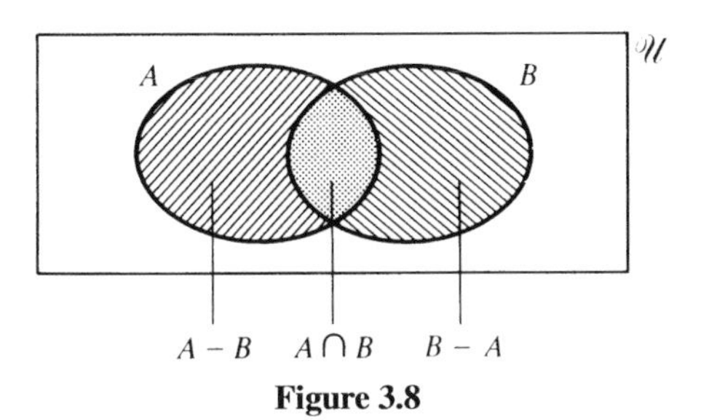

Figure 3.8

Therefore, by counting principle 2,

$$|A \cup B| = |A - B| + |A \cap B| + |B - A|. \tag{1}$$

The sets A and B can themselves be split into disjoint subsets $A - B$, $A \cap B$ and $B - A$, $A \cap B$ respectively. Thus

$$|A| = |A - B| + |A \cap B| \tag{2}$$

and

$$|B| = |B - A| + |A \cap B|. \tag{3}$$

It is now a simple exercise to combine equations (1), (2) and (3) to produce the desired result. □

The inclusion–exclusion principle is so called because to count the elements of $A \cup B$ we could have added the number of elements of A and the number of elements of B, in which case we have *included* the elements of $A \cap B$ twice: once as elements of A and once as elements of B. To obtain the correct number of elements in $A \cup B$, we would then need to *exclude* those in $A \cap B$ once, so that overall they are just counted once.

There are corresponding identities for more than two sets. The result for three sets is theorem 3.3, the proof of which we leave as an exercise.

Theorem 3.3

If A, B and C are finite sets, then

$$\begin{aligned} |A \cup B \cup C| = |A| + |B| + |C| - |A \cap B| - |B \cap C| \\ - |C \cap A| + |A \cap B \cap C|. \end{aligned}$$

Example 3.9

Each of the 100 students in the first year of Utopia University's Computer Science Department studies at least one of the subsidiary subjects: mathematics, electronics and accounting. Given that 65 study mathematics, 45 study electronics, 42 study accounting, 20 study mathematics and electronics, 25 study mathematics and accounting, and 15 study electronics and accounting, find the number who study:

(i) all three subsidiary subjects;
(ii) mathematics and electronics but not accounting;
(iii) only electronics as a subsidiary subject.

Solution

Let $\mathscr{U}$ = {students in the first year of Utopia's Computer Science Department}
M = {students studying mathematics}
E = {students studying electronics}
A = {students studying accounting}.

We are given the following information: $|\mathscr{U}| = 100$, $|M| = 65$, $|E| = 45$, $|A| = 42$, $|M \cap E| = 20$, $|M \cap A| = 25$, $|E \cap A| = 15$. Also, since every student takes at least one of three subjects as a subsidiary, $\mathscr{U} = M \cup E \cup A$.

Let $|M \cap E \cap A| = x$. Figure 3.9 shows the cardinalities of the various disjoint subsets of $\mathscr{U}$. These are calculated as follows, beginning with the innermost region representing $M \cap E \cap A$ and working outwards in stages.

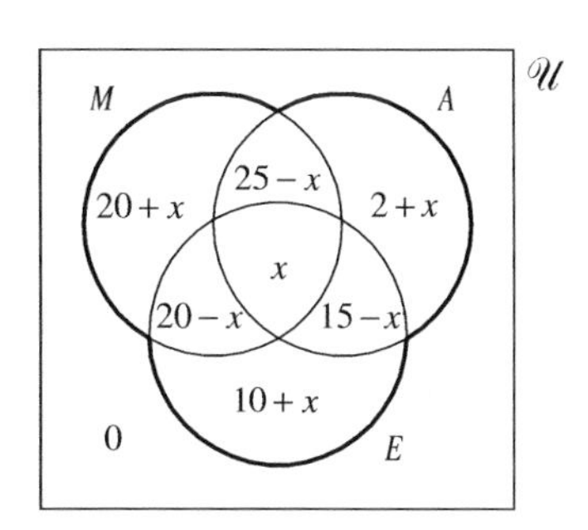

Figure 3.9

By Counting Principle 1,

$$|M \cap A| = |M \cap A \cap E| + |(M \cap A) - E|$$

so

$$|(M \cap A) - E| = |M \cap A| - |M \cap A \cap E| = 25 - x.$$

Similarly

$$|(M \cap E) - A| = |M \cap E| - |M \cap E \cap A| = 20 - x$$

and

$$|(A \cap E) - M| = |A \cap E| - |M \cap E \cap A| = 15 - x.$$

Now consider set M. By Counting Principle 2,

$$|M| = |M - (A \cup E)| + |(M \cap A) - E| + |(M \cap E) - A| + |M \cap E \cap A|$$

so

$$\begin{aligned} |M - (A \cup E)| &= |M| - |(M \cap A) - E| - |(M \cap E) - A| - |M \cap E \cap A| \\ &= 65 - (25 - x) - (20 - x) - x \\ &= 20 + x. \end{aligned}$$

Similarly

$$\begin{aligned} |A - (M \cup E)| &= |A| - |(A \cap M) - E| - |(A \cap E) - M| - |M \cap E \cap A| \\ &= 42 - (25 - x) - (15 - x) + x \\ &= 2 + x \end{aligned}$$

and

$$
\begin{aligned}
|E - (M \cup A)| &= |E| - |(E \cap M) - A| - |(E \cap A) - M| - |M \cap E \cap A| \\
&= 45 - (20 - x) - (15 - x) + x \\
&= 10 + x.
\end{aligned}
$$

Now, using Counting Principle 2 again, $|M \cup A \cup E| = 100$ is the sum of the cardinalities of its seven disjoint subsets, so:

$$
\begin{aligned}
& 100 = (20 + x) + (2 + x) + (10 + x) + (25 - x) \\
& \qquad + (20 - x) + (15 - x) + x \\
\Rightarrow \quad & 100 = 92 + x \\
\Rightarrow \quad & x = 8.
\end{aligned}
$$

We could now re-draw figure 3.9 showing the cardinality of each disjoint subset of $M \cup A \cup E$. However, this is not necessary to answer the three parts of the question.

(i) Eight students study all three subsidiary subjects.

(ii) The number of students who study mathematics and electronics but not accounting is $|(M \cap E) - A| = 20 - x = 20 - 8 = 12$.

(iii) The number of students who study only electronics as a subsidiary subject is $|E - (M \cup A)| = 10 + x = 10 + 8 = 18$.

3.5 The Algebra of Sets

From example 3.8.1 and exercise 3.3.7 above, it is clear that the intersection, union, complement (and hence difference) operations on sets are related to one another. For instance,

$$\overline{A \cap B} = \bar{A} \cup \bar{B}$$

for the sets defined in example 3.8.1. In fact, this equation holds for all sets.

We give below the basic identities connecting the operations of intersection, union and complement. Compare these with the laws for propositions given in §1.5. Given the connection between operations on sets and logical connectives between propositions, each of the set theory laws listed below can be derived from the corresponding logical equivalence between compound propositions. The following laws hold for all sets A, B and C.

Idempotent laws

$$A \cap A = A$$
$$A \cup A = A.$$

Commutative laws

$$A \cap B = B \cap A$$
$$A \cup B = B \cup A.$$

Associative laws

$$A \cap (B \cap C) = (A \cap B) \cap C$$
$$A \cup (B \cup C) = (A \cup B) \cup C.$$

Absorption laws

$$A \cap (A \cup B) = A$$
$$A \cup (A \cap B) = A.$$

Distributive laws

$$A \cap (B \cup C) = (A \cap B) \cup (A \cap C)$$
$$A \cup (B \cap C) = (A \cup B) \cap (A \cup C).$$

Involution law

$$\bar{\bar{A}} = A.$$

De Morgan's laws

$$(\overline{A \cup B}) = \bar{A} \cap \bar{B}$$
$$(\overline{A \cap B}) = \bar{A} \cup \bar{B}.$$

Identity laws

$$
\begin{aligned}
A \cup \varnothing &= A \\
A \cap \mathscr{U} &= A \\
A \cup \mathscr{U} &= \mathscr{U} \\
A \cap \varnothing &= \varnothing.
\end{aligned}
$$

Complement laws

$$
\begin{aligned}
A \cup \bar{A} &= \mathscr{U} \\
A \cap \bar{A} &= \varnothing \\
\bar{\varnothing} &= \mathscr{U} \\
\bar{\mathscr{U}} &= \varnothing.
\end{aligned}
$$

Although these laws can be derived from the corresponding equivalences between propositions, they are probably best *illustrated* using Venn-Euler diagrams. For example, the second of the distributive laws is illustrated by the Venn-Euler diagram in figure 3.10. The Venn-Euler diagram of figure 3.10(a) shows the set $A \cup (B \cap C)$. In figure 3.10(b), the two sets $A \cup B$ and $A \cup C$ are shaded differently, so the double shading represents their intersection $(A \cup B) \cap (A \cup C)$. The regions shaded in figure 3.10(a) and doubly shaded in figure 3.10(b) are the same, indicating that the two sets are equal.

The other laws may be illustrated similarly. For example, figure 3.11 explains the first of De Morgan's laws.

In figure 3.11(a) the complement of $A \cup B$ is shaded and in figure 3.11(b) $\bar{A}$ and $\bar{B}$ are shaded, the double shading representing $\bar{A} \cap \bar{B}$. The double-shaded area in (b) is the same as the shaded area in (a) indicating that the two sets represented are equal.

The Duality Principle

Just as a compound proposition involving the connectives $\wedge$, $\vee$ and negation has a dual proposition, so, too, does a statement about sets which involves $\cap$, $\cup$ and complement. The **dual** of such a statement is obtained by interchanging $\cap$ and $\cup$ everywhere and interchanging $\varnothing$ and $\mathscr{U}$ everywhere in the original statement. For example, the dual of

$$(A \cap \varnothing) \cup (B \cap \mathscr{U}) \cup \bar{B} = \mathscr{U}$$

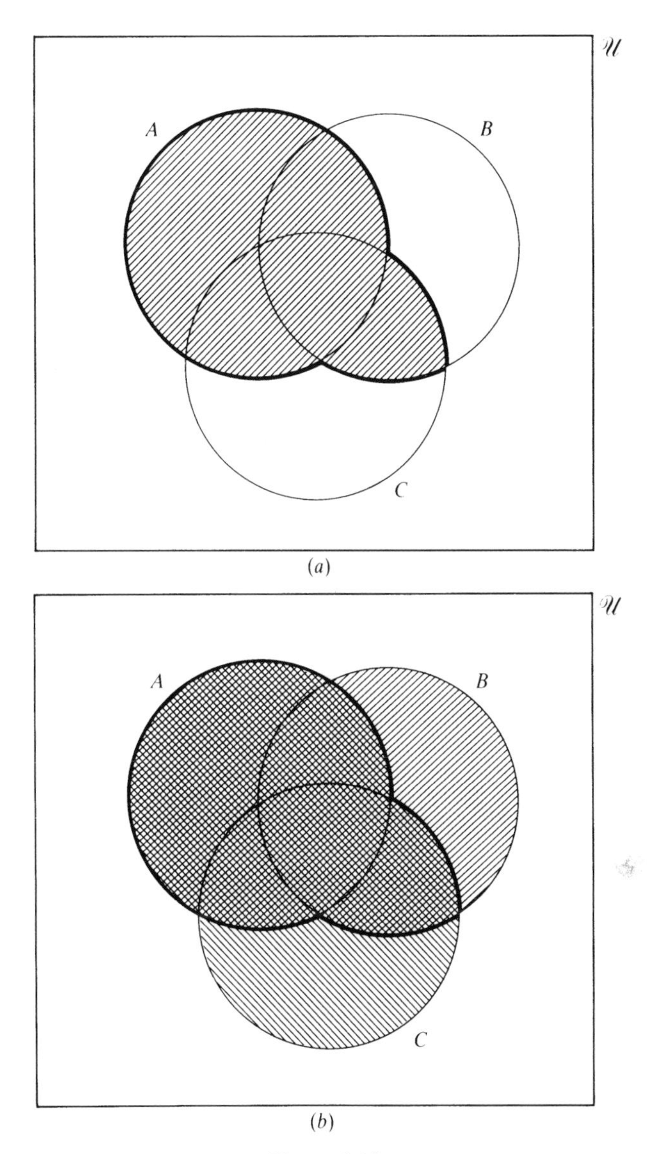

Figure 3.10

is

$$(A \cup \mathscr{U}) \cap (B \cup \varnothing) \cap \bar{B} = \varnothing.$$

For each of the laws of the algebra of sets, its dual is also a law. This suggests the following duality principle for sets which, although not a mathematical theorem, is extremely useful.

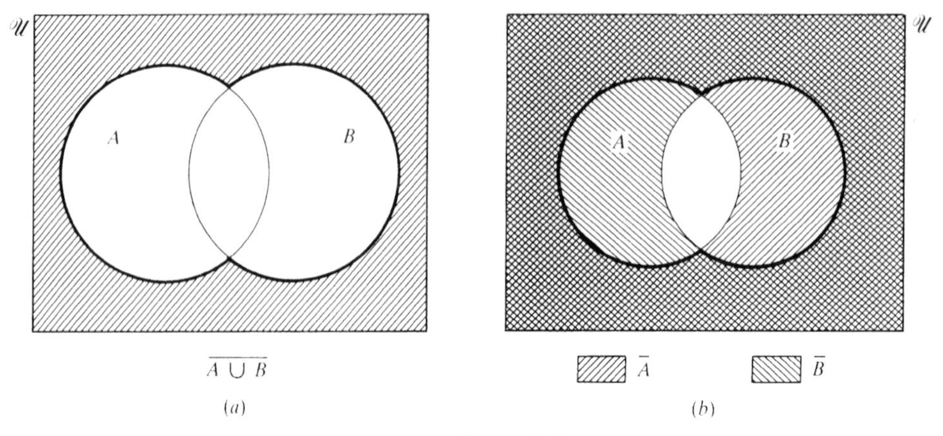

Figure 3.11

> **Duality Principle for Sets**
>
> If a statement about sets is true *for all sets* then its dual statement is necessarily true *for all sets* also.

Exercises 3.4

1. For each of the following set identities, draw a pair of Venn-Euler diagrams (as in figures 3.10 and 3.11) to illustrate the identity.

   (i) $A \cap (B \cup C) = (A \cap B) \cup (A \cap C)$
   (ii) $\overline{A - B} = B \cup \bar{A}$
   (iii) $\overline{A \cap B} = \bar{A} \cup \bar{B}$
   (iv) $(A - B) \cap C = (A \cap C) - B$
   (v) $(A - B) \cup (B - A) = (A \cup B) - (B \cap A)$.

2. For each of the following nine sets, draw a Venn-Euler diagram and shade the region corresponding to the set. Use your diagrams to identify which sets are equal.

(i)	$A \cup B$	(iv)	$A - (A \cap B)$	(vii)	$(A - B) \cup (B - A)$
(ii)	$\overline{A - B}$	(v)	$A \cap \bar{B}$	(viii)	$(A \cup B) - (A \cap B)$
(iii)	$\overline{A \cup B}$	(vi)	$B \cup \bar{A}$	(ix)	$\overline{A \cap B}$.

3. For each of the following four sets, draw a Venn-Euler diagram and shade the region corresponding to the set. Use your diagrams to identify which sets are equal.

(i) $(A \cap B) \cup (A \cap C)$ (iii) $(A \cap C) - B$
(ii) $(A - B) \cap C$ (iv) $A \cap (B \cup C)$.

4. The laws for the algebra of sets can sometimes be used to give proofs of set identities that are simpler than showing each set is a subset of the other (the method used in example 3.8.2(ii)). For example, the following is an alternative proof of the result in example 3.8.2(ii).

Proof

For all sets A, B and C:

$$
\begin{aligned}
A - (B \cap C) &= A \cap (\overline{B \cap C}) && \text{(definition of difference)} \\
&= A \cap (\bar{B} \cup \bar{C}) && \text{(De Morgan's law)} \\
&= (A \cap \bar{B}) \cup (A \cap \bar{C}) && \text{(distributive law)} \\
&= (A - B) \cup (B - C) && \text{(definition of difference).}
\end{aligned}
$$

In a similar way, prove each of the identities given in exercises 3.3.7.

5. The **symmetric difference** $A * B$ of sets A and B is defined by:

$$A * B = (A - B) \cup (B - A).$$

(i) Using the laws for the algebra of sets, show that, for every set A, $A * \varnothing = A$ and $A * A = \varnothing$.

(ii) Draw Venn-Euler diagrams to illustrate the identity

$$A \cap (B * C) = (A \cap B) * (A \cap C).$$

(This is called the **distributive law**: we say that intersection is **distributive over symmetric difference**.)

(iii) Find a counter-example to the proposition that, for all sets A, B and C,

$$A \cup (B * C) = (A \cup B) * (A \cup C).$$

(This shows that union is *not* distributive over symmetric difference.)

6. (i) Find sets A and B such that $A \in B$ and $A \subseteq B$.

(ii) Find sets A, B and C such that ($A \in B$ and $B \subseteq C$) and ($A \subseteq B$ and $B \in C$).

7. Write down the dual of each of the following statements:

(i) $\bar{A} \cap \bar{B} = (\overline{A \cup B})$
(ii) $A \cup (B \cap \mathscr{U}) = (A \cup \varnothing) \cup B$
(iii) $A \cap B = \varnothing$.

Note that if statement (iii) is true its dual may not also be true. Explain why this fact does *not* violate the principle of duality.

8. Use theorem 3.2, the counting principles and the algebra of sets to prove theorem 3.3.

9. Given that $|A| = 55$, $|B| = 40$, $|C| = 80$, $|A \cap B| = 20$, $|A \cap B \cap C| = 17$, $|B \cap C| = 24$, and $|A \cup C| = 100$, find:

(i) $|A \cap C|$
(ii) $|C - B|$
(iii) $|(B \cap C) - (A \cap B \cap C)|$.

Draw a Venn-Euler diagram and mark on it the cardinalities of the sets corresponding to each region of the diagram.

If $|\mathscr{U}| = 150$ find $|\overline{A \cup B \cup C}|$.

10. In a survey of 1000 households, 275 owned a home computer, 455 a video, 405 two cars, and 265 households owned neither a home computer, nor a video, nor two cars. Given that 145 households owned both a home computer and a video, 195 both a video and two cars, and 110 both two cars and a home computer, find the number of households surveyed which owned:

(i) a home computer, a video and two cars;
(ii) a video only;
(iii) two cars, a video but not a home computer;
(iv) a video, a home computer but not two cars.

11. In a certain village, there are three sports clubs: the soccer club, the rugby club and the cricket club. Everyone who belongs to the cricket club also belongs to the soccer club or rugby club (or both). The following additional information is known:

42 people belong to the soccer club;
45 people belong to the rugby club;

7 people belong to both the soccer and rugby clubs;
11 people belong to both the soccer and cricket clubs;
28 people belong to both the rugby and cricket clubs;
twice as many people belong only to the soccer club as belong only to the rugby club.

Find the number of people in the village who belong to

(i) all three clubs
(ii) the cricket club
(iii) only the soccer club.

3.6 Families of Sets

In section 3.3, we defined the intersection and union of a collection of n sets as follows:

$$\bigcap_{r=1}^{n} A_r = A_1 \cap A_2 \cap \cdots \cap A_n = \{x : x \in A_r \text{ for each } r = 1, 2, \ldots, n\}$$

and

$$\bigcup_{r=1}^{n} A_r = A_1 \cup A_2 \cup \cdots \cup A_n = \{x : x \in A_r \text{ for some } r = 1, 2, \ldots, n\}.$$

In this section we turn our attention to more general 'families' or 'collections' of sets which will include the case where there are infinitely many sets in the family. By a family or collection of sets, we really mean a *set* of sets, although the terms 'family of sets' or 'collection of sets' are both in widespread use and we shall use the three terms interchangeably. Before we can consider intersections and unions of arbitrary families of sets, we need first to describe carefully what we mean by such a family.

In the examples above, we have defined the intersection and union of the family (or set) of sets $\{A_1, A_2, \ldots, A_n\}$. In this family, the integers $1, 2, \ldots, n$ serve as labels to distinguish the various sets in the collection. In principle, any collection of labels would be suitable; for example, if we were to choose *Alice*, *Bob*, ..., *Nina* as labels, then we could write the family as $\{A_{\mathit{Alice}}, A_{\mathit{Bob}}, \ldots, A_{\mathit{Nina}}\}$. In practice, the labels $1, 2, \ldots, n$ are usually preferable. Whatever labels we choose form an **indexing set** or **labelling set** I for the collection. For the collection $\{A_1, A_2, \ldots, A_n\}$, the indexing set is $I = \{1, 2, \ldots, n\}$ and we can write the family as

$$\{A_i : i \in I\} = \{A_1, A_2, \ldots, A_n\}.$$

Using this idea of indexing set, we can define more general families of sets. For example, any collection of sets that has $\mathbb{Z}^+$ as the indexing set will contain infinitely many sets, one corresponding to each positive integer:

$$\{A_r : r \in \mathbb{Z}^+\} = \{A_1, A_2, A_3, \ldots\}.$$

If the set of real numbers $\mathbb{R}$ is the indexing set then the resulting family of sets $\{A_r : r \in \mathbb{R}\}$ also contains infinitely many sets, but this time we cannot list them even in an infinite list (see §5.5 for further details of the quantitative difference between the infinite sets $\mathbb{Z}$ and $\mathbb{R}$).

An arbitrary family of sets is of the form $\mathscr{F} = \{A_i : i \in I\}$ where I is *any* (indexing) set; in such a collection $\mathscr{F}$, there is exactly one set A_r for each element r of the indexing set I. Recall that the indexing set is just a collection of labels for the sets in the family $\mathscr{F}$. It is now straightforward to modify the definition given at the beginning of the section and define the intersection and union of the family $\mathscr{F}$ as follows:

$$\bigcap_{i \in I} A_i = \{x : x \in A_i \text{ for } \textit{all } i \in I\}$$

$$\bigcup_{i \in I} A_i = \{x : x \in A_i \text{ for } \textit{some } i \in I\}.$$

Examples 3.10

1. The definitions given above for intersection and union of arbitrary families of sets include as special cases our previous definitions for finite collections of sets. For example, let $I = \{1, 2\}$. A corresponding family of sets is $\{A_1, A_2\}$. Now

$$\bigcap_{i \in I} A_i = \{x : x \in A_i \text{ for } i = 1 \text{ and } i = 2\} = \{x : x \in A_1 \text{ and } x \in A_2\}$$
$$= A_1 \cap A_2$$
$$\bigcup_{i \in I} A_i = \{x : x \in A_i \text{ for } i = 1 \text{ or } i = 2\} = \{x : x \in A_1 \text{ or } x \in A_2\}$$
$$= A_1 \cup A_2.$$

So the definitions above agree with our previous definitions of intersection and union of two sets.

2. Let $I = \mathbb{Z}^+ = \{1, 2, 3, \ldots\}$, and for each $i \in \mathbb{Z}^+$ let $A_i = \{i\}$. Thus $A_1 = \{1\}$, $A_2 = \{2\}$, etc. Therefore

$$\bigcap_{i \in \mathbb{Z}^+} A_i = \varnothing \quad \text{and} \quad \bigcup_{i \in \mathbb{Z}^+} A_i = \{1, 2, 3, \ldots\} = \mathbb{Z}^+.$$

When the indexing set is $\mathbb{Z}^+$ we frequently write

$$\bigcap_{i=1}^{\infty} A_i \quad \text{for} \quad \bigcap_{i \in \mathbb{Z}^+} A_i \quad \text{and} \quad \bigcup_{i=1}^{\infty} A_i \quad \text{for} \quad \bigcup_{i \in \mathbb{Z}^+} A_i.$$

3. Let $I = \mathbb{Z}^+$ and for each $n \in \mathbb{Z}^+$ let $A_n = \{k \in \mathbb{Z} : k \leqslant n\}$. Thus:

$$\begin{aligned} A_1 &= \{k \in \mathbb{Z} : k \leqslant 1\} = \{\ldots, -3, -2, -1, 0, 1\} \\ A_2 &= \{k \in \mathbb{Z} : k \leqslant 2\} = \{\ldots, -3, -2, -1, 0, 1, 2\} \\ A_3 &= \{k \in \mathbb{Z} : k \leqslant 3\} = \{\ldots, -3, -2, -1, 0, 1, 2, 3\}, \text{ etc.} \end{aligned}$$

Then

$$\bigcap_{n=1}^{\infty} A_n = \{k \in \mathbb{Z} : k \leqslant n \text{ for all } n \in \mathbb{Z}^+\} = \{k \in \mathbb{Z} : k \leqslant 1\} = A_1.$$

Note that the family satisfies $A_1 \subseteq A_2 \subseteq A_3 \subseteq \cdots \subseteq A_n \subseteq A_{n+1} \subseteq \cdots$. Whenever this is the case, we have $\bigcap_{n=1}^{\infty} A_n = A_1$.

Now

$$\bigcup_{n=1}^{\infty} A_n = \{k \in \mathbb{Z} : k \leqslant n \text{ for some } n \in \mathbb{Z}^+\}.$$

Note that *every* integer k satisfies $k \leqslant n$ for some $n \in \mathbb{Z}^+$: if $k > 0$ we may take $n = k$, and if $k \leqslant 0$ we may take $n = 1$. Therefore every integer k belongs to the union of the family so

$$\bigcup_{n=1}^{\infty} A_n = \mathbb{Z}.$$

4. Let $I = \mathbb{R}$, the set of real numbers, and for each $m \in I$ let A_m be the set of points in the plane which lie on the line of gradient m which passes through the origin $(0, 0)$—figure 3.12. That is,

$$A_m = \{(x, y) : x \text{ and } y \text{ are real numbers and } y = mx\}.$$

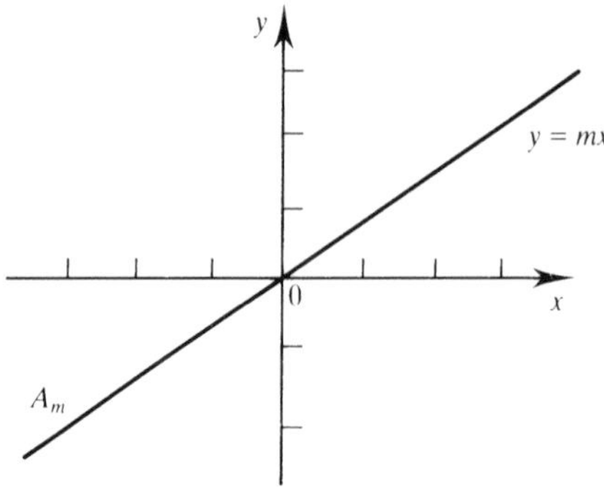

Figure 3.12

Note that in this case we cannot list the sets in the family $\{A_m : m \in \mathbb{R}\}$, even in an infinite list. This is because the real numbers themselves cannot be listed in an infinite list $x_1, x_2, x_3, \ldots$ (see §5.5).

Then

$$\bigcap_{m \in I} A_m = \{(0,0)\}$$

since the origin $(0,0)$ is the unique point common to all such lines.

The union

$$\bigcup_{m \in I} A_m$$

is the whole plane except the positive and negative parts of the y-axis. Points on the y-axis (except the origin) do not occur in the union because none of the lines A_m are vertical. The union of the sets A_m can also be defined by

$$\{(x, y) : x \text{ and } y \text{ are real numbers and } x \neq 0\} \cup \{(0,0)\}.$$

Power Set

Given any set A we can define the set consisting of all subsets of A. Called the 'power set of A', this is almost certainly the most widely used and important example of a family of sets.

Definition 3.3

Let A be any set. The **power set** of A, denoted $\mathscr{P}(A)$, is the set of all subsets of A:

$$\mathscr{P}(A) = \{B : B \subseteq A\}.$$

Notice that the power set of any set A contains $\varnothing$ and A since both are subsets of A. In particular the power set is necessarily non-empty.

The following theorem shows how the power set is related to subset, intersection and union.

Theorem 3.4

For all sets A and B:

(i) $A \subseteq B$ if and only if $\mathscr{P}(A) \subseteq \mathscr{P}(B)$.
(ii) $\mathscr{P}(A) \cap \mathscr{P}(B) = \mathscr{P}(A \cap B)$.
(iii) $\mathscr{P}(A) \cup \mathscr{P}(B) \subseteq \mathscr{P}(A \cup B)$.

Proof

We shall prove part (i) as an illustration; the proofs of parts (ii) and (iii) are left as exercises.

To prove the biconditional statement we prove the two conditional statements:

$$A \subseteq B \Rightarrow \mathscr{P}(A) \subseteq \mathscr{P}(B) \quad \text{and} \quad \mathscr{P}(A) \subseteq \mathscr{P}(B) \Rightarrow A \subseteq B.$$

Firstly, suppose $A \subseteq B$. We must show that $\mathscr{P}(A) \subseteq \mathscr{P}(B)$, so let $X \in \mathscr{P}(A)$. This means $X \subseteq A$. Since $A \subseteq B$, it follows from exercise 3.2.10(i) that $X \subseteq B$, which means that $X \in \mathscr{P}(B)$. Since $X \in \mathscr{P}(A)$ implies $X \in \mathscr{P}(B)$, we conclude that $\mathscr{P}(A) \subseteq \mathscr{P}(B)$, which completes the first half of the proof.

To prove the converse statement, suppose $\mathscr{P}(A) \subseteq \mathscr{P}(B)$. Since $A \in \mathscr{P}(A)$, it follows that $A \in \mathscr{P}(B)$. This means that $A \subseteq B$, which completes the proof. $\square$

Examples 3.11

1.
$$\begin{aligned}\mathscr{P}(\varnothing) &= \{\varnothing\}\\ \mathscr{P}\{a\} &= \{\varnothing, \{a\}\}\\ \mathscr{P}\{a,b\} &= \{\varnothing, \{a\}, \{b\}, \{a,b\}\}\\ \mathscr{P}\{a,b,c\} &= \{\varnothing, \{a\}, \{b\}, \{c\}, \{a,b\}, \{a,c\}, \{b,c\}, \{a,b,c\}\}.\end{aligned}$$

2. Let $A = \{1,2,3\}$ and $B = \{1,2\}$. Determine whether each of the following is true or false and give a brief justification.

(i) $B \in \mathscr{P}(A)$
(ii) $B \in A$
(iii) $A \in \mathscr{P}(A)$
(iv) $A \subseteq \mathscr{P}(A)$
(v) $B \subseteq \mathscr{P}(A)$
(vi) $\{\{1\}, B\} \subseteq \mathscr{P}(A)$
(vii) $\varnothing \in \mathscr{P}(A)$
(viii) $\varnothing \subseteq \mathscr{P}(A)$.

Solution

(i) True: B is a subset of A so B is an element of its power set.
(ii) False: B is a set but the elements of A are numbers, so B is not an element of A.
(iii) True: since $A \subseteq A$ it follows that $A \in \mathscr{P}(A)$. In fact, as noted above, this is the case for any set A.
(iv) False: the elements of A are numbers whereas the elements of $\mathscr{P}(A)$ are sets (namely subsets of A). Hence the elements of A cannot also be elements of $\mathscr{P}(A)$, so $A \not\subseteq \mathscr{P}(A)$.
(v) False: for the same reasons as given in part (iv).
(vi) True: $\{1\} \in \mathscr{P}(A)$ (since $\{1\} \subseteq A$) and $B \in \mathscr{P}(A)$ (part (i)) so each element of the set $\{\{1\}, B\}$ is also an element of $\mathscr{P}(A)$; hence $\{\{1\}, B\} \subseteq \mathscr{P}(A)$.
(vii) True: since $\varnothing \subseteq A$, we have $\varnothing \in \mathscr{P}(A)$.
(viii) True: $\varnothing \subseteq X$ for every set X and $\mathscr{P}(A)$ is certainly a set, so $\varnothing \subseteq \mathscr{P}(A)$.

3. Again we emphasize that great care should be taken to use $\in$ and $\subseteq$ correctly. For instance, if $a \in A$ then $\{a\} \subseteq A$ so $\{a\} \in \mathscr{P}(A)$. There is particular scope for confusion when x and $\{x\}$ are *both* elements of a set X.

Let $A = \{1, 2, \{1\}\}$. Then $1 \in A$ so $\{1\} \subseteq A$, and therefore $\{1\} \in \mathscr{P}(A)$. In this case $\{1\} \in A$ as well, so $\{\{1\}\} \in \mathscr{P}(A)$. In fact

$$\mathscr{P}(A) = \{\varnothing, \{1\}, \{2\}, \{\{1\}\}, \{1, 2\}, \{1, \{1\}\}, \{2, \{1\}\}, A\}.$$

Recall that 1, $\{1\}$, $\{\{1\}\}$ are all different. The first is a number, the second a set whose only element is a number, and the third a set whose only element is a set. Clearly we could continue in this way to produce an infinite sequence of different sets:

$$\{1\}, \{\{1\}\}, \{\{\{1\}\}\}, \ldots.$$

Each set in this sequence (except the first) could be defined as the set whose single element is the previous set in the sequence. More precisely, if we define

$$X_1 = \{1\} \quad \text{and} \quad X_{n+1} = \{X_n\} \quad \text{for } n = 1, 2, 3, \ldots$$

then the sequence $X_1, X_2, X_3, \ldots$ is identical with the sequence of sets above. As a final step, let

$$X = \bigcup_{n \in \mathbb{Z}^+} X_n = \bigcup_{n=1}^{\infty} X_n.$$

Note that X *is* a well defined set: given x we can decide definitely whether $x \in X$ or $x \notin X$. If x is of the form $\{\cdots\{1\}\cdots\}$, where a finite number of braces appear, then $x \in X$; otherwise $x \notin X$.

We could define this union X directly as follows:

$$X = \{x : x = \{1\} \text{ or } x = \{y\} \text{ where } y \in X\}.$$

This is an example of a **recursively defined set**—that is, one defined partially in terms of itself. Of course, we cannot define a set completely in terms of itself, which is why we also need $x = \{1\}$ as part of the definition.

The idea of recursion—defining something partially in terms of itself—is important in mathematics and computer science, both theoretically and practically. In computing, for example, many high-level programming languages allow procedures to call themselves—such procedures are called **recursive**.

The sets given in example 3.11.1 above suggest that, if A is finite and $|A| = n$ then $|\mathscr{P}(A)| = 2^n$. To prove this let A be the set $\{a_1, a_2, \ldots, a_n\}$. We can form a subset of A by considering each element a_i in turn and either including it or not in the subset. For each element there are two choices (either include it or don't) and the choice for each element is independent of the choices for the other elements, so there are 2^n choices altogether. Each of these 2^n choices gives a different subset and every subset of A can be obtained in this way. We have proved the following theorem (which can also be proved by mathematical induction).

Theorem 3.5

If $|A| = n$ then $|\mathscr{P}(A)| = 2^n$.

Some authors use 2^A to denote the power set: then theorem 3.5 takes the elegant form $|2^A| = 2^{|A|}$.

Partitions of a Set

It is sometimes important to divide a set into non-intersecting subsets. For instance, in §3.4, this device was frequently used when counting elements of sets. Such a division of a set into non-intersecting subsets is called a 'partition' of the set. It is closely related to the important notion of an equivalence relation on a set, which is introduced in the next chapter.

Definition 3.4

Let A be a set. A **partition** of A is a family (i.e. a *set*) $\{S_i : i \in I\}$ of non-empty subsets of A such that:

(i) $\bigcup_{i \in I} S_i = A$, and

(ii) $S_i \cap S_j = \varnothing$ if $i \neq j$, for all $i, j \in I$.

The first condition says that the sets S_i in the family 'fill out' all of A. The second condition says that, for any pair of sets S_i, S_j in the partition, these sets are disjoint. Whenever the second condition is satisfied we say that the sets S_i, $i \in I$, are **pairwise disjoint**. It is useful to visualize the elements S_i of the partition as non-overlapping 'blocks' which fit together to form A rather like the pieces of a jigsaw puzzle—see figure 3.13. Using this analogy, the first condition of definition 3.4 says that there are no missing pieces to the jigsaw puzzle and the second condition says that the pieces fit together 'snugly' with no overlaps between pieces. Clearly these are exactly the properties required of the pieces of a jigsaw puzzle.

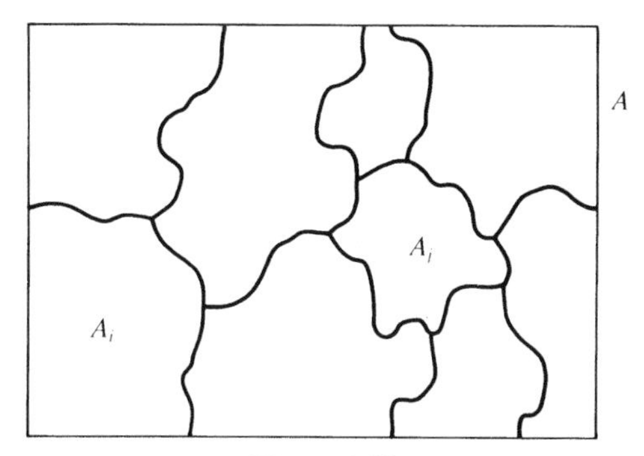

Figure 3.13

Perhaps it is worth pointing out that pairwise disjoint is a *stronger* condition than requiring the intersection of the whole family to be the empty set. For example, if $A = \{1, 2\}$, $B = \{2, 3\}$ and $C = \{3, 4\}$ then the family $\{A, B, C\}$ is *not* pairwise disjoint since $A \cap B \neq \varnothing$, for example. However, $A \cap B \cap C = \varnothing$, since there is no element common to *all three* sets.

Examples 3.12

1. $\{\{1\}, \{2, 3\}, \{4, 5, 6\}\}$ is a partition of $\{1, 2, 3, 4, 5, 6\}$.

2. Each of the following is a partition of $\mathbb{Z}$, the set of integers.

   (i) $\{\mathbb{Z}^-, \{0\}, \mathbb{Z}^+\}$, where $\mathbb{Z}^-$ and $\mathbb{Z}^+$ are the sets of negative and positive integers respectively.
   (ii) $\{\mathbb{E}, \mathbb{O}\}$, where $\mathbb{E} = \{\ldots, -4, -2, 0, 2, 4, 6, \ldots\}$, the set of even integers, and $\mathbb{O} = \{\ldots, -3, -1, 1, 3, 5, 7, \ldots\}$, the set of odd integers.
   (iii) $\{\{n\} : n \in \mathbb{Z}\}$.

   Clearly, for any set A we can form a partition in this way by taking the sets in the partition to be all the singleton subsets of A. (A **singleton set** is simply a set with only one element.)

3. For each real number α, let L_α be the set of points in the plane which lie on the vertical line through the point $(\alpha, 0)$:

$$L_\alpha = \{(x, y) : x = \alpha \text{ and } y \text{ is a real number}\}$$

$$= \{(\alpha, y) : y \in \mathbb{R}\}.$$

The family of these sets, $\{L_\alpha : \alpha \in \mathbb{R}\}$, is a partition of the plane: every point of the plane lies on one of the lines L_α and any two of the lines are disjoint.

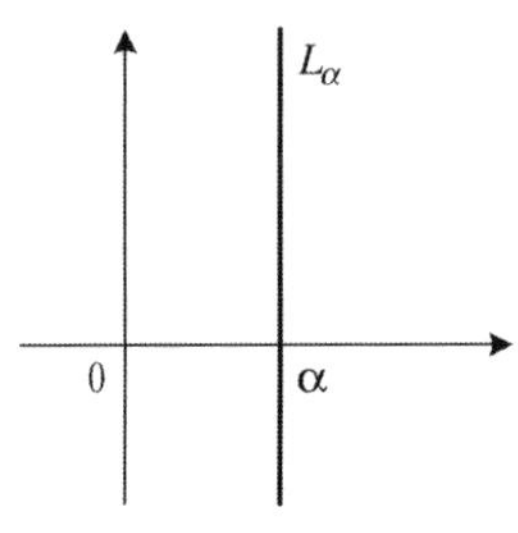

Figure 3.14

Exercises 3.5

1. List the elements of $\mathscr{P}(A)$ in the following cases:

   (i) $A = \{a, b, c, d\}$
   (ii) $A = \{\{1\}, \{1, 2\}\}$
   (iii) $A = \{\{1\}, \{1, 2\}, \{1, 2, 3\}\}$
   (iv) $A = \mathscr{P}\{1, 2\}$
   (v) $A = \mathscr{P}(\varnothing)$.

2. Let $A = \{1, 2, 3, 4, 5, 6, 7, 8, 9, 10\}$. Determine whether each of the following is a partition of A. If the set is *not* a partition, explain why not.

   (i) $\{1, 2, \{3, 4\}, \{5, 6\}, \{7, 8\}, \{9, 10\}\}$
   (ii) $\{\{1, 2\}, \{3, 4\}, \{5, 6\}, \{7, 8\}, \{9, 10\}\}$
   (iii) $\{\{1, 3, 5, 7, 9\}, \{2, 4, 8\}, \{10\}\}$
   (iv) $\{\{1, 5\}, \{2, 6, 10\}, \{3\}, \{4, 7, 9\}, \{8\}\}$
   (v) $\{\{2, 8, 10\}, \{1, 6\}, \{3, 4, 5\}, \{7, 8, 9\}\}$.

3. Which of the following are partitions of the set $\{2, 3, 7, 9, 10\}$?

   (i) $\{\{2, 3\}, \{3, 7, 9\}, \{10\}\}$
   (ii) $\{\{2, 10\}, \{3, 7\}, \{9\}\}$
   (iii) $\{\{2, 3, 4\}, \{7, 9, 10\}\}$
   (iv) $\{\{2\}, \{3\}, \{7\}, \{9\}, \{10\}\}$

(v) $\{2, 3, 7, 9, 10\}$
(vi) $\{\{2, 3, 7, 9, 10\}\}$
(vii) $\{\{10, 3\}, \{7, 2\}\}$
(viii) $\{\{2, 9, 10\}, \{3, 7\}, \varnothing\}$.

4. (i) How many partitions are there of the set $\{a, b, c, d\}$?
(ii) Find all the partitions, if any, of the empty set $\varnothing$.

5. Let $\{A_m : m \in \mathbb{R}\}$ be the family of sets defined in example 3.10.4—that is, A_m is the set of points in the plane lying on the line $y = mx$.

Is $\{A_m : m \in \mathbb{R}\}$ a partition of the plane? Explain your answer.

6. Which of the following families of sets are partitions of the set $\mathbb{Z}$ of integers? Explain your answers.

(i) $\{\{n, n + 1\} : n \in \mathbb{Z}\}$
(ii) $\{\{-n, n\} : n \in \mathbb{Z}^+\}$
(iii) $\{\{n, n^2, n^3\} : n \in \mathbb{Z}\}$
(iv) $\{\{2n : n \in \mathbb{Z}\}, \{2n + 1 : n \in \mathbb{Z}\}\}$.

7. Which of the following are partitions of $\mathbb{R}$, the set of real numbers? Explain your answers.

(i) $\{I_n : n \in \mathbb{Z}\}$, where $I_n = \{x \in \mathbb{R} : n \leqslant x \leqslant n + 1\}$.
(ii) $\{J_n : n \in \mathbb{Z}\}$, where $J_n = \{x \in \mathbb{R} : n \leqslant x < n + 1\}$.
(iii) $\{K_n : n \in \mathbb{Z}\}$, where $K_n = \{x \in \mathbb{R} : n < x < n + 1\}$.

8. Define a sequence of sets $X_0, X_1, X_2, \ldots$ by $X_0 = \varnothing$ and, for $n > 0$, $X_{n+1} = X_n \cup \{X_n\}$.

List the elements of X_1, X_2, and X_3.

What is the cardinality of X_n?

Give a recursive definition for the union

$$X = \bigcup_{n=0}^{\infty} X_n.$$

(This sequence of sets was invented/discovered in the 1920s by the mathematician, and later theoretical computer scientist, John von Neumann (1903–57)†. His idea was to start with only the empty set and

† To some extent this discovery of von Neumann's was anticipated some 20 years earlier by Bertrand Russell.

'create' the natural numbers. Von Neumann *defined* the natural number n to be the set X_n.)

9. Prove parts (ii) and (iii) of theorem 3.4.

   Find sets A and B such that $\mathscr{P}(A) \cup \mathscr{P}(B)$ is a *proper* subset of $\mathscr{P}(A \cup B)$.

10. Use the Principle of Mathematical Induction to prove theorem 3.5.

3.7 The Cartesian Product‡

The order in which the elements of a (finite) set are listed is immaterial; in particular, $\{x, y\} = \{y, x\}$. In some circumstances, however, order is significant. For instance, in coordinate geometry the points with coordinates $(1, 2)$ and $(2, 1)$, respectively, are distinct. We therefore wish to define, in the context of sets, something akin to the coordinates of points used in analytical geometry.

In order to deal with situations where order is important, we define the **ordered pair** (x, y) of objects x and y, to be such that

$$(x, y) = (x', y') \quad \text{if and only if } x = x' \text{ and } y = y'. \qquad (*)$$

With this definition it is clear that (x, y) and (y, x) are different (unless $x = y$), so the order is significant. It could be argued, with justification, that we have not really *defined* the ordered pair, but merely listed a *property* which we desire of it. Those who are concerned about the way we have plucked the ordered pair out of thin air, as it were, should note that (x, y) can be defined in terms of (unordered) sets considered earlier. (See exercise 3.6.1 for a way of doing this.) We have not formally defined the ordered pair in this way because the particular choice of definition (and there is more than one way of defining (x, y)) is unimportant. What *is* significant about the ordered pair is precisely the property (*) above.

We are now in a position to define the Cartesian product of two sets, a concept which is fundamental to several later chapters.

‡ Named after the French mathematician and philosopher René Descartes (1596–1650), the founder of analytical geometry.

Definition 3.5

The **Cartesian product**, $X \times Y$, of two sets X and Y is the set of all ordered pairs (x, y) where x belongs to X and y belongs to Y:

$$X \times Y = \{(x, y) : x \in X \text{ and } y \in Y\}.$$

When $X = Y$, it is usual to denote $X \times X$ by X^2. This is read as 'X two' and not 'X squared'.

Note that, if either X or Y (or both) is the empty set then $X \times Y$ is also the empty set. For example, if $X = \varnothing$ then there are no elements x to place in the first position of the ordered pair (x, y), so there are no ordered pairs in $X \times Y$.

If X and Y are both non-empty, then $X \times Y = Y \times X$ if and only if $X = Y$. The implication in one direction is obvious; if $X = Y$ then clearly $X \times Y = Y \times X$. For the converse, we prove its contrapositive: if $X \neq Y$ then $X \times Y \neq Y \times X$. Now, if $X \neq Y$ then either there exists an element x^* which belongs to X but not to Y, or there exists an element y^* which belongs to Y but not to X (or both). In the former case, choose any element $y \in Y$—we can make such a choice since we are assuming that Y in non-empty. Now the ordered pair (x^*, y) belongs to $X \times Y$, but does not belong to $Y \times X$ since $x^* \notin Y$. In the latter case we choose an element $x \in X$; then (x, y^*) belongs to $X \times Y$ but not to $Y \times X$, since in this case $y^* \notin X$. Therefore in either case we can find an element which belongs to $X \times Y$ but not to $Y \times X$, so the sets are not equal.

Examples 3.13

1. If $X = \{1, 2\}$ and $Y = \{a, b, c\}$ then

$$X \times Y = \{(1, a), (1, b), (1, c), (2, a), (2, b), (2, c)\}.$$

The elements of the sets X, Y and $X \times Y$ can be represented systematically on a single Venn-Euler diagram, as in figure 3.15.

2. If $X = Y = \mathbb{R}$, the set of real numbers, then $X \times Y = \mathbb{R} \times \mathbb{R} = \mathbb{R}^2$ which is the coordinate geometry representation of the (two-dimensional) plane. The corresponding diagram to figure 3.15 in this case is the plane with its usual rectangular coordinate axes. A point P in the plane is represented by an ordered pair (x, y) of real numbers—its coordinates.

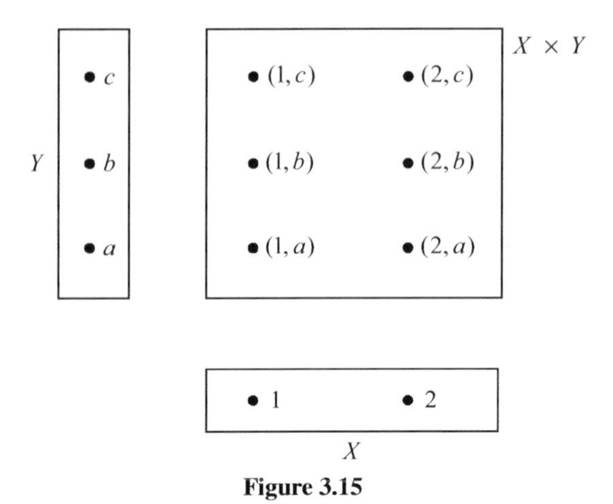

Figure 3.15

3. Let $X = \{\text{main courses offered by a certain restaurant}\}$ and $Y = \{\text{desserts offered by the same restaurant}\}$. The $X \times Y$ is the set of all (two-course) meals which can be ordered at the restaurant.

Diagrams such as figure 3.15 above and the coordinate geometry picture of the plane $\mathbb{R}^2 = \mathbb{R} \times \mathbb{R}$ are useful ways of visualizing Cartesian products. We can mimic these to obtain a pictorial way of representing an arbitrary Cartesian product $X \times Y$, given in figure 3.16. The sets X and Y are drawn as one-dimensional regions, rather than the usual two-dimensional ones in a Venn-Euler diagram. That is, X and Y are drawn as line segments, with elements belonging to them placed on the line segment. It is convenient to draw these lines perpendicular to one another with the line representing X horizontal. The Cartesian product is then represented as the rectangular region which lies above X and to the right of Y, and the ordered pair (x, y) is placed in this rectangle at the point vertically above x and horizontally to the right of y.

This type of diagram is useful for visualizing the intersections and unions of Cartesian products, and it also indicates other properties of the Cartesian product which are perhaps not so apparent from the ordered pair definition. For example, if we choose an element $x^* \in X$ and keep it fixed, then the set

$$\{x^*\} \times Y = \{(x^*, y) : y \in Y\}$$

is a 'copy' of Y in the sense that for every $y \in Y$ there corresponds one and only one element $(x^*, y) \in \{x^*\} \times Y$. This subset $\{x^*\} \times Y$ of $X \times Y$ can be visualized in figure 3.16 as the vertical line in $X \times Y$ which lies above the point in X representing the element x^*. We shall consider this kind of correspondence in more detail in chapter 5.

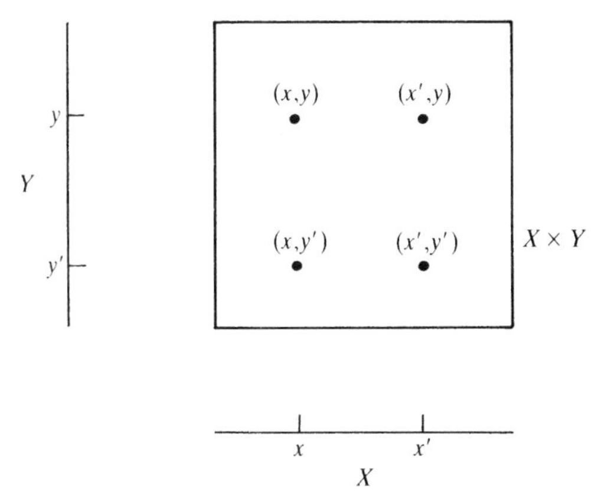

Figure 3.16

The ordered pair (x, y) may be generalized to an **ordered n-tuple** $(x_1, x_2, \ldots, x_n)$ with the property that

$$(x_1, x_2, \ldots, x_n) = (x_1', x_2', \ldots, x_n')$$

if and only if $x_1 = x_1', x_2 = x_2', \ldots, x_n = x_n'$.

Again we should note that the ordered n-tuple can be defined formally in terms of (unordered) sets. In particular, if ordered pairs have already been defined in terms of (unordered) sets (as indicated in exercise 3.6.1 below), then ordered n-tuples can be defined inductively using ordered pairs. (See exercise 3.6.7 for the details.)

ORDERED n-TUPLE

The Cartesian product of n sets is now a natural generalization of the case of two sets.

Definition 3.6

The **Cartesian product** of the sets $X_1, X_2, \ldots, X_n$ is

$$\begin{aligned} X_1 \times X_2 \times \cdots \times X_n \\ &= \{(x_1, x_2, \ldots, x_n) : x_1 \in X_1 \text{ and } x_2 \in X_2 \text{ and } \ldots \text{ and } x_n \in X_n\} \\ &= \{(x_1, x_2, \ldots, x_n) : x_i \in X_i \text{ for } i = 1, 2, \ldots, n\}. \end{aligned}$$

Again we write X^n (which is read 'X n' rather than 'X to the (power) n') in the case where $X_i = X$ for $i = 1, 2, \ldots, n$. For the general case, the Cartesian product $X_1 \times X_2 \times \cdots \times X_n$ is sometimes abbreviated

$$\mathop{\times}_{r=1}^{n} X_r.$$

Examples 3.14

1. If $A = \{1, 2\}$, $B = \{a, b\}$ and $C = \{\alpha, \beta\}$ then

$$\begin{aligned} A \times B \times C = \{&(1, a, \alpha), (1, a, \beta), (1, b, \alpha), (1, b, \beta), (2, a, \alpha), \\ &(2, a, \beta), (2, b, \alpha), (2, b, \beta)\}. \end{aligned}$$

   It is harder to picture the Cartesian product of three sets, $A \times B \times C$, in a diagram similar to figure 3.16 for two sets, but clearly its elements could be displayed in a three-dimensional region. Of course, for the Cartesian product of n sets, an n-dimensional region would be required, which is even more difficult to visualize!

2. As in the case of two sets, if any one (or more) of the sets X_r (for $r = 1, 2, \ldots, n$) is empty then so, too, is their Cartesian product

$$\mathop{\times}_{r=1}^{n} X_r.$$

   For instance, if $X_j = \varnothing$ then there is no element x_j to place in the jth position of the ordered n-tuple, so there can be no ordered n-tuples at all.

3. If $X_1 = X_2 = \cdots = X_n = \mathbb{R}$, then the Cartesian product $\mathbb{R}^n$ is the set of all n-tuples of real numbers $(x_x, x_2, \ldots, x_n)$. The set $\mathbb{R}^n$ is a coordinate representation of **real n-dimensional space**, which again is

somewhat (!) difficult to visualize. One of the reasons why ordered n-tuples are important is that they provide a framework for studying and understanding 'n-dimensional sets', whether in mathematics, computer science or elsewhere.

Of course, for the case $n = 3$, the set $\mathbb{R}^3$ is (or can be identified with) three-dimensional space familiar to those who have studied three-dimensional geometry.

4. We can extend example 3.13.3 by adding starters to the menu! Let $V = \{$starters offered by a certain restaurant$\}$ and, as before, $X = \{$main courses offered by the restaurant$\}$, $Y = \{$desserts offered by the restaurant$\}$. Then an ordered triple $(v, x, y) \in V \times X \times Y$ comprises a starter v, main course x and dessert y and so represents a three course meal. Therefore the Cartesian product $V \times X \times Y$ represents the set of all three course meals offered by the restaurant.

If X and Y are finite sets with $|X| = n$ and $|Y| = m$, then it is clear from the 'coordinate grid' diagram of $X \times Y$ (see figure 3.15) that the Cartesian product has nm elements. That is,

$$|X \times Y| = |X| \times |Y|.$$

This result clearly generalizes to the following for n sets, which may be proved formally using mathematical induction.

Theorem 3.6

If $X_1, X_2, \ldots, X_n$ are finite sets then

$$|X_1 \times X_2 \times \cdots \times X_n| = |X_1| \times |X_2| \times \cdots \times |X_n|.$$

We now turn to the question of how the Cartesian product operation behaves with respect to the other set theory operations such as intersection and union. Before we consider the general situation, let's look at two examples to see what is likely to happen in general.

Examples 3.15

1. Let $A = \{a, b, c, d\}$, $X = \{x, y, z\}$ and $Y = \{y, z, t\}$. Then

$$X \cap Y = \{y, z\}$$

so

$$A \times (X \cap Y) = \{(a, y), (a, z), (b, y), (b, z), (c, y), (c, z), (d, y), (d, z)\}.$$

Now

$$\begin{aligned} A \times X = \{&(a, x), (a, y), (a, z), (b, x), (b, y), (b, z), (c, x), \\ &(c, y), (c, z), (d, x), (d, y), (d, z)\}, \end{aligned}$$

and

$$\begin{aligned} A \times Y = \{&(a, y), (a, z), (a, t), (b, y), (b, z), (b, t), (c, y), \\ &(c, z), (c, t), (d, y), (d, z), (d, t)\}. \end{aligned}$$

Therefore

$$\begin{aligned} (A \times X) \cap (A \times Y) = \{&(a, y), (a, z), (b, y), (b, z), (c, y), (c, z), \\ &(d, y), (d, z)\}. \end{aligned}$$

Therefore, for the sets in *this* example,

$$A \times (X \cap Y) = (A \times X) \cap (A \times Y),$$

so we may wish to investigate whether this identity is true for *all* sets A, X and Y.

2. To investigate whether a similar identity may hold for unions, consider the sets $A = \{a, b\}$, $X = \{x, y\}$ and $Y = \{y, z\}$. Then $X \cup Y = \{x, y, z\}$, so

$$\begin{aligned} A \times (X \cup Y) &= \{(a, x), (a, y), (a, z), (b, x), (b, y), (b, z)\} \\ &= \{(a, x), (a, y), (b, x), (b, y)\} \\ &\qquad\qquad \cup \{(a, y), (a, z), (b, y), (b, z)\} \\ &= (A \times X) \cup (A \times Y). \end{aligned}$$

The results suggested by these examples do in fact hold for all sets A, X and Y. We list below identities which indicate how the Cartesian product behaves with respect to the intersection and union operations.

Theorem 3.7

(i) For all sets A, X and Y

$$A \times (X \cap Y) = (A \times X) \cap (A \times Y)$$

and

$$(X \cap Y) \times A = (X \times A) \cap (Y \times A).$$

(This says that the Cartesian product is **distributive** over intersection.)

(ii) For all sets A, X and Y

$$A \times (X \cup Y) = (A \times X) \cup (A \times Y)$$

and

$$(X \cup Y) \times A = (X \times A) \cup (Y \times A).$$

(This says that the Cartesian product is **distributive** over union.)

Proof

We shall prove the first identity in part (i) only—the others are left as exercises (3.6.9).

Let $(a, x) \in A \times (X \cap Y)$. By the definition of the Cartesian product, this means that $a \in A$ and $x \in (X \cap Y)$. Thus $x \in X$, so (a, x) belongs to $A \times X$; and $x \in Y$, so (a, x) belongs to $A \times Y$ as well. Therefore $(a, x) \in (A \times X) \cap (A \times Y)$, which proves that $A \times (X \cap Y) \subseteq (A \times X) \cap (A \times Y)$.

To prove the subset relation the other way round as well, let

$$(a, x) \in (A \times X) \cap (A \times Y).$$

Then $(a, x) \in (A \times X)$, so $a \in A$ and $x \in X$; and $(a, x) \in (A \times Y)$, so $a \in A$ and $x \in Y$. Therefore $a \in A$ and $x \in (X \cap Y)$ which means that the ordered pair (a, x) belongs to the Cartesian product $A \times (X \cap Y)$. Hence $(A \times X) \cap (A \times Y) \subseteq A \times (X \cap Y)$.

The conclusion that the sets $A \times (X \cap Y)$ and $(A \times X) \cap (A \times Y)$ are equal now follows, since each is a subset of the other. □

Figure 3.17 illustrates the identity proved above. The sets X and Y are both drawn as vertical line segments, which are kept distinct to avoid confusion over where one begins and the other ends. This means that it is more difficult to represent their intersection adequately—we have indicated $X \cap Y$ by a thickened line on both X and Y. The Cartesian products $A \times X$ and $A \times Y$ are shaded differently, the region of double shading representing $(A \times X) \cap (A \times Y)$. It is clear from the diagram that this doubly shaded region corresponds to the Cartesian product $A \times (X \cap Y)$.

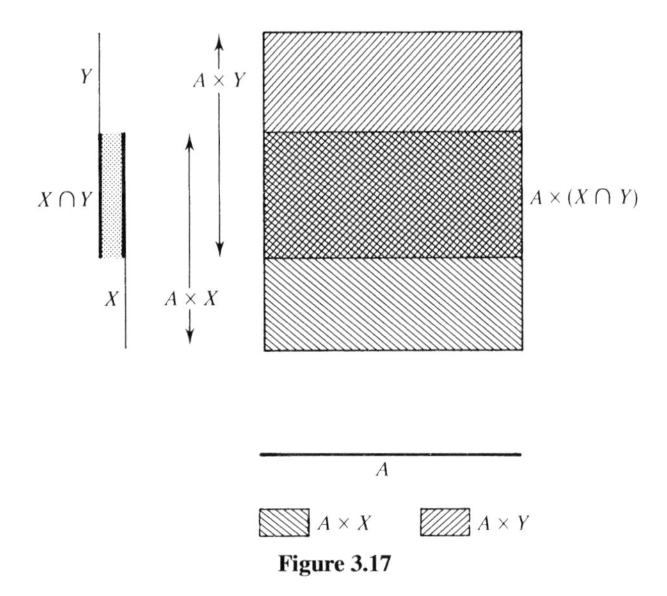

Figure 3.17

Finally, we state how the Cartesian product behaves with respect to the subset relationship. The proof of theorem 3.8 is left as an exercise (3.6.12). Before attempting to prove this, it is advisable to draw a 'coordinate grid' diagram to represent the situation.

Theorem 3.8

(i) For all sets A, B and X, $A \subseteq B$ implies $(A \times X) \subseteq (B \times X)$.

(ii) If X is *non-empty*, then $(A \times X) \subseteq (B \times X)$ implies $A \subseteq B$.

Exercises 3.6

1. (Kuratowski's definition† of the ordered pair.) If (x, y) is defined by $(x, y) = \{\{x\}, \{x, y\}\}$, show that

$$(x, y) = (a, b) \quad \text{if and only if } x = a \text{ and } y = b.$$

2. In each of the following cases list the elements of $X \times Y$, and draw a 'coordinate grid' diagram similar to figure 3.15:

   (i) $X = \{1, 2, 3, 4\}$ $Y = \{a, b\}$
   (ii) $X = \{1, 2\}$ $Y = \{a, b, c, d, e\}$
   (iii) $X = \{(1, 2)\}$ $Y = \{a, b, c, d, e\}$.

3. Let $A = \{1, 2, 3, 4\}$, $B = \{3, 4, 5\}$, $X = \{a, b\}$, $Y = \{b, c, d\}$. List the elements of each of the following sets.

   (i) $(A \cap B) \times (X \cap Y)$
   (ii) $(A \times X) \cap (B \times Y)$
   (iii) $(A \times Y) \cap (B \times X)$
   (iv) $(A \cap X) \times Y$
   (v) $(A \cap B) \times (X \cup Y)$
   (vi) $(A \times X) \cup (B \times Y)$.

4. In a simple library catalogue, each book has just four properties or attributes: title, author, class number, publication date. We assume that each book has a single author (co-authored books, such as this one, are listed only by their first named author) and that the class number is a positive decimal (for example, 314.25). Further, we assume that the library holds at most one copy of any book. Then each book in the library's stock corresponds to a unique quadruple of the form

$$(\textit{title}, \textit{author}, \textit{class number}, \textit{year of publication}).$$

   Let C (for 'collection') denote the set of all such quadruples corresponding to the books held in the library's collection. Then $C \subseteq T \times A \times \mathbb{R}^+ \times \mathbb{Z}$ where T is the set of all titles of books in the library's collection and A denotes the set of all authors of books in the library's collection. Informally, we can think of C as 'being' the set of all books in

† Named after the twentieth-century Polish mathematician Kazimierz Kuratowski, whose name is also associated with a theorem about planar graphs—see chapter 11.

the library's collection. Of course, this is not strictly correct since a book is not an ordered quadruple; however, there is an 'exact correspondence' between real books in the collection and quadruples in the set C. (The precise nature of such an 'exact correspondence' will be made clear in chapter 5.)

(i) Explain briefly why C is a *proper* subset of $T \times A \times \mathbb{R}^+ \times \mathbb{Z}$.

(ii) Let $D = \{n \in \mathbb{Z} : \text{there exists } (t, a, x, n) \in C\}$.

Describe in words what the set D represents in terms of the library's collection. What is the significance for the library of the smallest element of D?

(iii) Let $S = \{(t, a, x, n) \in C : a = \text{Shakespeare}\}$. Describe informally in words the set S.

(iv) Define formally (in a similar manner to the set S defined in part (iii)) the set of all books in C authored by 'Garnier'.

(v) Suppose $\{(t, a, x, n) \in C : x = 514.3\} = \varnothing$. What does this tell us about the library's collection?

(vi) Suppose $\{(t, a, x, n) \in C : t = \text{'Crime and Punishment'} \wedge a = \text{'Dostoyevsky'}\} \neq \varnothing$. What does this tell us about the library's collection?

Note: Representing objects by n-tuples corresponding to various attributes is a useful and extremely common way of organizing data. See sections 4.7 and 5.6 for a brief introduction to relational databases which are founded on the use of n-tuples in this way.

5. If $X \times Y = X \times Z$ does it *necessarily* follow that $Y = Z$? Explain your answer.

6. Let

$$[0,1] = \{x \in \mathbb{R} : 0 \leqslant x \leqslant 1\} \qquad (0,1) = \{x \in \mathbb{R} : 0 < x < 1\}$$
$$[0,1) = \{x \in \mathbb{R} : 0 \leqslant x < 1\} \qquad (0,1] = \{x \in \mathbb{R} : 0 < x \leqslant 1\}.$$

Describe (geometrically) each of the following sets:

(i) $[0,1] \times [0,1]$
(ii) $(0,1) \times (0,1)$
(iii) $[0,1) \times (0,1]$
(iv) $[0,1] \times (0,1)$.

7. (i) Defining the ordered triple (x, y, z) in terms of ordered pairs by

$$(x, y, z) = ((x, y), z)$$

show that $(x, y, z) = (a, b, c)$ if and only if $x = a$, $y = b$ and $z = c$.

(ii) If, for $n \geqslant 3$, ordered n-tuples are defined inductively by

$$(x_1, x_2, \ldots, x_n) = ((x_1, x_2, \ldots, x_{n-1}), x_n)$$

show that $(x_1, x_2, \ldots, x_n) = (y_1, y_2, \ldots, y_n)$ if and only if $x_i = y_i$ for each $i = 1, 2, \ldots, n$.

8. Let $A = \{a, b\}$ and $X = \{1, 2, 3\}$.

(i) List all the non-empty subsets of A and all the non-empty subsets of X.

(ii) List all the non-empty subsets of $A \times X$ which are of the form $B \times Y$ for some $B \subseteq A$ and some $Y \subseteq X$.

(iii) Write down a subset of $A \times X$ that is *not* of the form $B \times Y$ for some $B \subseteq A$ and some $Y \subseteq X$.

9. Prove the identities omitted from the proof of theorem 3.7. That is, for all sets A, X and Y:

(i) $(X \cap Y) \times A = (X \times A) \cap (Y \times A)$
(ii) $A \times (X \cup Y) = (A \times X) \cup (A \times Y)$
(iii) $(X \cup Y) \times A = (X \times A) \cup (Y \times A)$.

10. Using theorem 3.7 and the laws for the algebra of sets, show that, for all sets A, B, X and Y,

(i) $(A \cap B) \times (X \cap Y) = (A \times X) \cap (A \times Y) \cap (B \times X) \cap (B \times Y)$
(ii) $(A \cup B) \times (X \cup Y) = (A \times X) \cup (A \times Y) \cup (B \times X) \cup (B \times Y)$.

Draw diagrams to represent these identities. (Hint: for clarity in the diagram representing identity (ii), it is best to draw the sets A and B as if disjoint, and the sets X and Y also as if disjoint.)

11. (i) Prove that, for all sets A, B, X and Y,

$$\begin{aligned}(A \cap B) \times (X \cap Y) &= (A \times X) \cap (B \times Y) \\ &= (A \times Y) \cap (B \times X).\end{aligned}$$

Draw a diagram to represent each of these identities.

(ii) Find sets A, B, X and Y such that

$$(A \cup B) \times (X \cup Y) \neq (A \times X) \cup (B \times Y).$$

12. Prove theorem 3.8.

13. Prove that, for non-empty sets A, B, X and Y,

$$(A \times B) \subseteq (X \times Y) \quad \text{if and only if } A \subseteq X \text{ and } B \subseteq Y.$$

14. Prove each of the following identities, and draw diagrams to illustrate each:

(i) $(A - B) \times X = (A \times X) - (B \times X)$
(ii) $(A - B) \times (X - Y) = (A \times X) - [(A \times Y) \cup (B \times X)]$.

3.8 Types and Typed Set Theory

In software engineering, the notion of 'types' plays an important role in the various phases of software development: specification, design and implementation. Objects of different types behave differently and have different operations associated with them. In a software system managing a library, for example, objects classified as being of the type 'book' clearly have rather different properties than objects classified as being of the type 'borrower'. Similarly, in programming languages, variables need to be declared to be of type 'integer', 'real', 'string' and so forth, again because these types have different properties. In this section we will introduce types from a mathematical point of view and consider how we can formally define various operations on a type.

Consider the set of integers $\mathbb{Z}$. Various operations are defined on this set, such as addition, subtraction, multiplication and so on. In other words, given two integers n and m, we can define $n + m$, $n - m$, $n \times m$, etc. This is rather obvious, but note that the subset relation is not defined on the set of integers; if m and n are integers, then $n \subseteq m$ is meaningless. Similarly, the operations defined in chapter 1—conjunction, disjunction, implication, etc—are not defined on the set of integers; if n and m are integers then $n \wedge m$, $n \vee m$ and $n \rightarrow m$ are all meaningless.

Each operation defined on $\mathbb{Z}$ has a 'signature' which describes the 'inputs' and 'outputs' of the operation. The operations of addition, subtraction, multiplication each take two integers as 'input' and give a single integer as 'output'. For example, we could input 2 and 5 into the addition operation and obtain output 7; similarly inputting 2 and 5 into the subtraction or multiplication operation would give output -3 or 10 respectively. We say that each of these operations has signature

$$\mathit{Integer}, \mathit{Integer} \longrightarrow \mathit{Integer}$$

reflecting the fact that two integers are required for input and a single integer is the result of performing the operation.

Some integer operations take as their input a single integer. For example, the 'negation' operation or the 'square' operation both operate on a single integer. Input 3 into the 'negation' operation and the output is -3; similarly, input -2 and the output is $-(-2) = 2$. Or, for the 'square' operation, input 3 (or -3) and the output is 9. Each of these operations thus has signature

$$\mathit{Integer} \longrightarrow \mathit{Integer}.$$

We can now informally define the 'type' *Integer* to comprise the set of integers $\mathbb{Z}$ together with the operations that are defined on integers and their signatures. In general, a type T has a set of allowed values that variables of the type can take, together with a collection of operations in which variables of the type can participate as inputs or 'arguments'.

Other 'standard' types include the following.

Real	The type of the real numbers.
Boolean	The type of logical expressions (propositions and propositional functions).
String	The type of strings of characters (such as 'agxp nyt' or 'Hello Paul! How are you?').

Usually, (mathematical) operations have as input one or more arguments of the same type, but the output is frequently of a different type to the input(s). For example, if n and m are integers then $n + m$ and $n \leqslant m$ are both meaningful expressions but of a different nature: the value of $n + m$ is another integer but the value of $n \leqslant m$ is either 'true' or 'false'. More precisely, '$n \leqslant m$' is a proposition or propositional function (depending on whether the integers n and m are given specified values) and so is of type *Boolean*. Therefore the 'less than or equal' operation has signature

$$\mathit{Integer}, \mathit{Integer} \longrightarrow \mathit{Boolean}.$$

Examples 3.16

1. We summarize some (but not all) of the operations defined on the type *Integer*, together with their signatures.

$$\text{Addition} \qquad \_+\_ : \mathit{Integer}, \mathit{Integer} \rightarrow \mathit{Integer}.$$

The notation $\_+\_$ means that addition is an 'infix' operation where the sign of the operation comes in between the two integer 'arguments'. The two 'underscores' on either side of the addition sign represent placeholders which will be filled by the two integer input values.

Subtraction and multiplication have the same signature as addition.

$$\begin{array}{ll} \text{Subtraction} & \_-\_ : \mathit{Integer}, \mathit{Integer} \rightarrow \mathit{Integer} \\ \text{Multiplication} & \_\times\_ : \mathit{Integer}, \mathit{Integer} \rightarrow \mathit{Integer}. \end{array}$$

As noted above, negation takes a single integer as argument and returns the integer with the sign changed. For example, the negation of 2 is -2, the negation of -5 is 5, the negation of 0 is 0 ($=-0$). Negation is a 'prefix' operation because the operation sign precedes the input argument; it has the following signature.

$$\text{Negation} \qquad -\_ : \mathit{Integer} \rightarrow \mathit{Integer}.$$

Note that negation and subtraction are *different* operations because they have different signatures. Some people very sensibly use different words for the two operations: they say 'minus' for subtraction and 'negative' for negation.

$$\begin{array}{ll} 2-3 & \text{'two minus three'} \\ -3 & \text{'negative three'} \\ 2-(-3) & \text{'two minus negative three'.} \end{array}$$

Each of the order operations $<$ (less than), $\leqslant$ (less than or equal to), $>$ (greater than) and $\geqslant$ (greater than or equal to) are infix operations with the same signature. For example 'less than or equal to' has signature:

$$\_\leqslant\_ : \mathit{Integer}, \mathit{Integer} \rightarrow \mathcal{B}oolean.$$

The operation 'divides' means 'is a factor of' or 'goes exactly into'. For example, 6 divides 48 but 6 does not divide 15. What is the signature of the operation? It takes two integers as arguments, m and n say, and the result is either true or false. Hence the expression 'm divides n' has a

Boolean value (true or false) depending on whether m is or is not a factor of n. Therefore the operation has the following signature. (Note that $m|n$ is read as 'm divides n'.)

$$\text{Divides} \qquad \_|\_ : \mathit{Integer}, \mathit{Integer} \to \mathcal{Boolean}.$$

The 'absolute value' or 'modulus' operation, $|\_|$, takes a single integer as input and returns a non-negative integer as output. If the input value is greater than or equal to zero then the operation 'leaves it alone'; if the input value is negative then the operation returns the corresponding positive value. For example, $|3| = 3$, $|-3| = 3$, $|0| = 0$, etc. The absolute value operation has signature:

$$|\_| : \mathit{Integer} \to \mathit{Integer}.$$

2. Recall that $\mathcal{Boolean}$ is the type of logical expressions and we know that such expressions can have one of two values, T (true) or F (false). Therefore the set of values of the $\mathcal{Boolean}$ type is $\{\mathrm{T}, \mathrm{F}\}$. Some of the operations defined on the type $\mathcal{Boolean}$, together with their signatures are given below. The values returned by these operations are defined by the truth tables given in chapter 1.

Negation	$\neg\_ : \mathcal{Boolean} \to \mathcal{Boolean}$
Conjunction	$\_ \wedge \_ : \mathcal{Boolean}, \mathcal{Boolean} \to \mathcal{Boolean}$
Disjunction	$\_ \vee \_ : \mathcal{Boolean}, \mathcal{Boolean} \to \mathcal{Boolean}$
Exclusive disjunction	$\_ \veebar \_ : \mathcal{Boolean}, \mathcal{Boolean} \to \mathcal{Boolean}$
Conditional	$\_ \to \_ : \mathcal{Boolean}, \mathcal{Boolean} \to \mathcal{Boolean}$
Biconditional	$\_ \leftrightarrow \_ : \mathcal{Boolean}, \mathcal{Boolean} \to \mathcal{Boolean}.$

3. The type $\mathcal{Real}$ has the set of real numbers $\mathbb{R}$ as its set of values. Many of the operations defined on the type $\mathit{Integer}$ are also defined on the type $\mathcal{Real}$. Some of the operations defined on $\mathcal{Real}$ are given below, together with their signatures. We shall meet other operations on the type later.

Addition	$\_ + \_ : \mathcal{Real}, \mathcal{Real} \to \mathcal{Real}$
Subtraction	$\_ - \_ : \mathcal{Real}, \mathcal{Real} \to \mathcal{Real}$
Multiplication	$\_ \times \_ : \mathcal{Real}, \mathcal{Real} \to \mathcal{Real}$
Division	$\_/\_ : \mathcal{Real}, \mathcal{Real} \to \mathcal{Real}$
Negation	$-\_ : \mathcal{Real} \to \mathcal{Real}$
Less than or equal to	$\_ \leqslant \_ : \mathcal{Real}, \mathcal{Real} \to \mathcal{Boolean}$
Less than	$\_ < \_ : \mathcal{Real}, \mathcal{Real} \to \mathcal{Boolean}$

Greater than or equal to	$\_ \geqslant \_ : \mathit{Real}, \mathit{Real} \to \mathit{Boolean}$
Greater than	$\_ > \_ : \mathit{Real}, \mathit{Real} \to \mathit{Boolean}$
Square	$\_^2 : \mathit{Real} \to \mathit{Real}$
Square root	$\sqrt{\_} : \mathit{Real} \to \mathit{Real}$.

Typed Set Theory

Typed set theory is a more restricted version of set theory than the version considered in the previous sections of this chapter. In typed set theory, all the elements of a set are required to have the same type. Thus, for example, a set containing an element of type *Real* and an element of type *String* is not permitted. The notation for describing typed sets is slightly different from the notation we used previously for 'untyped' sets. If $\mathcal{T}$ is a type then the notation $x : \mathcal{T}$ means 'x is of type $\mathcal{T}$'. It is similar to $x \in A$ ('x belongs to A') but gives more information as it indicates the operations in which x may participate as an argument. We use the notation $\{x : \mathcal{T} \mid P(x)\}$ to define the set of all elements of type $\mathcal{T}$ for which the propositional function $P(x)$ is true. We read $\{x : \mathcal{T} \mid P(x)\}$ as 'the set of all x of type $\mathcal{T}$ such that $P(x)$ (is true)'. For example, $\{x : \mathit{Real} \mid x^2 = 2\} = \{-\sqrt{2}, \sqrt{2}\}$.

In typed set theory every element and every set is required to have a specified type. Any set whose elements are all of type $\mathcal{T}$ itself has type $\mathit{Set}[\mathcal{T}]$ which simply indicates that it is a set of 'things' of type $\mathcal{T}$. Thus a set of the form $\{x : \mathcal{T} \mid P(x)\}$ has type $\mathit{Set}[\mathcal{T}]$.

Examples 3.17

1. Let $A = \{n : \mathit{Integer} \mid -2 \leqslant n \leqslant 3\}$ and $B = \{n : \mathit{Real} \mid -2 \leqslant n \leqslant 3\}$. Then A has type $\mathit{Set}[\mathit{Integer}]$ and B has type $\mathit{Set}[\mathit{Real}]$. Also note that A is finite whereas B is infinite: $A = \{-2, -1, 0, 1, 2, 3\}$ but B contains all real numbers between -2 and 3 (inclusive) and it is impossible to list them.

2. We assume a type *Person* has been defined which is the type of all people, living or deceased. Although we shall not consider them here, we can imagine some of the operations defined on this type: *motherOf*, *age*, *gender*, *maritalStatus*, etc. Given this type, we can define various sets of people; for example:

   $A = \{x : \mathit{Person} \mid x$ is/was UK Prime Minister during

$$
\begin{aligned}
& \text{part of the period 1970–99}\} \\
&= \{\text{Blair, Callaghan, Heath, Major, Thatcher, Wilson}\} \\
B &= \{x : \mathit{Person} \mid x \text{ is/was US President during} \\
& \text{part of the period 1970–99}\} \\
&= \{\text{Bush, Carter, Clinton, Ford, Nixon, Reagan}\} \\
C &= \{x : \mathit{Person} \mid x \text{ was born on 29 February}\} \\
D &= \{x : \mathit{Person} \mid x \text{ has Spanish nationality}\}.
\end{aligned}
$$

Each of these sets has type $\mathit{Set}[\mathit{Person}]$.

3. Let $A = \{1, 2, 3\}$; then $A : \mathit{Set}[\mathit{Integer}]$. What is the type of $\mathscr{P}(A)$, the power set of A? Recall that the elements of $\mathscr{P}(A)$ are the subsets of A:

$$\mathscr{P}(A) = \{\varnothing, \{1\}, \{2\}, \{3\}, \{1,2\}, \{1,3\}, \{2,3\}, \{1,2,3\}\}.$$

The elements of $\mathscr{P}(A)$ have type $\mathit{Set}[\mathit{Integer}]$ so $\mathscr{P}(A)$ itself has type $\mathit{Set}[\mathit{Set}[\mathit{Integer}]]$ because it is a set of sets of integers.

4. Consider the informal definition of a set as 'the set of all cities in Canada.' In order to be able to define this as a *typed* set, we need to assume the existence of a type City, say, which is the type of cities. (We could imagine some of the operations that might be defined on this type: $\mathit{Population}(\_) : \mathit{City} \to \mathit{Integer}$, $\mathit{Mayor}(\_) : \mathit{City} \to \mathit{Person}$, $\mathit{Country}(\_) : \mathit{City} \to \mathit{Nation}$, and so on.) Provided City is a defined type, we can then define:

$$
\begin{aligned}
A &= \{x : \mathit{City} \mid x \text{ is in Canada}\} \\
&= \{\text{Ottawa, Montreal, Vancouver}, \ldots\} : \mathit{Set}[\mathit{City}].
\end{aligned}
$$

Operations on Typed Sets

The usual set theory operations—intersection, union, complement, and so on—are defined on typed sets. However, only sets of the *same* type can 'participate in' these operations. For example, if we were to attempt to form the union of a set of integers with a set of people, say, the result would not be a well formed typed set because its elements would not all be of the same type. For a fixed type $\mathcal{T}$, the signatures of the standard operations on the type $\mathit{Set}[\mathcal{T}]$ are given below. Intersection, union and difference are infix operations which take as arguments two sets of the same type and produce another set of the same type. Thus their signatures are the following.

Intersection $\quad \_ \cap \_ : \mathit{Set}[\mathcal{T}], \mathit{Set}[\mathcal{T}] \to \mathit{Set}[\mathcal{T}]$

Union	$\_ \cup \_ : \mathcal{S}et[\mathcal{T}], \mathcal{S}et[\mathcal{T}] \to \mathcal{S}et[\mathcal{T}]$
Difference	$\_ - \_ : \mathcal{S}et[\mathcal{T}], \mathcal{S}et[\mathcal{T}] \to \mathcal{S}et[\mathcal{T}]$.

Subset

If $A : \mathcal{S}et[\mathcal{T}]$ and $B : \mathcal{S}et[\mathcal{T}]$ then $A \subseteq B$ is either true or false. (Note that 'subset' behaves rather like 'less than or equal' in this respect.) Therefore 'subset' takes two sets of the same type as arguments and returns a Boolean expression so it has signature:

$$\_ \subseteq \_ : \mathcal{S}et[\mathcal{T}], \mathcal{S}et[\mathcal{T}] \to \mathcal{B}oolean.$$

Membership

For $x \in A$ to be defined, we require x and A to have appropriate types. More precisely, we require $x : \mathcal{T}$ and $A : \mathcal{S}et[\mathcal{T}]$ so that the types 'match'. Given this, what is the type of the statement $x \in A$? As with subset, '$A \subseteq B$', the expression '$x \in A$' is either true or false depending on whether or not x really is a member of the set A. Therefore set membership has signature

$$\_ \in \_ : \mathcal{T}, \mathcal{S}et[\mathcal{T}] \to \mathcal{B}oolean.$$

This means that the placeholder to the left of the membership symbol $\in$ can take a value of type $\mathcal{T}$ and the placeholder on the right takes a value of type $\mathcal{S}et[\mathcal{T}]$. Note that the set membership operation is unusual for a mathematical operation in that the types of its two inputs are necessarily different, $\mathcal{T}$ and $\mathcal{S}et[\mathcal{T}]$ respectively†.

Empty Set

What is the type of the empty set $\varnothing$? Given the signature of 'subset' defined above, we can only compare sets of the same type:

> writing $\varnothing \subseteq \{1, 2, 3\}$ implies that the empty set $\varnothing$ must be of type $\mathcal{S}et[\mathcal{I}nteger]$,
> writing $\varnothing \subseteq \{1.23, -19.857, \pi\}$ implies that $\varnothing$ is of type $\mathcal{S}et[\mathcal{R}eal]$,

† It is interesting to note that, in typed set theory, Russell's paradox disappears. (See exercise 3.2.12 for a discussion of Russell's paradox.) Indeed, in *Principia Mathematica*, their monumental work on the foundations of mathematics, Russell and Whitehead use a theory of types to avoid the paradox. The reason the paradox does not occur in typed set theory is that we are unable to form the set that gives rise to the difficulty. If $A : \mathcal{S}et[\mathcal{T}]$ then $A \notin A$, which is just shorthand for $\neg(A \in A)$, is not an allowed expression because it does not conform to the signature of $\notin$ (or $\in$). Hence we are unable to form the (typed) set R in exercise 3.2.12 that gives rise to the paradox.

writing $\varnothing \subseteq$ {Blair, Callaghan, Heath, Major, Thatcher, Wilson} implies that $\varnothing$ is of type $\mathcal{S}et[\mathcal{P}erson]$, etc.

There seems to be a difficulty here since we have stated that each set must have a unique type. The way around this problem is to define one empty set of *each* type. Thus there is an empty set of integers (containing no integers) which has type $\mathcal{S}et[\mathcal{I}nteger]$, an empty set of real numbers (containing no real numbers) which has type $\mathcal{S}et[\mathcal{R}eal]$, an empty set of people (containing no people) which has type $\mathcal{S}et[\mathcal{P}erson]$, and so on. We shall continue to use $\varnothing$ to denote each of these empty sets. It should usually be clear from the context which empty set is being represented by $\varnothing$.

Using a single notation to stand for several different concepts is called **overloading** the notation. Actually, we do this all the time in mathematics. For example, we use a single symbol $+$ to represent addition of integers, real numbers, matrices (see chapter 6), elements of an Abelian group (see chapter 8), and so forth. Similarly we use the symbol $-$ to represent both subtraction and negation of integers and of real numbers, and the difference of sets, etc. So using $\varnothing$ to denote the empty set of each type should not cause any difficulty.

Power Set

In example 3.17.3, we saw that if A is of type $\mathcal{S}et[\mathcal{I}nteger]$ then its power set $\mathscr{P}(A)$ is of type $\mathcal{S}et[\mathcal{S}et[\mathcal{I}nteger]]$. This generalizes to sets of any type. If a set A has type $\mathcal{S}et[\mathcal{T}]$ then any subset also has type $\mathcal{S}et[\mathcal{T}]$; therefore the *elements* of $\mathscr{P}(A)$ have type $\mathcal{S}et[\mathcal{T}]$ so $\mathscr{P}(A)$ itself has type $\mathcal{S}et[\mathcal{S}et[\mathcal{T}]]$. Since the power set operation takes a single set A as input and produces a single set $\mathscr{P}(A)$ as output, it has signature

$$\mathscr{P}(\_) : \mathcal{S}et[\mathcal{T}] \to \mathcal{S}et[\mathcal{S}et[\mathcal{T}]].$$

Cardinality

For finite sets†, the cardinality operation takes a set as argument and returns an integer value, namely the number of elements in the set. Hence cardinality has signature:

$$|\_| : \mathcal{S}et[\mathcal{T}] \to \mathcal{I}nteger.$$

† To include infinite sets, we would need to augment the $\mathcal{I}nteger$ type by adding a special symbol ∞ to produce a new type $\mathcal{I}nteger^*$ whose set of values is $\mathbb{Z} \cup \{\infty\}$. Then cardinality would have signature $|\_| : \mathcal{S}et[\mathcal{T}] \to \mathcal{I}nteger^*$.

Type Checking

Notice that, according to the signatures of $\cap$, $\cup$ and $-$ defined above, we can only form the union, intersection and difference of sets that are of the same type. Thus, for example, if $A : \mathit{Set}[\mathcal{S}]$ and $B : \mathit{Set}[\mathcal{T}]$ then $A \cap B$ is meaningless in typed set theory unless $\mathcal{S} = \mathcal{T}$. Similarly, if $x : \mathcal{S}$ and $A : \mathit{Set}[\mathcal{T}]$ then $x \in A$ is also meaningless unless $\mathcal{S} = \mathcal{T}$. (Actually, this is not quite true. As we shall see shortly, it is possible for $\mathcal{S}$ to be a 'subtype' of $\mathcal{T}$ and, in this case, $A \cap B$ and $x \in A$ are properly defined—see example 3.18.3.)

In fact, this phenomenon occurs in other situations in mathematics, even if we have not formally defined types. For example, if m and n are integers then the expression $(m \leqslant n) + 3$ is meaningless because the first argument of the addition operation is $m \leqslant n$ which is of type $\mathcal{Boolean}$ whereas addition requires two integers (or two reals) as arguments.

In each of these cases, there is an operation whose arguments do not match the signature of the operation. **Type checking** an expression means verifying that, for each operation in the expression, the types of its arguments agree with those specified by the signature of the operation. For example, if the expression includes $\cap$ then both its arguments must be sets of the same type; if it includes $+$ then both arguments must be integers (or both real numbers), and so on.

Examples 3.18

Suppose the following type declarations have been made:

$$
\begin{aligned}
&k, n, m : \mathit{Integer} \\
&x, y : \mathcal{Real} \\
&P, Q : \mathcal{Boolean} \\
&\text{Anne, Brian} : \mathcal{Person}.
\end{aligned}
$$

For each of the following statements or terms, decide whether it is meaningful (in other words, whether it 'type checks') and, if so, what is the type of the expression. (Assume the 'obvious' operations are defined on the type $\mathcal{Person}$.)

1. $n \geqslant m$.

   'Greater than or equal' is an infix operation that takes two integers as arguments, which is what we have here. Thus the expression type checks (i.e. it is meaningful). The expression has type $\mathcal{Boolean}$.

2. $(n \geqslant m) + k$.

This does not type check so is not meaningful. We noted above that $n \geqslant m$ has type $\mathcal{Boolean}$ but $+$ does not take a $\mathcal{Boolean}$ type as either argument.

3. $n + x$.

   Surprisingly, perhaps, this does not type check because addition (as given in examples 3.16) is defined between two integers or two real numbers. Addition either has signature $\_ + \_ : \mathit{Integer}, \mathit{Integer} \rightarrow \mathit{Integer}$ or $\_ + \_ : \mathit{Real}, \mathit{Real} \rightarrow \mathit{Real}$ but the given expression attempts to add an integer to a real.

   However $n+x$ clearly *ought* to be a meaningful expression—we have not previously had any difficulty in adding, say, 7 and 4.32! The way round this difficulty is to regard *Integer* as a **subtype** of *Real*. This means that we may substitute an integer value in any expression that requires a real argument. Clearly this can always be done: it amounts to regarding the integer 3, for example, as a real number 3.00. With this convention the expression $n + x$ is meaningful and has type *Real*.

4. Anne *IsOlderThan* Brian.

   Assuming that *IsOlderThan* is an infix operation with signature

   $$\_IsOlderThan\_ : \mathit{Person}, \mathit{Person} \rightarrow \mathit{Boolean}$$

   then the expression type checks and gives a *Boolean* result.

5. $n + \mathit{Age}(\text{Anne})$.

   This is meaningful provided *Age* has signature $\mathit{Person} \rightarrow \mathit{Integer}$ so that both n and *Age*(Anne) are of type *Integer*; then the expression has type *Integer*. (Alternatively, we could define *Age* to have signature $\mathit{Person} \rightarrow \mathit{Real}$, then the expression would also have type *Real*.)

6. $(x < y) \vee (P \rightarrow Q)$.

   This type checks as both $x < y$ and $P \rightarrow Q$ have type *Boolean*. The expression has type *Boolean*.

7. $\mathit{Age}(\text{Anne}) + 5 = \text{Brian}$.

   This is not meaningful since $\mathit{Age}(\text{Anne}) + 5 : \mathit{Integer}$ and Brian : *Person* but equality is only defined for values of the same type. Note that, however, $\mathit{Age}(\text{Anne}) + 5 = \mathit{Age}(\text{Brian})$ is meaningful and has type

$\mathcal{Boolean}$ because now both sides of the $=$ sign have the same type: *Integer*.

8. $(x + y) \leftrightarrow$ (Brian *IsSonOf* Anne).

We assume *IsSonOf* has signature $\_IsSonOf\_ : \mathcal{Person}, \mathcal{Person} \to \mathcal{Boolean}$. Then the given expression does not type check since $x + y : \mathcal{Real}$ and Brian *IsSonOf* Anne : $\mathcal{Boolean}$ but $\leftrightarrow$ requires both arguments to have type $\mathcal{Boolean}$.

Defining Operations: Preconditions and Postconditions

So far we have only defined the signature of various operations defined on *Integer*, $\mathcal{Boolean}$, $\mathcal{S}et[T]$ and so on. We have not defined the *behaviour* of any of the operations. This may not seem very important because we are all agreed what addition means for integers or intersection for sets and so on. However, being able to define precisely what an operation achieves is extremely important in software specification. To build a piece of software, it is vital to be able to define exactly what each component should do. Unfortunately, there are many examples of software failures because this precise specification stage has not been properly completed. We shall describe a way of specifying the behaviour of an operation using logical expressions as 'preconditions' and 'postconditions' for the operation. To keep the discussion as simple as possible, we shall restrict our examples to familiar mathematical operations.

A **precondition** is a condition that must be fulfilled before an operation can be invoked and a **postcondition** is a condition that is fulfilled as a result of the operation being invoked. We can think of the precondition and postcondition as defining a contract between the operation and any user of the operation. To satisfy the contract, the user is 'obliged' to supply values to the operation which satisfy the precondition; the operation is then 'required' to return a value which satisfies the postcondition. Note also that such a 'contract' only specifies *what* an operation should do and not *how* it should do it.

Examples 3.19

1. Consider the operation of division of real numbers. The operation takes two real numbers x and y as input and produces a single real number x/y as output. Therefore division has signature

$$\_/\_ : \mathcal{Real}, \mathcal{Real} \to \mathcal{Real}.$$

However, division by zero is meaningless so x/y is only defined when $y \neq 0$. If we are to 'feed in' two real numbers to the division operation, we had better ensure that the second of them is non-zero. This defines the precondition: $y \neq 0$.

Provided the precondition is satisfied, the result of performing the division operation on real numbers x, y is the real number which is 'x divided by y'. How can we define this number without simply asserting that it is x divided by y? For example, if we input π and 4, what is the property that the output number $\pi/4$ must satisfy? Suppose we have a 'division operation machine' that gives answers rounded to three decimal places; we input π and 4 and the machine returns the answer 0.792. Is the machine functioning correctly? The simplest test is to multiply the output by 4: $0.792 \times 4 = 3.168$ which is *not* the value of π correct to three decimal places. Therefore the division machine is faulty. In general, assuming that the operation of multiplication has been defined, the real number r that is the result of dividing x by y is defined by the equation $x = r \times y$. This equation is the postcondition.

We can now give the full description of the division operation. It has three parts: signature, precondition and postcondition.

$$\_/\_ : x : \mathcal{Real},\ y : \mathcal{Real} \to r : \mathcal{Real}$$

precondition $\quad y \neq 0$

postcondition $\quad x = r \times y.$

Note that we have extended the usual signature expression by adding labels for the input and output variables. This is so that we can refer to particular inputs and the output in the precondition and postcondition. For example, the precondition must state that it is the *second* of the two arguments that does not take a zero value. Hence we need to be able to distinguish between the two input variables.

This specification of division is the required contract between the operation and its user. The user is obliged to 'feed in' real numbers x and y satisfying the precondition $y \neq 0$; then the operation will keep its side of the contract by producing a real number r (which is the value of x/y) satisfying the postcondition: $x = r \times y$. The contract does not specify *how* the operation will calculate the value $r = x/y$—provided the result satisfies the postcondition, any method of calculation is acceptable. If we were required to implement the division operation as a software routine then the method of calculation would be important in terms of the speed of the operation and accuracy of the output. However, as far as the specification of the operation is concerned, these issues are not relevant.

2. Consider the 'square root' operation defined on real numbers. Since the operation takes a real number as input and returns a real number value, it has signature

$$\sqrt{\_} : \mathcal{Real} \to \mathcal{Real}.$$

What should be the precondition(s) and postcondition(s)? Imagine a square root machine as a 'black box'; we input a real number into the machine and out comes another real number.

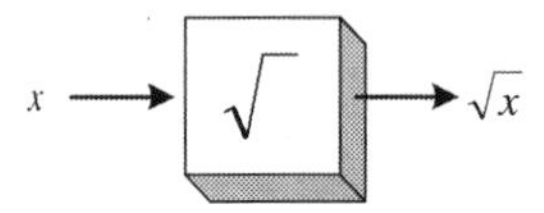

To determine the precondition, consider what real numbers we are allowed to 'feed into' the machine without 'breaking it' (we imagine that feeding in an illegal value is likely to break the machine). We cannot feed in a negative number since the square root of a negative real number is not defined (in the context of the type $\mathcal{Real}$). But this is the only restriction—any other real number is an allowed input. Therefore the precondition is $x \geqslant 0$ where x is the input value.

To determine the postcondition, suppose we feed in the value x and the real number r is the resulting output. What test(s) would need to be carried out on the output r in order to determine whether the machine was working properly? For example, suppose we feed in 7 and the machine (working to three decimal places, as before) outputs the answer 2.615. Is it working correctly? Since $2.615^2 = 2.615 \times 2.615 = 6.838$ to three decimal places (which is not equal to 7), we conclude the machine is not functioning properly.

In general, if we 'feed in' x (satisfying the precondition $x \geqslant 0$) the output r should satisfy $r^2 = x$ if the machine is working properly and this equation forms part of the postcondition. We are assuming here that the 'square' operation has been defined and properly specified—see example 3(i) below. If the square operation has not been specified then we would need to use the equation $r \times r = x$ in the postcondition in place of $r^2 = x$.

In fact there is another condition which must be satisfied. It is a common *but erroneous* belief that $\sqrt{x}$ means 'the positive or negative square root of x'. In fact, the symbol $\sqrt{\ }$ means 'the non-negative square root of'. For example, $\sqrt{4} = 2$ and *not* ± 2. (The equation $x^2 = 4$ has two solutions, namely $2 = \sqrt{4}$ and $-2 = -\sqrt{4}$, and these are frequently summarized as $\pm\sqrt{4}$ which is probably the cause of the error.) This means that there is another part of the postcondition which says that the output should be non-negative: $r \geqslant 0$.

Putting all the pieces together gives the following specification of the square root operation.

$$\sqrt{\_} : x : \mathcal{Real} \to r : \mathcal{Real}$$
$$\textbf{precondition} \quad x \geqslant 0$$
$$\textbf{postcondition} \quad r \geqslant 0 \wedge r^2 = x.$$

3. The following are some further specifications of operations given with somewhat briefer explanation. In general, each specification has a signature, precondition and postcondition although sometimes no precondition is required.

(i) square $\quad \_^2 : x : \mathcal{Real} \to r : \mathcal{Real}$
precondition There is no precondition since we are allowed to 'feed in' any real number to the 'square operation'
postcondition $r = x \times x$.

(ii) absolute value $\quad |\_| : x : \mathcal{Real} \to r : \mathcal{Real}$
precondition none (from now on we shall simply miss out the precondition part if there is no precondition)
postcondition $(x \geqslant 0 \to r = x) \wedge (x < 0 \to r = -x)$.

(iii) 'floor' or 'integer part' $\quad \lfloor \_ \rfloor : x : \mathcal{Real} \to n : \mathit{Integer}$.

The floor or integer part of a real number is the largest integer that is less than or equal to the given real number. For example, $\lfloor 8.74 \rfloor = 8$, $\lfloor \pi \rfloor = 3$, $\lfloor -2.38 \rfloor = -3$, $\lfloor 4 \rfloor = 4$, etc.

postcondition $(n \leqslant x) \wedge (n + 1 > x)$.

The postcondition uses $\leqslant$ and $>$ each with one *Integer* and one $\mathcal{Real}$ argument. We have defined these operations to have signature $\mathcal{Real}, \mathcal{Real} \to \mathcal{Real}$ (or $\mathit{Integer}, \mathit{Integer} \to \mathit{Integer}$). Therefore, in the postcondition, we are assuming that *Integer* is a subtype of $\mathcal{Real}$ and that, in each operation, we are substituting a value of type *Integer* for the first argument of the operation with signature $\mathcal{Real}, \mathcal{Real} \to \mathcal{Real}$.

(iv) intersection $\quad \_ \cap \_ : A : \mathit{Set}[\mathcal{T}],\ B : \mathit{Set}[\mathcal{T}] \to C : \mathit{Set}[\mathcal{T}]$
postcondition for all $x : \mathcal{T}$, $x \in C \leftrightarrow (x \in A \wedge x \in B)$.

The output set should be the intersection of the two input sets, $C = A \cap B$, and the postcondition defines this intersection. In this case, to capture the required defining property of $A \cap B$, we need to quantify over all elements of type $\mathcal{T}$.

(v) union $\_ \cup \_ : A : \mathit{Set}[\mathcal{T}],\ B : \mathit{Set}[\mathcal{T}] \to C : \mathit{Set}[\mathcal{T}]$
postcondition for all $x : \mathcal{T}, x \in C \leftrightarrow (x \in A \vee x \in B)$.

(vi) difference $\_ - \_ : A : \mathit{Set}[\mathcal{T}],\ B : \mathit{Set}[\mathcal{T}] \to C : \mathit{Set}[\mathcal{T}]$
postcondition for all $x : \mathcal{T}, x \in C \leftrightarrow (x \in A \wedge x \notin B)$.

(vii) empty set

It may seem odd to think of the empty set as an *operation* at all. However, for every type $\mathcal{T}$, there is an empty set of type $\mathit{Set}[\mathcal{T}]$. We can think of the empty set operation as delivering an empty set of type $\mathit{Set}[\mathcal{T}]$ 'automatically' without having first to receive an input. This means that there is no input type and hence no precondition. We can define:

empty set $\varnothing : \to C : \mathit{Set}[\mathcal{T}]$
postcondition for all $x : \mathcal{T}, \neg(x \in C)$.

The postcondition characterizes the empty set as the set that contains no element of type $\mathcal{T}$. Any operation like this that has no input type is called a **constant**.

(viii) subset $\_ \subseteq \_ : A : \mathit{Set}[\mathcal{T}],\ B : \mathit{Set}[\mathcal{T}] \to \mathcal{Boolean}$
postcondition for all $x : \mathcal{T}, A \subseteq B \leftrightarrow (x \in A \to x \in B)$.

(ix) set equality $\_ = \_ : A : \mathit{Set}[\mathcal{T}],\ B : \mathit{Set}[\mathcal{T}] \to \mathcal{Boolean}$
postcondition $A = B \leftrightarrow (A \subseteq B \wedge B \subseteq A)$.

(x) husband of $\mathit{HusbandOf}(\_) : p : \mathcal{Person} \to q : \mathcal{Person}$.

In the following specification of *HusbandOf* we assume the operations *IsFemale*, *IsMarried* and *IsMarriedTo* have clear meanings (and have been properly specified). See exercise 3.7.6 for the specification of further operations on the type $\mathcal{Person}$.

precondition $\mathit{IsFemale}(p) \wedge \mathit{IsMarried}(p)$
postcondition $\neg\mathit{IsFemale}(q) \wedge p\ \mathit{IsMarriedTo}\ q$.

Exercises 3.7

1. Suppose the following type declarations have been made.

$$k, m, n : \mathit{Integer} \qquad x, y, z : \mathcal{Real} \qquad P, Q : \mathcal{Boolean}.$$

Where necessary, assume that *Integer* is a subtype of *Real*; for example, the division operation for real numbers $\_/\_ : \mathcal{Real}, \mathcal{Real} \to \mathcal{Real}$ can take integer arguments (although the result will always be of type *Real*).

(a) State the type of each of the following terms.

(i)	$m + (k \times n)$	(iv)	$P \leftrightarrow Q$
(ii)	$x \leqslant (y - z)$	(v)	$P \vee (x \neq y)$
(iii)	$m - n = 2 \times k$	(vi)	$(n/k) + z$.

(b) Determine whether each of the following expressions is meaningful (that is, 'type checks').

(i)	$x \times (k/n)$	(iv)	$(n = m) \to (P \vee Q)$
(ii)	$P \wedge (x \geqslant y)$	(v)	$\neg(m < k)$
(iii)	$(P \wedge x) \geqslant y$	(vi)	$(\neg m) < k$.

2. Consider the type *Person* of all human beings (living and deceased). Various operations are to be defined on the type. Define the signature of each of the following operations. In some cases you will need to make choices regarding the meaning of the operation. You may assume the existence of other types that you require. For example, suppose *Name* is an operation which returns the family name of a person. You may assume the existence of the type *String*, which is the type of a person's family name, and then define Name to have the following signature:

$$\mathit{Name}(\_) : \mathcal{Person} \to \mathit{String}.$$

Operation	Comment
Height	Gives a person's height in metres in the form $\mathit{Height}(Jack) = 1.913$.
DateOfBirth	Assume that *Date* is a defined type.
YearOfBirth	
Age	Gives the age in years (i.e. age last birthday).
Mother	Gives a person's mother.
IsOlderThan	An infix operation; for particular p and q, p *IsOlderThan* q gives the truth value of 'p is older than q'.

CitizenOf	Gives the country of the person's nationality. Assume that $\mathcal{Nation}$ is a defined type. What happens if we allow dual (or multiple) nationality?
Children	Gives the set of children of a particular person.
IsTallerThan	
Qualifications	
Siblings	Gives the set of siblings of a person.

3. When quantifying over propositional functions, it is often necessary to specify the types of the variables. We use the following obvious generalization of the notation introduced in chapter 1.

   $\forall x : \mathcal{T}, P(x)$ means 'for all x of type $\mathcal{T}$, $P(x)$ (is true)'
   $\exists x : \mathcal{T}, P(x)$ means 'there exists (at least one) x of type $\mathcal{T}$ such that $P(x)$ (is true)'.

   Determine whether of not each of the following statements is true or false. (In some cases you will need to make assumptions about the operations involved—see question 2 above.)

   (i) $\exists n : \mathit{Integer}, n^2 = 2$
   (ii) $\exists x : \mathcal{Real}, x^2 = 2$
   (iii) $\forall n : \mathit{Integer}, n - 1 < n$
   (iv) $\forall x : \mathcal{Real}, \exists n : \mathit{Integer}, n > x$
   (v) $\forall x : \mathcal{Person}, \mathit{Mother}(x)\ \mathit{IsOlderThan}\ x$
   (vi) $\exists x : \mathcal{Real}, x^2 < 0$
   (vii) $\exists x : \mathcal{Person}, \mathit{IsQueen}(x)$
   (viii) $\exists x\, \exists y : \mathcal{Person}, \mathit{Age}(x) = \mathit{Age}(y)$
   (ix) $\forall n : \mathit{Integer}, \exists m : \mathit{Integer}, m > n$
   (x) $\forall n : \mathit{Integer}, (n < 0) \vee (n \geqslant 0)$.

4. Determine whether or not each of the following statements is true. Note that, for a statement to be true, it must *first* be meaningful (that is, it must type check).

   (i) $\forall n : \mathit{Integer}, n < n + 1$
   (ii) $\forall x : \mathcal{Person}, x \geqslant 0$
   (iii) $\forall n : \mathit{Integer}, n + 1 > 0$
   (iv) $\exists n : \mathit{Integer}, n + 1 > 0$
   (v) $\forall x : \mathcal{Real}, x^2 \geqslant 0$
   (vi) $\exists x : \mathcal{Real}, x^2 = 3$
   (vii) $\forall x : \mathcal{Person}, \exists n : \mathit{Integer}, \mathit{Age}(x) = n$
   (viii) $\exists n : \mathit{Integer}, \forall x : \mathcal{Person}, \mathit{Age}(x) = n$

(ix) $\exists n\, \exists m : \mathit{Integer},\ (n < m) \wedge (n^2 > m^2)$

(x) $\forall P\, \forall Q : \mathit{Boolean},\ (P \leftrightarrow Q) \vee (P \leftrightarrow \neg Q)$ (*Hint*: draw up a truth table)

(xi) $\forall x : \mathit{Real},\ \exists n : \mathit{Integer},\ n \geqslant x$

(xii) $\forall x : \mathit{Real},\ (x^2 < 0) \rightarrow (x < 0)$.

5. Suppose that, on the type *Integer*, the operations of addition and multiplication are given and are fully specified. Suppose also that there is an operation *IsPositive* with signature

$$\mathit{IsPositive}(\_) : \mathit{Integer} \rightarrow \mathit{Boolean}$$

such that $\mathit{IsPositive}(n)$ is true when n is greater than zero and false otherwise.

Write down complete specifications for each of the following operations. That is, write down preconditions (if any are necessary) and postconditions which define the operation. To begin with, you may *only* use addition, multiplication and *IsPositive* (together with equality). However, once an operation has been defined, it may then be used in subsequent specifications. In many cases there is more than one way of correctly specifying the operation.

(i)	subtraction	$\_ - \_ : \mathit{Integer}, \mathit{Integer} \rightarrow \mathit{Integer}$
(ii)	negation	$-\_ : \mathit{Integer} \rightarrow \mathit{Integer}$
(iii)	greater than	$\_ > \_ : \mathit{Integer}, \mathit{Integer} \rightarrow \mathit{Boolean}$
(iv)	is negative	$\mathit{IsNegative}(\_) : \mathit{Integer} \rightarrow \mathit{Boolean}$
(v)	less than	$\_ < \_ : \mathit{Integer}, \mathit{Integer} \rightarrow \mathit{Boolean}$
(vi)	reciprocal	$1/\_ : \mathit{Integer} \rightarrow \mathit{Real}$
(vii)	greater than or equal to	$\_ \geqslant \_ : \mathit{Integer}, \mathit{Integer} \rightarrow \mathit{Boolean}$
(viii)	less than or equal to	$\_ \leqslant \_ : \mathit{Integer}, \mathit{Integer} \rightarrow \mathit{Boolean}$
(ix)	even	$\mathit{IsEven}(\_) : \mathit{Integer} \rightarrow \mathit{Boolean}$
(x)	odd	$\mathit{IsOdd}(\_) : \mathit{Integer} \rightarrow \mathit{Boolean}$

(xi) mod $\_\mathrm{mod}\_ : \mathit{Integer}, \mathit{Integer} \rightarrow \mathit{Integer}$

The operation 'n mod k' gives the remainder when n is divided by k. For example, 2 mod 3 $=$ 2, 4 mod 3 $=$ 1, 38 mod 3 $=$ 2, 180 mod 3 $=$ 0, 52 mod 5 $=$ 2, 17 mod 7 $=$ 3, etc.

(xii) divides $\_|\_ : \mathit{Integer}, \mathit{Integer} \rightarrow \mathit{Boolean}$.

Recall that $n|m$ is true if n is a factor of m (that is, 'n goes exactly into m') and is false otherwise.

6. In this question, you may assume the type *Person* has the following operations already specified.

$IsMarried(p)$	p is married
$IsFemale(p)$	p is female
$IsChildOf(p, q)$	p is a child of q
$IsMarriedTo(p, q)$	p is married to q

(i) Write down the signatures of *IsMarried*, *IsFemale*, *IsChildOf* and *IsMarriedTo*.

(ii) The operation $WifeOf(p)$ is to be defined as returning the wife of p. Write down informal (in English) and formal preconditions and postconditions for *WifeOf*. In your formal version, you may use any of the operations above, but no others.

(iii) $Sons(p)$ is to be defined as returning the set (which may be empty) of sons of p : *Person*. Write down the signature, informal (in English) and formal preconditions and postconditions for *Sons*, again using only the operations above.

(iv) An operation f is defined on the set of people as follows:

signature	$f(\_) : p : \mathit{Person} \rightarrow q : \mathit{Person}$
precondition	none
postcondition	$(q = f(p)) \leftrightarrow (IsMale(q) \wedge IsChildOf(p, q))$

In ordinary English, describe what output the function f produces.

(v) Write a formal specification for the function *FatherInLaw*, that is to return a person's father-in-law (the father of his or her spouse).

7. A type *Pet* is to be defined as the type of all living domestic pets. In this question, assume the existence of the types *Integer*, *Real*, *Boolean* and *Person*.

(i) Assuming that every pet has one and only one (human) owner but a person may own more than one pet, write down the signature of each of the following operations:

(a) $OwnerOf(\_)$ this gives the owner of a pet
(b) $Owns(\_)$ this gives the pets owned by a person.

(ii) If a : *Pet*, what is the relationship between a and $Owns(OwnerOf(a))$?

(iii) The operation *hasPet* : $\mathcal{Person} \rightarrow \mathcal{Boolean}$ returns true if the person owns at least one pet and false otherwise. Give a formal specification of *hasPet* in terms of preconditions and/or postconditions.

(iv) An operation f has signature $a : \mathcal{Pet}, b : \mathcal{Pet} \rightarrow \mathcal{Boolean}$. It has no precondition but has the following postcondition:

postcondition
$$f(a,b) \leftrightarrow \exists p : \mathcal{Person}\, p = OwnerOf(a) \wedge p = OwnerOf(b).$$

Describe in simple terms the meaning of this operation f.

Chapter 4

Relations

4.1 Relations and Their Representations

The mathematical notion of a relation, like that of a set, is a very general one. It is one of the key concepts of mathematics and examples of relations occur throughout the subject. Three special types of relation are particularly important: functions, equivalence relations and order relations. Functions are the subject of the next chapter; equivalence and order relations are considered later in this chapter. We begin, though, with a look at the general concept of a relation and various ways of visualizing relations.

In §1.8 we considered two-place predicates such as 'is heavier than'. A two-place predicate requires two variables to convert it into a propositional function. For example, if H is the predicate 'is heavier than', then $H(x, y)$ denotes the propositional function 'x is heavier than y'. We can think of a two-variable propositional function as defining some kind of relationship between its two variables. Given objects a and b, the proposition $H(a, b)$ is true if and only if the objects are related in the appropriate way.

The first thing to note is that, in a two-variable propositional function $F(x, y)$, the order of the variables may be significant. For specific objects a and b, $F(a, b)$ and $F(b, a)$ may have different truth values. This is the case for the propositional functions 'x is heavier than y', 'x is the mother of y' or 'x is greater than y', for instance. Therefore, the set of objects for which $F(a, b)$ is a true proposition will be a set of ordered pairs. It is also important to realize that the two variables x and y may represent different *kinds* of object. For example, consider the propositional function $C(x, y)$: x is the capital city of y. Here x is the name of a city but y is the name of a country, so the set of ordered pairs (a, b) for which $C(a, b)$ is a true proposition is a subset of the Cartesian product $A \times B$, where $A = \{\text{cities}\}$ and $B = \{\text{countries}\}$.

The following mathematical definition of a 'relation' is surprisingly simple and very general. Some authors refer to this as a **binary relation** because it relates two objects. (There is a generalization of this which relates n objects—see exercise 4.1.11.)

Definition 4.1

Let A and B be sets. A **relation from A to B** (or **between A and B**) is a subset of the Cartesian product $A \times B$.

The first thing to notice is that a relation as we have defined it is a set; namely a set of ordered pairs. If R is a relation from A to B, we say that $a \in A$ is **related** to $b \in B$ if $(a, b) \in \mathsf{R}$. Thus the relation R itself is simply the set of all related pairs of elements. For the most part we shall adopt the commonly used notation and write $a \; \mathsf{R} \; b$ to denote 'a is related to b', and $a \not\mathsf{R} \; b$ to denote $(a, b) \notin \mathsf{R}$ or 'a is not related to b'. If $A = B$ it is also common to refer to R as a **relation on A**.

Examples 4.1

1. Let $A = \{\text{cities of the world}\}$, $B = \{\text{countries of the world}\}$ and $\mathsf{R} = \{(a, b) : a \text{ is the capital city of } b\}$. Thus $a \; \mathsf{R} \; b$ denotes 'a is the capital city of b'.

   Examples are: (Paris) R (France), (Moscow) R (Russia), (Tirana) R (Albania), etc.

   Also we have: (London) $\not\mathsf{R}$ (Zimbabwe), (Naples) $\not\mathsf{R}$ (Italy), (New York) $\not\mathsf{R}$ (United States), etc.

2. Let $A = B = \{1, 2, 3, 4, 5, 6\}$ and $\mathsf{R} = \{(a, b) : a \text{ divides } b\}$. Since A is a small finite set we can list the elements of the relation:

$$\mathsf{R} = \{(1,1), (1,2), (1,3), (1,4), (1,5), (1,6), (2,2), (2,4), (2,6), (3,3), (3,6), (4,4), (5,5), (6,6)\}.$$

   We represent R diagrammatically in figure 4.1 by plotting its elements on the coordinate grid diagram of the Cartesian product $A \times B = A^2$.

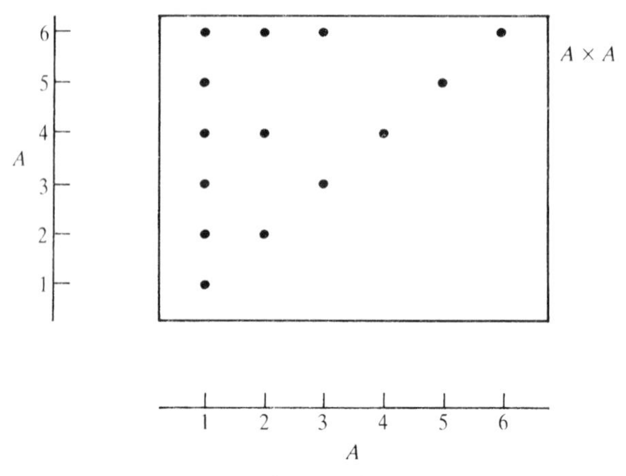

Figure 4.1

3. Let $A = B = \mathbb{Z}^+$, the set of positive integers, and let $a \, \mathsf{R} \, b$ denote 'a has the same parity as b'; that is, either a and b are both even or they are both odd. More precisely

$$\mathsf{R} = \{(a, b) : a - b \text{ is an integer multiple of } 2\}.$$

Then
$$\begin{array}{l} 1 \, \mathsf{R} \, 1, 1 \, \mathsf{R} \, 3, 1 \, \mathsf{R} \, 5, \ldots \\ 2 \, \mathsf{R} \, 2, 2 \, \mathsf{R} \, 4, 2 \, \mathsf{R} \, 6, \ldots \\ 3 \, \mathsf{R} \, 1, 3 \, \mathsf{R} \, 3, 3 \, \mathsf{R} \, 5, \ldots \\ 4 \, \mathsf{R} \, 2, 4 \, \mathsf{R} \, 4, 4 \, \mathsf{R} \, 6, \ldots \text{ etc.} \end{array}$$

A picture for this relation is figure 4.2, where again we have plotted the elements of R on the diagram for $A \times B$.

There are various ways of representing relations visually, particularly relations between finite sets. In figures 4.1 and 4.2, the elements of R are marked on the coordinate grid diagram of the Cartesian product $A \times B$. Diagrams such as these show clearly R as a subset of $A \times B$, but are not so good at showing additional properties of the relation.

An alternative for finite sets is to represent A and B as two side-by-side Venn-Euler diagrams with the elements arranged vertically. An arrow is drawn from $a \in A$ to $b \in B$ whenever $a \, \mathsf{R} \, b$. We refer to this as the **arrow diagram** of the relation. For example, the arrow diagram for the relation defined in example 4.1.2 above is given in figure 4.3.

Unfortunately figure 4.3 does not show very clearly at a glance which elements are related to which. For sets larger than $\{1, 2, 3, 4, 5, 6\}$ diagrams of this type

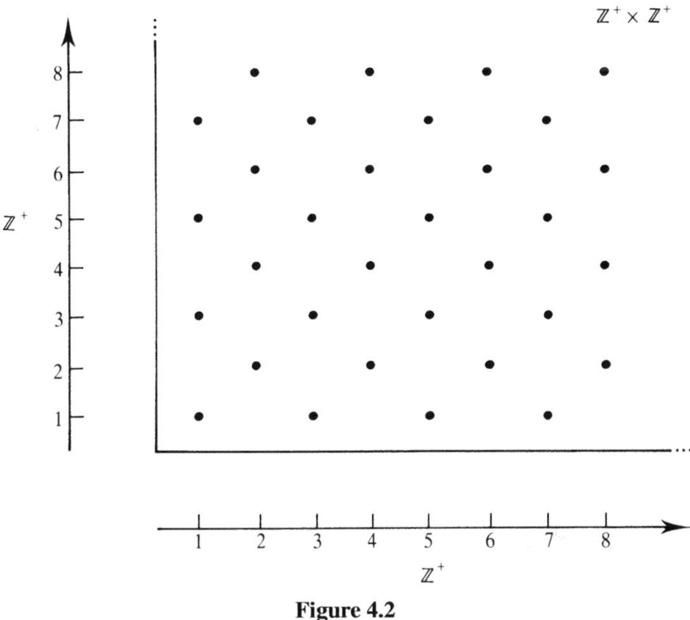

Figure 4.2

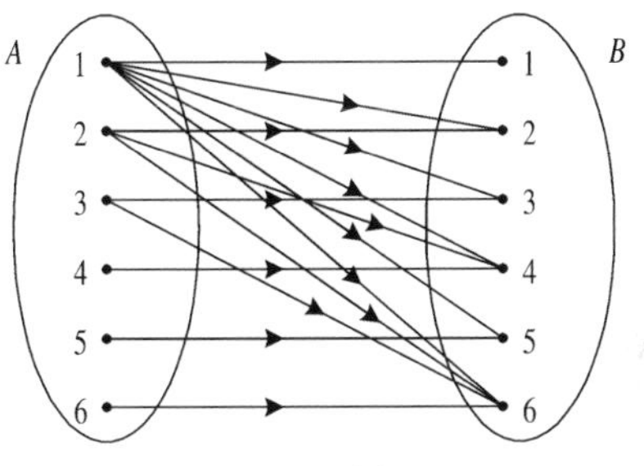

Figure 4.3

would become too cluttered to be of much use. However, for relations *on* a set (i.e. where $A = B$), there is a slight modification we can make which clarifies the diagram. Instead of listing the elements of A twice, once in each Venn-Euler diagram, we can represent each element of A once by a point in the plane. A directed arrow is still drawn from a to b if and only if $a \mathrel{\mathsf{R}} b$. The resulting diagram (see figure 4.4) is an example of a **directed graph**† or **digraph** and is called the **directed graph of the relation**. We shall study graphs and directed graphs in greater detail in chapters 10 and 11.

If two elements a and b are such that $a \mathrel{\mathsf{R}} b$ and $b \mathrel{\mathsf{R}} a$, we will usually connect

† More precisely, figure 4.4 is the *diagram* of a directed graph—see chapter 10.

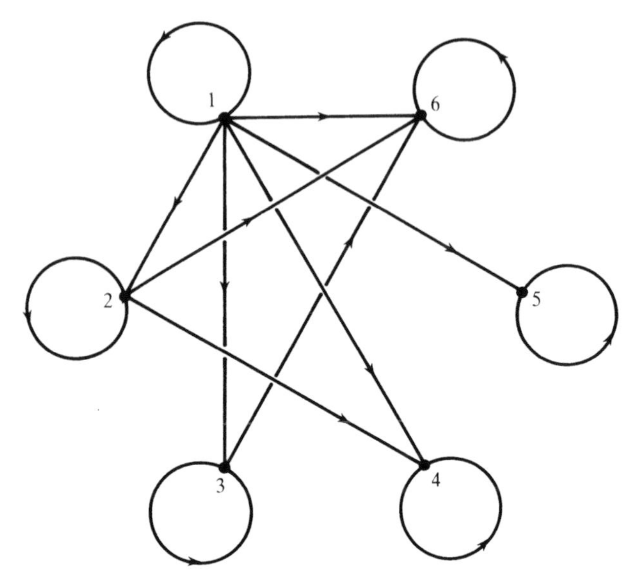

Figure 4.4

their points in the directed graph by a single bi-directional arrow, rather than two directed arrows. (See the diagram in exercise 4.1.5.)

A third way to represent a relation is by a 'binary matrix'. Let $A = \{a_1, a_2, \ldots, a_n\}$ and $B = \{b_1, b_2, \ldots, b_m\}$ be finite sets and let R be a relation from A to B. The **binary matrix of R** is a rectangular array of zeros and ones with n rows and m columns. The rows correspond to the elements of A (in the order listed above) and the columns correspond to the elements of B (again, in the order listed above). At the intersection of the ith row and jth column we place a one if $a_i \mathrel{\mathsf{R}} b_j$ or a zero if $a_i \not\mathrel{\mathsf{R}} b_j$. For example, the binary matrix representing the relation R on $A = \{1, 2, 3, 4, 5, 6\}$ given by $a \mathrel{\mathsf{R}} b$ if and only if a divides b (example 4.1.2) is the following:

$$
\begin{array}{c} \\ a_1 = 1 \\ a_2 = 2 \\ a_3 = 3 \\ a_4 = 4 \\ a_5 = 5 \\ a_6 = 6 \end{array}
\begin{array}{c} \begin{array}{cccccc} b_1 = 1 & b_2 = 2 & b_3 = 3 & b_4 = 4 & b_5 = 5 & b_6 = 6 \end{array} \\
\left(\begin{array}{cccccc}
1 & 1 & 1 & 1 & 1 & 1 \\
0 & 1 & 0 & 1 & 0 & 1 \\
0 & 0 & 1 & 0 & 0 & 1 \\
0 & 0 & 0 & 1 & 0 & 0 \\
0 & 0 & 0 & 0 & 1 & 0 \\
0 & 0 & 0 & 0 & 0 & 1
\end{array} \right) \end{array}.
$$

We have taken the elements of $A = B$ in increasing order so that row i represents the number i and column j represents the number j. The zero at the intersection of the fifth row and second column means that $a_5 \not\mathrel{\mathsf{R}} b_2$—that is, 5 does *not* divide 2.

Similarly the one at the intersection of the second row and fourth column means that $a_2 \mathsf{R} b_4$—that is, 2 divides 4.

Normally we will not write $a_1, \ldots, a_6$ to label the rows and $b_1, \ldots, b_6$ to label the columns as we have here, provided it is clearly understood which rows correspond to which elements of A and which columns correspond to which elements of B. Several of the properties of the relation R which we consider later can be deduced from properties of its binary matrix. Matrix algebra itself is the subject of chapter 6.

Note that the binary matrix of a relation on a finite set A is square—that is, it has an equal number of rows and columns. The number of rows or columns is of course $|A|$, the cardinality of A.

Relations and Types

We now consider briefly how typed sets introduced in section 3.8 fit in with the theory of relations. If A and B are typed sets and R is a relation from A to B, then R should also have a specified type. Before we can determine the type of R, we must first define the type of the Cartesian product $A \times B$ (because R is a subset of $A \times B$).

Given elements a of type $\mathcal{S}$ and b of type $\mathcal{T}$, we define the ordered pair (a, b) to have type $\mathcal{S} \times \mathcal{T}$. Symbolically,

$$a : \mathcal{S}, b : \mathcal{T} \rightarrow (a, b) : \mathcal{S} \times \mathcal{T}.$$

For example, if n : *Integer* and x : *Real* then (n, x) : *Integer* $\times$ *Real*. Now if A : *Set*$[\mathcal{S}]$ and B : *Set*$[\mathcal{T}]$ then their Cartesian product $A \times B$ is a set containing ordered pairs (a, b) of type $\mathcal{S} \times \mathcal{T}$. Therefore $A \times B$ has type *Set*$[\mathcal{S} \times \mathcal{T}]$. In symbols,

$$A : \mathit{Set}[\mathcal{S}], B : \mathit{Set}[\mathcal{T}] \rightarrow A \times B : \mathit{Set}[\mathcal{S} \times \mathcal{T}].$$

Now suppose that R is a relation from A to B. Then $\mathsf{R} \subseteq A \times B$ so R has the same type as $A \times B$, namely *Set*$[\mathcal{S} \times \mathcal{T}]$. To summarize:

if R is a relation from A : *Set*$[\mathcal{S}]$ to B : *Set*$[\mathcal{T}]$ then R : *Set*$[\mathcal{S} \times \mathcal{T}]$.

In examples 4.1:

the relation in example 1 has type *Set*[*City* $\times$ *Country*];
the relation in example 2 has type *Set*[*Integer* $\times$ *Integer*];
the relation in example 3 has type *Set*[*Integer* $\times$ *Integer*].

Exercises 4.1

1. For each of the following relations R on a set A, draw:

   (a) its coordinate grid diagram,
   (b) its directed graph, and
   (c) its binary matrix.

   (i) $A = \{1, 2, 3, 4, 5, 6, 7, 8\}$;
   $a \mathrel{\mathsf{R}} b$ if and only if $a < b$.

   (ii) $A = \{1, 2, 3, 4, 5, 6, 7, 8\}$;
   $a \mathrel{\mathsf{R}} b$ if and only if $a = b$.

   (iii) $A = \{1, 2, 3, 4, 5, 6, 7, 8\}$;
   $a \mathrel{\mathsf{R}} b$ if and only if $a \leqslant b$.

   (iv) $A = \{1, 2, 3, 4, 5, 6, 7, 8\}$;
   $a \mathrel{\mathsf{R}} b$ if and only if $a/b \in \mathbb{Z}$.

   (v) $A = \{a, b, c, d, e\}$;
   $\mathsf{R} = \{(a, b), (a, c), (a, e), (b, c), (c, a), (c, d), (d, e), (e, c), (e, d)\}$.

   (vi) $A = \{a, b, c, d, e, f\}$;
   $x \mathrel{\mathsf{R}} y$ if and only if x and y are both vowels or x and y are both consonants.

   (vii) $A = \{1, 2, 3, 4, 5, 6, 7, 8\}$;
   $a \mathrel{\mathsf{R}} b$ if and only if $a = 2b$.

   (viii) $A = \{1, 2, 3, 4, 5, 6, 7, 8\}$;
   $a \mathrel{\mathsf{R}} b$ if an only if $a = 2^n b$ for some $n = \mathbb{Z}^+$.

   (ix) $A = \mathscr{P}\{1, 2, 3\}$, the power set of $\{1, 2, 3\}$;
   $a \mathrel{\mathsf{R}} b$ if and only if $a \subseteq b$.

   (x) $A = \mathscr{P}\{1, 2, 3\}$, the power set of $\{1, 2, 3\}$;
   $a \mathrel{\mathsf{R}} b$ if and only if $a \subset b$.

2. The binary matrices M_{R} and M_{S} for two relations R and S respectively on the set $A = \{1, 2, 3, 4, 5\}$ are given below.

$$M_{\mathsf{R}} = \begin{pmatrix} 0 & 0 & 0 & 0 & 0 \\ 1 & 0 & 1 & 0 & 1 \\ 1 & 1 & 0 & 1 & 0 \\ 1 & 1 & 1 & 0 & 1 \\ 1 & 0 & 0 & 1 & 0 \end{pmatrix} \quad \text{and} \quad M_{\mathsf{S}} = \begin{pmatrix} 1 & 1 & 1 & 1 & 1 \\ 0 & 0 & 0 & 0 & 0 \\ 1 & 1 & 1 & 0 & 0 \\ 1 & 0 & 0 & 1 & 1 \\ 0 & 0 & 0 & 0 & 0 \end{pmatrix}.$$

(i) List the elements of R and S.
(ii) Draw the directed graphs of R and S.

3. A relation R on the set $A = \{a, b, c, d, e\}$ has the directed graph shown in the diagram below.

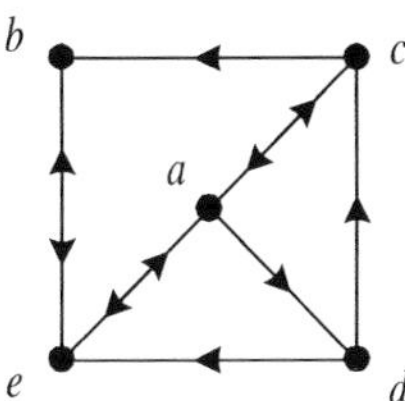

(i) List the elements of R.
(ii) Write down the binary matrix of R.

4. A relation R between the sets $A = \{1, 2, 3\}$ and $B = \mathscr{P}(A) = \{\varnothing, \{1\}, \{2\}, \{3\}, \{1, 2\}, \{2, 3\}, \{1, 3\}, \{1, 2, 3\}\}$ has the following binary matrix. (The row and columns of the matrix correspond to the elements of A and B as they are listed respectively.)

$$\begin{pmatrix} 0 & 1 & 0 & 0 & 1 & 0 & 1 & 1 \\ 0 & 0 & 1 & 0 & 1 & 1 & 0 & 1 \\ 0 & 0 & 0 & 1 & 0 & 1 & 1 & 1 \end{pmatrix}.$$

List the elements of R and define $a \mathrel{\mathsf{R}} b$ in words or symbols.

5. Each of the four football teams A, B, C, D in a mini-league plays every other team both at home and away. A relation R on the set $S = \{\mathrm{A, B, C, D}\}$ is defined by:

$X \mathrel{\mathsf{R}} Y$ if and only if X beat Y when X played at home.

The following diagram is the directed graph of R.

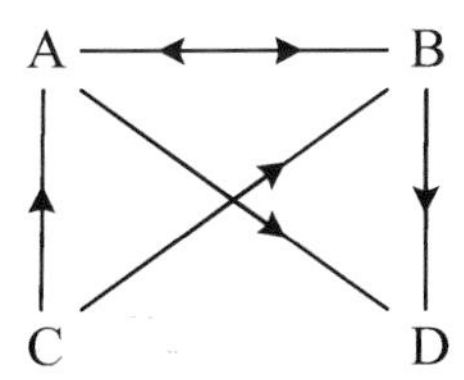

List the elements of R, and write down its binary matrix.

6. For each of the following relations R on the (ordered) set $A = \{a, b, c, d, e\}$, whose binary matrix is given, list the elements of R and draw its directed graph:

(i) $$\begin{pmatrix} 0 & 1 & 1 & 1 & 1 \\ 0 & 0 & 1 & 1 & 1 \\ 0 & 0 & 0 & 1 & 1 \\ 0 & 0 & 0 & 0 & 1 \\ 0 & 0 & 0 & 0 & 0 \end{pmatrix}$$

(ii) $$\begin{pmatrix} 1 & 0 & 0 & 0 & 0 \\ 1 & 1 & 0 & 0 & 0 \\ 1 & 1 & 1 & 0 & 0 \\ 1 & 1 & 1 & 1 & 0 \\ 1 & 1 & 1 & 1 & 1 \end{pmatrix}$$

(iii) $$\begin{pmatrix} 1 & 0 & 1 & 0 & 1 \\ 0 & 1 & 0 & 1 & 0 \\ 1 & 0 & 1 & 0 & 1 \\ 0 & 1 & 0 & 1 & 0 \\ 1 & 0 & 1 & 0 & 1 \end{pmatrix}.$$

What is the connection between the relations in parts (i) and (ii)?

7. Let $A = \{\text{rivers in the world}\}$, $B = \{\text{towns or cities in the world}\}$ and define a relation R from A to B by

$$a \mathsf{R} b \quad \text{if and only if } a \text{ flows through } b.$$

(i) Describe in words each of the following sets:

(a) $\{b \in B : (\text{Thames}) \mathsf{R} b\}$
(b) $\{a \in A : a \mathsf{R} (\text{London})\}$.

(ii) Describe in words each of the following propositions:

(a) $\neg(\exists x \in A, x \mathsf{R} (\text{Toronto}))$
(b) $\forall a \in A\ (a \mathsf{R} (\text{Washington, DC}) \rightarrow a = \text{Potomac})$.

(iii) Define each of the following sets symbolically.

(a) The set of rivers which flow through Paris.
(b) The set of rivers which flow through some town or city.

(iv) Write symbolically each of the following statements:

(a) All rivers flow through some town or city.
(b) All towns or cities have a river flowing through them.
(c) The River Nile flows through more than one town or city.

8. Let A be any (finite) set.

(i) The **identity relation** I_A on A is the relation of equality defined by: $a \mathsf{I}_A b$ if and only if $a = b$.

Describe (a) the directed graph and (b) the binary matrix of I_A.

(ii) The **universal relation** U_A on A is the relation defined by: $a \, \mathsf{U}_A \, b$ for all $a, b \in A$.

Describe (a) the directed graph and (b) the binary matrix of U_A.

9. (i) How many relations are there from $\{a, b, c\}$ to $\{1, 2, 3, 4\}$? (Do not try to list them all!)

   (ii) More generally, if $|A| = n$ and $|B| = m$, how many relations are there from A to B?

10. Given a relation R from set A to set B, its **inverse relation** R^{-1} is the relation from B to A defined by

$$x \, \mathsf{R}^{-1} \, y \quad \text{if and only if } y \, \mathsf{R} \, x.$$

   (a) Let $A = \{1, 2, 3, 4\}$ and let R be the relation on A defined by

$$R = \{(1,2), (1,4), (2,2), (2,3), (3,4), (4,3), (4,4)\}.$$

   (i) List the elements of R^{-1}.
   (ii) Draw the directed graphs of both R and R^{-1}.
   (iii) Write down the binary matrices of both R and R^{-1}.

   (b) Let R be a relation on a set A.

   (i) Describe the connection between the directed graphs of R and its inverse R^{-1}.

   (ii) Describe the connection between the binary matrices of R and its inverse R^{-1}.

11. An **n-ary relation** between sets $A_1, A_2, \ldots, A_n$ is defined to be a subset of the Cartesian product $A_1 \times A_2 \times \cdots \times A_n$. If $A = A_1 = A_2 = \cdots = A_n$ we refer to an n-ary relation **on** A. Let $A = \{1, 2, \ldots, 6\}$ and define a 3-ary relation R on A by $(x, y, z) \in \mathsf{R}$ if and only if $x < y$ and y divides z.

    List the elements of R.

12. Determine the type of each of the relations defined in questions 1, 4, 5 and 8 above.

13. (i) Let $A : \mathit{Set}[S]$, $B : \mathit{Set}[T]$ and let R be a relation from A to B. What is the type of the inverse relation R^{-1}?

(ii) Define the type of the relation R defined in question 7 above. Describe the inverse relation R^{-1} in words and define its type.

4.2 Properties of Relations

Up to now we have not justified our assertion, made at the beginning of this chapter, that relations are important in mathematics. Indeed, if all we were able to do with relations between sets were to draw diagrams to represent them, the concept of a relation would not be very significant. Its importance is mainly due to special kinds of relation which satisfy additional properties. The two of these which we shall study in this chapter—equivalence relations and order relations—are both relations *on* a set, so we look now at some of the properties which a relation on a set may have.

Definitions 4.2

Let R be a relation on set A. We say that R is:

(i) **reflexive** if and only if $a \mathrel{R} a$ for every $a \in A$;
(ii) **symmetric** if and only if $a \mathrel{R} b$ implies $b \mathrel{R} a$ for every $a, b, \in A$;
(iii) **anti-symmetric** if and only if $a \mathrel{R} b$ and $b \mathrel{R} a$ implies $a = b$ for every $a, b \in A$;
(iv) **transitive** if and only if $a \mathrel{R} b$ and $b \mathrel{R} c$ implies $a \mathrel{R} c$ for every $a, b, c, \in A$.

Note that to prove that a relation R on a set A satisfies one of these four properties, we need to show that the appropriate property is satisfied by an *arbitrary* element or elements of A. For example, to prove that R is symmetric, we need to show that $a \mathrel{R} b \to b \mathrel{R} a$ for arbitrary elements $a, b \in A$. However, to show that R does not satisfy one of the properties, we need to find a *particular* element or elements of A that show this. These particular elements are a counter-example to the property (see §2.3). For example, to show that R is not symmetric, we need to find particular elements $a, b \in A$ such that $a \mathrel{R} b$ but $b \not\mathrel{R} a$.

Examples 4.2

1. Let R be the relation on the set of real numbers defined by

$$x \mathsf{R} y \quad \text{if and only if } x \leqslant y.$$

Then:

(i) R is reflexive because $x \leqslant x$ for every $x \in \mathbb{R}$;
(ii) R is not symmetric because, for example, $1 \leqslant 2$ but $2 \nleqslant 1$, so $x \mathsf{R} y$ does not imply $y \mathsf{R} x$;
(iii) R is anti-symmetric: if $x \leqslant y$ and $y \leqslant x$ then it follows that $x = y$;
(iv) R is transitive because if $x \leqslant y$ and $y \leqslant z$ then it follows that $x \leqslant z$.

2. Let $A = \{a, b, c, d\}$ and $\mathsf{R} = \{(a, a), (a, b), (a, c), (b, a), (b, b), (b, c), (b, d), (d, d)\}$.

The relation R satisfies none of the properties of definition 4.2:

R is not reflexive since $c \not\mathsf{R} c$; therefore it is not true that $x \mathsf{R} x$ for every $x \in A$;
R is not symmetric since, for example, $a \mathsf{R} c$ but $c \not\mathsf{R} a$;
R is not anti-symmetric since $a \mathsf{R} b$ and $b \mathsf{R} a$ but $a \neq b$;
R is not transitive since $a \mathsf{R} b$ and $b \mathsf{R} d$ but $a \not\mathsf{R} d$.

3. Let $A = \mathbb{Z}^+ \times \mathbb{Z}^+$ and R be the relation on A defined by $(a, b) \mathsf{R} (c, d)$ if and only if $a + d = b + c$. Show that R is reflexive, symmetric and transitive, but not anti-symmetric.

Solution

For all positive integers a and $b, a + b = b + a$, so $(a, b) \mathsf{R} (a, b)$ for every $(a, b) \in A$, Therefore R is reflexive.

R is symmetric since if $(a, b) \mathsf{R} (c, d)$ then $a + d = b + c$ which implies that $c + b = d + a$, so $(c, d) \mathsf{R} (a, b)$.

To show that R is transitive, suppose $(a, b) \mathsf{R} (c, d)$ and $(c, d) \mathsf{R} (e, f)$. This means that

$$a + d = b + c \quad \text{and} \quad c + f = d + e.$$

Adding these equations gives

$$a + d + c + f = b + c + d + e$$

so

$$a + f = b + e$$

which means that $(a, b) \mathsf{R} (e, f)$.

Therefore $(a, b) \mathsf{R} (c, d)$ and $(c, d) \mathsf{R} (e, f)$ implies that $(a, b) \mathsf{R} (e, f)$, so R is transitive.

Finally, to show that R is not anti-symmetric we need to find a counter-example; that is, we need to find elements (a, b) and (c, d) of A such that $(a, b) \mathsf{R} (c, d)$ and $(c, d) \mathsf{R} (a, b)$ but $(a, b) \neq (c, d)$. Clearly the elements $(1, 2)$ and $(2, 3)$ will do.

We now consider how we can recognize whether a relation satisfies any of these properties given its directed graph or its binary matrix. Firstly, if R is a reflexive relation on a finite set A then $a \mathsf{R} a$ for every $a \in A$. This means that, in the directed graph of R, there is a directed arrow from every point to itself. The directed graph of the relation in example 4.1.2 (figure 4.4) has this property. In the binary matrix of a reflexive relation R, there is a one in every position along the diagonal which runs from the top left to the bottom right of the matrix. (This diagonal is called the 'leading diagonal' of the matrix.)

If R is symmetric then every arrow connecting different points in its directed graph is bidirectional. This is because an arrow from a to b means that $a \mathsf{R} b$; but if R is symmetric this implies that $b \mathsf{R} a$ as well, so the arrow must also go from b to a. The binary matrix of a symmetric relation is symmetric about its leading diagonal; whatever appears at the intersection of the ith row and jth column also appears at the intersection of the jth row and ith column.

For an anti-symmetric relation, the directed graph is such that there are no bidirectional arrows connecting different points. This is because for distinct elements a and b of A we cannot have both $a \mathsf{R} b$ and $b \mathsf{R} a$. The property satisfied by the binary matrix of an anti-symmetric relation is slightly less obvious. A relation being anti-symmetric means that, for distinct elements a and b, if $a \mathsf{R} b$ then $b \not\mathsf{R} a$ and if $b \mathsf{R} a$ then $a \not\mathsf{R} b$. Thus in the binary matrix, if there is a one at the intersection of the ith row and jth column (where $i \neq j$) then there must be a zero at the intersection of the jth row and ith column. However, since it is possible to have $a \not\mathsf{R} b$ and $b \not\mathsf{R} a$, there could be zeros in both of these positions.

The properties satisfied by the directed graph and binary matrix of a transitive relation are less obvious still. For the graph, however, we can describe the property reasonably easily. If three points are such that there are arrows from the first to the second and from the second to the third, then there must also be an arrow from the first to the third.

Examples 4.3

1. Consider the directed graph given in figure 4.5 of a relation R on the set $A = \{a, b, c, d, e\}$.

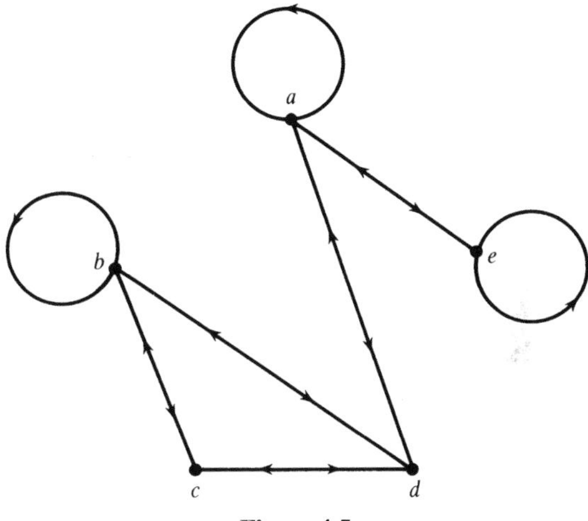

Figure 4.5

From this diagram we can see that:

(i) R is not reflexive, since there is no arrow from c to itself, for example.

(ii) R is symmetric, but not anti-symmetric, since every arrow connecting distinct points is bidirectional.

(iii) R is not transitive since, for instance, there are arrows from a to d, and from d to b, but not from a to b.

2. A relation R on a four-element set A has the following binary matrix:

$$\begin{pmatrix} 1 & 0 & 1 & 0 \\ 1 & 1 & 0 & 0 \\ 0 & 0 & 1 & 1 \\ 0 & 1 & 0 & 1 \end{pmatrix}.$$

Which of the properties of definitions 4.2 does R satisfy?

Solution

Firstly, it is clear that R is reflexive since there are only ones along the leading diagonal.

R is not symmetric because the matrix is not symmetric about the leading diagonal. For instance, there is a one in row 1, column 3, but a zero in row 3, column 1.

We have to look a bit harder to see that R is anti-symmetric; except for the leading diagonal wherever a one appears in row i, column j, a zero appears in row j, column i. Note that sometimes a zero appears in both these places; for example, in row 1, column 4 and row 4, column 1.

With some trial and error, we can also spot that R is not transitive. If we label the elements of the set a_1, a_2, a_3 and a_4, in that order, then $a_1 \mathrel{\mathsf{R}} a_3$ and $a_3 \mathrel{\mathsf{R}} a_4$ but $a_1 \not\mathrel{\mathsf{R}} a_4$. We leave it as an exercise to discover whether there are any other counter-examples to transitivity.

Exercises 4.2

1. For each of the binary relations on a set defined in exercises 4.1, determine whether the relation is:

   (i) reflexive;
   (ii) symmetric;
   (iii) anti-symmetric;
   (iv) transitive.

2. Let A be any set of living human beings. For each of the following relations on A, defined by a two-place predicate, determine which, if any, of the four properties of definitions 4.2 is/are satisfied:

   (i) 'is the mother of'.
   (ii) 'is the brother of'.
   (iii) 'is the sibling of'.
   (iv) 'is at least as tall as'.
   (v) 'is taller than'.
   (vi) 'is the same age (in years) as'.
   (vii) 'is the same gender as'.
   (viii) 'is an ancestor of'.
   (ix) 'is married to'. (Is this affected by whether the people come from a monogamous or polygamous society?)
   (x) 'is an acquaintance of'.

3. Let $A = \{a, b, c, d, e\}$. For each of the following relations R on

A, determine which of the four properties (reflexive, symmetric, anti-symmetric, transitive) are satisfied by the relation. Justify your answers.

(i) $\mathsf{R} = \{(a,a),(a,b),(a,c),(b,a),(b,b),(b,c),(c,a),(c,b),(c,c)\}$.
(ii) $\mathsf{R} = \{(a,a),(b,b),(c,c),(d,d),(e,e),(a,b),(b,c)\}$.
(iii) $\mathsf{R} = \{(a,a),(a,d),(b,b),(c,c),(d,d),(d,e),(e,a),(e,e)\}$.
(iv) $\mathsf{R} = \{(a,b),(b,c),(c,d),(d,e),(e,a)\}$.
(v) $\mathsf{R} = \{(a,b),(b,a),(b,d),(d,a),(c,e),(e,c),(e,e)\}$.

4. For each of the following relations, determine which of the four properties are satisfied by the relation. Justify your answers.

(i) $A = \{1,2,3,4,5,6,7,8\}$ $n\,\mathsf{R}\,m$ if and only if $n = 2^k m$ for some $k \in \mathbb{Z}$.
(ii) $A = \{1,2,3,4,5,6,7,8\}$ $n\,\mathsf{R}\,m$ if and only if $n \leqslant m$.
(iii) $A = \{1,2,3,4,5,6,7,8\}$ $n\,\mathsf{R}\,m$ if and only if $n \neq m$.
(iv) $A = \mathscr{P}(\{1,2,3\})$ $B\,\mathsf{R}\,C$ if and only if $B \subseteq C$.
(v) $A = \mathscr{P}(\{1,2,3\})$ $B\,\mathsf{R}\,C$ if and only if $|B| = |C|$.

5. Let A be any non-empty set and $\mathsf{R} = \varnothing$ be the **empty relation** on A. Which, if any, of the four properties defined in definitions 4.2 is/are satisfied by R?

If A itself is empty, which, if any, of the properties are satisfied?

6. Is it possible for a relation to be *both* symmetric and anti-symmetric? If so, how is it possible?

7. Let $\mathsf{R} = \{(a,a),(a,b),(a,c),(b,b),(b,c)\}$ be a relation on the set $\{a,b,c,d\}$. What is the minimum number of elements which need to be added to R in order that it becomes:

(i) reflexive;
(ii) symmetric;
(iii) anti-symmetric;
(iv) transitive?

8. Let $A = \{a,b,c,d\}$. For each of the following, define a relation R on A which satisfies the given properties. Try to keep the examples simple.

(i) R is reflexive and transitive but not symmetric.
(ii) R is both symmetric and anti-symmetric.
(iii) R is symmetric but not reflexive and not transitive.

9. Determine which, if any, of the four properties in definitions 4.2 is

satisfied by each of the following relations on the set $\mathbb{Z}^+$ of positive integers.

(i) $n \mathsf{R} m$ if and only if $n - m$ is a multiple of 3.
(ii) $n \mathsf{R} m$ if and only if $n = 3^k m$ for some $k \in \mathbb{Z}^+$.
(iii) $n \mathsf{R} m$ if and only if $n \neq m$.
(iv) $n \mathsf{R} m$ if and only if n/m is an integer.

10. Let A be the set of all lines in the plane $\mathbb{R}^2$. Which of our four properties is satisfied by the following relations?

(i) $l_1 \mathsf{R} l_2$ if and only if l_1 is parallel to l_2.
(ii) $l_1 \mathsf{R} l_2$ if and only if l_1 is perpendicular to l_2.
(iii) (For those readers who have studied elementary coordinate geometry.) $l_1 \mathsf{R} l_2$ if and only if the product of the gradients of l_1 and l_2 is equal to one.

11. (i) Given an example of a relation which is symmetric and transitive but not reflexive.

(ii) The following argument is sometimes given to show that a relation which is symmetric and transitive must also be reflexive.

Suppose R is a symmetric and transitive relation on a set A. Let a be any element of A. By the symmetric property, $a \mathsf{R} b$ implies $b \mathsf{R} a$. Since $a \mathsf{R} b$ and $b \mathsf{R} a$ we can deduce $a \mathsf{R} a$ from the transitive property. Because a was an arbitrary element, we have proved that $a \mathsf{R} a$ for every $a \in A$; therefore R is reflexive.

The example in (i) shows that this argument must be false. What is wrong with it?

12. Let A be any set of propositions not all of which have the same truth values. Define a relation R on A by:

$$p \mathsf{R} q \quad \text{if and only if } p \rightarrow q \text{ is true.}$$

Which of the four properties is satisfied by this relation?

13. For each of the four properties, reflexive, symmetric, anti-symmetric and transitive, if a relation R on a set A satisfies the property does its inverse relation R^{-1} necessarily satisfy the property as well? (The inverse relation is defined in exercise 4.1.10.)

Show also that R is symmetric if and only if $\mathsf{R} = \mathsf{R}^{-1}$.

4.3 Intersections and Unions of Relations

Since a relation R between A and B is simply a set—a subset of the Cartesian product $A \times B$—we can define intersections and unions of relations.

Let R and S be two relations from a set A to a set B. Both their intersection $\mathsf{R} \cap \mathsf{S}$ and their union $\mathsf{R} \cup \mathsf{S}$ are subsets of $A \times B$ also. That is, both the intersection and the union of two relations from A to B are also relations from A to B.

UNION OF RELATIONS

The situation is not quite so clear when R and S are relations between different pairs of sets. Suppose R is a relation from A to B, and S is a relation from C to D. Since R and S are both sets of ordered pairs, so, too, are their intersection and union. Thus $\mathsf{R} \cap \mathsf{S}$ and $\mathsf{R} \cup \mathsf{S}$ are both relations. However, it is not immediately apparent exactly *which* sets are related by $\mathsf{R} \cap \mathsf{S}$ and which sets are related by $\mathsf{R} \cup \mathsf{S}$. We leave consideration of this situation to exercise 4.3.3.

A natural question to ask is: if R and S are both relations from A to B, which (if any) of their properties are inherited by $\mathsf{R} \cap \mathsf{S}$ and $\mathsf{R} \cup \mathsf{S}$? Since the four properties of relations defined in the previous section are for relations on a set, we shall further restrict our attention to the case where R and S are both relations on the same set A.

Consider first the reflexive property. If R and S are *both* reflexive then $(a, a) \in \mathsf{R}$ and $(a, a) \in \mathsf{S}$ for every $a \in A$. Thus (a, a) belongs to $\mathsf{R} \cap \mathsf{S}$ and to $\mathsf{R} \cup \mathsf{S}$ for every $a \in A$, so the intersection and union of R and S are both also reflexive.

Secondly, suppose that R and S are both symmetric. Then so, too, are $\mathsf{R} \cap \mathsf{S}$ and $\mathsf{R} \cup \mathsf{S}$. We show this for the intersection only—the argument for the union is similar (see exercise 4.3.4(i)). Let $a, b \in A$ be such that $a(\mathsf{R} \cap \mathsf{S})b$ or, in set notation, $(a, b) \in \mathsf{R} \cap \mathsf{S}$. Then $a \,\mathsf{R}\, b$ and $a \,\mathsf{S}\, b$. Since R and S are both symmetric this implies $b \,\mathsf{R}\, a$ and $b \,\mathsf{S}\, a$, which means that $(b, a) \in \mathsf{R} \cap \mathsf{S}$. Thus $a(\mathsf{R} \cap \mathsf{S})b$ implies $b(\mathsf{R} \cap \mathsf{S})a$, so $\mathsf{R} \cap \mathsf{S}$ is symmetric.

The situation for anti-symmetry is more complicated. If R and S are both anti-symmetric, then an argument along the lines of that above for symmetry shows that the intersection $\mathsf{R} \cap \mathsf{S}$ is also anti-symmetric. However, the union need not be anti-symmetric. In order to demonstrate this we need to produce a counter-example; that is, an example of anti-symmetric relations R and S whose union $\mathsf{R} \cup \mathsf{S}$ is not anti-symmetric. A simple example is the following. Let $A = \{a, b\}$, $\mathsf{R} = \{(a, b)\}$ and $\mathsf{S} = \{(b, a)\}$. The relations R and S are both anti-symmetric since we never have x related to y, and y related to x for *different* elements x and y. However, the union $\mathsf{R} \cup \mathsf{S} = \{(a, b), (b, a)\}$ is not anti-symmetric because a is related to b, b is related to a, but a and b are not equal.

The situation for transitivity is similar to anti-symmetry: namely, if R and S are both transitive then $\mathsf{R} \cap \mathsf{S}$ is also transitive, but $\mathsf{R} \cup \mathsf{S}$ need not be. The proof that $\mathsf{R} \cap \mathsf{S}$ is transitive is similar to the proof for symmetry and we again leave it as an exercise (4.3.4(ii)). The simplest possible counter-example which shows that $\mathsf{R} \cup \mathsf{S}$ need not be transitive requires a three-element set A. (Why is this the simplest case?) Let $A = \{a, b, c\}$, $\mathsf{R} = \{(a, b)\}$ and $\mathsf{S} = \{(b, c)\}$. Then R and S are both transitive in a rather trivial way: we never have different elements x, y and z such that x is related to y and y is related to z, so the relations cannot fail to be transitive. (The transitive property is a conditional: *if* $x \,\mathsf{R}\, y$ and $y \,\mathsf{R}\, z$ *then* $x \,\mathsf{R}\, z$. Recall that a conditional proposition $p \to q$ is true whenever p is false. Thus if $x \,\mathsf{R}\, y$ and $y \,\mathsf{R}\, z$ is false for all $x, y, z \in A$ then R has the transitive property.) However, $\mathsf{R} \cup \mathsf{S} = \{(a, b), (b, c)\}$ is not transitive since a is related to b and b is related to c, but a is not related to c.

We summarize these considerations in the following theorem.

Theorem 4.1

Let R and S be two relations on the same set as A.

(i) If R and S are both reflexive then so, too, are $R \cap S$ and $R \cup S$.
(ii) If R and S are both symmetric then so, too, are $R \cap S$ and $R \cup S$.
(iii) If R and S are both anti-symmetric then so, too, is $R \cap S$ but $R \cup S$ need not be anti-symmetric.
(iv) If R and S are both transitive, then so, too, is $R \cap S$ but $R \cup S$ need not be transitive.

Exercises 4.3

1. Two relations R and S on the set $A = \{a, b, c, d\}$ are defined by:

$$R = \{(a, b), (a, c), (a, d), (b, b), (b, c), (c, a), (c, d)\}$$
$$S = \{(a, b), (a, c), (c, b), (c, d), (d, a)\}.$$

(i) Find $R \cap S$ and $R \cup S$.
(ii) Draw the directed graphs of R, S, $R \cap S$ and $R \cup S$.
(iii) Write down the binary matrices of R, S, $R \cap S$ and $R \cup S$.

2. Let R and S be relations on a set A.

(i) Explain how the directed graphs of $R \cap S$ and $R \cup S$ are related to the directed graphs of R and S.

(ii) Explain how the binary matrices of $R \cap S$ and $R \cup S$ are related to the binary matrices of R and S.

3. Let R_1 be a relation from A_1 to B_1 and let R_2 be a relation from A_2 to B_2. Show that $R_1 \cap R_2$ and $R_1 \cup R_2$ are both relations from $A_1 \cup A_2$ to $B_1 \cup B_2$.

4. Let R and S be relations on the same set A. Prove that:

(i) if R and S are both symmetric then so is $R \cup S$;
(ii) if R and S are both transitive then so is $R \cap S$.

Questions 5–11 refer to the composite of two relations which is defined as follows. Let R be a relation from A to B, and S be a relation from B

to C. The **composite** of R and S is the relation $\mathsf{S} \circ \mathsf{R}$ from A to C defined by $a(\mathsf{S} \circ \mathsf{R})c$ if and only if there exists an element $b \in B$ such that $a \,\mathsf{R}\, b$ and $b \,\mathsf{S}\, c$. This is illustrated in the following diagram.

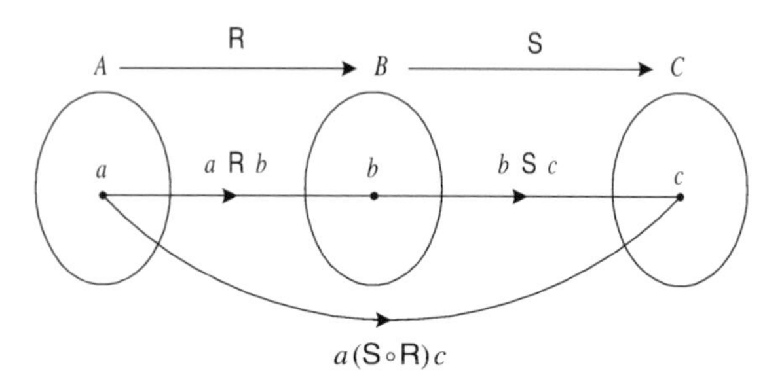

5. Let $A = \{1, 2, 3, 4\}$ and define two relations R and S on A by:

$$\mathsf{R} = \{(1,3), (2,2), (3,1), (3,4), (4,2)\}$$
$$\mathsf{S} = \{(1,2), (2,3), (3,4), (4,1)\}.$$

(i) List the elements of the relations $\mathsf{S} \circ \mathsf{R}$ and $\mathsf{R} \circ \mathsf{S}$.

(ii) List the elements of the relations R^{-1}, S^{-1}, $(\mathsf{S} \circ \mathsf{R})^{-1}$ and $(\mathsf{R} \circ \mathsf{S})^{-1}$.

(iii) List the elements of the relations $\mathsf{R}^{-1} \circ \mathsf{S}^{-1}$ and $\mathsf{S}^{-1} \circ \mathsf{R}^{-1}$.

(iv) What do you notice about the relations in parts (ii) and (iii)? Can you prove the general result that this suggests?

6. A relation R on the set $A = \{a, b, c, d, e, f, g, h\}$ has the following directed graph.

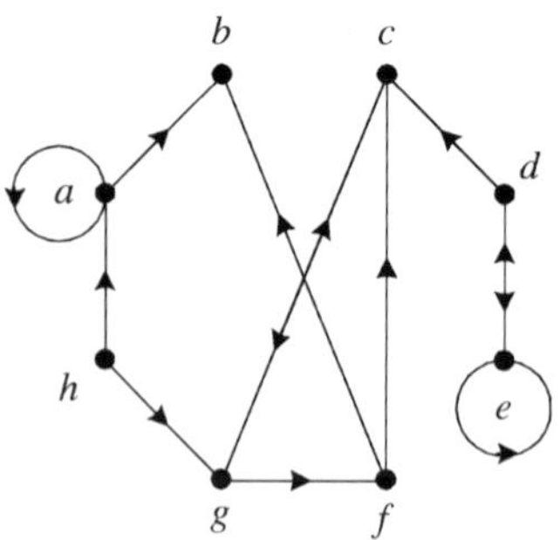

(i) List the elements of R.
(ii) List the elements of $\mathsf{R} \circ \mathsf{R}$.
(iii) Draw the directed graph of the relation $\mathsf{R} \circ \mathsf{R}$.

7. Let R and S be the relations on a set A of people defined by:

$$x \mathsf{R} y \quad \text{if and only if } x \text{ is the mother of } y;$$
$$x \mathsf{S} y \quad \text{if and only if } x \text{ is the father of } y.$$

Describe the relations (i) $\mathsf{S} \circ \mathsf{R}$, and (ii) $\mathsf{R} \circ \mathsf{S}$.

8. Let R be the relation on $\mathbb{Z}^+$ defined by

$$n \mathsf{R} m \quad \text{if and only if } m = n^2.$$

Describe the relation $\mathsf{R}^2 = \mathsf{R} \circ \mathsf{R}$ on $\mathbb{Z}^+$.

9. Let R be a relation from A to B and let I_A and I_B be the identity relations on A and B respectively. (See exercise 4.1.8.) Show that: (i) $\mathsf{R} \circ \mathsf{I}_A = \mathsf{R}$, and (ii) $\mathsf{I}_B \circ \mathsf{R} = \mathsf{R}$.

10. Let R be a relation from A to B, S a relation from B to C, and T a relation from C to D. Show that $(\mathsf{T} \circ \mathsf{S}) \circ \mathsf{R} = \mathsf{T} \circ (\mathsf{S} \circ \mathsf{R})$.

11. Let R be a relation from A to B and let S be a relation from B to C. Describe the relationship between the types of R, S and $\mathsf{S} \circ \mathsf{R}$.

4.4 Equivalence Relations and Partitions

One of the most important types of relation is an equivalence relation on a set. In this section we define the notion of an equivalence relation and explore the close connection between equivalence relations and partitions of a set.

Consider the relation R on the set of living people defined by: $x \mathsf{R} y$ if and only if x resides in the same country as y. Assuming each person is resident in only one country, the relation satisfies three obvious properties:

x resides in the same country as x; that is, R is reflexive;

if x resides in the same country as y, then y resides in the same country as x; that is, R is symmetric;

if x resides in the same country as y, and y resides in the same country as z, then x resides in the same country as z; that is, R is transitive.

Any given element x is related to everyone who lives in the same country as x and to no one else. Therefore the relation subdivides the set of living people into subsets according to their countries of residence. This is an example of an equivalence relation, which we now define formally.

Definition 4.3

A relation R on a set A is an **equivalence relation** if R is reflexive, symmetric and transitive.

Examples 4.4

1. Let $A = \mathbb{R}$, the set of real numbers, and define a relation R on A by

$$x \mathrel{\mathsf{R}} y \quad \text{if and only if } x^2 = y^2.$$

Then:

R is reflexive since $x^2 = x^2$ for every real number x;
R is symmetric since $x^2 = y^2$ implies $y^2 = x^2$;
R is transitive since $x^2 = y^2$ and $y^2 = z^2$ implies $x^2 = z^2$.

Therefore R is an equivalence relation.

2. Let $A = \mathbb{R}^2 - \{(0,0)\}$, the set of points in the plane except the origin†, and define a relation R on A by

$$(a,b) \mathrel{\mathsf{R}} (c,d) \quad \text{if and only if } (a,b) \text{ and } (c,d) \text{ both lie on the same straight line through the origin.}$$

Clearly R is both reflexive and symmetric. Also it is not difficult to see that R is transitive: if (a,b) and (c,d) both lie on the same straight line through the origin, and similarly (c,d) and (e,f) both lie on the same straight line through the origin, then so, too, do (a,b) and (e,f).

† For obvious reasons the set A is often referred to as the **punctured plane**.

Therefore R is an equivalence relation.

3. Let $A = \mathbb{Z}$, the set of integers, and define a relation R on A by

$$n \,\mathsf{R}\, m \quad \text{if and only if } n = 2^k m \text{ for some integer } k.$$

Show that R is an equivalence relation.

Solution

Firstly, R is reflexive since $n = 2^0 n$ for every integer n.

Secondly, if $n = 2^k m$ then $m = 2^{-k} n$ so $n \,\mathsf{R}\, m$ implies $m \,\mathsf{R}\, n$; therefore R is symmetric.

Thirdly, suppose $n \,\mathsf{R}\, m$ and $m \,\mathsf{R}\, p$; then there exist integers k and l such that $n = 2^k m$ and $m = 2^l p$. Combining these two equations gives $n = 2^k 2^l p = 2^{k+l} p$ where $k + l$ is an integer. Thus $n \,\mathsf{R}\, m$ and $m \,\mathsf{R}\, p$ implies $n \,\mathsf{R}\, p$ so R is transitive.

4. Consider the relation R defined on $\mathbb{Z}^+$ by

$$n \,\mathsf{R}\, m \quad \text{if and only if } n \text{ divides } m.$$

R is not an equivalence relation. To show this we only need to show that one of the three properties is not satisfied by R. Clearly R is not symmetric since, for example, 2 divides 4 but 4 does not divide 2.

(Note, however, that R is both reflexive and transitive. In fact R is also anti-symmetric because if n divides m and m divides n then $n = m$. Of course, these facts are not important in showing that R is not an equivalence relation.)

As we mentioned in chapter 3, there is a close connection between partitions of a set and equivalence relations on the set. Recall that a partition of a set A is a family of non-empty subsets which are pairwise disjoint and whose union is all of A (see definition 3.4). Suppose R is an equivalence relation on A. We can form subsets by grouping together in the same subset all elements which are related. We shall see that the properties of the equivalence relation guarantee that the subsets formed in this way form a partition of A. These subsets are called the 'equivalence classes' of the relation which we now define formally.

Definition 4.4

Let R be an equivalence relation on a set A, and let $x \in A$. The **equivalence class of x**, denoted $[x]$, is the set of all elements of A to which x is related:

$$[x] = \{y \in A : x \,\mathsf{R}\, y\}.$$

Note that, since R is symmetric, the equivalence class of x is also equal to $\{y \in A : y \,\mathsf{R}\, x\}$. In other words, the equivalence class of x can equally well be defined either as the set of elements that are related to x or as the set of elements to which x is related. Sometimes, if we need to emphasize the relation R, we refer to the **R-equivalence class of x** which we denote by $[x]_{\mathsf{R}}$.

Examples 4.5

1. Let R be the equivalence relation on $\mathbb{Z}^+$ defined in example 4.1.3 by $n \,\mathsf{R}\, m$ if and only if $n - m$ is divisible by 2. Then:

$$\begin{aligned}
[1] &= \{1, 3, 5, 7, 9, \ldots\} \\
[2] &= \{2, 4, 6, 8, 10, \ldots\} \\
[3] &= \{1, 3, 5, 7, 9, \ldots\} \\
[4] &= \{2, 4, 6, 8, 10, \ldots\} \text{ etc.}
\end{aligned}$$

   In this example there are clearly only two *different* equivalence classes—the sets of even and odd positive integers respectively. Note that these two equivalence classes form a partition of $\mathbb{Z}^+$.

2. Let R be the equivalence relation defined on the set of integers $\mathbb{Z}$ by $n \mathsf{R} m$ if and only if $n^2 = m^2$.

   For each integer n, only $n \,\mathsf{R}\, n$ and $n \,\mathsf{R}\, (-n)$ so the equivalence class of n contains two integers, namely n and its negative:

$$[n] = \{n, -n\}.$$

   There is one exception: since 0 equals its negative, the equivalence class of 0 contains only itself, $[0] = \{0\}$.

3. Let R be the relation on the punctured plane $\mathbb{R}^2 - \{(0,0)\}$ defined in example 4.4.2 by $(a,b) \,\mathsf{R}\, (c,d)$ if and only if (a,b) and (c,d) lie on the same straight line through the origin.

Let (x, y) be any point in $\mathbb{R}^2 - \{(0,0)\}$. The equivalence class of (x, y) is the set of all points (except the origin which is not an element of the punctured plane itself) which lie on the line through $(0, 0)$ and (x, y). In this case we can visualize the equivalence classes geometrically. There are infinitely many different classes in this example; one for each (punctured) line through the origin.

4. Let A be any non-empty set of people and define a relation R on A by $x \mathsf{R} y$ if and only if x is the same height as y (measured, let us say, to the nearest centimetre). Then R is an equivalence relation on A.

   For any person in the set, his or her equivalence class is the set of all people (in the set A) who are the same height.

It might seem from the definition that there are as many equivalence classes as there are elements of the set A. However, the above examples show that there are many fewer *distinct* (i.e. unequal) equivalence classes in general. This is because if two elements are related then their equivalence classes are equal. To see this, suppose that R is an equivalence relation on A and $x \mathsf{R} y$ for two elements x and y of A. We wish to show that $[x] = [y]$. Let $z \in [x]$; then $x \mathsf{R} z$ by definition. Since $x \mathsf{R} y$ and R is symmetric we know that $y \mathsf{R} x$ also. Thus $y \mathsf{R} x$ and $x \mathsf{R} z$; it follows that $y \mathsf{R} z$, by the transitivity of R, so $z \in [y]$. This shows that $[x] \subseteq [y]$. The proof that $[y] \subseteq [x]$ is similar, so we can conclude that $[x] = [y]$. The converse is also true; namely, if $[x] = [y]$ then x is related to y. This is very easy to prove: $y \in [y]$ since R is reflexive, so $y \in [x]$ which means $x \mathsf{R} y$ by definition. We have proved the following result.

Theorem 4.2

Let R be an equivalence relation on A and $x, y \in A$. Then $[x] = [y]$ if and only if $x \mathsf{R} y$.

The observation made in theorem 4.2 paves the way for us to prove the result mentioned above: the family of equivalence classes of an equivalence relation on a set form a partition of the set.

Theorem 4.3

Let R be an equivalence relation on a non-empty set A. The family of distinct R-equivalence classes is a partition of A.

Proof

A partition must be a family of *non-empty* sets—see definition 3.4. This is clearly satisfied here. Since R is reflexive, $x \in [x]$ for every $x \in A$. Therefore, every equivalence class is non-empty. This also shows that the first property of a partition is satisfied. Since $x \in [x]$ for every $x \in A$, it also follows that every element of A belongs to some equivalence class, namely its own. Therefore the union of all the equivalence classes contains all the elements of A.

The second property of a partition, that the members of the family of subsets should be pairwise disjoint, sometimes causes confusion because equivalence classes need not be disjoint according to theorem 4.2. The point here is that it is the family of *distinct* equivalence classes which we are considering†. So we must show that any two distinct classes are disjoint. In fact it is easier to prove the contrapositive: if two classes have elements in common (are not disjoint) then they are equal (are not distinct).

So suppose $[x] \cap [y] \neq \varnothing$. Then we may choose an element z in the intersection. Thus $x \mathrel{\mathsf{R}} z$ since $z \in [x]$ and $y \mathrel{\mathsf{R}} z$ since $z \in [y]$. The symmetric and transitive properties of R imply that $x \mathrel{\mathsf{R}} y$, so we conclude $[x] = [y]$ by theorem 4.2.

We have now checked both properties of a partition, so the proof is complete. □

Examples 4.6

1. Recall from example 3.19.3(iii) that the 'floor' or 'integer part' of a real number is the largest integer that is less than or equal to the given real number. For example,

$$\lfloor 2.4 \rfloor = 2 \qquad \lfloor -3.8 \rfloor = -4 \qquad \lfloor \sqrt{10} \rfloor = 3 \text{ etc.}$$

† From a theoretical point of view we do not need to emphasize that we are considering the family of distinct classes as this is taken care of automatically. Since the family of equivalence classes is a *set* of classes, our usual convention for sets, that we disregard any repeated elements, applies.

Let R be the relation on the set $\mathbb{R}$ of real numbers defined by $x \,\mathsf{R}\, y$ if and only if $\lfloor x \rfloor = \lfloor y \rfloor$. It is straightforward to check that R is an equivalence relation. Consider $\frac{1}{2} \in \mathbb{R}$: since $\lfloor \frac{1}{2} \rfloor = 0$, the equivalence class of $\frac{1}{2}$ is

$$[\tfrac{1}{2}] = \{x \in \mathbb{R} : \lfloor x \rfloor = 0\} = \{x \in \mathbb{R} : 0 \leqslant x < 1\}.$$

This set, called a half-open interval, is denoted $[0, 1)$. Similarly,

$$[2.4] = \{x \in \mathbb{R} : \lfloor x \rfloor = \lfloor 2.4 \rfloor = 2\} = \{x \in \mathbb{R} : 2 \leqslant x < 3\} = [2, 3).$$

In fact every equivalence class is a half-open interval of the form $[n, n+1)$ for some integer n, so the partition of $\mathbb{R}$ by equivalence classes is

$$\{[n, n+1) : n \in \mathbb{Z}\}.$$

2. A relation R is defined on the set $\mathbb{R}$ of real numbers by

$$x \,\mathsf{R}\, y \text{ if and only if } (x = 0 = y) \vee (xy > 0).$$

We leave it as an exercise to check that R is an equivalence relation. (This is not difficult although, since there are two cases in the definition of $x\mathsf{R}y$, the proof of each property requires consideration of cases.) What are the equivalence classes? Since the equivalence classes form a partition of $\mathbb{R}$, we can adopt the following simple strategy for finding all the equivalence classes:

(I) choose any $a \in \mathbb{R}$ and find its equivalence class $[a]$;
(II) choose $b \notin [a]$ and find $[b]$;
(III) choose $c \notin [a] \cup [b]$ and find $[c]$;
(IV) continue in this way, at each stage choosing $x \in \mathbb{R}$ that does not belong to any existing equivalence class, until it is no longer possible to choose such an x.

So we first select $a = 1$, say. Since $1 \neq 0$ we have $1 \,\mathsf{R}\, y \Leftrightarrow 1 \times y > 0 \Leftrightarrow y > 0$, so

$$[1] = \{y \in \mathbb{R} : y > 0\} = \mathbb{R}^+, \text{ the set of positive real numbers.}$$

Next we must choose $b \notin \mathbb{R}^+$; so let $b = 0$. Now $0 \,\mathsf{R}\, y \Leftrightarrow y = 0$ so the equivalence class is a singleton set

$$[0] = \{0\}.$$

Now we must select $c \notin \mathbb{R}^+ \cup \{0\}$; so let $c = -1$. Again $-1 \neq 0$ so $-1 \,\mathsf{R}\, y \Leftrightarrow -1 \times y > 0 \Leftrightarrow -y > 0 \Leftrightarrow y < 0$. Therefore

$$[-1] = \{y \in \mathbb{R} : y < 0\} = \mathbb{R}^-, \text{ the set of negative real numbers.}$$

Since every real number belongs to one of these equivalence classes $[1] = \mathbb{R}^+$, $[0] = \{0\}$ or $[-1] = \mathbb{R}^-$, we have found all the (distinct) classes. Hence the partition of $\mathbb{R}$ into equivalence classes is

$$\{\mathbb{R}^+, \{0\}, \mathbb{R}^-\}.$$

3. Define a relation R on $\mathscr{P}(\{1,2,3\})$, the power set of $\{1,2,3\}$, by $A \mathrel{\mathsf{R}} B$ if and only if $|A| = |B|$. For example, $\{1,2\} \mathrel{\mathsf{R}} \{2,3\}$, $\{1,2\} \not\mathrel{\mathsf{R}} \{3\}$, etc. It is easy to verify that R is an equivalence relation on $\mathscr{P}(\{1,2,3\})$—see exercise 4.2.4(v). For $A \subseteq \{1,2,3\}$, the equivalence class of A contains all those subsets of $\{1,2,3\}$ with the same cardinality. Hence the equivalence classes are:

$$\begin{aligned}
[\varnothing] &= \{\varnothing\} \\
[\{1\}] &= \{\{1\}, \{2\}, \{3\}\} \\
[\{1,2\}] &= \{\{1,2\}, \{1,3\}, \{2,3\}\} \\
[\{1,2,3\}] &= \{\{1,2,3\}\}.
\end{aligned}$$

More generally, let $\mathscr{F}$ be any family of finite sets and define a relation R on $\mathscr{F}$ by $A \mathrel{\mathsf{R}} B$ if and only if $|A| = |B|$. Then again R is an equivalence relation on $\mathscr{F}$. Given $A \in \mathscr{F}$, its equivalence class $[A]$ is the set of those sets in $\mathscr{F}$ with the same cardinality as A. Therefore the equivalence relation partitions $\mathscr{F}$ into subfamilies of sets with the same cardinality.

We have noted the connection between equivalence relations and partitions in one 'direction' only—from equivalence relations to partitions. That is, given an equivalence relation on a set we have defined a partition by equivalence classes. However, we can proceed in the other direction as well. If we are given a partition of a set we can use it to define an equivalence relation in such a way that the equivalence classes are the original subsets which make up the partition. This is easily done. Given a partition $\{S_i : i \in I\}$ of a set A, we define a relation R on A by

$x \mathrel{\mathsf{R}} y$ if and only if x and y belong to the same subset S_i of the partition.

The properties of the partition mean that the relation is well defined: given two elements x and y of the set, each belongs to precisely one of the sets S_i so we can determine with certainty whether $x \mathrel{\mathsf{R}} y$ or $x \not\mathrel{\mathsf{R}} y$. It should be clear that R is an equivalence relation. Every element of A belongs to the same subset as itself so R is reflexive. The statements that guarantee that R is symmetric and transitive are equally as trite!

The equivalence classes of this relation coincide with the original subsets of the partition. An equivalence class $[x]$ contains all those elements which belong to the same subset in the partition as x, which means that $[x]$ must be equal to the subset containing x. Thus every equivalence class is one of the 'original' subsets S_i.

We can summarize the preceding remarks in the following theorem.

Theorem 4.4

Let $\{S_i : i \in I\}$ be a partition of a set A. Then $x \mathsf{R} y$ if and only if $x, y \in S_i$ for some $i \in I$ defines an equivalence relation R on A whose equivalence classes are precisely the sets S_i in the partition.

Theorems 4.3 and 4.4 mean that we can pass freely back and forth between equivalence relations and partitions. In a sense they are two aspects of the same phenomenon.

Modulo Arithmetic

Let n be a fixed positive integer. A relation, called **congruence modulo n**, is defined on the set $\mathbb{Z}$ of integers by

$$a \equiv_n b \quad \text{if and only if } a - b = kn \text{ for some } k \in \mathbb{Z}.$$

Alternative notations for $a \equiv_n b$ are $a \equiv b \mod n$ or simply $a \equiv b$ if the value of n has already been established. If $a \equiv_n b$ we say that ***a* is congruent to *b* modulo n**. Thus a is congruent to b modulo n if n divides their difference or, equivalently, if a and b have the same remainder after division by n.

We leave it as a straightforward exercise to show that congruence modulo n is an equivalence relation—see exercise 4.4.10.

Example 4.7

Before considering the general case we look first at a specific example, that of congruence modulo 5. Since we are fixing $n = 5$ in this example, we write $a \equiv b$ as an abbreviation for $a \equiv_5 b$.

In this case $a \equiv b$ if and only if $a - b = 5k$ for some integer k; that is, if and only if there exists an integer k such that $a = 5k + b$. Therefore

$$[p] = \{q \in \mathbb{Z} : q = 5k + p, \text{ for some } k \in \mathbb{Z}\}.$$

All of the equivalence classes are infinite; some of them are listed below:

$$\begin{aligned}
[0] &= \{\ldots, -10, -5, 0, 5, 10, 15, \ldots\} \\
[1] &= \{\ldots, -9, -4, 1, 6, 11, 16, \ldots\} \\
[2] &= \{\ldots, -8, -3, 2, 7, 12, 17, \ldots\} \\
[3] &= \{\ldots, -7, -2, 3, 8, 13, 18, \ldots\} \\
[4] &= \{\ldots, -6, -1, 4, 9, 14, 19, \ldots\}.
\end{aligned}$$

Clearly these five are all the distinct equivalence classes, since every integer is contained in one of these. For instance

$$\cdots = [-8] = [-3] = [2] = [7] = [12] = \cdots$$

since

$$\cdots - 8 \equiv -3 \equiv 2 \equiv 7 \equiv 12 \cdots.$$

Returning to the general case, it can be shown that the relation of congruence modulo n partitions $\mathbb{Z}$ into the n distinct equivalence classes

$$[0], [1], [2], \ldots, [n-1]\dagger.$$

Let $\mathbb{Z}_n = \{[0], [1], [2], \ldots, [n-1]\}$ denote the set of equivalence classes. We can define the arithmetic operations of addition and multiplication on the set $\mathbb{Z}_n$ by

$$[a] +_n [b] = [a + b]$$

and

$$[a] \times_n [b] = [a.\, b].$$

This is not quite as straightforward as it might seem; hidden in these definitions is a potential problem. The crucial point is that a given equivalence class has many different 'names'. For example, in the case of congruence modulo 5, we saw that $[-8], [-3], [2], [7], [12]$, etc. are different notations for the *same* equivalence class. The potential problem with the definitions above of addition and multiplication on $\mathbb{Z}_n$ is that using different labels for the equivalence classes may produce different results.

† To emphasize the role of n, these classes are sometimes denoted $[0]_n, [1]_n, [2]_n, \ldots, [n-1]_n$.

Before we consider the general case, let us look again at the example of $n = 5$. In this case $[-8] = [2]$ and $[4] = [19]$ so we would hope that $[-8] +_5 [4]$ and $[2] +_5 [19]$ would define the same equivalence class, and $[-8] \times_5 [4]$ and $[2] \times_5 [19]$ would similarly define the same class. Now from the definition of addition on $\mathbb{Z}_n$, $[-8] +_5 [4] = [-4]$ and $[2] +_5 [19] = [21]$. However, all is well since $-4 \equiv_5 21$ so $[-4] = [21]$; the more convenient name for this particular class is $[1]$. Similarly $[-8] \times_5 [4] = [-32]$ and $[2] \times_5 [19] = [38]$, but again these two classes are equal since $-32 \equiv_5 38$.

Returning once more to the general case, to show that there really is no problem with our definitions of addition and multiplication on $\mathbb{Z}_n$, we need to prove:

$$\text{if } [a] = [a'] \text{ and } [b] = [b'] \text{ then } [a+b] = [a'+b'] \text{ and } [ab] = [a'b'].$$

The actual proof is not difficult and is left as an exercise—see exercise 4.4.11.

We now have a well defined 'arithmetic modulo n'. The arithmetic of these sets $\mathbb{Z}_n$ is important to mathematicians as they are all examples of a mathematical structure called a **ring**. In computer science the systems of arithmetic modulo 2, modulo 8 and modulo 16 have some importance.

Some of these 'finite arithmetics' as we may call them have some slightly unusual properties. For instance, the product of two non-zero elements (i.e. classes other than [0]) may sometimes be zero, $[0]$. In $\mathbb{Z}_6$, for example, $[3]_6 \times_6 [4]_6 = [0]_6$. We leave it as an investigation to determine for what values of n this can occur. (See exercise 4.4.12.)

We conclude this section with the tables for addition and multiplication in the set $\mathbb{Z}_5$.

Addition

$+_5$	[0]	[1]	[2]	[3]	[4]
[0]	[0]	[1]	[2]	[3]	[4]
[1]	[1]	[2]	[3]	[4]	[0]
[2]	[2]	[3]	[4]	[0]	[1]
[3]	[3]	[4]	[0]	[1]	[2]
[4]	[4]	[0]	[1]	[2]	[3]

Multiplication

$\times_5$	[0]	[1]	[2]	[3]	[4]
[0]	[0]	[0]	[0]	[0]	[0]
[1]	[0]	[1]	[2]	[3]	[4]
[2]	[0]	[2]	[4]	[1]	[3]
[3]	[0]	[3]	[1]	[4]	[2]
[4]	[0]	[4]	[3]	[2]	[1]

Exercises 4.4

1. A relation R on the set of integers $\mathbb{Z}$ is defined by $n \,\mathsf{R}\, m$ if and only if $|n| = |m|$. Show that R is an equivalence relation and determine the corresponding equivalence classes.

2. (i) A relation R on the set of real numbers $\mathbb{R}$ is defined by $x \,\mathsf{R}\, y$ if and only if $\lfloor 2x \rfloor = \lfloor 2y \rfloor$.

   (a) Verify that R is an equivalence relation.
   (b) Determine the equivalence classes of $1/4$ and $1/2$.
   (c) Describe the partition of $\mathbb{R}$ into the equivalence classes of R.

   (ii) An equivalence relation S is defined on $\mathbb{R}$ by $x \,\mathsf{S}\, y$ if and only if $\lfloor 3x \rfloor = \lfloor 3y \rfloor$. Determine the partition of $\mathbb{R}$ into the equivalence classes of S.

3. Let A be any non-empty set of people.

   (i) A relation R on A is defined by $P \,\mathsf{R}\, Q$ if and only if P and Q are the same age (in years). Show that R is an equivalence relation on A and describe the equivalence classes of R.

   (ii) A second relation S on A is defined by $P \,\mathsf{S}\, Q$ if and only if P and Q were born in the same country. Given that S is an equivalence relation on A, describe the equivalence classes of S.

4. Show that both the identity relation I_A and the universal relation U_A, as defined in exercise 4.1.8, are equivalence relations on a set A. What are the corresponding equivalence classes?

5. Verify that each of the following are equivalence relations on the plane $\mathbb{R}^2$ and describe the equivalence classes:

   (i) $(x_1, y_1) \,\mathsf{R}\, (x_2, y_2)$ if and only if $x_1 = x_2$.
   (ii) $(x_1, y_1) \,\mathsf{R}\, (x_2, y_2)$ if and only if $x_1 + y_1 = x_2 + y_2$.
   (iii) $(x_1, y_1) \,\mathsf{R}\, (x_2, y_2)$ if and only if $x_1^2 + y_1^2 = x_2^2 + y_2^2$.

6. A relation R on $\mathbb{Z}^+ \times \mathbb{Z}^+$ is defined by

$$(m, n) \,\mathsf{R}\, (p, q) \quad \text{if and only if } m + q = n + p.$$

   Show that R is an equivalence relation and describe the equivalence classes of $(1, 1)$, $(2, 1)$, $(3, 1)$, $(1, 2)$ and $(1, 3)$.

How are the set of equivalence classes and the set of integers related?

7. Verify that $x \mathsf{R} y$ if and only if $(x - y) \in \mathbb{Z}$ defines an equivalence relation on the set $\mathbb{Q}$ of rational numbers. Describe the equivalence classes of 2, $\frac{1}{4}$ and $-\frac{1}{4}$.

8. How many different equivalence relations are there on the sets (i) $\{a, b, c\}$, and (ii) $\{a, b, c, d\}$?

9. Let $A = \{n \in \mathbb{Z} : n \geqslant 2\} = \{2, 3, 4, 5, 6, \ldots\}$. For $n \in A$, let $P(n)$ denote the smallest prime number that divides n and let $Q(n)$ denote the largest prime number that divides n. For example:

$$P(14) = 2, P(15) = 3, P(16) = 2, P(17) = 17, P(18) = 2, \ldots$$
$$Q(14) = 7, Q(15) = 5, Q(16) = 2, Q(17) = 17, Q(18) = 3, \ldots.$$

(i) Show that

$$n \mathsf{R} m \quad \text{if and only if } P(n) = P(m)$$

defines an equivalence relation on A. List the first few elements of the equivalence classes of 2, 3 and 5.

(ii) It is given that

$$n \mathsf{S} m \quad \text{if and only if } Q(n) = Q(m)$$

also defines an equivalence relation on A. List the first few elements of the equivalence classes of 2, 3 and 5.

10. Show that, for a fixed positive integer n, the relation of congruence modulo n is an equivalence relation on $\mathbb{Z}$.

Show also that $a \equiv_n b$ if and only if a and b have the same remainder after division by n.

11. Show that addition and multiplication on $\mathbb{Z}_n$ is well defined. That is, prove:

$$\text{if} \quad [a] = [a'] \quad \text{and} \quad [b] = [b']$$

then

$$[a + b] = [a' + b'] \quad \text{and} \quad [ab] = [a'b'].$$

12. Draw up addition and multiplication tables for the set $\mathbb{Z}_n$ for $n = 3, 4, 6$ and 7.

For which of these values of n do there exist non-zero elements $[a]_n$ and $[b]_n$ such that $[a]_n \times_n [b]_n = [0]_n$?

What is the general condition on n such that there do *not* exist non-zero elements $[a]_n$ and $[b]_n$ in $\mathbb{Z}_n$ such that $[a]_n \times_n [b]_n = [0]_n$?

13. Let A be any set of propositions and define R by

$$p \,\mathsf{R}\, q \quad \text{if and only if } p \leftrightarrow q \text{ is true.}$$

Show that R is an equivalence relation on A. What are the equivalence classes?

14. A relation R on a set A is reflexive and satisfies the 'circular' property

$$\text{if } x \,\mathsf{R}\, y \text{ and } y \,\mathsf{R}\, z \text{ then } z \,\mathsf{R}\, x, \text{ for all } x, y, z, \in A.$$

Show that R is an equivalence relation on A.

15. Let R and S be equivalence relations on a set A. Show that $\mathsf{R} \circ \mathsf{S}$ is an equivalence relation on A if and only if $\mathsf{R} \circ \mathsf{S} = \mathsf{S} \circ \mathsf{R}$.

(The composite $\mathsf{R} \circ \mathsf{S}$ of two relations is defined in exercises 4.3.)

4.5 Order Relations

Many sets have a natural ordering of their elements. Probably the most familiar example is the set of real numbers ordered by 'magnitude'. We are used to statements such as $3 \leqslant \pi$, $-4 < -3$, $2 < \sqrt{8} < 3$, and $x^2 > 0$ for every non-zero $x \in \mathbb{R}$. Similarly, any family of sets is ordered by 'inclusion': if $A \subseteq B$ we may regard A as being 'smaller' than B. As a non-mathematical example, a set of people could be ordered by age or by height.

Unlike equivalence relations, there are various different types of order relation. Consider the relations on $\mathbb{R}$ defined by $x < y$ and $x \leqslant y$, respectively. These have different properties; for example, the latter is reflexive, but the former is not. There is another, perhaps less obvious, difference. Given any two real numbers x and y, at least one of the statements $x \leqslant y$ and $y \leqslant x$ is valid, but this is not true of the statements $x < y$ and $y < x$.

The most general order relation we consider is called a 'partial order' which we define as follows.

Definition 4.5

A **partial order** on a set is a relation which is reflexive†, anti-symmetric and transitive.

A set together with a partial order is called a **partially ordered set** or, somewhat less elegantly, a **poset**.

Examples 4.8

1. The relation R on the set of real numbers defined by $x \mathrel{\mathsf{R}} y$ if and only if $x \leqslant y$ is a partial order (see example 4.2.1).

   However, the relation S defined by $x \mathrel{\mathsf{S}} y$ if and only if $x < y$ is not a partial order, since it is not reflexive.

2. Let $\mathscr{F}$ be any family of sets and define a relation R on $\mathscr{F}$ by $A \mathrel{\mathsf{R}} B$ if and only if $A \subseteq B$. Every set is a subset of itself so R is reflexive. The anti-symmetry property is precisely the property we frequently use to prove that two sets are equal: see theorem 3.1. The transitivity of $\subseteq$ is dealt with in exercise 3.2.10(i). Therefore R is a partial order.

   Again note that the strict inclusion of subsets, $\subset$, is not a partial order because it is not reflexive.

3. The 'divisibility' relation on the set of positive integers $\mathbb{Z}^+$, defined by $n \mathrel{\mathsf{R}} m$ if and only if n divides m, is a partial order. (Note: n divides m is frequently written $n|m$.)

4. The relation on the set of English words defined by 'the word w_1 is related to the word w_2 if $w_1 = w_2$ or w_1 comes before w_2 in a dictionary' is a partial ordering. This is the usual alphabetical (or lexicographic) ordering of words.

5. A relation R is defined on $\mathbb{R}^2$ by

   $(x_1, y_1) \mathrel{\mathsf{R}} (x_2, y_2)$ if and only if *either* $x_1 < x_2$ *or* both $x_1 = x_2$ and $y_1 \leqslant y_2$.

† Some authors do not require a partial order to be reflexive, although it is much more common to include the condition.

Show that R is a partial order on $\mathbb{R}^2$.

Solution

Every $(x, y) \in \mathbb{R}^2$ is related to itself by the second part of the definition of $\mathsf{R} : x = x$ and $y \leqslant y$. Hence R is reflexive.

To prove the anti-symmetric and transitive properties, it helps to note that $(x_1, y_1)\,\mathsf{R}\,(x_2, y_2)$ implies $x_1 \leqslant x_2$.

Suppose $(x_1, y_1)\,\mathsf{R}\,(x_2, y_2)$ and $(x_2, y_2)\,\mathsf{R}\,(x_1, y_1)$. Then $x_1 \leqslant x_2$ and $x_2 \leqslant x_1$, so $x_1 = x_2$. This means that we must also have $y_1 \leqslant y_2$ and $y_2 \leqslant y_1$, so $y_1 = y_2$ as well. Hence $(x_1, y_2) = (x_2, y_2)$ which shows that R is anti-symmetric.

Finally, to prove that R is transitive, suppose that $(x_1, y_1)\,\mathsf{R}\,(x_2, y_2)$ and $(x_2, y_2)\,\mathsf{R}\,(x_3, y_3)$. Then $x_1 \leqslant x_2$ and $x_2 \leqslant x_3$. If $x_1 < x_2$ or $x_2 < x_3$ (or both), then $x_1 < x_3$ which means that $(x_1, y_1)\,\mathsf{R}\,(x_3, y_3)$. The other possibility is that $x_1 = x_2$ and $x_2 = x_3$. In this case we must have $y_1 \leqslant y_2$ and $y_2 \leqslant y_3$, since $(x_1, y_1)\,\mathsf{R}\,(x_2, y_2)$ and $(x_2, y_2)\,\mathsf{R}\,(x_3, y_3)$. Therefore $x_1 = x_3$ and $y_1 \leqslant y_3$ so again $(x_1, y_1)\,\mathsf{R}\,(x_3, y_3)$. In both cases we may conclude that $(x_1, y_1)\,\mathsf{R}\,(x_3, y_3)$ which shows that R is transitive.

This partial order may seem a little strange at first, It is, however, very similar to the alphabetical (lexicographic) ordering of words. To compare two ordered pairs, we first compare their initial elements; if these are unequal then we know how the ordered pairs are related. If the first elements are equal, then we need to look at the second elements of the ordered pairs to see how they are related. (Of course in the case of English words we may need to continue this process and consider the third letter, the fourth, etc, until we can order the words.)

There is also a geometric way of visualizing this partial order. Let P_1 and P_2 be the points in the plane with coordinates (x_1, y_1) and (x_2, y_2) respectively. Then $(x_1, y_1)\,\mathsf{R}\,(x_2, y_2)$ if and only if either P_1 is to the left of P_2 or the points are on the same vertical level and P_1 is below (or coincides with) P_2.

The following theorem says that any subset of a partially ordered set is automatically a partially ordered set. It gives a way of generating many more examples of partially ordered sets. The proof is straightforward and we leave it as an exercise.

Theorem 4.5

Let R be a partial order on a set A, and let B be any subset of A. Then $\mathsf{S} = \mathsf{R} \cap (B \times B)$ is a partial order on B.

Although the definition of the relation S looks somewhat technical, it is the obvious relation on B. For $b_1, b_2 \in B$ we have $b_1 \mathsf{S} b_2$ if and only if $b_1 \mathsf{R} b_2$. Therefore elements of B are related by S in exactly the same way as they are related by R, when we consider them as elements of A. This relation S is called the **restriction of R to B**, and we say that B **inherits** the relation S from the relation R on A.

Maximal and Minimal Elements

According to theorem 4.5, any subset of the real numbers is partially ordered by the relation $\leqslant$. Some sets of real numbers ordered in this way will have a greatest element and some will not, and similarly for the smallest or least element. For example, the set of integers has no greatest or least element, but the set of positive integers has a least element, namely 1, but no greatest element.

Clearly an finite subset of $\mathbb{R}$ will have both a greatest and a least element with respect to this order. An infinite subset of $\mathbb{R}$, however, may or may not have a greatest and/or least element. For example, the open interval

$$(0, 1) = \{x \in \mathbb{R} : 0 < x < 1\}$$

which contains an infinite number of elements, has no greatest or least element. However, its companion closed interval

$$[0, 1] = \{x \in \mathbb{R} : 0 \leqslant x \leqslant 1\}$$

which also contains an infinite number of elements, does have a greatest and a least element, namely 1 and 0 respectively.

We should not be led by the case of subsets of $\mathbb{R}$ to believe that *every* finite poset has a (single) greatest and a (single) least element. Consider the set of all *proper* subsets of $\{a, b, c\}$ ordered by inclusion $\subseteq$. The least element is $\varnothing$ but there is no (single) greatest element since there are three different two-element subsets of $\{a, b, c\}$. This example indicates that we should be more precise about our meaning of greatest and least elements. The following definition of greatest and least elements is the obvious one if we regard $a \mathsf{R} b$ to mean in some sense 'a is less than (or equal to) b'. The greatest element of a poset, for example, is then the element which is 'bigger' than *all* the other elements.

Definition 4.6

Let R be a partial order on a set A. The **greatest element** of A (if it exists) is the element α such that $a \mathsf{R} \alpha$ for every $a \in A$.

Similarly, the **least element** of A (if it exists) is the element β such that $\beta \mathsf{R} a$ for every $a \in A$.

Returning to the example of the proper subsets of $\{a, b, c\}$ ordered by inclusion, we can verify that there is no greatest element according to our definition. However, each of the two-element subsets can be regarded as the 'largest possible' in the sense that there are no subsets which are 'bigger' than these. We formalize this idea in the definition of 'maximal' elements.

Definition 4.7

Let A be a poset, with order relation R. An element x of A is **maximal** if, for every $a \in A$, $x \mathsf{R} a$ implies $x = a$.

Similarly, an element y is **minimal** if, for every $a \in A$, $a \mathsf{R} y$ implies $a = y$.

If we regard $a \mathsf{R} b$ as meaning 'a is less than or equal to b' in whatever sense, then an element is maximal if there is no 'greater' element in the set, i.e. the element is related only to itself. Similarly an element is minimal if there is no 'smaller' element in the set, i.e. no other element is related to it.

Examples 4.9

1. Consider again the proper subsets of $\{a, b, c\}$ ordered by inclusion. In this case there are three different maximal elements $\{a, b\}$, $\{b, c\}$ and $\{a, c\}$. There is a single minimal element, namely the least element $\varnothing$.

2. Let $A = \{2, 3, 4, 5, 6, 7, 8\}$, ordered by divisibility: $x \mathsf{R} y$ if and only if x divides y.

   There are four minimal elements, 2, 3, 5 and 7. If a divides 2, where $a \in A$, then $a = 2$; and similarly for 3, 5 and 7.

   The elements 5, 6, 7 and 8 are all maximal. For $a \in A$, if 5 divides a then $a = 5$; and similarly for 6, 7 and 8.

   Note that, with this ordering, A has no greatest or least element. Clearly the only candidates for a least element are the minimal elements, none of which is the least element. For example, since $2 \not\mathsf{R} 3$ it is not true that $2 \mathsf{R} a$ for *every* $a \in A$, so 2 is not the least element. Also $3 \not\mathsf{R} 2$, $5 \not\mathsf{R} 2$ and $7 \not\mathsf{R} 2$ so neither 3 nor 5 nor 7 is the least element. Similar remarks apply to the maximal elements, so there is no greatest element.

We have seen that a partially ordered set may have several minimal and/or maximal elements. It can, however, have at most one greatest element and at most one least element. That is, if a poset A has a greatest element α, then α is unique; and similarly for a least element β. (We have, in fact, been tacitly assuming this by referring to *the* greatest and least elements.) It is easy to see, for example, that A has at most one least element: suppose β and β' are two least elements. Then $\beta \mathsf{R} \beta'$ since β is a least element, and $\beta' \mathsf{R} \beta$, since β' is a least element. Therefore $\beta = \beta'$ (by anti-symmetry), so there is only one least element. The same kind of argument clearly works for the greatest element as well.

The following theorem clarifies the connection between least and minimal elements and between greatest and maximal elements.

Theorem 4.6

Let A be a poset with partial order relation R.

If A has a greatest element α, then α is maximal and there are no other maximal elements.

Similarly, if A has a least element β, then β is minimal and there are no other minimal elements.

Proof

We prove the proposition for the greatest element only; the proof for the least element is similar.

Let α be the greatest element and suppose $\alpha \mathsf{R} a$ where $a \in A$. Since α is the greatest element we also know that $a \mathsf{R} \alpha$. Therefore $a = \alpha$, by the anti-symmetric property, so α is a maximal element.

Suppose, now, that x is a maximal element. Since α is the greatest element, we have $x \mathsf{R} \alpha$. By the maximal property of x this implies $x = \alpha$, so α is the only maximal element. □

We have seen that, in a partially ordered set, there may be elements a and b such that neither $a \mathsf{R} b$ nor $b \mathsf{R} a$†. For our most familiar order relation, $\leqslant$ on $\mathbb{R}$, this cannot occur. A partial order such as this, where every pair of elements is related (at least one way round), is called a 'total order'.

Definition 4.8

A **total order** (or **linear order**) on a set A is a partial order R which satisfies the following **dichotomy law**.

For every pair $a, b \in A$, either $a \mathsf{R} b$ or $b \mathsf{R} a$ (or both).

Note that there is a certain amount of redundancy in the definition of a total order in that the reflexive condition (which is included in the statement that R is a partial order) follows from the dichotomy law. This is because if we let $b = a$ then this last condition implies $a \mathsf{R} a$ for every $a \in A$. Thus a total order could be defined slightly more efficiently as a relation which is anti-symmetric, transitive and satisfies the dichotomy law.

Examples 4.10

1. The relation $\leqslant$ on $\mathbb{R}$ is a total order. Any subset of a totally ordered set is also totally ordered by the same relation (exercise: prove this). Thus the relation $\leqslant$ is a total order on any set of real numbers.

† Historically, orders where this cannot occur were studied before partial orders. The term *partial* order was therefore required to emphasize the possibility that two elements need not be related.

2. The alphabetical ordering of English words is a total order, as is the related order on $\mathbb{R}^2$ given in example 4.8.5.

3. Let $\mathscr{F}$ be any family of finite sets such that no two sets have the same cardinality and R the relation on $\mathscr{F}$ defined by $A \mathrel{\mathsf{R}} B$ if and only if $|A| \leqslant |B|$. (Exercise: verify that this is a partial order.)

   That this is a total order essentially follows from the fact that $|A|$ is an integer and $\leqslant$ is a total order on the set of integers.

4. Let R be the relation on $\mathbb{R}^2$ defined by $(x_1, y_1) \mathrel{\mathsf{R}} (x_2, y_2)$ if and only if $x_1 \leqslant x_2$ and $y_1 \leqslant y_2$. Again, we leave it as an exercise (4.5.3(i)) to show that R is a partial order. It is not, however, a total order because, for example, $(0, 1)$ and $(1, 0)$ are not related.

5. Let $A = \{1, 2, 3, 4, 6, 12\}$, the set of factors of 12, ordered by divisibility. A is not a totally ordered set because, for example, 2 and 3 are not related.

   However, A does have subsets which are totally ordered by the inherited relation of divisibility. For example, the subsets $\{1, 2, 4, 12\}$ and $\{1, 3, 6, 12\}$—and of course any subsets of *these*—are totally ordered.

Subsets such as those in the last example are sufficiently important to be given a name: a subset of a partially ordered set which is totally ordered by the inherited relation is called a **chain**. Note that a chain may be finite, as in the previous example, or infinite in length; for example, the set $\{1, 2, 4, 8, \ldots, 2^k, \ldots\}$ is a chain in $\mathbb{Z}^+$ ordered by divisibility. Of course, in a totally ordered set, every non-empty subset is a chain.

Exercises 4.5

1. Verify that the divisibility relation, $n \mathsf{R} m$ if and only if n divides m, $n|m$, is a partial order on the set of positive integers. What is the least element?

2. A class of students who have been studying relations has proposed (incorrectly) that each of the following relations R on set A is a partial order. For each relation, determine which property or properties (reflexive, anti-symmetric, transitive) the relation fails to satisfy.

(i) $A = \mathscr{P}(\{1, 2, 3\})$; $B \mathsf{R} C$ if and only if $B \subset C$.

(ii) $A = \mathscr{P}(\{1, 2, 3\})$; $B \mathsf{R} C$ if and only if $|B| \leqslant |C|$.

(iii) $A = \mathbb{Z}$; $n \mathsf{R} m$ if and only if $n^2 \leqslant m^2$.

(iv) $A = \mathbb{R} \times \mathbb{R}$; $(x_1, x_2) \mathsf{R} (y_1, y_2)$ if and only if $x_1 \leqslant y_1$.

(v) A is any non-empty set of people no two of whom are both the same age and the same height; $P \mathsf{R} Q$ if and only if $(age(P) \leqslant age(Q)) \vee (height(P) \leqslant height(Q))$.

(vi) $A = \mathbb{Z} \times \mathbb{Z}$; $(n, m) \mathsf{R} (p, q)$ if and only if $n \leqslant p \vee m \leqslant q$.

3. (i) Show that the relation R on the plane $\mathbb{R}^2$ defined by

$$(x_1, y_1) \mathsf{R} (x_2, y_2) \quad \text{if and only if } x_1 \leqslant x_2 \text{ and } y_1 \leqslant y_2$$

is a partial order.

(ii) More generally, show that if R is a partial order on a set A then the relation $\mathsf{R} \times \mathsf{R}$ defined by

$$(x_1, y_1)(\mathsf{R} \times \mathsf{R})(x_2, y_2) \quad \text{if and only if } x_1 \mathsf{R}\, x_2 \text{ and } y_1 \mathsf{R}\, y_2$$

is a partial order on the Cartesian product $A \times A$.

4. Prove theorem 4.5.

5. Show that, if $\mathscr{F}$ is any family of finite sets such that no two sets have the same cardinality, then the relation R defined on $\mathscr{F}$ by

$$A \mathsf{R} B \quad \text{if and only if } |A| \leqslant |B|$$

is a total order. (See example 4.10.3.)

Describe the maximal and minimal elements.

6. Let A be a set of people. Under what circumstances does the relation defined by

$$x \mathsf{R} y \quad \text{if and only if } x \text{ is younger than or the same age as } y$$

define a partial order on A? (Assume, say, that age is measured to the nearest day.)

In the situation where R is a partial order, show that it is in fact a total order, and describe the greatest and least elements.

7. Let $\mathscr{F}$ be a non-empty family of finite sets. A relation R is defined on $\mathscr{F}$ by:

$$A \mathsf{R} B \quad \text{if and only if } A = B \text{ or } |A| < |B|.$$

   (i) Show that R is a partial order on $\mathscr{F}$. Describe the maximal and minimal elements.

   (ii) Is R a total order in the case where $\mathscr{F} = \mathscr{P}(\{1, 2, 3\})$? Explain your answer.

8. Show that

$$n \mathsf{R} m \quad \text{if and only if } n = 2^k m \text{ for some } k \in \mathbb{Z}$$

defines an equivalence relation on $\mathbb{Z}^+$ but

$$n \mathsf{S} m \quad \text{if and only if } n = 2^k m \text{ for some } k \in \mathbb{N}$$

defines a partial order relation on $\mathbb{Z}^+$.

9. A **strict order relation** on a set A is a transitive relation which satisfies the following **trichotomy law**.

For every $a, b \in A$ exactly one of the following three conditions hold:

$$a \mathsf{R} b, \qquad b \mathsf{R} a, \qquad a = b.$$

   (i) Show that the strict inequality relation $<$ is a strict order on $\mathbb{R}$.

   (ii) More generally, show that if R is a total order on A then the relation R^*, defined by $x \mathsf{R}^* y$ if and only if $x \mathsf{R} y$ and $x \neq y$, is a strict order on A.

10. Let R be a strict order relation on A. Show that the relation $\mathsf{R}^\#$ defined by

$$x \mathsf{R}^\# y \quad \text{if and only if either } x \mathsf{R} y \text{ or } x = y$$

is a total order on A. (Compare with question 9(ii) above.)

11. Let A be a poset with order relation R, and let $a_1, a_2, \ldots, a_n$ be elements of A such that $a_1 \mathsf{R} a_2, a_2 \mathsf{R} a_3, \ldots, a_{n-1} \mathsf{R} a_n, a_n \mathsf{R} a_1$.

Show that $a_1 = a_2 = \cdots = a_n$.

12. Prove that the inverse relation R^{-1} of a partial order R is a partial order. (See exercise 4.1.10 for the definition of R^{-1}.) Prove also that an element

is maximal with respect to R if and only if it is minimal with respect to R^{-1}.

If R is a total order, is R^{-1} necessarily a total order as well?

If R is a strict order, is R^{-1} necessarily a strict order as well?

13. A total order R on a set A is said to be a **well ordering** if every non-empty subset of A has a least element with respect to R.

(i) Show that every total order on a *finite* set A is a well ordering.

(ii) Find an example of an infinite set with a well ordering.

(iii) Show that the usual (total) order relation $\leqslant$ is not a well ordering on either the set $\mathbb{Z}$ of integers or on the set $\mathbb{R}^+$ of positive real numbers.

4.6 Hasse Diagrams

Consider the set $\{1, 2, 3, 4, 5, 6\}$ ordered by divisibility. The directed graph of the relation was given in figure 4.4 above. Although the diagram is not very complicated, it is apparent that for sets much larger than this the directed graph would become too complex to be of much use. Since the partial order is reflexive and transitive, we can obtain much the same information in a modified version of the directed graph, called a 'Hasse diagram'.

Let A be a finite set partially ordered by the relation R. We say that b **covers** a if $a \mathrel{\mathsf{R}} b$ and there is no element c such that $a \mathrel{\mathsf{R}} c$ and $c \mathrel{\mathsf{R}} b$. More formally, b covers a if $a \mathrel{\mathsf{R}} b$ and, for all $x \in A$, $a \mathrel{\mathsf{R}} x$ and $x \mathrel{\mathsf{R}} b$ implies either $a = x$ or $x = b$.

The **Hasse diagram** of a finite poset can now be defined as follows. The elements of the set are represented as points in the plane and the points representing a and b are joined by a rising line if and only if b covers a.

The Hasse diagram for $\{1, 2, 3, 4, 5, 6\}$ ordered by divisibility is given in figure 4.6 which is clearer and less complicated than its directed graph (figure 4.4). We can still reconstruct the relation R from the Hasse diagram, given that we know that R is a partial order. Every element is certainly related to itself since R is reflexive. If $a \neq b$ then $a \mathrel{\mathsf{R}} b$ if and only if there is a sequence of rising lines connecting a to b. (We hope this is self-evident: a formal proof of this last

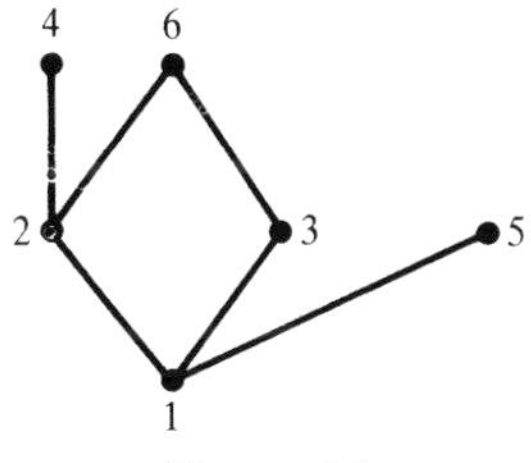

Figure 4.6

statement uses the transitivity of R and mathematical induction.) For example, from figure 4.6 we can see that 1 R 6 either because there are rising lines from 1 to 2 and from 2 to 6, or because there are rising lines from 1 to 3 and from 3 to 6.

Examples 4.11

1. The Hasse diagram for $\mathcal{P}\{a, b, c\}$ ordered by inclusion is given in figure 4.7. We leave it as an exercise to draw the directed graph of this relation as a comparison.

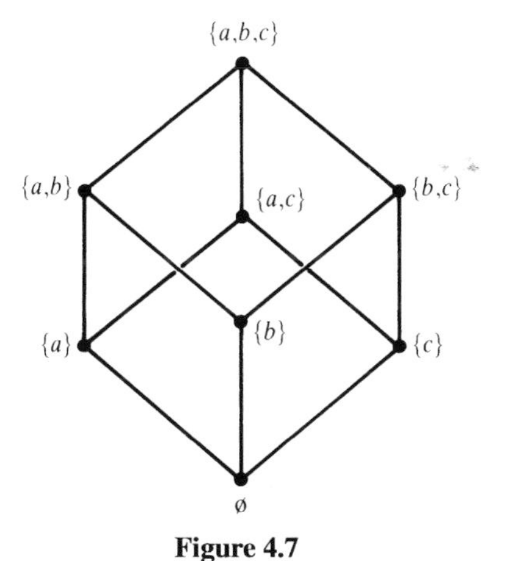

Figure 4.7

2. A partial order R on a set A has the Hasse diagram in figure 4.8. List the elements of R.

Solution

Since R is a partial order, every element of the set is related to itself.

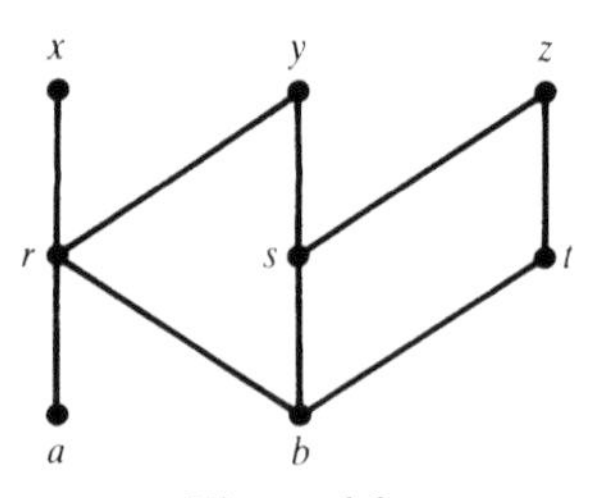

Figure 4.8

For each element p of the set, we can find all elements q (different from p) such that $p \mathsf{R} q$ by following the lines upwards from the point p. For example, beginning at the point b we find:

$$b \mathsf{R} b, b \mathsf{R} r, b \mathsf{R} x, b \mathsf{R} y, b \mathsf{R} s, b \mathsf{R} z \text{ and } b \mathsf{R} t.$$

Following this procedure for each element in turn enables us to list the elements of R:

$$\begin{aligned}\mathsf{R} = \{&(a,a),(a,r),(a,x),(a,y),(b,b),(b,r),(b,s),(b,t),(b,x),\\&(b,y),(b,z),(r,r),(r,x),(r,y),(s,s),(s,y),(s,z),(t,t),\\&(t,z),(x,x),(y,y),(z,z)\}.\end{aligned}$$

3. Draw a Hasse diagram of the partial order relation R on $A = \{a,b,c,p,q,x,y\}$ given by

$$\begin{aligned}\mathsf{R} = \{&(a,a),(b,b),(c,c),(p,p),(q,q),(x,x),(y,y),(a,p),(b,q),\\&(c,q),(x,a),(x,b),(x,p),(x,q),(y,b),(y,c),(y,q)\}.\end{aligned}$$

Solution

First note that p and q are maximal elements since the only ordered pairs in R of the form $(p, \_)$ or $(q, \_)$ are (p,p) and (q,q). Similarly x and y are minimal elements since the only ordered pairs in R of the form $(\_, x)$ or $(\_, y)$ are (x,x) and (y,y). The element a is neither maximal nor minimal since $x \mathsf{R} a$ and $a \mathsf{R} p$; similarly neither b nor c is maximal or minimal. Hence we may arrange the points representing the elements of A as shown in figure 4.9(a) with the maximal elements at the top and the minimal elements at the bottom of the diagram.

Now a covers x since $x \mathsf{R} a$ but there is no element $t \in A$ such that $x \mathsf{R} t$ and $t \mathsf{R} a$; hence we join the points corresponding to x and a. Similarly p covers a so we join the corresponding points. However, p does not cover x because $x \mathsf{R} a$ and $a \mathsf{R} p$. Continuing in this way, we obtain the Hasse diagram shown in figure 4.9(b).

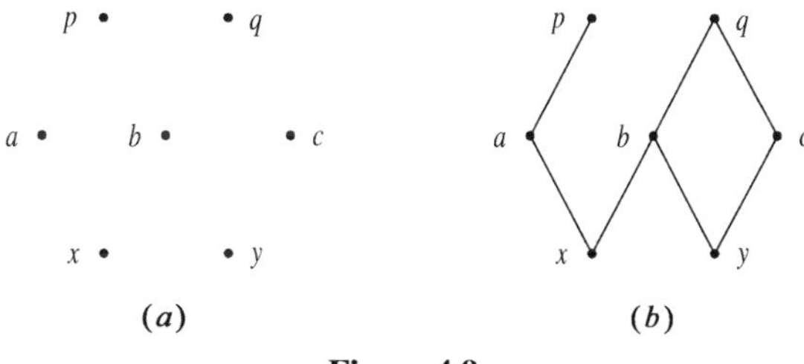

Figure 4.9

4. Show that neither of the configurations in figure 4.10 can occur anywhere in the Hasse diagram of a poset.

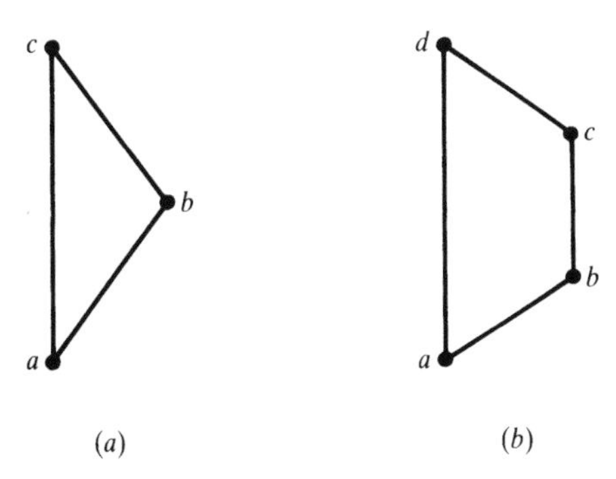

Figure 4.10

Solution

In (a) the line joining a to c should not occur because $a \mathsf{R} b$ and $b \mathsf{R} c$ which means that c does *not* cover a. Of course, a is related to c by the transitive property.

The configuration in figure 4.10(b) cannot occur for a similar reason. The right-hand part of the diagram implies that $a \mathsf{R} b$ and $b \mathsf{R} d$ (assuming transitivity), so d does not cover a. The line joining a to d should therefore be deleted.

5. Let $A = \{a_1, a_2, \ldots, a_n\}$ be a finite set with a *total* order R. Then every pair of elements is related, so given $x, y \in A$ either we can get from x to y or we can get from y to x by a sequence of rising lines in the Hasse diagram.

This means that, in the Hasse diagram, the elements are arranged in a single vertical line as in figure 4.11. This diagram explains why a total order is sometimes called a linear order.

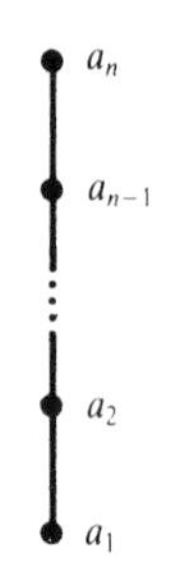

Figure 4.11

The Hasse diagram of a finite poset shows clearly the greatest and least elements (if these exist) as well as the maximal and minimal elements.

The least element has the property that every element can be reached from it by a sequence of rising lines. For example, the empty set in figure 4.7 has this property. The greatest element has the corresponding property that it can be reached from every element by a sequence of rising lines. In figure 4.7, the set $\{a, b, c\}$ has this property.

Minimal elements are those which have no lines rising *to* them. They usually occur at the bottom of the Hasse diagram with only rising lines coming from them, although a minimal element could be represented by an isolated point, connected to no lines at all if the element is related to nothing but itself. The dual property for maximal elements is that they have no rising lines *from* them. They usually appear at the top of the diagram, but again could be represented by isolated points. For example, in figure 4.8, the minimal elements are a and b, and the maximal ones are x, y and z.

A consideration of the possibilities for the Hasse diagram of a finite poset suggests that there must be at least one minimal and at least one maximal element. This we now prove.

Theorem 4.7

Let A be a finite non-empty poset. Then A contains at least one minimal element and if there is only one it is the least element.

Similarly A must contain at least one maximal element and if there is only one it is the greatest element.

Proof

As usual, we shall prove the first part only since the proof of the second is similar.

Choose any element a_1 of A. If a_1 is minimal we are finished. Otherwise there exists an element a_2 such that $a_2 \mathsf{R} a_1$. Either a_2 is minimal or there exists a_3 such that $a_3 \mathsf{R} a_2$. Since A is finite this sequence of elements $a_1, a_2, a_3, \ldots$ must terminate at some element a_k which must therefore be minimal.

Now suppose that A has a *unique* minimal element, β say. Let a_1 be any element of A different from β. Then a_1 is not minimal so, by the first part of the proof, there exists a sequence of elements $a_1, a_2, a_3, \ldots$, with each related to the previous one, which must terminate at the minimal element β. Therefore $\beta \mathsf{R} a_1$, for every $a_1 \in A$, so β is the least element. □

Finally, we note that it is easy to identify chains—totally ordered subsets—from the Hasse diagram of a poset. In the diagram a chain is seen as any part which resembles figure 4.11. That is, a chain is a portion of the diagram consisting of a single line with no branches. From figure 4.8 we can identify the chains; they are the following subsets of $\{a, b, r, s, t, x, y, z\}$:

$$\{a, r, x\}, \{a, r, y\}, \{b, r, x\}, \{b, r, y\}, \{b, s, y\}, \{b, s, z\}, \{b, t, z\}$$

and, of course, any subsets of these.

Exercises 4.6

1. Draw Hasse diagrams for each of the following sets under the divisibility relation: $n \mathsf{R} m$ if and only if n divides m:

   (i) $\{1, 2, 3, 4, 6, 12\}$ (ii) $\{1, 2, 4, 5, 10, 20\}$
   (iii) $\{1, 2, 4, 8, 16, 32\}$ (iv) $\{1, 2, 3, 5, 6, 10, 15, 30\}$.

   In each case identify the longest chain(s) (i.e. the chain(s) with the greatest number of elements).

2. The Hasse diagram of a partial order R on the set $\{a, b, c, d, e, f, g, h, i\}$ is given in the diagram below. List the elements of R and identify the

maximal and minimal elements.

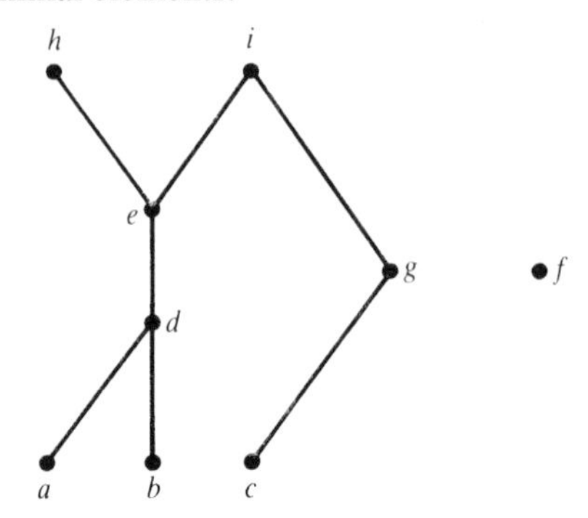

3. Let A be a poset with three elements. How many different kinds of Hasse diagrams of A are possible? Hence find the total number of different partial orders that can be defined on a set with three elements.

   Repeat for a four-element poset.

4. Draw the Hasse diagram for the set of non-empty proper subsets of $\{a, b, c, d\}$ ordered by inclusion. Identify the maximal and minimal elements and the chain(s) of greatest length.

5. Let $A = \{0, 1, 2\} \times \{2, 5, 8\}$
   $= \{(0,2), (0,5), (0,8), (1,2), (1,5), (1,8), (2,2), (2,5), (2,8)\}$.

   A partial order relation R on A is defined by

   $$(a, b) \mathrel{\mathsf{R}} (c, d) \qquad \text{if and only if } (a + b) \text{ divides } (c + d).$$

   (i) Draw a Hasse diagram for the poset A.
   (ii) What are the maximal and minimal elements of the poset A? Does A have a greatest and/or a least element?

6. Let S be the set of non-empty subsets of $\{a, b, c\}$. A partial order relation R on S is defined by

   $$A \mathrel{\mathsf{R}} B \qquad \text{if and only if either } (A = \{a\} \text{ and } a \notin B) \text{ or } (A \subseteq B).$$

   (i) Draw a Hasse diagram for the poset S.
   (ii) What are the maximal and minimal elements of the poset S? Does S have a greatest and/or a least element?

7. Let $A = \{n \in \mathbb{Z} : 2 \leqslant n \leqslant 12\}$. A partial order relation R on A is defined by

   $$a \mathrel{\mathsf{R}} b \quad \text{if and only if either } (a \text{ divides } b) \text{ or } (a \text{ is prime and } a < b).$$

(i) Draw a Hasse diagram for the poset A.
(ii) Identify the least element and the maximal elements.

8. A partial order relation R on $A = \{a, b, c, d, e, f, g\}$ has the directed graph given below. Draw its Hasse diagram.

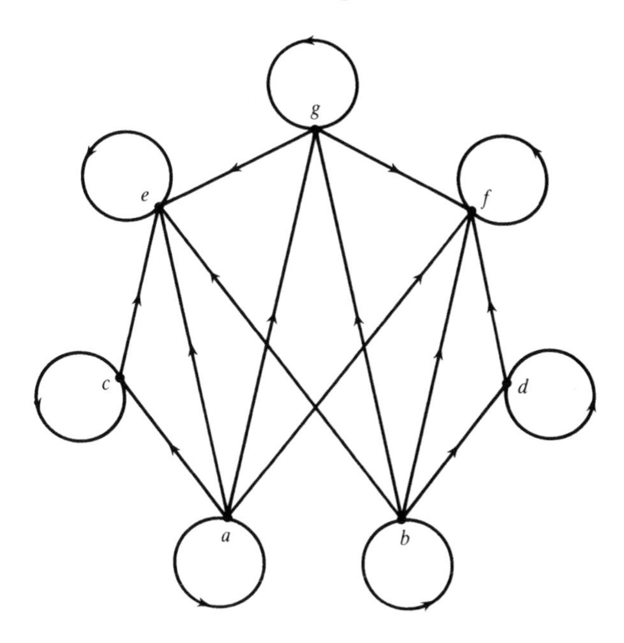

4.7 Application: Relational Databases

The advent of relatively cheap computers has made ours the 'information age'—large quantities of information are routinely stored, retrieved and manipulated electronically. A computer system designed to store and handle information is called a **database system**. The software which controls the manipulation of the stored data is called a **database management system** or **DBMS**.

There are many different ways of representing data; for example, as lists, tables, diagrams of various kinds, and so on. Any representation of data inevitably imposes some kind of structure on them. All database management systems are designed assuming that data have a particular type of structure and the way that a DBMS manipulates the stored data depends on a theoretical model of the data themselves. Thus database management systems can be classified into various types, the most common being relational, network (or 'CODASYL') and hierarchical. More recently, object-oriented and deductive (or logical) database systems have been developed to handle more sophisticated data structures and relationships between data.

In this section we shall consider briefly relational database systems which are based on the mathematics of relations. The relational model was first proposed by E F Codd in a paper in 1970. Although the idea was initially greeted with some scepticism, its potential was soon appreciated and most new database systems developed in recent years have been relational. Networks and hierarchical systems remain important largely because significant quantities of data are still stored in such systems.

Almost invariably a data item comprises, and is classified into, several parts. For instance, an entry in an address book might be classified in one of the following ways:

1. Name, address, telephone number.
2. Family name, first name, address, telephone number.
3. Family name, first name, street number and name, town/city, county/state, postcode/ZIP code, telephone area code, telephone number.

Each part of a data item is called an 'attribute'. We shall always assume that data items have a specified set of attributes and each attribute has a specified type. Thus each data item itself has a defined type, sometimes called the 'record type' of the item. Each of the three classifications above defines a different record type. In the first classification, suppose that attribute 'name' has type String, attribute 'address' has type $\mathcal{A}\mathit{ddress}$ (a defined type which may have both $\mathit{Integer}$ and String components) and attribute 'telephone number' has type $\mathit{Integer}$. Then any corresponding data item has type $\mathit{String} \times \mathcal{A}\mathit{ddress} \times \mathit{Integer}$ (see §4.1). A collection of data items all of the same type is called a 'table' or, sometimes, a 'relation'. A table defined in this way (a set of items all of which have the same type) is said to be in **first normal form** and we shall assume all our sets of data items have this property. Note that, since a table is a set of items, the order in which they appear is not significant. It is also worth noting that a table is conceptual and need not correspond to any actual file stored on a computer disk or other media. Any useful collection of data items of the same type may be defined as a table.

FIRST NORMAL FORM

Definition 4.9

Data are classified into components (or 'headings') called **attributes** (or **fields**). Each attribute has a specified type and this defines the type of a data item, sometimes called its **record type**. A **record instance** is an actual data item of a particular type and a **table** is a set of record instances of the same type.

A detailed discussion of suitable guidelines for choosing record types for particular kinds of data is beyond the scope of this book. However, it should be clear that a record type with many attributes is more flexible than one with fewer attributes. This is illustrated in the following example, which we shall develop further below.

Example 4.12

A charity, *Goodworks*, wishes to set up a database holding information regarding its donors, their names, addresses, and telephone numbers, the details of their contributions, and so on.

Goodworks initially decides to set up the database with a single table and to classify the data using five attributes with the following names and types.

DONOR_NAME : *String*

$$\text{DONOR\_ADDRESS} : \mathcal{Address}$$
$$\text{DONOR\_TELEPHONE} : \mathit{Integer}$$
$$\text{DONATION\_AMOUNT} : \mathcal{Currency}$$
$$\text{DONATION\_DATE} : \mathcal{Date}$$

Hence record instances have type $\mathcal{String} \times \mathcal{Address} \times \mathit{Integer} \times \mathcal{Currency} \times \mathcal{Date}$. Table 4.1 shows five (fictitious) record instances showing part of the table (in the sense of definition 4.9) that comprises *Goodworks*' database of donors.

Table 4.1

DONOR_NAME	DONOR_ADDRESS	DONOR_ TELEPHONE	DONATION_ AMOUNT	DONATION_ DATE
Smith, A	33 New Street Great Oldtown XP3 9NJ	9612-3993	£100	June 1996
Smith, A	33 New Street Great Oldtown XP3 9NJ	9612-3993	£250	Dec. 2000
Smith, A	33 New Street Great Oldtown XP3 9NJ	9612-3993	£150	July 2001
Thomas, N	2A Oaks Road Suburbia Bigcity BC3 5NR	2468-9753	£500	May 1994
Thomas, N	2A Oaks Road Suburbia Bigcity BC3 5NR	2468-9753	£350	Oct. 1998

Since a donor's complete address is labelled by a single attribute, it would be difficult to extract geographical information from this table. For instance, suppose *Goodworks* is launching a special campaign in 'Bigcity' and wishes to write to all past donors who live there. Obtaining a list of all such donors would be a difficult task from this table because the donor's town or city has not been defined as a separate attribute. The required list of past donors whose address is in Bigcity could be obtained more easily if the single attribute DONOR_ADDRESS were replaced, say, by the following set of attributes: STREET, CITY, POSTCODE. (Here we may assume that STREET contains all of the address coming before the town or city name.) It would then be relatively straightforward to pick out the record instances whose CITY attribute has the value 'Bigcity'.

This example illustrates an important guideline for defining attributes: each potentially useful individual piece of information in a record instance should be specified by an attribute. Of course this is not a precise rule for defining attributes because the meaning of 'potentially useful individual piece of information' will depend on the context of the data usage. In the example above, if geographical information is of no interest (and is never likely to be of any interest), it may be perfectly acceptable to specify the donors' addresses using a single attribute.

In the relational model of databases, all data is held in tables. The columns of a table are headed by attribute names; the types of these attributes define the record type of the table. Each row of a table represents a record instance of the given type.

In general, suppose a table R has n attributes $A_1, A_2, \dots, A_n$. Informally, we say that R has 'attribute set' or 'attribute type' $(A_1, A_2, \dots, A_n)$. Each attribute A_i has a particular type $\mathcal{T}_i$ and a corresponding set of data entries. In example 4.12, the attribute DONOR_NAME has type *String* and the corresponding set of data entries would be the set of all the names of past donors to *Goodworks*. Similarly the set of all donations made corresponds to the attribute DONATION_AMOUNT, which has type *Currency* and so on. Let X_i denote the set corresponding to the attribute A_i; thus A_i denotes the name of the attribute and X_i the set of values attained by the attribute. The sets X_i are time dependent and may change as new record instances are added to or existing record instances are deleted from the table.

With this notation, a given record instance is an n-tuple $(x_1, x_2, \dots, x_n)$ where each x_i belongs to the set X_i, corresponding to the attribute A_i, and has type $\mathcal{T}_i$. This ensures that all record instances have the same type $\mathcal{T}_1 \times \mathcal{T}_2 \times \cdots \times \mathcal{T}_n$. A table is a collection of record instances of the same type; that is, a set of n-tuples $(x_1, x_2, \dots, x_n)$. Recall that the set of *all* n-tuples $(x_1, x_2, \dots, x_n)$, where $x_i \in X_i$ for $i = 1, 2, \dots, n$, is the Cartesian product $X_1 \times X_2 \times \cdots \times X_n$. Therefore a table R is just a subset of this Cartesian product $\mathsf{R} \subseteq X_1 \times X_2 \times \cdots \times X_n$. This is the definition of an n-ary relation between the sets $X_1, X_2, \dots, X_n$. To summarize: in the relational model of database system:

each table is just an n-ary relation, $R \subseteq X_1 \times X_2 \times \cdots \times X_n$;
if each x_i has type $\mathcal{T}_i$ then each record instance has type $\mathcal{T}_1 \times \mathcal{T}_2 \times \cdots \times \mathcal{T}_n$;
each X_i has type $\mathit{Set}[\mathcal{T}_i]$ and the table R has type $\mathit{Set}[\mathcal{T}_1 \times \mathcal{T}_2 \times \cdots \times \mathcal{T}_n]$.

Example 4.13

For future reference, we label the attributes in example 4.12 as follows:

A_1 : DONOR_NAME

A_2 : DONOR_ADDRESS

A_3 : DONOR_TELEPHONE

A_4 : DONATION_AMOUNT

A_5 : DONATION_DATE.

For each attribute A_i we suppose that there is a set X_i of individual data items for the corresponding attribute. Then a record instance is a 5-tuple $(x_1, x_2, x_3, x_4, x_5)$ where $x_i \in X_i$ for $i = 1, \ldots, 5$.

With this record type there is likely to be a certain amount of duplication of information. For instance, a donor's name, address and telephone number are recorded for every donation he or she makes. Apart from being a wasteful use of the storage medium, this can cause problems in updating the record file. Suppose, for example, that Ms A Smith, who has made three donations, moves to a new address. To update the table, the new address would need to be changed in each of her three record instances.

For these reasons it may be more sensible to split the data into two separate tables, one with attributes

A_1, A_2, A_3; DONOR_NAME, DONOR_ADDRESS, DONOR_TELEPHONE

and the other with the attributes

A_1, A_4, A_5; DONOR_NAME, DONATION_AMOUNT, DONATION_DATE.

In this way much of the duplication of information in the original database is avoided and most updating tasks are achieved more simply. The database would now consist of two (related) tables, one a subset of $X_1 \times X_2 \times X_3$ and the other a subset of $X_1 \times X_4 \times X_5$. Of course the attribute DONOR_NAME serves to link the two tables.

Table 4.2 FILE_ONE

DONOR_NAME	DONOR_ADDRESS	DONOR_TELEPHONE
Smith, A	33 New Street Great Oldtown XP3 9NJ	9612-3993
Thomas, N	2A Oaks Road Suburbia Bigcity BC3 5NR	2468-9753

Tables 4.2 and 4.3 show how the information contained in table 4.1 would be split into these two new tables, which we call FILE_ONE and FILE_TWO. Individual record instances in FILE_ONE have type *String* × *Address* × *Integer*

Table 4.3 FILE_TWO

DONOR_NAME	DONATION_AMOUNT	DONATION_DATE
Smith, A	£100	June 1996
Smith, A	£250	Dec. 2000
Smith, A	£150	July 2001
Thomas, N	£500	May 1994
Thomas, N	£350	Oct. 1998

so FILE_ONE itself has type $\mathcal{Set}\,[\mathcal{String} \times \mathcal{Address} \times \mathcal{Integer}]$. Similarly record instances in FILE_TWO have type $\mathcal{String} \times \mathcal{Currency} \times \mathcal{Date}$ so FILE_TWO itself has type $\mathcal{Set}\,[\mathcal{String} \times \mathcal{Currency} \times \mathcal{Date}]$.

Example 4.13 indicates why it may be desirable to organize a database into several related tables. This poses the question of how we can specify the relationship between the different tables in the database and, in particular, how we can access related record instances from different tables. Organizing *Goodworks*' database into two related tables, as suggested above, should not prevent us from obtaining record instances with all five attributes, A_1, A_2, A_3, A_4, A_5, as listed in table 4.1. Before we turn to a consideration of how this can be achieved, we are now in a position to define formally a relational database.

Definition 4.10

Let $A_1, A_2, \ldots, A_n$ be a collection of attributes and suppose with each A_i there is associated a set X_i of data items. Each data item associated with the attribute A_i has type $\mathcal{T}_i$, so X_i has type $\mathcal{Set}\,[\mathcal{T}_i]$.

A **relational database** with attributes $A_1, A_2, \ldots, A_n$ is a collection (or set) of relations, each of which is a relation between some (possibly all) of the sets X_i $(i = 1, 2, \ldots, n)$. Each relation $\mathsf{R} \subseteq X_{i_1} \times X_{i_2} \times \cdots \times X_{i_m}$ is called a **table**; its elements are called **record instances** and each record instance has type $\mathcal{T}_{i_1} \times \mathcal{T}_{i_2} \times \cdots \times \mathcal{T}_{i_m}$.

Example 4.14

According to the formal definition, the (modified) *Goodworks* database defined in example 4.13 consists of two 3-ary relations, one a relation between X_1, X_2, X_3

and the other a relation between X_1, X_4, X_5. The database thus contains two tables.

Record instances in a table can be accessed by a 'key'. This is simply a set of attributes whose values uniquely specify a record instance, but no proper subset has this property of uniquely specifying record instances. In other words, specifying the values of the attributes in the key determines the values of the other attributes not belonging to the key, but specifying the values of only a subset of these attributes would not necessarily determine the values of the other attributes.

In practice there may be several possible choices of key. A set of attributes which could serve as a key is called a **candidate key**. One of these is selected to be used as the actual key—it is called the **primary key**. In other words, the candidate keys are the potential keys and the primary key is the one actually chosen to act as key.

In the *Goodworks* database, {DONOR_NAME} is a candidate key to FILE_ONE, with attributes (DONOR_NAME, DONOR_ADDRESS, DONOR_TELEPHONE), provided no two donors have the same name. If this is the case, each record instance can be identified solely by the donor's name. If, on the other hand, there were two different donors with (exactly) the same name, then {DONOR_NAME} would not be a key, but the attribute set {DONOR_NAME, DONOR_TELEPHONE} might well be a key.

We now turn to the kind of information which can be obtained from a relational database. Two basic types of operation which can be performed are the extraction of a list of all record instances which satisfy a certain set of criteria and the creation of new tables (relations) from the existing ones in the database. We consider five fundamental operations which can be combined together to provide most of the classes of information commonly required by database users.

Selection

The process of **selection** lists all record instances from a table which satisfy a given set of criteria. The following commands are examples of selection.

1. List all name and address records for customers whose address is in City X.
2. List all customers whose bank account is overdrawn and whose overdraft exceeds their agreed limit.
3. List the names and occupations of all students who graduated in 2000.

In order to be able execute these instructions the corresponding attributes need to have been defined. For example, as we have already explained, for command 1 above to be executable, the city of a customer's address needs to be a separate attribute. Similarly, account balance and overdraft limit both need to be attributes of the appropriate table in order to be able to execute command 2. Likewise, year of graduation must be one of the attributes if command 3 is to be performed.

Let R be a table with attributes $A_1, A_2, \ldots, A_m$ and let a_i be a specified value for the attribute A_i. We wish to select all those record instances in R whose attribute A_i has the value a_i. This can be described mathematically as follows.

The table is an m-ary relation between the sets $X_1, \ldots, X_m$; that is, $\mathsf{R} \subseteq (X_1 \times X_2 \times \cdots \times X_m)$. The selection process is nothing other than defining the subset of R consisting of all m-tuples $(x_1, \ldots, x_m)$ whose x_i entry is the specified value a_i:

$$\{(x_1, \ldots, x_m) \in \mathsf{R} : x_i = a_i\}.$$

Selecting from R all record instances with the property that several attributes have certain specified values also corresponds mathematically to defining a subset of R. For instance, to list all record instances whose attribute A_i has value a_i, A_j has value a_j and A_k has value a_k, we need to define the following subset of R:

$$\{(x_1, \ldots, x_m) \in \mathsf{R} : x_i = a_i, x_j = a_j \text{ and } x_k = a_k\}.$$

We can regard the selection process as defining new tables, namely subsets of given tables in the database. These new tables would probably have only temporary existence; they would not be added to the collection of tables which constitute the (theoretical model of the) database. It should be noted that a new table obtained by selection has the same record type as the original table.

Selection can be described simply as follows. The new table is obtained by picking out those rows (record instances) which have the corresponding attribute values. Since entire rows are selected to form the new table, it is clear that it must have the same record type as the original table.

For example, the command 'SELECT ALL DONATIONS MADE AFTER DECEMBER 1996', when performed on the part of the record file represented in table 4.3, picks out the second, third and fifth rows, as indicated in table 4.4. Actually, this operation is a slight generalization of that defined above. We have implicitly assumed that the set of dates corresponding to the DONATION_DATE attribute are ordered in the obvious way and we are selecting all those record instances whose DONATION_DATE value is *greater* than some specified value, namely December 1996.

Table 4.4

DONOR_NAME	DONATION_AMOUNT	DONATION_DATE
Smith, A	£100	June 1996
Smith, A	**£250**	**Dec. 2000**
Smith, A	**£150**	**July 2001**
Thomas, N	£500	May 1994
Thomas, N	**£350**	**Oct. 1998**

Projection

Whereas selection picks out certain rows from a table, the next operation we describe, 'projection', picks out certain columns. Since the columns correspond to attributes, it is clear that the resulting table has fewer attributes than the original.

We can formally describe **projection** as follows. Let R be a table with attributes $A_1, \ldots, A_p$ and let $B_1, \ldots, B_q$ be attributes with $q \leqslant p$ such that each attribute B_i is also an attribute of R; that is, each $B_i = A_j$ for some j. Projection defines a new table with attributes $B_1, \ldots, B_q$ whose record instances comprise the B_i attributes of each of the record instances of R.

Example 4.15

Consider again the *Goodworks* FILE_TWO with attributes $(A_1, A_2, A_3) = $ DONOR_NAME, DONATION_AMOUNT, DONATION_DATE), part of which is represented in table 4.3. Projection onto the attributes $(A_1, A_2) = $ (DONOR_NAME, DONATION_AMOUNT) produces the new table whose record entries consist only of the name and donation amount attribute values. Thus projection onto A_1, A_2 'forgets' the donation date. This is illustrated in table 4.5. Note that only the first two columns of table 4.3 have been selected, but that the new table has the same number of rows (record instances) as the original.

Mathematically, the projection operation, like selection, produces a new relation. Since this is most easily described in terms of certain naturally defined functions on Cartesian products, we shall leave the mathematical description to the next chapter which deals with functions: see §5.6.

Table 4.5

DONOR_NAME	DONATION_AMOUNT
Smith, A	£100
Smith, A	£250
Smith, A	£150
Thomas, N	£500
Thomas, N	£350

Natural Join

Suppose *Goodworks* has organized its database into the two tables, FILE_ONE and FILE_TWO, as described in example 4.13. How is it then possible to obtain a list of, say, all donor names, telephone numbers and donation amounts? The problem is that the donor telephone numbers and the donation amounts are held in different tables. We need a method of joining the tables together to produce a new table with all three required attributes: DONOR_NAME, DONOR_TELEPHONE and DONATION_AMOUNT. Since the two tables also contain respectively DONOR_ADDRESS and DONATION_DATE, the result of the 'joining' will produce a table whose record type also includes these attributes. This is not a problem, however, since we can then project the joined table onto the required record type. In fact joining the two tables will produce the original five-attribute table introduced in example 4.12.

The 'natural join' has the following mathematical basis. Suppose R and S are tables with attributes $A_1, \ldots, A_p, B_1, \ldots, B_q$ and $A_1, \ldots, A_p, C_1, \ldots, C_r$. Note that we allow the possibility that $p = 0$ which represents the case where R and S have no attributes in common. Their **natural join** is a new table with attributes $A_1, \ldots, A_p, B_1, \ldots, B_q, C_1, \ldots, C_r$. The record instances which comprise the natural join are all the $(p+q+r)$-tuples $(x_1, \ldots, x_p, y_1, \ldots, y_q, z_1, \ldots, z_r)$ with the property that $(x_1, \ldots, x_p, y_1, \ldots, y_q) \in \mathsf{R}$ and $(x_1, \ldots, x_p, z_1, \ldots, z_r) \in \mathsf{S}$. Note that for simplicity we have listed the common attributes in R and S at the beginning of the record type. In practice, this need not be the case, but to describe the more general situation is notationally more complex. In set notation the natural join of R and S is

$$\{(x_1, \ldots, x_p, y_1, \ldots, y_q, z_1, \ldots, z_p) : (x_1, \ldots, x_p, y_1, \ldots, y_q) \in \mathsf{R} \\ \text{and } (x_1, \ldots, x_p, z_1, \ldots, z_r) \in \mathsf{S}\}.$$

Example 4.16

To obtain a list of all *Goodworks* donors, their addresses and their donations, we first need to join the tables FILE_ONE with attribute type (A_1, A_2, A_3) and FILE_TWO with attribute type (A_1, A_4, A_5). This produces the table with attribute type $(A_1, A_2, A_3, A_4, A_5)$ shown in table 4.1. Then we project this joined table onto (A_1, A_2, A_4). The resulting table is shown below in table 4.6.

Table 4.6

DONOR_NAME	DONOR_ADDRESS	DONATION_AMOUNT
Smith, A	33 New Street Great Oldtown XP3 9NJ	£100
Smith, A	33 New Street Great Oldtown XP3 9NJ	£250
Smith, A	33 New Street Great Oldtown XP3 9NJ	£150
Thomas, N	2A Oaks Road Suburbia Bigcity BC3 5NR	£500
Thomas, N	2A Oaks Road Suburbia Bigcity BC3 5NR	£350

Union and Difference

Given two tables R and S *of the same record type* their **union** and **difference** are both simply the usual (typed) set theory union $R \cup S$ and difference $R - S$ respectively. Thus $R \cup S$ is the table which contains all record instances in R or in S (but does not list the repeats twice). It corresponds to pasting the table representing S under that representing R and then deleting the repeated rows, if any. The difference $R - S$ is the table which contains all the record instances in R which do not appear in S. The need for R and S to have the same set of attributes is evident in both cases.

Exercises 4.7

The exercises refer to the following relational database of a fictitious college which contains information concerning its students, their current courses, etc.

Attributes: $A_1 =$ ID_NUMBER
$A_2 =$ STUDENT_NAME
$A_3 =$ DATE_OF_BIRTH
$A_4 =$ DATE_OF_ENTRY
$A_5 =$ MAJOR_DISCIPLINE
$B_1 =$ CURRENT_COURSE_#1
$B_2 =$ CURRENT_COURSE_#2
$B_3 =$ CURRENT_COURSE_#3
$B_4 =$ CURRENT_COURSE_#4.

Tables:

PERSONAL	Attributes: (A_1, A_2, A_3, A_4)
DISCIPLINE	Attributes: (A_1, A_2, A_5)
CURRENT_COURSE	Attributes: $(A_1, B_1, B_2, B_3, B_4)$.

Parts of three tables, PERSONAL, DISCIPLINE and CURRENT_COURSE, are given below.

PERSONAL

M1452	Adams, K	23/06/71	1990
F3286	Johnson, D	15/12/69	1989
F5419	Kirby, F	29/07/63	1990
M3415	Singer, R	03/10/71	1989
F0278	Williams, L	19/03/70	1989

DISCIPLINE

M1452	Adams, K	CompSci
F3286	Johnson, D	Psyc
F5419	Kirby, F	Math/Econ
M3415	Singer, R	Hist
F0278	Williams, L	CompSci/Math

CURRENT_COURSE

M1452	Comp100	Math150	Bus_105	Econ110
F3286	Psyc250	Psyc280	Psyc281	Soc_200
F5419	Math100	Math150	Econ110	Econ120
M3415	Hist210	Hist220	Lit_200	Stat120
F0278	Comp210	Comp230	Math205	Math215

1. List the tables which result from performing each of the following operations.

   (i) Select from PERSONAL those record instances who entered the college in 1989.

   (ii) Project PERSONAL onto (A_2, A_3).

   (iii) Perform the natural join of PERSONAL and DISCIPLINE.

   (iv) Perform the natural join of PERSONAL and CURRENT_COURSE and then project the result onto $(A_2, B_1, B_2, B_3, B_4)$.

   (v) Using * as a 'wildcard' which can represent any number, select from PERSONAL those record instances whose A_1 attribute is F**** and project the result onto (A_1, A_2).

2. Select from CURENT_COURSE those students with B_3 attribute value equal to 'Econ110'.

   Why does this selection not list all students who are currently taking course Econ110? Explain how to obtain a list from CURRENT_COURSE of those students currently taking Econ110.

3. Perform the natural join of PERSONAL and DISCIPLINE and then perform the natural join of the result with CURRENT_COURSE.

   Perform the natural join of PERSONAL with the result of performing the natural join of DISCIPLINE and CURRENT_COURSE.

   Is the natural join operation associative in general? Justify your answer.

4. Explain how the following lists of information can be obtained using the operations described in the text. (Where necessary use * as a 'wildcard' which can stand for any character or number—see question 1(v) above for an example.)

   (i) A list of student names and current courses.

   (ii) A list of student ID numbers and names for those students who entered the college in 1990.

   (iii) A list of student names and current courses for those students whose major disciplines include CompSci.

   (iv) A list of names, major disciplines and current courses of all students.

(v) A list of ID numbers, dates of entry and current courses of those students born in 1971.

Chapter 5

Functions

5.1 Definitions and Examples

In this chapter we consider another of the central concepts of modern mathematics, that of a function or mapping. Although functions have been used in mathematics for several centuries, it is only comparatively recently that a rigorous and generally accepted definition of the concept has emerged. When historians come to write the history of mathematics in the second half of the twentieth century, the rise in importance of functions of various kinds will almost certainly be one of their major themes.

Like many of the concepts which we deal with in this book, that of a function is both simple and very general. Instead of giving the definition immediately, we shall begin with a notion with which you may very well be familiar from your previous studies—that of a (real) variable. Traditionally labelled x, a variable is often associated with expressions such as

$$x^2 + 4x - 7, \quad 1/(x+1)^3, \quad \sin x, \quad \log x, \quad \text{etc.}$$

Expressions like these are frequently denoted $f(x)$ and called 'a function of (the variable) x'. In such cases there is generally the assumption (which is often only implicit) that the variable x refers to an 'arbitrary' real number, although it may be subject to some restrictions, such as it must be positive. For us this idea of a function is both too restrictive and somewhat incomplete, although it does point towards a simpler and more general definition. The essence of the examples above is that we can calculate (in principle, at least) the value of the expression for any (allowed) value of the variable x. More important than an expression itself is the fact that it provides a 'rule' for calculating its value given any value of x. Two different expressions $f(x)$ and $g(x)$ may give the same values for all real numbers x, and we would regard the two expressions as defining the same function. A simple example of this is provided by the expressions $f(x) = x^2 + 4x - 7$ and $g(x) = (x + 2)^2 - 11$.

In this book we will need to use functions where the 'variable' is not a real number, nor even a number, but an element of some given set A. Thus it may also be somewhat misleading to refer to a rule for 'calculating' the value of an expression. With these points in mind, the following is a reasonable working definition.

Working Definition

Let A and B be two sets. A **function f from A to B**, written $f : A \to B$, is a rule which associates to each $a \in A$ a unique element $f(a) \in B$.

This is a very general definition, which includes the examples above as well as many non-numerical examples. It is quite common to visualize the function rule as being encapsulated in a 'function machine'. This is a 'black box', illustrated below, which has the property that if an element $a \in A$ is fed into the machine, it produces as output the associated element $f(a) \in B$.

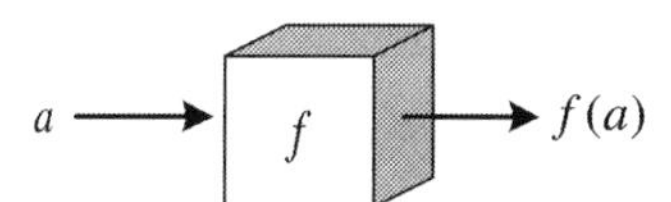

Examples 5.1

1. Let $A = \{a, b, c, d, e\}$, $B = \{\alpha, \beta, \gamma, \delta\}$ and define a function $f : A \to B$ by $f(a) = \beta$, $f(b) = \alpha$, $f(c) = f(d) = f(e) = \delta$.

   An arrow diagram such as figure 5.1 is a useful way of visualizing a function like this, where the sets A and B are finite. The sets are represented as regions of the plane and an arrow is drawn from each element of A to its associated element of B. (Compare this with the arrow diagram of a relation—figure 4.3.)

2. The expression $x^2 + 4x - 7$ referred to above is *not* a function on its own according to the working definition because the sets A and B have not

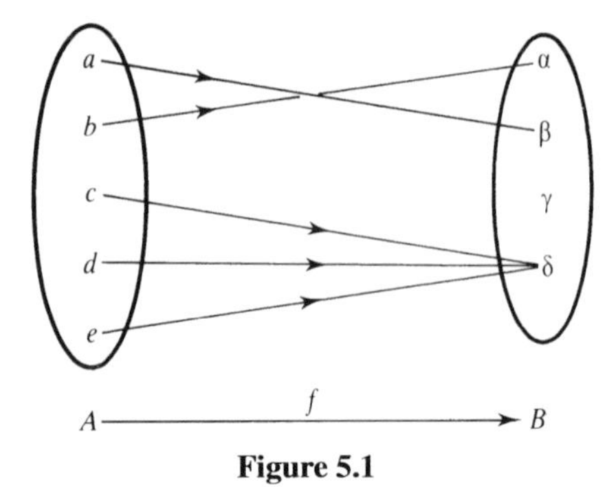

Figure 5.1

been specified. However, the expression can be used to define a function $f : \mathbb{R} \to \mathbb{R}$ which associates to each real number x the real number $f(x) = x^2 + 4x - 7$.

For example, 2 is associated with $f(2) = 2^2 + 4 \times 2 - 7 = 5$, -4 is associated with $f(-4) = (-4)^2 + 4 \times (-4) - 7 = -7$, and so on.

Note: we emphasize that a function is more than just the rule of association—we should always specify the two sets involved as well.

3. The second expression given above, $1/(x+1)^3$, cannot be used to define a function from $\mathbb{R}$ to $\mathbb{R}$. This is because $1/(x+1)^3$ is not defined for every real number x. When x has the value -1, the expression is undefined. (Division by zero is not allowed: 1/0 is meaningless.)

   However, since -1 is the only 'troublesome' element of $\mathbb{R}$ in this respect, there is a function

$$f : \mathbb{R} - \{-1\} \to \mathbb{R} \quad \text{defined by} \quad f(x) = 1/(x+1)^3.$$

   (Recall that $\mathbb{R} - \{-1\} = \{x \in \mathbb{R} : x \neq -1\}$.) The function f associates 1 with $f(1) = 1/2^3 = 1/8$, -4 with $f(-4) = 1/(-3)^3 = -1/27$, etc.

   An alternative approach is to view the association $f(x) = 1/(x+1)^3$ as defining a 'partial function' $\mathbb{R} \to \mathbb{R}$. In a partial function, $f(a)$ need not be defined for *every* $a \in A$. See exercise 5.1.12 for details.

   This example underlines the importance of the sets A and B in our working definition. We have chosen the largest possible subset of $\mathbb{R}$ for the set A, but we could have been more restrictive and defined a different function using the same expression, say $g : \mathbb{R}^+ \to \mathbb{R}$, $g(x) = 1/(x+1)^3$.

4. Another non-numerical example is the following. Let $A = \{$living human beings$\}$ and $B = \{$human beings, living or dead$\}$. A function f from A

to B could be defined by associating to each person his or her mother. Symbolically,

$$f : A \to B, \quad f(p) = \text{the mother of } p.$$

One of the drawbacks with our working definition is that it begs an important question: what do we mean by a rule? Intuitively, a rule in the sense of the working definition is some method of specification whereby, given any $a \in A$, the element $f(a) \in B$ can be determined, at least in principle. If A is a finite set this could be achieved by tabulating its elements alongside their associated elements of B.

For instance, the function defined in example 5.1.1 could be tabulated as follows.

A	B
a	β
b	α
c	δ
d	δ
e	δ

A more concise way of doing the same thing would be to list the pairs $(a, f(a))$ for each element a of A. The list (a, β), (b, α), (c, δ), (d, δ), (e, δ) completely specifies the rule of association which defines the function from A to B. Since each member of this list is an ordered pair, we are simply defining a subset of the Cartesian product $A \times B$.

In this example we have used the given rule of association to define the subset of $A \times B$ consisting of all pairs $(a, f(a))$ where $a \in A$. Changing our perspective slightly, we can regard a subset of the Cartesian product as defining a rule of association itself. In other words specifying a rule is nothing more or less than specifying a subset of $A \times B$. This leads us then to our formal definition of a function.

Definition 5.1

Let A and B be sets. A **function f from A to B**, written $f : A \to B$, is a subset $f \subseteq (A \times B)$ which satisfies:

($*$) for each $a \in A$ there exists a unique $b \in B$ such that $(a, b) \in f$.

The set A is called the **domain**, and the set B the **codomain**, of f. If $(a, b) \in f$ the element $b \in B$ is called the **image** of $a \in A$ and is written $b = f(a)$, or $f : a \mapsto b$†.

A function is also called a **mapping** or a **transformation**.

The condition ($*$) on the subset f of $A \times B$ corresponds to the condition in the working definition that *each* $a \in A$ is associated with a *unique* element $f(a) \in B$.

Recall from chapter 4 that a relation from a set A to a set B is a subset of $A \times B$. According to definition 5.1, therefore, a function $f : A \to B$ is just a special kind of relation from A to B—one which satisfies the property ($*$).

Definition 5.2

Two functions $f : A \to B$ and $g : A' \to B'$ are **equal** if:

(i) $A = A'$,
(ii) $B = B'$, and
(iii) $f(a) = g(a)$ for all elements a belonging to $A = A'$.

Although this is the most common (and probably the most useful) definition of equality of functions, it is slightly at odds with what we might expect. Since the functions $f : A \to B$ and $g : A' \to B'$ are both sets (subsets of $A \times B$ and $A' \times B'$, respectively), we ought perhaps to define them to be equal if they contain the same elements (see §3.1). Now f contains an ordered pair (a, b) for every $a \in A$ (and similarly for g). Therefore, if the sets f and g contain the same elements, it follows that $A = A'$ and $f(a) = g(a)$ for every $a \in A = A'$. However,

† Note that the barred arrow, $\mapsto$, is used exclusively for denoting the image of an element to avoid confusion with the function itself, $f : A \to B$.

the second condition of definition 5.2 is not implied by the equality of the sets f and g. Suppose, for example, that *as sets* $f = g = \{(a,1),(b,2),(c,3)\}$. Then $A = A' = \{a,b,c\}$, $f(a) = g(a) = 1$, $f(b) = g(b) = 2$ and $f(c) = g(c) = 3$. However, all that can be said about the codomains, B and B', is that they must both contain the elements 1, 2 and 3; they need not be equal.

We shall see later that it is highly desirable to impose the additional condition for equality of two functions that their codomains be equal.

Examples 5.2

1. The function $f : A \to B$ defined informally in example 5.1.1 can now be defined formally as the set $f \subseteq A \times B$, where

$$f = \{(a,\beta),(b,\alpha),(c,\delta),(d,\delta),(e,\delta)\}.$$

2. Similarly the function $f : \mathbb{R} \to \mathbb{R}$ of example 5.1.2 is defined formally as the set

$$f = \{(x,y) \in \mathbb{R} \times \mathbb{R} : y = x^2 + 4x - 7\}.$$

3. Again in the same way, the function $f : A \to B$ given in example 5.1.4 is defined as the set

$$f = \{(a,b) \in A \times B : b \text{ is the mother of } a\}$$

where $A = \{\text{living humans}\}$ and $B = \{\text{humans, living or dead}\}$.

4. Other familiar functions can be defined in this manner. For instance, the 'square' and 'cube' functions f and g from $\mathbb{R}$ to $\mathbb{R}$ are respectively defined as the sets

$$f = \{(x,y) \in \mathbb{R} \times \mathbb{R} : y = x^2\} \quad \text{and} \quad g = \{(x,y) \in \mathbb{R} \times \mathbb{R} : y = x^3\}.$$

5. Let A be any set. The **identity function** $id_A : A \to A$ is defined by

$$id_A = \{(x,x) : x \in A\}.$$

The identity function is simply the identity relation on A—see exercise 4.1.8(i). Less formally we could write the identity function as $id_A : A \to A$, $id_A(x) = x$, for all $x \in A$.

6. We might attempt to define the 'square root' function $\mathbb{R} \to \mathbb{R}$ as the set

$$f = \{(x,y) \in \mathbb{R} \times \mathbb{R} : x = y^2\}.$$

The reason for defining this set is, of course, that if $y = \sqrt{x}$ then $y^2 = x$. However, it should be emphasized that this subset of $\mathbb{R} \times \mathbb{R}$ is *not* a function; it fails condition $(*)$ of definition 5.1 on two counts. (It is, of course, a relation from $\mathbb{R}$ to $\mathbb{R}$.)

Firstly, it is not true that for *each* $x \in \mathbb{R}$ there exists an element $y \in \mathbb{R}$ such that $(x, y) \in f$. If $x = -1$ for instance, there is no y with the required property (namely that $y^2 = -1$).

Secondly, even when there does exist the required element y it is (usually) *not unique*. Consider $x = 4$ for example. In this case there are two corresponding elements (x, y) of f such that $y^2 = 4$; namely, $(4, 2)$ and $(4, -2)$.

7. In some instances it may be difficult or even impossible in practice to compute $f(a)$ for some elements a of the domain of a function f. For example, define a function $f : \mathbb{Z}^+ \to \{0, 1, 2, 3, 4, 5, 6, 7, 8, 9\}$ by $f(n) =$ the digit in the nth decimal place in the expansion of π.

   More formally,

$$f = \{(n, m) : m = \text{the digit in the } n\text{th decimal place in the expansion of } \pi\}.$$

   Although the value of π has been calculated to many million decimal places†, for very large values of n it may still be impractical to calculate $f(n)$. For example, what is $f(10^{10})$ or $f(10^{20})$?

The informal description of a function as a rule which associates $f(x)$ to x is too appealing to drop altogether and we shall continue to use it. Thus we shall frequently use expressions like 'the function $f : A \to B$ defined by $b = f(a)$, or $a \mapsto f(a)$'. You should be able to reinterpret this in terms of the formal definition, if necessary.

Diagrams such as figure 5.1 above will continue to be useful visual aids, even when the sets involved are arbitrary. Although it seems very obvious and natural to us now, the 'arrow notation' for a function is comparatively recent. It only became widely used after the development of category theory beginning in the late 1940s.

† You may wonder why anyone should ever be interested in computing several million decimal places of *any* number! However, there are some interesting questions concerning the randomness or otherwise of the distribution of digits in the decimal expansion of π; the actual expansion has been calculated in order to provide evidence for or against various possible answers to these questions. Also, calculations such as these are used to test the performance of high powered 'super computers'.

You are almost certainly familiar with the notion of the graph of a function $f : \mathbb{R} \to \mathbb{R}$. This is the curve drawn in the plane $\mathbb{R}^2 = \mathbb{R} \times \mathbb{R}$ consisting of all the points (x, y) such that $y = f(x)$. However, according to definition 5.1, this set of points *is* the function f itself. In other words the graph of $f : \mathbb{R} \to \mathbb{R}$ is just a pictorial representation of the set f. From our point of view, there is little distinction to be made between the function itself and its graph.

It should be noted, however, that not every curve in the (x, y)-plane is the graph of some function $f : A \to \mathbb{R}$, where $A \subseteq \mathbb{R}$. A circle is a simple example of a curve which is not the graph of a function. Consider, for instance, the circle centred at the origin $(0, 0)$ with radius 1; its equation is $x^2 + y^2 = 1$ (figure 5.2). For each value of x (strictly) between -1 and 1, there correspond two values of y. For example, if $x = \frac{1}{2}$ the corresponding values of y are $\sqrt{3}/2$ and $-\sqrt{3}/2$. Therefore the condition $(*)$ of definition 5.1 is violated.

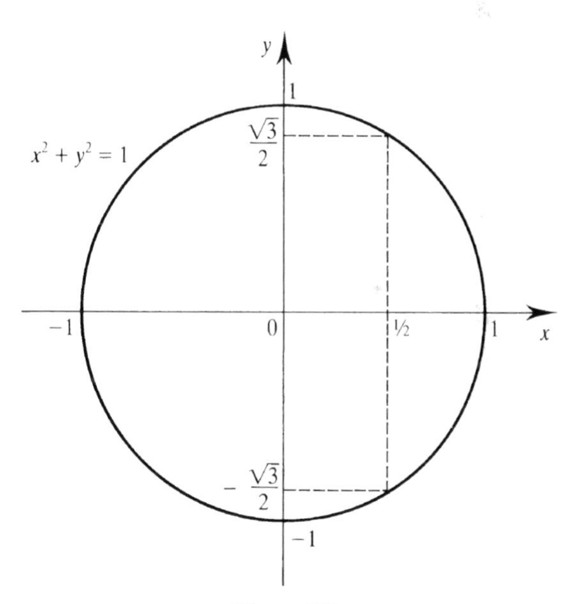

Figure 5.2

Given a curve in the (x, y)-plane, it is easy to see whether it is the graph of some function $f : A \to \mathbb{R}$. Given $a \in A$, there exists a unique $y \in \mathbb{R}$ such that $y = f(a)$ if and only if the vertical line through $x = a$ meets the curve exactly once. The 'problem' with the example of the circle above is that the vertical line through $x = \frac{1}{2}$, for instance, meets the curve twice. This leads us to the following 'test for functionhood'.

Vertical line test

A curve in the (x, y)-plane is the graph of some function $f : A \to \mathbb{R}$ where $A \subseteq \mathbb{R}$ if and only if the following condition is satisfied.

(#) Every vertical line in the plane meets the curve at most once.

If the condition (#) is satisfied, then the domain A of the function f is the set of points $a \in \mathbb{R}$ such that the vertical line through a meets the curve.

There are two features of the definition of a function which sometimes cause confusion, both of which are illustrated by the function $f : A \to B$ defined in example 5.1.1. The first is that two (or more) elements of the domain may have the same image in the codomain. In the case of our function we have $f(c) = f(d) = f(e) = \delta$. Secondly, not every element of the codomain need necessarily be the image of some element of the domain. Again, for our function f, there is no $x \in A$ such that $f(x) = \gamma$, so γ is not the image of any element of the domain.

Whether or not either of these actually occurs for a given function is easily observed in the 'arrow diagram' of a function—see figure 5.3.

We shall consider these points in more detail in §5.3. For now, the second point leads us to make the next definition.

Definition 5.3

Let $f : A \to B$ be a function. The **image set of *f*** (or **range† of *f***) is the set

$$im(f) = \{b \in B : (a, b) \in f \text{ for some } a \in A\}.$$

Note that $im(f)$ is a subset of B, the codomain of f; it should not be confused with $f(a)$, the image of an element $a \in A$. The image of an element (of A) is an element (of B), but the image set of the function is a set, namely the set of all the images of elements of the domain:

$$im(f) = \{f(a) : a \in A\}.$$

† Unfortunately the term 'range' is used differently by different authors. Some use it as we have and others use it to mean codomain. For this reason we shall avoid using the word.

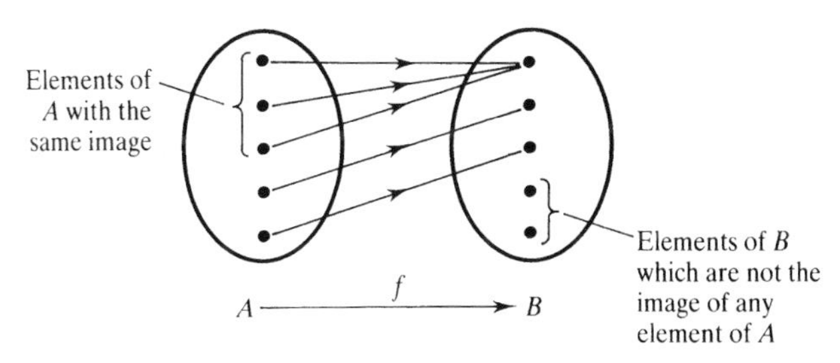

Figure 5.3

For the function defined in example 5.1.1, $im(f) = \{\alpha, \beta, \delta\}$. Thus $im(f)$ may be a proper subset of the codomain of f. Figure 5.4 should help you understand the definition of $im(f)$.

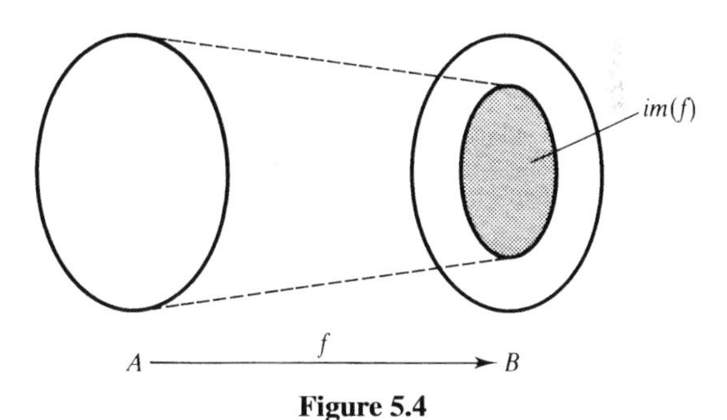

Figure 5.4

Examples 5.3

1. The image set of $f : \mathbb{Z}^+ \to \mathbb{Z}^+, n \mapsto 2^n$ is

$$\{2^n : n \in \mathbb{Z}^+\} = \{2, 4, 8, 16, 32, \ldots\}.$$

This function can be represented visually by a modified version of the 'arrow diagram'—see figure 5.5.

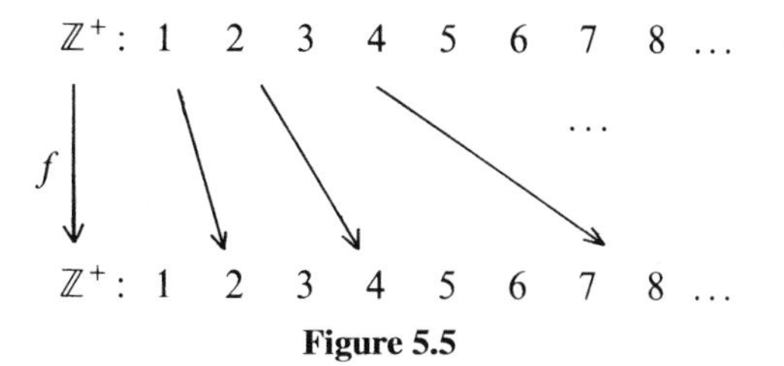

Figure 5.5

Similarly, the image set of $f : \mathbb{Z}^+ \to \mathbb{R}$, $n \mapsto 2^{-n}$ is

$$\{2^{-n} : n \in \mathbb{Z}^+\} = \{1/2, 1/4, 1/8, 1/16, 1/32, \ldots\}.$$

2. Let f and g be the square and cube functions $\mathbb{R} \to \mathbb{R}$ defined by $f(x) = x^2$ and $g(x) = x^3$ respectively. (The formal definitions are given in example 5.2.4.) Then

$$im(f) = \{y \in \mathbb{R} : y = x^2 \text{ for some } x \in \mathbb{R}\} = \{x^2 : x \in \mathbb{R}\}.$$

We show that $im(f) = \mathbb{R}^+ \cup \{0\} = \{y \in \mathbb{R} : y \geqslant 0\}$, by proving that $im(f) \subseteq \mathbb{R}^+ \cup \{0\}$ and $\mathbb{R}^+ \cup \{0\} \subseteq im(f)$. (Recall that this is frequently how we prove two sets are equal—see theorem 3.1.)

Let $y \in im(f)$. Then, by definition, $y = f(x) = x^2$ for some $x \in \mathbb{R}$, so $y \geqslant 0$. Hence $im(f) \subseteq \mathbb{R}^+ \cup \{0\}$.

Now let $y \in \mathbb{R}^+ \cup \{0\}$. To show $y \in im(f)$, we need to find a real number x such that $f(x) = y$. Since $y \geqslant 0$, its square root is a real number. So let $x = \sqrt{y} \in \mathbb{R}$. Then

$$f(x) = f(\sqrt{y}) = (\sqrt{y})^2 = y$$

so $y \in im(f)$. Hence $\mathbb{R}^+ \cup \{0\} \subseteq im(f)$.

Since $im(f) \subseteq \mathbb{R}^+ \cup \{0\}$ and $\mathbb{R}^+ \cup \{0\} \subseteq im(f)$, we conclude that $im(f) = \mathbb{R}^+ \cup \{0\}$.

For the cube function,

$$im(g) = \{y \in \mathbb{R} : y = x^3 \text{ for some } x \in \mathbb{R}\} = \{x^3 : x \in \mathbb{R}\}.$$

In this case, however, $im(g) = \mathbb{R}$. Since $im(g)$ is clearly a subset of $\mathbb{R}$, we need to show that $\mathbb{R} \subseteq im(g)$. Note that $\sqrt[3]{y} \in \mathbb{R}$ for every real number y. Therefore, given $y \in \mathbb{R}$, let $x = \sqrt[3]{y}$; then

$$g(x) = g(\sqrt[3]{y}) = (\sqrt[3]{y})^3 = y$$

so $y \in im(g)$. Hence $im(g) \subseteq \mathbb{R}$, so we conclude $im(g) = \mathbb{R}$.

3. Find the image set of the function $f : \mathbb{R} \to \mathbb{R}$ defined by

$$f(x) = \frac{3x}{x^2 + 1}.$$

Solution

By definition $y \in im(f)$ if and only if

$$y = \frac{3x}{x^2 + 1} \quad \text{for some } x \in \mathbb{R}.$$

Now this equation is equivalent to

$$yx^2 + y = 3x$$

or

$$yx^2 - 3x + y = 0.$$

Regarding this as a quadratic equation in x and using the quadratic formula, we have, provided $y \neq 0$,

$$x = \frac{3 \pm \sqrt{9 - 4y^2}}{2y}.$$

In order that this has a real solution we require $y \neq 0$ and

$$9 - 4y^2 \geqslant 0.$$

Hence

$$y^2 \leqslant 9/4 \quad (\text{and } y \neq 0)$$

which means

$$-3/2 \leqslant y \leqslant 3/2 \quad (\text{and } y \neq 0).$$

Therefore, provided $-3/2 \leqslant y \leqslant 3/2$, $y \neq 0$, there exists a real number x such that $y = f(x)$. The value $y = 0$ is a special case, but clearly $f(0) = 0$, so $0 \in im(f)$.

Hence

$$im(f) = [-3/2, 3/2] = \{y \in \mathbb{R} : -3/2 \leqslant y \leqslant 3/2\}.$$

Finding the image set of a function $f : A \to \mathbb{R}$, where A is a subset of $\mathbb{R}$, involves determining the real numbers y such that the equation $y = f(x)$ has a solution for some $x \in A$. This was the method adopted in the last example; we were able to find $im(f)$ because the equation $y = f(x)$ had a fairly simple form. In general, however, it may be rather more difficult to find $im(f)$ in such cases.

If we are given (or can determine) the graph of $f : A \to \mathbb{R}$, then the image set of the function can be found in a simple geometric way. For each element a of A, its image $f(a)$ can be determined from the graph by drawing a vertical line through

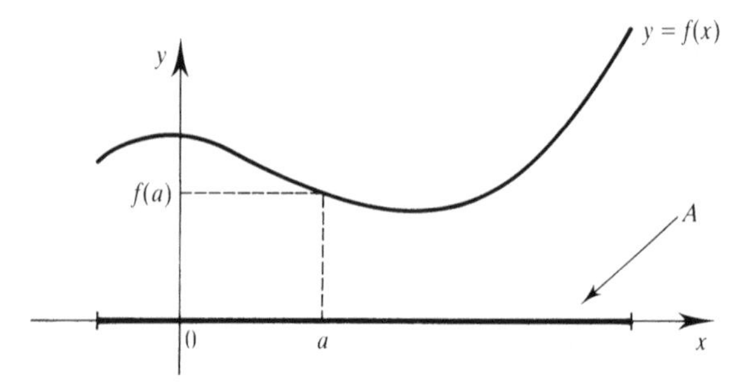

Figure 5.6

a until it meets the graph and then drawing a horizontal line from the point on the graph to the y-axis—see figure 5.6.

We now see that the image set of f is the set of points on the y-axis which arise from the graph in this way. In other words, the image set of f is the set of points on the y-axis such that the horizontal line through the point meets the graph of f in *at least* one point. Of course if the horizontal line through y meets the graph more than once, there is more than one element of A which has image equal to y.

Example 5.4

Let

$$f : \mathbb{R} \to \mathbb{R}, \quad x \mapsto \frac{3x}{x^2 + 1}$$

be the function considered in example 5.3.3. We have given its graph in figure 5.7, from which it is easy to see that $im(f) = [-3/2, 3/2]$.

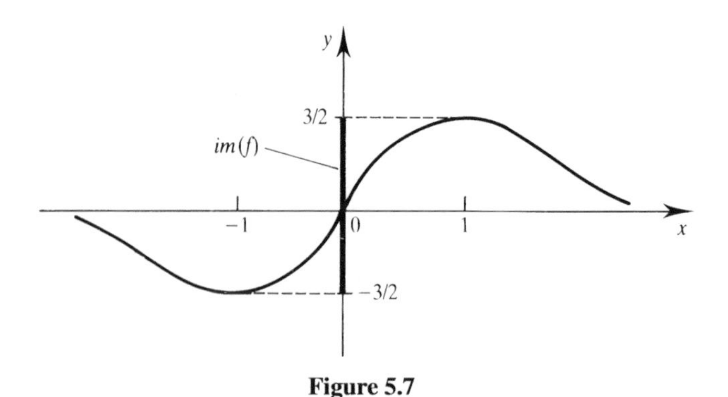

Figure 5.7

Functions and Types

We now consider briefly how functions fit in the theory of typed sets introduced in §3.8. In fact, since a function is a special kind of relation, we have essentially defined the type of a function in chapter 4. Suppose A and B are typed sets, where $A : \mathit{Set}[\mathcal{S}]$ and $B : \mathit{Set}[\mathcal{T}]$ and suppose $f : A \to B$ is a function. According to the formal definition 5.1, $f \subseteq A \times B$ so f has type $\mathit{Set}[\mathcal{S} \times \mathcal{T}]$. Thus, for example, a function $f : \mathbb{Z} \to \mathbb{Z}$ has type $\mathit{Set}[\mathit{Integer} \times \mathit{Integer}]$, a function $g : \mathbb{R} \to \mathbb{R}$ has type $\mathit{Set}[\mathcal{Real} \times \mathcal{Real}]$, a function $h : \mathbb{R} \to \mathcal{P}(\mathbb{Z})$ has type $\mathit{Set}[\mathcal{Real} \times \mathit{Set}[\mathit{Integer}]]$, and so on.

Exercises 5.1

1. Three functions f, g and h are defined as follows.

$$f : \mathbb{R} \to \mathbb{R} \qquad f(x) = x^2 - 5$$

$$g : \mathbb{Z} \to \mathbb{R} \qquad g(x) = \frac{5x}{x^2 - 2}$$

$$h : \mathbb{R} \to \mathbb{Z} \qquad h(x) = \lfloor x \rfloor. \qquad \text{(See example 4.6.1.)}$$

Find the value of each of the following.

(i)	$f(3)$	(v)	$h(-3.7)$
(ii)	$g(3)$	(vi)	$g(h(3.7))$
(iii)	$h(3.7)$	(vii)	$f(a+1)$
(iv)	$f(\sqrt{2})$	(viii)	$g(a^2)$.

2. Let $A = \{1, 2, 3, 4\}$. Two functions $f, g : \mathcal{P}(A) \to \mathcal{P}(A)$ are defined by $f(X) = A - X$ and $g(X) = X \cup \{1\}$. Find the value of each of the following.

(i)	$f(\{1, 2\})$	(v)	$g(\{1, 2\})$
(ii)	$f(\{4\})$	(vi)	$g(\{4\})$
(iii)	$f(A)$	(vii)	$g(A)$
(iv)	$f(\varnothing)$	(viii)	$g(\varnothing)$.

3. Which of the following subsets of $\mathbb{Z} \times \mathbb{Z}$ are functions $\mathbb{Z} \to \mathbb{Z}$? Justify your answers.

(i) $\{(n, 2n) : n \in \mathbb{Z}\}$
(ii) $\{(2n, n) : n \in \mathbb{Z}\}$

(iii) $\{(n, n^3) : n \in \mathbb{Z}\}$
(iv) $\{(n^3, n) : n \in \mathbb{Z}\}$
(v) $\{(n, n+4) : n \in \mathbb{Z}\}$
(vi) $\{(n+4, n) : n \in \mathbb{Z}\}$
(vii) $\{(n, 2^n) : n \in \mathbb{Z}\}$
(viii) $\{(n, m) : n \in \mathbb{Z}$ and $m = a^n$ for some $a \in \mathbb{Z}\}$.

4. Which of the following subsets f of $A \times B$ are functions $A \to B$? Justify your answers.

(i) $A = B = \{$human beings, living or dead$\}$
$f = \{(a, b) \in A \times B : a$ is a parent of $b\}$

(ii) $A = B = \{$human beings, living or dead$\}$
$f = \{(a, b) \in A \times B : b$ is a parent of $a\}$

(iii) $A = \{$countries of the world$\}$, $B = \{$cities of the world$\}$
$f = \{(a, b) \in A \times B : b$ is the capital city of $a\}$

(iv) $A = B = \{$living human beings$\}$
$f = \{(a, b) \in A \times B : b$ is married to $a\}$.

5. Let A be any non-empty set and $\mathscr{P}(A)$ its power set. Which of the following subsets of $A \times \mathscr{P}(A)$ are functions $A \to \mathscr{P}(A)$? Justify your answers.

(i) $f = \{(a, B) : a \in B\}$
(ii) $f = \{(a, B) : B = \{a\}\}$
(iii) $f = \{(a, B) : B \neq \varnothing\}$
(iv) $f = \{(a, B) : B \cup \{a\} = A\}$.

6. Let $A = \{1, 2, 3, 4, 5, 6, 7, 8, 9\}$. For each of the following functions $f : A \to A$ defined informally:

(a) list the images $f(1), f(2), \ldots, f(9)$,
(b) write down the image set of the function,
(c) list the elements of f as a subset of $A \times A$.

(i) $f(x) =$ the larger of x and 4 (and $f(4) = 4$)
(ii) $f(x) =$ the smaller of $x + 4$ and 9 (and $f(5) = 9$)
(iii) $f(x) =$ the smallest prime number which (exactly) divides $x + 1$
(iv) $f(x) = |x - 3| + 1$
(v) $f(x) = |2x - 9|$
(vi) $f(x) = \dfrac{x^2 + x}{x + 1}$

(vii) $f(x) = \begin{cases} 4 & \text{if } x \leqslant 5 \\ 3 & \text{if } x > 5 \end{cases}$

(viii) $f(x) = \begin{cases} 2 & \text{if } x^2 \leqslant 2^x \\ 3 & \text{if } x^2 > 2^x. \end{cases}$

7. Determine the image set of each of the following functions:

(i) $f : \mathbb{R} \to \mathbb{R}, x \mapsto x^2 + 2$
(ii) $f : \mathbb{R} \to \mathbb{R}, x \mapsto (x + 2)^2$
(iii) $f : \mathbb{R} \to \mathbb{R}, x \mapsto 1/(x^2 + 2)$
(iv) $f : \mathbb{R} \to \mathbb{R}, x \mapsto x^4$
(v) $f : \mathbb{R} \to \mathbb{R}, x \mapsto (x + 2)/(x^2 + 5)$
(vi) $f : \mathbb{R} \to \mathbb{R}, x \mapsto \sqrt{x^2 + 1}$.

8. Describe the image set of each of the following functions.

(i) $A = \mathscr{P}\{a, b, c, d\}$, the power set of $\{a, b, c, d\}$
$f : A \to \mathbb{Z}$, $f(C) = |C|$.
(ii) $f : \mathbb{Z} \to \mathbb{Z}$, $f(n) = n^2$.
(iii) $A = \{\text{countries of the world}\}$, $B = \{\text{cities of the world}\}$
$f : A \to B$, $f(X) =$ the capital city of X.
(iv) $A = \mathscr{P}\{a, b, c, d\}$, $f : A \to A$, $f(X) = X \cap \{a\}$.
(v) $A = \mathscr{P}\{a, b, c, d\}$, $f : A \to A$, $f(X) = X \cup \{a\}$.

9. Determine the type of each of the functions defined in questions 1, 2, 6, 7 and 8 above.

10. Let $f : A \to B$ be a function and C a subset of A. The **image of C** is the set denoted by $f(C) = \{f(c) : c \in C\}$. Thus $f(C)$ is the set of all images of elements of C; in particular $f(A) = im(f)$ and, if $a \in A$, $f\{a\} = \{f(a)\}$. (See the diagram below.)

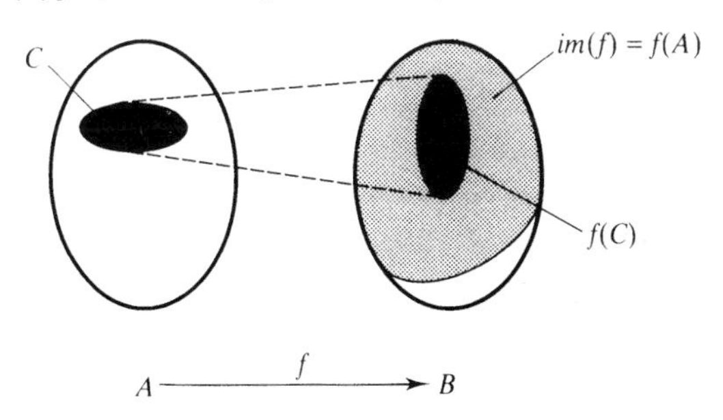

Determine $f(C)$ in each of the following cases.

(i) $f : \mathbb{R} \to \mathbb{R}$, $f(x) = x^2$; $C = [-3, 2] = \{x \in \mathbb{R} : -3 \leqslant x \leqslant 2\}$.

(ii) $f : \mathbb{R} \to \mathbb{R}$, $f(x) = 2/x$; $C = (0, 8] = \{x \in \mathbb{R} : 0 < x \leqslant 8\}$.
(iii) $f : \mathbb{Z} \to \mathbb{Z}$, $f(x) = 2^x$; $C = \{n \in \mathbb{Z} : -1 \leqslant n \leqslant 6\}$.
(iv) $f : \mathbb{R}^2 \to \mathbb{R}$, $f(x, y) = x^2 + y^2$;
$C = [-2, 3] \times [-1, 2] = \{(x, y) \in \mathbb{R}^2 : -2 \leqslant x \leqslant 3$ and $-1 \leqslant y \leqslant 2\}$.
(v) $f : \{\text{English words}\} \to \mathbb{Z}^+$, $f(w) =$ the number of letters in w;
$C = \{\text{mathematics, is, a, fascinating, subject}\}$.
(vi) $f : A \to B$ is any function; $C = \varnothing$.

11. Let $f : A \to B$ be a function and D a subset of B. The **inverse image of D** is the set $f^{-1}(D) = \{a \in A : f(a) \in D\}$. Thus $f^{-1}(D)$ is the set of all elements of A whose image lies in D. Note that f^{-1} is not necessarily a function so that the inverse image is a different concept from that of the image defined in the previous question. (See the diagram below.)

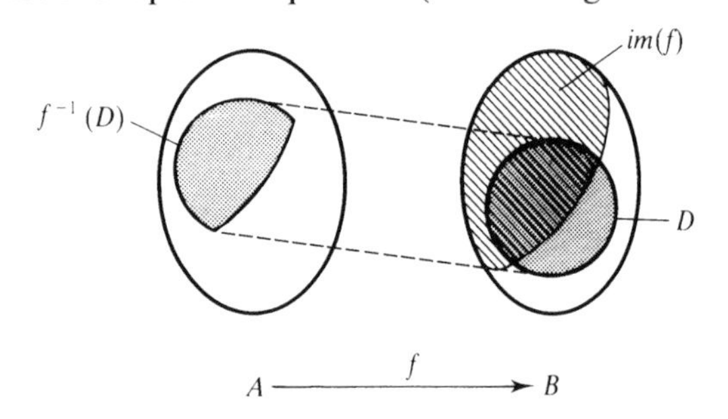

Find $f^{-1}(D)$ in each of the following cases.

(i) $f : \mathbb{R} \to \mathbb{R}$, $f(x) = x^2$; $D = [4, 9] = \{x \in \mathbb{R} : 4 \leqslant x \leqslant 9\}$.
(ii) $f : \mathbb{R} \to \mathbb{R}$, $f(x) = x^2$; $D = [-9, -4]$
$= \{x \in \mathbb{R} : -9 \leqslant x \leqslant -4\}$.
(iii) $f : \mathbb{R} \to \mathbb{R}$, $f(x) = x^2$; $D = [-4, 9] = \{x \in \mathbb{R} : -4 \leqslant x \leqslant 9\}$.
(iv) $f : \mathbb{Z} \to \mathbb{R}$, $f(x) = 2^x$; $D = \{n \in \mathbb{Z} : 0 \leqslant n \leqslant 10\}$.
(v) $f : \mathbb{R}^2 \to \mathbb{R}$, $f(x, y) = x^2 + y^2$; $D = [0, 1]$.
(vi) $f : A \to B$ is any function; $D = \varnothing$.
(vii) $f : A \to B$ is any function; $D = B$.

12. A **partial function** f from A to B is a 'function' in which $f(a)$ is not defined for *every* $a \in A$. A partial function is sometimes denoted $f : A \nrightarrow B$. In example 5.1.3, the rule $f(x) = 1/(x+1)^3$ defines a partial function $f : \mathbb{R} \nrightarrow \mathbb{R}$ because $f(-1)$ is not defined. For a partial function $f : A \nrightarrow B$, the set A is called the **source** of f and the set of elements of A for which $f(a)$ is defined, $\{a \in A : f(a) \text{ is defined}\}$, is the **domain** of f.

Sometimes, to emphasize that a function is *not* partial; i.e. $f(a)$ is defined for all $a \in A$, we say f is a **total function**. So a total function is what we have previously called simply a function and we will always use the unqualified term 'function' to mean total function.

Explain why each of the following is a partial function and determine the domain in each case.

(i) $f : \{1, 2, 3, 4, 5, 6, 7, 8, 9\} \nrightarrow \{1, 2, 3, 4, 5, 6, 7, 8, 9\}$, $f(n) = n + 3$.
(ii) $f : \{1, 2, 3, 4, 5, 6, 7, 8, 9\} \nrightarrow \{1, 2, 3, 4, 5, 6, 7, 8, 9\}$, $f(n) = 2n - 5$.
(iii) $f : \{1, 2, 3, 4, 5, 6, 7, 8, 9\} \nrightarrow \{1, 2, 3, 4, 5, 6, 7, 8, 9\}$, $f(n) = n^2$.
(iv) $f : \{1, 2, 3, 4, 5, 6, 7, 8, 9\} \nrightarrow \{1, 2, 3, 4, 5, 6, 7, 8, 9\}$, $f(n) = \sqrt{n}$.
(v) $f : \mathbb{R} \nrightarrow \mathbb{R}$, $f(x) = 1/x$.
(vi) $f : \mathbb{Z} \nrightarrow \mathbb{Z}$, $f(n) = 1/n$.
(vii) $f : \mathbb{Z} \nrightarrow \mathbb{Z}$, $f(n) = n/4$.
(viii) $f : \mathbb{Z} \nrightarrow \mathbb{Z}$, $f(n) = 4/n$.
(ix) $f : \mathbb{Z}^+ \times \mathbb{Z}^+ \nrightarrow \mathbb{Z}^+$, $f(n, m) = n/m$.
(x) $f : \mathbb{Z} \nrightarrow \mathbb{Z}$, $f(n) = 2^n$.

13. Classify each of the following as (a) a total function, (b) a partial function or (c) not a function (either partial or total). Give brief reasons for your answers.

(i) $f : \mathbb{Z} \to \mathbb{Z}$, $f(n) = n/2$.
(ii) $A = \{\text{countries of the world}\}$;
$f : A \to A$, $f(X) =$ countries sharing a border with X.
(iii) $A = \{\text{countries of the world}\}$;
$f : A \to \mathscr{P}(A)$, $f(X) =$ the *set* of countries sharing a border with X.
(iv) $f : \mathbb{R} \to \mathbb{R}$, $f(x) = \sqrt{x}$.
(v) $f : \mathbb{R} \to \mathbb{R}$, $f(x) = \pm\sqrt{x}$.
(vi) $f : \mathbb{R}^+ \to \mathbb{R}^+$, $f(x) = \sqrt{x}$.

14. Let A and B be finite sets such that $|A| = n$ and $|B| = m$. How many different functions are there from A to B?

15. Let $f : A \to B$ be a function. Define a relation R on its domain A by:

$$x \,\mathsf{R}\, y \quad \text{if and only if } f(x) = f(y).$$

Show that R is an equivalence relation on A, and describe the equivalence classes.

16. Let $f : A \to B$ be a function. Under what circumstances is

$$g = \{(b, a) : (a, b) \in f\}$$

a function $B \to A$? (This question is considered in §5.4.)

17. (i) If R is an equivalence relation on a set A, is R necessarily a function $A \to A$? Justify your answer.

(ii) If R is a partial order relation on a set A, is R necessarily a function $A \to A$? Justify your answer.

5.2 Composite Functions

Let $f : A \to B$ and $g : B \to C$ be functions. If x is an element of A then $y = f(x)$ belongs to B. Therefore $g(y) = g(f(x))$ is an element of C. We can use the association $x \mapsto g(f(x))$ to define a function from A to C, called the **composite of f and g**, denoted $g \circ f$†. The composite $g \circ f$ can be represented by the diagram in figure 5.8(i).

Alternatively, if we think of the functions f and g being represented by function machines, then the composite $g \circ f$ has a function machine that is obtained by connecting the output of f to the input of g. This is represented by figure 5.8(ii).

Example 5.5

Let $A = \{a, b, c, d, e\}$, $B = \{\alpha, \beta, \gamma, \delta\}$ and $C = \{1, 2, 3, 4, 5, 6\}$. Let $f : A \to B$ be the function defined in example 5.1.1 and $g : B \to C$ be the function defined by

$$\alpha \mapsto 3, \quad \beta \mapsto 5, \quad \gamma \mapsto 1, \quad \delta \mapsto 5.$$

Then the composite function, $g \circ f : A \to C$, is given by

$$a \mapsto 5, \quad b \mapsto 3, \quad c \mapsto 5, \quad d \mapsto 5, \quad e \mapsto 5.$$

† Thus the composite $g \circ f$ is the function 'f followed by g'. This is an instance where notation can cause some confusion; the function f is written after g but 'acts' before it. Some authors avoid this 'problem' with the notation by writing the function on the right; that is, they write xf instead of $f(x)$. Written in this notation $g(f(x))$ becomes xfg, so the composite function 'f followed by g' is denoted fg.

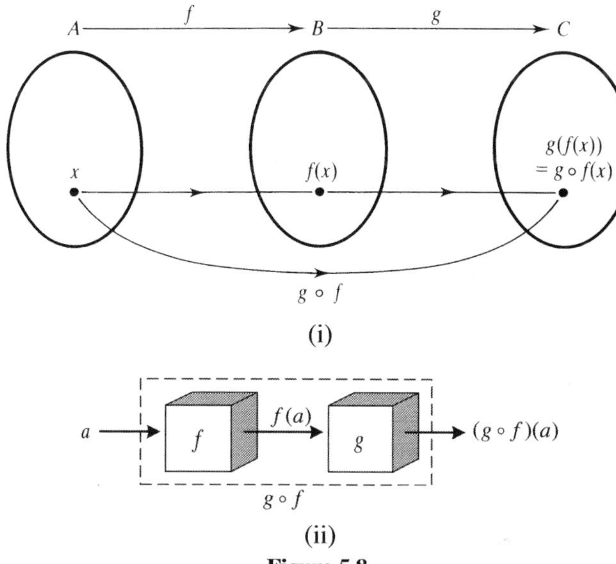

Figure 5.8

This example is illustrated by figure 5.9. The diagram shows very clearly that the composite function $g \circ f$ is 'f followed by g'.

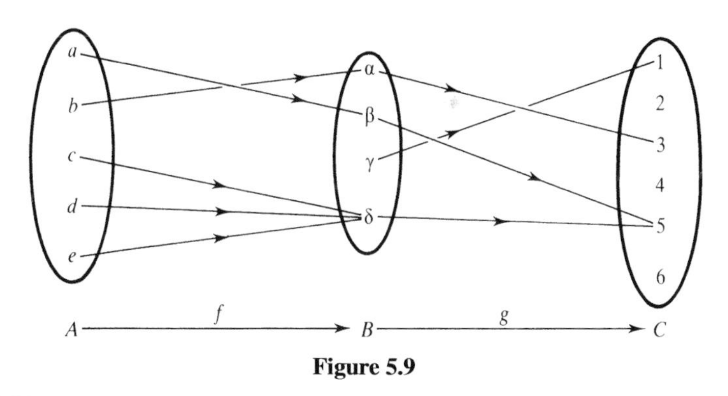

Figure 5.9

So far we have considered only the informal definition of $g \circ f$. However, we should be able to make the notion precise in terms of our Cartesian product definition of a function.

According to the formal definition, the function $g \circ f$ should be the subset of the Cartesian product $A \times C$ consisting of all those elements (x, z) such that $z = g \circ f(x)$. If we let $y = f(x) \in B$ then $(x, y) \in f$ and $(y, z) \in g$. Therefore we may formally define the composite function as follows.

Definition 5.4

Let $f : A \to B$ and $g : B \to C$ be functions. The **composite function** $g \circ f : A \to C$ is

$$g \circ f = \{(x, z) \in A \times C : (x, y) \in f \text{ and } (y, z) \in g \text{ for some } y \in B\}.$$

Note that the composite $g \circ f$ of two arbitrary functions may not exist. In definition 5.4, the domain of g equals the codomain of f. It is usual to define the composite function only when the sets 'match up' in this way. However, this is slightly more restrictive than is strictly necessary and we can widen the conditions under which $g \circ f$ is defined as follows. Let $f : A \to B$ and $g : B' \to C$ be functions and let $a \in A$. In order that $g(f(a))$ be defined, we require that $f(a)$ belong to B', the domain of g. Now to define $g \circ f$ it is necessary (and sufficient) that $g(f(a))$ be defined for all $a \in A$. Hence $g \circ f$ is defined if and only if the image set of f is a subset of the domain of g. Of course this condition is satisfied if $B = B'$, which is the condition given in definition 5.4. (Figure 5.11 below may help you to visualize the situation described here.)

Examples 5.6

1. The formal definition of the composite function $g \circ f$ in example 5.5 is

$$g \circ f = \{(a, 5), (b, 3), (c, 5), (d, 5), (e, 5)\}.$$

2. Let f and g be the functions $\mathbb{R} \to \mathbb{R}$ defined by $f(x) = x + 2$ and $g(x) = 1/(x^2 + 1)$ respectively.

   Then $g \circ f : \mathbb{R} \to \mathbb{R}$ is defined by

$$\begin{aligned} g \circ f(x) &= g(f(x)) \\ &= g(x + 2) \\ &= \frac{1}{(x + 2)^2 + 1} \\ &= \frac{1}{x^2 + 4x + 5}. \end{aligned}$$

   Similarly,

$$f \circ g(x) = f(g(x))$$

$$= f(1/(x^2+1))$$
$$= \frac{1}{x^2+1} + 2$$
$$= \frac{2x^2+3}{x^2+1}.$$

This example illustrates that, in general, $f \circ g \neq g \circ f$. Of course, given two functions f and g, it is quite possible for $g \circ f$ to be defined but $f \circ g$ not to be defined. (See exercise 5.2.7.)

3. Three functions, f, g and h, are defined by $f : \mathbb{Z}^+ \to \mathbb{R}$, $f(x) = \dfrac{2}{x+1}$, $g : \mathbb{Z} \to \mathbb{Z}$, $g(x) = x^2 + 3$ and $h : \mathbb{R} \to \mathbb{R}$, $h(x) = 3x + 2$.

   Determine which of the following composite functions are defined.

   (i) $g \circ f$
   (ii) $f \circ g$
   (iii) $h \circ f$
   (iv) $f \circ h$
   (v) $g \circ h$
   (vi) $h \circ g$.

Solution

(i) Since $f(2) = 2/3$, we have that $2/3 \in im(f)$ but $2/3 \notin \mathbb{Z}$. Therefore $im(f) \nsubseteq \mathbb{Z}$, so $g \circ f$ is not defined.
(ii) For all $x \in \mathbb{Z}$, $g(x) = x^2 + 3 \geqslant 3$ (since $x^2 \geqslant 0$). Hence the image set of g is a subset of $\mathbb{Z}^+$, the domain of f. Therefore $f \circ g$ is defined.
(iii) Since $im(f) \subseteq \mathbb{R}$, $h \circ f$ is defined.
(iv) The composite $f \circ h$ is not defined: $h(1/2) = 7/2 \notin \mathbb{Z}^+$ so $im(h) \nsubseteq \mathbb{Z}^+$.
(v) The same reasoning as in part (iv) shows that $g \circ h$ is not defined.
(vi) Since $im(g) \subseteq \mathbb{Z} \subseteq \mathbb{R}$, it follows that the composite $h \circ g$ is defined.

4. Consider $f : \mathbb{R} \to (\mathbb{R}^+ \cup \{0\})$, $f(x) = x^2$ and $g : (\mathbb{R}^+ \cup \{0\}) \to \mathbb{R}$, $g(x) = \sqrt{x}$. Determine the composite functions $g \circ f$ and $f \circ g$.

Solution

The function $g \circ f : \mathbb{R} \to \mathbb{R}$ is $g \circ f(x) = g(x^2) = \sqrt{x^2}$. Note that $\sqrt{x^2}$ is positive (or zero). If x is positive (or zero), then $\sqrt{x^2}$ is just x. However, if x is negative, then $\sqrt{x^2} = -x$. For instance, if $x = -2$, then $\sqrt{x^2} = \sqrt{4} = 2 = -(-2)$.

In other words,

$$g \circ f(x) = \begin{cases} x & \text{if } x \geqslant 0 \\ -x & \text{if } x < 0. \end{cases}$$

This is called the **modulus function**, and is denoted $x \mapsto |x|$. Its graph is given in figure 5.10.

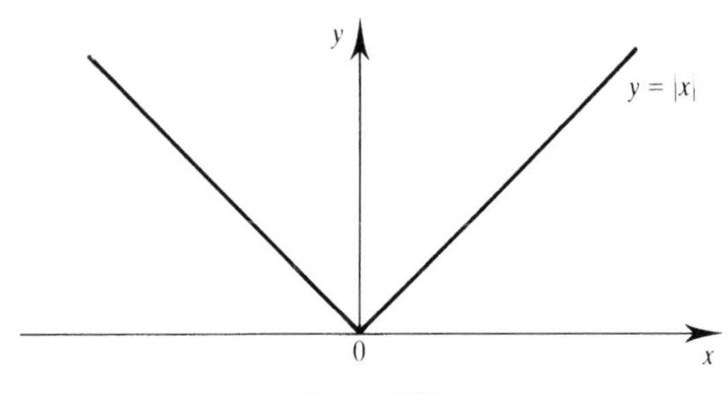

Figure 5.10

The function $f \circ g : (\mathbb{R}^+ \cup \{0\}) \to (\mathbb{R}^+ \cup \{0\})$ is $f \circ g(x) = (\sqrt{x})^2$. Since x is positive (or zero) here, $(\sqrt{x})^2$ is just x itself. In other words, $f \circ g$ is the identity function on $\mathbb{R}^+ \cup \{0\}$.

Theorem 5.1

Let $f : A \to B$ and $g : B \to C$ be functions. Then

$$\mathit{im}(g \circ f) \subseteq \mathit{im}(g).$$

Proof

Let $c \in \mathit{im}(g \circ f)$. Then there exists $a \in A$ such that $(g \circ f)(a) = g(f(a)) = c$. Now let $b = f(a) \in B$; then $g(b) = c$, so $c \in \mathit{im}(g)$. Therefore $\mathit{im}(g \circ f) \subseteq \mathit{im}(g)$. □

Theorem 5.1 is probably best visualized by figure 5.11.

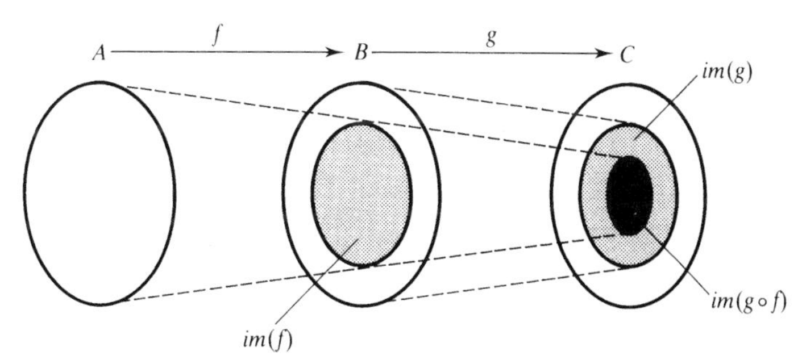

Figure 5.11

Exercises 5.2

1. Let f, g and h be the functions defined in exercise 5.1.1:

$$f : \mathbb{R} \to \mathbb{R} \qquad f(x) = x^2 - 5$$
$$g : \mathbb{Z} \to \mathbb{R} \qquad g(x) = \frac{5x}{x^2 - 2}$$
$$h : \mathbb{R} \to \mathbb{Z} \qquad h(x) = \lfloor x \rfloor.$$

Find the value of each of the following.

(i)	$(f \circ f)(2)$	(v)	$(h \circ g)(3)$
(ii)	$(g \circ h)(2.5)$	(vi)	$(h \circ f)(1.5)$
(iii)	$(f \circ g)(2)$	(vii)	$(f \circ h)(1.5)$
(iv)	$(h \circ h)(3.7)$	(viii)	$(g \circ h)(2)$.

2. Let A = {*Anna Karenina*, *Crime and Punishment*, *Sons and Lovers*, *War and Peace*}, B = {*Dostoyevsky*, *Lawrence*, *Tolstoy*, *Zola*}, and C = {*America*, *England*, *France*, *Russia*}.

Define two functions $f : A \to B$ and $g : B \to C$ by

f : *Anna Karenina* $\mapsto$ *Tolstoy*
f : *Crime and Punishment* $\mapsto$ *Dostoyevsky*
f : *Sons and Lovers* $\mapsto$ *Lawrence*
f : *War and Peace* $\mapsto$ *Tolstoy*

g : *Dostoyevsky* $\mapsto$ *Russia*
g : *Lawrence* $\mapsto$ *England*

$$g: \qquad \textit{Tolstoy} \mapsto \textit{Russia}$$
$$g: \qquad \textit{Zola} \mapsto \textit{France}.$$

(i) Define the composite function $g \circ f$ in a similar way, and draw a diagram to represent the composite function.

(ii) Write down (in words) rules which define each of the functions f, g and $g \circ f$.

3. Let $f, g : \mathbb{R} \to \mathbb{R}$ be defined by $f(x) = 4x - 1$ and $g(x) = x^2 + 1$. Find:

(i)	$f(2)$	(vi)	$(g \circ g)(2)$
(ii)	$g(2)$	(vii)	$(f \circ g \circ f)(3)$
(iii)	$(g \circ f)(2)$	(viii)	$(g \circ f \circ g)(3)$
(iv)	$(f \circ g)(2)$	(ix)	$(g \circ f)(x)$
(v)	$(f \circ f)(2)$	(x)	$(f \circ g)(x)$.

4. Let f, g and h be functions $\mathbb{R} \to \mathbb{R}$ defined respectively by

$$f(x) = 2x + 1, \quad g(x) = 1/(x^2 + 1), \quad \text{and} \quad h(x) = \sqrt{x^2 + 1}.$$

Find expressions for each of the following:

(i)	$(g \circ f)(1)$	(vi)	$(g \circ f)(x)$
(ii)	$(f \circ g)(1)$	(vii)	$(g \circ h)(x)$
(iii)	$(g \circ h)(2)$	(viii)	$(f \circ f)(x)$
(iv)	$(h \circ f)(3)$	(ix)	$((f \circ g) \circ h)(x)$
(v)	$(f \circ g)(x)$	(x)	$(f \circ (g \circ h))(x)$.

5. Let $A = \{\text{humans, living or dead}\}$ and let f and g be the functions $A \to A$ defined by $f(x) =$ the father of x and $g(x) =$ the mother of x, respectively.

Describe the composite functions $f \circ f$, $f \circ g$, $g \circ f$ and $g \circ g$.

6. Let $f : A \to B$ be any function. Show that $f \circ id_A = f$ and $id_B \circ f = f$.

7. Let $f : A \to B$ and $g : C \to D$ be two functions. What are the most general conditions under which both composites $g \circ f$ and $f \circ g$ can be defined?

8. (**Associativity of composition.**) Let $f : A \to B$, $g : B \to C$, $h : C \to D$ be functions. Explain why $(h \circ g) \circ f = h \circ (g \circ f)$. (Hint: find expressions for $((h \circ g) \circ f)(x)$ and $(h \circ (g \circ f))(x)$.)

9. Let $A = \{a, b, c\}$. Define a function $f : A \to A$, which is not the identity function on A, such that $f \circ f = f$.

10. Let $f : \mathbb{R} \to \mathbb{R}$ be the function $f(x) = \lfloor x \rfloor$, where $\lfloor x \rfloor$ is the largest integer less than or equal to x (see example 4.6.1).

    (i) Show that $f \circ f = f$.
    (ii) Show that $f(x+k) = f(x) + k$ for all $x \in \mathbb{R}$ if and only if $k \in \mathbb{Z}$.
    (iii) For what values of x is $f(2x) = 2f(x)$?

11. Given $f : \mathbb{R} \to \mathbb{R}$, $x \mapsto (2x+1)$, define $f^{[n]} : \mathbb{R} \to \mathbb{R}$ inductively by

$$f^{[1]} = f, \quad f^{[n]} = f^{[n-1]} \circ f \quad \text{for } n > 1.$$

    Prove that $f^{[n]}(x) = 2^n x + (2^n - 1)$.

12. Consider the function $f : \mathbb{Z}^+ \to \mathbb{Z}^+$, $x \mapsto x + 2$.

    (i) Show that there are infinitely many different functions $g : \mathbb{Z}^+ \to \mathbb{Z}^+$ such that $g \circ f = id_{\mathbb{Z}^+}$.
    (ii) Show that there is no function $h : \mathbb{Z}^+ \to \mathbb{Z}^+$ such that $f \circ h = id_{\mathbb{Z}^+}$.
    (iii) Evaluate $f^{[n]}(x)$. (See exercise 5.2.11 above.)

13. In each of the following, define the composite function $g \circ f$:

    (i) $$f : \mathbb{R} \to \mathbb{R}, \quad f(x) = \begin{cases} x^2 + x & \text{if } x \geqslant 0 \\ 1/x & \text{if } x < 0 \end{cases}$$

    $$g : \mathbb{R} \to \mathbb{R}, \quad g(x) = \begin{cases} \sqrt{x+1} & \text{if } x \geqslant 0 \\ 1/x & \text{if } x < 0 \end{cases}$$

    (ii) $$f : \mathbb{R} \to \mathbb{R}, \quad f(x) = \begin{cases} x - 2 & \text{if } x \geqslant 1 \\ x^3 & \text{if } x < 1 \end{cases}$$

    $$g : \mathbb{R} \to \mathbb{R}, \quad g(x) = \begin{cases} (x+4)/3 & \text{if } x \geqslant 0 \\ |x+1| & \text{if } x < 0. \end{cases}$$

14. Let C be a subset of A. The function $i_C : C \to A$, $c \mapsto c$, is called the **inclusion of C in A**. (Thus i_C is the same as the identity function on C *except* that its codomain is A.)

    (i) Define i_C as a subset of $C \times A$.

    Now let $f : A \to B$ be a function. The function $f|_C : C \to B$, $c \mapsto f(c)$, is called the **restriction of f to C**. (Thus $f|_C$ is similar to f except that it has domain C.)

    (ii) Show that $f|_C = f \cap (C \times B)$.
    (iii) Using the informal definitions, show that $f|_C = f \circ i_C$.

15. Let $f : A \to B$ and $g : C \to D$ be functions. If either f or g is a partial function or if the image set of f is not a subset of the domain of g, $im(f) \not\subseteq C$, then their composite $g \circ f$ may be a partial function (see exercise 5.1.12). For each of the following pairs of functions, determine whether f, g and the composite $g \circ f$ are partial or total functions. If $g \circ f$ is partial, determine its domain.

    (i) $f : \{1, 2, 3, 4, 5\} \to \{1, 2, 3, 4, 5, 6, 7, 8, 9, 10\}$, $f(x) = 2x$
    $g : \{1, 2, 3, 4, 5\} \to \{1, 2, 3, 4, 5, 6, 7, 8, 9, 10\}$, $g(x) = x + 3$

    (ii) $f : \mathbb{Z} \to \mathbb{Z}$, $f(x) = 2x$; $g : \mathbb{Z} \to \mathbb{Z}$, $g(x) = x + 3$

    (iii) $f : \mathbb{Z} \to \mathbb{R}$, $f(x) = \dfrac{1}{x^2 + 1}$; $g : \mathbb{Z} \to \mathbb{Z}$, $g(x) = 2x + 3$

    (iv) $f : \mathbb{R} \to \mathbb{R}$, $f(x) = \dfrac{1}{x^2 + 1}$; $g : \mathbb{R} \to \mathbb{R}$, $g(x) = \sqrt{x}$

    (v) $f : \mathbb{Z} \to \mathbb{Z}$, $f(x) = \dfrac{x}{2}$; $g : \mathbb{R} \to \mathbb{R}$, $g(x) = 2x$

    (vi) $f : \mathbb{R} \to \mathbb{R}$, $f(x) = 2x$; $g : \mathbb{Z} \to \mathbb{Z}$, $g(x) = \dfrac{x}{2}$

    (vii) $f : \mathbb{Z} \to \mathbb{Z}$, $f(x) = 2x$; $g : \mathbb{Z} \to \mathbb{Z}$, $g(x) = \dfrac{x}{2}$

    (viii) $f : \mathbb{R} \to \mathbb{R}$, $f(x) = x^2$; $g : \mathbb{R} \to \mathbb{R}$, $g(x) = \sqrt{1 - x}$.

16. Determine the type of each of the functions (f, g and $g \circ f$) defined in question 15.

5.3 Injections and Surjections

In this section we consider two special kinds of functions: 'injections' and 'surjections'. Recall from §5.1 that a function $f : A \to B$ can be such that:

(i) different elements of the domain may have the same image in the codomain;

(ii) there may be elements of the codomain which are not the image of any element of the domain.

Both these possibilities are exhibited by the square function $f : \mathbb{R} \to \mathbb{R}$, $f(x) = x^2$. For instance, both 2 and -2 have the same image (namely 4) and any negative real number does not belong to the image of f (because $x^2 \geqslant 0$ for all real numbers x).

A function where the first possibility does *not* occur is called 'injective' and a function where the second possibility does *not* occur is called 'surjective'. The two cases are represented in figure 5.12. The function $f : \{a, b, c, d\} \to \{\alpha, \beta, \gamma, \delta, \varepsilon\}$ illustrated in figure 5.12(a) is injective but not surjective. It is injective because no two elements of the domain have the same image; it is not surjective because the element δ of the codomain is not the image of any element of the domain. On the other hand, the function $g : \{a, b, c, d, e\} \to \{\alpha, \beta, \gamma, \delta\}$, illustrated in figure 5.12(b), is surjective but not injective. It is surjective because every element of the codomain is the image of at least one element of the domain; it is not injective because the two elements a and b of the domain have the same image.

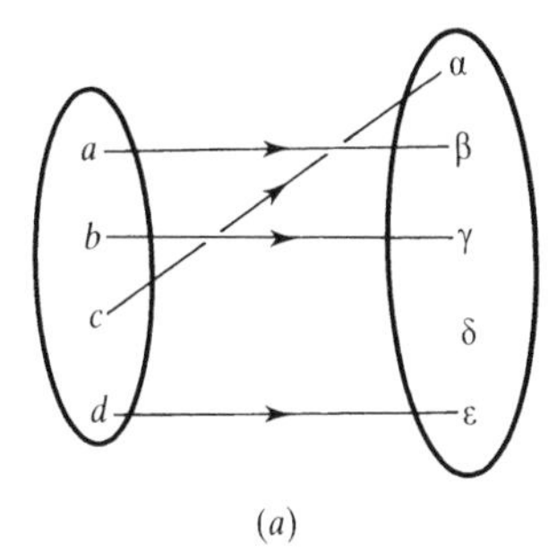

(a)

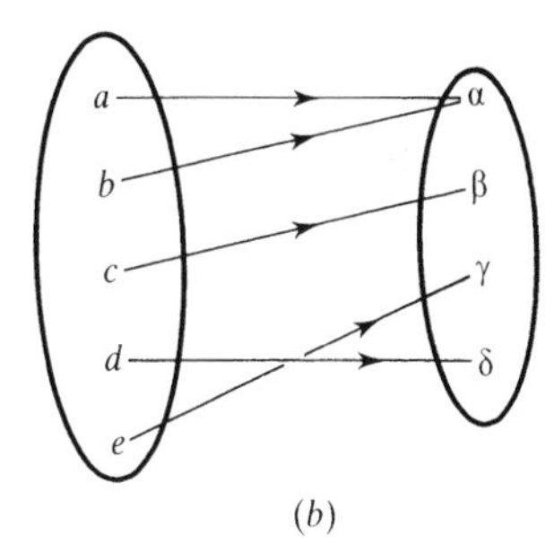

(b)

Figure 5.12

The following are the formal definitions of injective and surjective, given in terms of the Cartesian product definition of a function.

Definition 5.5

Let $f : A \to B$ be a function.

(i) We say that f is **injective** or is **an injection**† if the following is satisfied for all elements $a, a' \in A$:

$$\text{if } (a, b), (a', b') \in f \text{ and } a \neq a' \text{ then } b \neq b'.$$

(ii) We say that f is **surjective** or is **a surjection**† if for every $b \in B$ there exists $a \in A$ such that $(a, b) \in f$.

† Some authors use the term **'one-to-one function'** for an injective function and **'onto function'** for a surjective function.

Written less formally, a function f is injective if, for all $a, a' \in A$,

$$\text{if } a \neq a' \text{ then } f(a) \neq f(a').$$

However, to *prove* that a given function is injective it is generally easier to use the equivalent contrapositive statement. That is, for all $a, a' \in A$,

$$\text{if } f(a) = f(a') \text{ then } a = a'.$$

Of course, to show a function is *not* injective we need to find a counter-example to the general condition. In other words, we need to find two different elements a and a' of A which have the same image, $f(a) = f(a')$.

The second part of the definition can be rephrased simply to say that $f : A \to B$ is surjective if its image set equals its codomain, i.e. $im(f) = B$.

INJECTIVE FUNCTION

Examples 5.7

1. We have seen that $f : \mathbb{R} \to \mathbb{R}$, $x \mapsto x^2$, is neither injective nor surjective.

2. Let $f : \{1, 2, 3, 4, 5\} \to \{1, 2, 3, 4, 5, 6\}$,
   be defined by $1 \mapsto 4, 2 \mapsto 6, 3 \mapsto 1, 4 \mapsto 3, 5 \mapsto 5$,
   and let $g : \{2, 4, 6, 8, 10, 12\} \to \{2, 3, 5, 7, 11\}$,
   be defined by $2 \mapsto 11, 4 \mapsto 2, 6 \mapsto 5, 8 \mapsto 3, 10 \mapsto 5, 12 \mapsto 7$.

The function f is injective because each element of the domain has a different image. In other words, the following situation does *not* occur in the arrow diagram of f.

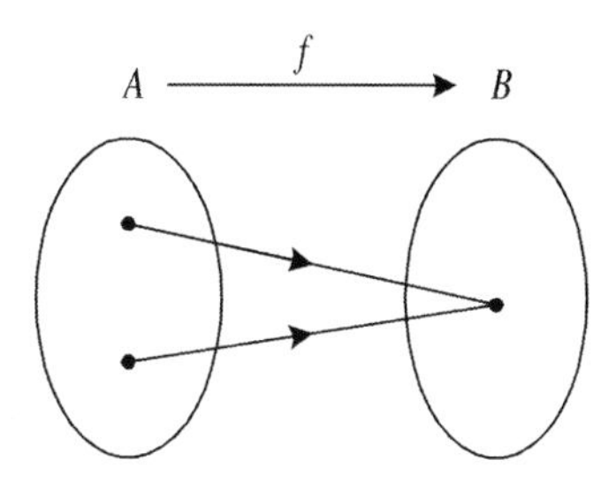

The element 2 in the codomain is not the image of any element of the domain, $2 \notin im(f)$. Therefore f is not surjective.

Now consider g. There do exist two different elements of the domain with the same image, $g(6) = 5 = g(10)$, so g is not injective. Since each element of the codomain is 'hit by an arrow'—$2 = g(4)$, $3 = g(8)$, $5 = g(6)$, $7 = g(12)$ and $11 = g(2)$—it follows that g is surjective.

3. Consider $f : \mathbb{R} \to \mathbb{R}$ defined by $f(x) = 3x - 7$. Show that f is both injective and surjective.

Solution

To show that f is an injection we prove that, for all real numbers x and y, $f(x) = f(y)$ implies $x = y$.

Now

$$\begin{aligned} & f(x) = f(y) \\ \Rightarrow \quad & 3x - 7 = 3y - 7 \\ \Rightarrow \quad & 3x = 3y \\ \Rightarrow \quad & x = y \end{aligned}$$

so f is injective.

To show that f is a surjection, let y be any element of the codomain $\mathbb{R}$. We need to find $x \in R$ such that $f(x) = y$. Let $x = (y + 7)/3$. Then $x \in \mathbb{R}$ and

$$\begin{aligned} f(x) &= f((y + 7)/3) \\ &= 3 \times \frac{y + 7}{3} - 7 \end{aligned}$$

$$= y + 7 - 7$$
$$= y$$

so f is surjective.

In the proof of surjectivity above, we seem to have plucked the value $x = (y+7)/3$ 'out of thin air' as it were. In fact, to discover that $(y+7)/3$ is the appropriate value of x, we let $y = f(x)$ and solve this to find x, as follows:

$$y = 3x - 7 \quad \Rightarrow \quad y + 7 = 3x \quad \Rightarrow \quad x = (y+7)/3.$$

However, this process of 'working backwards' to discover the appropriate value of x to consider is not strictly part of the proof of the surjectivity of f.

Clearly this proof can be generalized to show that any **linear function** $f : \mathbb{R} \to \mathbb{R}$, $f(x) = ax + b$ (where a and b are fixed real numbers with $a \neq 0$), is both injective and surjective.

4. Let $A = \{\text{countries of the world}\}$, $B = \{\text{cities of the world}\}$ and define $f : A \to B$ by $f(x) =$ the capital city of X (exercise 5.1.4(iii)). Then f is injective since (we suppose) different countries have different capital cities. Since there are cities which are not the capitals of any countries, f is not surjective. For example, New York $\notin im(f)$, Birmingham $\notin im(f)$, etc.

   With the same sets A and B define $g : B \to A$ by $g(C) =$ the country to which C belongs. This function is a surjection because (again we suppose) every country contains at least one city within its borders. However, g is not injective as several cities may belong to the same country. For example, f(Paris) $= f$(Nice) $=$ France, f(Ottawa) $= f$(Vancouver) $=$ Canada, etc.

5. Let X and Y be non-empty sets and $X \times Y$ their Cartesian product. The functions

$$p_1 : X \times Y \to X, p_1(x, y) = x \quad \text{and} \quad p_2 : X \times Y \to Y, p_2(x, y) = y$$

   are called the **natural projections** of $X \times Y$ onto X and of $X \times Y$ onto Y respectively. Both are clearly surjective and, provided X and Y are not singleton sets, neither is injective.

Consider a function $f : A \to B$ where A and B are subsets of $\mathbb{R}$. Just as we can 'read off' $im(f)$ from the graph of the function, we can also tell from the graph whether or not the function is injective or surjective.

Suppose that f is *not* injective. Then there are two elements a_1 and a_2 in A such that $f(a_1) = f(a_2) = b$, say. This means that the horizontal line at height b meets the graph at points corresponding to $x = a_1$ and $x = a_2$ on the x-axis. This situation is illustrated in figure 5.13.

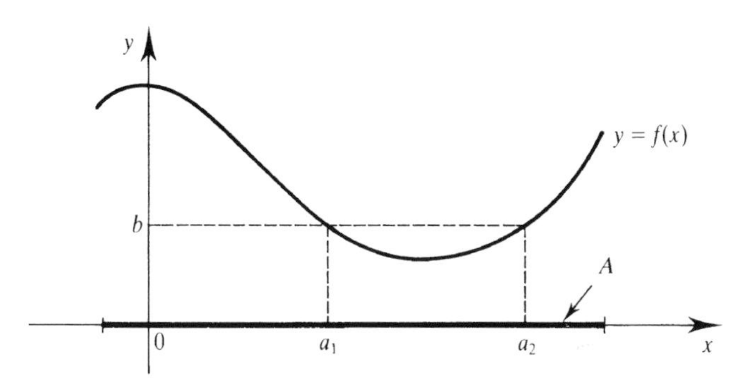

Figure 5.13

If, on the other hand, f is injective then this situation never occurs. In other words, a horizontal line through any point $b \in B$ on the y-axis will not meet the graph in *more* than one point.

We saw in §5.2 that $im(f)$ is represented by the region of the y-axis consisting of those points such that a horizontal line through a point meets the graph somewhere. Therefore f is surjective (i.e. $im(f) = B$) if and only if every horizontal line through a point of B meets the graph at least once.

These considerations are summarized in the following theorem.

Theorem 5.2

Let $f : A \to B$ be a function, where A and B are subsets of $\mathbb{R}$. Then:

(i) f is injective if and only if every horizontal line through a point of B on the y-axis meets the graph of f *at most* once;

(ii) f is surjective if and only if every horizontal line through a point of B on the y-axis meets the graph of f *at least* once.

Example 5.8

The graphs of four functions $A \to B$ are given below. Determine whether each function is injective and/or surjective.

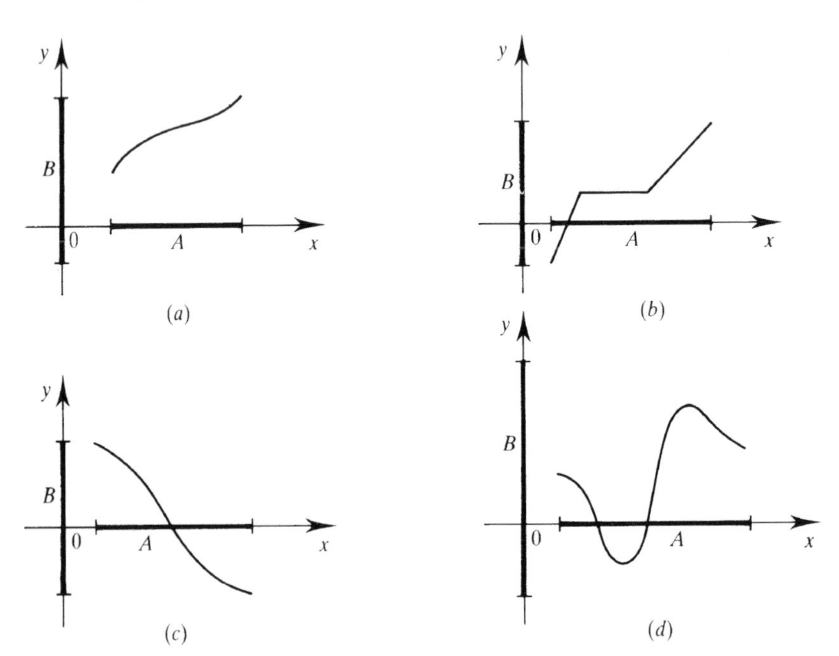

Solution

The function with graph (a) is injective since each horizontal line drawn through B meets the graph at most once. It is not surjective because, for instance, a horizontal line through any negative element of B does not meet the graph at all.

Graph (b) is the graph of a surjective but not injective function. Every horizontal line through B meets the graph somewhere, but the horizontal line at the same height as the horizontal portion of the graph meets the graph more than once—it meets it in infinitely many points in fact.

Similar arguments show that the function represented by graph (c) is both injective and surjective, and the function represented by graph (d) is neither injective nor surjective.

The examples above illustrate that the injective and surjective properties are independent of one another. A function may be injective but not surjective, surjective but not injective, both or neither.

Theorem 5.3

Let $f : A \to B$ and $G : B \to C$ be two functions.

(i) If f and g are both injective then so, too, is the composite $g \circ f$.

(ii) If f and g are both surjective then so, too, is the composite $g \circ f$.

Proof

(i) Suppose f and g are injections. Let $a, a' \in A$, $b = f(a)$ and $b' = f(a')$. Then

$$\begin{aligned}
& g \circ f(a) = g \circ f(a') & \\
\Rightarrow\quad & g(f(a)) = g(f(a')) & \\
\Rightarrow\quad & g(b) = g(b') & \\
\Rightarrow\quad & b = b' & \text{(since } g \text{ is injective)} \\
\Rightarrow\quad & f(a) = f(a') & \text{(since } f(a) = b,\ f(a') = b'\text{)} \\
\Rightarrow\quad & a = a' & \text{(since } f \text{ is injective).}
\end{aligned}$$

Hence $g \circ f$ is an injection.

(ii) Suppose f and g are surjections and let $c \in C$. Since g is surjective, there exists $b \in B$ such that $g(b) = c$, and since f is surjective, there exists $a \in A$ such that $f(a) = b$. Therefore there exists $a \in A$ such that

$$g \circ f(a) = g(f(a)) = g(b) = c$$

so $g \circ f$ is surjective. □

It is reasonable to ask whether the converse of each part of theorem 5.3 is also true: if $g \circ f$ is injective (surjective), does it follow that f and g are necessarily injective (surjective)? The answer to both questions is ‘no’, as the following example shows.

Example 5.9

Let $A = \{a_1, a_2\}$, $B = \{b_1, b_2, b_3\}$ and $C = \{c_1, c_2\}$, and define

$$f : A \to B \qquad \text{by} \qquad f(a_1) = b_1,\ f(a_2) = b_2$$

and

$$g : B \to C \qquad \text{by} \qquad g(b_1) = c_1,\ g(b_2) = g(b_3) = c_2.$$

Figure 5.14 illustrates these functions.

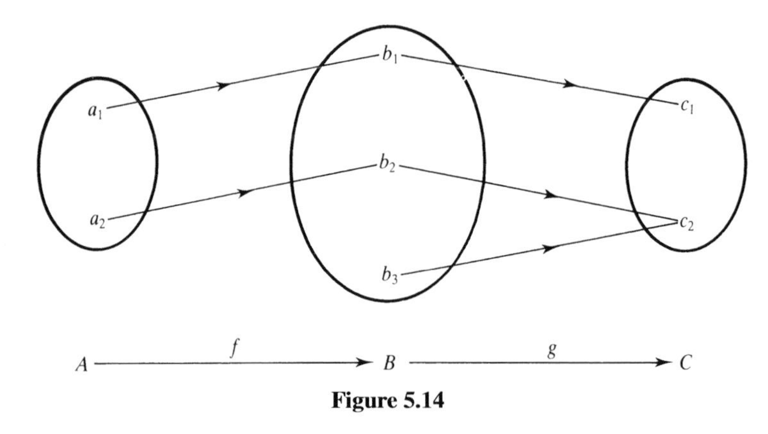

Figure 5.14

Clearly f is injective but not surjective and g is surjective but not injective. However, the composite function $g \circ f : A \to C$, which is given by

$$g \circ f(a_1) = c_1 \quad \text{and} \quad g \circ f(a_2) = c_2$$

is both injective and surjective.

This example suggests the following which is a partial converse to theorem 5.3.

Theorem 5.4

Let $f : A \to B$ and $g : B \to C$ be two functions:

(i) If the composite $g \circ f$ is injective then so, too, is f.

(ii) If the composite $g \circ f$ is surjective then so, too, is g.

Proof

In both cases we prove the contrapositive statement.

(i) The contrapositive is: if f is not injective then $g \circ f$ is not injective. Suppose that f is not injective. Then there exist $a, a' \in A$ such that $a \neq a'$ but $f(a) = f(a')$. Hence $g \circ f(a) = g \circ f(a')$ as well, so the composite function is also not injective.

(ii) The contrapositive statement here is: if g is not surjective then $g \circ f$ is not surjective. Suppose that g is not surjective. Then *im*(g) is a proper subset of C. Since *im*$(g \circ f) \subseteq$ *im*(g) (theorem 5.1), it follows that *im*$(g \circ f)$ is also a proper subset of C, so $g \circ f$ is not surjective either. □

The existence of an injection or a surjection from one set to another also has implications for the cardinalities of the sets concerned. Suppose $A = \{a_1, a_2, \ldots, a_n\}$ and $B = \{b_1, b_2, \ldots, b_m\}$ are finite sets and $f : A \to B$ is an injection. Then, assuming we have not listed any element of A twice, the elements $f(a_1), f(a_2), \ldots, f(a_n)$ are all different, so B contains at least n elements. Now suppose instead that $f : A \to B$ is a surjection. Then the list of elements $f(a_1), f(a_2), \ldots, f(a_n)$ must include every element of B at least once (but may contain repeats), so B contains at most n elements. We have proved the following theorem.

Theorem 5.5

Let $f : A \to B$ be a function between finite sets.

(i) If f is injective then $|A| \leqslant |B|$.

(ii) If f is surjective then $|A| \geqslant |B|$.

Exercises 5.3

1. For each of the following functions F determine whether or not F is (a) injective, (b) surjective. Justify your answers.

   (i) $F : \{a, b, c, d, e, f\} \to \{a, b, c, d, e, f\}$,
   $a \mapsto f, b \mapsto b, c \mapsto d, d \mapsto e, e \mapsto b, f \mapsto c$.

(ii) $F : \{a, b, c, d, e, f\} \to \{a, b, c, d, e, f\}$,
$a \mapsto f, b \mapsto e, c \mapsto d, d \mapsto c, e \mapsto b, f \mapsto a$.

(iii) $F : \{a, b, c, d, e\} \to \{a, b, c, d, e, f, g\}$,
$a \mapsto b, b \mapsto e, c \mapsto f, d \mapsto c, e \mapsto a$.

(iv) $F : \{a, b, c, d, e, f, g\} \to \{a, b, c, d, e\}$,
$a \mapsto e, b \mapsto c, c \mapsto d, d \mapsto a, e \mapsto d, f \mapsto e, g \mapsto a$.

(v) $F : \{a, b, c, d, e, f\} \to \{a, b, c, d, e, f, g\}$,
$a \mapsto b, b \mapsto e, c \mapsto d, d \mapsto b, e \mapsto a, f \mapsto g$.

2. For each of the following functions f determine whether or not f is (a) injective, (b) surjective. Justify your answers. **Hint**: in some cases it may help to evaluate $f(n)$ for a few values of n.

(i) $f : \mathbb{Z} \to \mathbb{Z}, f(n) = n - 6$.
(ii) $f : \mathbb{Z} \to \mathbb{Z}, f(n) = 3n - 5$.
(iii) $f : \mathbb{Z} \to \mathbb{Z}, f(n) = n^2$.
(iv) $f : \mathbb{Z} \to \mathbb{Z}, f(n) = n^3$.
(v) $f : \mathbb{Z} \to \mathbb{Z}, f(n) = n^2 + n$.
(vi) $f : \mathbb{Z} \to \mathbb{Z}, f(n) = (-1)^n$.
(vii) $f : \mathbb{Z} \to \mathbb{Z}, f(n) = n + (-1)^n$.

(viii) $f : \mathbb{Z} \to \mathbb{Z}, f(n) = \begin{cases} n & \text{if } n \geqslant 0 \\ n - 1 & \text{if } n < 0. \end{cases}$

(ix) $f : \mathbb{Z} \to \mathbb{Z}, f(n) = \begin{cases} n & \text{if } n \text{ is even} \\ \dfrac{n+1}{2} & \text{if } n \text{ is odd.} \end{cases}$

(x) $f : \mathbb{Z} \to \mathbb{Z}, f(n) = \begin{cases} n - 1 & \text{if } n \text{ is even} \\ n + 1 & \text{if } n \text{ is odd.} \end{cases}$

3. Each of the following is the graph of a function $A \to B$ (where A and B are subsets of $\mathbb{R}$). Determine whether or not each function is: (a) injective; (b) surjective.

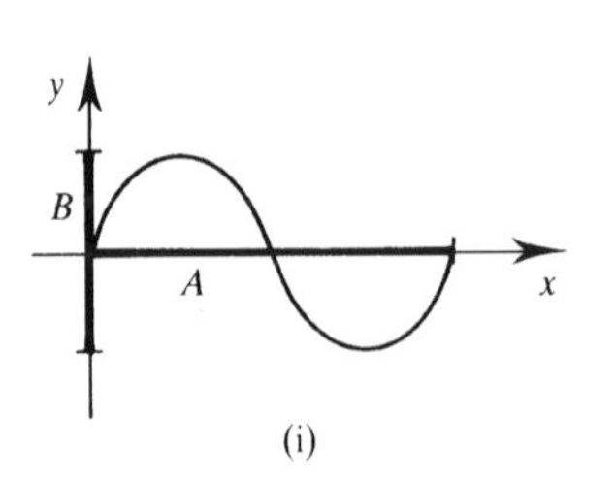

(i)

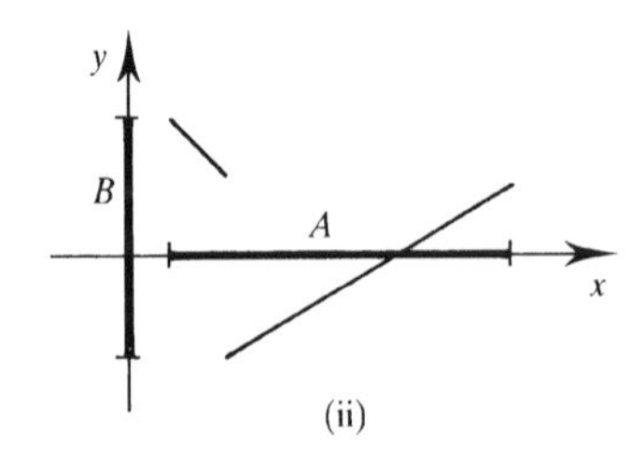

(ii)

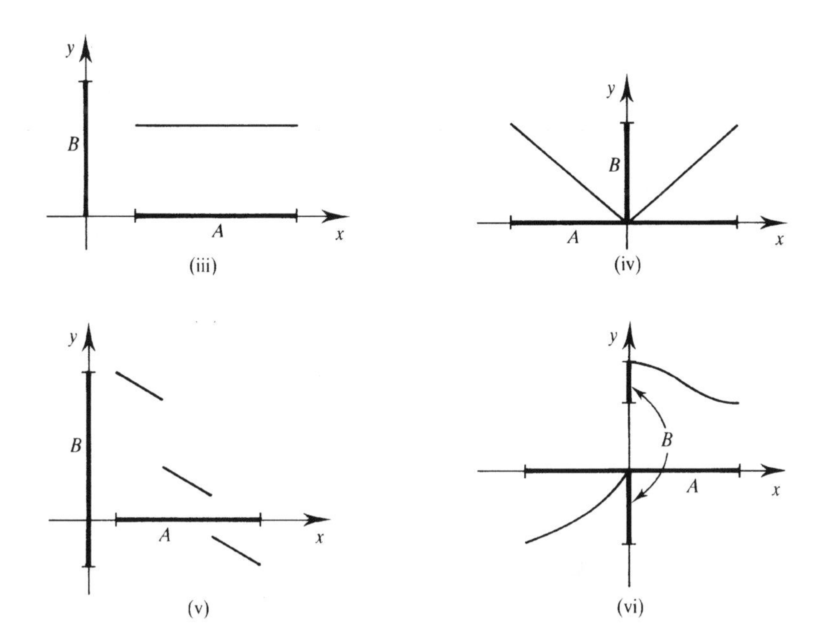

4. Determine whether or not each of the following functions is (a) injective, (b) surjective. Justify your answers.

(i) $A = \{1, 2, 3\}$, $B = \mathscr{P}(A)$; $f : A \to B$, $f(x) = \{x\}$.
(ii) $A =$ any non-empty set, $B = \mathscr{P}(A)$; $f : A \to B$, $f(x) = \{x\}$.
(iii) $A = \{1, 2, 3, 4\}$, $B = \mathscr{P}(A)$; $f : B \to B$, $f(X) = X \cap \{1, 2\}$.
(iv) $A = \{1, 2, 3, 4\}$, $B = \mathscr{P}(A)$; $f : B \to B$, $f(X) = X \cup \{1, 2\}$.
(v) $A = \{1, 2, 3, 4\}$, $B = \mathscr{P}(A)$; $f : B \to B$, $f(X) = A - X$.
(vi) $A = \{1, 2, 3, 4\}$, $B = \mathscr{P}(A)$; $f : B \to A$, $f(X) =$ smallest element in X.

5. Determine whether or not each of the following functions is (a) injective, (b) surjective. Justify your answers.

(i) $A = B = \mathbb{Z}_5 = \{[0], [1], [2], [3], [4]\}$, $f([n]) = [n^2]$.
(ii) $A = B = \mathbb{Z}_5 = \{[0], [1], [2], [3], [4]\}$, $f([n]) = [n^3]$.
(iii) $A = B = \mathbb{Z}_5 = \{[0], [1], [2], [3], [4]\}$, $f([n]) = [2n + 3]$.
(iv) $A = B = \mathbb{Z}_5 = \{[0], [1], [2], [3], [4]\}$, $f([n]) = [5n + 3]$.
(v) $A = B = \mathbb{Z}_6 = \{[0], [1], [2], [3], [4], [5]\}$, $f([n]) = [n^2]$.
(vi) $A = B = \mathbb{Z}_6 = \{[0], [1], [2], [3], [4], [5]\}$, $f([n]) = [n^3]$.
(vii) $A = B = \mathbb{Z}_6 = \{[0], [1], [2], [3], [4], [5]\}$, $f([n]) = [2n + 3]$.
(viii) $A = B = \mathbb{Z}_6 = \{[0], [1], [2], [3], [4], [5]\}$, $f([n]) = [5n + 3]$.

6. Determine whether each of the following real-valued functions is injective, surjective, both or neither.

(i) $f : \mathbb{R} \to \mathbb{R}$, $f(x) = x^2 + 4$.
(ii) $f : \mathbb{R} - \{1\} \to \mathbb{R} - \{1\}$, $f(x) = x/(x-1)$.
(iii) $f : \mathbb{R} \to \mathbb{R}$, $f(x) = 2^x$.
(iv) $f : \mathbb{R} \to \mathbb{R}$, $f(x) = |x|$.
(v) $f : \mathbb{R} \to \mathbb{R}^+ \cup \{0\}$, $f(x) = x + |x|$.
(vi) $f : \mathbb{R}^2 \to \mathbb{R}$, $f(x, y) = xy$.
(vii) $f : \mathbb{R} \to \mathbb{R}^2$, $f(x) = (x, x^2)$.
(viii) $f : \mathbb{R}^2 \to \mathbb{R}^2$, $f(x, y) = (x + y, x - y)$.
(ix) $f : \mathbb{R}^2 \to \mathbb{R}^2$, $f(x, y) = (x + y, x^2 + y^2)$.
(x) $f : \mathbb{R}^2 \to \mathbb{R}^2$, $f(x, y) = (x - y, x^2 - y^2)$.

7. For each of the following functions $f : A \to B$:

(a) Determine what conditions, if any, must be placed on the sets A and/or B to ensure that f is injective.

(b) Determine what conditions, if any, must be placed on the sets A and/or B to ensure that f is surjective.

(i) A is a non-empty set of people, $B = \{n \in \mathbb{Z} : 0 \leqslant n \leqslant 100\}$
$f : A \to B$, $f(p) =$ age last birthday of p.

(ii) A is a non-empty set of cities, B is a non-empty set of countries
$f : A \to B$, $f(X) =$ country containing city X.

(iii) A is a non-empty set of countries, B is a non-empty set of cities
$f : A \to B$, $f(X) =$ city with the largest population (in thousands) in country X.

(iv) $A = \{n \in \mathbb{Z} : a \leqslant n \leqslant b\}$, $B = \{n \in \mathbb{Z} : c \leqslant n \leqslant d\}$
$f : A \to B$, $f(n) = n$.

(v) $A = \{n \in \mathbb{Z} : a \leqslant n \leqslant b\}$, $B = \{n \in \mathbb{Z} : c \leqslant n \leqslant d\}$
$f : A \to B$, $f(n) = n + 10$.

8. Let $f : A \to B$ be a function and let C_1, C_2 be subsets of A. Prove that:

(i) $f(C_1 \cup C_2) = f(C_1) \cup f(C_2)$
(ii) $f(C_1 \cap C_2) \subseteq f(C_1) \cap f(C_2)$
(iii) f is injective if and only if, for all subsets C_1, C_2 of A,
$f(C_1 \cap C_2) = f(C_1) \cap f(C_2)$.

(See exercise 5.1.10 for the definition of $f(C)$ where $C \subseteq A$.)

9. Let $f : A \to B$ be a function. Prove each of the following.

   (i) For all subsets C of A, $C \subseteq f^{-1}(f(C))$.
   (ii) If f is injective then $C = f^{-1}(f(C))$ for all subsets C of A.
   (iii) If $C = f^{-1}(f(C))$ for all subsets C of A then f is injective.
   (iv) If f is surjective then $f(f^{-1}(D)) = D$ for all subsets D of B.

   (See exercise 5.1.11 for the definition of $f^{-1}(D)$ where $D \subseteq B$.)

10. Let A_1 and A_2 be non-empty sets and $f_1 : A_1 \to B_1$ and $f_2 : A_2 \to B_2$ be two functions. Define $F = (f_1 \times f_2) : (A_1 \times A_2) \to (B_1 \times B_2)$ by $F(a_1, a_2) = (f_1(a_1), f_2(a_2))$.

    Prove that:

    (i) F is injective if and only if f_1 and f_2 are both injective;
    (ii) F is surjective if and only if f_1 and f_2 are both surjective.

11. Let A and B be sets and $f : A \to B$ a function. Define $\mathscr{P}_f : \mathscr{P}(A) \to \mathscr{P}(B)$ by $(\mathscr{P}_f)(C) = f(C)$, where $C \subseteq A$.

    Prove that:

    (i) if f is injective then $\mathscr{P}_f$ is injective;
    (ii) if f is surjective then $\mathscr{P}_f$ is surjective.

    Are the converse statements true?

12. Let $f : A \to B$ be a function and let C be a subset of A.

    (i) Show that if f is injective then so, too, is the restriction $f|_C$. (See exercise 5.2.14.)
    (ii) Under what conditions is $f|_C$ surjective?

13. Let $X_1 \times X_2 \times \cdots \times X_n$ be the Cartesian product of non-empty sets $X_1, X_2, \ldots, X_n$.

    (i) Show that, for each $i = 1, 2, \ldots, n$, the natural projection

    $$p_i : X_1 \times X_2 \times \cdots \times X_n \to X_i, \quad (x_1, x_2, \ldots, x_n) \mapsto x_i$$

    is surjective.

    (ii) If one of the sets X_j is empty, is the natural projection still surjective?

(iii) Let $\{j_1, j_2, \ldots, j_m\}$ be a set of positive integers such that $1 \leqslant j_1 < j_2 < \cdots < j_m \leqslant n$. Show that the natural projection

$$X_1 \times X_2 \times \cdots \times X_n \to X_{j_1} \times X_{j_2} \times \cdots \times X_{j_m}$$

defined by

$$(x_1, x_2, \ldots, x_n) \mapsto (x_{j_1}, x_{j_2}, \ldots, x_{j_m})$$

is a surjection.

5.4 Bijections and Inverse Functions

In the previous sections we defined two special kinds of functions: injections and surjections. Functions which are both injective and surjective have interesting and important properties; they are the subjects of this section.

Definition 5.6

A function $f : A \to B$ is **bijective** or is **a bijection** if it is both injective and surjective.

The terms **one-to-one correspondence** and **one-to-one onto function** are also used for 'bijection'.

Examples 5.10

1. In example 5.7.3 we proved that the function $f : \mathbb{R} \to \mathbb{R}$ defined by $f(x) = 3x - 7$ is a bijection.

2. Show that $f : \mathbb{R}^+ \cup \{0\} \to \mathbb{R}^+ \cup \{0\}$, $x \mapsto x^2$, is a bijection.

Solution

If x_1 and x_2 are both non-negative then $x_1^2 = x_2^2$ implies $x_1 = x_2$, so f is injective. If $y \geqslant 0$ then $\sqrt{y}$ is also a non-negative real number and $f(\sqrt{y}) = y$, so f is surjective.

Note that this example underlines again the importance of the domain and codomain in the definition of a function. We have already seen in example 5.7.1 that $f : \mathbb{R} \to \mathbb{R}$, $x \mapsto x^2$, is neither injective nor surjective. Thus the properties of a function depend crucially on its domain and codomain as well as the 'rule of association'. In particular, the statement 'the function $f(x) = x^2$ is bijective' is ambiguous at best and meaningless at worst.

3. Let $\mathbb{E}^+ = \{2n : n \in \mathbb{Z}^+\}$ denote the set of even positive integers and consider the function $f : \mathbb{Z}^+ \to \mathbb{E}^+$ defined by $f(n) = 2n$.

   Now $f(n) = f(n') \Rightarrow 2n = 2n' \Rightarrow n = n'$, so f is injective, and if $m \in \mathbb{E}^+$ then $n = m/2 \in \mathbb{Z}^+$ and $f(n) = m$, so f is surjective. Therefore f is a bijection.

4. Let f be the function $\mathbb{R}^2 \to \mathbb{R}^2$ defined by $f(x, y) = (2x - 3y, x - 2y)$. Show that $f \circ f = id_{\mathbb{R}^2}$ and deduce that f is a bijection.

Solution

Let $(x, y) \in \mathbb{R}^2$. Then

$$\begin{aligned}(f \circ f)(x, y) &= f(2x - 3y, x - 2y) \\ &= (2(2x - 3y) - 3(x - 2y), (2x - 3y) - 2(x - 2y)) \\ &= (x, y).\end{aligned}$$

Therefore $f \circ f = id_{\mathbb{R}^2}$. We can use this property to prove that f is both injective and surjective. Let $(x, y), (x', y') \in \mathbb{R}^2$. Then:

$$\begin{aligned} & f(x, y) = f(x', y') \\ \Rightarrow \quad & f(f(x, y)) = f(f(x', y')) \\ \Rightarrow \quad & (f \circ f)(x, y) = (f \circ f)(x', y') \\ \Rightarrow \quad & (x, y) = (x', y') \end{aligned}$$

so f is injective.

To show that f is surjective, let $(a, b) \in \mathbb{R}^2$ and define $(x, y) = f(a, b)$. Then

$$f(x, y) = f(f(a, b)) = (f \circ f)(a, b) = (a, b)$$

so f is surjective.

The properties of injections and surjections given in theorems 5.2, 5.3 and 5.5 immediately imply the following results.

Theorem 5.6

Let $f : A \to B$ be a function where A and B are subsets of $\mathbb{R}$. Then f is bijective if and only if every horizontal line through a point of B meets the graph of f exactly once.

Theorem 5.7

(i) The composite of two bijections is a bijection.

(ii) If $f : A \to B$ is a bijection, where A and B are finite sets, then $|A| = |B|$.

Note that the converse of (i) is false—if a composite function $g \circ f$ is bijective, it does *not* follow that both f and g need be bijective. A counter-example to this is provided by the functions in example 5.9. Theorem 5.4 gives the most general result in the reverse direction—if $g \circ f$ is a bijection then f is injective and g is surjective.

There is a kind of converse to (ii). If A and B are finite sets with the same cardinality, then there exists a bijection from A to B which can easily be defined as follows. Suppose $|A| = |B| = n$; list the elements of A and B respectively as $\{a_1, \ldots, a_n\}$ and $\{b_1, \ldots, b_n\}$. Then a bijection f is given by $f(a_i) = b_i$ for $i = 1, \ldots, n$. Clearly if $n \geqslant 2$ there is more than one choice of bijection $A \to B$; in fact there are $n! = n(n-1)(n-2)\ldots 2.1$ different such bijections—see exercise 5.4.7.

Theorem 5.7(ii) implies that there is no bijection from a finite set to a proper subset. By contrast, example 5.10.3 gives a bijection from the (infinite) set of positive integers to a proper subset—the (infinite) set $\mathbb{E}^+$ of even positive integers. In fact every infinite set has the property that there exists a bijection from itself to some proper subset. This property can therefore be used to characterize infinite sets without having to refer to numbers of elements: a set A is infinite if and only if there exists a proper subset B and a bijection $A \to B$.

The argument used in example 5.10.4 to prove that f is bijective generalizes to any function $f : A \to A$ such that $f \circ f = id_A$. We state this as a theorem.

Theorem 5.8

Let $f : A \to A$ be a function such that $f \circ f = id_A$. Then f is a bijection.

Proof

Suppose $f \circ f = id_A$. Let $a, b \in A$. Then:

$$\begin{aligned} & f(a) = f(b) \\ \Rightarrow \quad & f(f(a)) = f(f(b)) \\ \Rightarrow \quad & (f \circ f)(a) = (f \circ f)(b) \\ \Rightarrow \quad & a = b \end{aligned}$$

so f is injective.

Let $c \in A$ and define $a = f(c)$. Then

$$f(a) = f(f(c)) = (f \circ f)(c) = c$$

so f is surjective. □

We now turn to the question raised in exercise 5.1.16: given a function $f : A \to B$, under what circumstances does $g = \{(b, a) : (a, b) \in f\}$ define a function? We can think of g as 'reversing the arrows' in the arrow diagram of f: if $b = f(a)$ then $a = g(b)$. See figure 5.15.

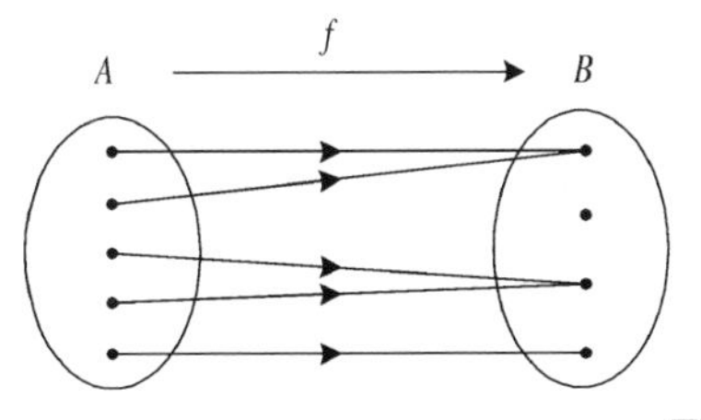

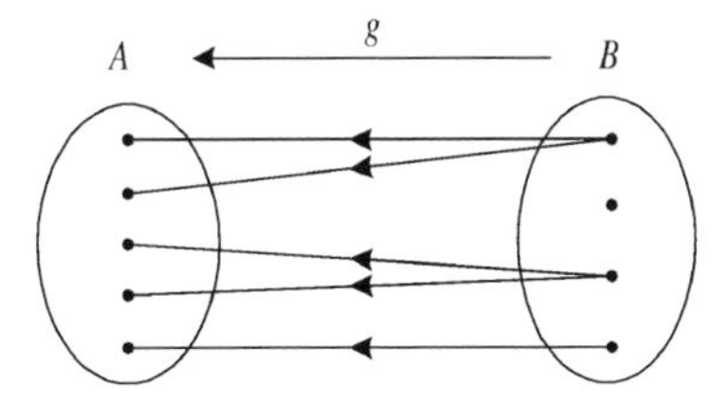

Figure 5.15

Example 5.2.6 indicates that this will not in general define a function. The square function f from $\mathbb{R}$ to $\mathbb{R}$ is formally defined as $\{(x, y) \in \mathbb{R} \times \mathbb{R} : y = x^2\}$. However, the relation $g = \{(y, x) \in \mathbb{R} \times \mathbb{R} : y = x^2\}$ is not a function as we explained in example 5.2.6.

Returning to the general situation, let $f : A \to B$ be a function and define the relation $g = \{(b, a) : (a, b) \in f\}$. Now, according to definition 5.1, g is a function if for each $b \in B$ there exists a unique $a \in A$ such that $(b, a) \in g$ or, equivalently, $(a, b) \in f$. The existence of *some* $a \in A$ with the required property for each element of B is precisely the requirement that f is surjective. Furthermore, an element $a \in A$ such that $(a, b) \in f$ is *unique* if and only if f is injective. This is because the existence of two elements $a, a' \in A$ such that $(a, b), (a', b) \in f$ is equivalent to $f(a) = f(a')$ for two different elements of A. These arguments can be summarized as follows.

Theorem 5.9

Let $f : A \to B$ be a function. The relation $g = \{(b, a) \in B \times A : (a, b) \in f\}$ is a function from B to A if and only if f is a bijection.

Definition 5.7

If $f : A \to B$ is a bijection then the function $g : B \to A$ defined by $g(b) = a$ if and only if $f(a) = b$ is called the **inverse function** of f and is denoted f^{-1}.

Theorem 5.10

Let $f : A \to B$ be a bijection and let $f^{-1} : B \to A$ be its inverse. Then $f^{-1} \circ f = id_A$ and $f \circ f^{-1} = id_B$.

Proof

Let $a \in A$ and suppose $b = f(a)$. Then $a = f^{-1}(b)$, so

$$(f^{-1} \circ f)(a) = f^{-1}(f(a)) = f^{-1}(b) = a.$$

Hence $f^{-1} \circ f = id_A$.

Now let $b' \in B$ and suppose $a' = f^{-1}(b')$. Then $b' = f(a')$, so

$$(f \circ f^{-1})(b') = f(f^{-1}(b')) = f(a') = b'.$$

Hence $f \circ f^{-1} = id_B$. □

Examples 5.11

1. For any set A the identity function on A, id_A, is its own inverse function.

2. We can now define the 'square root' function as the inverse of the bijection $f : \mathbb{R}^+ \cup \{0\} \to \mathbb{R}^+ \cup \{0\}$, $x \mapsto x^2$ defined in example 5.10.2. Since, for non-negative real numbers, $y = x^2$ if and only if $x = \sqrt{y}$, the inverse function is

$$f^{-1} : \mathbb{R}^+ \cup \{0\} \to \mathbb{R}^+ \cup \{0\} \text{ defined by } f^{-1}(y) = \sqrt{y}.$$

3. Let $f : \mathbb{R} \to \mathbb{R}$ be defined by $f(x) = 5x + 8$. Show that f is a bijection and find its inverse.

Solution

If we can find the inverse function f^{-1} then f must be a bijection, by theorem 5.9. To find f^{-1} we simply use its definition: if $y = f(x)$ then $x = f^{-1}(y)$.

Now

$$
\begin{aligned}
& & y &= f(x) \\
&\Rightarrow & y &= 5x + 8 \\
&\Rightarrow & y - 8 &= 5x \\
&\Rightarrow & (y-8)/5 &= x \\
&\Rightarrow & x &= f^{-1}(y) = (y-8)/5.
\end{aligned}
$$

Therefore the inverse function is $f^{-1} : \mathbb{R} \to \mathbb{R}$, $f^{-1}(y) = (y-8)/5$.

4. In example 5.10.4, we showed that $f : \mathbb{R}^2 \to \mathbb{R}^2$, $f(x,y) = (2x - 3y, x - 2y)$ is a bijection. Find its inverse.

Solution

In example 5.10.4 we showed that $f \circ f = id_{\mathbb{R}^2}$. It follows from theorem 5.10 that $f^{-1} = f$, so

$$f^{-1} : \mathbb{R}^2 \to \mathbb{R}^2, \quad f^{-1}(x,y) = (2x - 3y, x - 2y).$$

5. Show that $f : \mathbb{R} - \{1\} \to \mathbb{R} - \{2\}$ defined by

$$f(x) = \frac{2x}{x-1}$$

is bijective and find its inverse.

Solution

Again we show that f is a bijection by finding its inverse. We let $y = f(x)$ and solve for x to find $x = f^{-1}(y)$.

Now

$$\begin{aligned}
& & y &= \frac{2x}{x-1} \\
&\Rightarrow & y(x-1) &= 2x \\
&\Rightarrow & yx - 2x &= y \\
&\Rightarrow & x(y-2) &= y \\
&\Rightarrow & x &= \frac{y}{y-2}.
\end{aligned}$$

Therefore we define a function

$$g : \mathbb{R} - \{2\} \to \mathbb{R} - \{1\}, \quad g(y) = \frac{y}{y-2}.$$

It is a routine matter to check that $(g \circ f)(x) = x$ for all $x \in \mathbb{R} - \{1\}$ and $(f \circ g)(y) = y$ for all $y \in \mathbb{R} - \{2\}$. Therefore f is bijective and $f^{-1} = g$.

Exercises 5.4

1. Determine which (if any) of the following functions $\mathbb{Z} \to \mathbb{Z}$ is bijective and, for each bijection, find its inverse.

   (i) $f : \mathbb{Z} \to \mathbb{Z}$, $f(n) = n - 17$.
   (ii) $f : \mathbb{Z} \to \mathbb{Z}$, $f(n) = 2n + 8$.
   (iii) $f : \mathbb{Z} \to \mathbb{Z}$, $f(n) = (n-1)(n+3)$.
   (iv) $f : \mathbb{Z} \to \mathbb{Z}$, $f(n) = n + 5$.
   (v) $f : \mathbb{Z} \to \mathbb{Z}$, $f(n) = \begin{cases} n-1 & \text{if } n \text{ is even} \\ n+1 & \text{if } n \text{ is odd.} \end{cases}$

2. (i) Let $f : \mathbb{Z}_5 \to \mathbb{Z}_5$, $f([n]) = [3n + 1]$ and $g : \mathbb{Z}_5 \to \mathbb{Z}_5$, $g([n]) = [2n + 3]$. Find $g \circ f$ and $f \circ g$. Deduce that f is a bijection and identify f^{-1}.

   (ii) Let $f : \mathbb{Z}_5 \to \mathbb{Z}_5$ be given by $f([n]) = [4n + 3]$. Find $f \circ f$. What can you deduce about f?

   (iii) Let $f : \mathbb{Z}_5 \to \mathbb{Z}_5$ be given by $f([n]) = [n^3]$. Evaluate $f([0])$, $f([1])$, $f([2])$, $f([3])$ and $f([4])$. Hence determine f^{-1}.

3. Show that each of the following is a bijection and find its inverse.

   (i) $f : \mathbb{R} \to \mathbb{R}$, $f(x) = \dfrac{5x + 3}{8}$.

   (ii) $f : \mathbb{R} - \{-1\} \to \mathbb{R} - \{3\}$, $f(x) = \dfrac{3x}{x + 1}$.

   (iii) $f : [1, 3] \to [-2, 2]$, $f(x) = 2x - 4$.

   (iv) $f : \mathbb{R}^+ \to (0, 1)$, $f(x) = \dfrac{1}{x + 1}$.

   (v) $f : \mathbb{R}^2 \to \mathbb{R}^2$, $f(x, y) = (y, x)$.

   (vi) $f : \mathbb{R}^2 \to \mathbb{R}^2$, $f(x, y) = (2x - 1, 5y + 3)$.

   (vii) $f : \mathbb{R}^2 \to \mathbb{R}^2$, $f(x, y) = (2x - y, x - 2y)$.

   (viii) $f : \mathbb{R} \to \mathbb{R}$, $f(x) = (2x + 3)^3$.

   (ix) $f : \mathbb{Z}^+ \to \mathbb{Z}$, $f(n) = \begin{cases} n/2 & \text{if } n \text{ is even} \\ (1 - n)/2 & \text{if } n \text{ is odd.} \end{cases}$

   (x) $f : \mathbb{Z}^+ \times \{0, 1\} \to \mathbb{Z}$, $f(n, m) = \begin{cases} n - 1 & \text{if } m = 0 \\ -n & \text{if } m = 1. \end{cases}$

4. (i) Let A be any (non-empty) set and let C be a subset of A. A function δ_C is defined by

$$\delta_C : A \to \{0, 1\}, \quad \delta_C(a) = \begin{cases} 0 & \text{if } a \notin C \\ 1 & \text{if } a \in C. \end{cases}$$

   (a) Let $A = \{a, b, c, d, e\}$ and $C = \{b, d, e\}$. Evaluate $\delta_C(x)$ for each $x \in A$.

(b) Under what circumstances is δ_C injective?
(c) Under what circumstances is δ_C surjective?

(ii) Let $A = \mathscr{P}\{a,b\}$ and $B = \{0,1\} \times \{0,1\} = \{(0,0),(0,1),(1,0),(1,1)\}$. A function $f : A \to B$ is defined by $f(C) = (\delta_C(a), \delta_C(b))$. Show that f is a bijection.

(iii) Suppose $|X| = n$. How can the function f defined in part (ii) be generalized to a bijection $\mathscr{P}(X) \to \{0,1\}^n$?

5. For each of the following pairs, A and B, of subsets of $\mathbb{R}$, find an explicit bijection $f : A \to B$.

(i) $A = [0,1]$, $B = [1,3]$.
(ii) $A = (0,1)$, $B = \mathbb{R}^+$.
(iii) $A = \mathbb{R}^+$, $B = \mathbb{R}$.
(iv) $A = (0,1)$, $B = \mathbb{R}$.
(v) $A = \mathbb{Z}$, $B = \mathbb{Z}^+$.

6. Let f and g be functions $\mathbb{R} \to \mathbb{R}$ and $k \in \mathbb{R}$. The functions $f+g$, $f * g$ and $kf : \mathbb{R} \to \mathbb{R}$ are defined respectively by

$$(f+g)(x) = f(x) + g(x)$$
$$(f * g)(x) = f(x)g(x)$$
$$(kf)(x) = kf(x).$$

(i) Prove that, if $k \neq 0$, then kf is a bijection if and only if f is a bijection.

(ii) Find bijections f and g such that neither $f+g$ nor $f * g$ is a bijection.

7. (i) Prove that if $|A| = n$ then there are $n! = n(n-1)(n-2)\dots 2.1$ different bijections $A \to A$.

(ii) Suppose $|A| = n$ and $|B| = m$, where $n \leqslant m$. Show that there are $m!/(m-n)! = m(m-1)(m-2)\dots(m-n+1)$ different injections $A \to B$.

(iii) Counting the number of surjections $A \to B$ in the case where $|A| = n \geqslant m = |B|$ is much harder. The number of such surjections is $S(n,m) \times m!$, where $S(n,m)$ is a so-called 'Stirling number of the second kind'. (Unfortunately, there is no easy formula for $S(n,m)$. Stirling numbers crop up in various counting problems such as this.)

In some special cases, however, the number of surjections $A \to B$ can be identified. Show that:

(a) if $m = 1$, there is only one surjection $A \to B$;
(b) if $m = 2$, there are $2^n - 2$ surjections $A \to B$;
(c) if $m = n-1$, there are $\frac{1}{2}n(n-1) \times m!$ surjections $A \to B$.

8. Let A and B be finite sets with the same number of elements and let $f : A \to B$ be a function. Prove that f is injective if and only if f is surjective.

   (Note: this result is useful if we need to show that a given function $f : A \to B$ is a bijection where $|A| = |B|$. All we are required to do is either show that f is an injection or show that f is a surjection.)

9. Let $f : A \to B$ be a function. Show that f is bijective if and only if $f(A - C) = B - f(C)$, for every subset C of A.

10. (If you are not familiar with the theory of matrices, you should read chapter 6 before attempting this question.)

    Let $X = \{\mathbf{x} : \mathbf{x} \text{ is a } 2 \times 1 \text{ column matrix/vector}\}$. Thus X is just $\mathbb{R}^2$ in another notation.

    Let A be a 2×2 matrix, and define $f : X \to X$ by $f(\mathbf{x}) = A\mathbf{x}$.

    Show that f is a bijection if and only if A is non-singular.

11. If $f : A \to B$ is injective but not surjective then defining the inverse $f^{-1} : B \to A$ by $f^{-1}(b) = a$ if and only if $b = f(a)$ defines a partial function. This is illustrated in the diagrams below.

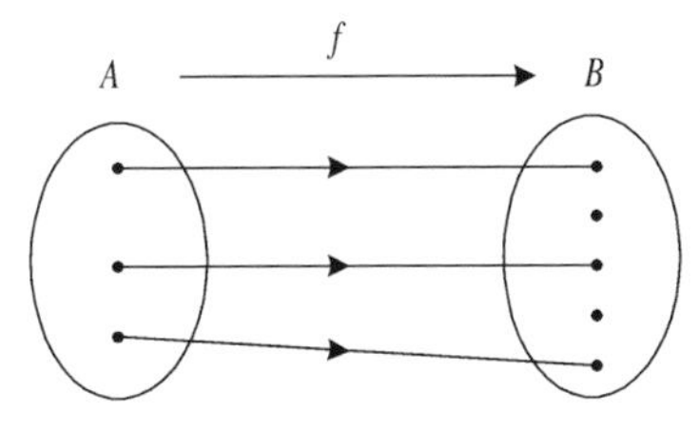

f is injective but not surjective

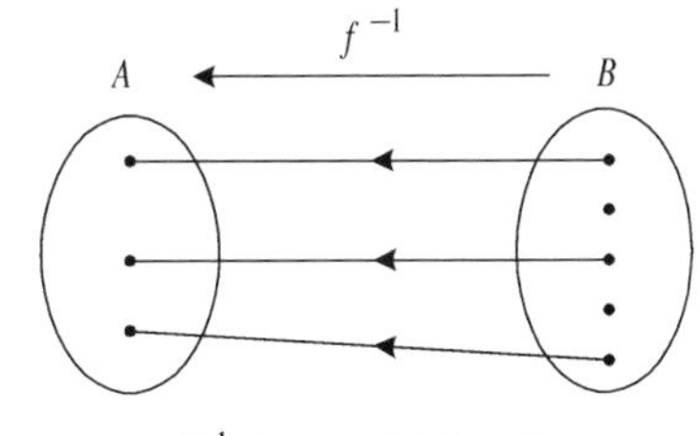

f^{-1} is a partial function

Determine the inverse of each of the following functions. In each case, state whether the inverse is a partial function or a total function; if the inverse function is partial, state its domain.

(i) $f : \{1, 2, 3, 4, 5\} \to \{1, 2, 3, 4, 5, 6, 7, 8\}$, $f : x \mapsto x + 2$.

(ii) $f : \{1, 2, 3, 4, 5\} \to \{0, 1, 2, 3, 4\}$, $f : x \mapsto \dfrac{x^2 - 1}{x + 1}$.

(iii) $f : \mathbb{R} \to \mathbb{R}$, $f : x \mapsto 4x - 8$.

(iv) $f : \mathbb{Z} \to \mathbb{Z}$, $f : x \mapsto 2x + 1$.

(v) $f : \mathbb{Z} \to \mathbb{Z}^+$, $f(n) = \begin{cases} n^2 + 1 & \text{if } n < 0 \\ n^2 & \text{if } n \geqslant 0. \end{cases}$

5.5 More on Cardinality

In this section we introduce briefly a theory of cardinality due to Cantor† which enables infinite sets of different cardinality to be defined. This theory, which is in essence very simple, caused great controversy in the mathematical community when it was introduced by Cantor in the 1870s and 1880s. As the material in this section is not used elsewhere in the book, it may safely be omitted. However, the ideas that Cantor introduced are indeed of great importance to mathematics and we hope you will find them stimulating. As Hallett (1984) says: 'Cantor was the founder of the mathematical theory of the infinite, and so one might with justice call him the founder of modern mathematics'.

With the advertisement over, we now turn to the theory. The starting point is theorem 5.7(ii): if there exists a bijection between finite sets then those sets must have the same cardinality. Since the notion of bijection is a purely set-theoretic one, it does not require the sets involved to be finite. We can therefore use theorem 5.7(ii) to *define* cardinality for infinite sets. More precisely, it defines the notion of two sets (finite or infinite) having the *same* cardinality.

Definition 5.8

Two (finite or infinite) sets A and B are said to have the **same cardinality**, written $|A| = |B|$, if there exists a bijection $f : A \to B$.

† Georg Cantor was born in St Petersburg in 1845, but spent most of his life in Germany. He was the first person to provide a satisfactory theory of the infinite. One of the fiercest critics of Cantor's theory was his former teacher, Leopold Kronecker, who Cantor believed was responsible for his failure to be appointed professor at the University of Berlin. Possibly the attacks of Kronecker and others led to the nervous breakdowns Cantor suffered. Although he also received praise from contemporaries, notably David Hilbert, Cantor was plagued by self-doubt and eventually died in 1918 in a mental institution.

Example 5.10.3 shows that there is a bijection between the sets of positive integers $\mathbb{Z}^+$ and positive even integers $\mathbb{E}^+$; therefore the two sets have the same cardinality. Thus an infinite set may have the same cardinality as a proper subset! In fact, as we remarked in §5.4, this property characterizes infinite sets. Any set which has the same cardinality as $\mathbb{Z}^+$ is said to have cardinality $\aleph_0$. (This is read as 'aleph nought' or 'aleph zero'; the symbol $\aleph$ is the first letter of the Hebrew alphabet). Thus $|A| = \aleph_0$ if there exists a bijection $\mathbb{Z}^+ \to A$.

Any set with cardinality $\aleph_0$ is said to be **countably infinite**. The reason for the terminology is the following. Suppose $|A| = \aleph_0$. Then, by definition, there is a bijection $f : \mathbb{Z}^+ \to A$. If we denote $f(n) \in A$ by a_n, we can regard the bijection f as 'listing' or 'counting' the elements of A as $a_1, a_2, \ldots, a_n, \ldots$. Since this listing or counting of the elements of A is an infinite process, we say that A is countably infinite.

Examples 5.12

1. The set $\mathbb{P}$ of prime numbers is countably infinite. In chapter 2 we presented Euclid's proof that $\mathbb{P}$ is an infinite set. We can define the required bijection $f : \mathbb{Z}^+ \to \mathbb{P}$ as follows. List the elements of $\mathbb{Z}^+$ and $\mathbb{P}$ in increasing order:

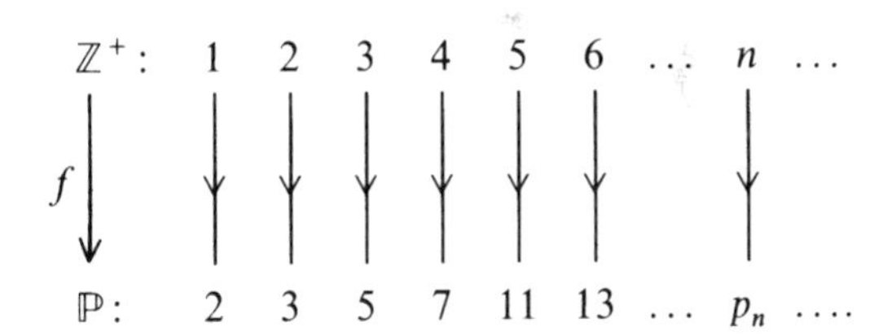

   Now define $f : \mathbb{Z}^+ \to \mathbb{P}$ by $f(n) = p_n$, the nth prime number in the list. Although there is no (known) formula for $f(n)$, we can in principle find $f(n)$ for any positive integer n. Thus our description of f does define a function, which is clearly bijective. Hence $|\mathbb{P}| = \aleph_0$.

2. The set $\mathbb{Q} = \{p/q : p, q \in \mathbb{Z} \text{ and } q \neq 0\}$ of rational numbers also has cardinality $\aleph_0$. (In view of the fact that between any two integers there are infinitely many rationals, this result is, at first sight, rather surprising.)

   We need to define a bijection $\mathbb{Z}^+ \to \mathbb{Q}$; again our definition is descriptive. Firstly, note that we can list the rational numbers in a two-dimensional array as follows. (Ignore the arrows for the present.)

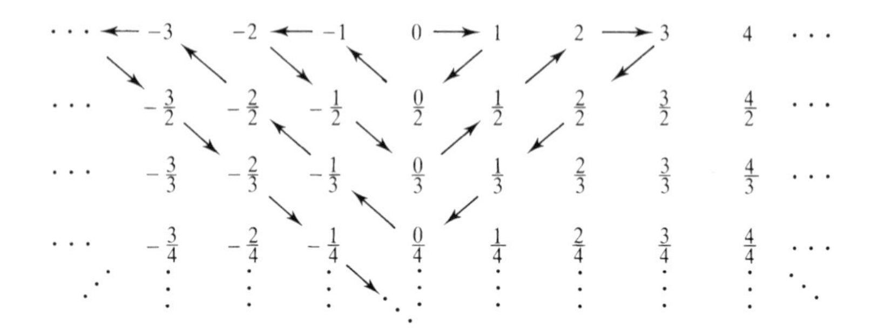

This array clearly includes every rational number, but it also has many repeats. For example $0 = 0/2 = 0/3 = \cdots$; $1/2 = 2/4 = 3/6 = \cdots$; etc.

To define a bijection $f : \mathbb{Z}^+ \to \mathbb{Q}$, begin with $f(1) = 0$ and follow the arrows around the array, ignoring any elements of $\mathbb{Q}$ which have already been encountered. Now define $f(n)$ to be the nth rational obtained in this way.

Thus we have $f(1) = 0$, $f(2) = 1$, $f(3) = -1$, $f(4) = -2$, $f(5) = -1/2$, $f(6) = 1/2$, $f(7) = 2$, $f(8) = 3$, $f(9) = 1/3$, $f(10) = -1/3$, $f(11) = -3$, etc.

As in the previous example, it would be virtually impossible to define a formula for $f(n)$. However, our description does define a function $f : \mathbb{Z}^+ \to \mathbb{Q}$ which is bijective because, in defining f, we skip over those rational numbers which we have already encountered.

Therefore $|\mathbb{Q}| = \aleph_0$.

At this point you may be beginning to think that all infinite sets have cardinality $\aleph_0$. In fact, this is not the case. Using his now-famous 'diagonal argument', Cantor proved that the set of real numbers is not countably infinite.

Theorem 5.11

The set $\mathbb{R}$ is not countably infinite.

Proof

The proof is by contradiction.

We first represent each real number x as a (non-terminating) decimal numeral $x_0.x_1x_2\ldots x_n\ldots$. Here x_0 is an integer and, for $i \geqslant 1$, each x_i is an integer in the range $0 \leqslant x_i \leqslant 9$. If the usual decimal of x terminates we simply add a whole string of zeros to the end.

For example, $37\frac{1}{4} = 37.250\,00\ldots$, $\pi = 3.141\,59\ldots$, and so on. There is a problem with this, however—the decimal numeral for x is not necessarily unique due to the possibility of recurring nines. For instance, $0.5000\ldots = 0.4999\ldots$. If we adopt the convention that the decimal numeral is not to end in recurring nines, then every $x \in \mathbb{R}$ does have a unique decimal expansion.

Now suppose there does exist a bijection $f : \mathbb{Z}^+ \to \mathbb{R}$. Thus we can list the elements of $\mathbb{R}$ in some infinite list:

$$\begin{aligned} f(1) &= a_0.a_1a_2a_3a_4\ldots \\ f(2) &= b_0.b_1b_2b_3b_4\ldots \\ f(3) &= c_0.c_1c_2c_3c_4\ldots \\ \vdots \quad & \quad \vdots \end{aligned}$$

Of course, we soon run out of letters of the alphabet, but that does not alter the principle. To obtain a contradiction, we are going to define a real number x (between 0 and 1) which is not equal to $f(n)$ for any $n \in \mathbb{Z}^+$. The existence of such an x shows that f is not surjective.

In order to define x, we specify its decimal numeral $0.x_1x_2x_3\ldots$ as follows. If the nth decimal place of $f(n)$ is 4, we define $x_n = 3$; if the nth decimal place of $f(n)$ is not equal to 4, we define $x_n = 4$. The choice of 3 and 4 here is more or less arbitrary. What is important is that x_n is defined to be different from the nth decimal place of $f(n)$.

To clarify the definition of x, suppose that the first few elements in out list are:

$$\begin{aligned} f(1) &= 3.\mathbf{2}178\ldots \\ f(2) &= -1.6\mathbf{4}22\ldots \\ f(3) &= 13.01\mathbf{8}7\ldots \\ f(4) &= -0.987\mathbf{6}\ldots. \end{aligned}$$

In this case the decimal numeral of x would begin 0.4344.... (The second decimal place of x has the value 3 because the second decimal place of $f(2)$ is 4.) Note that the nth decimal places of $f(n)$ and x are different in each case.

This defines a real number x. Now x is not equal to $f(n)$, for any $n \in \mathbb{Z}^+$, because it differs from $f(n)$ at least in the nth decimal place. (Of course, x and $f(n)$ will usually differ in many other decimal places as well.) Therefore x does not appear anywhere in the list, which contradicts the surjectivity of f.

Hence there is no bijection $f : \mathbb{Z}^+ \to \mathbb{R}$. □

The cardinality of $\mathbb{R}$ is usually denoted c, which stands for 'continuum'. We now know that there are at least two different infinite cardinalities, $\aleph_0$ and c.

In fact Cantor's diagonal argument, given in the proof of the previous theorem, can be modified to show that, for any set A, the cardinality of A differs from the cardinality of its power set $\mathscr{P}(A)$; see exercise 5.5.2. We are already familiar with this in the case of finite sets, for which $|\mathscr{P}(A)| = 2^{|A|}$ (theorem 3.5).

Knowing this, Cantor was able to determine an infinite sequence of different infinite cardinalities:

$$|\mathbb{Z}^+|,\ |\mathscr{P}(\mathbb{Z}^+)|,\ |\mathscr{P}(\mathscr{P}(\mathbb{Z}^+))|,\ |\mathscr{P}(\mathscr{P}(\mathscr{P}(\mathbb{Z}^+)))|, \ldots .$$

Although it is beyond the scope of the present chapter, it is possible to show that c is the second cardinality in this sequence, in other words, that the sets $\mathbb{R}$ and $\mathscr{P}(\mathbb{Z}^+)$ have the same cardinality. An interesting question is whether there is any cardinality 'in between' $\aleph_0$ and c. To be more precise, extending theorem 5.5 (i) to arbitrary sets, we can define an order relation on cardinalities by

$\alpha \leqslant \beta$ if and only if there exists sets A and B such that $|A| = \alpha$, $|B| = \beta$ and there exists an injection $A \to B$.

Then the question is: does there exist a set S such that

$$\aleph_0 < |S| < c\,?$$

Cantor believed, although he was unable to prove, that the answer to the question is 'no'. This conjecture—that there is no such set S—became known as **Cantor's continuum hypothesis**, and its proof (or disproof) was much sought†. Eventually, in 1938,the Austrian logician Kurt Gödel showed that it is impossible to prove the

† In 1900, David Hilbert, whom many regard as the leading mathematician of his day, addressed the Second International Congress of Mathematicians in Paris. In order to anticipate the future development of mathematics, Hilbert outlined 23 problems whose solutions would make significant progress in the subject. In the first of these problems, Hilbert asked for a resolution of the continuum hypothesis. (Hilbert himself had tried unsuccessfully to prove Cantor's hypothesis.)

Hilbert was one of the earliest champions of Cantor's work on the infinite. At the height of the controversy surrounding Cantor's theory, he wrote: 'No one shall drive us from the paradise which Cantor has created for us'.

continuum hypothesis is *false* using the usual axioms about sets. Much later, in 1963, the American Paul Cohen showed that the continuum hypothesis cannot be proved to be *true* either using the usual axioms about sets. Therefore the truth or falsity of the continuum hypothesis is *undecidable* in axiomatic set theory. One can therefore choose whether to assume its truth or its falsity. More precisely, we can choose to include the continuum hypothesis or its negation as an additional axiom of set theory. This seems somewhat paradoxical. It means that there are two different versions of set theory—one where the continuum hypothesis is 'true' and one where it is 'false'. (In fact, it is now known that there are other statements of this type so there are many different versions of set theory! Fortunately they only differ in rather esoteric aspects—the properties of sets developed in chapter 3 are common to all the different set theories.)

Having defined a hierarchy of infinite cardinalities, Cantor set about defining arithmetic operations for them. That is, Cantor defined addition, multiplication and exponentiation of (infinite) cardinalities. This might sound like an extremely difficult task, but in fact the definitions are very obvious! Recall the following facts about finite sets A and B.

1. If $A \cap B = \varnothing$ then $|A \cup B| = |A| + |B|$. (Counting principle 1, p 94.)
2. $|A \times B| = |A| \times |B|$. (See theorem 3.6.)
3. If $C = \{\text{functions } A \to B\}$ then $|C| = |B|^{|A|}$. (Exercise 5.1.14.)

Cantor simply—and boldly!—used these three facts to *define* addition, multiplication and exponentiation for arbitrary cardinalities.

Definition 5.9

Let A and B be (finite or infinite) sets and let $|A| = \alpha$ and $|B| = \beta$. Then

(i) $\alpha + \beta = |A \cup B|$, provided $A \cap B = \varnothing$;
(ii) $\alpha\beta = |A \times B|$;
(iii) $\beta^\alpha = |C|$, where C is the set of all functions $A \to B$.

We need to check that these definitions make sense. For addition to be well defined, for example, we need to show that $\alpha + \beta$ depends only on α and β and not on the particular sets A and B used in the definition. That is, if $|A| = |A'|$ and $|B| = |B'|$, where $A \cap B$ and $A' \cap B'$ are both empty, then $|A \cup B| = |A' \cup B'|$. Similar remarks apply to the definition of $\alpha\beta$ and β^α. The proofs of these necessary facts are beyond the scope of this book, and we shall have to be content with assuming them.

Provided we accept that these definitions are indeed well defined, some interesting results can be obtained. We conclude the section with a small selection of these.

Examples 5.13

1. $\aleph_0 + \aleph_0 = \aleph_0$.

Proof

We have seen in example 5.10.3 that the set of even positive integers $\mathbb{E}^+$ has cardinality $\aleph_0$. A similar argument shows that $\mathbb{O}^+$, the set of odd positive integers, also has the cardinality $\aleph_0$. Since $\mathbb{E}^+ \cap \mathbb{O}^+ = \varnothing$,

$$\aleph_0 + \aleph_0 = |\mathbb{E}^+ \cup \mathbb{O}^+| = |\mathbb{Z}^+| = \aleph_0.$$

2. $\aleph_0 \aleph_0 = \aleph_0$.

Proof

Using a similar argument to the one given in example 5.12.2 (which showed $|\mathbb{Q}| = \aleph_0$), we can prove that $|\mathbb{Z}^+ \times \mathbb{Z}^+| = \aleph_0$, from which the result follows by definition.

3. $2^{\aleph_0} = \aleph_1$.

Proof

Recall that $\aleph_1 = |\mathscr{P}(\mathbb{Z}^+)|$ and $2^{\aleph_0} = |C|$, where C is the set of functions $\mathbb{Z}^+ \to A$, for some set A with two elements. Since we can choose A to be any set with two elements, we might as well take $A = \{0, 1\}$.

Given $X \subseteq \mathbb{Z}^+$ define a function $f_X : \mathbb{Z}^+ \to \{0, 1\}$ by $f_X(n) = 0$ if $n \notin X$ and $f_X(n) = 1$ if $n \in X$. We can now define a function $F : \mathscr{P}(\mathbb{Z}^+) \to C$ by $F(X) = f_X$.

To prove $2^{\aleph_0} = \aleph_1$, we need to show that F is a bijection. This is most easily done by defining its inverse. So we define $G : C \to \mathscr{P}(\mathbb{Z}^+)$ as follows. If $f : \mathbb{Z}^+ \to \{0, 1\}$ is a function, define a subset $X_f = \{n \in \mathbb{Z}^+ : f(n) = 1\} \subseteq \mathbb{Z}^+$; now define $G(f) = X_f$.

A little thought should convince you that $G = F^{-1}$. Hence F is a bijection, so $2^{\aleph_0} = \aleph_1$.

Exercises 5.5

1. Determine the cardinality of each of the following sets:

   (i) $\{n \in \mathbb{Z} : n \geqslant 10^6\}$
   (ii) $(0, 1) = \{x \in \mathbb{R} : 0 < x < 1\}$
   (iii) $\mathbb{Z}^+ \times \{0, 1\}$
   (iv) $\mathbb{Z}^+ \times \mathbb{Z}^+ \times \mathbb{Z}^+$.

2. Use modification of Cantor's diagonal argument (theorem 5.11) to show that there is no bijection $A \to \mathscr{P}(A)$, for any set A.

   (Hint: suppose that there is a bijection $f : A \to \mathscr{P}(A)$. Then, for every $x \in A$, $f(x)$ is a subset of A; call it A_x. Now consider the subset B of A defined by $B = \{x \in A : x \notin A_x\}$. Show, by contradiction, that B is not the image of any $x \in A$.)

3. Show that, for any set A, $|\mathscr{P}(A)| = 2^{|A|}$. (Hint: see example 5.13.3.)

4. Prove each of the following.

   (i) $\aleph_0 + k = \aleph_0$, for any $k \in \mathbb{Z}^+$.
   (ii) $(\aleph_0)^2 = \aleph_0$. (Note that this is *not* the same as example 5.13.2, which shows $\aleph_0\aleph_0 = \aleph_0$. By definition 5.9(iii), $(\aleph_0)^2$ is the cardinality of the set of functions $\{0, 1\} \to \mathbb{Z}^+$.)

5. Show that $\alpha^2 = \alpha\alpha$ for any cardinality α. (See the remark in exercise 5.5.4(ii) above.)

5.6 Databases: Functional Dependence and Normal Forms

In §4.7 we introduced some of the basic concepts of and operations on relational databases. The purpose of this section is to use some of the ideas of this chapter to develop further the database concepts. In particular we consider the notion of functional dependence and normal forms as applied to relational databases.

Firstly, however, we reconsider from the point of view of functions the operation of projection introduced in §4.7. The following example shows that the projection operation on tables can be viewed as a function.

Example 5.14

Let A_1, A_2, A_3, A_4 be attributes and let R be a table with attribute type (A_1, A_2, A_3, A_4). Recall that this means that $\mathsf{R} \subseteq (X_1 \times X_2 \times X_3 \times X_4)$ where each X_i is the set of data items corresponding to the attribute A_i. The elements $(x_1, x_2, x_3, x_4) \in \mathsf{R}$ are called record instances.

Projection of R onto (A_2, A_4) produces a new table, S say, with attribute type (A_2, A_4) whose record instances consist of just the (A_2, A_4) values of the record instances of R. (Recall that a table is an abstract concept and need not correspond to a file stored on any medium. Thus projection does not produce a new file stored on disk or tape.) The table S is

$$\mathsf{S} = \{(a_2, a_4) : x_2 = a_2 \text{ and } x_4 = a_4 \text{ for some } (x_1, x_2, x_3, x_4) \in \mathsf{R}\}.$$

Recall from exercise 5.3.13 that there is a natural projection function

$$p : X_1 \times X_2 \times X_3 \times X_4 \to X_2 \times X_4, \quad p(x_1, x_2, x_3, x_4) = (x_2, x_4).$$

It should be clear that the table S is simply $p(\mathsf{R})$, the image of R under the natural projection p.

The general situation is conceptually no more difficult than the previous example. However, the general case is more complicated to describe because the notation is necessarily more complex. Let R be a table of attribute type $(A_1, A_2, \ldots, A_n)$. Let $I = \{i_1, i_2, \ldots, i_k\}$ be a set of indices such that $1 \leqslant i_1 < i_2 < \cdots < i_k \leqslant n$. The natural projection function with index set I is the function

$$p_I : X_1 \times X_2 \times \cdots \times X_n \to X_{i_1} \times X_{i_2} \times \cdots \times X_{i_k}$$

defined by $p_I(x_1, x_2, \ldots, x_n) = (x_{i_1}, x_{i_2}, \ldots, x_{i_k})$. Projection of the table R onto attribute type $(A_{i_1}, A_{i_2}, \ldots, A_{i_k})$ defines the new table $\mathsf{S} = p_I(\mathsf{R})$. The record instances of S are the k-tuples $(x_{i_1}, x_{i_2}, \ldots, x_{i_k})$ obtained from the n-tuples of R by deleting those x_j where j does not belong to the index set I. In example 5.14, the index set is $I = \{2, 4\}$.

Functional Dependence

The idea of functional dependence between attributes is of fundamental importance to relational database theory in general and the definition of various normal forms in particular. As the terminology suggests, functional dependence between attributes is closely connected with the concept of a function. Informally, in a table R, an attribute A_j is said to be 'functionally dependent' on an attribute A_i if record instances with the same A_i value but different A_j values never occur. This means that, as far as the table R is concerned, specifying a value of A_i uniquely determines a value of A_j. In this situation we also say that A_i 'functionally determines' A_j.

FUNCTIONAL DEPENDENCE

The precise connection between functional dependence and the concept of a function can be described as follows. Let A_i and A_j be attributes of a table R and let $I = \{i, j\}$. For convenience we suppose $i < j$. Natural projection of R onto attribute set $\{A_i, A_j\}$ produces the relation $\mathsf{S} = p_I(\mathsf{R})$ comprising all ordered pairs (a_i, a_j) such that there is some element $(x_1, \ldots, x_n) \in \mathsf{R}$ with $x_i = a_i$ and $x_j = a_j$. Thus S is a subset of the Cartesian product $X_i \times X_j$. The attribute A_i **functionally determines** the attribute A_j if and only if the relation S is a function $X_i \to X_j$. Briefly, A_i functionally determines A_j (and A_j is **functionally dependent** on A_i) if projection onto $\{A_i, A_j\}$ defines a relation which is a function $X_i \to X_j$.

Example 5.15

Let $X_1 = \{\alpha_1, \alpha_2, \alpha_3\}$, $X_2 = \{\beta_1, \beta_2, \beta_3, \beta_4\}$, $X_3 = \{\gamma_1, \gamma_2\}$ and $X_4 = \{\delta_1, \delta_2, \delta_3, \delta_4, \delta_5\}$. A table R of type $\{A_1, A_2, A_3, A_4\}$ is defined below.

A_1	A_2	A_3	A_4
α_1	β_1	γ_2	δ_1
α_1	β_2	γ_2	δ_3
α_1	β_3	γ_2	δ_5
α_2	β_4	γ_1	δ_1
α_3	β_3	γ_2	δ_2
α_3	β_1	γ_2	δ_4
α_3	β_2	γ_2	δ_4

Projecting R onto the attribute set $\{A_1, A_2\}$ gives the relation

$$\mathsf{S} = \{(\alpha_1, \beta_1), (\alpha_1, \beta_2), (\alpha_1, \beta_3), (\alpha_2, \beta_4), (\alpha_3, \beta_3), (\alpha_3, \beta_1), (\alpha_3, \beta_2)\}.$$

The relation S is not a function $X_1 \to X_2$ since, for example, $\alpha_1 \mathrel{\mathsf{S}} \beta_1$ and $\alpha_1 \mathrel{\mathsf{S}} \beta_2$. Therefore the attribute A_2 is not functionally dependent on attribute A_1. Specifying the value of A_1 to be α_1, for example, does not determine the value of A_2—it could be β_1, β_2 or β_3.

Now project R onto attribute set $\{A_1, A_3\}$. This produces the relation $\mathsf{T} = \{(\alpha_1, \gamma_2), (\alpha_2, \gamma_1), (\alpha_3, \gamma_2)\}$. Now T is a function $X_1 \to X_3$ given in function notation by

$$\alpha_1 \mapsto \gamma_2 \qquad \alpha_2 \mapsto \gamma_1 \qquad \alpha_3 \mapsto \gamma_2.$$

Therefore A_3 is functionally dependent on A_1.

We leave it as an exercise (5.6.1) to determine all of the remaining functional dependences (if any) in R.

The notion of functional dependence readily generalizes to sets of attributes but again is notationally more complicated to describe. Informally, a set of attributes $\{A_{j_1}, \ldots, A_{j_m}\}$ is functionally dependent on a set of attributes $\{A_{i_1}, \ldots, A_{i_k}\}$ if specifying the values of each of the attributes $A_{i_1}, \ldots, A_{i_k}$ uniquely determines the values of each of the attributes $A_{j_1}, \ldots, A_{j_m}$.

For the relation R defined in example 5.15, it is not too difficult to see that $\{A_1, A_2\}$ functionally determines $\{A_4\}$. Projecting R onto $\{A_1, A_2, A_4\}$ produces the relation

$$\{(\alpha_1, \beta_1, \delta_1), (\alpha_1, \beta_2, \delta_3), (\alpha_1, \beta_3, \delta_5), (\alpha_2, \beta_4, \delta_1), (\alpha_3, \beta_3, \delta_2), (\alpha_3, \beta_1, \delta_4), (\alpha_3, \beta_2, \delta_4)\}.$$

Therefore, specifying the values of A_1 *and* A_2 determines the value of A_4. We may regard this relation as defining a function

$$\{(\alpha_1,\beta_1),(\alpha_1,\beta_2),(\alpha_1,\beta_3),(\alpha_2,\beta_4),(\alpha_3,\beta_3),(\alpha_3,\beta_1),(\alpha_3,\beta_2)\} \to X_4$$

given by

$$(\alpha_1,\beta_1) \mapsto \delta_1 \qquad (\alpha_1,\beta_2) \mapsto \delta_3 \qquad (\alpha_1,\beta_3) \mapsto \delta_5 \qquad (\alpha_2,\beta_4) \mapsto \delta_1$$
$$(\alpha_3,\beta_3) \mapsto \delta_2 \qquad (\alpha_3,\beta_1) \mapsto \delta_4 \qquad (\alpha_3,\beta_2) \mapsto \delta_4.$$

Note that the domain of this function,

$$\{(\alpha_1,\beta_1),(\alpha_1,\beta_2),(\alpha_1,\beta_3),(\alpha_2,\beta_4),(\alpha_3,\beta_3),(\alpha_3,\beta_1),(\alpha_3,\beta_2)\},$$

is the projection of R onto $\{A_1, A_2\}$ and the codomain X_4 is the projection of R onto $\{A_4\}$.

To give a formal definition of functional dependence between sets of attributes, let $I = \{i_1, i_2, \ldots, i_k\}$ and $J = \{j_1, j_2, \ldots, j_m\}$ be *disjoint* sets of indices. For convenience we suppose $i_1 < i_2 < \cdots < i_k < j_1 < j_2 < \cdots < j_m$. Then $I \cup J = \{i_1, \ldots, i_k, j_1, \ldots, j_m\}$ where the indices are written in increasing order. Let $\mathsf{S}_{I\cup J} = p_{I\cup J}(\mathsf{R})$ be the projection of R onto the set of attributes $\boldsymbol{A_{I\cup J}} = \{A_{i_1}, \ldots, A_{i_k}, A_{j_1}, \ldots, A_{j_m}\}$. Then $\boldsymbol{A_J} = \{A_{j_1}, \ldots, A_{j_m}\}$ is **functionally dependent** on $\boldsymbol{A_I} = \{A_{i_1}, \ldots, A_{i_k}\}$ if the set $\mathsf{S}_{I\cup J}$ is a function from $\mathsf{S}_I = p_I(\mathsf{R})$ to $\mathsf{S}_J = p_J(\mathsf{R})$. Again we also say that $\boldsymbol{A_I}$ **functionally determines** $\boldsymbol{A_J}$ in this situation. In the example above $I = \{1, 2\}$ and $J = \{4\}$.

What this formal definition means is that for every k-tuple $(a_{i_1}, \ldots, a_{i_k}) \in \mathsf{S}_I = p_I(\mathsf{R})$ there exists a unique m-tuple $(a_{j_1}, \ldots, a_{j_m}) \in \mathsf{S}_J = p_J(\mathsf{R})$ such that the $(k+m)$-tuple $(a_{i_1}, \ldots, a_{i_k}, a_{j_1}, \ldots, a_{j_m})$ belongs to $\mathsf{S}_{I\cup J}$. In other words, within the table R, the values of the attributes $A_{i_1}, \ldots, A_{i_k}$ uniquely determine the values of the attributes $A_{j_1}, \ldots, A_{j_m}$.

The notation needed to describe this general situation may obscure the general concept being described. Hopefully, a further example will help to clarify the situation.

Example 5.16

A relation R of type $(A_1, A_2, A_3, A_4, A_5)$ is defined by the following table.

A_1	A_2	A_3	A_4	A_5
α_1	β_1	γ_1	δ_1	ε_1
α_1	β_1	γ_1	δ_2	ε_1
α_1	β_2	γ_2	δ_1	ε_1
α_1	β_3	γ_1	δ_2	ε_2
α_1	β_3	γ_1	δ_1	ε_2
α_2	β_1	γ_3	δ_1	ε_3
α_3	β_1	γ_2	δ_1	ε_1
α_3	β_1	γ_2	δ_2	ε_1
α_3	β_4	γ_4	δ_2	ε_3

It is not too difficult to see that no single attribute is functionally dependent on any other single attribute. For instance, A_2 is not functionally dependent on A_1 since in rows 2 and 3 there are record instances with the same A_1 value (namely α_1) but different A_2 values (β_1 and β_2 respectively). From the point of view of functions, the projection of R onto $\{A_1, A_2\}$ is the relation

$$\mathsf{S} = \{(\alpha_1, \beta_1), (\alpha_1, \beta_2), (\alpha_1, \beta_3), (\alpha_2, \beta_1), (\alpha_3, \beta_1), (\alpha_3, \beta_4)\}.$$

The set S is not a function from $X_1 = \{\alpha_1, \alpha_2, \alpha_3\}$ to $X_2 = \{\beta_1, \beta_2, \beta_3, \beta_4\}$ because there are ordered pairs with the same first element but different second elements.

Using similar arguments, it is not too difficult to see that no single attribute functionally determines any other single attribute. An equivalent way of viewing this is to say that no projection of R onto a *pair* of attributes produces a function.

However, the pair of attributes $\{A_1, A_2\}$ functionally determines the pair of attributes $\{A_3, A_5\}$. (This implies that $\{A_1, A_2\}$ functionally determines $\{A_3\}$ and also that $\{A_1, A_2\}$ functionally determines $\{A_5\}$. Why?) To see that $\{A_1, A_2\}$ functionally determines $\{A_3, A_5\}$, consider the projection of R onto $\{A_1, A_2, A_3, A_5\}$ which gives the following relation.

A_1	A_2	A_3	A_5
α_1	β_1	γ_1	ε_1
α_1	β_2	γ_2	ε_1
α_1	β_3	γ_1	ε_2
α_2	β_1	γ_3	ε_3
α_3	β_1	γ_2	ε_1
α_3	β_4	γ_4	ε_3

If we let p_{ij} denote the projection of R onto $\{A_i, A_j\}$ then the set defined by this table defines a function from

$$p_{12}(\mathsf{R}) = \{(\alpha_1, \beta_1), (\alpha_1, \beta_2), (\alpha_1, \beta_3), (\alpha_2, \beta_1), (\alpha_3, \beta_1), (\alpha_3, \beta_4)\}$$

to

$$p_{35}(\mathsf{R}) = \{(\gamma_1, \varepsilon_1), (\gamma_2, \varepsilon_1), (\gamma_1, \varepsilon_2), (\gamma_3, \varepsilon_3), (\gamma_4, \varepsilon_3)\}.$$

In the usual notation for a function, we can write this function $p_{12}(\mathsf{R}) \to p_{35}(\mathsf{R})$ as follows:

$$\begin{array}{lll} (\alpha_1, \beta_1) \mapsto (\gamma_1, \varepsilon_1) & (\alpha_1, \beta_2) \mapsto (\gamma_2, \varepsilon_1) & (\alpha_1, \beta_3) \mapsto (\gamma_1, \varepsilon_2) \\ (\alpha_2, \beta_1) \mapsto (\gamma_3, \varepsilon_3) & (\alpha_3, \beta_1) \mapsto (\gamma_2, \varepsilon_1) & (\alpha_3, \beta_4) \mapsto (\gamma_4, \varepsilon_3). \end{array}$$

Hence $\{A_1, A_2\}$ functionally determines $\{A_3, A_5\}$. Note that this function is not bijective. This means that $\{A_3, A_5\}$ does *not* functionally determine $\{A_1, A_2\}$.

Note also that projection onto $\{A_1, A_2, A_3, A_5\}$ does *not* define a function between the 'complete' Cartesian products $X_1 \times X_2 \to X_3 \times X_5$. This is because there exist ordered pairs—(α_2, β_2), for example—which belong to $X_1 \times X_2$ but do not belong to $p_{12}(\mathsf{R})$. In the terminology of exercise 5.1.12, projection onto $\{A_1, A_2, A_3, A_5\}$ defines a partial function $X_1 \times X_2 \to X_3 \times X_5$.

The notion of a key can also be defined quite succinctly using functional dependence. Recall that a candidate key is a set of attributes whose values uniquely specify a record instance but no proper subset of the candidate key has this property. For this to be the case, specifying the values of the attributes in the candidate key uniquely determines the values of the attributes not in the candidate key. In other words a candidate key is a set of attributes which functionally determines each attribute of the table but no proper subset of the candidate key functionally determines each attribute. It will be useful to distinguish between those attributes in a table R which belong to some candidate key and those which do not. An attribute of R which does belong to a candidate key is called **prime** and an attribute which belongs to no candidate key is called **non-prime**.

Normal Forms

The various normal forms for tables are designed to avoid problems of redundancy and inconsistency in the data. A given table will almost inevitably be updated during its lifetime in the database, usually many times. Record instances will be modified or deleted and new record instances will be added to the table. In order to be able to update tables as appropriate and to prevent this causing anomalies in the data, it is necessary to avoid certain kinds of functional dependence between the various attributes of the tables.

To understand what kind of updating anomalies can occur consider the following example.

Example 5.17

A company, which manufactures components, stores the data relating to its prices, customers and the customers' orders in a single table of attribute type (ORDER_#, PART_#, CUSTOMER_NAME, CUSTOMER_ADDRESS, PART_PRICE, QUANTITY, DATE). Whenever an order arrives at the company, it is assigned an order number and the information it contains is entered as one or more record instances. We suppose that a customer may order several different components at a time so that a single order number may have associated with it several different part numbers. Given the attribute type of the company's table, this means that a given customer order may be recorded as several different record instances in the table, one for each part ordered. Thus {ORDER_#} alone would not be a candidate key for the table, but the pair {ORDER_#, PART_#} would be a candidate key. We suppose that {ORDER_#, PART_#} has been chosen as the primary key for the table. This is indicated by printing the key attributes in bold type, so that we denote the type of this record file by (**ORDER_#**, **PART_#**, CUSTOMER_NAME, CUSTOMER_ADDRESS, PART_PRICE, QUANTITY, DATE). In fact, {ORDER_#, PART_#} is the only candidate key for this record file so there is no choice to be made.

It should be clear that the company is not being very sensible in recording data referring to its products, its customers and the orders which it receives in a single table. There are several problems with tables which contain data referring to different kinds of entities. (We are not suggesting that this scenario is realistic. It merely serves to highlight some of the problems which can arise and which the normal forms are designed to eliminate.)

The first problem is the considerable amount of duplication of the data being stored. For instance the component prices and the customers' names and addresses will be recorded many times in different record instances which is clearly wasteful of the storage medium and causes updating problems. When a customer's address or the price of a particular part changes, the company has a choice of whether or not to update all the relevant record instances. Clearly such updating may involve a considerable amount of work as these details need to be changed in many different record instances. If, however, the company chooses not to update, then an inconsistency will result with, for example, the same component having different prices in different record instances or the same customer having different addresses in different record instances. Again there are clearly problems associated with this choice. What would result if, for instance, the company wished to produce a list of its customers' names and addresses? Projection onto {CUSTOMER_NAME, CUSTOMER_ADDRESS} would produce a table where the same customer appears with more than one address. Similarly, projection onto {PART_#, PART_PRICE} would produce a less than helpful table.

There are other perhaps less obvious, but no less serious, problems with the linking of customer-related and product-related information in a single table. If the company produces a new product, it has no means of including information in the table about the part number or price for the new item until one of its customers places an order for the particular component. Similarly, if for whatever reason the information concerning a particular order is deleted from the table, then so too are the data relating to a part number and price. If the record instance being deleted is the last to contain information about a particular component, then that information will be lost completely from the table.

As we have said, it is precisely to avoid the kinds of updating problems indicated in example 5.17 that the different normal forms for tables have been introduced. Since any table in a relational database is, we suppose, in first normal form (see §4.7), we begin by defining the second normal form.

Definition 5.10

A table R is in **second normal form** if no non-prime attribute is functionally dependent on a proper subset of any candidate key.

We should note that this definition is not universally accepted. Some authors define normal forms in terms of the chosen primary key rather than in terms of all possible candidate keys. Thus they define a prime attribute to be one which appears in the primary key and a table to be in second normal form if no non-prime attribute is functionally dependent on a proper subset of the primary key.

The definition of a candidate key says that no proper subset of it functionally determines *every* attribute. However, it may happen that a proper subset of a key functionally determines *some* attribute and it is precisely this situation that the second normal form rules out. Of course, a table is automatically in second normal form if the only candidate keys are single attributes. We can think of the second normal form as eliminating 'partial dependences'. Every non-prime

SECOND NORMAL FORM

attribute must be functionally dependent on a complete candidate key and not any 'part' of it.

Example 5.18

Consider the table introduced in example 5.17. This is *not* in second normal form. Recall that the only candidate key is {ORDER_#, PART_#}. The non-prime attribute PART_PRICE is functionally dependent on a proper subset of the key, namely {PART_#}. (We are assuming that the part price does not vary from order to order or from customer to customer.) Similarly the attributes CUSTOMER_NAME and CUSTOMER_ADDRESS are functionally dependent on the proper subset {ORDER_#} of the key.

In order to 'normalize' we need to split the table into more than one table. (**Normalizing** a table means replacing it with tables which are in the appropriate normal form.) One way of obtaining tables in second normal form is to divide the original table into three new tables of attribute types (**ORDER_#**, CUSTOMER_NAME, CUSTOMER_ADDRESS, DATE), (**ORDER_#**, **PART_#**, QUANTITY) and (**PART_#**, PART_PRICE) respectively. The primary key for each table is indicated in bold type. (Again in each case there is, in fact, only one candidate key.) It should be clear that each of these tables is in second normal form. The first and third have single-attribute keys and must therefore be in second normal form. For the second table, the quantity ordered clearly depends on *both* the order number and the part number so it, too, is in second normal form.

One of the problems indicated in example 5.17 has not been resolved in example 5.18 by splitting the table into three. Since the customer-related information is still linked with order number in the first of the new tables, a particular customer's name and address will be duplicated many times. If a customer changes address, this information will have to be changed in many different record instances. The problem here is essentially that there remains what is often called a 'transitive dependence' in the table. The attribute ORDER_# functionally determines CUSTOMER_NAME and the attribute CUSTOMER_NAME functionally determines CUSTOMER_ADDRESS (assuming that each customer has only one address.) The third normal form is designed to eliminate such 'hidden' or 'transitive' dependences as these.

Definition 5.11

A table R is in **third normal form** if

(i) it is in second normal form, and
(ii) whenever a non-prime attribute is functionally dependent on a set of attributes, the set of attributes contains a candidate key as a subset.

THIRD NORMAL FORM

Suppose $\{A_i : i \in I\}$ is a set of attributes which does not contain a candidate key. For a table in third normal form the set $\{A_i : i \in I\}$ does not functionally determine any attribute A_j. (Note that we suppose here that $A_j \notin \{A_i : i \in I\}$—the definition of functional dependence applies to disjoint sets of attributes.) If $\{A_i : i \in I\}$ does functionally determine A_j then the table contains the following 'transitive dependence': any key K functionally determines $\{A_i : i \in I\}$ (since K functionally determines *every* attribute) and $\{A_i : i \in I\}$ functionally determines A_j (and, of course, K functionally determines A_j). Thus the third normal form rules out this kind of 'transitive dependence'.

It should be noted, however, that it is only this kind of transitive dependence which is ruled out by the third normal form. A table in third normal form can contain some transitive dependence if it has more then one candidate key. Suppose K_1 and K_2 are candidate keys for a table R and suppose A_j is an attribute which does not belong to $K_1 \cup K_2$. Then K_1 functionally determines K_2 (since K_1 is a key), K_2 functionally determines A_j (since K_2 is a key) and, of course, K_1 functionally determines A_j. Writing $X \dashrightarrow Y$ to stand for 'X functionally determines Y' we have:

$$K_1 \dashrightarrow K_2 \dashrightarrow A_j \quad \text{and} \quad K_1 \dashrightarrow A_j.$$

Example 5.19

Consider the table above with attribute type (**ORDER_#**, CUSTOMER_NAME, CUSTOMER_ADDRESS, DATE) and primary key {ORDER_#} introduced in example 5.18. (Recall that {ORDER_#} is the unique candidate key.) This is not in third normal form. The attribute CUSTOMER_ADDRESS is functionally dependent on {CUSTOMER_NAME} which does not contain the key as a subset. This table could be 'normalized' into third normal form by splitting it into two tables, one of attribute type (**ORDER_#**, CUSTOMER_NAME, DATE) and the other of attribute type (**CUSTOMER_NAME**, CUSTOMER_ADDRESS) with the keys indicated in bold type as usual.

The remaining two tables in example 5.18—those of attribute types (**ORDER_#**, **PART_#**, QUANTITY) and (**PART_#**, PART_PRICE) respectively—are in third normal form. Thus to normalize into third normal form the original table of attribute type (**ORDER_#**, **PART_#**, CUSTOMER_NAME, CUSTOMER_ADDRESS, PART_PRICE, QUANTITY, DATE) defined in example 5.17, we could split it into four tables with the following attribute types:

(**ORDER_#**, CUSTOMER_NAME, DATE)

(**CUSTOMER_NAME**, CUSTOMER_ADDRESS)

(**ORDER_#**, **PART_#**, QUANTITY)

(**PART_#**, PART_PRICE).

Each of these tables is in third normal form. They suffer from none of the updating problems mentioned in example 5.17. Using the operations, such as selection, projection, and natural join, introduced in §4.7, we can still create tables with attribute sets such as (ORDER_#, CUSTOMER_NAME, CUSTOMER_ADDRESS, DATE) or (ORDER_#, CUSTOMER_NAME, PART_#, PART_PRICE, QUANTITY).

We should mention in conclusion that this is not the end of the story as far as normal forms are concerned. It is generally regarded as a 'rule of thumb', but not a rigidly applied principle, that a database designer should aim for a collection of tables which are in third normal form. There are, however, 'higher' normal forms—Boyce–Codd normal form and the fourth and fifth normal forms. Boyce–Codd normal form is similar to but slightly more restrictive than the third normal form. The fourth and fifth normal forms are designed to eliminate certain kinds of so-called 'multidependences' amongst the attributes. A consideration of these normal forms (and when and why a database designer should aim for 'higher normalization') is beyond the scope of our brief excursion into relational database theory. The interested reader should consult a more specialized book. (A selection of possible titles for further reading is given in the list of references.)

Exercises 5.6

1. This question refers to the table defined in example 5.15.

   (i) Determine the functional dependences between single attributes.
   (ii) Find all the candidate keys for the table.
   (iii) Is the table in (a) second, or (b) third normal form?

2. This question refers to the table R defined in example 5.16.

   (i) Show that $K = \{A_1, A_2, A_4\}$ is a candidate key for R.
   (ii) Is R in (a) second, or (b) third normal form?

3. This question refers to the tables PERSONAL, DISCIPLINE and CURRENT_COURSE introduced in exercises 4.7 (but *not* the parts of the tables displayed there). In each case $\{A_1\} = \{\text{ID\_NUMBER}\}$ is the only candidate key to the table.

   (i) Assuming that no two students have exactly the same name, are these tables in third normal form?

(ii) If two students do have exactly the same name, are these tables in third normal form?

4. Determine whether each of the following tables is in third normal form. You may assume that there is a unique candidate key in each case. If the table is not in third normal form, split the data into two or more tables which are in third normal form. (Where appropriate, state any assumptions you need to make about various functional dependences between attributes or sets of attributes.)

   (i) A table stores employee-related information and has attribute type {**EMPLOYEE_#**, EMPLOYEE_NAME, DEPARTMENT_#, JOB_DESCRIPTION, WORK_LOCATION}. You may assume that each employee works for only one department and each department is located in only one place.

   (ii) A travel agency records information about its customers' flight bookings in a table of attribute type {**PASSENGER_NAME**, **FLIGHT_#**, AIRLINE, DATE, EMBARKATION, DESTINATION, CLASS}.

   (iii) A hospital uses three tables to maintain its patient-related information. The name and attribute type of each table is given below.

   PATIENT_HISTORY:
   {**PATIENT_#**, PATIENT_NAME, **ADMISSION_DATE**, DISCHARGE _DATE, CONDITION}.

   PATIENT_CURRENT:
   {**PATIENT_#**, PATIENT_NAME, ADMISSION_DATE, CONDITION, CONSULTANT_#, CONSULTANT_NAME, CONSULTANT_PHONE, WARD_#}.

   TREATMENT_CURRENT:
   {**PATIENT_#**, **DRUG**, QUANTITY, DAILY_COST}.

5. Recast the definition of functional dependence given on page 281 in terms of partial functions. (See exercise 5.1.12 for the definition of a partial function.)

Chapter 6

Matrix Algebra

6.1 Introduction

It is often convenient to present certain kinds of data in tabular form. For example, the following table shows, in a concise way, the percentage marks obtained by five students (denoted by A–E) in three examinations.

	Examination		
Student	Discrete maths	Data structures	Operating systems
A	72	68	60
B	35	48	42
C	61	60	76
D	84	82	90
E	53	62	51

Here, the data values (the examination marks) are positioned in the table according to two variable factors: student and examination subject.

A two-dimensional rectangular array of numbers, such as that in the table above, is called a **matrix** (plural—**matrices**) and the numbers which constitute the matrix are called its **elements** (or **entries**). A matrix may have elements which are other than numbers, for example sets or variables. We define the **dimension** (or **order**) of a matrix to be the number of rows and columns (in that order) which it contains.

Thus a matrix having m rows and n columns is said to have dimension $m \times n$ (read as 'm by n'). Since it has five rows and three columns, the matrix above has dimension 5×3.

Whilst a matrix may be presented as a table with row and column headings, it may also be shown as a rectangular array of elements enclosed in brackets. In this form, the matrix above would be presented as:

$$\begin{pmatrix} 72 & 68 & 60 \\ 35 & 48 & 42 \\ 61 & 60 & 76 \\ 84 & 82 & 90 \\ 53 & 62 & 51 \end{pmatrix}.$$

Example 6.1

State the dimension of each of the following matrices.

(i) $\begin{pmatrix} 2 & 4 \\ -1 & 5 \end{pmatrix}$

(ii) $\begin{pmatrix} -2 & 14 & 1 \\ 0 & 3 & -2 \end{pmatrix}$

(iii) $\begin{pmatrix} 3 & 1 & 4 & 2 & -9 \end{pmatrix}$

(iv) $\begin{pmatrix} 1 \\ 5 \\ 7 \end{pmatrix}$.

Solution

(i) This matrix has two rows and two columns. Its dimension is therefore 2×2.

(ii) The dimension of this matrix 2×3.

(iii) This is a 1×5 matrix.

(iv) This matrix has dimension 3×1.

It is conventional to refer to a matrix using an upper-case letter. For instance:

$$A = \begin{pmatrix} 2 & 1 \\ -4 & 3 \\ 7 & 0 \end{pmatrix} \qquad B = \begin{pmatrix} 1 & 0 \\ 3 & -4 \end{pmatrix}.$$

The elements within the matrix are usually denoted by the corresponding lowercase letter to which two subscripts are attached. The first identifies the row within which the element lies and the second its column. For example, a_{31} denotes the element in matrix A lying in the third row and first column. For the matrix A above, $a_{31} = 7$. Similarly $a_{22} = 3$, $a_{12} = 1$, etc. For the matrix B, $b_{12} = 0$, $b_{22} = -4$, and so on.

In general, a_{ij} denotes the element in matrix A occupying the ith row and jth column. This element is often referred to as the $\boldsymbol{(i, j)}$**-entry** or the $\boldsymbol{(i, j)}$**-element** of A. An alternative notation is to denote the matrix A by $[a_{ij}]$ where $A = [a_{ij}]$ refers to the matrix A whose (i, j)-entry is a_{ij}. Note the important distinction between $[a_{ij}]$ and a_{ij}. The former refers to the whole matrix whereas the latter, for different values of i and j, denotes the individual elements within the matrix.

Equal Matrices

Matrices are **equal** if they are identical in every respect—that is, they have the same dimension and the same elements in the same positions. Put more formally: let $A = [a_{ij}]$ and $B = [b_{ij}]$; then $A = B$ if and only if A and B have the same dimension and $a_{ij} = b_{ij}$ for all values of i and j.

Example 6.2

Find all the values of a, b, c and d if

$$\begin{pmatrix} 3a & 2b \\ c & -d \end{pmatrix} = \begin{pmatrix} 10 & 6 \\ -4 & 1 \end{pmatrix}.$$

Solution

Since the condition that the matrices have the same dimension is satisfied (both are 2×2 matrices), all we must ensure is that corresponding elements are equal. Therefore

$$3a = 10 \qquad 2b = 6 \qquad c = -4 \qquad -d = 1.$$

Solving these simple equations gives

$$a = \tfrac{10}{3} \qquad b = 3 \qquad c = -4 \qquad d = -1.$$

6.2 Some Special Matrices

It is convenient to distinguish several different 'families' of matrices. A list of these together with their defining characteristics is given below.

A **square matrix** is one having the same number of rows as it has columns. The following are examples of square matrices:

$$\begin{pmatrix} 2 & 3 \\ -1 & 4 \end{pmatrix} \quad \begin{pmatrix} 3 & 2 & 0 \\ 4 & 5 & -1 \\ 2 & -3 & 7 \end{pmatrix}.$$

A **column matrix** (or **column vector**) is a matrix having only one column. Examples are:

$$\begin{pmatrix} 2 \\ 1 \end{pmatrix} \quad \begin{pmatrix} -1 \\ 4 \\ 2 \\ 10 \end{pmatrix}.$$

A **row matrix** (or **row vector**) is a matrix having only one row, for example:

$$\begin{pmatrix} 7 & 1 & -2 \end{pmatrix} \qquad \begin{pmatrix} 3 & 2 & 1 & 0 & -1 \end{pmatrix}.$$

Row and column vectors are often (but not always) denoted by lower-case bold letters. In handwriting, lower case letters with a line or tilde underneath (the printer's notation for bold print) are used. Thus we may write:

$$\boldsymbol{u} = \begin{pmatrix} 2 & 4 & 3 & -6 \end{pmatrix} \quad \text{or} \quad \underset{\sim}{u} = \begin{pmatrix} 2 & 4 & 3 & -6 \end{pmatrix}.$$

A row vector is sometimes written with its elements separated by commas, for example $\boldsymbol{u} = (2, 4, 6, -2, 2)$.

A **zero matrix** (or **null matrix**) is one where every element is zero. So for a zero matrix A, $a_{ij} = 0$ for all values of i and j. Since different zero matrices exist for every possible dimension it is usual to denote a zero matrix by $O_{m\times n}$ where $m \times n$ is the dimension of the matrix. For example,

$$O_{2\times 3} = \begin{pmatrix} 0 & 0 & 0 \\ 0 & 0 & 0 \end{pmatrix} \quad \text{and} \quad O_{2\times 2} = \begin{pmatrix} 0 & 0 \\ 0 & 0 \end{pmatrix}.$$

A **diagonal matrix** is a square matrix where all of the elements are zero except possibly those occupying the positions diagonally from the top-left corner to the bottom right corner. In any square matrix, these elements constitute what is termed the **leading diagonal** or **principal diagonal**. Thus $A = [a_{ij}]$ is a diagonal matrix if $a_{ij} = 0$ for $i \neq j$. Examples of diagonal matrices are:

$$\begin{pmatrix} 3 & 0 \\ 0 & 2 \end{pmatrix} \quad \begin{pmatrix} -2 & 0 & 0 \\ 0 & 1 & 0 \\ 0 & 0 & 4 \end{pmatrix} \quad \begin{pmatrix} 5 & 0 & 0 \\ 0 & -4 & 0 \\ 0 & 0 & 0 \end{pmatrix}.$$

An **identity matrix** (or **unit matrix**) is a diagonal matrix whose leading diagonal elements are all 1. Note that this means that an identity matrix is necessarily square. Thus if $A = [a_{ij}]$ is an identity matrix, then $a_{ij} = 0$ for $i \neq j$ and $a_{ii} = 1$. An identity matrix is often denoted by I when its dimension is clear from the context or irrelevant to the discussion. When it is necessary to distinguish identity matrices of different dimensions, we denote the identity matrix of dimension $n \times n$ by I_n. For example,

$$I_2 = \begin{pmatrix} 1 & 0 \\ 0 & 1 \end{pmatrix} \quad \text{and} \quad I_4 = \begin{pmatrix} 1 & 0 & 0 & 0 \\ 0 & 1 & 0 & 0 \\ 0 & 0 & 1 & 0 \\ 0 & 0 & 0 & 1 \end{pmatrix}.$$

A matrix $A = [a_{ij}]$ is **symmetric** if it is square and $a_{ij} = a_{ji}$ for all values of i and j. (This means that the 'symmetry' of a symmetric matrix is about the leading diagonal.) The following are symmetric matrices:

$$\begin{pmatrix} 3 & -1 \\ -1 & 2 \end{pmatrix} \quad \begin{pmatrix} 5 & -1 & 3 & 7 \\ -1 & -9 & 2 & 5 \\ 3 & 2 & 6 & 0 \\ 7 & 5 & 0 & 2 \end{pmatrix}.$$

Exercises 6.1

1. Give the truth values of the following propositions:

   (i) $\{A : A \text{ is an identity matrix}\} \subset \{B : B \text{ is a symmetric matrix}\}$.
   (ii) $\{B : B \text{ is a symmetric matrix}\} \subset \{C : C \text{ is a diagonal matrix}\}$.
   (iii) $\begin{pmatrix} 1 & 1 \\ 1 & 1 \end{pmatrix} \in \{A : A \text{ is an identity matrix}\}$.
   (iv) $\{D : D \text{ is a square matrix}\} \subset \{C : C \text{ is a diagonal matrix}\}$.
   (v) $\{A : A \text{ is an identity matrix}\} \subset \{C : C \text{ is a diagonal matrix}\}$.

2. Write the 3×2 matrices A, B and C defined as follows:

$$A = [a_{ij}] \qquad B = [b_{ij}] \qquad C = [c_{ij}]$$

   where $a_{ij} = i - j$, $b_{ij} = i - 2j$, $c_{ij} = 4i + 3j$.

3. Write down the matrix A which has dimension 4×4, is symmetric and has the following properties: $a_{ii} = i^2$, $a_{13} = a_{24} = 0$, $a_{14} = 3$, $a_{12} = a_{23} = a_{11} + a_{22}$, $a_{34} = a_{23} - a_{14}$.

4. Give an example of a matrix which is both a row matrix and a column matrix.

5. If

$$A = [a_{ij}] = \begin{pmatrix} x+y & 10 \\ 2x-y & 4 \end{pmatrix}$$

find x and y if $a_{11} = a_{22}$ and $a_{12} = \frac{1}{2}a_{21}$.

6. The **transpose** of a matrix A (denoted by A^{T}) is the matrix obtained by interchanging the rows and columns of A. For example, if

$$A = \begin{pmatrix} 2 & 3 & 1 \\ -2 & 4 & 3 \end{pmatrix} \quad \text{then} \quad A^{\mathrm{T}} = \begin{pmatrix} 2 & -2 \\ 3 & 4 \\ 1 & 3 \end{pmatrix}.$$

In general, if $A = [a_{ij}]$, then $A^{\mathrm{T}} = [a_{ji}]$, i.e. the (i, j)-entry of A^{T} is a_{ji}, the (j, i)-entry of A.

(i) Write down the transpose of each of the following matrices:

$$\text{(a)} \begin{pmatrix} -1 \\ 2 \\ 3 \end{pmatrix} \quad \text{(b)} \begin{pmatrix} 4 & 0 & -1 \\ 3 & 1 & 3 \end{pmatrix} \quad \text{(c)} \begin{pmatrix} -1 & 6 & 3 & 4 \\ 0 & 7 & -1 & 3 \end{pmatrix}.$$

(ii) Prove that A is a symmetric matrix if and only if $A = A^{\mathrm{T}}$. (This is often given as the definition of a symmetric matrix.)

6.3 Operations on Matrices

Multiplication of a Matrix by a Scalar

A **scalar** is simply a real number; for instance, 6, -2, 3 and 0.672 are all scalars. To multiply a matrix by a scalar, we multiply every element of the matrix by that number. For instance, if A is the matrix given by

$$A = \begin{pmatrix} 2 & 3 & -2 \\ 1 & -4 & 6 \end{pmatrix}$$

then multiplication of A by the scalar 3 gives

$$3 \times A = 3A = \begin{pmatrix} 6 & 9 & -6 \\ 3 & -12 & 18 \end{pmatrix}.$$

We say that multiplication by a scalar is 'commutative' (see definition 8.3 on page 364); i.e. $k \times A = A \times k = kA$ for any matrix A and scalar k.

The formal definition of multiplication of a matrix by a scalar is given below.

Definition 6.1

If $A = [a_{ij}]$ is any matrix and k is a scalar, then the product kA is the matrix given by $kA = [ka_{ij}]$, i.e. the (i, j)-entry of kA is k times the (i, j)-entry of A.

Note that multiplication of a matrix by a scalar -1 results in another matrix whose elements are the same as those for A but with the opposite sign. For instance, for the matrix A given above

$$-1A = \begin{pmatrix} -2 & -3 & 2 \\ -1 & 4 & -6 \end{pmatrix}.$$

In general if $A = [a_{ij}]$, then $-1A = [-a_{ij}]$. We normally denote the matrix $-1A$ by $-A$.

Addition of Matrices

Unlike real numbers, it is not always possible to add two matrices. Only matrices having the same dimension can be added and, where this condition is satisfied, we say that the matrices are **conformable for addition**. Given two matrices A and B which have the same dimension, the result of adding A and B is simply the matrix whose elements are the sums of the elements in corresponding positions of A and B. For instance, if

$$A = \begin{pmatrix} 3 & -1 \\ 2 & 7 \\ 5 & 4 \end{pmatrix} \quad \text{and} \quad B = \begin{pmatrix} 4 & 2 \\ -1 & 1 \\ -6 & 0 \end{pmatrix}$$

then

$$
\begin{aligned}
A + B &= \begin{pmatrix} 3 & -1 \\ 2 & 7 \\ 5 & 4 \end{pmatrix} + \begin{pmatrix} 4 & 2 \\ -1 & 1 \\ -6 & 0 \end{pmatrix} \\
&= \begin{pmatrix} 3+4 & -1+2 \\ 2+(-1) & 7+1 \\ 5+(-6) & 4+0 \end{pmatrix} \\
&= \begin{pmatrix} 7 & 1 \\ 1 & 8 \\ -1 & 4 \end{pmatrix}.
\end{aligned}
$$

We can define matrix addition in general as follows.

Definition 6.2

Let $A = [a_{ij}]$ and $B = [b_{ij}]$ be matrices of the same dimension. Then $A + B = C$ where $C = [c_{ij}]$ and $c_{ij} = a_{ij} + b_{ij}$ for all values of i and j.

Note that, for any $m \times n$ matrix A, we have $A + O_{m\times n} = A$ and also $A + (-A) = O_{m\times n}$. Because of the latter property, the matrix $-A$ is sometimes referred to as the **additive inverse** of A. (See chapter 8 for a detailed explanation of the concept of inverses.)

We can use our definition of scalar multiplication together with the definition of matrix addition to define matrix subtraction. We have already noted that $-A$ means $-1A$, i.e. the result of multiplying the matrix A by -1. So $B - A$ is just the sum of B and $-A$, i.e. the sum of B an the additive inverse of A. This can be found according to the rules of matrix addition provided that the condition that A and B have the same dimension is satisfied.

Examples 6.3

1. If

$$A = \begin{pmatrix} -1 & 0 & 2 \\ 4 & 5 & -3 \end{pmatrix} \quad \text{and} \quad B = \begin{pmatrix} 2 & 4 & 7 \\ 1 & -6 & -1 \end{pmatrix}$$

find $A - B$.

Solution

$$
\begin{aligned}
A - B &= A + (-B) \\
&= \begin{pmatrix} -1 & 0 & 2 \\ 4 & 5 & -3 \end{pmatrix} + \begin{pmatrix} -2 & -4 & -7 \\ -1 & 6 & 1 \end{pmatrix} \\
&= \begin{pmatrix} -3 & -4 & -5 \\ 3 & 11 & -2 \end{pmatrix}.
\end{aligned}
$$

2. If

$$
A = \begin{pmatrix} 2 & 7 \\ 1 & 4 \\ 3 & -2 \end{pmatrix} \quad \text{and} \quad B = \begin{pmatrix} -1 & 0 \\ 0 & 4 \\ -3 & 1 \end{pmatrix}
$$

find $3A - 2B$.

Solution

$$
\begin{aligned}
3A - 2B &= 3A + 2(-B) \\
&= 3\begin{pmatrix} 2 & 7 \\ 1 & 4 \\ 3 & -2 \end{pmatrix} + 2\begin{pmatrix} 1 & 0 \\ 0 & -4 \\ 3 & -1 \end{pmatrix} \\
&= \begin{pmatrix} 6 & 21 \\ 3 & 12 \\ 9 & -6 \end{pmatrix} + \begin{pmatrix} 2 & 0 \\ 0 & -8 \\ 6 & -2 \end{pmatrix} \\
&= \begin{pmatrix} 8 & 21 \\ 3 & 4 \\ 15 & -8 \end{pmatrix}.
\end{aligned}
$$

(We could also have written $3A - 2B = 3A + (-2)B$ and obtained the same result.)

It is a simple matter to show that matrix addition and scalar multiplication have the following properties (provided, of course, that the appropriate matrices have the same dimension and are therefore conformable for addition).

(a) $A + B = B + A$ (commutative law).
(b) $A + (B + C) = (A + B) + C$ (associative law).
(c) $k(A + B) = kA + kB$ (i.e. multiplication by a scalar is distributive over matrix addition).
(d) $k(lA) = (kl)A$, where k and l are scalars.

Multiplication of Matrices

Given the way in which we have defined matrix addition, you might expect that matrix multiplication is carried out in an analogous way—by multiplying the elements in corresponding positions in matrices with the same dimension. However, this is not the case. If it were, then matrix algebra would be worthy of little attention since the rules governing it would be very much like the familiar algebra of real numbers. Matrix multiplication lacks the intuitive appeal of matrix addition and, for the time being, we will concentrate on how matrices are multiplied rather than attempting a justification as to why they are multiplied in this way. However, be assured that matrix multiplication does have significant applications in mathematics and elsewhere. We shall consider one such application in the next chapter when we deal with solving systems of linear equations. Matrix multiplication is also used in chapter 10.

We first consider the simplest case of multiplication of two matrices—that of multiplying a row matrix by a column matrix. For the moment, we assume that the following two conditions must be satisfied:

(a) each matrix has the same number of elements;

(b) the row matrix is placed to the left of the column matrix in forming the product.

Thus, if

$$\boldsymbol{u} = (u_1\, u_2\, \ldots\, u_n) \quad \text{and} \quad \boldsymbol{v} = \begin{pmatrix} v_1 \\ v_2 \\ \vdots \\ v_n \end{pmatrix}$$

then the product $\boldsymbol{uv}$ is defined as follows:

$$\begin{aligned} \boldsymbol{uv} &= (u_1\, u_2\, \ldots\, u_n) \begin{pmatrix} v_1 \\ v_2 \\ \vdots \\ v_n \end{pmatrix} \\ &= (u_1 v_1 + u_2 v_2 + \cdots + u_n v_n) \end{aligned}$$

i.e. $\boldsymbol{uv}$ is the 1×1 matrix whose single element is calculated by multiplying together the first elements in $\boldsymbol{u}$ and $\boldsymbol{v}$, the second elements in $\boldsymbol{u}$ and $\boldsymbol{v}$, etc and summing the results.

Example 6.4

Where possible calculate the matrix product $\boldsymbol{uv}$ in each of the following cases:

(i) $\boldsymbol{u} = \begin{pmatrix} 4 & 2 & 3 \end{pmatrix} \qquad \boldsymbol{v} = \begin{pmatrix} 1 \\ 0 \\ 5 \end{pmatrix}$

(ii) $\boldsymbol{u} = \begin{pmatrix} -1 & 2 & 4 & 3 \end{pmatrix} \qquad \boldsymbol{v} = \begin{pmatrix} -2 \\ 1 \\ 7 \end{pmatrix}$

(iii) $\boldsymbol{u} = \begin{pmatrix} -1 & 2 & 4 & -2 \end{pmatrix} \qquad \boldsymbol{v} = \begin{pmatrix} 3 \\ -1 \\ 2 \\ 2 \end{pmatrix}.$

Solution

(i) Here $\boldsymbol{u}$ is a row matrix, $\boldsymbol{v}$ is a column matrix and both have the same number of elements. Therefore the product $\boldsymbol{uv}$ exists and

$$\begin{aligned} \boldsymbol{uv} &= \begin{pmatrix} 4 & 2 & 3 \end{pmatrix} \begin{pmatrix} 1 \\ 0 \\ 5 \end{pmatrix} \\ &= (4 \times 1 + 2 \times 0 + 3 \times 5) \\ &= (19). \end{aligned}$$

(ii) Since $\boldsymbol{u}$ and $\boldsymbol{v}$ do not have the same number of elements, we cannot form the product $\boldsymbol{uv}$.

(iii) $$\begin{aligned} \boldsymbol{uv} &= \begin{pmatrix} -1 & 2 & 4 & -2 \end{pmatrix} \begin{pmatrix} 3 \\ -1 \\ 2 \\ 2 \end{pmatrix} \\ &= ((-1) \times 3 + 2 \times (-1) + 4 \times 2 + (-2) \times 2) \\ &= (-1). \end{aligned}$$

Matrix multiplication in general consists of repeated applications of the operation we have just described for the multiplication of a row matrix by a column matrix. When multiplying two matrices A and B to form the product AB, we repeatedly 'multiply' a row of A by a column of B as we did in the examples above. We perform this operation on every combination of a row from A together with a column from B. Beginning with the first row of A we multiply it by each column of B in turn beginning with the first. The scalars which result are the elements comprising the first row of the product matrix. We then repeat the process with the second row of A together with each column of B giving the second row of AB. The process continues until the final row of A has been multiplied by each column of B.

The process is illustrated below.

$$\begin{pmatrix} \boxed{a_{11} \quad a_{12} \quad \cdots \quad a_{1n}} \\ a_{21} \quad a_{22} \quad \cdots \quad a_{2n} \\ \vdots \quad \vdots \quad \ddots \quad \vdots \\ a_{m1} \quad a_{m2} \quad \cdots \quad a_{mn} \end{pmatrix} \begin{pmatrix} \begin{array}{|c|} \hline b_{11} \\ b_{21} \\ \vdots \\ b_{n1} \\ \hline \end{array} & \begin{array}{c} b_{12} \\ b_{22} \\ \vdots \\ b_{n2} \end{array} & \begin{array}{c} \cdots \\ \cdots \\ \ddots \\ \cdots \end{array} & \begin{array}{c} b_{1r} \\ b_{2r} \\ \vdots \\ b_{nr} \end{array} \end{pmatrix}$$

1st row of A 1st column of B

$$= \begin{pmatrix} \boxed{c_{11}} & & \\ & & \\ & & \end{pmatrix}$$

$(1,1)$-entry of AB

$$\begin{pmatrix} \boxed{a_{11} \quad a_{12} \quad \cdots \quad a_{1n}} \\ a_{21} \quad a_{22} \quad \cdots \quad a_{2n} \\ \vdots \quad \vdots \quad \ddots \quad \vdots \\ a_{m1} \quad a_{m2} \quad \cdots \quad a_{mn} \end{pmatrix} \begin{pmatrix} \begin{array}{c} b_{11} \\ b_{21} \\ \vdots \\ b_{n1} \end{array} & \begin{array}{|c|} \hline b_{12} \\ b_{22} \\ \vdots \\ b_{n2} \\ \hline \end{array} & \begin{array}{c} \cdots \\ \cdots \\ \ddots \\ \cdots \end{array} & \begin{array}{c} b_{1r} \\ b_{2r} \\ \vdots \\ b_{nr} \end{array} \end{pmatrix}$$

1st row of A 2nd column of B

$$= \begin{pmatrix} c_{11} & \boxed{c_{12}} & \\ & & \\ & & \end{pmatrix}$$

$(1,2)$-entry of AB.

In general:

$$\begin{pmatrix} a_{11} \quad a_{12} \quad \cdots \quad a_{1n} \\ \vdots \quad \vdots \quad \quad \vdots \\ \boxed{a_{i1} \quad a_{i2} \quad \cdots \quad a_{in}} \\ \vdots \quad \vdots \quad \quad \vdots \\ a_{m1} \quad a_{m2} \quad \cdots \quad a_{mn} \end{pmatrix} \begin{pmatrix} \begin{array}{c} b_{11} \\ b_{21} \\ \vdots \\ b_{n1} \end{array} & \begin{array}{c} \cdots \\ \cdots \\ \\ \cdots \end{array} & \begin{array}{|c|} \hline b_{j1} \\ b_{j2} \\ \cdots \\ b_{jn} \\ \hline \end{array} & \begin{array}{c} \cdots \\ \cdots \\ \\ \cdots \end{array} & \begin{array}{c} b_{1r} \\ b_{2r} \\ \vdots \\ b_{nr} \end{array} \end{pmatrix}$$

ith row of A jth column of B

$$= \begin{pmatrix} & & \vdots & & \\ & \cdots & \boxed{c_{ij}} & \cdots & \\ & & \vdots & & \end{pmatrix}$$

(i, j)-entry of AB.

Note that the sequence of operations which we have described requires that A must have the same number of elements in a row as B has in a column or, equivalently, A must have the same number of columns as B has rows. Also the (i, j)-entry in AB is the result of multiplying the ith row of A by the jth column of B.

Example 6.5

Given the matrices

$$A = \begin{pmatrix} 3 & 1 & -2 \\ 2 & 4 & 3 \end{pmatrix} \quad \text{and} \quad B = \begin{pmatrix} 2 & 1 & 1 \\ -3 & 4 & -1 \\ 1 & 3 & 0 \end{pmatrix}$$

calculate $C = AB$.

Solution

Since A has the same number of elements in a row as B has in a column, the product AB exists. We perform our 'row times column' operation taking each row of A with each column of B.

We take the first row of A and the first column of B:

$$\begin{pmatrix} \boxed{3 \quad 1 \quad -2} \\ 2 \quad 4 \quad 3 \end{pmatrix} \begin{pmatrix} \boxed{\begin{matrix} 2 \\ -3 \\ 1 \end{matrix}} & \begin{matrix} 1 \\ 4 \\ 3 \end{matrix} & \begin{matrix} 1 \\ -1 \\ 0 \end{matrix} \end{pmatrix}.$$

We perform our multiplication operation and get:

$$3 \times 2 + 1 \times (-3) + (-2) \times 1 = 1.$$

This scalar is c_{11}, the $(1, 1)$ entry of C:

$$\begin{pmatrix} \boxed{1} & & \\ & & \end{pmatrix}.$$

Repeating this operation with the first row of A and second column of B we obtain c_{12}, where

$$c_{12} = 3 \times 1 + 1 \times 4 + (-2) \times 3 = 1.$$

Continuing in this way we calculate systematically all the elements in C. Thus

$$\begin{aligned}C &= \begin{pmatrix} 3 & 1 & -2 \\ 2 & 4 & 3 \end{pmatrix} \begin{pmatrix} 2 & 1 & 1 \\ -3 & 4 & -1 \\ 1 & 3 & 0 \end{pmatrix} \\ &= \begin{pmatrix} 3.2 + 1.(-3) + (-2).1 & 3.1 + 1.4 + (-2).3 & 3.1 + 1.(-1) + (-2).0 \\ 2.2 + 4.(-3) + 3.1 & 2.1 + 4.4 + 3.3 & 2.1 + 4.(-1) + 3.0 \end{pmatrix} \\ &= \begin{pmatrix} 1 & 1 & 2 \\ -5 & 27 & -2 \end{pmatrix}.\end{aligned}$$

We have already said that, for the matrix product AB to exist, the number of columns of A must be the same as the number of rows of B. Where this condition is satisfied, what will be the dimension of the product AB? Each row of A, when multiplied by all the columns of B, produces a row of AB. So AB must have the same number of rows as A. Further, given a row of A, the 'row times column' operation is performed on each column of B. So the number of elements within a row of AB must be the same as the number of columns of B. In other words, the number of columns of AB is the same as the number of columns of B.

We can state this more formally as follows:

> If A has dimension $m \times n$ and B has dimension $p \times q$, then the matrix product AB exists if and only if $n = p$. The dimension of AB is then $m \times q$.

Example 6.6

For each of the following pairs of matrices A and B, state whether the matrix product AB exists. If it does exist, state the dimension of AB.

(i) $A = \begin{pmatrix} 2 & 3 \\ -1 & 2 \\ 5 & 4 \end{pmatrix}$ $\quad B = \begin{pmatrix} 4 & 1 & 2 \\ 3 & 2 & 1 \end{pmatrix}$

(ii) $A = \begin{pmatrix} -1 & 0 \\ 0 & 0 \end{pmatrix}$ $\quad B = \begin{pmatrix} 4 & 0 \\ 1 & 2 \\ 3 & -1 \end{pmatrix}$

(iii) $A = \begin{pmatrix} 1 & 4 & 2 & -5 \end{pmatrix}$ $\quad B = \begin{pmatrix} 2 & 0 \\ 4 & 1 \\ 3 & -3 \\ -1 & -1 \end{pmatrix}.$

Solution

(i) The matrix A has dimension 3×2 and B has dimension 2×3. Since A has the same number of columns as B has rows, AB exists and has dimension 3×3. Evaluation of the product gives

$$AB = \begin{pmatrix} 17 & 8 & 7 \\ 2 & 3 & 0 \\ 32 & 13 & 14 \end{pmatrix}.$$

(ii) The matrices A and B have dimensions 2×2 and 3×2 respectively. Matrix A has two columns but B has three rows and hence the product AB does not exist (although BA does).

(iii) The dimension of A is 1×4 and that of B is 4×2. Thus AB exists and has dimension 1×2. Evaluation of the product gives

$$AB = \begin{pmatrix} 29 & 3 \end{pmatrix}.$$

We now give a formal definition of matrix multiplication. The notation is a little cumbersome but it should be clear that it summarizes exactly those steps detailed in the examples above.

Definition 6.3

If $A = [a_{ij}]$ is an $m \times n$ matrix and $B = [b_{ij}]$ is an $n \times r$ matrix, then $AB = C$ where $C = [c_{ij}]$ has dimension $m \times r$ and

$$\begin{aligned} c_{ij} &= a_{i1}b_{1j} + a_{i2}b_{2j} + \cdots + a_{in}b_{nj} \\ &= \sum_{k=1}^{n} a_{ik}b_{kj}. \end{aligned}$$

It is important to note that, in general, $AB \neq BA$. We say that matrix multiplication is 'not commutative' (see definition 8.3). In fact the dimensions of A and B could well be such that, although one of these two products exists, the other does not (see example 6.6(ii)). Even if A and B are such that AB and BA both exist, in general these two products may not be equal. This means 'multiply

A by B' is ambiguous when A and B are matrices since it is not clear which of the two products AB or BA is intended. Where the product AB is required, we say that B is to be **pre-multiplied** by A (or that A is to be **post-multiplied** by B). If the product BA is required, we say that A is to be pre-multiplied by B (or that B is to be post-multiplied by A).

Matrix multiplication has the following properties provided, of course, that the appropriate matrix sums and products exist. It is easy to demonstrate that these properties hold for specific matrices but their proof for general matrices is a tedious and lengthy exercise.

(a) $(AB)C = A(BC)$ (associative law);

(b) $A(B+C) = AB + AC$, $(B+C)D = BD + CD$ (distributive law);

(c) $k(AB) = (kA)B = A(kB)$ where k is a scalar.

Note that property (a) means that we can write this matrix product as ABC without fear of ambiguity.

Exercises 6.2

1. If

$$A = \begin{pmatrix} 2 & -3 \\ 0 & -1 \end{pmatrix} \quad \text{and} \quad B = \begin{pmatrix} 4 & -1 \\ 2 & -1 \end{pmatrix}$$

find:

(i)	$A + 2B$	(ii)	$3A - 6B$	(iii)	AB
(iv)	A^2 (i.e. AA)	(v)	BA	(vi)	$A(BA)$
(vii)	$(AB)A$	(viii)	$A(A - B)$	(ix)	$A^{\mathrm{T}}B$
(x)	$(AB)^{\mathrm{T}}$	(xi)	$B^{\mathrm{T}}A^{\mathrm{T}}$.		

(See exercise 6.1.6 for the definition of A^{T}, the transpose of A.)

2. If

$$A = \begin{pmatrix} 1 & 3 & 2 \end{pmatrix} \quad B = \begin{pmatrix} 2 & 4 \\ 3 & 3 \\ -1 & 0 \end{pmatrix} \quad C = \begin{pmatrix} 2 & 3 & -1 \\ 4 & 1 & -1 \end{pmatrix}$$

find the following, if they exist. If any matrix product does not exist, explain why it does not.

(i) AB (ii) BA (iii) AC (iv) CA (v) BC (vi) CB (vii) ABC.

3. If
$$A = \begin{pmatrix} 0 & 1 \\ 2 & 0 \end{pmatrix}$$
find a matrix B such that $AB = BA$.

4. Let A be any 2×2 matrix. Show that $AI_2 = I_2A = A$, where I_2 is the 2×2 identity matrix.

5. Let A be any $n \times n$ matrix. Show that $AI_n = I_nA = A$. (Use the definition of matrix multiplication.)

6. Let A be any $m \times n$ matrix. Show that $AI_n = I_mA = A$.

7. If
$$A = \begin{pmatrix} 1 & -3 & 2 \\ 2 & 1 & -3 \\ 4 & -3 & -1 \end{pmatrix} \quad B = \begin{pmatrix} 1 & 4 & 1 \\ 2 & 1 & 1 \\ 1 & -2 & 1 \end{pmatrix} \quad C = \begin{pmatrix} 2 & 1 & -1 \\ 3 & -2 & -1 \\ 2 & -5 & -1 \end{pmatrix}$$
show that $AB = AC$. (But note that $B \neq C$, i.e. the 'cancellation law' does not hold for matrix multiplication in general.)

8. Let A, B and C be matrices such that A has dimension $m \times n$, B has dimension $p \times q$ and C has dimension $r \times s$. For each of the following, write down the condition(s) for the product to exist and state the dimension of the product:

(i) ABC (ii) CBA (iii) $(A + B)C$.

9. Find a counter-example to show that, if A and B are matrices and $AB = O$ (where O is the zero matrix of appropriate dimension), then we cannot conclude that either A or B is a zero matrix.

10. If $A = [a_{ij}]$ is a diagonal matrix of dimension $n \times n$ (so that $a_{ij} = 0$ if $i \neq j$) and $B = [b_{ij}]$ is another diagonal matrix of dimension $n \times n$, find the products AB and BA. Use your result to write down the products AB and BA where
$$A = \begin{pmatrix} 2 & 0 & 0 \\ 0 & 3 & 0 \\ 0 & 0 & -3 \end{pmatrix} \quad B = \begin{pmatrix} -4 & 0 & 0 \\ 0 & -1 & 0 \\ 0 & 0 & 5 \end{pmatrix}.$$
If $A = [a_{ij}]$ is a diagonal matrix, write down the matrix A^2. What would you expect the result to be for A^n? Prove this result by mathematical induction.

11. In exercise 6.1.6 we defined A^{T}, the transpose of matrix A. Prove each of the following properties of the transpose:

    (i) $(A^{\mathrm{T}})^{\mathrm{T}} = A$
    (ii) $(A+B)^{\mathrm{T}} = A^{\mathrm{T}} + B^{\mathrm{T}}$
    (iii) $(AB)^{\mathrm{T}} = B^{\mathrm{T}}A^{\mathrm{T}}$.

12. Show that, if A is a square matrix, then $A + A^{\mathrm{T}}$ is a symmetric matrix. (Recall that a matrix B is symmetric if and only if $B = B^{\mathrm{T}}$—see exercise 6.1.6.)

6.4 Elementary Matrices

An **elementary matrix** is one which can be obtained from an identity matrix by performing only one of the following operations on that identity matrix:

(R1) interchanging two rows;
(R2) multiplying the elements of one row by a non-zero real number;
(R3) adding to the elements of one row, any multiple of the corresponding elements of another row;
(C1) interchanging two columns;
(C2) multiplying the elements of one column by a non-zero real number;
(C3) adding to the elements of one column any multiple of the corresponding elements of another column.

Notice that (R1), (R2) and (R3) describe operations which are applied to rows of the identity matrix whilst (C1), (C2) and (C3) describe operations applied to its columns. The operations (R1), (R2) and (R3), when applied to any matrix (not necessarily an identity matrix), are called **elementary row operations** (or **elementary row transformations**). Operations (C1), (C2) and (C3) are called **elementary column operations** (or **elementary column transformations**).

An elementary matrix is always square since it is obtained from another square matrix (an identity matrix) by one of the operations described above, none of which alter the dimension of the matrix.

ELEMENTARY ROW OPERATIONS

Example 6.7

State whether or not each of the following is an elementary matrix:

(i) $\begin{pmatrix} 7 & 3 & 2 \\ 1 & 4 & 1 \end{pmatrix}$ (ii) $\begin{pmatrix} 0 & 0 & 1 \\ 0 & 1 & 0 \\ 1 & 0 & 0 \end{pmatrix}$ (iii) $\begin{pmatrix} 1 & 5 \\ 0 & 1 \end{pmatrix}$

(iv) $\begin{pmatrix} 4 & 0 \\ 0 & -3 \end{pmatrix}$ (v) $\begin{pmatrix} 5 & 0 & 0 \\ 0 & 1 & 0 \\ 0 & 1 & 1 \end{pmatrix}$.

Solution

(i) This matrix is not square and therefore cannot be an elementary matrix.

(ii) Starting from the identity matrix

$$I_3 = \begin{pmatrix} 1 & 0 & 0 \\ 0 & 1 & 0 \\ 0 & 0 & 1 \end{pmatrix}$$

the matrix

$$\begin{pmatrix} 0 & 0 & 1 \\ 0 & 1 & 0 \\ 1 & 0 & 0 \end{pmatrix}$$

can be obtained by interchanging the first and third rows (or by interchanging the first and third columns). Hence it is an elementary matrix.

(iii) The matrix

$$\begin{pmatrix} 1 & 5 \\ 0 & 1 \end{pmatrix}$$

can be obtained from

$$I_2 = \begin{pmatrix} 1 & 0 \\ 0 & 1 \end{pmatrix}$$

by adding five times the second row to the first row (or by adding five times the first column to the second). In either case this involves one elementary row or column operation and hence

$$\begin{pmatrix} 1 & 5 \\ 0 & 1 \end{pmatrix}$$

is an elementary matrix.

(iv) In order to obtain

$$\begin{pmatrix} 4 & 0 \\ 0 & -3 \end{pmatrix}$$

from I_2 we would need to perform two elementary row or column operations: either multiply the first row by 4 and the second by -3 or multiply the first column by 4 and the second by -3. No single elementary row or column operation produces the required result and hence

$$\begin{pmatrix} 4 & 0 \\ 0 & -3 \end{pmatrix}$$

is not an elementary matrix.

(v) The matrix

$$\begin{pmatrix} 5 & 0 & 0 \\ 0 & 1 & 0 \\ 0 & 1 & 1 \end{pmatrix}$$

cannot be obtained from I_3 by means of any single elementary row or column operation and hence it is not an elementary matrix.

In the examples above, each of the elementary matrices could be formed in two ways—either by a single elementary row operation or by a single elementary column operation. It is not difficult to see that this is the case for all elementary matrices.

Note that any identity matrix I_n is itself an elementary matrix since it can be regarded as being derived from I_n by multiplying any row or column by 1.

What is interesting about elementary matrices is their effect upon another matrix of appropriate dimension when the two are multiplied together. Consider the elementary matrix

$$E = \begin{pmatrix} 3 & 0 & 0 \\ 0 & 1 & 0 \\ 0 & 0 & 1 \end{pmatrix}$$

obtained from I_3 by multiplying the first row by 3. Observe what happens when we multiply this matrix by another matrix, say,

$$A = \begin{pmatrix} 2 & -3 \\ 1 & 2 \\ 4 & 6 \end{pmatrix}$$

so that the elementary matrix is on the left (i.e. we pre-multiply A by the elementary matrix):

$$EA = \begin{pmatrix} 3 & 0 & 0 \\ 0 & 1 & 0 \\ 0 & 0 & 1 \end{pmatrix} \begin{pmatrix} 2 & -3 \\ 1 & 2 \\ 4 & 6 \end{pmatrix} = \begin{pmatrix} 6 & -9 \\ 1 & 2 \\ 4 & 6 \end{pmatrix}.$$

The product matrix is simply the matrix obtained from A by multiplying the first row by 3. In this case pre-multiplication of A by the elementary matrix E effects the same elementary row operation on A as was necessary on I_3 to produce the elementary matrix itself.

The following matrices can be obtained from I_3 by multiplying a row by a non-zero constant k:

$$\begin{pmatrix} k & 0 & 0 \\ 0 & 1 & 0 \\ 0 & 0 & 1 \end{pmatrix} \quad \begin{pmatrix} 1 & 0 & 0 \\ 0 & k & 0 \\ 0 & 0 & 1 \end{pmatrix} \quad \begin{pmatrix} 1 & 0 & 0 \\ 0 & 1 & 0 \\ 0 & 0 & k \end{pmatrix}.$$

It is easy to show that, if one of these elementary matrices is multiplied by any $3 \times n$ matrix A, with the elementary matrix on the left of the product, then the result is the matrix A with the corresponding row multiplied by k.

Does this result generalize to elementary matrices obtained by other elementary row operations? Consider the following elementary matrix E, obtained from I_4 by interchanging rows 1 and 2:

$$\begin{pmatrix} 0 & 1 & 0 & 0 \\ 1 & 0 & 0 & 0 \\ 0 & 0 & 1 & 0 \\ 0 & 0 & 0 & 1 \end{pmatrix}.$$

If

$$A = \begin{pmatrix} 7 & 4 & 2 & -1 \\ 3 & 1 & 2 & 0 \\ -3 & 6 & -3 & 2 \\ 1 & -2 & 4 & 3 \end{pmatrix}$$

then

$$\begin{aligned} EA &= \begin{pmatrix} 0 & 1 & 0 & 0 \\ 1 & 0 & 0 & 0 \\ 0 & 0 & 1 & 0 \\ 0 & 0 & 0 & 1 \end{pmatrix} \begin{pmatrix} 7 & 4 & 2 & -1 \\ 3 & 1 & 2 & 0 \\ -3 & 6 & -3 & 2 \\ 1 & -2 & 4 & 3 \end{pmatrix} \\ &= \begin{pmatrix} 3 & 1 & 2 & 0 \\ 7 & 4 & 2 & -1 \\ -3 & 6 & -3 & 2 \\ 1 & -2 & 4 & 3 \end{pmatrix}. \end{aligned}$$

The effect of pre-multiplication by the elementary matrix is to interchange rows 1 and 2 of A, the same elementary row operation by which the elementary matrix was obtained from the appropriate identity matrix.

In a similar way we can check the effect of an elementary matrix formed by the addition of k times one row to another when multiplied by a matrix A of appropriate dimension (with the elementary matrix on the left). The product is again a matrix which is the result of applying the same elementary row operation to A.

We now generalize these results in a useful theorem, which we state without proof.

Theorem 6.1

(i) Consider an elementary matrix E formed from I_n by an elementary row operation. If A is any matrix of dimension $n \times m$, the matrix product EA is the matrix resulting from performing the same elementary row operation on A.

(ii) Consider an elementary matrix F formed from I_n by an elementary column operation. If A is any matrix of dimension $m \times n$, the matrix product AF is the matrix resulting from performing the same elementary column operation on A.

The theorem confirms that elementary row operations on a matrix can be effected by pre-multiplication by the appropriate elementary matrix. It also states that we can effect elementary column operations by post-multiplication by the elementary matrix formed from the appropriate identity matrix by the same column operation.

As you read this chapter and the next, you will no doubt notice that we tend to neglect elementary column operations in favour of elementary row operations. There is no particular reason for this other than convention. The uses to which we put row operations could equally well be served by column operations.

Examples 6.8

1. Given

$$A = \begin{pmatrix} 3 & 4 & 2 \\ -1 & 5 & 4 \\ 6 & 1 & 8 \end{pmatrix}$$

find a matrix E so that the product EA is given by

$$EA = \begin{pmatrix} 3 & 4 & 2 \\ 11 & 7 & 20 \\ 6 & 1 & 8 \end{pmatrix}.$$

Solution

We note that the product EA is the result of the following elementary row operation on A: add twice the third row to the second. Therefore E is the elementary matrix obtained from I_3 by this same elementary row operation. Therefore

$$E = \begin{pmatrix} 1 & 0 & 0 \\ 0 & 1 & 2 \\ 0 & 0 & 1 \end{pmatrix}.$$

We can check that E is the required matrix:

$$\begin{aligned} EA &= \begin{pmatrix} 1 & 0 & 0 \\ 0 & 1 & 2 \\ 0 & 0 & 1 \end{pmatrix} \begin{pmatrix} 3 & 4 & 2 \\ -1 & 5 & 4 \\ 6 & 1 & 8 \end{pmatrix} \\ &= \begin{pmatrix} 3 & 4 & 2 \\ 11 & 7 & 20 \\ 6 & 1 & 8 \end{pmatrix}. \end{aligned}$$

2. Given

$$A = \begin{pmatrix} 2 & 1 \\ 3 & -2 \\ 0 & -4 \end{pmatrix}$$

find F so that

$$AF = \begin{pmatrix} 1 & 2 \\ -2 & 3 \\ -4 & 0 \end{pmatrix}.$$

Solution

The product AF is the result of the elementary column operation 'interchange columns 1 and 2' on the matrix A. Therefore F is the elementary matrix obtained from I_2 by interchanging the first and second columns. Thus

$$F = \begin{pmatrix} 0 & 1 \\ 1 & 0 \end{pmatrix}.$$

Checking that AF gives the required matrix:

$$\begin{aligned} AF &= \begin{pmatrix} 2 & 1 \\ 3 & -2 \\ 0 & -4 \end{pmatrix} \begin{pmatrix} 0 & 1 \\ 1 & 0 \end{pmatrix} \\ &= \begin{pmatrix} 1 & 2 \\ -2 & 3 \\ -4 & 0 \end{pmatrix}. \end{aligned}$$

3. Given

$$A = \begin{pmatrix} 3 & 0 & -4 \\ -4 & 2 & 1 \\ -2 & -1 & 5 \end{pmatrix} \quad \text{and} \quad B = \begin{pmatrix} -4 & 2 & 1 \\ -6 & 0 & 8 \\ -2 & -1 & 5 \end{pmatrix}$$

find a matrix P such that $PA = B$.

Solution

The matrix B is not the result of any one elementary row operation on the matrix A so P is not an elementary matrix. However, B can be considered as the result of applying the following two elementary row operations to A: (i) interchange rows 1 and 2, then (ii) multiply (the new) row 2 by -2.

Pre-multiplication of A by the elementary matrix

$$E_1 = \begin{pmatrix} 0 & 1 & 0 \\ 1 & 0 & 0 \\ 0 & 0 & 1 \end{pmatrix}$$

will effect the interchange of the first and second rows. Pre-multiplication of the result, namely E_1A, by the elementary matrix

$$E_2 = \begin{pmatrix} 1 & 0 & 0 \\ 0 & -2 & 0 \\ 0 & 0 & 1 \end{pmatrix}$$

will have the effect of multiplying the second row of E_1A by -2. Thus

$$E_2(E_1A) = \begin{pmatrix} -4 & 2 & 1 \\ -6 & 0 & 8 \\ -2 & -1 & 5 \end{pmatrix}.$$

Using the associative property of matrix multiplication, we have

$$E_2(E_1A) = (E_2E_1)A$$

so that

$$\begin{aligned} P &= E_2E_1 \\ &= \begin{pmatrix} 1 & 0 & 0 \\ 0 & -2 & 0 \\ 0 & 0 & 1 \end{pmatrix} \begin{pmatrix} 0 & 1 & 0 \\ 1 & 0 & 0 \\ 0 & 0 & 1 \end{pmatrix} \\ &= \begin{pmatrix} 0 & 1 & 0 \\ -2 & 0 & 0 \\ 0 & 0 & 1 \end{pmatrix}. \end{aligned}$$

Note that the result PA could also have been obtained by the following two elementary row operations on A: (i) multiply row 1 by -2, then (ii) interchange rows 1 and 2.

In this case the corresponding elementary matrices are

$$E_3 = \begin{pmatrix} -2 & 0 & 0 \\ 0 & 1 & 0 \\ 0 & 0 & 1 \end{pmatrix} \quad \text{and} \quad E_4 = \begin{pmatrix} 0 & 1 & 0 \\ 1 & 0 & 0 \\ 0 & 0 & 1 \end{pmatrix}$$

and therefore

$$P = E_4E_3 = \begin{pmatrix} 0 & 1 & 0 \\ 1 & 0 & 0 \\ 0 & 0 & 1 \end{pmatrix} \begin{pmatrix} -2 & 0 & 0 \\ 0 & 1 & 0 \\ 0 & 0 & 1 \end{pmatrix}$$
$$= \begin{pmatrix} 0 & 1 & 0 \\ -2 & 0 & 0 \\ 0 & 0 & 1 \end{pmatrix}$$

as before.

In these examples we have seen how one matrix may be obtained from another by a sequence of one or more elementary row operations and how this is equivalent to pre-multiplication of the matrix by the appropriate elementary matrices. In general, if a matrix B can be obtained from matrix A by a finite sequence of elementary row operations, we say that B is **row-equivalent** to A and we write $A \sim B$.

Exercises 6.3

1. State whether or not each of the following is an elementary matrix. For each elementary matrix state the alternative elementary row and elementary column operations by which it is formed from the appropriate identity matrix.

   (i) $\begin{pmatrix} 1 & 1 \\ 0 & 1 \end{pmatrix}$ (ii) $\begin{pmatrix} 0 & 0 & 0 & 1 \\ 0 & 0 & 1 & 0 \\ 0 & 1 & 0 & 0 \\ 1 & 0 & 0 & 0 \end{pmatrix}$ (iii) $\begin{pmatrix} 1 & 0 & -3 & 0 \\ 0 & 1 & 0 & 0 \\ 0 & 0 & 1 & 0 \\ 0 & 0 & 0 & 1 \end{pmatrix}$

   (iv) $\begin{pmatrix} 1 & 0 \\ 0 & 4 \end{pmatrix}$ (v) $\begin{pmatrix} -1 & 0 \\ 0 & -1 \end{pmatrix}$ (vi) $\begin{pmatrix} 1 & 0 & 0 \\ 0 & 1 & 0 \\ 0 & 0 & 0 \end{pmatrix}$

   (vii) $\begin{pmatrix} 1 & 0 & 0 \\ 0 & 1 & 0 \\ 0 & 0 & 1 \end{pmatrix}$.

2. If

$$A = \begin{pmatrix} 2 & -4 & 1 \\ -1 & 2 & 3 \\ 0 & 1 & -2 \end{pmatrix} \quad \text{and} \quad B = \begin{pmatrix} 0 & 1 & -2 \\ -1 & 2 & 3 \\ 2 & -4 & 1 \end{pmatrix}$$

(i) find an elementary matrix E_1 such that $E_1 A = B$
(ii) find an elementary matrix E_2 such that $E_2 B = A$
(iii) evaluate $E_1 E_2$ and $E_2 E_1$.

3. If

$$A = \begin{pmatrix} 1 & -1 \\ 3 & 2 \\ -4 & 4 \end{pmatrix} \quad \text{and} \quad B = \begin{pmatrix} 1 & -1 \\ 5 & 0 \\ -4 & 4 \end{pmatrix}$$

(i) find an elementary matrix E_1 such that $A = E_1 B$
(ii) find an elementary matrix E_2 such that $E_2 A = B$
(iii) evaluate $E_1 E_2$ and $E_2 E_1$.

4. If

$$A = \begin{pmatrix} 2 & 4 & -2 \\ 1 & 3 & 7 \\ -1 & 1 & 1 \end{pmatrix} \quad \text{and} \quad B = \begin{pmatrix} 2 & -8 & -2 \\ 1 & -6 & 7 \\ -1 & -2 & 1 \end{pmatrix}$$

(i) find a matrix F_1 such that $AF_1 = B$
(ii) find a matrix F_2 such that $BF_2 = A$
(iii) evaluate $F_1 F_2$ and $F_2 F_1$.

5. If

$$A = \begin{pmatrix} 2 & -1 & 2 \\ 7 & 0 & 4 \\ 1 & 6 & 2 \end{pmatrix} \quad \text{and} \quad B = \begin{pmatrix} 7 & 0 & 4 \\ 2 & -1 & 2 \\ 1 & 6 & 2 \end{pmatrix}$$

find an elementary matrix E_1 so that $E_1 A = B$.

If

$$C = \begin{pmatrix} 3\frac{1}{2} & 0 & 2 \\ 2 & -1 & 2 \\ 1 & 6 & 2 \end{pmatrix}$$

find an elementary matrix E_2 such that $E_2 B = C$. Hence find a matrix Q such that $QA = C$.

6. If

$$A = \begin{pmatrix} 2 & 4 \\ 1 & 4 \end{pmatrix}$$

find elementary matrices E_1, E_2 and E_3 so that $E_3 E_2 E_1 A = I_2$.

7. Is multiplication of elementary matrices commutative? Why or why not?

8. Show that the relation defined by $A \mathsf{R} B$ if and only if A is row-equivalent to B is an equivalence relation on the set of $m \times n$ matrices.

6.5 The Inverse of a Matrix

Consider the matrices

$$A = \begin{pmatrix} 1 & -2 \\ -1 & 3 \end{pmatrix} \quad \text{and} \quad B = \begin{pmatrix} 3 & 2 \\ 1 & 1 \end{pmatrix}.$$

Then

$$AB = \begin{pmatrix} 1 & -2 \\ -1 & 3 \end{pmatrix}\begin{pmatrix} 3 & 2 \\ 1 & 1 \end{pmatrix} = \begin{pmatrix} 1 & 0 \\ 0 & 1 \end{pmatrix}$$

the 2×2 identity matrix. Also

$$BA = \begin{pmatrix} 3 & 2 \\ 1 & 1 \end{pmatrix}\begin{pmatrix} 1 & -2 \\ -1 & 3 \end{pmatrix} = \begin{pmatrix} 1 & 0 \\ 0 & 1 \end{pmatrix}.$$

Here we have a matrix B which, when multiplied by A on the left or on the right, gives the 2×2 identity matrix. The matrix B is said to be the 'inverse' (or 'multiplicative inverse', to give it its full name) of A and we write $B = A^{-1}$ where A^{-1} denotes the inverse of A.

(Remember that in §6.3 we referred to $-A$ as the additive inverse of the matrix A. The word 'inverse' is therefore ambiguous. However, in matrix algebra, the term is conventionally taken to refer to the multiplicative inverse rather than the additive inverse. If the latter is intended, then its full title must be used.)

We now give a formal definition of the inverse of a matrix.

Definition 6.4

Let A be a square matrix of dimension $n \times n$. A matrix B such that $AB = BA = I_n$ is called the **inverse** of A and we write $B = A^{-1}$.

Note that the following points are implied in the definition.

(a) The inverse is defined only for a square matrix.

(b) Since A is a square matrix, B must be a square matrix with the same dimension as A, so that both the products AB and BA exist.

(c) If B is the inverse of A, then A is the inverse of B. Hence the inverse of A^{-1} is A, i.e. $(A^{-1})^{-1} = A$.

It is not clear whether the inverse of a square matrix can always be found. As we shall see, some square matrices do not have an inverse. Such matrices are called **singular matrices**. Those matrices which do have an inverse are called **non-singular matrices** (or **invertible matrices**).

Although matrix multiplication is not in general commutative, it can be shown that for square matrices, if $AB = I$, then $BA = I$, so that to establish that B is the inverse of A only one of the two products needs to be evaluated.

Whilst the definition tells us what the inverse of a matrix is, it does not give us any idea as to how to find it for a given matrix, nor even how to discover whether a matrix has an inverse. There are a number of methods for finding the inverse of a square matrix where it exists. We shall develop a method based on elementary matrices. To do this we shall need some simple theorems.

In exercise 6.2.5 we showed that if I_n is the $n \times n$ identity matrix and A is any $n \times n$ matrix then $AI_n = I_nA = A$. We can use this result to prove the following important theorem which guarantees that, where the inverse of a square matrix exists, that inverse is unique. (Actually we have been assuming that this is the case by referring to 'the' inverse of a square matrix.)

Theorem 6.2

A non-singular matrix has only one inverse, i.e. if B and C are both inverses of a matrix A, then $B = C$.

Proof

Suppose that B and C are both inverses of A.

Since B is an inverse of A, we have

$$AB = BA = I.$$

The matrix C is also an inverse of A so that

$$AC = CA = I.$$

We multiply both sides of the equation $BA = I$ on the right by C. This gives

$$\begin{aligned}
& (BA)C = IC & \\
\Rightarrow \quad & B(AC) = C & \text{(since matrix multiplication is associative)} \\
\Rightarrow \quad & BI = C & \text{(since } AC = I\text{)}
\end{aligned}$$

$$\Rightarrow \qquad B = C. \qquad \square$$

The next theorem guarantees the existence of inverses for all elementary matrices.

Theorem 6.3

Every elementary matrix has an inverse (i.e. is non-singular) and the inverse of an elementary matrix is also an elementary matrix.

Proof

This theorem follows from the fact that, for every elementary row operation, there is another elementary row operation which 'undoes' its effect. Suppose we have a matrix A to which we apply an elementary row operation which results in matrix B. Then there is another elementary row operation which, when applied to B, results in A. These two elementary row operations are said to be **inverses** of each other. For instance, if we perform the operation 'multiply row 2 by k' ($k \neq 0$) on matrix A to produce matrix B, then the operation 'multiply row 2 by $1/k$' performed on B results in A. For the elementary row operation 'add k times row i to row j' the inverse elementary row operation is 'add $(-k)$ times row i to row j'. The elementary row operation 'interchange two rows' is its own inverse.

Since each elementary row operation corresponds to pre-multiplication by an elementary matrix, it follows from the above that, for every elementary matrix, there is another elementary matrix which 'reverses' the elementary row operation performed by the first.

Let E_1 be an elementary matrix obtained from I_n by applying an elementary row operation. Applying the inverse of this elementary row operation to I_n produces another elementary matrix E_2. Since applying an elementary row operation followed by its inverse has no net effect $E_2 E_1 I_n = I_n$, and therefore $E_2 E_1 = I_n$.

Thus E_2 is the inverse of E_1 (and E_1 is the inverse of E_2). $\square$

Example 6.9

Find the inverse of the elementary matrix

$$E_1 = \begin{pmatrix} 1 & 0 & 0 \\ 0 & 1 & 0 \\ 0 & -3 & 1 \end{pmatrix}.$$

Proof

The elementary matrix E_1 performs the elementary row operation 'add (-3) times the second row to the third row'. The inverse operation is 'add 3 times the second row to the third row' which corresponds to the elementary matrix

$$E_2 = \begin{pmatrix} 1 & 0 & 0 \\ 0 & 1 & 0 \\ 0 & 3 & 1 \end{pmatrix}$$

and E_2 is the inverse of E_1. We can check this result by confirming that $E_2E_1 = I_3$ (or $E_1E_2 = I_3$).

We need to prove one more simple theorem before we can establish a method for finding the inverse of a non-singular square matrix.

Theorem 6.4

If A and B are non-singular matrices of dimension $n \times n$, then AB is non-singular and $(AB)^{-1} = B^{-1}A^{-1}$.

Proof

Consider the matrix product $(AB)(B^{-1}A^{-1})$. Using the associative property of matrix multiplication (twice), we have:

$$\begin{aligned}(AB)(B^{-1}A^{-1}) &= A(BB^{-1})A^{-1} \\ &= AI_nA^{-1} \\ &= AA^{-1} \\ &= I_n.\end{aligned}$$

Hence $B^{-1}A^{-1}$ is the inverse of AB, i.e. $(AB)^{-1} = B^{-1}A^{-1}$. □

This theorem can be extended to cover the inverse of the product of any finite number of non-singular matrices to give the result

$$(A_1A_2 \ldots A_n)^{-1} = {A_n}^{-1}{A_{n-1}}^{-1} \ldots {A_1}^{-1}$$

(see exercise 6.4.1).

We are now in a position to devise a method for finding the inverse of a non-singular matrix.

Suppose that the $n \times n$ matrix A is row-equivalent to I_n, i.e. using a finite number of elementary row operations on A, we can obtain I_n. This means that pre-multiplication of A by a finite number of elementary matrices, one for each elementary row operation, results in I_n. Thus we have:

$$(E_m E_{m-1} \ldots E_2 E_1)A = I_n$$

so that

$$\begin{aligned} A^{-1} &= E_m E_{m-1} \ldots E_1 \\ &= (E_m E_{m-1} \ldots E_1) I_n. \end{aligned}$$

Thus whatever elementary row operations are necessary to reduce A to the identity matrix, the same operations performed in the same order on the appropriate identity matrix will result in the inverse of A.

Notice that A will have an inverse provided that A is row-equivalent to I_n. If A cannot be reduced to I_n by a finite sequence of elementary row operations then our method breaks down because we cannot write the matrix equation

$$(E_m E_{m-1} \ldots E_2 E_1)A = I_n.$$

Although we shall not prove it, it is the case that where A is not row-equivalent to I_n then A has no inverse. We state this formally in the following theorem.

Theorem 6.5

If A is an $n \times n$ matrix, A^{-1} exists if and only if A is row-equivalent to I_n.

Example 6.10

1. Find the inverse of

$$A = \begin{pmatrix} 2 & 4 \\ 1 & 4 \end{pmatrix}$$

(see exercise 6.3.6).

Solution

We perform a sequence of elementary row operations on A with the object of reducing this matrix to I_2. We then perform the same sequence of operations on I_2. The result is A^{-1}.

The operations on A are shown below. The elementary row operation used at each stage and the row to which it is applied are shown in square brackets after the matrix which results. We denote the ith row by R_i so that, for instance, $R_2 \to (R_2 - R_1)$ indicates that the second row has been transformed by having the first row subtracted from it:

$$\begin{aligned} A &= \begin{pmatrix} 2 & 4 \\ 1 & 4 \end{pmatrix} \\ &\sim \begin{pmatrix} 1 & 0 \\ 1 & 4 \end{pmatrix} \qquad [R_1 \to (R_1 - R_2)] \\ &\sim \begin{pmatrix} 1 & 0 \\ 0 & 4 \end{pmatrix} \qquad [R_2 \to (R_2 - R_1)] \\ &\sim \begin{pmatrix} 1 & 0 \\ 0 & 1 \end{pmatrix} \qquad [R_2 \to (R_2 \div 4)]. \end{aligned}$$

These three elementary row operations 'reduce' A to I_2. We now perform these operations, in the same order, on I_2:

$$\begin{aligned} I_2 &= \begin{pmatrix} 1 & 0 \\ 0 & 1 \end{pmatrix} \\ &\sim \begin{pmatrix} 1 & -1 \\ 0 & 1 \end{pmatrix} \qquad [R_1 \to (R_1 - R_2)] \\ &\sim \begin{pmatrix} 1 & -1 \\ -1 & 2 \end{pmatrix} \qquad [R_2 \to (R_2 - R_1)] \\ &\sim \begin{pmatrix} 1 & -1 \\ -\frac{1}{4} & \frac{1}{2} \end{pmatrix} \qquad [R_2 \to (R_2 \div 4)]. \end{aligned}$$

Thus

$$A^{-1} = \begin{pmatrix} 1 & -1 \\ -\frac{1}{4} & \frac{1}{2} \end{pmatrix}.$$

We can check this:

$$\begin{aligned} AA^{-1} &= \begin{pmatrix} 2 & 4 \\ 1 & 4 \end{pmatrix} \begin{pmatrix} 1 & -1 \\ -\frac{1}{4} & \frac{1}{2} \end{pmatrix} \\ &= \begin{pmatrix} 1 & 0 \\ 0 & 1 \end{pmatrix}. \end{aligned}$$

Alternatively

$$A^{-1}A = \begin{pmatrix} 1 & -1 \\ -\frac{1}{4} & \frac{1}{2} \end{pmatrix} \begin{pmatrix} 2 & 4 \\ 1 & 4 \end{pmatrix}$$
$$= \begin{pmatrix} 1 & 0 \\ 0 & 1 \end{pmatrix}.$$

So we have $AA^{-1} = A^{-1}A = I_2$ as required.

2. Find the inverse of

$$A = \begin{pmatrix} 6 & 2 \\ 4 & 1 \end{pmatrix}.$$

Solution

Rather than perform row operations on A followed by the same sequence on I_2, we may as well do both simultaneously. The usual way of presenting this is to write the matrices A and I_2 side by side thus:

$$\left(\begin{array}{cc|cc} 6 & 2 & 1 & 0 \\ 4 & 1 & 0 & 1 \end{array}\right).$$

This matrix, denoted by $(A\, I_2)$, is an example of a **partitioned matrix**. It is a 2×4 matrix partitioned into two 2×2 blocks and each block is termed a **submatrix** of the partition. We now perform elementary row operations on $(A\, I_2)$ until we obtain $(I_2\, A^{-1})$:

$$\begin{aligned}
(A\, I_2) &= \left(\begin{array}{cc|cc} 6 & 2 & 1 & 0 \\ 4 & 1 & 0 & 1 \end{array}\right) \\
&\sim \left(\begin{array}{cc|cc} -2 & 0 & 1 & -2 \\ 4 & 1 & 0 & 1 \end{array}\right) && [R_1 \to (R_1 - 2R_2)] \\
&\sim \left(\begin{array}{cc|cc} 1 & 0 & -\frac{1}{2} & 1 \\ 4 & 1 & 0 & 1 \end{array}\right) && [R_1 \to (R_1 \div (-2))] \\
&\sim \left(\begin{array}{cc|cc} 1 & 0 & -\frac{1}{2} & 1 \\ 0 & 1 & 2 & -3 \end{array}\right) && [R_2 \to (R_2 - 4R_1)].
\end{aligned}$$

Hence

$$A^{-1} = \begin{pmatrix} -\frac{1}{2} & 1 \\ 2 & -3 \end{pmatrix}.$$

We can check that $AA^{-1} = I_2$ (or $A^{-1}A = I_2$) as in the last example.

In general, for a non-singular matrix A, there will be many different sequences of elementary row operations which will reduce A to I_n and it does not matter which of these we use. Any sequence of elementary row operations which reduces A to I_n will, when applied to I_n, result in the inverse of A. It is useful, however, to develop a systematic method of utilizing elementary row operations which will reduce a square matrix to the appropriate identity matrix. Provided that the matrix concerned is row-equivalent to I_n the following steps will always reduce an $n \times n$ matrix to I_n.

1. Obtain a one in the top left-hand corner of the matrix A either by

   (a) dividing (or multiplying) the first row by a suitable constant, or
   (b) if the top left-hand element is zero, interchanging the first row with another row which has a non-zero element as its first entry and then performing step (a).

2. Subtract a suitable multiple of the first row from every other row so as to obtain zero in every first column entry apart from row 1.

The first column now has zeros in every row except the first where the element is a one. We now work on the second column with the object of obtaining a one in the second row and zeros elsewhere.

3. Divide (or multiply) the second row by a suitable constant so as to produce a one in the second column. If this is not possible because the $(2, 2)$-entry is zero, interchange row 2 with a row below it which does not have zero in the second column and then divide (or multiply) by a suitable constant.

4. Subtract a suitable multiple of the second row from every other row so that the second column consists of zeros apart from a one in row 2.

The second column now has zeros in every row except the second where the element is a one.

The process continues in the same way operating on each column in turn so as to produce a one on the leading diagonal and zeros everywhere else. Applying these steps to the partitioned matrix $(A\,I)$ finally results in $(I\,A^{-1})$ so long as A is non-singular. A flowchart for this algorithm is given in figure 6.1.

We illustrate these steps in the examples below.

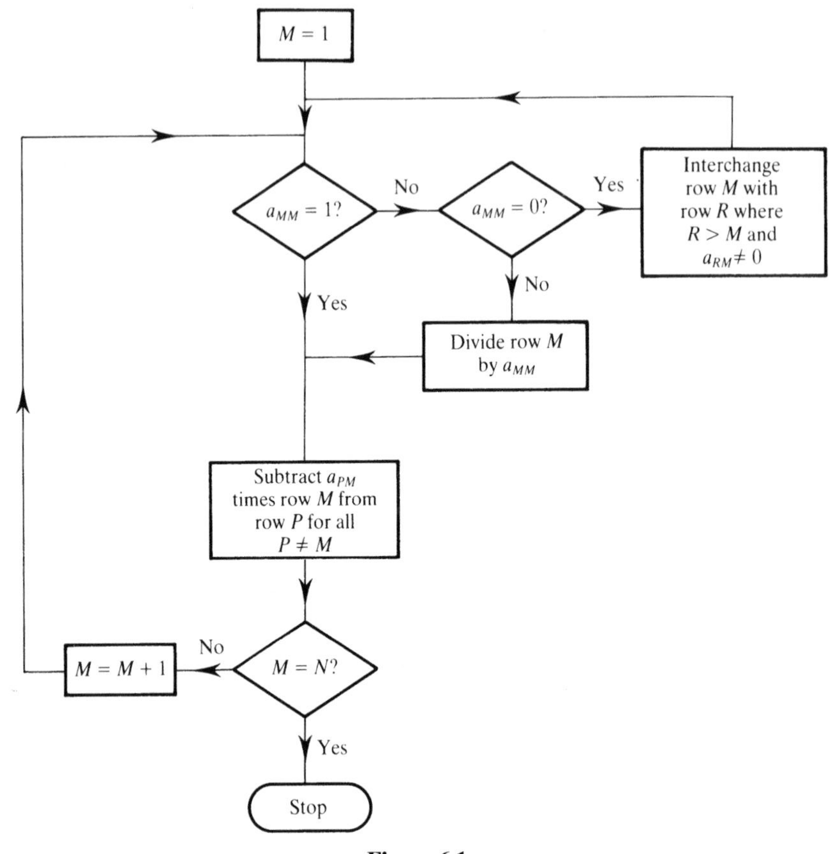

Figure 6.1

Examples 6.11

1. Find the inverse of

$$A = \begin{pmatrix} 2 & 2 & -6 \\ -1 & 1 & 2 \\ -3 & 5 & 3 \end{pmatrix}.$$

Solution

$$(A\,I_3) = \left(\begin{array}{rrr|rrr} 2 & 2 & -6 & 1 & 0 & 0 \\ -1 & 1 & 2 & 0 & 1 & 0 \\ -3 & 5 & 3 & 0 & 0 & 1 \end{array}\right)$$

$$\sim \left(\begin{array}{rrr|rrr} 1 & 1 & -3 & \frac{1}{2} & 0 & 0 \\ -1 & 1 & 2 & 0 & 1 & 0 \\ -3 & 5 & 3 & 0 & 0 & 1 \end{array}\right) \qquad [R_1 \to (R_1 \div 2)]$$

$$\sim \left(\begin{array}{rrr|rrr} 1 & 1 & -3 & \frac{1}{2} & 0 & 0 \\ 0 & 2 & -1 & \frac{1}{2} & 1 & 0 \\ 0 & 8 & -6 & \frac{3}{2} & 0 & 1 \end{array}\right) \qquad \begin{array}{l} [R_2 \to (R_2 + R_1)] \\ [R_3 \to (R_3 + 3R_1)] \end{array}$$

$$\sim \left(\begin{array}{rrr|rrr} 1 & 1 & -3 & \frac{1}{2} & 0 & 0 \\ 0 & 1 & -\frac{1}{2} & \frac{1}{4} & \frac{1}{2} & 0 \\ 0 & 8 & -6 & \frac{3}{2} & 0 & 1 \end{array}\right) \qquad [R_2 \to (R_2 \div 2)]$$

$$\sim \left(\begin{array}{rrr|rrr} 1 & 0 & -\frac{5}{2} & \frac{1}{4} & -\frac{1}{2} & 0 \\ 0 & 1 & -\frac{1}{2} & \frac{1}{4} & \frac{1}{2} & 0 \\ 0 & 0 & -2 & -\frac{1}{2} & -4 & 1 \end{array}\right) \qquad \begin{array}{l} [R_1 \to (R_1 - R_2)] \\ [R_3 \to (R_3 - 8R_2)] \end{array}$$

$$\sim \left(\begin{array}{rrr|rrr} 1 & 0 & -\frac{5}{2} & \frac{1}{4} & -\frac{1}{2} & 0 \\ 0 & 1 & -\frac{1}{2} & \frac{1}{4} & \frac{1}{2} & 0 \\ 0 & 0 & 1 & \frac{1}{4} & 2 & -\frac{1}{2} \end{array}\right) \qquad [R_3 \to (R_3 \div (-2))]$$

$$\sim \left(\begin{array}{rrr|rrr} 1 & 0 & 0 & \frac{7}{8} & \frac{9}{2} & -\frac{5}{4} \\ 0 & 1 & 0 & \frac{3}{8} & \frac{3}{2} & -\frac{1}{4} \\ 0 & 0 & 1 & \frac{1}{4} & 2 & -\frac{1}{2} \end{array}\right) \qquad \begin{array}{l} [R_1 \to (R_1 + \frac{5}{2}R_3)] \\ [R_2 \to (R_2 + \frac{1}{2}R_3)]. \end{array}$$

Thus

$$A^{-1} = \begin{pmatrix} \frac{7}{8} & \frac{9}{2} & -\frac{5}{4} \\ \frac{3}{8} & \frac{3}{2} & -\frac{1}{4} \\ \frac{1}{4} & 2 & -\frac{1}{2} \end{pmatrix}.$$

We can confirm the result by checking that $AA^{-1} = I_3$ (or $A^{-1}A = I_3$). Since it is only too easy to make errors in the process of finding a matrix inverse, it is always wise to do this.

2. Find the inverse of

$$A = \begin{pmatrix} 1 & 1 & 1 \\ 2 & -1 & 3 \\ 4 & 1 & 5 \end{pmatrix}.$$

Solution

$$(A\, I_3) = \left(\begin{array}{rrr|rrr} 1 & 1 & 1 & 1 & 0 & 0 \\ 2 & -1 & 3 & 0 & 1 & 0 \\ 4 & 1 & 5 & 0 & 0 & 1 \end{array}\right)$$

$$\sim \left(\begin{array}{ccc|ccc} 1 & 1 & 1 & 1 & 0 & 0 \\ 0 & -3 & 1 & -2 & 1 & 0 \\ 0 & -3 & 1 & -4 & 0 & 1 \end{array}\right) \qquad \begin{array}{l} [R_2 \to (R_2 - 2R_1)] \\ [R_3 \to (R_3 - 4R_1)] \end{array}$$

$$\sim \left(\begin{array}{ccc|ccc} 1 & 1 & 1 & 1 & 0 & 0 \\ 0 & 1 & -\frac{1}{3} & \frac{2}{3} & -\frac{1}{3} & 0 \\ 0 & -3 & 1 & -4 & 0 & 1 \end{array}\right) \qquad [R_2 \to (R_2 \div (-3))]$$

$$\sim \left(\begin{array}{ccc|ccc} 1 & 0 & \frac{4}{3} & \frac{1}{3} & \frac{1}{3} & 0 \\ 0 & 1 & -\frac{1}{3} & \frac{2}{3} & -\frac{1}{3} & 0 \\ 0 & 0 & 0 & -2 & -1 & 1 \end{array}\right) \qquad \begin{array}{l} [R_1 \to (R_1 - R_2)] \\ \\ [R_3 \to (R_3 + 3R_2)]. \end{array}$$

No further sequence of elementary row operations will complete the conversion of the matrix A to I_3. The matrix A is not row-equivalent to I_3. Therefore, by theorem 6.5, A does not have an inverse and is a singular matrix.

You are probably beginning to realize that finding the inverse of even a 3×3 matrix can be a tedious and error-prone activity. Using this method for inverting a 10×10 matrix is not for the faint-hearted! Fortunately, computer programs are available which will invert matrices of considerable dimension, e.g. Derive, Maple, Mathematica, etc.

Exercises 6.4

1. Prove by mathematical induction that

$$(A_1 A_2 \ldots A_n)^{-1} = A_n{}^{-1} A_{n-1}{}^{-1} \ldots A_2{}^{-1} A_1{}^{-1}.$$

2. Find the inverse (if it exists) of each of the following matrices:

(i) $\begin{pmatrix} 5 & -3 \\ -3 & 2 \end{pmatrix}$ (ii) $\begin{pmatrix} -4 & 8 \\ -1 & 3 \end{pmatrix}$ (iii) $\begin{pmatrix} 4 & 2 \\ 6 & 3 \end{pmatrix}$ (iv) $\begin{pmatrix} 1 & 1 & 0 \\ 1 & 0 & 0 \\ 1 & 1 & 1 \end{pmatrix}$

(v) $\begin{pmatrix} 1 & 2 & 3 \\ 4 & 5 & 6 \\ 7 & 8 & 9 \end{pmatrix}$ (vi) $\begin{pmatrix} 2 & 3 & -2 \\ 1 & 2 & -1 \\ -2 & 1 & 0 \end{pmatrix}$ (vii) $\begin{pmatrix} 4 & 2 & 1 \\ 3 & -1 & 0 \\ 2 & -4 & 2 \end{pmatrix}$

(viii) $\begin{pmatrix} 0 & 2 & 2 \\ 2 & 2 & 0 \\ 1 & 4 & 3 \end{pmatrix}$.

3. If A, B and C are square matrices of the same dimension, prove that, if A is non-singular and $AB = AC$, then $B = C$. Show that the matrix A in exercise 6.2.7 is singular so that the above result does not necessarily hold for A.

4. (i) Show that if A and B are square matrices of the same dimension and A is non-singular, then

$$(A^{-1}BA)^2 = A^{-1}B^2A.$$

(ii) Let

$$A = \begin{pmatrix} 2 & 1 \\ 3 & 1 \end{pmatrix} \quad \text{and} \quad B = \begin{pmatrix} 4 & -2 \\ 3 & -1 \end{pmatrix}.$$

Calculate $A^{-1}BA$. Using the result in (i), calculate $A^{-1}B^2A$ and hence $A^{-1}B^4A$.

(iii) Prove that $(A^{-1}BA)^n = A^{-1}B^nA$ for all positive integers n.

5. Let

$$A = \begin{pmatrix} a & b \\ c & d \end{pmatrix}.$$

Show that, if $ad - bc \neq 0$, then

$$A^{-1} = \frac{1}{ad - bc}\begin{pmatrix} d & -b \\ -c & a \end{pmatrix}.$$

Use this result to write down the inverses of

(i) $\begin{pmatrix} 2 & -3 \\ 4 & 1 \end{pmatrix}$ (ii) $\begin{pmatrix} 3 & 7 \\ 1 & 1 \end{pmatrix}$.

6. Suppose that A is a non-singular diagonal matrix of dimension $n \times n$ with (non-zero) diagonal elements a_{ii} $(i = 1, \ldots, n)$. Determine A^{-1}.

Hence write down the inverse of

$$\begin{pmatrix} \frac{4}{7} & 0 & 0 \\ 0 & -\frac{5}{3} & 0 \\ 0 & 0 & \frac{4}{9} \end{pmatrix}.$$

7. A matrix which is its own inverse is said to be **involutary**, i.e. A is an involutary matrix if $A^2 = I$. Prove that an $n \times n$ matrix A is involutary if and only if $(I_n - A)(I_n + A) = O_{n \times n}$.

8. Suppose that partitioned matrices A and B are as follows:

$$A = \left(\begin{array}{cc|ccc} 4 & 2 & 2 & 1 & 3 \\ 1 & -3 & 0 & -2 & 3 \end{array}\right)$$

and

$$B = \left(\begin{array}{cc|c} 2 & -1 & 4 \\ 0 & 3 & -1 \\ \hline 3 & -2 & 0 \\ -1 & 4 & -2 \\ 1 & 0 & 1 \end{array}\right)$$

so that

$$A = (A_1\ A_2) \quad \text{and} \quad B = \begin{pmatrix} B_1 & B_2 \\ B_3 & B_4 \end{pmatrix}.$$

Evaluate AB and show that

$$AB = (A_1B_1 + A_2B_3 \quad A_1B_2 + A_2B_4).$$

Show also that

$$B^{\mathrm{T}} = \begin{pmatrix} {B_1}^{\mathrm{T}} & {B_3}^{\mathrm{T}} \\ {B_2}^{\mathrm{T}} & {B_4}^{\mathrm{T}} \end{pmatrix}.$$

Chapter 7

Systems of Linear Equations

7.1 Introduction

A **linear equation** in n **variables** (or **unknowns**) $x_1, x_2, \ldots, x_n$ is one which can be expressed in the form

$$a_1x_1 + a_2x_2 + \cdots + a_nx_n = b \qquad (1)$$

where $a_1, a_2, \ldots, a_n$ and b are real numbers. The constants $a_1, a_2, \ldots, a_n$ are called the **coefficients** of the variables $x_1, x_2, \ldots, x_n$ respectively.

Examples of linear equations are:

$$\begin{aligned} 3x_1 + 2x_2 - 5x_3 &= 7, \\ 3y &= 7x - 4z + 2, \\ 5x_1 - 2x_3 &= 4x_2 + 3x_4. \end{aligned}$$

A linear equation which is expressed in the form of (1) above, with the variables on the left-hand side of the equation and the constant term on the right, is said to be written in **standard form**. Of the examples above, only the first is in standard form.

A **solution** of the linear equation

$$a_1x_1 + a_2x_2 + \cdots + a_nx_n = b$$

is an ordered n-tuple $(c_1, c_2, \ldots, c_n)$ such that $x_1 = c_1$, $x_2 = c_2, \ldots, x_n = c_n$ satisfies the equation, i.e. such that $a_1c_1 + a_2c_2 + \cdots + a_nc_n = b$.

For the simplest linear equation, $ax = b$, there are three possibilities concerning solutions.

1. If $a \neq 0$, the equation has the single solution $x = b/a$.
2. If $a = 0$ and $b \neq 0$, then the equation has no solution.
3. If $a = 0$ and $b = 0$, then any real number is a solution of the equation, i.e. the equation has an infinite number of solutions.

For the general linear equation (1) with more than one variable, we have two possibilities. If $a_1 = a_2 = \cdots = a_n = 0$, $b \neq 0$, the equation will have no solution. Otherwise a single linear equation in more than one variable will have an infinite number of solutions. For example, if we take the linear equation

$$3x - 2y + z = 4$$

some solutions are:

$$\begin{array}{llll} x = 0 & y = 0 & z = 4 & \text{i.e. } (0, 0, 4), \\ x = 1 & y = \frac{1}{2} & z = 2 & \text{i.e. } (1, \frac{1}{2}, 2), \\ x = 2 & y = 1 & z = 0 & \text{i.e. } (2, 1, 0). \end{array}$$

The **solution set** of a linear equation is the set of all possible solutions. For a linear equation in n variables, this is a set of ordered n-tuples. Apart from the trivial case referred to above, for a linear equation with two or more variables this will be an infinite set.

We can interpret linear equations in two and three variables geometrically. The solution set of the equation $a_1x + a_2y = b$ (where a_1 and a_2 are not both zero) is represented by all points on a line in two-dimensional space, $\mathbb{R}^2$.

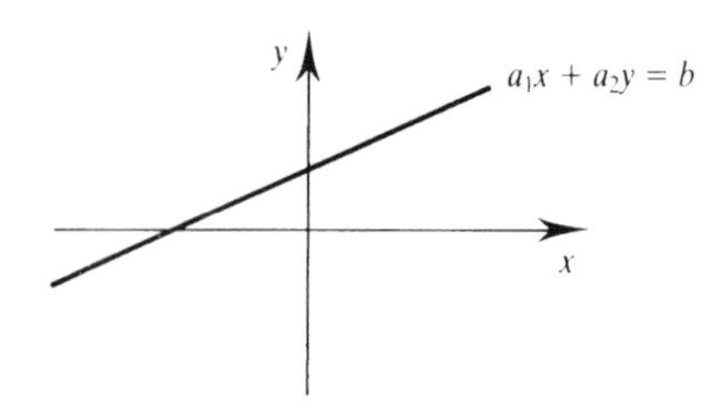

The general linear equation in three variables $a_1x + a_2y + a_3z = b$ has a solution set which, so long as it is not empty, defines a plane in three-dimensional space, $\mathbb{R}^3$.

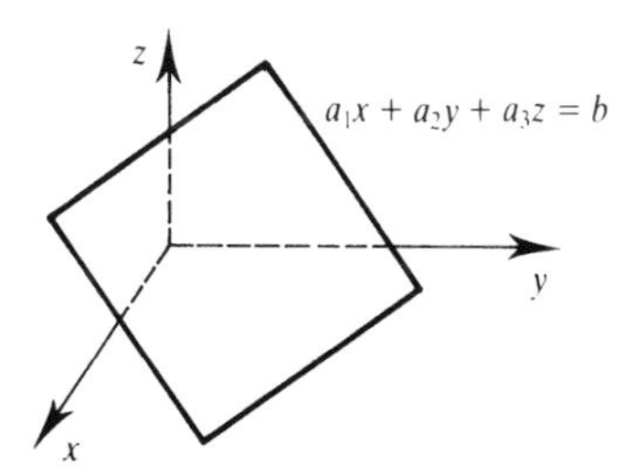

A **system of linear equations** is simply several linear equations involving the same variables. In general a system consisting of m linear equations in n variables can be written:

$$\begin{array}{ccccccccc} a_{11}x_1 & + & a_{12}x_2 & + & \cdots & + & a_{1n}x_n & = & b_1 \\ a_{21}x_1 & + & a_{22}x_2 & + & \cdots & + & a_{2n}x_n & = & b_2 \\ \vdots & & \vdots & & & & \vdots & & \vdots \\ a_{m1}x_1 & + & a_{m2}x_2 & + & \cdots & + & a_{mn}x_n & = & b_m. \end{array}$$

If $b_1 = b_2 = \cdots = b_m = 0$, then the system of linear equations is said to be **homogeneous**. If any of these constants is not zero, then the system is said to be **non-homogeneous**. For example,

$$\begin{aligned} 6x + 2y - z &= 7 \\ 4x + \ \ y + z &= 2 \end{aligned}$$

is a non-homogeneous system of two linear equations in the three variables x, y and z. The system

$$\begin{aligned} 2x_1 - x_2 + x_3 - x_4 &= 0 \\ x_2 &= 2x_1 - 4x_3 - x_4 \\ 3x_1 + 5x_2 - 2x_3 &= 0 \end{aligned}$$

is recognizable as a homogeneous system of three linear equations in the four variables x_1, x_2, x_3 and x_4 once the second equation has been written in standard form:

$$2x_1 - x_2 - 4x_3 - x_4 = 0.$$

A **solution of a system of linear equations** is an ordered n-tuple defining values of the variables which satisfy each equation in the system. For instance, the system

$$\begin{aligned} 3x - 2y + z &= -3 \\ x + \ \ y + z &= \ \ 5 \\ x - 2y - z &= -9 \\ y + z &= \ \ 6 \end{aligned}$$

has a solution $(-1, 2, 4)$.

In general, just as for the single linear equation $ax = b$, a system of linear equations may have none, one or many solutions. A system which has no solution is called **inconsistent**. A system which has one or many solutions is called **consistent**.

A convenient way to represent a system of linear equations is in matrix form. Consider the following general system of m equations in the n variables $x_1, x_2, \ldots, x_n$:

$$\begin{array}{ccccccc} a_{11}x_1 & + & a_{12}x_2 & + \cdots + & a_{1n}x_n & = & b_1 \\ a_{21}x_1 & + & a_{22}x_2 & + \cdots + & a_{2n}x_n & = & b_2 \\ \vdots & & \vdots & & \vdots & & \vdots \\ a_{m1}x_1 & + & a_{m2}x_2 & + \cdots + & a_{mn}x_n & = & b_m. \end{array}$$

This system can be represented by the equivalent matrix equation:

$$\begin{pmatrix} a_{11} & a_{12} & \cdots & a_{1n} \\ a_{12} & a_{22} & \cdots & a_{2n} \\ \vdots & \vdots & & \vdots \\ a_{m1} & a_{m2} & \cdots & a_{mn} \end{pmatrix} \begin{pmatrix} x_1 \\ x_2 \\ \vdots \\ x_n \end{pmatrix} = \begin{pmatrix} b_1 \\ b_2 \\ \vdots \\ b_m \end{pmatrix}.$$

Multiplying together the matrices on the left-hand side of the equation gives a matrix of dimension $m \times 1$ whose elements are the left-hand side of the system of equations. Equating the elements of this matrix with those in the matrix on the right-hand side of the matrix equation gives each of the m equations in the system.

If we let

$$A = [a_{ij}] \qquad \boldsymbol{x} = \begin{pmatrix} x_1 \\ x_2 \\ \vdots \\ x_n \end{pmatrix} \qquad \text{and} \quad \boldsymbol{b} = \begin{pmatrix} b_1 \\ b_2 \\ \vdots \\ b_m \end{pmatrix}$$

then we can write the matrix equation as

$$A\boldsymbol{x} = \boldsymbol{b}.$$

The matrix A is often referred to as the **matrix of coefficients**. A solution of the system corresponds to any column matrix $\boldsymbol{x}$ which satisfies the matrix equation. The elements of this column matrix are values of $x_1, x_2, \ldots, x_n$ which satisfy all the equations in the system.

Example 7.1

Write the following system of linear equations in matrix form:

$$\begin{aligned} 3x_1 + 2x_2 - x_3 &= 7 \\ x_1 - 3x_2 - 2x_3 &= -5 \\ 2x_1 + x_2 &= 4 \\ 6x_2 + 7x_3 &= 12. \end{aligned}$$

Solution

The equivalent matrix equation is

$$\begin{pmatrix} 3 & 2 & -1 \\ 1 & -3 & -2 \\ 2 & 1 & 0 \\ 0 & 6 & 7 \end{pmatrix} \begin{pmatrix} x_1 \\ x_2 \\ x_3 \end{pmatrix} = \begin{pmatrix} 7 \\ -5 \\ 4 \\ 12 \end{pmatrix}.$$

Thus the system can be written in the form $A\boldsymbol{x} = \boldsymbol{b}$, where

$$A = \begin{pmatrix} 3 & 2 & -1 \\ 1 & -3 & -2 \\ 2 & 1 & 0 \\ 0 & 6 & 7 \end{pmatrix} \quad \boldsymbol{x} = \begin{pmatrix} x_1 \\ x_2 \\ x_3 \end{pmatrix} \quad \boldsymbol{b} = \begin{pmatrix} 7 \\ -5 \\ 4 \\ 12 \end{pmatrix}.$$

That

$$\boldsymbol{x} = \begin{pmatrix} 1 \\ 2 \\ 0 \end{pmatrix}$$

(i.e. $x_1 = 1$, $x_2 = 2$, $x_3 = 0$) is a solution of the system can be checked by substitution into the matrix equation.

We saw earlier that the linear equation $ax = b$ has none, one or infinitely many solutions. The situation is exactly the same for any system of linear equations. We can justify this assertion for a system of two equations in two variables:

$$\begin{aligned} a_{11}x + a_{12}y &= b_1 \\ a_{21}x + a_{22}y &= b_2. \end{aligned}$$

As long as the coefficients of x and y are not both zero, each of these equations represents a line in the (x, y)-plane and the solution of the system can be interpreted as a point or points which are common to *both* lines. There are three possibilities.

(a) The lines are parallel and do not meet at all. In this case the system has no solution and is inconsistent. This situation is illustrated in the diagram below.

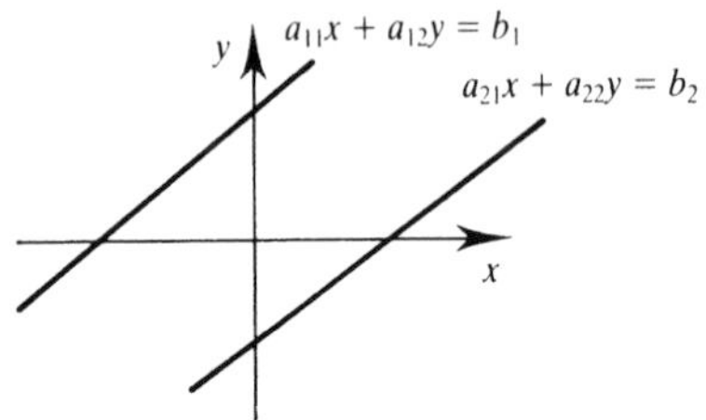

(b) The lines cross and therefore have one point in common. In this case the system has one solution.

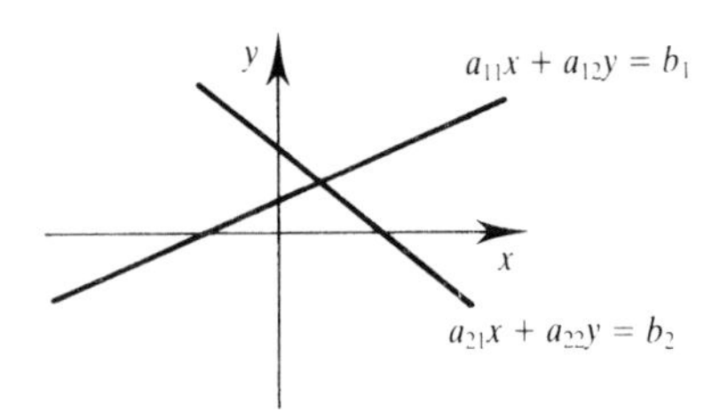

(c) The lines coincide so that all points on one line are common to the other. The system then has an infinite number of solutions.

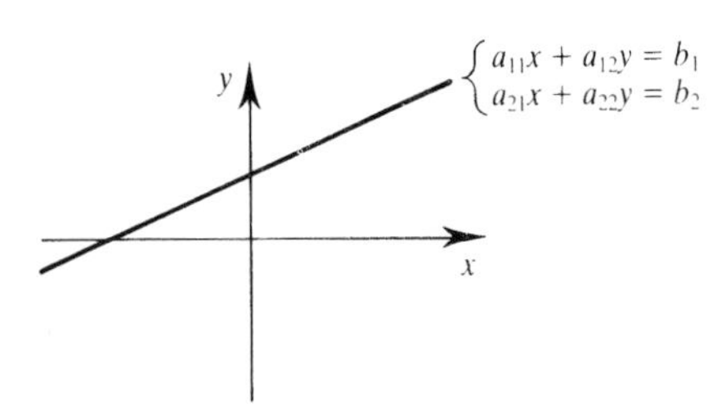

It is clearly not possible for two lines to have more than one point in common unless they have all points in common.

We now prove this result for the general system of linear equations. We show that, for a consistent set of equations, either there is one solution or there are infinitely many solutions.

Theorem 7.1

Every system of linear equations has no solution, one solution or infinitely many solutions.

Proof

We shall come across examples of each of the three cases during the course of this chapter so, to prove the theorem, we must show that there are no other possibilities. This means that we must show that a system of linear equations having more then one solution cannot have a finite number of solutions.

Suppose that the system of equations represented in matrix form by

$$A\boldsymbol{x} = \boldsymbol{b}$$

has two different solutions $\boldsymbol{x} = \boldsymbol{u}$ and $\boldsymbol{x} = \boldsymbol{v}$ so that

$$A\boldsymbol{u} = \boldsymbol{b} \quad \text{and} \quad A\boldsymbol{v} = \boldsymbol{b}.$$

Consider the matrix $t\boldsymbol{u} + (1-t)\boldsymbol{v}$ where t is any real number. Now

$$\begin{aligned} A[t\boldsymbol{u} + (1-t)\boldsymbol{v}] &= tA\boldsymbol{u} + (1-t)A\boldsymbol{v} \\ &= t\boldsymbol{b} + (1-t)\boldsymbol{b} \\ &= \boldsymbol{b}. \end{aligned}$$

This shows that, if $\boldsymbol{u}$ and $\boldsymbol{v}$ are solutions to the system, then so is $t\boldsymbol{u} + (1-t)\boldsymbol{v}$ for any value of t. This proves the result, i.e. that a system having more than one solution has an infinite number of solutions. □

We now consider some methods for solving systems of linear equations.

7.2 Matrix Inverse Method

Suppose that we have a system of m linear equations in n variables expressed in matrix form $A\boldsymbol{x} = \boldsymbol{b}$. Now consider the case where $m = n$, i.e. we have the same number of equations as variables. The matrix A is then square and of dimension

$n \times n$. If A is non-singular so that A^{-1} exists, we can pre-multiply the matrix equation above by A^{-1} to give:

$$\begin{aligned} & A^{-1}A\boldsymbol{x} = A^{-1}\boldsymbol{b} \\ \Rightarrow \quad & I_n\boldsymbol{x} = A^{-1}\boldsymbol{b} \\ \Rightarrow \quad & \boldsymbol{x} = A^{-1}\boldsymbol{b}. \end{aligned}$$

Equating the elements in the matrices on each side of the last equation gives the values of $x_1, x_2, \ldots, x_n$ which satisfy all the equations in the system, i.e. a solution to the system of equations.

Example 7.2

Solve the system of equations

$$\begin{aligned} 2x + 2y - 6z &= 4 \\ -x + y + 2z &= 3 \\ -3x + 5y + 3z &= -1. \end{aligned}$$

Solution

The system of equations can be written in matrix form

$$A\boldsymbol{x} = \boldsymbol{b}$$

where

$$A = \begin{pmatrix} 2 & 2 & -6 \\ -1 & 1 & 2 \\ -3 & 5 & 3 \end{pmatrix} \quad \boldsymbol{x} = \begin{pmatrix} x \\ y \\ z \end{pmatrix} \quad \text{and} \quad \boldsymbol{b} = \begin{pmatrix} 4 \\ 3 \\ -1 \end{pmatrix}.$$

If A is non-singular, a solution of the system is given by

$$\boldsymbol{x} = A^{-1}\boldsymbol{b}.$$

In example 6.10.1 we found that

$$A^{-1} = \begin{pmatrix} \frac{7}{8} & \frac{9}{2} & -\frac{5}{4} \\ \frac{3}{8} & \frac{3}{2} & -\frac{1}{4} \\ \frac{1}{4} & 2 & -\frac{1}{2} \end{pmatrix}$$

so that $\boldsymbol{x} = A^{-1}\boldsymbol{b}$ is equivalent to

$$\begin{pmatrix} x \\ y \\ z \end{pmatrix} = \begin{pmatrix} \frac{7}{8} & \frac{9}{2} & -\frac{5}{4} \\ \frac{3}{8} & \frac{3}{2} & -\frac{1}{4} \\ \frac{1}{4} & 2 & -\frac{1}{2} \end{pmatrix} \begin{pmatrix} 4 \\ 3 \\ -1 \end{pmatrix}.$$

Multiplying the matrices on the right-hand side we have

$$\begin{pmatrix} x \\ y \\ z \end{pmatrix} = \begin{pmatrix} \frac{146}{8} \\ \frac{50}{8} \\ \frac{15}{2} \end{pmatrix}$$

and a solution of the system is

$$x = \frac{146}{8} = \frac{73}{4},\ y = \frac{50}{8} = \frac{25}{4},\ z = \frac{15}{2},$$

i.e. $(x, y, z) = \left(\frac{73}{4}, \frac{25}{4}, \frac{15}{2}\right)$.

This can be confirmed by substituting these values of x, y and z into each equation and checking that they satisfy every equation in the system.

We have already shown that a system of linear equations may have none, one or infinitely many solutions. It is reasonable therefore to ask whether the solution obtained by the method above is the only solution or whether it is just one of an infinite number still to be found. It is not difficult to prove that, when A is non-singular, the solution $\boldsymbol{x} = A^{-1}\boldsymbol{b}$ is the only solution. This we do in the following theorem.

Theorem 7.2

If A is a non-singular square matrix, then the system of linear equations represented by $A\boldsymbol{x} = \boldsymbol{b}$ has a unique solution given by $\boldsymbol{x} = A^{-1}\boldsymbol{b}$.

Proof

Suppose that $\boldsymbol{x}$ and $\boldsymbol{y}$ are both solutions so that

$$A\boldsymbol{x} = \boldsymbol{b} \quad \text{and} \quad A\boldsymbol{y} = \boldsymbol{b}.$$

We have

$$
\begin{aligned}
& & A\boldsymbol{x} &= A\boldsymbol{y} \\
&\Rightarrow & A^{-1}A\boldsymbol{x} &= A^{-1}A\boldsymbol{y} \\
&\Rightarrow & \boldsymbol{x} &= \boldsymbol{y}.
\end{aligned}
$$

We have already seen that $\boldsymbol{x} = A^{-1}\boldsymbol{b}$ is a solution and we can therefore conclude that this is the unique solution of the equation. □

Example 7.3

Solve the homogeneous system of equations

$$
\begin{aligned}
2x + 2y - 6z &= 0 \\
-x + y + 2z &= 0 \\
-3x + 5y + 3z &= 0.
\end{aligned}
$$

Solution

In matrix form the system of equations is given by $A\boldsymbol{x} = \boldsymbol{b}$ where

$$
A = \begin{pmatrix} 2 & 2 & -6 \\ -1 & 1 & 2 \\ -3 & 5 & 3 \end{pmatrix} \quad \boldsymbol{x} = \begin{pmatrix} x \\ y \\ z \end{pmatrix} \quad \boldsymbol{b} = \begin{pmatrix} 0 \\ 0 \\ 0 \end{pmatrix}.
$$

The solution is $\boldsymbol{x} = A^{-1}\boldsymbol{b}$ where A^{-1} is as in the last example. So we have

$$
\begin{aligned}
\boldsymbol{x} &= \begin{pmatrix} \frac{7}{8} & \frac{9}{2} & -\frac{5}{4} \\ \frac{3}{8} & \frac{3}{2} & -\frac{1}{4} \\ \frac{1}{4} & 2 & -\frac{1}{2} \end{pmatrix} \begin{pmatrix} 0 \\ 0 \\ 0 \end{pmatrix} \\
&= \begin{pmatrix} 0 \\ 0 \\ 0 \end{pmatrix}.
\end{aligned}
$$

Thus the solution of the system is $(0, 0, 0)$.

This result is fairly obvious when we look at the original set of equations. It is clear that any set of homogeneous equations will always have a solution where all variables are zero. This solution is called the **trivial solution**. For a homogeneous system of n equations in n variables where the matrix of coefficients is non-singular, the implication of theorem 7.2 is that the *only* solution is the trivial one.

We now have a method for solving a system of n linear equations in n variables so long as A, the matrix of coefficients, is non-singular. However, the method is computationally inefficient because, as we have already noted, inverting a matrix can involve a lot of arithmetic operations. Also, what happens if A is not square or is square but singular? In either of these cases, the method fails.

The method based on finding the inverse of the matrix of coefficients, although useful, has obvious limitations. In the next section we consider another method which can be applied to any system of linear equations and which therefore does not suffer from the disadvantages of the method described above.

Exercises 7.1

1. Find the inverse of the matrix

$$A = \begin{pmatrix} 2 & -1 \\ 1 & 2 \end{pmatrix}.$$

Hence solve the system of equations

$$\begin{aligned} 2x - y &= 6 \\ x + 2y &= 8. \end{aligned}$$

2. Find the inverse of the matrix

$$A = \begin{pmatrix} 2 & 2 & 1 \\ 1 & -1 & -1 \\ 1 & 3 & 3 \end{pmatrix}.$$

Hence solve the system of equations

$$\begin{aligned} 2x + 2y + z &= 4 \\ x - y - z &= 1 \\ x + 3y + 3z &= 1. \end{aligned}$$

Write down the solution of the system

$$\begin{aligned} 2x + 2y + z &= 0 \\ x - y - z &= 0 \\ x + 3y + 3z &= 0. \end{aligned}$$

3. Solve the system of linear equations

$$\begin{aligned} 3x + 2y + z &= 4 \\ x - y + 2z &= 8 \\ 6x - 3y - z &= -4. \end{aligned}$$

7.3 Gauss†–Jordan‡ Elimination

In order to explain the method we need two definitions.

Definition 7.1

A matrix is said to be in **row echelon form** if all the following are true.

(a) The first (i.e. furthest to the left) non-zero element in every row is a 1.
(b) In each row the 'leading 1' is further to the right than the leading 1 in any preceding row.
(c) Every row of zeros is below all non-zero rows.

(The first non-zero element in any row of a matrix is termed the **leading element** of that row.)

Examples of matrices in row echelon form are:

$$\begin{pmatrix} 1 & 3 & -2 & 5 \\ 0 & 0 & 1 & 4 \\ 0 & 0 & 0 & 1 \\ 0 & 0 & 0 & 0 \end{pmatrix} \quad \begin{pmatrix} 1 & -1 & 0 \\ 0 & 1 & 2 \\ 0 & 0 & 1 \end{pmatrix} \quad \begin{pmatrix} 0 & 1 & -1 & 4 \\ 0 & 0 & 1 & -2 \\ 0 & 0 & 0 & 0 \end{pmatrix}.$$

† Carl Friedrich Gauss (1777–1855), a German, is considered by many to be the greatest mathematician of all time. He was a child prodigy and, when he was 20, gave the first proof of the fundamental theorem of algebra. Much of his extensive research he never published. As a result, a great deal of the mathematics published in the mid nineteenth century was subsequently found to have been discovered earlier by Gauss.
‡ Wilhelm Jordan (1842–99) was a surveyor and professor of geodesy at the Karlsruhe Technical College in Germany. However, this method is often mistakenly attributed to Camille Jordan (1838–1921), an engineer who published material in many branches of mathematics, particularly group theory.

GAUSS–JORDAN ELIMINATION

The following matrices are not in row echelon form, the first because the leading element in the second row is not a one and the second because the leading 1 in the second row is to the left of the leading 1 in the first row. The third matrix fails because a row of zeros (the third row) is above a non-zero row.

$$\begin{pmatrix} 1 & 1 & 3 \\ 0 & 2 & 0 \\ 0 & 0 & 1 \end{pmatrix} \quad \begin{pmatrix} 0 & 1 & 2 \\ 1 & 2 & -1 \\ 0 & 0 & 1 \end{pmatrix} \quad \begin{pmatrix} 1 & 0 & 0 & 3 \\ 0 & 0 & 1 & -2 \\ 0 & 0 & 0 & 0 \\ 0 & 0 & 0 & 1 \end{pmatrix}.$$

Definition 7.2

A matrix is said to be in **reduced row echelon form** if it is in row echelon form and every column which contains a leading 1 contains zeros for all its other elements.

ROW ECHELON FORM

REDUCED ROW ECHELON FORM

Examples of reduced row echelon matrices are:

$$\begin{pmatrix} 1 & 0 & 0 & -4 \\ 0 & 1 & 0 & 6 \\ 0 & 0 & 1 & 5 \end{pmatrix} \quad \begin{pmatrix} 1 & 0 & 0 \\ 0 & 1 & 0 \\ 0 & 0 & 1 \end{pmatrix} \quad \begin{pmatrix} 0 & 1 & -1 & 0 & 0 \\ 0 & 0 & 0 & 1 & 0 \\ 0 & 0 & 0 & 0 & 1 \\ 0 & 0 & 0 & 0 & 0 \end{pmatrix}.$$

You are probably familiar with the method used in the example below to solve a system of two linear equations in two variables.

Example 7.4

Solve the system of equations

$$\begin{aligned} 2x - 2y &= 6 \\ 4x + y &= 7. \end{aligned}$$

Solution

We label the equations (e_1) and (e_2) so that we can refer to them:

$$2x - 2y = 6 \qquad (e_1)$$
$$4x + y = 7. \qquad (e_2)$$

We can eliminate x from (e_2) by subtracting from it twice (e_1). This gives the two equations

$$2x - 2y = 6 \qquad (e_1)$$
$$5y = -5. \qquad (e_2 - 2e_1)$$

Dividing the second of these equations by 5 gives $y = -1$. Substitution of $y = -1$ into (e_1) gives $2x = 4$ so that $x = 2$.

This method essentially consists of applying certain 'allowed' operations to the equations within the system to produce another system having the same solution but which is easier to solve. Two or more systems of equations having the same solution(s) are called **equivalent** systems. The following operations on a system of linear equations are 'permitted' in that they produce another system which is equivalent:

(a) interchanging two equations;
(b) multiplying (or dividing) one equation by a non-zero constant;
(c) adding to one equation a multiple of another equation.

These operations are precisely the elementary row operations (§6.4) but applied to the equations in a system rather than the rows of a matrix.

Now suppose that we have the following linear system to solve:

$$\begin{aligned} x + y + z &= 2 \\ 2x - 2y - z &= 2 \\ 3x + y - 2z &= -2. \end{aligned}$$

We represent the system by the following partitioned matrix:

$$\left(\begin{array}{rrr|r} 1 & 1 & 1 & 2 \\ 2 & -2 & -1 & 2 \\ 3 & 1 & -2 & -2 \end{array}\right).$$

This is the matrix of coefficients which we denoted by A but with an extra column consisting of the constant terms on the right-hand sides of the equations. It is called the **augmented matrix** (or the **augmented matrix of coefficients**) and is denoted by $(A\ \boldsymbol{b})$. Each equation in the system can be reconstructed from the augmented matrix.

Each permitted operation on the equations of the system corresponds to an elementary row operation on the augmented matrix. Each elementary row operation produces a row-equivalent matrix representing an equivalent system of equations. In other words, row-equivalent augmented matrices represent equivalent systems of equations. Now suppose that we can reduce the matrix to reduced row echelon form using elementary row operations. The result might be something like

$$\left(\begin{array}{rrr|r} 1 & 0 & 0 & a \\ 0 & 1 & 0 & b \\ 0 & 0 & 1 & c \end{array}\right)$$

where a, b and c are constants. This matrix represents the system $x = a$, $y = b$, $z = c$, i.e. the solution of the original system.

This suggests a useful method for solving a system of linear equations. We write down the augmented matrix and, by applying elementary row operations to it, we reduce it (if possible) to reduced row echelon form. The solution(s) of the system can then be read off from this matrix. This process is called **Gauss–Jordan elimination**. Note that the method does not depend on being able to invert any matrix nor on the number of equations or variables. Hence it is general enough to apply to any system of linear equations.

Example 7.5

Solve the following system of linear equations:

$$\begin{aligned} x + \ y + \ z &= \ \ 2 \\ 2x - 2y - \ z &= \ \ 2 \\ 3x + \ y - 2z &= -2. \end{aligned}$$

Solution

Starting from the augmented matrix we reduce it by elementary row operations as follows:

$$\left(\begin{array}{ccc|c} 1 & 1 & 1 & 2 \\ 2 & -2 & -1 & 2 \\ 3 & 1 & -2 & -2 \end{array}\right) \quad \text{(augmented matrix)}$$

$$\sim \left(\begin{array}{ccc|c} 1 & 1 & 1 & 2 \\ 0 & -4 & -3 & -2 \\ 0 & -2 & -5 & -8 \end{array}\right) \quad \begin{array}{l} [R_2 \to (R_2 - 2R_1)] \\ [R_3 \to (R_3 - 3R_1)] \end{array}$$

$$\sim \left(\begin{array}{ccc|c} 1 & 1 & 1 & 2 \\ 0 & 1 & \frac{3}{4} & \frac{1}{2} \\ 0 & -2 & -5 & -8 \end{array}\right) \quad [R_2 \to (R_2 \div (-4))]$$

$$\sim \left(\begin{array}{ccc|c} 1 & 0 & \frac{1}{4} & \frac{3}{2} \\ 0 & 1 & \frac{3}{4} & \frac{1}{2} \\ 0 & 0 & -\frac{7}{2} & -7 \end{array}\right) \quad \begin{array}{l} [R_1 \to (R_1 - R_2)] \\ \\ [R_3 \to (R_3 + 2R_2)] \end{array}$$

$$\sim \left(\begin{array}{ccc|c} 1 & 0 & \frac{1}{4} & \frac{3}{2} \\ 0 & 1 & \frac{3}{4} & \frac{1}{2} \\ 0 & 0 & 1 & 2 \end{array}\right) \quad [R_3 \to (R_3 \div (-\tfrac{7}{2}))]$$

$$\sim \left(\begin{array}{ccc|c} 1 & 0 & 0 & 1 \\ 0 & 1 & 0 & -1 \\ 0 & 0 & 1 & 2 \end{array}\right) \quad \begin{array}{l} [R_1 \to (R_1 - \tfrac{1}{4}R_3)] \\ [R_2 \to (R_2 - \tfrac{3}{4}R_3)]. \end{array}$$

Thus the solution is $x = 1$, $y = -1$, $z = 2$. These can be checked by substitution into the three equations.

The systematic steps required to reduce the augmented matrix to reduced row echelon form are exactly those which we employed to find the inverse of a matrix A when we used elementary row operations to convert the partitioned matrix $(A\ I)$ to $(I\ A^{-1})$. The sequence of steps is given on page 325 and illustrated in the flowchart in figure 6.1. In fact, if A (the matrix of coefficients) is a square non-singular matrix, the process of Gauss–Jordan elimination will inevitably result in a reduced row echelon form which consists of the appropriate identity matrix with an extra column on the right-hand side.

We now see what happens if A does not have an inverse, either because (a) A is a square singular matrix, or (b) A is not a square matrix.

Example 7.6

Solve the system of linear equations

$$\begin{aligned} 2x + 2y + z &= 4 \\ x - \ \ y - z &= 2 \\ 3x + \ \ y \qquad &= 6. \end{aligned}$$

Solution

$$\left(\begin{array}{ccc|c} 2 & 2 & 1 & 4 \\ 1 & -1 & -1 & 2 \\ 3 & 1 & 0 & 6 \end{array}\right) \quad \text{(augmented matrix)}$$

$$\sim \left(\begin{array}{ccc|c} 1 & 1 & \frac{1}{2} & 2 \\ 1 & -1 & -1 & 2 \\ 3 & 1 & 0 & 6 \end{array}\right) \quad [R_1 \to (R_1 \div 2)]$$

$$\sim \left(\begin{array}{ccc|c} 1 & 1 & \frac{1}{2} & 2 \\ 0 & -2 & -\frac{3}{2} & 0 \\ 0 & -2 & -\frac{3}{2} & 0 \end{array}\right) \quad \begin{array}{l} [R_2 \to (R_2 - R_1)] \\ [R_3 \to (R_3 - 3R_1)] \end{array}$$

$$\sim \left(\begin{array}{ccc|c} 1 & 1 & \frac{1}{2} & 2 \\ 0 & 1 & \frac{3}{4} & 0 \\ 0 & -2 & -\frac{3}{2} & 0 \end{array}\right) \quad [R_2 \to (R_2 \div (-2))]$$

$$\sim \left(\begin{array}{ccc|c} 1 & 0 & -\frac{1}{4} & 2 \\ 0 & 1 & \frac{3}{4} & 0 \\ 0 & 0 & 0 & 0 \end{array}\right) \quad \begin{array}{l} [R_1 \to (R_1 - R_2)] \\ \\ [R_3 \to (R_3 + 2R_2)]. \end{array}$$

This is now in reduced row echelon form and represents the system of equations

$$\begin{aligned} x - \tfrac{1}{4}z &= 2 \\ y + \tfrac{3}{4}z &= 0. \end{aligned}$$

The last equation which would normally be expected to give us the value for z, corresponds to $0z = 0$. This is satisfied by any value of z and therefore z can be chosen arbitrarily. Writing $z = t$, where t is a parameter which can take any real value, we have

$$\begin{aligned} x &= 2 + \tfrac{1}{4}t \\ y &= \ \ -\tfrac{3}{4}t \end{aligned}$$

and the system has an infinite number of solutions corresponding to all possible values of the parameter t. We can write the solution set as

$$\begin{aligned}&\left\{(x,y,z) : x = 2 + \tfrac{1}{4}t, y = -\tfrac{3}{4}t, z = t, \text{where } t \in \mathbb{R}\right\}\\ &\quad = \left\{\left(2 + \tfrac{1}{4}t, -\tfrac{3}{4}t, t\right) : t \in \mathbb{R}\right\}.\end{aligned}$$

In the example above, we can see from the reduced row echelon matrix that A, the matrix of coefficients, is singular since it has not been possible to reduce that part of the augmented matrix corresponding to A to the identity matrix. (Remember that any non-singular square matrix is row-equivalent to the appropriate identity matrix.) However, it does not matter that A is singular; the method has still enabled us to solve the system.

Example 7.7

Solve the linear system

$$\begin{aligned}2x_1 - x_2 + 5x_3 + 3x_4 &= 5\\ x_1 + x_2 + 4x_3 + 3x_4 &= 7\\ x_1 \qquad\;\; + 3x_3 + 2x_4 &= 4\\ x_2 + x_3 + x_4 &= 3.\end{aligned}$$

Solution

$$\left(\begin{array}{cccc|c} 2 & -1 & 5 & 3 & 5\\ 1 & 1 & 4 & 3 & 7\\ 1 & 0 & 3 & 2 & 4\\ 0 & 1 & 1 & 1 & 3\end{array}\right) \qquad \text{(augmented matrix)}$$

$$\sim \left(\begin{array}{cccc|c} 1 & -\frac{1}{2} & \frac{5}{2} & \frac{3}{2} & \frac{5}{2}\\ 1 & 1 & 4 & 3 & 7\\ 1 & 0 & 3 & 2 & 4\\ 0 & 1 & 1 & 1 & 3\end{array}\right) \qquad [R_1 \to (R_1 \div 2)]$$

$$\sim \left(\begin{array}{cccc|c} 1 & -\frac{1}{2} & \frac{5}{2} & \frac{3}{2} & \frac{5}{2}\\ 0 & \frac{3}{2} & \frac{3}{2} & \frac{3}{2} & \frac{9}{2}\\ 0 & \frac{1}{2} & \frac{1}{2} & \frac{1}{2} & \frac{3}{2}\\ 0 & 1 & 1 & 1 & 3\end{array}\right) \qquad \begin{array}{l}[R_2 \to (R_2 - R_1)]\\ [R_3 \to (R_3 - R_1)]\end{array}$$

$$\sim \left(\begin{array}{cccc|c} 1 & -\frac{1}{2} & \frac{5}{2} & \frac{3}{2} & \frac{5}{2} \\ 0 & 1 & 1 & 1 & 3 \\ 0 & \frac{1}{2} & \frac{1}{2} & \frac{1}{2} & \frac{3}{2} \\ 0 & 1 & 1 & 1 & 3 \end{array}\right) \qquad [R_2 \to (R_2 \times \tfrac{2}{3})]$$

$$\sim \left(\begin{array}{cccc|c} 1 & 0 & 3 & 2 & 4 \\ 0 & 1 & 1 & 1 & 3 \\ 0 & 0 & 0 & 0 & 0 \\ 0 & 0 & 0 & 0 & 0 \end{array}\right) \qquad \begin{array}{l} [R_1 \to (R_1 + \frac{1}{2}R_2)] \\ \\ [R_3 \to (R_3 - \frac{1}{2}R_2)] \\ [R_4 \to (R_4 - R_2)]. \end{array}$$

This is equivalent to the system

$$\begin{aligned} x_1 + 3x_3 + 2x_4 &= 4 \\ x_2 + x_3 + x_4 &= 3. \end{aligned}$$

In this example two variables, x_3 and x_4, can be assigned arbitrary values. Writing $x_3 = t$, $x_4 = u$, we have

$$\begin{aligned} x_1 &= 4 - 3t - 2u \\ x_2 &= 3 - t - u. \end{aligned}$$

As in the last example, the system has an infinite number of solutions and the solution set can be written

$$\begin{aligned} &\{(x_1, x_2, x_3, x_4) : x_1 = 4 - 3t - 2u, x_2 = 3 - t - u, x_3 = t, x_4 = u; t, u \in \mathbb{R}\} \\ &\qquad = \{(4 - 3t - 2u, 3 - t - u, t, u) : t, u \in \mathbb{R}\}. \end{aligned}$$

Example 7.8

Solve the linear system

$$\begin{aligned} x + 2y - 3z &= 0 \\ 3x - y + z &= 3 \\ 2x - 3y + 4z &= 1. \end{aligned}$$

Solution

$$\left(\begin{array}{ccc|c} 1 & 2 & -3 & 0 \\ 3 & -1 & 1 & 3 \\ 2 & -3 & 4 & 1 \end{array}\right) \qquad \text{(augmented matrix)}$$

$$\sim \left(\begin{array}{ccc|c} 1 & 2 & -3 & 0 \\ 0 & -7 & 10 & 3 \\ 0 & -7 & 10 & 1 \end{array}\right) \qquad \begin{array}{l} [R_2 \to (R_2 - 3R_1)] \\ [R_3 \to (R_3 - 2R_1)] \end{array}$$

$$\sim \left(\begin{array}{ccc|c} 1 & 2 & -3 & 0 \\ 0 & 1 & -\frac{10}{7} & -\frac{3}{7} \\ 0 & -7 & 10 & 1 \end{array}\right) \qquad [R_2 \to (R_2 \div (-7))]$$

$$\sim \left(\begin{array}{ccc|c} 1 & 0 & -\frac{1}{7} & \frac{6}{7} \\ 0 & 1 & -\frac{10}{7} & -\frac{3}{7} \\ 0 & 0 & 0 & -2 \end{array}\right) \qquad \begin{array}{l} [R_1 \to (R_1 - 2R_2)] \\ \\ [R_3 \to (R_3 + 7R_2)] \end{array}$$

$$\sim \left(\begin{array}{ccc|c} 1 & 0 & -\frac{1}{7} & \frac{6}{7} \\ 0 & 1 & -\frac{10}{7} & -\frac{3}{7} \\ 0 & 0 & 0 & 1 \end{array}\right) \qquad [R_3 \to (R_3 \div (-2))]$$

$$\sim \left(\begin{array}{ccc|c} 1 & 0 & -\frac{1}{7} & 0 \\ 0 & 1 & -\frac{10}{7} & 0 \\ 0 & 0 & 0 & 1 \end{array}\right) \qquad \begin{array}{l} [R_1 \to (R_1 - \frac{6}{7}R_3)] \\ [R_2 \to (R_2 + \frac{3}{7}R_3)]. \end{array}$$

The bottom row of the reduced row echelon matrix represents the equation $0x + 0y + 0z = 1$, which clearly has no solution. Hence the system of equations is inconsistent. (Note that A is again a singular square matrix.)

In each of the last three examples, A, the matrix of coefficients, was a square singular matrix and so any process of applying elementary row operations to the augmented matrix will not reduce that part of the matrix to an identity matrix. Hence a system of equations where A is singular will not have a unique solution. Depending on the form of the reduced row echelon matrix, such a system has either an infinite number of solutions or no solution at all.

The method of Gauss–Jordan elimination can be used just as easily in the case where A is not a square matrix, i.e. where there are more equations than variables or more variables than equations. We consider each of these cases in the two examples below.

Example 7.9

Solve the system of equations

$$\begin{aligned} 2x - 3y &= -4 \\ x + 2y &= 5 \\ -4x + 6y &= 8. \end{aligned}$$

Solution

$$
\begin{aligned}
&\left(\begin{array}{cc|c} 2 & -3 & -4 \\ 1 & 2 & 5 \\ -4 & 6 & 8 \end{array}\right) && \text{(augmented matrix)} \\
\sim &\left(\begin{array}{cc|c} 1 & -\frac{3}{2} & -2 \\ 1 & 2 & 5 \\ -4 & 6 & 8 \end{array}\right) && [R_1 \to (R_1 \div 2)] \\
\sim &\left(\begin{array}{cc|c} 1 & -\frac{3}{2} & -2 \\ 0 & \frac{7}{2} & 7 \\ 0 & 0 & 0 \end{array}\right) && \begin{array}{l}[R_2 \to (R_2 - R_1)] \\ [R_3 \to (R_3 + 4R_1)]\end{array} \\
\sim &\left(\begin{array}{cc|c} 1 & -\frac{3}{2} & -2 \\ 0 & 1 & 2 \\ 0 & 0 & 0 \end{array}\right) && [R_2 \to (R_2 \times \tfrac{2}{7})] \\
\sim &\left(\begin{array}{cc|c} 1 & 0 & 1 \\ 0 & 1 & 2 \\ 0 & 0 & 0 \end{array}\right) && [R_1 \to (R_1 + \tfrac{3}{2}R_2)].
\end{aligned}
$$

Thus we have $x = 1$, $y = 2$, the last row of the matrix providing no information other than $0x + 0y = 0$!

Where there are more equations than variables, there may be none, one or an infinite number of solutions. However, if the number of variables exceeds the number of equations, then there will be none or an infinite number of solutions. The following example shows why this is the case.

Example 7.10

Solve the linear system

$$
\begin{aligned}
3x - 2y + z &= -4 \\
x + y + 2z &= 2.
\end{aligned}
$$

Solution

$$
\left(\begin{array}{ccc|c} 3 & -2 & 1 & -4 \\ 1 & 1 & 2 & 2 \end{array}\right) \quad \text{(augmented matrix)}
$$

$$\sim \left(\begin{array}{ccc|c} 1 & -\frac{2}{3} & \frac{1}{3} & -\frac{4}{3} \\ 1 & 1 & 2 & 2 \end{array}\right) \qquad [R_1 \rightarrow (R_1 \div 3)]$$

$$\sim \left(\begin{array}{ccc|c} 1 & -\frac{2}{3} & \frac{1}{3} & -\frac{4}{3} \\ 0 & \frac{5}{3} & \frac{5}{3} & \frac{10}{3} \end{array}\right) \qquad [R_2 \rightarrow (R_2 - R_1)]$$

$$\sim \left(\begin{array}{ccc|c} 1 & -\frac{2}{3} & \frac{1}{3} & -\frac{4}{3} \\ 0 & 1 & 1 & 2 \end{array}\right) \qquad [R_2 \rightarrow (R_2 \times \tfrac{3}{5})]$$

$$\sim \left(\begin{array}{ccc|c} 1 & 0 & 1 & 0 \\ 0 & 1 & 1 & 2 \end{array}\right) \qquad [R_1 \rightarrow (R_1 + \tfrac{2}{3}R_2)].$$

An equivalent system is therefore given by

$$\begin{aligned} x + z &= 0 \\ y + z &= 2. \end{aligned}$$

Writing $z = t$, we have $x = -t$, $y = 2 - t$ and the system has an infinite number of solutions because t can be chosen arbitrarily.

If we consider the possible final forms of the reduced row echelon matrix when we have more variables than equations, it is clear that such a system cannot have a unique solution. The system will either be inconsistent or have an infinite number of solutions. When we interpret the two equations in example 7.10 geometrically, we can see why this is the case. The two equations represent planes in three-dimensional space. There are three possibilities:

(a) the planes intersect in a line and there are an infinite number of solutions which can be expressed in terms of one parameter;

(b) the equations represent the *same* plane and there are an infinite number of solutions which can be expressed in terms of two parameters;

(c) the planes are parallel and do not intersect at all, in which case there are no solutions.

A final word is in order about homogeneous systems of linear equations. We have seen that such a system always has at least one solution—the trivial solution where all the variables are zero. However, a homogeneous system is just a special case of a general system of linear equations and the results that we have just deduced for linear systems apply no less to homogeneous linear systems. For instance, we have stated that if A, the matrix of coefficients, is a square singular matrix, then the system has no solution or an infinite number of solutions. Applying this result to homogeneous systems, we deduce that if A is a singular square matrix, the homogeneous system must have an infinite number of solutions since we know that it has at least one.

We give below a table summarizing the number of solutions which a system of linear equations may have for the four possible states of A, the matrix of coefficients.

	Homogeneous linear system	Non-homogeneous linear system
A is $m \times m$, non-singular	Trivial solution only	Unique solution
A is $m \times m$, singular	Infinite number of solutions	None or infinite number of solutions
A is $m \times m$, $m > n$	One or infinite number of solutions	None, one or infinite number of solutions
A is $m \times n$, $m < n$	Infinite number of solutions	None or infinite number of solutions

There are methods for determining from the augmented matrix $(A\ \boldsymbol{b})$ exactly how many solutions a linear system of equations has, but these are beyond the scope of this book.

Exercises 7.2

Use Gauss–Jordan elimination to solve the following systems of linear equations.

1.
$$\begin{aligned} x - y - 2z &= 2 \\ 2x + y - z &= 7 \\ x - 3y + 2z &= -12. \end{aligned}$$

Hence write down the solution of

$$\begin{aligned} x - y - 2z &= 0 \\ 2x + y - z &= 0 \\ x - 3y + 2z &= 0. \end{aligned}$$

2.
$$\begin{aligned} 2x_1 + 7x_2 + 3x_3 &= 14 \\ x_1 + 5x_2 + 3x_3 &= 13 \\ x_1 + 4x_2 + 2x_3 &= 8. \end{aligned}$$

3.
$$\begin{aligned} 2x_1 + x_2 - x_3 &= 2 \\ 4x_1 - x_2 - x_3 &= -2 \\ 3x_1 + 3x_2 - 2x_3 &= 6. \end{aligned}$$

4.
$$\begin{aligned} x_1 - x_2 + 2x_3 &= 0 \\ 2x_1 + 3x_2 - x_3 &= 0. \end{aligned}$$

5.
$$\begin{aligned} x_1 + 3x_2 + x_3 \qquad &= 1 \\ x_1 + 2x_2 + x_3 + x_4 &= 5 \\ - x_2 \qquad + x_4 &= 4 \\ x_1 + x_2 + x_3 + 2x_4 &= 9. \end{aligned}$$

6.
$$\begin{aligned} x - y + z &= 0 \\ 2x - y \qquad &= 0 \\ 3x + 2y - 7z &= 0. \end{aligned}$$

7.
$$\begin{aligned} 2x_1 + 4x_2 - x_3 + x_4 &= 2 \\ 3x_1 - x_2 \qquad + 2x_4 &= 3 \\ x_1 + 2x_2 + 3x_3 - x_4 &= 5. \end{aligned}$$

7.4 Gaussian Elimination

In Gauss–Jordan elimination the object is to reduce the augmented matrix of coefficients to reduced row echelon form by applying elementary row operations. Then any solution can be 'read off' directly from the final matrix.

In practice it is not necessary to reduce the augmented matrix right down to reduced row echelon form. If a matrix in row echelon form is obtained, the solutions to the system (if any) can be calculated easily. The steps necessary to obtain the row echelon form of the matrix are the same as those required to obtain the reduced row echelon form except that, having obtained a 'leading 1' in any row, suitable multiples of that row are subtracted only from the rows *below* it in order to obtain zeros below the leading 1 in any column. The method, which is called **Gaussian elimination** is illustrated in the examples below.

GAUSSIAN ELIMINATION

Example 7.11

Use Gaussian elimination to solve the following system of linear equations:

$$\begin{aligned} x + \ y + \ z &= \ \ 2 \\ 2x - 2y - \ z &= \ \ 2 \\ 3x + \ y - 2z &= -2. \end{aligned}$$

(In example 7.5 we solved this system using Gauss–Jordan elimination.)

Solution

$$\left(\begin{array}{ccc|c} 1 & 1 & 1 & 2 \\ 2 & -2 & -1 & 2 \\ 3 & 1 & -2 & -2 \end{array}\right) \quad \text{(augmented matrix)}$$

$$\sim \left(\begin{array}{ccc|c} 1 & 1 & 1 & 2 \\ 0 & -4 & -3 & -2 \\ 0 & -2 & -5 & -8 \end{array}\right) \quad \begin{array}{l} [R_2 \to (R_2 - 2R_1)] \\ [R_3 \to (R_3 - 3R_1)] \end{array}$$

$$\sim \left(\begin{array}{ccc|c} 1 & 1 & 1 & 2 \\ 0 & 1 & \frac{3}{4} & \frac{1}{2} \\ 0 & -2 & -5 & -8 \end{array}\right) \quad [R_2 \to (R_2 \times (-\tfrac{1}{4}))]$$

$$\sim \left(\begin{array}{ccc|c} 1 & 1 & 1 & 2 \\ 0 & 1 & \frac{3}{4} & \frac{1}{2} \\ 0 & 0 & -\frac{7}{2} & -7 \end{array}\right) \qquad [R_3 \to (R_3 + 2R_2)]$$

$$\sim \left(\begin{array}{ccc|c} 1 & 1 & 1 & 2 \\ 0 & 1 & \frac{3}{4} & \frac{1}{2} \\ 0 & 0 & 1 & 2 \end{array}\right) \qquad [R_3 \to (R_3 \times (-\tfrac{2}{7}))].$$

This gives the system of equations

$$\begin{aligned} x + y + \phantom{\tfrac{3}{4}}z &= 2 && \text{(i)} \\ y + \tfrac{3}{4}z &= \tfrac{1}{2} && \text{(ii)} \\ z &= 2 && \text{(iii)} \end{aligned}$$

which is equivalent to the original system.

From equation (iii) we have $z = 2$. Substitution of this value into equation (ii) gives

$$y + \tfrac{3}{2} = \tfrac{1}{2} \quad \text{so that} \quad y = -1.$$

Substitution of these values for y and z into equation (i) gives

$$x - 1 + 2 = 2 \quad \text{so that} \quad x = 1.$$

Thus the solution of the system is $(1, -1, 2)$.

Example 7.12

Use Gaussian elimination to solve the following system of linear equations:

$$\begin{aligned} 2x + 2y + z &= 4 \\ x - y - z &= 2 \\ 3x + y \phantom{{}+z} &= 6. \end{aligned}$$

(In example 7.6 we solved this system using Gauss–Jordan elimination.)

Solution

$$\left(\begin{array}{ccc|c} 2 & 2 & 1 & 4 \\ 1 & -1 & -1 & 2 \\ 3 & 1 & 0 & 6 \end{array}\right) \qquad \text{(augmented matrix)}$$

$$\sim \left(\begin{array}{ccc|c} 1 & 1 & \frac{1}{2} & 2 \\ 1 & -1 & -1 & 2 \\ 3 & 1 & 0 & 6 \end{array}\right) \qquad [R_1 \to (R_1 \div 2)]$$

$$\sim \left(\begin{array}{ccc|c} 1 & 1 & \frac{1}{2} & 2 \\ 0 & -2 & -\frac{3}{2} & 0 \\ 0 & -2 & -\frac{3}{2} & 0 \end{array}\right) \qquad \begin{array}{l} [R_2 \to (R_2 - R_1)] \\ [R_3 \to (R_3 - 3R_1)] \end{array}$$

$$\sim \left(\begin{array}{ccc|c} 1 & 1 & \frac{1}{2} & 2 \\ 0 & 1 & \frac{3}{4} & 0 \\ 0 & -2 & -\frac{3}{2} & 0 \end{array}\right) \qquad [R_2 \to (R_2 \div (-2))]$$

$$\sim \left(\begin{array}{ccc|c} 1 & 1 & \frac{1}{2} & 2 \\ 0 & 1 & \frac{3}{4} & 0 \\ 0 & 0 & 0 & 0 \end{array}\right) \qquad [R_3 \to (R_3 + 2R_2)].$$

This represents the system of equations

$$\begin{aligned} x + y + \tfrac{1}{2}z &= 2 \\ y + \tfrac{3}{4}z &= 0 \\ 0z &= 0. \end{aligned}$$

Writing $z = t$, we have

$$y + \tfrac{3}{4}t = 0$$

(from the second equation) so that

$$y = -\tfrac{3}{4}t.$$

Also

$$x - \tfrac{3}{4}t + \tfrac{1}{2}t = 2$$

(from the first equation) so that

$$x = 2 + \tfrac{1}{4}t.$$

In a consistent set of equations, Gaussian elimination results in a final matrix from which the value of at least one of the variables can be read off directly or assigned a parameter if the number of solutions is infinite. The values of the other variables must be calculated by a systematic method of 'back-substitution'. This usually involves fewer arithmetic operations than are involved in the extra elementary row operations necessary to obtain the reduced row echelon form rather than the row echelon form. For this reason, Gaussian elimination is usually preferred to Gauss–Jordan elimination as a method for solving systems of linear equations.

Exercises 7.3

1. Find the solution(s) (if any) of the following systems of linear equations using Gauss–Jordan elimination:

(i)
$$\begin{aligned} 3x - y + z &= 10 \\ x + y - z &= -2 \\ -x + 2y + 2z &= 0 \end{aligned}$$

(ii)
$$\begin{aligned} 2x_1 + x_2 + 8x_3 &= 14 \\ x_2 + 2x_3 &= 6 \\ x_1 + 3x_2 + 5x_3 &= 10 \end{aligned}$$

(iii)
$$\begin{aligned} x + 2y - 4z &= 4 \\ x + 3y - 6z &= 7 \\ 2x + 3y - 5z &= 9 \end{aligned}$$

(iv)
$$\begin{aligned} 3x + 2y - z &= 4 \\ 4x - 2y + 7z &= 3 \\ x + 4y - 2z &= 3 \end{aligned}$$

(v)
$$\begin{aligned} x + y + z &= 0 \\ -2x - y + z &= 0 \\ 3x + 2y + 2z &= 0. \end{aligned}$$

2. Find the solution(s), if any, of the following systems of linear equations, using Gaussian elimination:

(i)
$$\begin{aligned} x_1 + x_2 + 4x_3 &= 2 \\ 4x_1 + 3x_2 + 15x_3 &= 0 \\ 2x_1 + x_2 + 7x_3 &= -4 \end{aligned}$$

(ii)
$$\begin{aligned} x + y - z &= 0 \\ 2x + y + 2z &= 4 \\ x - 2y + z &= 8 \end{aligned}$$

(iii)
$$\begin{aligned} x + y - z &= 8 \\ y - 3z &= 2 \\ 2x + y + z &= 14 \end{aligned}$$

(iv)
$$\begin{aligned} x + 2y + z &= 0 \\ 3x - y - z &= 0 \\ 2x - 3y + 2z &= 0. \end{aligned}$$

3. Find the solution(s), if any, of the following systems of linear equations:

(i)
$$\begin{aligned} x_1 + x_2 - 3x_3 &= 3 \\ 2x_1 - 3x_2 - x_3 &= -9 \end{aligned}$$

(ii)
$$\begin{aligned} x + y &= 0 \\ 2x + y &= 1 \\ x - 2y &= 8 \end{aligned}$$

(iii)
$$\begin{aligned} 2x - y - z &= 1 \\ x - 2y + 3z &= 7 \\ -x + y - z &= 2 \end{aligned}$$

(iv)
$$\begin{aligned} x_1 + x_2 + x_3 &= 3 \\ 2x_1 - x_2 - x_3 &= 0 \\ -x_1 + 3x_2 - 2x_3 &= 0 \\ x_2 - 2x_3 &= -1 \end{aligned}$$

(v)
$$\begin{aligned} x + y - z &= 0 \\ 3x - 2y + 2z &= 0 \\ 2x - y + z &= 0 \end{aligned}$$

(vi)
$$\begin{aligned} 3x_1 - 2x_2 + x_3 &= 7 \\ 2x_1 - 2x_2 + 3x_3 &= -1 \\ -2x_1 + 4x_3 &= -1. \end{aligned}$$

Chapter 8

Algebraic Structures

8.1 Binary Operations and Their Properties

Very often in mathematics we are interested in combining the elements of some set. We have come across many such examples in earlier chapters of this book. In chapter 1 propositions were combined to form new propositions using logical connectives. The operations of union and intersection of sets introduced in chapter 3 each combine two sets to give a third denoted by $A \cup B$ and $A \cap B$ respectively. In chapter 5 we looked at composition of functions. Given functions f and g such that the image of f is a subset of the domain of g, we defined the composite function $g \circ f$. Other examples are the addition and multiplication of matrices and the familiar arithmetic operations of addition, subtraction, multiplication and division of real numbers.

The essential feature of each of these examples is a rule which allows two members of a specified set to be combined. For our purposes in this chapter we shall require that the rule must provide a means of combining *any* two elements and the result must itself be a member of the set. A rule which satisfies these criteria is called a 'binary operation'.

Of the examples given above, whether or not a particular rule is a binary operation depends critically upon the set in question. For example, addition is a binary operation on the set of positive integers; any two positive integers can be added and the result is also a positive integer. On the other hand subtraction is not a binary operation on this set because, given any two positive integers m and n, the result $m - n$ is not always a positive integer. Subtraction *is* a binary operation on the set of all integers, however.

For some binary operations the order of combining two elements matters and for some it does not. For instance, for the set of integers, it is always the case that $m + n = n + m$. This is not so for subtraction where in general the results of $m - n$ and $n - m$ are different. For this reason a binary operation must be viewed as acting, not just on a pair of elements of a set, but on an *ordered* pair.

In summary, a binary operation has two essential ingredients: a set and a rule for combining any ordered pair of its elements so that the result is also a member of the set.

Definition 8.1(a)

A **binary operation** $*$ on a non-empty set S is a rule for combining any two elements $x, y \in S$ to give an element $z \in S$ where z is denoted by $x * y$.

Notice from the definition that a binary operation is simply a function which assigns an element of S to every ordered pair of elements (x, y) where x and y belong to S. The set of these ordered pairs is, of course, the Cartesian product $S \times S$. This leads us to an alternative, rather more succinct, definition given below.

Definition 8.1(b)

A **binary operation** on a non-empty set S is a function $f : S \times S \to S$. If x and y are elements of S, we denote $f(x, y)$ by $x * y$.

The condition that $x * y$ must belong to S (or equivalently that the codomain of the function f is S) is called the **closure** property of the binary operation and, when this condition holds, we say that S is **closed** under the operation $*$. (In some texts, closure is not a required property of a binary operation, which is then defined as a function $f : S \times S \to T$, where S and T are non-empty sets and normally $S \subseteq T$.)

Examples 8.1

1. The operations of addition, subtraction and multiplication are each binary operations on $\mathbb{Z}$, the set of integers. Division is not a binary operation on $\mathbb{Z}$ since, for instance, $3 \div 4$ does not result in a member of $\mathbb{Z}$, i.e $\mathbb{Z}$ is not closed under division.

2. Addition, subtraction and multiplication are all binary operations on $\mathbb{Q}$, the set of rational numbers. Division is still not a binary operation on $\mathbb{Q}$ because $x \div 0$ is not defined for any $x \in \mathbb{Q}$.

3. If $S = \mathscr{P}(A)$, the power set of a set A (i.e. the set of all subsets of A), then the operations denoted by $\cup$ and $\cap$ are each binary operations on S.

4. A binary operation on a finite set may be defined using a table showing the result of applying that operation to any ordered pair of elements of the set. For example, if $S = \{a, b, c, d\}$, we can define a binary operation $*$ on S by the following table.

$*$	a	b	c	d
a	a	b	c	d
b	d	c	a	b
c	c	b	a	a
d	d	b	c	a

The convention for interpreting the table is that $b * d$, for example, is defined to be the element at the intersection of the row labelled 'b' and the column headed by 'd', so that $b*d = b$. Similarly $c*d = a$, $d*c = c$, $c*c = a$, and so on. A table which defines a binary operation in this way is called a **Cayley table**†.

We now consider some definitions which will enable us to distinguish the properties of certain binary operations. The first relates to the fact that a binary operation $*$ combines *pairs* of elements of a set, so that there are two different readings which we could give to the expression $a * b * c$. We could interpret it as $(a * b) * c$, i.e. combine a with b first and then combine the result with c. Alternatively, we could perform the operation in the order $a * (b * c)$, combining

† Named after Arthur Cayley (1821–95), the English mathematician who, during his professorship at Cambridge University, successfully brought about a change in the regulations so that women could be admitted to the university.

a with the result of $b * c$. For some binary operations, for example subtraction on the real numbers, the two interpretations give different results. For others, such as addition, the method of grouping the three elements makes no difference. Binary operations which have the latter property are termed 'associative'.

Definition 8.2

A binary operation $*$ on a set S is said to be **associative** if, for all $x, y, z \in S$,

$$(x * y) * z = x * (y * z).$$

For a binary operation which is not associative, an expression involving the combination of more than two elements must include brackets to indicate which elements are to be combined first. For an associative binary operation we can write $x * y * z$ without fear of ambiguity.

Recall that we have defined a binary operation on an ordered pair (x, y). Implicit in the definition is that, if x and y are elements of a set, $x * y$ and $y * x$ are also elements of the set. However, they may not be equal because the ordered pairs (x, y) and (y, x) are not equal (unless $y = x$). Binary operations, such as addition of real numbers, for which $x * y = y * x$ are said to be 'commutative'.

Definition 8.3

A binary operation $*$ on a set S is said to be **commutative** if, for all $x, y \in S$,

$$x * y = y * x.$$

For certain binary operations there is an element within the set which is neutral in the sense that, when it is combined with any member of the set, it leaves that element unchanged. For the real numbers under addition, zero has this property: $x + 0 = 0 + x = x$ for all real numbers x. Such an element, if it exists, is called an 'identity element'.

Definition 8.4

Let $*$ be a binary operation on a set S. An element $e \in S$ with the property that

$$x * e = e * x = x$$

for all $x \in S$ is called an **identity element** (or just an **identity**) for the operation $*$.

Notice that for e to be an identity element, *both* the equations $x * e = x$ and $e * x = x$ must be satisfied for all elements x of the set S. The element 0 is not an identity for the integers under subtraction because, if $x \in \mathbb{Z}$, $x - 0 = x$ but $0 - x = -x$.

Our final property relates only to binary operations for which an identity element exists.

Definition 8.5

Let $*$ be a binary operation on the set S and suppose that there is an identity element $e \in S$. Let x be an element of S. An **inverse** of x is an element $y \in S$ such that

$$x * y = y * x = e.$$

Where the inverse is unique, we write $y = x^{-1}$. (Note that in this case, x is also the inverse of y, i.e. $x = y^{-1}$.)

The fact that we write x^{-1} as a generic symbol for the inverse of x may be a little confusing given the more familiar interpretation of x^{-1} as meaning $1/x$ $(x \neq 0)$. (In fact $1/x$ *is* the inverse of x if S is the set of non-zero real numbers and $*$ is the binary operation of multiplication.) We must therefore be careful to interpret x^{-1} as the inverse of the element x with respect to the binary operation currently under consideration.

Examples 8.2

1. The binary operation of addition on $\mathbb{Z}$ is associative and commutative. The identity element is 0 since $0 \in \mathbb{Z}$ and

$$x + 0 = 0 + x = x$$

for all $x \in \mathbb{Z}$. The inverse of any element x is $-x$ since

$$x + (-x) = (-x) + x = 0$$

and for any $x \in \mathbb{Z}$, $(-x) \in \mathbb{Z}$. Hence for the set of integers under addition we can write $x^{-1} = -x$.

2. Multiplication on $\mathbb{Z}$ is also associative and commutative. The identity element is 1. Do any elements of $\mathbb{Z}$ have inverses? Consider the element 4, for example. To find the inverse of 4 we must find an element $a \in \mathbb{Z}$ such that

$$4 \times a = a \times 4 = 1.$$

The only number a which satisfies these equations is $\frac{1}{4}$ which does not belong to $\mathbb{Z}$. Thus 4 does not have an inverse. The only elements of $\mathbb{Z}$ which do have inverses under multiplication are 1 and -1. Since each of these is its own inverse, we say that these elements are **self-inverse**.

For multiplication on $\mathbb{Q}$ (the set of rational numbers), all elements except 0 have inverses. The inverse of any non-zero element x is $1/x$ and we can therefore write the familiar $x^{-1} = 1/x$.

3. Let $S = \mathscr{P}(A)$, the power set of a set A. We saw in chapter 3 that the binary operation of union of sets is both associative and commutative. The identity element is $\varnothing$ since $\varnothing \in S$ and

$$X \cup \varnothing = \varnothing \cup X = X$$

for any $X \in S$. The only member of S which has an inverse is $\varnothing$, which is self-inverse. (The power set $\mathscr{P}(A)$ together with set intersection is considered in exercise 8.1.4.)

4. The binary operation defined by the Cayley table in example 8.1.4 is not associative since, for instance

$$(b * d) * a = b * a = d$$

but

$$b * (d * a) = b * d = b.$$

Neither is the operation commutative, e.g. $b * a \neq a * b$. That no identity element exists can be readily verified from the table (see exercise 8.1.3) and, since there is no identity, there can be no inverses.

In the examples of binary operations described above, if an identity existed at all, there was only one element of the set which satisfied the necessary criteria. This is no accident. An identity element may or may not exist for a given set and binary operation but, where it does exist, it is unique. (In fact, we have been anticipating this result by referring to *the* identity.)

Theorem 8.1

Let $*$ be a binary operation on a set S. If an identity element exists, then it is unique.

Proof

Let e_1 and e_2 be identity elements in S under the operation $*$. Since e_2 is an identity,

$$e_1 * e_2 = e_2 * e_1 = e_1.$$

But e_1 is also an identity, so

$$e_2 * e_1 = e_1 * e_2 = e_2.$$

This establishes that $e_1 = e_2$, so the identity element is unique. □

We can also show that, for an associative binary operation, the inverse of an element, where it exists is unique.

Theorem 8.2

Let $*$ be an associative binary operation on a set S which has identity element e under $*$. For any element which has an inverse, the inverse is unique.

Proof

Suppose that an element $x \in S$ has inverses y and z so that

$$y * x = x * y = e$$
$$z * x = x * z = e.$$

Now

$$\begin{aligned} y &= y * e \\ &= y * (x * z) \\ &= (y * x) * z \quad \text{(by associativity)} \\ &= e * z \\ &= z. \end{aligned}$$

Hence the inverse of x is unique. □

Notice that associativity of the binary operation was essential to the proof of this theorem. If the binary operation is not associative, the uniqueness of any inverse cannot be guaranteed (see exercise 8.1.9).

Exercises 8.1

1. In each of the examples below, state whether $x * y$ defines a binary operation on the set S given. If it does not, explain why not.

   (i) $x * y = x - y$, $S = \mathbb{R}^+$.
   (ii) $x * y = z$ where $z < x + y$, $S = \mathbb{Z}$.
   (iii) $x * y = x^y$, $S = \mathbb{R}^+$.
   (iv) $x * y$ = the least common multiple of x and y, $S = \{1, 2, 3, 4, 6, 8, 12, 24\}$ (the set of divisors of 24).
   (v) $x * y$ = the greatest common factor of x and y, S as defined in (iv).
   (vi) $x * y = x + y$, $S = \{\text{all matrices}\}$.

2. Consider the binary operation of subtraction on the set of real numbers.

   (a) Is the operation associative?
   (b) Is the operation commutative?
   (c) Does an identity element exist and, if so, which elements have inverses?

3. Suppose that S is a finite set and a binary operation is defined on S by a Cayley table (as in example 8.1.4). How can you tell from the Cayley table:

   (i) whether $*$ is commutative on S?
   (ii) whether an identity exists?

4. Consider $S = \mathscr{P}(A)$, the set of all subsets of a set A, together with the binary operation of intersection, $\cap$.

   (i) Is $\cap$ commutative on S?
   (ii) What is the identity element?
   (iii) Which elements, if any, have inverses? What are their inverses?

5. Let S be a set together with a binary operation $*$. Suppose that an identity element exists and, for all $x, y, z \in S$,

$$x * (y * z) = (x * z) * y.$$

   Show that $*$ is commutative and associative.

6. How many distinct binary operations can be defined on a set with

   (i) two elements,
   (ii) three elements,
   (iii) four elements,
   (iv) n elements?

7. Let $S = \{a, b, c\}$ and $*$ be a commutative binary operation on S. Let a be the identity element and suppose that every element has a unique inverse. Draw the Cayley tables for all the binary operations which satisfy these criteria. Are any of these operations associative?

8. Let $S = \mathscr{P}(A)$, the power set of a set A. Let

$$X * Y = (X - Y) \cup (Y - X)$$

   for all $X, Y \in S$. (This operation is called the symmetric difference of X and Y; see exercise 3.4.5.)

   (i) Show that $*$ is a binary operation on S.
   (ii) Is $*$ commutative?
   (iii) Is $*$ associative?
   (iv) Is there an identity element? If so, what is it?
   (v) If there is an identity element, what is the inverse of an element $X \in S$?

9. The following is the Cayley table for a binary operation $*$ on the set $\{a, b, c, d\}$. Note that a is the identity element and that c and d are both inverses for b.

$*$	a	b	c	d
a	a	b	c	d
b	b	d	a	a
c	c	a	b	d
d	d	a	b	c

Show that $*$ is not associative on $\{a, b, c, d\}$ (cf. theorem 8.2).

8.2 Algebraic Structures

An **algebraic structure** consists of one or more sets together with one or more operations which enable members of the sets to be combined in some way. What is important about a particular algebraic structure is that many of its properties are predictable from the characteristics of the operation or operations involved. This means that we can classify algebraic structures into families whose members have many features in common. Identification of a given algebraic structure as belonging to a particular family of structures allows us to conclude that it has the properties characteristic of all members of the family. To illustrate the point: you may know nothing about a lory. However, if you are told that it is a type of parrot, then you may reasonably assume that it has amongst its attributes all those which are characteristic of parrots. So it is with algebraic structures. If a particular structure can be identified as a 'group' then it can be assumed to have all the properties characteristic of groups.

The algebraic structures with which we shall concern ourselves here are those which consist of a single set S together with a single binary operation for combining members of the set. We shall denote such a structure by $(S, *)$ to emphasize that the structure has two essential components—a set and a binary operation on that set. Properties of the binary operation provide the axioms defining the different families by which these structures are classified.

Semigroups

For our first class of algebraic structures we require of the binary operation only that it be associative. Algebraic structures with this property are called 'semigroups'.

Definition 8.6

Let S be a non-empty set and let $*$ be a binary operation defined on S. The structure $(S, *)$ is a **semigroup** if the operation $*$ is associative on S, i.e. if, for every $x, y, z \in S$,

$$(x * y) * z = x * (y * z).$$

If the operation is also commutative, then the structure $(S, *)$ is called an **abelian**† (or **commutative**) **semigroup**.

Examples 8.3

1. The structures $(\mathbb{N}, +)$, $(\mathbb{N}, \times)$, $(\mathbb{R}, +)$, $(\mathbb{R}, \times)$ are all abelian semigroups.

2. Let A denote a non-empty set of symbols. Such a set is called an **alphabet**. Some examples of alphabets are:

   (a) $A = \{\alpha, \beta, \gamma, \delta, \phi, \pi\}$
   (b) $A = \{a, b, c, d, \ldots, x, y, z\}$
   (c) $A = \{\times, +, -, \div, /, £, \$, \%, \&, \mathrm{q}\}$.

   Given an alphabet A, we define a **string** (or **word**) **over** $\boldsymbol{A}$ to be a finite ordered sequence of symbols from A. The **length** of a string is the number of symbols which it contains. Thus, if $A = \{a, b, c, d\}$, then $abbc$, $dcad$ and $abcd$ are all strings of length 4. The string $bccadaa$ has length 7, and so on.

   Now suppose that A is an alphabet and consider A^*, the set of all strings over A. (Note that A^* is an infinite set.) We define the operation of **concatenation** on the elements of A^* as follows. If x and y are two elements of A^* (i.e. two strings over the alphabet A), then the concatenation of x and y, denoted by $x * y$, is the string obtained by juxtaposing x and y so that x is on the left and y on the right. Thus for the set A^* of strings over the set $A = \{a, b, c, d\}$, we have, for example,

$$abd * cabc = abdcabc$$
$$baaa * ccbabb = baaaccbabb.$$

† Named after Niels Henrik Abel (1802–29), a Norwegian mathematician who contributed to the theory of equations and infinite series. A year after his premature death from tuberculosis, he was honoured with the award of the Grand Prize in Mathematics by the Royal Academy of France.

For any given alphabet A, the operation of concatenation on A^* is a binary operation and it is clear from the definition that this operation is associative. The structure $(A^*, *)$, where $*$ represents concatenation, is therefore a semigroup. It is called the **free semigroup generated by** A.

Note that concatenation is commutative only when A has just a single element.

3. Suppose that $S = \{a, b, c\}$ and a binary operation is defined on S by the following Cayley table.

$*$	a	b	c
a	a	b	c
b	a	b	c
c	a	b	c

The structure $(S, *)$ is in fact a semigroup, but to check that this is so we have to show that

$$(x * y) * z = x * (y * z)$$

for all $x, y, z \in S$. Often when a binary operation is defined by a Cayley table, establishing associativity involves checking that this equation holds for all possible choices of x, y and z. This can be a long and arduous process! However, in this case, notice that

$$x * y = y$$

for all $x, y \in S$. Hence

$$\begin{aligned}(x * y) * z &= y * z = z \\ x * (y * z) &= x * z = z\end{aligned}$$

for all $x, y, z \in S$, so $*$ is associative. For this structure we can save ourselves the trouble of testing all 27 equations.

Monoids

The single restriction on the binary operation of semigroups does not give them enough structure for many interesting properties to emerge. So for our next family of algebraic structures we add a second condition to that of associativity—the existence of an identity. (Remember that theorem 8.1 guarantees that there can be only one identity.) Algebraic structures having these two properties are called 'monoids'.

Definition 8.7

A **monoid** is a semigroup $(S, *)$ which has an identity element.

If $*$ is also commutative, the monoid is called an **abelian monoid** (or **commutative monoid**).

Examples 8.4

1. $(\mathbb{Z}^+, \times)$ is an abelian monoid with identity element 1. The structure $(\mathbb{Z}^+, +)$ is not a monoid because there is no identity element $(0 \notin \mathbb{Z}^+)$.

   The structures $(\mathbb{Z}, \times)$ and $(\mathbb{Z}, +)$ are each abelian monoids with identity elements 1 and 0 respectively.

2. In example 8.3.2 we defined the operation of concatenation on strings of symbols. Suppose we add the **empty string** (i.e. string containing no symbols) to the set A^*. Denoting the empty string by λ, we have

$$x * \lambda = \lambda * x = x$$

   for all $x \in A^* \cup \{\lambda\}$. The structure $(A^* \cup \{\lambda\}, *)$ is a monoid and it is called the **free monoid generated by** $\boldsymbol{A}$.

3. The structure $(S, *)$ defined in example 8.3.3 is not a monoid since there is no identity element.

4. If $S = \mathscr{P}(A)$, where A is any set, then $(S, \cup)$ is an abelian monoid with identity element $\varnothing$. Also $(S, \cap)$ is an abelian monoid with identity element A.

Groups

Many of the most important and interesting examples of algebraic structures involving a single binary operation satisfy a third condition in addition to the two defining a monoid. This is that each element of the set has an inverse element with respect to the operation. Adding this condition to those for a monoid defines the class of algebraic structures known generically as 'groups'.

Definition 8.8

A **group** is a monoid $(S, *)$ in which every element has an inverse, i.e. the pair $(S, *)$ satisfies the following three conditions:

(G1) $*$ is associative on S;
(G2) an identity element exists;
(G3) every element of S has an inverse.

Predictably, a group in which the binary operation is commutative is called an **abelian group** (or **commutative group**).

Remember that we proved (theorem 8.2) that, for a set with an associative binary operation, the inverse of any element is unique. When applied to a group $(S, *)$, the theorem guarantees the existence of a unique inverse for every element of S.

Examples 8.5

1. The structure $(\mathbb{Z}, +)$ is a group. The identity element is 0 and the inverse of any $z \in \mathbb{Z}$ is $-z$. Since addition is commutative, $(\mathbb{Z}, +)$ is an abelian group.

2. The structure $(\mathbb{R}^+, \times)$ is an abelian group with identity element 1. The inverse of x is $1/x$.

3. The monoid $(A^* \cup \{\lambda\}, *)$ defined in example 8.4.2 is not a group because, for any non-empty string x, we cannot find another string y so that

$$x * y = y * x = \lambda$$

   where λ is the empty string. Thus no element in the set $A^* \cup \{\lambda\}$, other than λ itself, has an inverse under concatenation. (Exercise 8.2.12 shows how a group can be defined from an alphabet.)

4. Consider $\mathbb{Z}$ together with the binary operation defined by

$$x * y = x + y + 1.$$

   Is the structure $(\mathbb{Z}, *)$ a group?

Testing first for associativity: for any $x, y, z \in \mathbb{Z}$ we have

$$\begin{aligned}(x * y) * z &= (x + y + 1) * z \\ &= (x + y + 1) + z + 1 \\ &= x + y + z + 2\end{aligned}$$

and

$$\begin{aligned}x * (y * z) &= x * (y + z + 1) \\ &= x + (y + z + 1) + 1 \\ &= x + y + z + 2.\end{aligned}$$

Thus $*$ is associative on $\mathbb{Z}$.

Is there an identity element? If so, the identity e must satisfy

$$e * x = x * e = x$$

for any $x \in \mathbb{Z}$.

Now

$$\begin{aligned} & e * x \quad \text{and} \quad x * e = x \\ \Leftrightarrow \quad & x + e + 1 = x \\ \Leftrightarrow \quad & e = -1.\end{aligned}$$

Since $-1 \in \mathbb{Z}$ and $x * (-1) = (-1) * x = x$ for all $x \in \mathbb{Z}$, -1 is the identity element under operation $*$.

What about inverses? For $x, y \in \mathbb{Z}$

$$\begin{aligned} & x * y = e \quad \text{and} \quad y * x = e \\ \Leftrightarrow \quad & x + y + 1 = -1 \\ \Leftrightarrow \quad & y = -2 - x.\end{aligned}$$

For every $x \in \mathbb{Z}$, $(-2 - x) \in \mathbb{Z}$, so every element has an inverse.

Since (G1), (G2) and (G3) are satisfied, $(\mathbb{Z}, *)$ is a group.

5. In chapter 4 we looked at modulo arithmetic. For fixed integer n, we defined the equivalence relation 'congruence modulo n' on the set $\mathbb{Z}$ of integers:

$$a \equiv_n b \quad \text{if and only if } a - b = kn \text{ for some } k \in \mathbb{Z}.$$

We found that this relation partitioned $\mathbb{Z}$ into the set of equivalence classes

$$\mathbb{Z}_n = \{[0], [1], [2], \ldots, [n-1]\}.$$

Consider $n = 5$ (see example 4.7). The table for $+_5$ addition modulo 5 (given on page 185), is the following.

$+_5$	[0]	[1]	[2]	[3]	[4]
[0]	[0]	[1]	[2]	[3]	[4]
[1]	[1]	[2]	[3]	[4]	[0]
[2]	[2]	[3]	[4]	[0]	[1]
[3]	[3]	[4]	[0]	[1]	[2]
[4]	[4]	[0]	[1]	[2]	[3]

Is the set $\mathbb{Z}_5$ together with addition modulo 5 a group?

That the operation is associative follows from the associativity of ordinary addition of integers. If we do not appeal to this property, we are faced with no alternative but to test all possible equations of the form

$$(x +_5 y) +_5 z = x +_5 (y +_5 z)$$

for all $x, y, z \in \mathbb{Z}_5$. (How many such equations are there?)

From the table above we can see that the identity element is [0] and that every element has an inverse. For example, $[1]^{-1} = [4]$, $[3]^{-1} = [2]$.

Hence $\mathbb{Z}_5$ with addition modulo 5 is a group. (See exercise 8.2.4 for consideration of the group properties of $\mathbb{Z}_5$ under multiplication modulo 5.)

Exercises 8.2

1. Show that the set of all 2×2 matrices with real elements together with the binary operation of matrix addition is a group. Why is this set together with matrix multiplication not a group?

   Show that matrix multiplication on the set of all 2×2 non-singular matrices is a binary operation. Prove that the set of all 2×2 non-singular matrices forms a group under matrix multiplication.

2. If $*$ is a binary operation on a set S, then an element $x \in S$ is said to be **idempotent** if $x * x = x$. Prove that a group has only one idempotent element.

3. Show that the set

$$\mathbb{Z}_6 = \{[0], [1], [2], [3], [4], [5]\}$$

together with addition modulo 6 (denoted by $+_6$) is a group. Is $\mathbb{Z}_6$ together with multiplication modulo 6 (denoted by $\times_6$) a group?

4. Show that the set

$$\mathbb{Z}_5 = \{[0], [1], [2], [3], [4]\}$$

under multiplication modulo 5 is not a group but that

$$\mathbb{Z}_5 - \{[0]\} = \{[1], [2], [3], [4]\}$$

is a group under this operation.

Is $\mathbb{Z}_4 - \{[0]\}$ under multiplication modulo 4 a group?

Under what circumstances will the set $\mathbb{Z}_n - \{[0]\}$ under multiplication modulo n be a group? (Cf. exercise 4.4.12.)

5. Let $P = \{p \in \mathbb{Z}^+ : \; p \text{ is prime and } p \leqslant 13\}$. A binary operation $*$ is defined on P by

$$p * q = \text{greatest prime divisor of } p + q - 2.$$

Construct a Cayley table for P under the operation $*$ and show that P has an identity element with respect to $*$. Is $(P, *)$ a group? Justify your answer.

6. Let S be a non empty set and $*$ a binary operation defined by

$$x * y = x$$

for all $x, y \in S$. Show that $(S, *)$ is a semigroup. Is $(S, *)$ a monoid? Why or why not?

7. Suppose that the binary operations $*$ and $\circ$ are defined on the sets S and T respectively and that $(S, *)$ and $(T, \circ)$ are both groups. Define the operation . on the Cartesian product $S \times T$ as follows:

$$(s_1, t_1).(s_2, t_2) = (s_1 * s_2, t_1 \circ t_2)$$

for all $s_1, s_2 \in S$ and $t_1, t_2 \in T$.

Show that . is a binary operation on $S \times T$ and that $(S \times T, .)$ is a group. What is the inverse of a typical element (s, t) of $S \times T$? (The algebraic structure $(S \times T, .)$ is called the **external direct product** of $(S, *)$ and $(T, \circ)$. In this exercise you are required to show that the external direct product of two groups is itself a group.)

8. Consider the structure $(\mathbb{N}, *)$ where $*$ is the binary operation defined by

$$x * y = \begin{cases} x & \text{if } x \geqslant y \\ y & \text{if } x < y \end{cases}$$

where $x, y \in \mathbb{N}$.

Show that $(\mathbb{N}, *)$ is a semigroup. Is $(\mathbb{N}, *)$ a monoid? Why or why not?

Define the binary operation $\circ$ on $\mathbb{N}$ by

$$x \circ y = \begin{cases} x & \text{if } x \leqslant y \\ y & \text{if } x > y. \end{cases}$$

Is $(\mathbb{N}, \circ)$ a semigroup? Is $(\mathbb{N}, \circ)$ a monoid?

9. Consider the set of 2×2 matrices of the form

$$\begin{pmatrix} a & 0 \\ 0 & b \end{pmatrix}$$

where $a, b \in \mathbb{R}$, together with the binary operation of matrix multiplication. Is this structure

(a) a semigroup,
(b) a monoid,
(c) a group?

10. Show that the set of all 2×2 matrices of the form

$$\begin{pmatrix} 1 & n \\ 0 & 1 \end{pmatrix}$$

where $n \in \mathbb{Z}$ is a group under the operation of matrix multiplication. What is the identity? What is the inverse of

$$\begin{pmatrix} 1 & 4 \\ 0 & 1 \end{pmatrix}?$$

11. Let M denote the set of real 2×2 matrices of the form

$$\begin{pmatrix} x & y \\ -y & x \end{pmatrix}$$

where x and y are not *both* zero. Show that M is a group under the operation of matrix multiplication.

12. Let A be a finite alphabet and let $\bar{A}$ be the set of symbols of the form $\bar{a}$ where $a \in A$, i.e. $\bar{A} = \{\bar{a} : a \in A\}$. Let $B = A \cup \bar{A}$ and let $F(A)$ be the subset of $B^* \cup \{\lambda\}$ consisting of those strings which do not contain pairs of symbols of the form $a\bar{a}$ or $\bar{a}a$. Define the binary operation $*$ on $F(A)$ to be concatenation of strings followed by the successive removal of all substrings of the form $a\bar{a}$ or $\bar{a}a$. For example

$$ab * \bar{b}ca = ab\bar{b}ca = aca$$
$$db\bar{a} * a\bar{b}c\bar{d}a = db\bar{a}a\bar{b}c\bar{d}a = db\bar{b}c\bar{d}a = dc\bar{d}a.$$

Assuming that the operation is associative, show that $(F(A), *)$ is a group. This is called the **free group generated by A**.

8.3 More about Groups

We now concentrate our attention on groups, the most important of our three algebraic structures and historically the first to be studied abstractly.

Of the three families of structures which we consider in this chapter, the class of groups is the most widely studied, has the most interesting structure and is the most extensively applied. In addition to its significance within mathematics itself, group theory has applications in fields as diverse as physics, chemistry and linguistics. In the last section of this chapter we look at how groups are utilized in coding theory.

The foundations of group theory were laid in the nineteenth century by the French mathematician Galois†. The subject is now a well developed component of

† Evariste Galois (1811–32) was born in Paris and had a short but eventful life. He twice failed the entrance examination to the L'Ecole Polytechnique although, in his late teens, he made discoveries which contributed significantly to the theory of equations. His political activities led to a six-month spell in prison and, shortly after his release, he was killed in a duel. Although not recognized in his lifetime, Galois is now regarded as one of the greatest of mathematical geniuses

abstract algebra and many books are devoted exclusively to the subject. We shall be able to do no more than prove some basic theorems about groups and look at some important examples of groups. In the following sections, we shall also look at some relations amongst groups themselves.

In this section and those which follow, we shall adopt the convention of omitting the symbol $*$ when writing expressions involving an unspecified binary operation. We shall only include this symbol where to omit it results in an ambiguous expression, for instance when we need to distinguish between two binary operations. Instead of $x * y$ we shall write xy. We also define 'powers' of x as follows. If $n \in \mathbb{Z}^+$,

$$x^n = \underbrace{x * x * \cdots * x}_{n \text{ terms}}$$

and if $n \in \mathbb{Z}^-$

$$x^n = (x^{-1})^{|n|} = \underbrace{x^{-1} * x^{-1} * \cdots * x^{-1}}_{|n| \text{ terms}}.$$

Predictably, we shall define $x^0 = e$, the identity element.

This 'multiplicative notation' has the advantage of convenience and brevity but the disadvantage that, for those of us who have studied any algebra, xy is already established in our minds as meaning 'x multiplied by y'. We must therefore be careful not to make assumptions which may be true for the operation of multiplication but not necessarily so for the binary operation under consideration. For example, we cannot assume $xy = yx$ unless the binary operation is known to be commutative. We must also be careful with the 'laws of indices'. It is not difficult to show that the following hold for the elements of a group:

$$\begin{array}{llll} & (x^{-1})^n = (x^n)^{-1} & = x^{-n} & \text{for all } n \in \mathbb{Z} \\ & x^m x^n = x^{m+n} & = x^n x^m & \text{for all } m, n \in \mathbb{Z} \\ & (x^m)^n = x^{mn} & = (x^n)^m & \text{for all } m, n \in \mathbb{Z}. \\ \text{However,} & (xy)^n = x^n y^n & & \text{for all } n \in \mathbb{Z} \textit{ only} \text{ for a} \\ & & & \text{commutative binary operation.} \end{array}$$

Where the binary operation is addition, it is usual to adopt the notation normally associated with that operation. The inverse of an element x is denoted by $-x$ and

$$\underbrace{x + x + \cdots + x}_{n \text{ terms}}$$

is written $n.x$ or nx. For an additive group, the analogues of the 'laws of indices' listed above are

$$\begin{aligned}
n(-x) &= -(nx) &&= (-n)x && \text{for all } n \in \mathbb{Z} \\
mx + nx &= (m+n)x &&= nx + mx && \text{for all } m, n \in \mathbb{Z} \\
n(mx) &= (nm)x &&= m(nx) && \text{for all } m, n \in \mathbb{Z}
\end{aligned}$$

and $$n(x + y) = nx + ny \qquad \text{for all } n \in \mathbb{Z} \text{ since addition is commutative.}$$

We shall denote a group by $(G, *)$ rather than $(S, *)$ in order to emphasize that we are referring to a group rather than some other algebraic structure.

Perhaps the most obvious characteristic of any group $(G, *)$ is its 'size', that is the number of elements in the underlying set G. This is termed the 'order' of the group $(G, *)$.

Definition 8.9

The **order** of a group $(G, *)$ is the cardinality of the set G. It is denoted by $|G|$ (see definition 3.1).

We now prove some useful theorems about the properties of groups.

Theorem 8.3

If $(G, *)$ is a group, then the left and right cancellation laws hold; that is, if $a, x, y \in G$, then

(a) $ax = ay$ implies that $x = y$ (left cancellation law), and
(b) $xa = ya$ implies that $x = y$ (right cancellation law).

Proof

Suppose that $ax = ay$.

Since $(G, *)$ is a group, then the element a has an inverse a^{-1}. 'Multiplying' on the left by a^{-1} gives

$$
\begin{aligned}
& a^{-1}(ax) = a^{-1}(ay) \\
\Rightarrow \quad & (a^{-1}a)x = (a^{-1}a)y \quad \text{(by associativity)} \\
\Rightarrow \quad & ex = ey \quad \text{(where } e \text{ is the identity)} \\
\Rightarrow \quad & x = y.
\end{aligned}
$$

We have proved that the left cancellation law holds in a group. A similar proof establishes that the right cancellation law also holds. □

These cancellation laws, as they apply to addition and multiplication of non-zero real numbers, are a familiar feature of elementary algebra. For example, from the equation $3x = 3y$ we can deduce that $x = y$. We can make the same deduction from the equation $x + 2 = y + 2$.

The next theorem also has a familiar application in elementary algebra. The linear equations $a + x = b$ and $ax = b$ have unique solutions for x as long as $a \neq 0$. (If $a = 0$, the first equation has a unique solution but the second does not.) The need to solve equations such as these arises frequently and we might therefore ask, given a binary operation $*$, under what circumstances does the 'linear' equation $a * x = b$ have a unique solution? That such an equation does not always have a unique solution is easy enough to demonstrate. Consider, for example, the binary operation defined in example 8.1.4. The equation $c * x = a$ has two solutions and the equation $c * x = d$ has none. For the members of the group, however, we can prove that every such equation has a unique solution.

Theorem 8.4

If $(G, *)$ is a group and $a, b \in G$, then

(a) the equation $ax = b$ has a unique solution $x = a^{-1}b$, and
(b) the equation $ya = b$ has a unique solution $y = ba^{-1}$.

Proof

(a) Suppose we have $ax = b$.

Pre-multiplying this equation by a^{-1} gives

$$
\begin{aligned}
& & a^{-1}(ax) &= a^{-1}b \\
&\Rightarrow & (a^{-1}a)x &= a^{-1}b \\
&\Rightarrow & ex &= a^{-1}b \\
&\Rightarrow & x &= a^{-1}b.
\end{aligned}
$$

Thus $x = a^{-1}b$ is a solution of the equation. We must now show that this is the only solution.

Suppose that x_1 and x_2 are both solutions of $ax = b$. Then we have

$$
\begin{aligned}
& & ax_1 &= ax_2 \\
&\Rightarrow & x_1 &= x_2 \quad \text{(by the left cancellation law).}
\end{aligned}
$$

Hence $x = a^{-1}b$ is the unique solution.

The proof of (b) is similar. □

A useful consequence of each of these two theorems is their implication for the Cayley table of a group $(G, *)$ with a finite number of elements. The second theorem guarantees that every element appears exactly once in every row and column. To see why this is so, consider an arbitrary element $a \in G$. Any element $g \in G$ appears in the row corresponding to a if the equation $ax = g$ has a solution for some $x \in G$. In this case g is in the column corresponding to x as shown in figure 8.1.

*	…	x	…
⋮		⋮	
a	…	g	…
⋮		⋮	

Figure 8.1

Theorem 8.4 states that this equation has a *unique* solution for each $g \in G$. Hence every element of G appears just once in the row corresponding to the element a and, since a was chosen arbitrarily, we can deduce that every element of G appears exactly once in every row. A similar argument can be used to show that each element appears exactly once in every column. Theorem 8.3 can also be used to establish this result which we summarize below.

Theorem 8.5

If $(G, *)$ is a finite group (i.e. one with finite order), its Cayley table is such that every element of G appears once and only once in every row and column.

Since we have not established the truth of the converse statement, we cannot use this property of the Cayley table to show that $(S, *)$ is a group although the fact that this criterion is not satisfied is often useful in proving that a structure is *not* a group. In fact, if the Cayley table for a binary operation on a finite set S is such that there is an identity element and every element of the set appears once and only once in every row and column, then $(S, *)$ is a group if and only if $*$ is an associative operation (see exercise 8.3.3). However, as we have seen, establishing associativity for a binary operation defined by a Cayley table can be a tedious process (see example 8.3.3).

We now turn our attention to some important families of groups.

8.4 Some Families of Groups

Cyclic Groups

Consider the group defined by the following Cayley table.

$*$	e	a	b	c
e	e	a	b	c
a	a	b	c	e
b	b	c	e	a
c	c	e	a	b

For this group we have $a^1 = a$, $a^2 = b$, $a^3 = c$, $a^4 = e$, from which we can deduce that every element of $\{e, a, b, c\}$ can be written in the form a^n for some integer n. For any given element, this representation is not unique. For instance, we could write $b = a^2 = a^6 = a^{-2}$ and so on. In fact there are an infinite number of ways of representing each element of the set as a 'power' of a. The point is that every element of $\{e, a, b, c\}$ can be written as a^n for *some* integer n and, where this is the case, we say that a is a 'generator' of the group.

CYCLIC GROUPS

It is reasonable to ask whether any other element is also a generator of the group. We can confirm that the element c is a generator but that b is not because $b^n = e$ if n is an even integer and $b^n = b$ if n is odd. A group which has at least one generator is said to be 'cyclic'.

Definition 8.10

A group $(G, *)$ is said to be **cyclic** if there exists an element $a \in G$ such that, for each $g \in G, g = a^n$ for some $n \in \mathbb{Z}$. The group $(G, *)$ is said to be **generated** by a and a is called a **generator** of $(G, *)$.

In the notation for an additive group (one where the binary operation is addition), the element a is a generator if, for all $g \in G$, $g = na$ for some integer n. A cyclic group is necessarily abelian because, given $g_1, g_2 \in G$, we have $g_1 = a^r$ and

$g_2 = a^s$ for some $r, s \in \mathbb{Z}$ so that

$$\begin{aligned} g_1 g_2 &= a^r a^s \\ &= a^{r+s} \\ &= a^{s+r} \\ &= g_2 g_1. \end{aligned}$$

Examples 8.6

1. Show that the group $(\mathbb{Z}, +)$ is cyclic with generator 1.

Solution

The identity element is 0 and the inverse of the element 1 is -1. For any element $n \in \mathbb{Z}$ where $n > 0$ we have

$$\begin{aligned} n &= \underbrace{1 + 1 + \cdots + 1}_{\longleftarrow n \text{ terms} \longrightarrow} \\ &= n.1. \end{aligned}$$

If $n < 0$,

$$\begin{aligned} n &= \underbrace{(-1) + (-1) + \cdots + (-1)}_{\longleftarrow |n| \text{ terms} \longrightarrow} \\ &= |n|.(-1) \\ &= n.1. \end{aligned}$$

If $n = 0$ then

$$n = 0.1 = n.1.$$

Hence $(\mathbb{Z}, +)$ is a cyclic group and 1 is a generator.

(A similar line of argument will show that -1 is also a generator of the group.)

2. Show that $\mathbb{Z}_7 = \{[0], [1], [2], [3], [4], [5], [6]\}$ together with addition modulo 7 is a cyclic group with generator $[2]$.

Solution

$$
\begin{aligned}
1.[2] &= [2] \\
2.[2] &= [2] +_7 [2] = [4] \\
3.[2] &= [4] +_7 [2] = [6] \\
4.[2] &= [6] +_7 [2] = [1] \\
5.[2] &= [1] +_7 [2] = [3] \\
6.[2] &= [3] +_7 [2] = [5] \\
7.[2] &= [5] +_7 [2] = [0].
\end{aligned}
$$

Hence every element of $\mathbb{Z}_7$ can be written as $n.[2]$ for some integer n and so $(\mathbb{Z}_7, +_7)$ is a cyclic group with generator $[2]$.

It is easy to verify that all elements of $\mathbb{Z}_7$ except $[0]$ are generators of the group $(\mathbb{Z}_7, +_7)$.

Dihedral Groups

Consider an equilateral triangle with vertices numbered 1,2 and 3 positioned as shown in the diagram below.

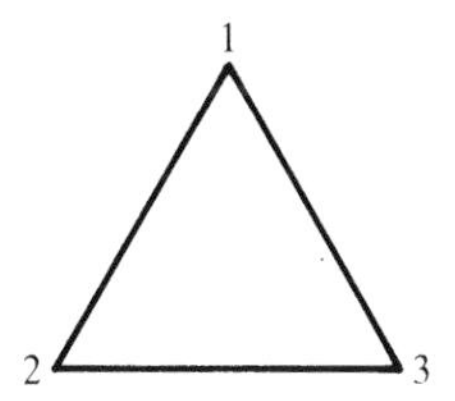

Now consider all the possible transformations of this triangle which result in an interchange of the positions of the vertices. For instance, if the triangle is rotated anti-clockwise through 120° about its 'centre', we obtain:

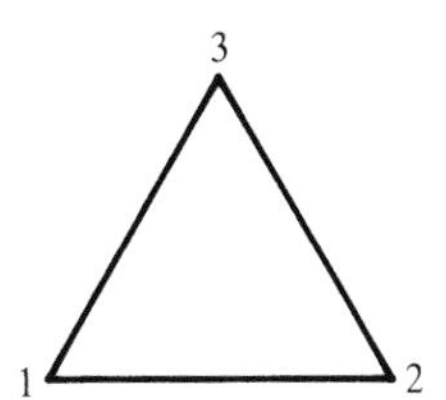

Reflection of the triangle in the line joining the uppermost vertex to the midpoint of the opposite side gives:

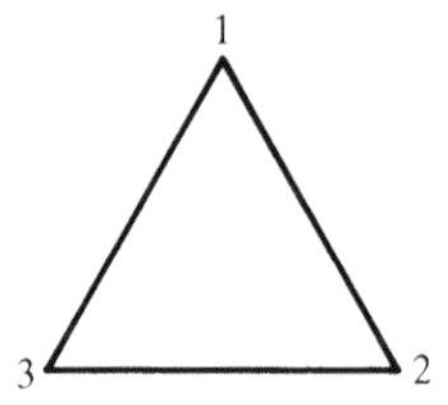

The set of all these transformations is called the **set of symmetries of the equilateral triangle**. There are six such symmetries, three involving rotations and three involving reflections in the lines L_1, L_2 and L_3 as shown in figure 8.2.

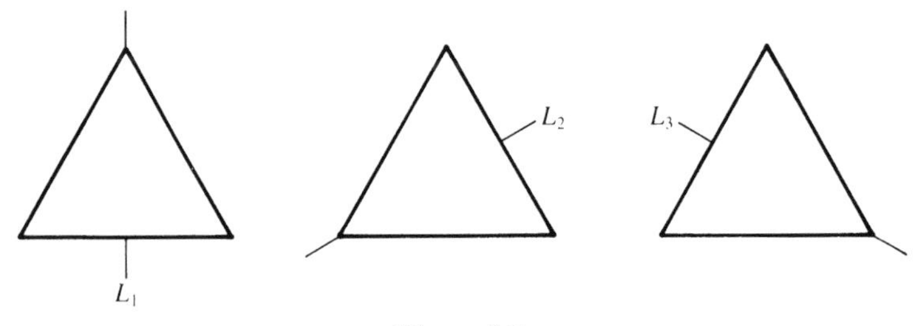

Figure 8.2

(Note that these lines are fixed in the plane and do not move when the triangle is rotated or reflected.)

Table 8.1 gives the position of the vertices of the triangle after each of the transformations has been effected, given the starting position indicated.

Consider the set $T = \{r_0, r_1, r_2, m_1, m_2, m_3\}$ and the operation $*$ where $a * b = ab$ means 'perform transformation a followed by transformation b'. Thus $r_1 * m_1$ means 'rotate the triangle through $120°$ anti-clockwise and then reflect the result in L_1'. Figure 8.3 shows the result of combining these two transformations.

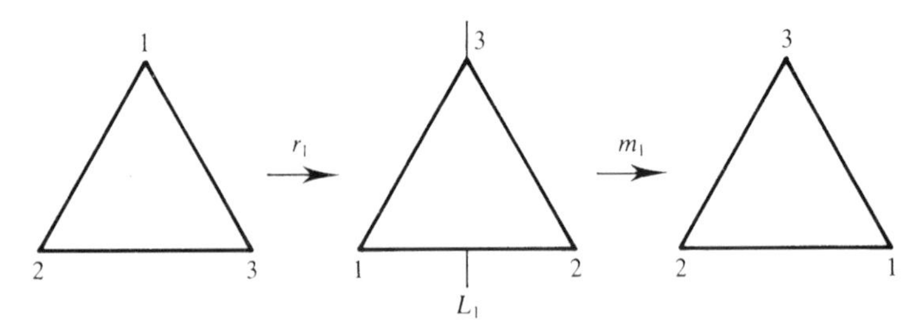

Figure 8.3

Table 8.1

Symmetry	Result of transformation
r_0: rotation through $0°$ anti-clockwise	1 / 2 3 → 1 / 2 3
r_1: rotation through $120°$ anti-clockwise	1 / 2 3 → 3 / 1 2
r_2: rotation through $240°$ anti-clockwise	1 / 2 3 → 2 / 3 1
m_1: reflection in L_1	1 / 2 3 → 1 / 3 2
m_2: reflection in L_2	1 / 2 3 → 3 / 2 1
m_3: reflection in L_3	1 / 2 3 → 2 / 1 3

The result is equivalent to the single transformation m_2 and we can write $r_1 m_1 = m_2$. The operation $*$ is not commutative since, for example $m_1 r_1 = m_3$; see figure 8.4.

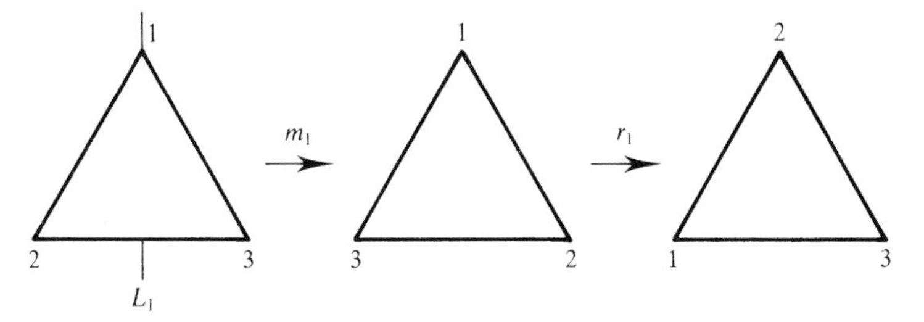

Figure 8.4

The Cayley table for the set T under the operation $*$ is given below.

$*$	r_0	r_1	r_2	m_1	m_2	m_3
r_0	r_0	r_1	r_2	m_1	m_2	m_3
r_1	r_1	r_2	r_0	m_2	m_3	m_1
r_2	r_2	r_0	r_1	m_3	m_1	m_2
m_1	m_1	m_3	m_2	r_0	r_2	r_1
m_2	m_2	m_1	m_3	r_1	r_0	r_2
m_3	m_3	m_2	m_1	r_2	r_1	r_0

It is clear that $*$ is a binary operation on T and we can show that $(T, *)$ is a non-abelian group. The identity is r_0 and each element has an inverse. We have the usual problem with associativity. However, since each transformation can be regarded as a function mapping the triangular region of the plane to itself, the operation $*$ is then composition of functions which we have already shown to be associative (see exercise 5.2.8).

The group $(T, *)$ is often denoted by D_3 (the operation being understood as that of combining transformations). It is referred to as the **group of symmetries of the equilateral triangle** or the **dihedral group of degree 3**.

A similar group of symmetries exists for any regular polygon. The dihedral group of degree n is the group of symmetries of a regular n-sided polygon. It has $2n$ elements and is denoted by D_n.

Groups of Permutations

Definition 8.11

Suppose that S is a non-empty set. A **permutation** of S is a bijection from S to S.

The usual way of defining a specific bijection would be to show the effect of the mapping on every element of S. For example, if $S = \{1, 2, 3, 4\}$, we could define a bijection p_1 by

$$p_1(1) = 2 \qquad p_1(2) = 4 \qquad p_1(3) = 3 \qquad p_1(4) = 1.$$

A more convenient way of representing p_1 is by using an array in which the elements of S occupy the first row and their corresponding images the second row. For the bijection p_1 defined above, we write

$$\begin{aligned} p_1 &= \begin{bmatrix} 1 & 2 & 3 & 4 \\ p_1(1) & p_1(2) & p_1(3) & p_1(4) \end{bmatrix} \\ &= \begin{bmatrix} 1 & 2 & 3 & 4 \\ 2 & 4 & 3 & 1 \end{bmatrix}. \end{aligned}$$

In the same way, we might define another permutation p_2 by

$$p_2 = \begin{bmatrix} 1 & 2 & 3 & 4 \\ 2 & 3 & 4 & 1 \end{bmatrix}.$$

This is equivalent to:

$$p_2(1) = 2 \qquad p_2(2) = 3 \qquad p_2(3) = 4 \qquad p_2(4) = 1.$$

Note that the order in which the elements of S are listed in the first row is immaterial. What is important is that below each element is its image under the appropriate bijection. Thus we could equally well write

$$p_1 = \begin{bmatrix} 2 & 1 & 4 & 3 \\ 4 & 2 & 1 & 3 \end{bmatrix} \quad \text{or} \quad p_2 = \begin{bmatrix} 4 & 3 & 2 & 1 \\ 1 & 4 & 3 & 2 \end{bmatrix}.$$

Consider now the set $A = \{1, 2, 3\}$ and let S_3 be the set of all permutations of A. (We use the notation S_3 for this set to emphasize that it is the set of permutations of a set with three elements.) It is not difficult to establish (see exercise 5.4.7) that S_3 has six elements $p_1, p_2, \ldots, p_6$ defined as follows:

$$p_1 = \begin{bmatrix} 1 & 2 & 3 \\ 1 & 2 & 3 \end{bmatrix} \quad p_2 = \begin{bmatrix} 1 & 2 & 3 \\ 2 & 3 & 1 \end{bmatrix} \quad p_3 = \begin{bmatrix} 1 & 2 & 3 \\ 3 & 1 & 2 \end{bmatrix}$$
$$p_4 = \begin{bmatrix} 1 & 2 & 3 \\ 1 & 3 & 2 \end{bmatrix} \quad p_5 = \begin{bmatrix} 1 & 2 & 3 \\ 3 & 2 & 1 \end{bmatrix} \quad p_6 = \begin{bmatrix} 1 & 2 & 3 \\ 2 & 1 & 3 \end{bmatrix}.$$

There is a natural binary operation which can be defined on S_3, that of composition of functions. Thus $p_i p_j$ $(p_i, p_j \in S_3)$ denotes composition of the bijections p_i and p_j in the order p_i *followed by* p_j. (This notation is convenient for our purpose but it is at odds with our usual notation for composition of functions. Remember that for functions f_1 and f_2, $(f_1 \circ f_2)(x)$ is interpreted as $f_1[f_2(x)]$, i.e. perform f_2 followed by f_1. Thus if $a \in A$,

$$\begin{aligned} (p_i p_j)(a) &= p_j[p_i(a)] \\ &= (p_j \circ p_i)(a).) \end{aligned}$$

The operation is clearly a binary operation since the composition of bijections on S is itself a bijection on S (see theorem 5.7(i)). Consider for example $p_3 p_5$. In array form we write

$$p_3 p_5 = \begin{bmatrix} 1 & 2 & 3 \\ 3 & 1 & 2 \end{bmatrix} \begin{bmatrix} 1 & 2 & 3 \\ 3 & 2 & 1 \end{bmatrix}.$$

To obtain the array representing the bijection $p_3 p_5$ we must find the effect of the bijection on each member of A. Take the element 1 for instance. From the array for p_3 we see that $1 \mapsto 3$. The array for p_5 gives $3 \mapsto 1$. Therefore under the bijection $p_3 p_5$ the image of 1 is 1. We show this below:

$$p_3 p_5 = \begin{bmatrix} 1 & 2 & 3 \\ 3 & 1 & 2 \end{bmatrix} \begin{bmatrix} 1 & 2 & 3 \\ 3 & 2 & 1 \end{bmatrix}$$
$$= \begin{bmatrix} 1 & 2 & 3 \\ 1 & ? & ? \end{bmatrix}.$$

Repeating this process with the remaining elements of A we have

$$p_3 p_5 = \begin{bmatrix} 1 & 2 & 3 \\ 3 & 1 & 2 \end{bmatrix} \begin{bmatrix} 1 & 2 & 3 \\ 3 & 2 & 1 \end{bmatrix}$$
$$= \begin{bmatrix} 1 & 2 & 3 \\ 1 & 3 & 2 \end{bmatrix}.$$

This is the array representing p_4 and so we can write

$$p_3 p_5 = p_4.$$

Completing the Cayley table for $(S_3, *)$ gives:

$*$	p_1	p_2	p_3	p_4	p_5	p_6
p_1	p_1	p_2	p_3	p_4	p_5	p_6
p_2	p_2	p_3	p_1	p_5	p_6	p_4
p_3	p_3	p_1	p_2	p_6	p_4	p_5
p_4	p_4	p_6	p_5	p_1	p_3	p_2
p_5	p_5	p_4	p_6	p_2	p_1	p_3
p_6	p_6	p_5	p_4	p_3	p_2	p_1

That the structure $(S_3, *)$ is a non-abelian group can easily be verified. The identity is p_1 and inverses are given by ${p_1}^{-1} = p_1$, ${p_2}^{-1} = p_3$, ${p_3}^{-1} = p_2$, ${p_4}^{-1} = p_4$, ${p_5}^{-1} = p_5$, ${p_6}^{-1} = p_6$. Associativity follows from the associativity of composition of functions. The set A, on which the bijections were defined, has three elements. The set of permutations of A, denoted by S_3, has six elements.

If $S = \{1, 2, \ldots, n\}$, so that $|S| = n$, then the set of permutations, S_n, would have $n(n-1)(n-2)\ldots 1 = n!$ elements. This is so because, in defining a bijection from S to S, the first element of S can be mapped to any one of the n elements of S, the second element of S to any one of the remaining $n-1$ elements, and so on. This gives $n!$ possible bijections in all (see exercise 5.4.7). For any positive integer n, $(S_n, *)$, where $*$ denotes composition of bijections, is a group called the **symmetric group of degree n**. It is usually referred to simply as S_n, the operation being understood as that of composition of bijections.

Exercises 8.3

1. Show that for any group $(G, *)$,

$$(ab)^{-1} = b^{-1}a^{-1}$$

for all $a, b \in G$. (This is sometimes known as the 'shoes and socks' theorem. Can you suggest why?)

Deduce that, if $a \in G$, $(a^{-1})^n = (a^n)^{-1}$ for all $n \in \mathbb{Z}$.

(Note that theorem 6.4 is the 'shoes and socks' theorem applied to the group of non-singular $n \times n$ matrices under multiplication.)

2. The following is part of the Cayley table of a finite group. Complete the table.

$*$	e	p	q	r	s	t
e	e	p	q	r	s	t
p	p	q	e	s		
q	q					
r	r	t		e		p
s	s					
t	t					

3. The binary operation $*$ is defined on the set $S = \{e, a, b, c, d\}$ by the following Cayley table.

$*$	e	a	b	c	d
e	e	a	b	c	d
a	a	e	d	b	c
b	b	c	e	d	a
c	c	d	a	e	b
d	d	b	c	a	e

Use this table to show that the converse of theorem 8.5 does not hold.

4. Consider a (non-square) rectangle with vertices numbered 1, 2, 3 and 4 positioned as shown in the diagram below.

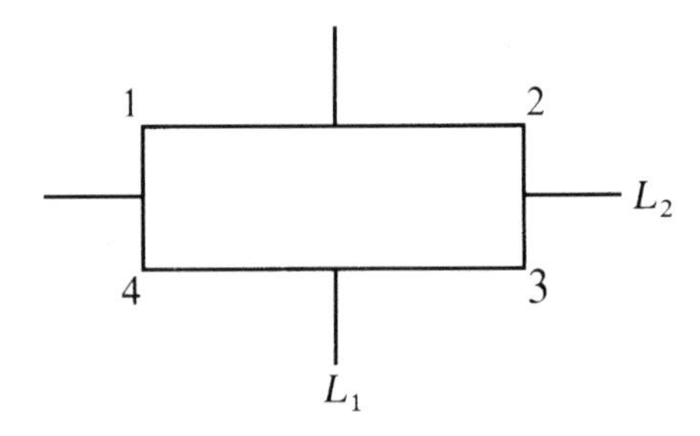

The rectangle has four symmetries:

r_0: rotation through 0° about the centre
r_1: rotation through 180° about the centre
m_1: reflection in the line L_1
m_2: reflection in the line L_2.

Draw the Cayley table for the composition of these transformations and show that the set $\{r_0, r_1, m_1, m_2\}$ together with this binary operation is a group. (This group is known as the **Klein four-group**.)

5. Let $(G, *)$ be a finite group with order n. Show that, for every element $g \in G$, there exists an integer $m \leqslant n$ such that $g^m = e$.

6. Draw up the Cayley table for D_4, the symmetries of a square under composition of transformations. Establish that the group properties hold for D_4.

7. Prove that a cyclic group with only one generator cannot have more than two elements.

8. Show that the group $(\mathbb{Z}_6, +_6)$ is cyclic and identify all its generators.

9. Show that the set of rotational symmetries of an equilateral triangle $\{r_0, r_1, r_2\}$ is a group under composition of rotations. Is this a cyclic group? If so, what are the generators?

10. Let $(G, *)$ be a group. Show that if $x^2 = e$ for all $x \in G$, then $(G, *)$ is abelian.

11. We have shown (exercise 8.2.10) that the set of all 2×2 matrices of the form

$$\begin{pmatrix} 1 & n \\ 0 & 1 \end{pmatrix}$$

where $n \in \mathbb{Z}$, is a group under matrix multiplication. Show that this is a cyclic group.

12. Find all the subsets of $\mathbb{Z}_{10}$ which form a group under the operation of $\times_{10}$. Identify the generators of any of these groups which are cyclic.

13. (a) For each of the following values of n, find the largest subset of $\mathbb{Z}_n$ which forms a group under $\times_n$.

    (i) $n = 6$;
    (ii) $n = 7$;
    (iii) $n = 8$;
    (iv) $n = 9$.

    (b) Given a set S such that $S \subseteq \mathbb{Z}_n$, how should the elements of S be chosen so that $(S, \times_n)$ is a group with the greatest possible order?

14. (a) Let $C_5 = \{e, g, g^2, g^3, g^4\}$ be the cyclic group of order 5 (so that $g^5 = e$). Which elements of C_5 generate the group?

    (b) Repeat (a) for the cyclic groups C_6 and C_9.

    (c) Generalize the results of (a) and (b). In other words, which of the elements of the cyclic group of order n, $C_n = \{e, g, g^2, g^3, \ldots, g^{n-1}\}$ generate the group?

8.5 Substructures

We have shown that the set of symmetries of an equilateral triangle $T = \{r_0, r_1, r_2, m_1, m_2, m_3\}$ is a group under composition of transformations. In exercise 8.3.9 we saw that the subset $\{r_0, r_1, r_2\}$ is also a group under the same binary operation. We have come across other examples of 'a group within a group'. For instance, $(\mathbb{Z}, +)$ and $(\mathbb{R}, +)$ are each groups and $\mathbb{Z}$ is a subset of $\mathbb{R}$. Where one group is contained within another, we refer to the former as a 'subgroup' of the latter.

Definition 8.12

Let $(G, *)$ be a group. If $H \subseteq G$ and $(H, *)$ is itself a group, we say that $(H, *)$ is a **subgroup** of $(G, *)$ and we write $(H, *) \leqslant (G, *)$.

Note that, in order to be a subgroup, the subset H must be a group under the *same* binary operation as that defined for the group $(G, *)$.

Every group $(G, *)$ with two or more elements has at least two subgroups. Since $G \subseteq G$, $(G, *)$ is a subgroup of itself. Also, if e is the identity element, $\{e\} \subseteq G$ and $(\{e\}, *)$ is a group and is therefore a subgroup of $(G, *)$. These two subgroups are called **improper** (or **trivial**) **subgroups**. All other subgroups (if any exist) are called **proper subgroups**.

Examples 8.7

1. It can be readily verified that the set

$$\mathbb{Z}_7 - \{[0]\} = \{[1], [2], [3], [4], [5], [6]\}$$

is a group under multiplication modulo 7 (see exercise 8.2.4). The Cayley table for the set $\{[1], [2], [4]\}$ under multiplication modulo 7 is given below.

$\times_7$	[1]	[2]	[4]
[1]	[1]	[2]	[4]
[2]	[2]	[4]	[1]
[4]	[4]	[1]	[2]

From the table we can see that $\{[1], [2], [4]\}$ is also a group under multiplication modulo 7 and is therefore a subgroup of $(\mathbb{Z}_7 - \{[0]\}, \times_7)$.

2. We denote by C_n the group of rotations of a regular n-sided polygon under composition of rotations. (The group considered in exercise 8.3.9 is C_3.) This group is cyclic and, for all positive integers n, C_n is a subgroup of D_n, the dihedral group of degree n.

Given a group $(G, *)$ and a set H where $H \subseteq G$, it is useful to have a set of criteria for determining whether $(H, *)$ is a subgroup of $(G, *)$. The following theorem provides a set of such criteria. The proof is simple and is therefore left as an exercise (8.4.8).

Theorem 8.6 (Subgroup test)

If $(G, *)$ is a group and H is a non-empty subset of G, then $(H, *)$ is a group if and only if

(a) $ab \in H$ for all $a, b \in H$ (i.e. H is closed under $*$), and
(b) for all $a \in H$, $a^{-1} \in H$.

The theorem states that, if $H \neq \varnothing$ and $H \subseteq G$, to establish that $(H, *)$ is a subgroup of $(G, *)$, we need only ensure that H is closed under $*$ and that the inverse of every element of H is also a member of H.

In fact, if H is a *finite* non-empty subset of G, all that is necessary to establish that $(H, *)$ is a subgroup of $(G, *)$ is to show that H is closed under the operation $*$. The second condition, that every element belonging to H has an inverse which belongs to H, is automatically satisfied. Since this is less obvious than the result of theorem 8.6 we give a proof.

Theorem 8.7 (Finite subgroup test)

Let $(G, *)$ be group and $H \subseteq G$, where H is finite and non-empty. If H is closed under $*$, then $(H, *)$ is a subgroup of $(G, *)$.

Proof

We are given that $ab \in H$ for all $a, b \in H$. To apply the result of theorem 8.6 we must show that $a^{-1} \in H$ for all $a \in H$.

Now, if $a \in H$, then $a^n \in H$ for all $n \in \mathbb{Z}^+$ by the closure property. Since H is a finite set, this apparently infinite collection of terms must contain some duplicates. In particular $a^r = a^s$ for some $r, s \in \mathbb{Z}^+$. Without loss of generality we can assume that $r > s$ and, since $a^{r-s} \in H$, we can write this equation as

$$a^s a^{r-s} = a^s.$$

Applying the left cancellation law (these elements belong to the group $(G, *)$), we have

$$a^{r-s} = e.$$

(This establishes that the identity, e, belongs to H.)

Since $r > s, r - s - 1 \geqslant 0$ so $a^{r-s-1} \in H$ and

$$\begin{aligned} aa^{r-s-1} &= a^{r-s} \\ &= e. \end{aligned}$$

Thus the inverse of a is a^{r-s-1}. This shows that every element of H has an inverse in H and the theorem is proved. □

A third test is given for subgroups in exercise 8.4.9.

The subgroup tests provide a useful means of proving that a particular structure is a group. If $H \subseteq G$ where $(G, *)$ is known to be group, then to prove that $(H, *)$ is a group it is sufficient to show that the appropriate subgroup conditions apply. Example 8.8.1 illustrates this.

Examples 8.8

1. Consider the set

$$A = \left\{ \begin{pmatrix} 1 & n \\ 0 & 1 \end{pmatrix} : n \in \mathbb{Z} \right\}$$

under the operation $*$ of matrix multiplication. Now A is a non-empty subset of the set of all 2×2 non-singular matrices and we have shown this to be a group under matrix multiplication (exercise 8.2.1). To show that $(A, *)$ is a group we simply apply theorem 8.6.

The set A is closed under matrix multiplication since

$$\begin{pmatrix} 1 & n \\ 0 & 1 \end{pmatrix}\begin{pmatrix} 1 & m \\ 0 & 1 \end{pmatrix} = \begin{pmatrix} 1 & m+n \\ 0 & 1 \end{pmatrix} \quad \text{for all } m, n \in \mathbb{Z}.$$

The inverse of

$$\begin{pmatrix} 1 & n \\ 0 & 1 \end{pmatrix} \quad \text{is} \quad \begin{pmatrix} 1 & -n \\ 0 & 1 \end{pmatrix}$$

which is an element of A.

Hence $(A, *)$ is a subgroup of the group of all 2×2 non-singular matrices under multiplication and so is a group.

2. We have already established that the structure $(\mathbb{R}^+, \times)$ is a group. The structure $(\mathbb{Q}^+, \times)$ is a subgroup of $(\mathbb{R}^+, \times)$ because:

   (a) $\mathbb{Q}^+ \subseteq \mathbb{R}^+$ and $\mathbb{Q}^+$ is non-empty;
   (b) for any $a, b \in \mathbb{Q}^+$, $ab \in \mathbb{Q}^+$, i.e. $\mathbb{Q}^+$ is closed under multiplication;
   (c) for any $a \in \mathbb{Q}^+$, $a^{-1} = 1/a \in \mathbb{Q}^+$.

Given any element $a \in G$, where $(G, *)$ is a group, it is reasonable to ask what is the smallest subgroup which contains a. By 'smallest' we mean the subgroup which is contained within any other subgroup of which the element a is a member. Clearly if the subgroup contains a, by the closure property it must contain $a^2, a^3, \ldots$, i.e. it must contain all positive powers of a. The identity, a^0, must also be included and, since the subgroup contains a, it must also contain a^{-1}, the inverse of a. Applying the closure property again, the subgroup must also contain the following: $(a^{-1})^2 = a^{-2}$, $(a^{-1})^3 = a^{-3}, \ldots$. To summarize, any subgroup $(G, *)$ which contains the element a must contain at least all elements of the form a^n where n is an integer. (These elements may not be distinct. Indeed, if G is finite they certainly will not be.) This is the essence of the proof of the following theorem.

Theorem 8.8

Let $(G, *)$ be a group and let $a \in G$. Let $H = \{a^n : n \in \mathbb{Z}\}$. Then $(H, *)$ is a subgroup of $(G, *)$ and, if $(H', *)$ is any other subgroup containing a, then $H \subseteq H'$.

The group $(H, *)$ is called the **cyclic subgroup of $(G, *)$ generated by a**. Note that $(H, *)$ may not be a proper subgroup of $(G, *)$. If $a = e$, the identity, then $H = \{e\}$ so that $(H, *)$ is an improper subgroup of $(G, *)$. Also if $(G, *)$ is cyclic and a is a generator then $H = G$ and $(H, *)$ is again an improper subgroup.

Example 8.9

Consider the group $(\mathbb{Z}_6, +_6)$. Find the cyclic subgroup generated by the element [2].

Solution

The subgroup must contain all elements of the form $n \times [2]$ where n is an integer:

$$\begin{aligned}
n &= 0; & 0 \times [2] &= [0], \text{ the identity element} \\
n &= 1; & 1 \times [2] &= [2] \\
n &= -1; & -1 \times [2] &= [4] \\
n &= 2; & 2 \times [2] &= [2] +_6 [2] = [4] \\
n &= -2; & -2 \times [2] &= 2(-1 \times [2]) = [4] +_6 [4] = [2].
\end{aligned}$$

It is clear that for all integers n, $n \times [2]$ gives one of $[0], [2], [4]$. Thus $[2]$ generates the subgroup $(\{[0], [2], [4]\}, +_6)$ and this is the smallest subgroup containing $[2]$.

In a similar way we can verify that [3] generates the subgroup $(\{[0], [3]\}, +_6)$ whilst [1] and [5] generate the group $(\mathbb{Z}_6, +_6)$ itself. The element [4] generates the same subgroup as does [2].

The 'powers' of an element a of a group $(G, *)$ may all be distinct, i.e. $a^m \neq a^n$ for any integers m, n where $m \neq n$. On the other hand there may be distinct integers m and n such that $a^m = a^n$. In this case we have $a^{m-n} = e$ and there is a power of a which gives the identity. The smallest positive value of r such that $a^r = e$ is called the 'order' of the element a.

Definition 8.13

If $(G, *)$ is a group with identity element e, the **order** of an element $a \in G$ is the least positive integer r such that $a^r = e$. If no such integer exists then a is said to be of **infinite order**. If the order of a is n we write $|a| = n$.

If $(G, *)$ is a finite group then the powers of any element $a \in g$ cannot be distinct and hence every element has finite order.

Example 8.10

Find the order of each element of the group $(G, *)$ defined by the following table.

$*$	e	a	b	c
e	e	a	b	c
a	a	e	c	b
b	b	c	e	a
c	c	b	a	e

Solution

Clearly the order of the identity element of any group is 1. (In fact, the identity element is the *only* element with order 1.) Since $a^2 = e$, the order of a is 2, i.e. $|a| = 2$. Also $b^2 = e$ and $c^2 = e$ so that the orders of b and c are also 2.

The definitions for a subsemigroup and submonoid are similar to that for a subgroup and we include them here for completeness.

Definition 8.14

Let $(S, *)$ be a semigroup and let $T \subseteq S$, where $T \neq \varnothing$. The structure $(T, *)$ is a **subsemigroup** of $(S, *)$ if $(T, *)$ is itself a semigroup.

Given a non-empty set T where $T \subseteq S$ and $(S, *)$ is a semigroup, the only criterion necessary to establish that $(T, *)$ is a semigroup is that T be closed under the operation $*$ so that $*$ is a binary operation on T. If this is so, $*$ will be an associative binary operation because it 'inherits' this property from the semigroup $(S, *)$.

Definition 8.15

Let $(S, *)$ be a monoid with identity e. If $T \subseteq S$ and $(T, *)$ is itself a monoid with identity e, then $(T, *)$ is a **submonoid** of $(S, *)$.

To test whether $(T, *)$ is a submonoid, we therefore need to establish that three criteria are satisfied:

(a) $T \subseteq S$;
(b) T is closed under $*$;
(c) T contains the identity element, e.

Examples 8.11

1. The structure $(\mathbb{Z}^+, +)$ is a semigroup. If $\mathbb{E}^+ = \{2, 4, 6, \ldots\}$ then $(\mathbb{E}^+, +)$ is a subsemigroup of $(\mathbb{Z}^+, +)$ since $\mathbb{E}^+ \subseteq \mathbb{Z}^+$ and $\mathbb{E}^+$ is closed under addition.

2. The structure $(\mathbb{Z}^+, \times)$ is a monoid with identity element 1. If $\mathbb{O}^+ = \{1, 3, 5, \ldots\}$ then $(\mathbb{O}^+, \times)$ is a submonoid of $(\mathbb{Z}^+, \times)$.

3. Let $A = \{a, b\}$. Consider $(A^*, *)$, the free semigroup generated by A (see example 8.3.2). Let $X = \{x : x \in A^*$ and x has a as its first symbol$\}$. X is clearly closed under the operation of concatenation since, if two strings having a as their first symbol are concatenated, the resulting string will also have a as its first symbol. Hence $(X, *)$ is a subsemigroup of $(A^*, *)$. Note that $(X, *)$ is not a submonoid of $(A^* \cup \{\lambda\}, *)$ (the free monoid generated by A), since λ is not a member of X.

Exercises 8.4

1. Let $(M, *)$ be an abelian monoid. Show that the set of idempotent elements of M is a submonoid under $*$. (The element $x \in M$ is idempotent if $x^2 = x$.)

2. Find all the proper subgroups of each of the following groups:

   (i) $(\mathbb{Z}_7, +_7)$

(ii) $(\mathbb{Z}_8, +_8)$
(iii) $(\mathbb{Z}_{10}, +_{10})$
(iv) $(\mathbb{Z}_{12}, +_{12})$.

3. Show that:

   (i) the set $\{3z : z \in \mathbb{Z}\}$ together with addition forms a subgroup of $(\mathbb{Z}, +)$;
   (ii) the set $\{nz : z \in \mathbb{Z}\}$ together with addition forms a subgroup of $(\mathbb{Z}, +)$ for any integer n.

4. Find all the proper subgroups of $(S_3, *)$, the group of permutations of a set with three elements (see §8.4).

5. Determine whether or not $(\{[0], [3], [6]\}, +_9)$ is a subgroup of $(\mathbb{Z}_9, +_9)$.

6. Find all the cyclic subgroups of D_4, the dihedral group of degree 4. Find also a non-cyclic proper subgroup of D_4. (See exercise 8.3.6.)

7. Given a group $(G, *)$, the **centre** is defined to be the set $\{a \in G : ag = ga$ for all $g \in G\}$, i.e. the subset of G containing all elements which commute with every element of G.

   (i) Show that the centre is a subgroup of $(G, *)$.
   (ii) Find the centre of D_3, the dihedral group of degree 3.
   (iii) Find the centre of D_4, the dihedral group of degree 4.

8. Prove theorem 8.6.

9. Let $(G, *)$ be a group and let $H \subseteq G$ where $H \neq \varnothing$. Prove that $(H, *)$ is a subgroup of $(G, *)$ if and only if $ab^{-1} \in H$ for all $a, b \in H$.

10. Prove that, if $(H, *)$ and $(K, *)$ are both subgroups of the group $(G, *)$, then so is $(H \cap K, *)$. Is $(H \cup K, *)$ necessarily a subgroup of $(G, *)$? Justify your answer.

11. Consider the set $\mathbb{Z}_7 - \{[0]\} = \{[1], [2], [3], [4], [5], [6]\}$ under multiplication modulo 7. Find all the proper subgroups of the group $(\mathbb{Z}_7 - \{[0]\}, \times_7)$.

12. Consider the set $T = \{A, B, C, D\}$ where

$$A = \begin{pmatrix} 1 & 0 \\ 0 & 1 \end{pmatrix} \qquad B = \begin{pmatrix} 0 & 1 \\ 1 & 0 \end{pmatrix}$$

$$C = \begin{pmatrix} 0 & -1 \\ -1 & 0 \end{pmatrix} \quad D = \begin{pmatrix} -1 & 0 \\ 0 & -1 \end{pmatrix}.$$

Show that matrix multiplication is a binary operation on this set and hence that T together with this operation is a subgroup of the set of all non-singular 2×2 matrices under multiplication.

13. Prove that every subgroup of a cyclic group is also cyclic.

14. The following is a well-known result in group theory, whose proof is beyond the scope of the current chapter.

Lagrange's theorem.† Let G be a finite group and let H be a subgroup of G. Then the order of H is a factor of the order of G.

Use Lagrange's theorem to prove that, if G is a group with order n and $g \in G$ then $g^n = e$.

8.6 Morphisms

Isomorphism

In §8.4 we considered examples of three important families of groups—cyclic groups, dihedral groups and groups of permutations. The Cayley tables for the dihedral group D_3 and for S_3, the group of permutations of a set with three elements, are reproduced below.

D_3

$*$	r_0	r_1	r_2	m_1	m_2	m_3
r_0	r_0	r_1	r_2	m_1	m_2	m_3
r_1	r_1	r_2	r_0	m_2	m_3	m_1
r_2	r_2	r_0	r_1	m_3	m_1	m_2
m_1	m_1	m_3	m_2	r_0	r_2	r_1
m_2	m_2	m_1	m_3	r_1	r_0	r_2
m_3	m_3	m_2	m_1	r_2	r_1	r_0

S_3

$*$	p_1	p_2	p_3	p_4	p_5	p_6
p_1	p_1	p_2	p_3	p_4	p_5	p_6
p_2	p_2	p_3	p_1	p_5	p_6	p_4
p_3	p_3	p_1	p_2	p_6	p_4	p_5
p_4	p_4	p_6	p_5	p_1	p_3	p_2
p_5	p_5	p_4	p_6	p_2	p_1	p_3
p_6	p_6	p_5	p_4	p_3	p_2	p_1

† Named after the Italian-born mathematician Joseph-Louis Lagrange (1736-1813). Lagrange lived before the development of abstract groups. He actually proved a result about polynomials which was later recognised to be a special case of this theorem.

Comparison of these tables leads to the rather surprising observation that the two are identical apart from the labelling of the elements. Wherever r_2 appears in the first table, p_3 appears in the second; wherever m_3 is positioned in the first, p_6 is found in the second, and so on. Had we called the transformations $p_1, p_2, \ldots, p_6$ instead of $r_0, r_1, r_2, m_1, m_2, m_3$ respectively, the two tables would have looked identical. When two finite groups are related in this way we say that they are 'isomorphic'.

It is important to appreciate that being isomorphic does not mean that groups are 'equal'. In our example, the two sets, however their elements are labelled, are different and the two binary operations are not the same. However, there is clearly a very close relationship between isomorphic groups in that their structure is the same even if the elements are not and we must somehow describe this relationship in mathematical terms.

What we mean by saying that the Cayley tables are 'identical apart from the naming of the elements' is that there exists a one-to-one correspondence between the elements of D_3 and the elements of S_3 so that corresponding elements occupy the same positions in their respective tables. This one-to-one correspondence is a bijective function which has the property of preserving the group structure. Such a function is called an 'isomorphism'. Put more formally: given two groups $(G, *)$ and $(G', \circ)$, an isomorphism is a bijective function $f : G \to G'$ which is such that the image of $g_1 * g_2$ is that element of G' which is the result of the operation $\circ$ applied to the images of g_1 and g_2. It is this important property of the isomorphism which ensures that the structures of isomorphic groups are the same.

We summarize these ideas in figure 8.5 and in formal definition which applies not only to finite groups but also to infinite ones.

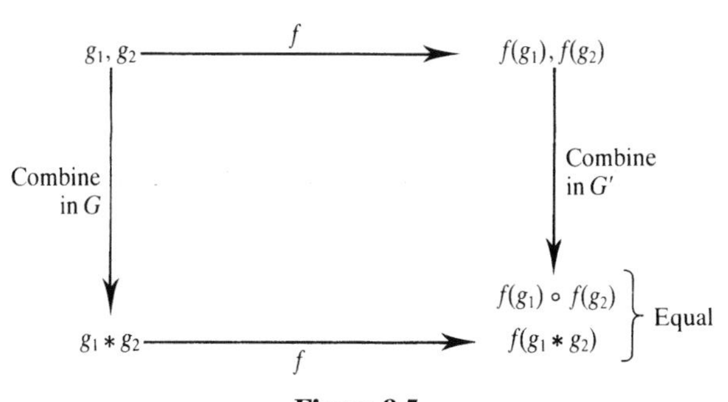

Figure 8.5

Definition 8.16

An **isomorphism** from the group $(G, *)$ to the group $(G', \circ)$ is a bijective function $f : G \to G'$ such that

$$f(g_1 * g_2) = f(g_1) \circ f(g_2)$$

for all $g_1, g_2 \in G$. If such a function exists, we say that $(G, *)$ is **isomorphic** to $(G', \circ)$ and we write $(G, *) \cong (G', \circ)$.

An isomorphism from D_3 to S_3 is defined by $f : r_0 \mapsto p_1$, $r_1 \mapsto p_2$, $r_2 \mapsto p_3$, $m_1 \mapsto p_4$, $m_2 \mapsto p_5$, $m_3 \mapsto p_6$.

More generally, we might ask whether the group of all permutations of a set with n elements is isomorphic with the dihedral group of degree n. The answer is no, because the orders of these groups are not equal for $n > 3$, so that there does not exist a bijection $D_n \to S_n$. For the dihedral group of degree n, $|D_n| = 2n$ whereas $|S_n| = n!$. Not every permutation of n elements corresponds to a symmetry of an n-sided regular polygon.

Examples 8.12

1. Consider the groups $(\mathbb{R}, +)$ and $(\mathbb{R}^+, \times)$. Show that the function $f : \mathbb{R} \to \mathbb{R}^+$, where $f(x) = 2^x$, defines an isomorphism from $(\mathbb{R}, +)$ to $(\mathbb{R}^+, \times)$.

Solution

We have to show two things:

(a) that f is a bijection;
(b) that, if $x, y, \in \mathbb{R}$, then $f(x + y) = f(x) \times f(y)$.

Perhaps the easiest way to confirm that f is a bijection is to plot the graph of $y = f(x)$ for $x \in \mathbb{R}$. This is given in figure 8.6.

Since any horizontal line through the positive part of the y-axis meets the graph exactly once, by theorem 5.6, the function f is bijective.

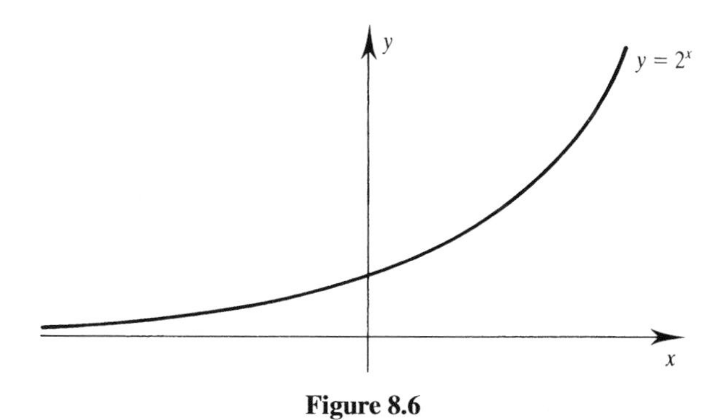

Figure 8.6

Also

$$\begin{aligned} f(x+y) &= 2^{x+y} \\ &= 2^x \times 2^y \\ &= f(x) \times f(y). \end{aligned}$$

We have shown that f is an isomorphism from $(\mathbb{R}, +)$ to $(\mathbb{R}^+, \times)$.

(Note that $f : \mathbb{R} \to \mathbb{R}^+$ where $f(x) = a^x$ is an isomorphism from $(\mathbb{R}, +)$ to $(\mathbb{R}^+, \times)$ for any $a \in \mathbb{R}^+$, $a \neq 1$.)

2. Show that the groups $(\mathbb{Z}_4, +_4)$ and $(\mathbb{Z}_5 - \{[0]\}, \times_5)$ are isomorphic.

Solution

The Cayley tables for each of these groups are given below.

$+_4$	[0]	[1]	[2]	[3]
[0]	[0]	[1]	[2]	[3]
[1]	[1]	[2]	[3]	[0]
[2]	[2]	[3]	[0]	[1]
[3]	[3]	[0]	[1]	[2]

$\times_5$	[1]	[2]	[3]	[4]
[1]	[1]	[2]	[3]	[4]
[2]	[2]	[4]	[1]	[3]
[3]	[3]	[1]	[4]	[2]
[4]	[4]	[3]	[2]	[1]

Our task is to find an isomorphism between the two groups—that is, we must find a bijection $f : \mathbb{Z}_4 \to \mathbb{Z}_5 - \{[0]\}$ such that

$$f(x +_4 y) = f(x) \times_5 f(y) \quad \text{for all } x, y \in \mathbb{Z}_4.$$

Comparison of the Cayley tables and a little trial and error reveals that there are two bijections which have the required properties. These are:

$$f : \mathbb{Z}_4 \to \mathbb{Z}_5 - \{[0]\} \qquad f([0]) = [1], f([1]) = [2],$$
$$f([2]) = [4], f([3]) = [3]$$

and

$$g : \mathbb{Z}_4 \to \mathbb{Z}_5 - \{[0]\} \qquad g([0]) = [1], g([1]) = [3],$$
$$g([2]) = [4], g([3]) = [2].$$

Both of these functions map the first Cayley table to the second and hence both are isomorphisms from $(\mathbb{Z}_4, +_4)$ to $(\mathbb{Z}_5 - \{[0]\}, \times_5)$.

These examples illustrate the fact that there may be more than one isomorphism between isomorphic groups. However, to establish that two groups are isomorphic, all that is necessary is to find one such function.

Determining whether or not two groups are isomorphic necessarily involves a certain amount of trial and error and can therefore be time consuming, especially if the order of the groups is large. A certain amount of guesswork can be eliminated by using known properties of isomorphic groups. Some of these are listed in the theorem below.

Theorem 8.9

If $f : G_1 \to G_2$ is an isomorphism between the groups $(G_1, *)$ and $(G_2, \circ)$ then:

(1) if e is the identity in $(G_1, *)$, $f(e)$ is the identity in $(G_2, \circ)$;

(2) $(G_1, *)$ is abelian if and only if $(G_2, \circ)$ is abelian;

(3) if a^{-1} is the inverse of a in $(G_1, *)$, then $f(a^{-1})$ is the inverse of $f(a)$ in $(G_2, \circ)$, i.e. $f(a^{-1}) = [f(a)]^{-1}$;

(4) the inverse function $f^{-1} : G_2 \to G_1$ defines an isomorphism from $(G_2, \circ)$ to $(G_1, *)$;

(5) if $(H_1, *)$ is a subgroup of $(G_1, *)$, then $(H_2, \circ)$ where $H_2 = \{f(a) : a \in H_1\}$ is a subgroup of $(G_2, \circ)$ and $(H_1, *) \cong (H_2, \circ)$;

(6) $(G_1, *)$ is cyclic if and only if $(G_2, \circ)$ is cyclic;

(7) if $a \in G_1$ then $|a| = |f(a)|$.

It is not difficult to prove that the properties listed apply to isomorphic groups and that if any one of them fails, then the two groups in question are not isomorphic. Therefore, to show that two groups are *not* isomorphic we look for a group-theoretic property which holds for one but not for the other. To show that two groups *are* isomorphic, however, it is not enough to show that they have common properties. We must actually find an isomorphism. These points are summarized in the following 'isomorphism principle'.

Isomorphism principle

To show that two groups are isomorphic, an isomorphism from one to the other must be found; to show that two groups are not isomorphic, a group-theoretic property must be found which one group has but the other does not.

In attempting to establish whether or not two groups are isomorphic, the properties listed above can be useful in determining which bijections are potential isomorphisms and which are not. Particularly helpful in this respect is the property concerning orders of elements. Given two groups $(G_1, *)$ and $(G_2, \circ)$ a useful first step in the search for an isomorphism is to write down the orders of the elements of each. At this stage we can at least see whether a bijective function with order-preserving properties is possible. If it is not then we can deduce immediately that the groups are not isomorphic. If it is possible to define such a function then, in our search for an isomorphism, some bijections can be eliminated by the fact that the orders of each element and its image must be equal. If an order-preserving bijective function exists this is not sufficient for us to conclude that the groups are isomorphic, since the property $f(x*y) = f(x) \circ f(y)$ must also be satisfied. However, order-preserving bijections are the only possible candidates for isomorphisms.

Examples 8.13

1. The group $(G, *)$ is defined as in example 8.10. Determine whether this group is isomorphic with $(\mathbb{Z}_4, +_4)$.

Solution

The Cayley table for $(\mathbb{Z}_4, +_4)$ is the following.

$+_4$	$[0]$	$[1]$	$[2]$	$[3]$
$[0]$	$[0]$	$[1]$	$[2]$	$[3]$
$[1]$	$[1]$	$[2]$	$[3]$	$[0]$
$[2]$	$[2]$	$[3]$	$[0]$	$[1]$
$[3]$	$[3]$	$[0]$	$[1]$	$[2]$

For this group the orders of the elements are given by $|[0]| = 1$, $|[1]| = 4$, $|[2]| = 2$, $|[3]| = 4$.

In example 8.10 we found that the orders of the elements of $(G, *)$ were as follows: $|e| = 1$, $|a| = 2$, $|b| = 2$, $|c| = 2$.

We can see immediately that there is no bijection possible where the order of each element is the same as that of its image. Hence the groups are not isomorphic.

2. Show that the groups $(\mathbb{Z}_4, +_4)$ and $(\{[1], [3], [7], [9]\}, \times_{10})$ are isomorphic.

Solution

The Cayley table for $\{[1], [3], [7], [9]\}$ under multiplication modulo 10 is given below.

$\times_{10}$	$[1]$	$[3]$	$[7]$	$[9]$
$[1]$	$[1]$	$[3]$	$[7]$	$[9]$
$[3]$	$[3]$	$[9]$	$[1]$	$[7]$
$[7]$	$[7]$	$[1]$	$[9]$	$[3]$
$[9]$	$[9]$	$[7]$	$[3]$	$[1]$

We first write down the orders of the elements of each group so that we can decide which bijective functions are possible isomorphisms. We saw in the example above that, for $(\mathbb{Z}_4, +_4)$, the orders of the elements are: $|[0]| = 1$, $|[1]| = 4$, $|[2]| = 2$, $|[3]| = 4$.

For the second group, orders of elements are as follows: $|[1]| = 1$, $|[3]| = 4$, $|[7]| = 4$, $|[9]| = 2$.

From these we can deduce that any isomorphism $f : \mathbb{Z}_4 \to \{[1], [3], [7], [9]\}$ must be such that $f([0]) = [1]$, and $f([2]) = [9]$. For the other two elements of $\mathbb{Z}_4$

there are two possibilities $f([1]) = [3]$ and $f([3]) = [7]$ or $f([1]) = [7]$ and $f([3]) = [2]$. Hence there are two isomorphism candidates, the functions f and g defined below.

$$f : \mathbb{Z}_4 \to \{[1],[3],[7],[9]\} \qquad f([0]) = [1], f([1]) = [3],$$
$$f([2]) = [9], f([3]) = [7].$$

$$f : \mathbb{Z}_4 \to \{[1],[3],[7],[9]\} \qquad f([0]) = [1], f([1]) = [7],$$
$$f([2]) = [9], f([3]) = [3].$$

In fact each of these functions is an isomorphism but this must be established as before.

3. Determine whether the groups D_3 and $(\mathbb{Z}_6, +_6)$ are isomorphic.

Solution

Applying theorem 8.9(2) we can say immediately that these groups are not isomorphic since $(\mathbb{Z}_6, +_6)$ is abelian whereas D_3 is not.

Isomorphisms are defined in exactly the same way between monoids and semigroups. Indeed, since any group is automatically a monoid and a semigroup, all of the group isomorphisms considered above could be regarded as isomorphisms between two monoids or between two semigroups. In the example below we consider an isomorphism between two monoids which are not groups.

Example 8.14

Show that, if $A = \{a, b\}$, then the monoids $(\mathscr{P}(A), \cup)$ and $(\mathscr{P}(A), \cap)$ are isomorphic.

Solution

If $A = \{a, b\}$ then $\mathscr{P}(A) = \{\varnothing, \{a\}, \{b\}, \{a, b\}\}$.

The Cayley tables for $(\mathscr{P}(A), \cup)$ and $(\mathscr{P}(A), \cap)$ are given below.

$\cup$	$\varnothing$	$\{a\}$	$\{b\}$	$\{a,b\}$
$\varnothing$	$\varnothing$	$\{a\}$	$\{b\}$	$\{a,b\}$
$\{a\}$	$\{a\}$	$\{a\}$	$\{a,b\}$	$\{a,b\}$
$\{b\}$	$\{b\}$	$\{a,b\}$	$\{b\}$	$\{a,b\}$
$\{a,b\}$	$\{a,b\}$	$\{a,b\}$	$\{a,b\}$	$\{a,b\}$

$\cap$	$\varnothing$	$\{a\}$	$\{b\}$	$\{a,b\}$
$\varnothing$	$\varnothing$	$\varnothing$	$\varnothing$	$\varnothing$
$\{a\}$	$\varnothing$	$\{a\}$	$\varnothing$	$\{a\}$
$\{b\}$	$\varnothing$	$\varnothing$	$\{b\}$	$\{b\}$
$\{a,b\}$	$\varnothing$	$\{a\}$	$\{b\}$	$\{a,b\}$

A bijective function which maps the first table to the second is

$$f : \mathscr{P}(A) \to \mathscr{P}(A) \qquad \begin{array}{l} f(\varnothing) = \{a,b\}, f(\{a,b\}) = \varnothing, \\ f(\{a\}) = \{a\}, f(\{b\}) = \{b\}. \end{array}$$

Thus the two monoids are isomorphic.

An alternative, and perhaps more natural, isomorphism is given by $g(X) = \bar{X}$ for all $X \in \mathscr{P}(A)$.

Morphisms

For two groups $(G, *)$ and $(G', \circ)$ to be isomorphic, we need to be able to define a function $f : G \to G'$ which is bijective and which also preserves the structure of the group. Dropping the bijective condition defines a more general concept of structure-preserving function called a 'morphism' (or 'homomorphism'). Morphisms are defined not only between pairs of groups but between any two algebraic structures of the types considered in this chapter.

Definition 8.17

Given two algebraic structures $(A, *)$ and $(B, \circ)$, a **morphism** from $(A, *)$ to $(B, \circ)$ is a function $f : A \to B$ such that

$$f(a_1 * a_2) = f(a_1) \circ f(a_2)$$

for all $a_1, a_2 \in A$.

A morphism need not be surjective so that there may be elements of B which are not the image of any element of A. If it is surjective, a morphism is called an **epimorphism**. Similarly a morphism need not be injective so that there may be elements of B which are the image of more than one element of A. An injective morphism is called a **monomorphism**. An isomorphism is a morphism which is both surjective and injective.

What is important about morphisms between two algebraic structures $(A, *)$ and $(B, \circ)$ is that many of the properties of A under the operation $*$ are preserved in the image set $f(A)$ under the operation $\circ$. In particular, if $(A, *)$ is a member of a particular family of structures then so too is $(f(A), \circ)$. This is stated in the following theorem the proof of which is left as an exercise (8.5.2).

Theorem 8.10

Let $(A, *)$ and $(B, \circ)$ be algebraic structures and let $f : A \to B$ be a morphism.

(a) If $(A, *)$ is a semigroup then so too is $(f(A), \circ)$.
(b) If $(A, *)$ is a monoid then so too is $(f(A), \circ)$.
(c) If $(A, *)$ is a group then so too is $(f(A), \circ)$.

We saw that when there exists an isomorphism between two algebraic structures $(A, *)$ and $(B, \circ)$, these structures could be regarded as 'essentially the same'. Dropping the bijective condition means that some of the likenesses between the two structures are lost. For a bijective morphism, $(B, \circ)$ can be regarded as a perfect copy of $(A, *)$ whereas if the morphism lacks the bijective property $(f(A), \circ)$ lacks some of the details of $(A, *)$.

The following theorem lists some of the general properties of all morphisms.

Theorem 8.11

Let $(A, *)$ and $(B, \circ)$ be algebraic structures and let $f : A \to B$ be a morphism from $(A, *)$ to $(B, \circ)$. Then:

(1) if e is the identity in $(A, *)$, $f(e)$ is the identity in $(f(A), \circ)$;
(2) if $(A, *)$ is abelian then $(f(A), \circ)$ is abelian;
(3) if a^{-1} is the inverse of a in $(A, *)$ then $f(a^{-1})$ is the inverse of $f(a)$ in $(f(A), \circ)$, i.e. $f(a^{-1}) = [f(a)]^{-1}$;
(4) if $(A, *)$ is a cyclic group then so too is $(f(A), \circ)$.

The structure $(f(A), \circ)$ is called the **morphic image** of $(A, *)$. If the morphism is surjective then $(f(A), \circ)$ can be replaced by $(B, \circ)$ in each of the statements above. An analogy which is sometimes used to illustrate the relationship between a structure and its morphic image is that between a colour photograph and the person appearing in it. Some characteristics of the individual can be obtained from the photograph, for example the colour of their hair. Others, such as their height and weight, cannot be determined.

Examples 8.15

1. Consider the groups $(\mathbb{Z}, +)$ and $(\{[0], [1]\}, +_2)$. Show that the function $f : \mathbb{Z} \to \{[0], [1]\}$ where

$$f(x) = \begin{cases} [0] & \text{if } x \text{ is even,} \\ [1] & \text{if } x \text{ is odd} \end{cases}$$

defines a morphism from $(\mathbb{Z}, +)$ to $(\{[0], [1]\}, +_2)$.

Solution

We must show that, for all $a, b \in \mathbb{Z}$

$$f(a + b) = f(a) +_2 f(b).$$

Clearly the values of each side of this equation depend on whether a and b are odd or even. All the four possible cases are considered in the table below. The last two columns show that, in each case, the equation above is satisfied and hence that f defines a morphism from $(\mathbb{Z}, +)$ to $(\{[0], [1]\}, +_2)$.

a	b	$a + b$	$f(a)$	$f(b)$	$f(a + b)$	$f(a) +_2 f(b)$
Even	Even	Even	[0]	[0]	[0]	[0]
Even	Odd	Odd	[0]	[1]	[1]	[1]
Odd	Even	Odd	[1]	[0]	[1]	[1]
Odd	Odd	Even	[1]	[1]	[0]	[0]

Note that f is surjective (and is therefore an epimorphism) but is not injective.

2. Consider the group $(\mathbb{Z}, +)$. Let $f : \mathbb{Z} \to \mathbb{Z}$ be defined by $f(x) = 2x$. Show that f is a morphism from $(\mathbb{Z}, +)$ to $(\mathbb{Z}, +)$.

Solution

Here we must show that $f(x+y) = f(x) + f(y)$ for all $x, y \in \mathbb{Z}$.

We have

$$\begin{aligned} f(x+y) &= 2(x+y) \\ &= 2x + 2y \\ &= f(x) + f(y). \end{aligned}$$

Hence f is a morphism. The function is injective (i.e. is a monomorphism) but is not surjective. The image set is the set $\mathbb{E}$ of even integers (including zero) and we can confirm the result of theorem 8.10(c). The set $\mathbb{E}$ is a group under the operation of addition.

The following is an example of a morphism between monoids.

Example 8.16

Consider the alphabet $A = \{a, b, c\}$ and let $(A^* \cup \{\lambda\}, *)$ be the free monoid generated by A (see example 8.4.2). A function $f : A^* \cup \{\lambda\} \to \mathbb{N}$ is defined by

$$f(x) = \text{the length of the string } x.$$

Show that f is a morphism from $(A^* \cup \{\lambda\}, *)$ to $(\mathbb{N}, +)$.

Solution

Again we must show that

$$f(x * y) = f(x) + f(y)$$

where x and y are strings and $*$ represents concatenation of strings. Since the number of symbols in two concatenated strings is clearly equal to the sum of the number of symbols in each string, the equation holds and f defines a morphism from $(A^* \cup \{\lambda\}, *)$ to $(\mathbb{N}, +)$. This function is surjective (but not injective) and the morphism is therefore an epimorphism.

Exercises 8.5

1. Prove theorem 8.9.

2. Prove theorem 8.10.

3. Let A be the set of all 2×2 matrices of the form

$$\begin{pmatrix} 1 & n \\ 0 & 1 \end{pmatrix}$$

where $n \in \mathbb{Z}$.

(In example 8.8.1 we showed that A is a group under matrix multiplication.) Show that the function $f : A \to \mathbb{Z}$ defined by

$$f\left[\begin{pmatrix} 1 & n \\ 0 & 1 \end{pmatrix}\right] = n$$

is an isomorphism from $(A, *)$ (where $*$ denotes matrix multiplication) to $(\mathbb{Z}, +)$.

4. Which of the following functions $f : \mathbb{R} \to \mathbb{R}$ defines a morphism from $(\mathbb{R}, +)$ to $(\mathbb{R}, +)$?

(i) $f(x) = x - 3$,
(ii) $f(x) = 5x$,
(iii) $f(x) = x^2$,
(iv) $f(x) = x/2$,
(v) $f(x) = |x|$,
(vi) $f(x) = 2^x$.

Which morphisms are isomorphisms? (An isomorphism from a group to itself is called an **automorphism**.)

5. Which of the following functions $f : \mathbb{R} \to \mathbb{R} - \{0\}$ are morphisms from $(\mathbb{R}, +)$ to $(\mathbb{R} - \{0\}, \times)$? Which morphisms are isomorphisms?

(i) $f(x) = 2^x$,
(ii) $f(x) = 5$,
(iii) $f(x) = 1$,
(iv) $f(x) = 3^{-x}$.

6. Prove that, if $f : A \to B$ defines a morphism from the algebraic structure $(A, *)$ to the structure $(B, \circ)$, then $f(e_A) = e_B$ where e_A is the identity in

$(A, *)$ and e_B is the identity in $(f(A), \circ)$ (assuming that these identities exist).

7. Let $T = \{A, B, C, D\}$ where

$$A = \begin{pmatrix} 1 & 0 \\ 0 & 1 \end{pmatrix} \qquad B = \begin{pmatrix} 0 & 1 \\ -1 & 0 \end{pmatrix}$$

$$C = \begin{pmatrix} -1 & 0 \\ 0 & -1 \end{pmatrix} \quad D = \begin{pmatrix} 0 & -1 \\ 1 & 0 \end{pmatrix}.$$

Show that T is a group under matrix multiplication. Show that this group is isomorphic to $(\mathbb{Z}_5 - \{[0]\}, \times_5)$.

8. Let $(G, *)$ be a group and let g be a particular element of G. Show that the function $f : G \to G$, where $f(x) = g^{-1}xg$, defines an isomorphism from $(G, *)$ to itself. (This is called an **inner automorphism**.)

9. (i) Let $\mathbb{R}^*$ be the set of non-zero real numbers. Show that the function $f : \mathbb{R}^* \to \mathbb{R}^*$ defined by $f(x) = x^2$ is a morphism from the group $(\mathbb{R}^*, \times)$ to itself.

   (ii) Let $(G, *)$ be a group. Show that the function $f : G \to G$ defined by $f(x) = x^2$ is a morphism from $(G, *)$ to itself if and only if $(G, *)$ is abelian.

10. Let $(G_1, *)$, $(G_2, \circ)$ and $(G_3, \cdot)$ be groups. Let $f : G_1 \to G_2$ and $g : G_2 \to G_3$ define morphisms from $(G_1, *)$ to $(G_2, \circ)$ and from $(G_2, \circ)$ to $(G_3, \cdot)$ respectively. Show that $g \circ f : G_1 \to G_3$ defines a morphism from $(G_1, *)$ to $(G_3, \cdot)$.

11. Let $(G, *)$ be a group and let a function $f : G \to G$ be defined by $f(x) = x^{-1}$ where x^{-1} denotes the inverse of x with respect to the operation $*$. Show that f is an isomorphism from $(G, *)$ to itself if and only if G is abelian.

12. Let $A = \{a\}$. Show that:

   (i) the free semigroup generated by A is isomorphic to $(\mathbb{Z}^+, +)$;
   (ii) the free monoid generated by A is isomorphic to $(\mathbb{N}, +)$;
   (iii) the free group generated by A is isomorphic to $(\mathbb{Z}, +)$.

13. Show that the set of matrices $T = \{A, B, C, D\}$ defined in exercise 8.4.12 is isomorphic to the set of symmetries of a rectangle under composition of transformations. (It is another manifestation of the Klein four-group.)

14. Let $(G_1, *)$ and $(G_2, \circ)$ be two groups with identities e_1 and e_2 respectively. The **kernel** of a morphism $f : G_1 \to G_2$, denoted by $\mathit{ker}\, f$, is the set $\{g \in G_1 : f(g) = e_2\}$, i.e. the set of elements of G_1 which map to the identity in G_2.

(a) For each pair of groups and the morphism given, find the kernel of the morphism:

(i) $(\mathbb{Z}, +)$, $(\mathbb{Z}, +)$; $f : \mathbb{Z} \to \mathbb{Z}$ where $f(x) = 7x$;
(ii) $(\mathbb{R}, +)$, $(\mathbb{R}, +)$; $f : \mathbb{Z} \to \mathbb{Z}$ where $f(x) = 7x$;
(iii) $(\mathbb{R}, +)$, $(\mathbb{R}, +)$; $f : \mathbb{R} \to \mathbb{R}$ where $f(x) = 0$;
(iv) $(\mathbb{R}, +)$, $(\mathbb{R}^+, \times)$; $f : \mathbb{R} \to \mathbb{R}^+$ where $f(x) = 2^x$;
(v) $(\mathbb{Z}, +)$, $(\mathbb{Z}_6, +_6)$; $f : \mathbb{Z} \to \mathbb{Z}_6$ where $f(x) = [x]$ (the modulo 6 equivalence class of x);
(vi) $(\mathbb{Z}_6, +_6)$, $(\mathbb{Z}_6, +_6)$; $f : \mathbb{Z}_6 \to \mathbb{Z}_6$ where $f([x]) = [2x]$.

(b) Show that:

(i) f is a monomorphism if and only if $\mathit{ker}\, f = \{e_1\}$;
(ii) $(\mathit{ker}\, f, *)$ is a subgroup of $(G_1, *)$;
(iii) if $x \in \mathit{ker}\, f$ and $g \in G_1$, then $g^{-1}xg \in \mathit{ker}\, f$. (A subgroup N, which is such that if $x \in N$ and $g \in G$ then $g^{-1}xg \in N$, is called a **normal subgroup** of G.)

8.7 Group Codes

Many of the applications of modern technology involve the communication of data from one point to another. The two points may be relatively close to each other, as in the case of data transfer from one memory location to another in a computer. On the other hand, telecommunication via satellite involves the transmission of data over many thousands of miles. In either case, however, the essential features of the system are the same. There is a communication channel along which data are transmitted where, ideally, the data received at one end of the channel are identical to those sent at the other.

For our purposes we shall assume that all relevant data can be represented by a string of digits each of which is either zero or one, i.e. a word over the alphabet $\{0, 1\}$. We shall refer to such words as **binary words** and their digits as **bits**.

However much we would like our transmission system to be completely reliable, it is inevitable that faults will develop from time to time and that there will be interference (known as 'noise') from external sources. These may cause an error in transmission so that a received word is different from that transmitted. It is important therefore to be able to detect when a received word is in error and, if possible, to determine what was the word actually sent. If the latter is not possible, then at least the detection of an error could lead to a request for the data to be retransmitted.

We shall make the following assumptions about transmission of errors.

(a) They take the form of the conversion of 1 to 0 or 0 to 1 in one or more of the bits which comprise the transmitted word.
(b) The conversions of 1 to 0 and 0 to 1 are equally likely.
(c) Errors in individual bits occur independently of each other.
(d) An error is equally likely in each of the bits which comprise the transmitted word.
(e) For $n < m$, n errors are more likely than m errors so that for an incorrectly transmitted word, the most likely number of errors is one.

We now consider some examples to illustrate the essential features of error detection and correction.

Suppose that words to be transmitted through a communication channel are all the members of the set of binary words of length 3. We denote this set B^3, i.e. $B^3 = \{000, 001, 010, 100, 110, 101, 011, 111\}$. Suppose that the word 010 is transmitted and that an error occurs in the third digit so that the word received is 011. There is no way of detecting this error because 011 is a member of the set of words which we might expect to receive. Further, if we cannot detect an error, there is certainly no chance of our correcting it. This example highlights one property which is essential if we are to detect errors at all—an incorrectly transmitted word must not be a member of the set of words which we are expecting to receive.

The words in the example above are in a sense 'too close together'. Any error results in another member of the set. Suppose instead that words for transmission are members of the set $\{111, 100, 001, 010\}$. In this case an error in the third digit of the transmitted word 010 will be detected because when we receive 011 we know that this could not have been transmitted. However, even though we know that an error has occurred, we cannot determine where it is. On the assumption that one error is the most probable, the word transmitted is equally likely to be 111, 001 or 010. We can detect the single error but we cannot correct it. Note that two errors cannot be detected because errors in any two digits would result in another member of the set.

The members of the set $\{111, 100, 001, 010\}$ are still too close together for even a single error to be corrected. Suppose that we transmit only members of the set $\{000, 111\}$. Now errors in one or two digits can be detected and, if a single error occurs, we can correct it. For example, if we receive 011 we will assume that the word closest to it, 111, was transmitted. However, two errors cannot be corrected. If 000 undergoes two transmission errors and is received as 011, we shall, on the assumption of one error, incorrectly assume that the word sent was 111. For this set of words, we can detect two errors but we cannot correct them.

The examples above illustrate that error detection is easier than error correction. However, both depend upon the words in the set of possible transmitted words being sufficiently different from one another. The following definition gives a means by which we can measure the difference between two individual words.

Definition 8.18

Let x and y be binary words of length n. The **distance** (or **Hamming distance**†) between $\boldsymbol{x}$ and $\boldsymbol{y}$, denoted by $d(\boldsymbol{x}, \boldsymbol{y})$, is the number of digits in which $\boldsymbol{x}$ and $\boldsymbol{y}$ differ.

For example, if $\boldsymbol{x} = 001101$ and $\boldsymbol{y} = 111110$, the two words have different first, second, fifth and sixth digits. Hence $d(\boldsymbol{x}, \boldsymbol{y}) = 4$.

It is easy enough to show that the distance has the following properties for all binary words $\boldsymbol{x}$, $\boldsymbol{y}$ and $\boldsymbol{z}$ of length n:

(a) $d(\boldsymbol{x}, \boldsymbol{y}) \geqslant 0$,
(b) $d(\boldsymbol{x}, \boldsymbol{y}) = 0$ if and only if $\boldsymbol{x} = \boldsymbol{y}$,
(c) $d(\boldsymbol{x}, \boldsymbol{y}) = d(\boldsymbol{y}, \boldsymbol{x})$,
(d) $d(\boldsymbol{x}, \boldsymbol{z}) \leqslant d(\boldsymbol{x}, \boldsymbol{y}) + d(\boldsymbol{y}, \boldsymbol{z})$.

(Any function $d : X \times X \to \mathbb{R}^+ \cup \{0\}$ having these properties is called a **metric** on the set X. Hence distance is a metric on the set B^n.)

For successful error detection and correction it is desirable that the distance between individual words in the set of possible transmitted words be as large as possible.

† Named after Richard Hamming who pioneered the field of error detection and correction in transmitted data in the 1950s.

INFORMATION BITS

In practice the detection and correction of errors are carried out by **coding** words before transmission. Generally this involves adding one or more bits to the end of the word. These are called **check digits** and they act as checks on the validity of some or all of the digits of the received word. Thus any transmitted binary word of length n consists of m digits called **information bits** which carry the information to be sent and $r = n - m$ check digits which provide the means for detecting and correcting errors. Denoting by B^n the set of all binary words of length n, we can view the coding mechanism as a function $E : B^m \to B^n$. Such a function is called an **encoding function** and the members of its image set are called **codewords**. Since each codeword must correspond to a unique word B^m, an encoding function must be injective. For each encoding function there is a **decoding function** $D : B^n \to B^m \cup \{\text{'error'}\}$ which maps a codeword $\boldsymbol{y} \in B^n$ to $\boldsymbol{x} \in B^m$, where $E(\boldsymbol{x}) = \boldsymbol{y}$. Since $m < n$ the set of codewords is a proper subset of B^n so that there are elements of the domain of D which are not codewords. If a word $\boldsymbol{w}'$ is received which falls into this category then $D(\boldsymbol{w}') = D(\boldsymbol{w})$ where $\boldsymbol{w}$ is the codeword 'nearest to' $\boldsymbol{w}'$ in the sense that it differs from $\boldsymbol{w}'$ in the fewest digits. This is called 'nearest-neighbour decoding'. If the 'nearest neighbour' is not unique, then $D(\boldsymbol{w}') =$ 'error'. If the data cannot be retransmitted, then one of the set of nearest neighbours may be chosen arbitrarily and decoded.

A coding/decoding procedure which consists of an encoding function $E : B^m \to B^n$ and a decoding function $D : B^n \to B^m \cup \{\text{'error'}\}$ is called an $\boldsymbol{(m, n)}$ **block code**. Such a code is said to be **systematic** if, given $\boldsymbol{x} \in B^m$, the first m digits of $E(x)$ are, in the same order, those of $\boldsymbol{x}$ itself. In this brief introduction to coding theory we shall restrict our discussion to systematic block codes.

The simplest encoding function involves the addition of a single digit to the end of a word where the digit is chosen to make the number of ones in the codeword even. Such a code is an $(m, m+1)$ systematic block code and it is called an **even parity check code**. The encoding function $E : B^m \to B^{m+1}$ is such that, for example, if $m = 4$, $E(0011) = 00110$ and $E(1000) = 10001$. Odd parity check codes can also be used. For an even parity check code using one check digit, one error can be detected because it will result in an odd number of ones. However, the error cannot be corrected because it is not possible to tell which digit is at fault.

A code in which any combination of k or fewer errors can be detected is said to be **k-error detecting** and a code in which any combination of k or fewer errors can be corrected is called **k-error correcting**. Even and odd parity check codes are 1-error detecting and 0-error correcting.

We have seen how the ability to detect and correct errors is dependent upon the distance between codewords. For codes involving check digits, the distance between each pair of codewords is not necessarily the same so that the factor determining the error-detecting and error-correcting capabilities of the code is the minimum of all the distances between pairs of codewords. The **minimum distance** of a code is defined to be the minimum of all the distances between distinct pairs of codewords.

The following two theorems give criteria for determining the capability for error detection in a code.

Theorem 8.12

A code is k-error detecting if and only if the minimum distance is at least $k + 1$.

Proof

Any number of errors in a codeword can be detected so long as they do not result in another codeword. If the minimum distance between codewords is $k + 1$, then any number of errors fewer than $k + 1$ will not result in a codeword and so will be detected. Hence k or fewer errors can be detected and so the code is k-error detecting. A similar line of argument establishes the converse. □

Theorem 8.13

A code is k-error correcting if and only if the minimum distance is at least $2k+1$.

Proof

We first prove that, if a code is k-error correcting, then the minimum distance between any pair of codewords is at least $2k+1$. We use proof by contradiction: we assume that the code is k-error correcting and that there exists a pair of words whose distance is less than $2k+1$. Denoting these by $\boldsymbol{w}$ and $\boldsymbol{x}$, we have

$$d(\boldsymbol{w}, \boldsymbol{x}) \leqslant 2k.$$

However, since the code is k-error correcting, we can certainly detect k errors, so, by the last theorem,

$$d(\boldsymbol{w}, \boldsymbol{x}) \geqslant k+1.$$

So we have

$$k+1 \leqslant d(\boldsymbol{w}, \boldsymbol{x}) \leqslant 2k$$

i.e. $\boldsymbol{w}$ and $\boldsymbol{x}$ differ in at least $k+1$ digits and at most $2k$ digits.

Now suppose that $\boldsymbol{w}$ undergoes k transmission errors and is received as $\boldsymbol{w}'$ so that $d(\boldsymbol{w}, \boldsymbol{w}') = k$. Suppose also that each of these k errors occurs in one of the digits in which $\boldsymbol{w}$ and $\boldsymbol{x}$ differ. This means that $\boldsymbol{w}'$ and $\boldsymbol{x}$ differ by at most $2k - k = k$ digits, so

$$d(\boldsymbol{w}', \boldsymbol{x}) \leqslant k$$

and $\boldsymbol{w}'$ is at least as close to $\boldsymbol{x}$ as it is to $\boldsymbol{w}$. Hence $\boldsymbol{w}'$ cannot be correctly decoded. Thus the code is not k-error correcting and we have a contradiction.

We now prove that, if the minimum distance is $2k+1$, then the code is k-error correcting. Suppose that a codeword $\boldsymbol{w}$ is transmitted with k or fewer errors and is received as $\boldsymbol{w}'$. For any other codeword $\boldsymbol{x}$ we have

$$d(\boldsymbol{w}, \boldsymbol{w}') + d(\boldsymbol{w}', \boldsymbol{x}) \geqslant d(\boldsymbol{w}, \boldsymbol{x})$$

so

$$d(\boldsymbol{w}', \boldsymbol{x}) \geqslant d(\boldsymbol{w}, \boldsymbol{x}) - d(\boldsymbol{w}, \boldsymbol{w}').$$

But

$$d(\boldsymbol{w}, \boldsymbol{x}) \geqslant 2k+1 \quad \text{and} \quad d(\boldsymbol{w}, \boldsymbol{w}') \leqslant k$$

so that

$$d(\boldsymbol{w}', \boldsymbol{x}) \geqslant (2k+1) - (k)$$

and hence

$$d(\boldsymbol{w}', \boldsymbol{x}) \geqslant k+1.$$

From this we can conclude that the distance between the received word $\boldsymbol{w}'$ and any other codeword is greater than the distance between $\boldsymbol{w}'$ and $\boldsymbol{w}$ so that $\boldsymbol{w}'$ will be correctly decoded as $\boldsymbol{w}$. □

Example 8.17

Consider the encoding function $E : B^2 \to B^6$ defined as follows:

$$E(00) = 001000 \qquad E(01) = 010100$$
$$E(10) = 100010 \qquad E(11) = 110001.$$

Find how many errors the code can detect and correct.

Solution

The distances between pairs of codewords are given below:

$$d(001000, 010100) = 3 \qquad d(001000, 100010) = 3$$
$$d(001000, 110001) = 4 \qquad d(010100, 100010) = 4$$
$$d(010100, 110001) = 3 \qquad d(100010, 110001) = 3.$$

The minimum distance is three and so the code is k-error detecting where $k+1 = 3$. The code is therefore 2-error detecting.

The code is k-error correcting where $2k+1 = 3$. This gives $k = 1$ and so only one error can be corrected.

In order to appreciate the importance of groups in coding theory we need to define a binary operation on the set of n bit words, B^n. We do this as follows.

Definition 8.19

Let $\boldsymbol{x}$ and $\boldsymbol{y}$ be codewords of length n such that the ith digit of $\boldsymbol{x}$ is x_i and the ith digit of $\boldsymbol{y}$ is y_i. The **sum** of $\boldsymbol{x}$ and $\boldsymbol{y}$, denoted $\boldsymbol{x} \oplus \boldsymbol{y}$, is the n bit word whose ith digit is $x_i +_2 y_i$, where $+_2$ denotes addition modulo 2.

Thus the sum of two codewords is obtained by applying modulo 2 addition to corresponding bits. For example

$$1011001 \oplus 1000111 = 0011110$$

and

$$111001 \oplus 110011 = 001010.$$

Note that the modulo 2 sum of two bits is 0 if the bits are the same and 1 if they are different. The distance between two words $\boldsymbol{x}$ and $\boldsymbol{y}$ could therefore be defined as the number of ones in $\boldsymbol{x} \oplus \boldsymbol{y}$.

Definition 8.20

The **weight** of a word $\boldsymbol{x}$, denoted by $w(\boldsymbol{x})$, is the number of ones which it contains.

For example, $w(101101) = 4$ and $w(011110111) = 7$.

The distance between two n bit binary words $\boldsymbol{x}$ and $\boldsymbol{y}$ is given by

$$d(\boldsymbol{x}, \boldsymbol{y}) = w(\boldsymbol{x} \oplus \boldsymbol{y}).$$

A code for which the set of codewords is a group under the operation $\oplus$ is called a **group code**. It is a simple matter to show that the set B^n of all n bit binary words is a group under this operation (see exercise 8.6.1). However, this is not a particularly useful set of codewords because, as we saw earlier, the words are too close together for any error detection to be possible. We have shown that the error-detecting and error-correcting capabilities of a code can be determined from the minimum distance between codewords. For an arbitrary code, determining this minimum distance involves comparing the distance between all possible pairs of codewords—a daunting prospect if the number of codewords is large! For a group code, however, we can show that the minimum distance is equal to the minimum weight of all non-zero codewords.

Theorem 8.14

The minimum distance of a group code is the minimum weight of all non-zero codewords.

Proof

Let n be the minimum weight of all non-zero codewords so that there exists a codeword $\boldsymbol{z}$ such that $w(\boldsymbol{z}) = n$ and, for any other codeword $\boldsymbol{x}, w(\boldsymbol{x}) \geqslant n$.

Suppose that d is the minimum distance of the code so that there exist distinct codewords $\boldsymbol{v}$ and $\boldsymbol{w}$ such that

$$d(\boldsymbol{v}, \boldsymbol{w}) = d$$

that is,

$$w(\boldsymbol{v} \oplus \boldsymbol{w}) = d.$$

Now $\boldsymbol{v} \oplus \boldsymbol{w}$ is also a codeword by the closure property of the binary operation $\oplus$ and we therefore have

$$w(\boldsymbol{v} \oplus \boldsymbol{w}) \geqslant n$$

so that

$$d \geqslant n.$$

Denoting by $\mathbf{0}$ the word whose bits are all zero, we have

$$\boldsymbol{x} \oplus \boldsymbol{x} = \mathbf{0}$$

for any codeword $\boldsymbol{x}$. Now $\mathbf{0}$ is the identity under $\oplus$ and so is a codeword.

Therefore

$$d(\mathbf{0}, \boldsymbol{z}) \geqslant d.$$

But

$$\begin{aligned} d(\mathbf{0}, \boldsymbol{z}) &= w(\boldsymbol{z}) \\ &= n \end{aligned}$$

so that

$$n \geqslant d.$$

We have shown that $d \geqslant n$ and also that $n \geqslant d$, and we can therefore conclude that $n = d$. □

An encoding function $E : B^m \to B^n$ $(n > m)$ which encodes a word by appending check digits can most easily be described using an $m \times n$ matrix G whose entries are zeros and ones. Such a matrix is known as a **generator matrix** for the code. To encode an m bit word, we view that word as a $1 \times m$ matrix and post-multiply this row matrix by the matrix G with all additions and multiplications carried out modulo 2. For a systematic code we require that the first m bits of the codeword are the same as the m bits of the word to be encoded. In this case it is necessary that the first m columns of G constitute the identity matrix I_m.

Example 8.18

Consider the generator matrix

$$G = \begin{pmatrix} 1 & 0 & 0 & 1 & 0 & 1 \\ 0 & 1 & 0 & 1 & 1 & 0 \\ 0 & 0 & 1 & 0 & 1 & 1 \end{pmatrix}.$$

An encoding function $E : B^3 \to B^6$ is defined by $E(\boldsymbol{x}) = \boldsymbol{x}G$ for any $\boldsymbol{x} \in B^3$. For instance

$$E(011) = \begin{pmatrix} 0 & 1 & 1 \end{pmatrix} \begin{pmatrix} 1 & 0 & 0 & 1 & 0 & 1 \\ 0 & 1 & 0 & 1 & 1 & 0 \\ 0 & 0 & 1 & 0 & 1 & 1 \end{pmatrix} = \begin{pmatrix} 0 & 1 & 1 & 1 & 0 & 1 \end{pmatrix}$$

so that 011 is encoded as 011101. As another example, consider

$$E(100) = \begin{pmatrix} 1 & 0 & 0 \end{pmatrix} \begin{pmatrix} 1 & 0 & 0 & 1 & 0 & 1 \\ 0 & 1 & 0 & 1 & 1 & 0 \\ 0 & 0 & 1 & 0 & 1 & 1 \end{pmatrix} = \begin{pmatrix} 1 & 0 & 0 & 1 & 0 & 1 \end{pmatrix}.$$

In general, for any three-bit word with digits x_1, x_2 and x_3

$$\begin{aligned} E(x_1 x_2 x_3) &= \begin{pmatrix} x_1 & x_2 & x_3 \end{pmatrix} \begin{pmatrix} 1 & 0 & 0 & 1 & 0 & 1 \\ 0 & 1 & 0 & 1 & 1 & 0 \\ 0 & 0 & 1 & 0 & 1 & 1 \end{pmatrix} \\ &= \begin{pmatrix} x_1 & x_2 & x_3 & x_1 +_2 x_2 & x_2 +_2 x_3 & x_1 +_2 x_3 \end{pmatrix}. \end{aligned}$$

If $x_1 x_2 x_3$ encodes as $w_1 w_2 w_3 w_4 w_5 w_6$ we have

$$\begin{array}{lll} w_1 = x_1 & w_2 = x_2 & w_3 = x_3 \\ w_4 = x_1 +_2 x_2 & w_5 = x_2 +_2 x_3 & w_6 = x_1 +_2 x_3 \end{array}$$

so that

$$w_4 = w_1 +_2 w_2 \qquad w_5 = w_2 +_2 w_3 \qquad w_6 = w_1 +_2 w_3.$$

Thus the last three digits of the codeword act as parity checks on different pairs of information bits. An error in any one of the six bits of a codeword will uniquely determine which of these three equations is not satisfied. For example, for an error in the fourth bit, the first equation alone will not hold; an error in the first bit will result in the first and third equations not holding.

Since $0 +_2 0 = 1 +_2 1 = 0$, the three equations above can be written

$$\begin{array}{llllllll} w_1 & +_2 w_2 & & +_2 w_4 & & & = 0 \\ & w_2 & +_2 w_3 & & +_2 w_5 & & = 0 \\ w_1 & & +_2 w_3 & & & +_2 w_6 & = 0 \end{array}$$

which is equivalent to the matrix equation

$$\begin{pmatrix} 1 & 1 & 0 & 1 & 0 & 0 \\ 0 & 1 & 1 & 0 & 1 & 0 \\ 1 & 0 & 1 & 0 & 0 & 1 \end{pmatrix} \begin{pmatrix} w_1 \\ w_2 \\ w_3 \\ w_4 \\ w_5 \\ w_6 \end{pmatrix} = \begin{pmatrix} 0 \\ 0 \\ 0 \end{pmatrix}.$$

The matrix

$$H = \begin{pmatrix} 1 & 1 & 0 & 1 & 0 & 0 \\ 0 & 1 & 1 & 0 & 1 & 0 \\ 1 & 0 & 1 & 0 & 0 & 1 \end{pmatrix}$$

is called a **parity check matrix**. For any (correctly transmitted) codeword $\boldsymbol{w}$ the equation

$$H\boldsymbol{w}^{\mathrm{T}} = \begin{pmatrix} 0 \\ 0 \\ 0 \end{pmatrix}$$

is always satisfied.

Notice the relationship between the generator matrix G and H, the parity check matrix. The matrix G can be regarded as the partitioned matrix $(I_3\ F)$ where

$$F = \begin{pmatrix} 1 & 0 & 1 \\ 1 & 1 & 0 \\ 0 & 1 & 1 \end{pmatrix}.$$

The matrix H is the partitioned matrix $(F^{\mathrm{T}}\ I_3)$ where F^{T} denotes the transpose of F.

We generalize this result in the following theorem.

Theorem 8.15

Let G be an $m \times n$ generator matrix such that $G = (I_m\ F)$ where F is an $(m \times r)$ matrix and $m = n - r$. Let an encoding function $E : B^m \to B^n$ be defined by $E(\boldsymbol{x}) = \boldsymbol{x}G$ for any $\boldsymbol{x} \in B^m$. Then for any codeword $\boldsymbol{w} \in B^n$,

$$H\boldsymbol{w}^{\mathrm{T}} = O_{r\times 1} \quad \text{where} \quad H = (F^{\mathrm{T}}\ I_r).$$

Proof

If $\boldsymbol{w}$ is a codeword, then $\boldsymbol{w} = \boldsymbol{x}G$ for some $\boldsymbol{x} \in B^m$.

$$\begin{aligned} H(\boldsymbol{x}G)^{\mathrm{T}} &= H(G^{\mathrm{T}}\boldsymbol{x}^{\mathrm{T}}) && \text{(see exercise 6.2.11 (iii))} \\ &= (HG^{\mathrm{T}})\boldsymbol{x}^{\mathrm{T}} \\ &= (F^{\mathrm{T}}\, I_r)\begin{pmatrix} I_m \\ F^{\mathrm{T}} \end{pmatrix}\boldsymbol{x}^{\mathrm{T}} \\ &= (F^{\mathrm{T}} +_2 F^{\mathrm{T}})\boldsymbol{x}^{\mathrm{T}} && \text{(see exercise 6.4.9)} \\ &= O_{r\times m}\boldsymbol{x}^{\mathrm{T}} \\ &= O_{r\times 1}. \end{aligned}$$

■

The converse of this theorem also holds (see exercise 8.6.9).

Example 8.19

The generator matrix

$$G = \begin{pmatrix} 1 & 0 & 0 & 0 & 1 & 1 & 1 \\ 0 & 1 & 0 & 0 & 0 & 1 & 1 \\ 0 & 0 & 1 & 0 & 1 & 1 & 0 \\ 0 & 0 & 0 & 1 & 0 & 0 & 1 \end{pmatrix}$$

defines an encoding function $E : B^4 \to B^7$. For example,

$$\begin{aligned} E(1101) &= \begin{pmatrix} 1 & 1 & 0 & 1 \end{pmatrix}\begin{pmatrix} 1 & 0 & 0 & 0 & 1 & 1 & 1 \\ 0 & 1 & 0 & 0 & 0 & 1 & 1 \\ 0 & 0 & 1 & 0 & 1 & 1 & 0 \\ 0 & 0 & 0 & 1 & 0 & 0 & 1 \end{pmatrix} \\ &= \begin{pmatrix} 1 & 1 & 0 & 1 & 1 & 0 & 1 \end{pmatrix}. \end{aligned}$$

Now $G = (I_4\, F)$ where

$$F = \begin{pmatrix} 1 & 1 & 1 \\ 0 & 1 & 1 \\ 1 & 1 & 0 \\ 0 & 0 & 1 \end{pmatrix}$$

so that the parity check matrix corresponding to G is given by

$$\begin{aligned} H &= (F^{\mathrm{T}}\, I_3) \\ &= \begin{pmatrix} 1 & 0 & 1 & 0 & 1 & 0 & 0 \\ 1 & 1 & 1 & 0 & 0 & 1 & 0 \\ 1 & 1 & 0 & 1 & 0 & 0 & 1 \end{pmatrix}. \end{aligned}$$

For the codeword 1101101 we have

$$H(\,1\ \ 1\ \ 0\ \ 1\ \ 1\ \ 0\ \ 1\,)^{\mathrm{T}} = \begin{pmatrix} 1 & 0 & 1 & 0 & 1 & 0 & 0 \\ 1 & 1 & 1 & 0 & 0 & 1 & 0 \\ 1 & 1 & 0 & 1 & 0 & 0 & 1 \end{pmatrix} \begin{pmatrix} 1 \\ 1 \\ 0 \\ 1 \\ 1 \\ 0 \\ 1 \end{pmatrix} = \begin{pmatrix} 0 \\ 0 \\ 0 \end{pmatrix}$$

as expected.

We have seen that for any positive integer n, the set of all elements of B^n is a group under bit-wise addition modulo 2 with identity $O_{1\times n}$. We now show that the set of all codewords where the encoding function is defined by a generator matrix is a group under this operation.

Theorem 8.16

Let $E : B^m \to B^n$ be an encoding function such that $E(\boldsymbol{x}) = \boldsymbol{x}G$ where G is a generator matrix. Then the set of codewords $E(B^m)$ is a group under bit-wise addition modulo 2.

Proof

Given $\boldsymbol{x}_1, \boldsymbol{x}_2 \in B^m$, we have

$$\begin{aligned} E(\boldsymbol{x}_1 \oplus \boldsymbol{x}_2) &= (\boldsymbol{x}_1 \oplus \boldsymbol{x}_2)G \\ &= \boldsymbol{x}_1 G \oplus \boldsymbol{x}_2 G \\ &= E(\boldsymbol{x}_1) \oplus E(\boldsymbol{x}_2). \end{aligned}$$

Therefore E is a morphism from $(B^m, \oplus)$ to $(B^n, \oplus)$. Since $(B^m, \oplus)$ is a group, we can conclude that $(E(B^m), \oplus)$ is a group (see theorem 8.10). (It is in fact a subgroup of $(B^n, \oplus)$.) □

We noted earlier that the error-detecting and error-correcting capabilities of a code depend on the minimum distance of the code. For a group code, the minimum distance is the minimum weight of a non-zero codeword. We now use these results to show how to tell from the parity check matrix how many errors can be detected or corrected.

Suppose that H is an $r \times n$ matrix with columns denoted by $\boldsymbol{h}_1, \boldsymbol{h}_2, \ldots, \boldsymbol{h}_n$, and suppose that for k of these columns the sum of the elements in the corresponding rows is zero. We will denote these columns by $\boldsymbol{h}_{i_1}, \boldsymbol{h}_{i_2}, \ldots, \boldsymbol{h}_{i_k}$. Now the n digit word $\boldsymbol{w}$ which has ones as its $i_1, i_2, \ldots, i_k$ digits and zeros elsewhere must be such that $H\boldsymbol{w}^{\mathrm{T}} = \mathbf{0}$ and so is a codeword. To illustrate why this is so, consider the following matrix H where

$$H = \begin{pmatrix} h_{11} & h_{12} & h_{13} & h_{14} & h_{15} \\ h_{21} & h_{22} & h_{23} & h_{24} & h_{25} \\ h_{31} & h_{32} & h_{33} & h_{34} & h_{35} \end{pmatrix}.$$

Suppose that the first, third and fourth columns of H sum to zero so that

$$\begin{aligned} h_{11} +_2 h_{13} +_2 h_{14} &= 0 \\ h_{21} +_2 h_{23} +_2 h_{24} &= 0 \\ h_{31} +_2 h_{33} +_2 h_{34} &= 0. \end{aligned}$$

Now consider a word $\boldsymbol{w}$ with digits w_1, w_2, w_3, w_4, w_5 where $w_1 = w_3 = w_4 = 1$ and $w_2 = w_5 = 0$. Then

$$\begin{aligned} H\boldsymbol{w}^{\mathrm{T}} &= \begin{pmatrix} h_{11}w_1 +_2 h_{12}w_2 +_2 h_{13}w_3 +_2 h_{14}w_4 +_2 h_{15}w_5 \\ h_{21}w_1 +_2 h_{22}w_2 +_2 h_{23}w_3 +_2 h_{24}w_4 +_2 h_{25}w_5 \\ h_{31}w_1 +_2 h_{32}w_2 +_2 h_{33}w_3 +_2 h_{34}w_4 +_2 h_{35}w_5 \end{pmatrix} \\ &= \begin{pmatrix} h_{11}w_1 +_2 h_{13}w_3 +_2 h_{14}w_4 \\ h_{21}w_1 +_2 h_{23}w_3 +_2 h_{24}w_4 \\ h_{31}w_1 +_2 h_{33}w_3 +_2 h_{34}w_4 \end{pmatrix} \qquad \text{since } w_2 = w_5 = 0 \\ &= \begin{pmatrix} h_{11} +_2 h_{13} +_2 h_{14} \\ h_{21} +_2 h_{23} +_2 h_{24} \\ h_{31} +_2 h_{33} +_2 h_{34} \end{pmatrix} \qquad \text{since } w_1 = w_3 = w_4 = 1 \\ &= \begin{pmatrix} 0 \\ 0 \\ 0 \end{pmatrix}. \end{aligned}$$

Hence 10110 is a codeword.

The converse is also true—if a codeword has ones only in its $i_1, i_2, \ldots, i_k$ positions then $i_1, i_2, \ldots, i_k$ columns of H must sum to zero.

This result enables us to determine the minimum weight of a code defined by a generator matrix G or, equivalently, by a parity check matrix H. It is simply the minimum number of columns of H which sum to zero. Since such a code is a group code, the minimum weight is equal to the minimum distance and from this we can determine the error-detecting and error-correcting capabilities of the code.

Example 8.20

Suppose a group code is defined by the encoding function $E : B^4 \to B^7$ where $E(\boldsymbol{x}) = \boldsymbol{x}G$ for any $\boldsymbol{x} \in B^4$ and G is the generator matrix

$$G = \begin{pmatrix} 1 & 0 & 0 & 0 & 1 & 1 & 1 \\ 0 & 1 & 0 & 0 & 0 & 1 & 1 \\ 0 & 0 & 1 & 0 & 1 & 1 & 0 \\ 0 & 0 & 0 & 1 & 1 & 0 & 1 \end{pmatrix}.$$

The parity check matrix corresponding to G is

$$H = \begin{pmatrix} 1 & 0 & 1 & 1 & 1 & 0 & 0 \\ 1 & 1 & 1 & 0 & 0 & 1 & 0 \\ 1 & 1 & 0 & 1 & 0 & 0 & 1 \end{pmatrix}.$$

To find the minimum weight of the code we must find the minimum number of columns of H which sum to zero. For two columns to sum to zero the entries in each must be identical. Since there is no column of zeros (in which case the minimum weight of the code would be one) and no identical columns (from which we would conclude that the minimum weight was two), the minimum weight is at least three. That it *is* three can be confirmed by adding columns 1, 2 and 5, although these are not the only three columns which sum to zero. The minimum weight of this code is three and hence the minimum distance between codewords is three. From this we deduce that the code is 2-error detecting and 1-error correcting.

Given a systematic code defined by a parity check (or generator) matrix, decoding a received word $\boldsymbol{w}$ involves the calculation of $H\boldsymbol{w}^{\mathrm{T}}$. This quantity is called the **syndrome** of $\boldsymbol{w}$. If the syndrome has entries which are all zero, then we may reasonably conclude that the word was correctly transmitted and decoding involves the selection of the first m information bits.

What if the syndrome has one or more entries which are not zero? In this case we know that at least one transmission error has occurred. Suppose that there is one error and it is in the ith digit. Denoting the received word by $\boldsymbol{w}_{\mathrm{r}}$ and the transmitted word by $\boldsymbol{w}_{\mathrm{t}}$, these two words differ in that $\boldsymbol{w}_{\mathrm{r}}$ is $\boldsymbol{w}_{\mathrm{t}}$ with 1 added (modulo 2, of course) to its ith digit. If we define $\boldsymbol{e}$ to be the binary word with all digits zero except for the ith, then $\boldsymbol{e}$ is called the **error pattern** and $\boldsymbol{w}_{\mathrm{r}} = \boldsymbol{w}_{\mathrm{t}} \oplus \boldsymbol{e}$.

Definition 8.21

If an n bit word $\boldsymbol{w}_{\mathrm{t}}$ is transmitted and an n bit word $\boldsymbol{w}_{\mathrm{r}}$ is received, the error pattern is the binary word $\boldsymbol{e}$ with digits $e_1, e_2, \ldots, e_n$ where

$$e_i = \begin{cases} 0 & \text{if the } i\text{th digits of } \boldsymbol{w}_{\mathrm{r}} \text{ and } \boldsymbol{w}_{\mathrm{t}} \text{ are the same} \\ 1 & \text{if the } i\text{th digits of } \boldsymbol{w}_{\mathrm{r}} \text{ and } \boldsymbol{w}_{\mathrm{t}} \text{ are different.} \end{cases}$$

For a transmission error in only the ith digit of $\boldsymbol{w}_{\mathrm{t}}$, the syndrome $H\boldsymbol{w}_{\mathrm{r}}{}^{\mathrm{T}}$ is given by

$$\begin{aligned} H\boldsymbol{w}_{\mathrm{r}}{}^{\mathrm{T}} &= H(\boldsymbol{w}_{\mathrm{t}} \oplus \boldsymbol{e})^{\mathrm{T}} \\ &= H(\boldsymbol{w}_{\mathrm{t}}{}^{\mathrm{T}} \oplus \boldsymbol{e}^{\mathrm{T}}) \\ &= H\boldsymbol{w}_{\mathrm{t}}{}^{\mathrm{T}} \oplus H\boldsymbol{e}^{\mathrm{T}} \\ &= H\boldsymbol{e}^{\mathrm{T}} & \text{(since } \boldsymbol{w}_{\mathrm{t}} \text{ is a codeword)} \\ &= i\text{th column of } H. \end{aligned}$$

So for a single error the syndrome tells us exactly which digit is in error.

Example 8.21

Suppose we have an encoding function $E : B^3 \rightarrow B^6$ with parity check matrix given by

$$H = \begin{pmatrix} 1 & 1 & 0 & 1 & 0 & 0 \\ 1 & 1 & 1 & 0 & 1 & 0 \\ 1 & 0 & 1 & 0 & 0 & 1 \end{pmatrix}.$$

Suppose that a word 100001 is received. What is most likely to be the word which was transmitted?

Solution

We first compute the syndrome

$$H\boldsymbol{w}^{\mathrm{T}} = \begin{pmatrix} 1 & 1 & 0 & 1 & 0 & 0 \\ 1 & 1 & 1 & 0 & 1 & 0 \\ 1 & 0 & 1 & 0 & 0 & 1 \end{pmatrix} \begin{pmatrix} 1 \\ 0 \\ 0 \\ 0 \\ 0 \\ 1 \end{pmatrix} = \begin{pmatrix} 1 \\ 1 \\ 0 \end{pmatrix}.$$

Since Hw^{T} is not zero, 100001 is not a codeword and could not have been transmitted. The result is in fact the second column of H from which we conclude that there is an error in the second digit of the received word and that 110001 was actually transmitted. This is then decoded as 110.

Suppose that the syndrome is neither zero nor a column of H. In this case we conclude that there is an error in more than one digit and for a single-error-correcting code the received word cannot be decoded reliably.

In this brief introduction to coding theory we have been concerned mainly with codes which are single error correcting. These are important because, for an incorrectly transmitted word, one error is more likely than several. Where the possibility of multiple errors is not small (e.g. in transmission from a spacecraft) codes which have more sophisticated error-detecting and error-correcting capabilities are used. These also involve theory from the realm of abstract algebra.

Exercises 8.6

1. Show that the set B^n is a group under operation $\oplus$ of bit-wise addition modulo 2.

2. An encoding function $E : B^3 \to B^9$ is defined by $E(x_1x_2x_3) = x_1x_2x_3x_1x_2x_3x_1x_2x_3$. (This is an example of a **triple-repetition block code**.) What is the generator matrix for this code? What is the maximum number of errors which the code will (i) detect and (ii) correct?

3. An encoding function $E : B^3 \to B^6$ is defined by the generator matrix

$$G = \begin{pmatrix} 1 & 0 & 0 & 1 & 0 & 0 \\ 0 & 1 & 0 & 1 & 1 & 0 \\ 0 & 0 & 1 & 0 & 1 & 1 \end{pmatrix}$$

so that $E(\boldsymbol{x}) = \boldsymbol{x}G$ for any $\boldsymbol{x} \in B^3$.

Find a parity check matrix. The words listed below are received at the end of a communication channel. For each one calculate the syndrome and indicate whether the word is likely to have been correctly transmitted:

(i) 111001 (ii) 101011 (iii) 001011 (iv) 101101 (v) 011111.

4. An encoding function $E : B^2 \to B^5$ is defined as follows:

$$E(00) = 00000 \qquad E(01) = 01111$$
$$E(10) = 10101 \qquad E(11) = 11010.$$

Show that this is a group code. Find the minimum distance of the code and hence the maximum number of errors which the code can (i) detect and (ii) correct.

5. A parity check matrix for a systematic (m, n) block code is given by

$$H = \begin{pmatrix} 1 & 0 & 1 & 1 & 1 & 0 & 0 \\ 1 & 1 & 1 & 0 & 0 & 1 & 0 \\ 0 & 1 & 1 & 1 & 0 & 0 & 1 \end{pmatrix}.$$

What are the values of m and n? Find the corresponding generator matrix.

6. Find a generator matrix for the $(m, m+1)$ even parity check code. What is the corresponding parity check matrix for this code?

7. Given the parity check matrix

$$H = \begin{pmatrix} 1 & 1 & 0 & 1 & 0 & 0 \\ 0 & 1 & 1 & 0 & 1 & 0 \\ 1 & 0 & 1 & 0 & 0 & 1 \end{pmatrix}$$

calculate the syndrome for each of the following received words and indicate which are likely to have been correctly transmitted. For those which contain an error, find, if possible, the word which was likely to have been transmitted:

(i) 011001 (ii) 111000 (iii) 001100 (iv) 111110.

8. Consider an encoding function $E : B^m \to B^n$ defined by an $m \times n$ generator matrix G. Let $r = n - m$ be the number of check digits in each codeword. Show that, for fixed r, the largest number of information bits for a single-error-correcting code is $2^r - r - 1$.

For given r ($r \in \mathbb{Z}^+, r > 1$) a $(2^r - r - 1, 2^r - 1)$ single-error-correcting block code is called a **Hamming code**. Write down parity check matrices for the $(1, 3)$ and $(4, 7)$ Hamming codes. (Note that these matrices are unique only up to a reordering of the first $2^r - r - 1$ columns.)

9. State the converse of theorem 8.15. (The converse is true but the proof is beyond the scope of this book.)

Chapter 9

Introduction to Number Theory

Number theory is concerned with the properties of the integers; it is sometimes referred to simply as 'arithmetic'. As such, it deals with some of the most familiar mathematical objects. Despite this, number theory is certainly not trivial. There are many deep and beautiful results in number theory. Gauss was said to have described mathematics as the 'Queen of the Sciences' and number theory as the 'Queen of Mathematics'.

Number theory has sometimes been thought of as an archetypal branch of pure mathematics—very interesting in its own right, but with little practical use outside mathematics. Indeed, G H Hardy who was a number theorist and a leading mathematician of the first half of the 20th century, wrote in his book, *A Mathematician's Apology*, 'I have never done anything 'useful'. No discovery of mine has made, or is likely to make, directly or indirectly, for good or ill, the least difference to the amenity of the world.'

However, results in elementary number theory have been used to devise encryption algorithms that are very secure. The particular 'public key' properties of these algorithms make them ideal for encryption of data sent via the internet and, in so doing, enable e-commerce to be viable. Hardy also claimed that pure mathematics does no harm to the world: 'no one has yet discovered any warlike purpose to be served by the theory of numbers . . . '. But with the arrival of encryption systems based on number theory, the factorisation of some large positive integers is now considered a military secret. The point here is that applications of even the 'purest' part of mathematics can arise in unexpected places. We will consider public key encryption algorithms based on number theory in section 9.5.

9.1 Divisibility

We begin with the simple process of dividing one integer by another. Early on in school mathematics, we learn to 'divide and take remainders'. For example, we learn something that we might phrase as '23 divided by 5 goes 4 times with a remainder of 3'. This is simply saying that

$$23 = 4 \times 5 + 3.$$

Here we say that dividing 23 by 5 gives *quotient* 4 and *remainder* 3.

We can do this for any positive integers a and b, a result known as the Division Algorithm .

Theorem 9.1 (The Division Algorithm)

Let $a, b \in \mathbb{Z}^+$. Then there exist unique $q, r \in \mathbb{N}$ such that

$$a = qb + r \quad \text{and} \quad 0 \leqslant r < b.$$

Here q is called the **quotient** and r is called the **remainder.**

Proof

We first prove the existence of the positive integers q and r. If $a < b$ then we may write $a = 0 \times b + a$ where $0 \leqslant a < b$ and the result follows with $q = 0$ and $r = a$. Now suppose that $a \geqslant b$.

Consider the set

$$S = \{a - nb : n \in \mathbb{Z}^+\} = \{a - b, a - 2b, a - 3b, \ldots\}.$$

In the case above where $a = 23$ and $b = 5$ the set S is

$$S = \{18, 13, 8, 3, -2, -7, \ldots\}.$$

It is clear that S always contains some non-negative integers; for example, the first listed element $a - b \geqslant 0$. Therefore S contains a *least* non-negative element which is of the form $a - qb$ for some $q \in \mathbb{Z}^+$. Call this least non-negative element r. Then $r = a - qb$ so $a = qb + r$ where $r \geqslant 0$ and $q, r \in \mathbb{N}$.

We are not quite finished with the existence part as we still need to show that $r < b$. Suppose that $r \geqslant b$. Then $a - (q+1)b < a - qb = r$ and

$$a - (q+1)b = a - qb - b = r - b \geqslant 0.$$

Hence $a - (q+1)b$ belongs to S, is non-negative and is smaller than r, contradicting the fact that r was the *least* non-negative element of S. Therefore $r < b$, as required. This establishes the existence of $q, r \in \mathbb{N}$.

Now we turn to the uniqueness of q and r. Suppose that

$$a = qb + r \quad \text{and} \quad 0 \leqslant r < b$$

and

$$a = q'b + r' \quad \text{and} \quad 0 \leqslant r' < b.$$

Subtracting gives

$$0 = (q - q')b + (r - r')$$

so

$$r - r' = (q' - q)b.$$

Suppose that $q' \neq q$. Then without loss of generality we may suppose that $q' > q$. Then we have

$$r - r' = (q' - q)b > b.$$

This is impossible since both r and r' lie between 0 and $b - 1$ inclusive so the greatest value of their difference is $b - 1$. Hence their difference cannot be greater than b. Therefore $q' = q$ from which it also follows that $r' = r$. Hence q and r are unique. □

Examples 9.1

1. With $a = 37$ and $b = 8$ we have $37 = 4 \times 8 + 5$ so $q = 4$ and $r = 5$.

   Although we would not use this to find the quotient and remainder, the set S used in the proof of the Division Algorithm is

   $$S = \{37 - 8n : n \in \mathbb{Z}^+\} = \{29, 21, 13, 5, -3, -11, \ldots\}.$$

   The smallest non-negative value in this set is the 4th element $5 = 37 - 4 \times 8$, so $q = 4$ and $r = 5$.

2. In fact the division algorithm is also valid when the integer a is negative but in this case the quotient q will also be negative. For example, if

$a = -37$ and $b = 5$ we have $-37 = -8 \times 5 + 3$ so $q = -8$ and $r = 3$.

When the remainder is zero in theorem 9.1, we say that b is a *divisor* or *factor* of a. Thus, for example, the divisors of 36 are: $1, 2, 3, 4, 6, 9, 12, 18$ and 36. In general, finding all the divisors of a large number is very time consuming and somewhat tedious.

Definition 9.1

Let $a, b \in \mathbb{Z}^+$. If $a = qb$ for some $q \in \mathbb{Z}^+$ (that is, if $r = 0$ in theorem 9.1) then we say that a is a **multiple** of b. In this case we also say that b is a **factor** of a or b is a **divisor** of a. We also say that b **divides** a, which is denoted $b|a$.

In chapter 4, we claimed that the divisibility relation is a partial order on $\mathbb{Z}^+$. See exercise 4.5.1 on page 195. This is captured in the second part of theorem 9.2 below. The first part says that divisibility is 'preserved' by taking linear combinations. The proof of the first part is left as an exercise; for the proof of the second part, see the solution to exercise 4.5.1 on page 723.

Theorem 9.2

Let $a, b, c \in \mathbb{Z}^+$.

(a) If $c|a$ and $c|b$ then $c|(ma + nb)$ for all $m, n \in \mathbb{Z}^+$.

(b) The divisibility relation on $\mathbb{Z}^+$ satisfies the following conditions.

Reflexive:	$a|a$ for all $a \in \mathbb{Z}^+$.
Anti-symmetric:	For all $a, b \in \mathbb{Z}^+$, if $a|b$ and $b|a$ then $a = b$.
Transitive:	For all $a, b, c \in \mathbb{Z}^+$, if $a|b$ and $b|c$ then $a|c$.

The notion of the *greatest common divisor* of two positive integers is probably a familiar one. For example, consider 36 and 30. Now 36 has divisors 1, 2, 3, 4, 6, 12, 18 and 36 and 30 has divisors 1, 2, 3, 5, 6, 10, 15 and 30. Hence the divisors they have in common are 1, 2, 3 and 6 so their greatest common divisor is 6.

Definitions 9.2

Let $a, b, c, d \in \mathbb{Z}^+$.
If $d|a$ and $d|b$ then we say that d is a **common divisor** (or **common factor**) of a and b.
If d is a common divisor of a and b and, for all other common divisors c, $c \leqslant d$, then d is called the **greatest common divisor** (or **greatest common factor**) of a and b, written $d = \gcd(a, b)$.

Example 9.2

Let $a = 24$ and $b = 44$. Then a has divisors $1, 2, 3, 4, 6, 8, 12, 24$ and b has divisors $1, 2, 4, 11, 22, 44$ so $\gcd(24, 44) = 4$.

For large integers a and b, finding their greatest common divisor by listing the divisors of each and selecting the largest is time consuming and tedious. Fortunately, there is a more efficient method, called *Euclid's algorithm* which is based on the following simple result.

Theorem 9.3

Let $a, b \in \mathbb{Z}^+$. If $a = qb + r$ where $q, r \in \mathbb{N}$ and $0 \leqslant r < b$ (as in the Division Algorithm) then $\gcd(a, b) = \gcd(b, r)$.

Proof

By theorem 9.2 (a), any common divisor of b and r also divides $a = qb + r$. Similarly, any common divisor of a and b also divides $r = a - qb$. Therefore the pair a, b and the pair b, r have exactly the same common divisors and therefore these pairs have the same greatest common divisor. □

Before we describe Euclid's greatest common divisor algorithm in general we work through an example to show how we can apply the Division Algorithm repeatedly to find the greatest common divisor of two integers.

Example 9.3

We will find the greatest common divisor of 2016 and 273 by repeated application of the division algorithm (theorem 9.1) together with theorem 9.3.

Applying the division algorithm to $a = 2016$ and $b = 273$ gives

$$2016 = 7 \times 273 + 105.$$

By theorem 9.3, $\gcd(2016, 273) = \gcd(273, 105)$ so applying the division algorithm once has considerably reduced the size of the numbers involved.

Next we apply the division algorithm to 273 and the remainder found at the previous stage, 105:

$$273 = 2 \times 105 + 63.$$

Hence, applying theorem 9.3 again,

$$\gcd(2016, 273) = \gcd(273, 105) = \gcd(105, 63).$$

Applying the division algorithm again with 105 and the remainder found at the previous stage, 63:

$$105 = 1 \times 63 + 42$$

so

$$\gcd(2016, 273) = \gcd(273, 105) = \gcd(105, 63) = \gcd(63, 42).$$

Continuing in the same way repeatedly applying the division algorithm we soon obtain a zero remainder:

$$\begin{aligned} 63 &= 1 \times 42 + 21 \\ 42 &= 2 \times 21 + \mathbf{0} \end{aligned}$$

From theorem 9.3,

$$\begin{aligned} \gcd(2016, 273) &= \gcd(273, 105) \\ &= \gcd(105, 63) \\ &= \gcd(63, 42) \\ &= \gcd(42, 21) = 21. \end{aligned}$$

In the previous example, we repeatedly applied the division algorithm where the numbers 'a' and 'b' at one stage are replaced by 'b'and the remainder 'r' at the next stage. We continue until a remainder of zero is obtained. (It is clear that the sequence of remainders must 'reach' zero since each remainder is non-negative and strictly less than its predecessor: $r_k \geqslant 0$ and $r_k < r_{k-1}$.) Then the greatest common divisor of the original pair of numbers is the last non-zero remainder.

This is Euclid's greatest common divisor algorithm, which we summarise in the following theorem. Figure 9.1 describes Euclid's algorithm using a flowchart.

Theorem 9.4 (Euclid's greatest common divisor algorithm)

Let $a, b \in \mathbb{Z}^+$ and suppose $b < a$. Repeatedly apply the division algorithm as follows until $r_n = 0$:

$$\begin{aligned}
a &= q_1 b + r_1 && \text{where} \quad 0 \leqslant r_1 < b \\
b &= q_2 r_1 + r_2 && \text{where} \quad 0 \leqslant r_2 < r_1 \\
r_1 &= q_3 r_2 + r_3 && \text{where} \quad 0 \leqslant r_3 < r_2 \\
&\;\;\vdots \\
r_{n-3} &= q_{n-1} r_{n-2} + r_{n-1} && \text{where} \quad 0 \leqslant r_{n-1} < r_{n-2} \\
r_{n-2} &= q_n r_{n-1} + r_n && \text{where} \quad r_n = 0.
\end{aligned}$$

Then $\gcd(a, b) = r_{n-1}$.

Example 9.4

Usually, when working through Euclid's algorithm, all the applications of the division algorithm are set out in one go. For example, using Euclid's greatest common divisor algorithm to calculate $d = \gcd(15\,300, 3510)$ we would normally proceed as follows.

$$\begin{aligned}
15\,300 &= 4 \times 3510 + 1260 \\
3510 &= 2 \times 1260 + 990 \\
1260 &= 1 \times 990 + 270 \\
990 &= 3 \times 270 + 180 \\
270 &= 1 \times 180 + 90 \\
180 &= 2 \times 90 + 0.
\end{aligned}$$

Hence $\gcd(15\,300, 3510) = 90$.

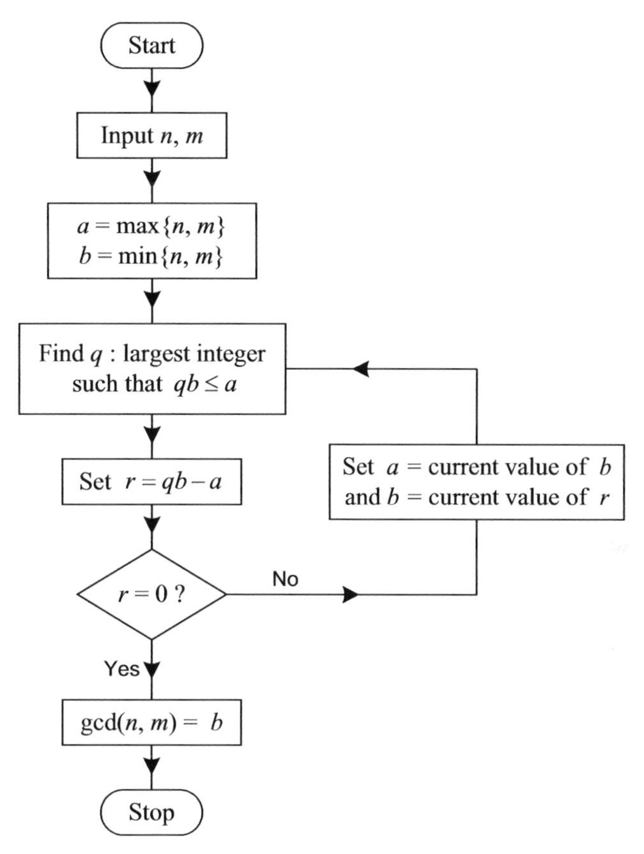

Figure 9.1 Euclid's gcd algorithm

Euclid's algorithm actually gives us something extra. By working 'backwards' through the algorithm, we can express $\gcd(a, b)$ as a linear combination of a and b. We illustrate this using the working through of the algorithm given in example 9.3 above.

Example 9.5

In example 9.3 above, we found $\gcd(2016, 273) = 21$.

First we re-write the penultimate application of the division algorithm, $63 = 1 \times 42 + 21$ making the gcd 21 the subject:

$$21 = 63 - 1 \times 42.$$

Now, from the previous application of the division algorithm, $105 = 1 \times 63 + 42$, we can replace 42 with $105 - 1 \times 63$ and simplify:

$$21 = 63 - 1 \times (105 - 1 \times 63) = 2 \times 63 - 105.$$

Next use the previous application of the division algorithm, $273 = 2 \times 105 + 63$, to replace 63 with $273 - 2 \times 105$ and simplify:

$$21 = 2 \times (273 - 2 \times 105) - 105 = 2 \times 273 - 5 \times 105.$$

Finally, the first step in Euclid's algorithm, $2016 = 7 \times 273 + 105$ allows us to replace 105:

$$21 = 2 \times 273 - 5 \times (2016 - 7 \times 273) = 37 \times 273 - 5 \times 2016.$$

Hence by working 'backwards' through Euclid's algorithm, successively replacing the remainders and simplifying, we have found

$$21 = 37 \times 273 - 5 \times 2016.$$

Clearly the process described in the previous example can always be carried through so that $\gcd(a, b)$ can always be expressed in the form $ma + nb$ for some integers m and n. This result is known as Bézout's identity. Note that in Bézout's identity, one of the integers m and n will be positive and the other negative—why is this?

Theorem 9.5 (Bézout's identity)

Let $a, b \in \mathbb{Z}^+$. Then there exist integers m and n such that

$$\gcd(a, b) = ma + nb.$$

Example 9.6

Express $\gcd(15\,300, 3510)$ in the form $m \times 15\,300 + n \times 3510$ where $m, n \in \mathbb{Z}$.

Solution

In example 9.4, we calculated $\gcd(15\,300, 3510) = 90$.

Working backwards through the application of Euclid's algorithm given in example 9.4, successively replacing remainders and simplifying, gives the following.

$$\begin{aligned}
90 &= 270 - 180 \\
&= 270 - (990 - 3 \times 270) \\
&= -990 + 4 \times 270 \\
&= -990 + 4 \times (1260 - 990) \\
&= 4 \times 1260 - 5 \times 990 \\
&= 4 \times 1260 - 5 \times (3510 - 2 \times 1260) \\
&= -5 \times 3510 + 14 \times 1260 \\
&= -5 \times 3510 + 14 \times (15\,300 - 4 \times 3510) \\
&= 14 \times 15\,300 - 61 \times 3510.
\end{aligned}$$

Hence $\gcd(15\,300, 3510) = 90 = 14 \times 15\,300 - 61 \times 3510$.

Bézout's identity says that $\gcd(a, b)$ can be expressed as an 'integer linear combination' of a and b. An obvious question to ask is: what other integers can be expressed in this way as $ma + nb$ for some integers m and n? Clearly any multiple of the greatest common divisor can be written in this way. The next theorem says that, in fact, *only* multiples of $\gcd(a, b)$ can be expressed as an 'integer linear combination' of a and b.

Theorem 9.6

Let $a, b \in \mathbb{Z}^+$ and let $d = \gcd(a, b)$.
An integer c can be expressed in the form $c = sa + tb$ for some $s, t \in \mathbb{Z}$ if and only if d divides c.
Hence $d = \gcd(a, b)$ is the smallest positive integer that can be expressed as $sa + tb$ where $s, t \in \mathbb{Z}$.

Proof

Let $a, b \in \mathbb{Z}^+$ and let $d = \gcd(a, b)$.

Suppose that $c = sa + tb$ for some $s, t \in \mathbb{Z}$. Since d divides both a and b, it follows from theorem 9.2 (a) that d divides c, $d|c$.

Conversely suppose d divides c. Then $c = kd$ for some integer k. From theorem 9.5 we know that d can be written as $d = ma + nb$ for some $m, n \in \mathbb{Z}$. Hence $c = kd = k(ma + nb) = (km)a + (kb)d$ where $ka, kb \in \mathbb{Z}$, as required.

Finally, since d divides every positive integer in the set $\{sa + tb : s, t \in \mathbb{Z}\}$, it follows that d is the smallest positive integer in the set. □

Definition 9.3

Two integers a and b are **coprime** (or **relatively prime**) if $\gcd(a, b) = 1$.

Examples 9.7

1. Consider 42 and 55. Since 42 has factors 1, 2, 3, 6, 7, 14, 21, 42 and 55 has factors 1, 5, 11, 55, their only common divisor is 1, Hence 42 and 55 are coprime.

2. The integers 2016 and 273 are not coprime. Since $2016 = 3 \times 672$ and $273 = 3 \times 91$, it follows that 3 is a common divisor of 2016 and 273 so their greatest common divisor is not 1.

   In fact, using Euclid's algorithm, we found in example 9.3 that $\gcd(2016, 273) = 21$.

As an important consequence of theorems 9.5 and 9.6 we have the following.

Theorem 9.7

Let $a, b \in \mathbb{Z}^+$. Then a and b are coprime if and only if there exist integers m and n such that

$$ma + nb = 1.$$

Exercises 9.1

1. (i) Use Euclid's greatest common divisor algorithm to calculate:

   (a) $\gcd(21\,600, 2970)$

   (b) $\gcd(1485, 1745)$

   (c) $\gcd(851, 2679)$

   (d) $\gcd(13\,376, 7980)$.

   (ii) Use the results of part (i) to express:

   (a) $\gcd(21\,600, 2970)$ in the form $21\,600m + 2970n$ where $m, n \in \mathbb{Z}$;

   (b) $\gcd(1485, 1745)$ in the form $1485m + 1745n$ where $m, n \in \mathbb{Z}$;

   (c) $\gcd(851, 2679)$ in the form $851m + 2679n$ where $m, n \in \mathbb{Z}$;

   (d) $\gcd(13\,376, 7980)$ in the form $13\,376m + 7980n$ where $m, n \in \mathbb{Z}$.

2. Prove each of the following where $a, b, c, d, m \in \mathbb{Z}^+$.

   (i) If $a|b$ and $b|c$ then $a|c$.

   (ii) If $a|b$ and $c|d$ then $ac|bd$.

   (iii) $a|b$ if and only if $ma|mb$.

3. Prove theorem 9.2 (a).

4. Show that, in Bézout's identity (theorem 9.5), the integers n and m are not uniquely determined by a and b. Specifically, using the result of example 9.5, find integers $s \neq 37$ and $t \neq -5$ such that

$$21 = 273s + 2016t.$$

5. Let a and b be positive integers.

   Show that *every* integer c can be expressed as $c = sa + tb$ for some integers s and t if and only if a and b are coprime.

6. Using the results of exercise 1 (i) (d) and theorem 9.6, determine which, if any, of the following integers can be expressed as $13\,376s + 7980t$ for some integers s and t:

$$226, \ 228, \ 230.$$

7. Determine which of the following pairs of integers are coprime:

   (i) 57 and 45,

   (ii) 54 and 101,

   (iii) 112 and 117.

8. We can extend the definition of greatest common divisor of two positive integers to a finite set of positive integers: $\gcd(a_1, a_2, \ldots, a_n)$ is the smallest positive integer d such that $d|a_1, d|a_2, \ldots, d|a_n$.

   (i) It can be shown that

   $$\gcd(a_1, a_2, \ldots, a_n) = \gcd(\gcd(a_1, a_2), a_3, \ldots, a_n).$$

   In particular, we have $\gcd(a, b, c) = \gcd(\gcd(a, b), c)$.

   Use this result to calculate:

   (a) $\gcd(63, 165, 297)$,

   (b) $\gcd(1092, 1155, 2002)$,

   (c) $\gcd(72, 48, 108, 54)$.

   (ii) Generalise theorem 9.2 (a) to the case of k integers, $a_1, a_2, \ldots, a_k$. That is, suppose that $c|a_1$, $c|a_2$, ..., $c|a_k$. Prove that, for any integers $n_1, n_2, \ldots, n_k$,

   $$c|(n_1a_1 + n_2a_2 + \ldots + n_ka_k).$$

9. Let $(a_n) = (1, 1, 2, 3, 5, 8, 13, 21 \ldots)$ be the Fibonacci sequence defined by $a_1 = a_2 = 1$ and, for $n \geqslant 3$, $a_n = a_{n-1} + a_{n-2}$.

   (i) Apply Euclid's greatest common divisor algorithm to calculate $\gcd(21, 13)$.

   (ii) Describe what happens in general when Euclid's greatest common divisor algorithm is used to calculate $\gcd(a_n, a_{n-1})$.

(iii) Use the result of part (i) to find integers r and s such that $1 = 21r + 13s$.

(iv) (Harder) Use the result of part (iii) to conjecture a general result which expresses $1 = ra_n + sa_{n-1}$ for some integers r and s.

Prove your conjecture using mathematical induction.

10. Let a and b be consecutive positive integers. Show that a and b are coprime.

11. Let a and b be coprime positive integers. Using theorem 9.7, prove that, for all positive integers c, if $a|c$ and $b|c$ then $ab|c$.

Show that this result is false if a and b are not coprime.

9.2 Prime Numbers

We begin with what is probably a familiar definition.

Definition 9.4

An integer p is **prime** if $p > 1$ and the only divisors of p are 1 and p itself. An integer $n > 1$ that is not prime is called **composite**.

Note that 1 is *not* prime; 2 is the smallest prime and all other primes are odd. The first few primes are: $2, 3, 5, 7, 11, 13, 17, 19, \ldots$. Later in this section we will find all the primes less than 100.

In two important ways, prime numbers play an analogous role in number theory to that played by atoms in molecular chemistry. Firstly, primes are 'indivisible' in the sense that they cannot be 'broken down' into the product of smaller positive integers. Secondly, as we shall see, primes are the 'building blocks' of the positive integers in the sense that every positive integer (except 1) can expressed as a product of primes.

The fact that primes are 'indivisible' has certain consequences for the divisibility relation that we now explore by considering two (related) questions. Suppose p is prime.

1. Which positive integers a are coprime with p?

2. If p divides a product of integers $a \times b$, what may we deduce?

To explore the first question, let $p = 7$. It is clear that, if a is a multiple of 7, then a and 7 are not coprime. What about those integers greater than 1 that are *not* divisible by 7:

$$2, 3, 4, 5, 6, 8, 9, 10, 11, 12, 13, 15, \ldots .$$

Each of these is coprime with 7. This would suggest that an integer $a > 1$ that is not a multiple of p is coprime with p.

To answer the second question, again let $p = 7$. Now 7 divides 126 and we can express 126 as a product of two integers in a number of ways:

$$126 = 2 \times 63 = 3 \times 42 = 6 \times 21 = 9 \times 14 = 18 \times 7.$$

In each of these ways of writing 126 as a product, 7 divides one of the factors. Similarly, $7|1155$ and however we write 1155 as $a \times b$ where $a, b \in \mathbb{Z}^+$, we find that 7 divides one of the terms. For example, $1155 = 15 \times 77$ and $7|77$; $1155 = 55 \times 21$ and $7|21$, and so forth.

These considerations are formalised in the following theorem.

Theorem 9.8

Let p be prime.

(a) For all $a \in \mathbb{Z}^+$, either $p|a$ or a and p are coprime.

(b) For all $a, b \in \mathbb{Z}^+$, if $p|ab$ then $p|a$ or $p|b$.

Proof

(a) Consider the greatest common divisor of a and p. By definition $\gcd(a, p)$ is a positive integer that divides p (as well as a). Since p is prime its only divisors are 1 and p, so the only possibilities for the greatest common divisor are $\gcd(a, p) = 1$ or $\gcd(a, p) = p$.

If $\gcd(a, p) = p$ then $p|a$ since, by definition, the greatest common divisor $\gcd(a, p)$ divides a.

If $\gcd(a, p) = 1$ than a and p are coprime, again by definition.

(b) Let $p|ab$.

Suppose that p does not divide a. Then we need to show that $p|b$.

By part (a), we have $\gcd(a, p) = 1$. Therefore, from theorem 9.7, there exist $m, n \in \mathbb{Z}$ such that $ma + np = 1$. Multiplying this equation by b gives

$$b = mab + npb$$

and we consider each of the terms in this expression. By our assumption $p|ab$ so that $p|mab$. Clearly $p|npb$. Therefore p divides each term so p divides $mab + npb = b$, as required. □

We claimed above that the prime numbers form the 'building blocks' from which all integers can be 'constructed' using multiplication in the sense that every integer $n > 1$ can be written as a product of primes. To see this, let n be an integer. If n is not prime then it has a factor a such that $1 < a < n$ so can be expressed as $n = ab$. Then consider a and b in turn. Each is either prime or can itself be factorised. Continuing in this way we eventually reach only prime factors. This is illustrated in the following examples.

Examples 9.8

1. Consider 120. First we may write $120 = 10 \times 12$. Then each of the factors also factorises, $10 = 2 \times 5$ and $12 = 3 \times 4$, so $120 = (2 \times 5) \times (3 \times 4)$. The only composite number in this expression is $4 = 2 \times 2$ so we obtain $120 = 2 \times 5 \times 3 \times 2 \times 2 = 2^3 \times 3 \times 5$.

   Note that it does not matter *how* we complete the factorisation. For example, the following is an alternative way of achieving the (same) prime factorisation of 120:

   $120 = 3 \times 40 = 3 \times 2 \times 20 = 3 \times 2 \times 5 \times 4 = 3 \times 2 \times 5 \times 2 \times 2 = 2^3 \times 3 \times 5.$

2. Repeating the same process with 675 gives

$$675 = 5 \times 135 = 5 \times 5 \times 27 = 5^2 \times 3^3.$$

The statement that this can always be done in an 'essentially' unique way is called the *Fundamental Theorem of Arithmetic*. By 'essentially unique' we mean unique apart from the order in which the primes are written.

Theorem 9.9 (The Fundamental Theorem of Arithmetic)

Every integer $n > 1$ has a factorisation

$$n = p_1^{e_1} p_2^{e_2} \dots p_k^{e_k}$$

where $p_1, p_2, \dots, p_k$ are distinct primes and $e_1, e_2, \dots, e_k$ are positive integers.
This factorisation is essentially unique; that is, it is unique up to the ordering of the primes.

The fundamental theorem gives a way of calculating the greatest common divisor, as illustrated in the following example.

Example 9.9

Find $\gcd(1350, 2205)$.

First we find the prime factorisation of 1350 and 2205.

$$1350 = 5 \times 270 = 5 \times 27 \times 10 = 5 \times 3^3 \times 2 \times 5 = 2 \times 3^3 \times 5^2$$

$$2205 = 5 \times 441 = 5 \times 3 \times 147 = 5 \times 3 \times 3 \times 49 = 3^2 \times 5 \times 7^2.$$

The greatest common divisor is then the product of those prime powers common to both expressions,

$$\gcd(1350, 2205) = 3^2 \times 5 = 45.$$

In other words, for each prime that is common to both factorisations, we chose the smaller of its powers in the two expressions. For example, 3^3 is a divisor of 1350 and 3^2 is a divisor of 2205. Hence the largest power of 3 that is a common divisor is 3^2.

Fermat Numbers and Mersenne Numbers

Numbers of the form $F_n = 2^{2^n} + 1$ are called **Fermat numbers.** Fermat conjectured that F_n is prime for all $n > 0$. The first few Fermat numbers are:

$$\begin{array}{lllll} F_0 & = & 2^{2^0}+1 & = & 2^1+1=3 & \text{which is prime,} \\ F_1 & = & 2^{2^1}+1 & = & 2^2+1=5 & \text{which is prime,} \\ F_2 & = & 2^{2^2}+1 & = & 2^4+1=17 & \text{which is prime,} \\ F_3 & = & 2^{2^3}+1 & = & 2^8+1=257 & \text{which is prime,} \\ F_4 & = & 2^{2^4}+1 & = & 2^{16}+1=65537 & \text{which is prime.} \end{array}$$

In 1732 Euler showed that the next Fermat number

$$F_5 = 2^{2^5} = 2^{32} + 1 = 4294967297 = 641 \times 6700417$$

is composite, thus proving that Fermat's conjecture is false.

However integers of the form $2^n - 1$ have been considered as possible candidates for (large) prime numbers. We show that a necessary condition for $2^n - 1$ to be prime is that n itself is prime. In other words, for all $n \in \mathbb{Z}^+$,

$$2^n - 1 \text{ is prime } \Rightarrow n \text{ is prime.}$$

This is most easily proved by proving the contrapositive:

$$n \text{ is not prime } \Rightarrow 2^n - 1 \text{ is not prime.}$$

Suppose that n is not prime. Then $n = rs$ where $1 < r < n$ and $1 < s < n$. Hence $2^n = 2^{rs} = (2^r)^s$. We know from exercise 2.3.4 that, for $x \neq 1$,

$$1 + x + x^2 + \ldots + x^{n-1} = \frac{x^n - 1}{x - 1}.$$

Using this result with $x = 2^r$ and $n = s$ gives

$$1 + 2^r + (2^r)^2 + \ldots + (2^r)^{s-1} = \frac{(2^r)^s - 1}{2^r - 1} = \frac{2^n - 1}{2^r - 1}$$

so, multiplying both sides by $2^r - 1$, gives

$$(2^r - 1)(1 + 2^r + (2^r)^2 + \ldots + (2^r)^{s-1}) = 2^n - 1.$$

This last equation expresses $2^n - 1$ as the product of two integers, namely $2^r - 1$ and $1 + 2^r + (2^r)^2 + \ldots + (2^r)^{s-1}$. Hence $2^n - 1$ is not prime, as required.

Integers of the form $M_p = 2^p - 1$ where p is prime are called **Mersenne numbers**.

We consider whether the converse of the result above is true. In other words, if p is prime, is it necessarily the case that $M_p = 2^p - 1$ is prime? The first few Mersenne numbers are:

$$\begin{array}{lllll} M_2 & = & 2^2 - 1 & = & 3 & \text{which is prime,} \\ M_3 & = & 2^3 - 1 & = & 7 & \text{which is prime,} \\ M_5 & = & 2^5 - 1 & = & 31 & \text{which is prime,} \\ M_7 & = & 2^7 - 1 & = & 127 & \text{which is prime.} \end{array}$$

However $M_{11} = 2^{11} - 1 = 2047 = 23 \times 89$, so the converse of the statement is false; in other words, not all Mersenne numbers are prime. However, over 40 Mersenne primes have been discovered. For example the 39th Mersenne prime is $M_{13466917} \approx 10^{4053946}$; hence, this prime number has over 4 million digits!

There is a collaborative project of volunteers called GIMPS—the Great Internet Mersenne Prime Search—that is dedicated to finding Mersenne primes. Mersenne primes with over ten million digits are now known.

Primality Testing and Factorisation

There are two practical questions associated with prime factorisation.

(a) How do we determine whether or not a given integer n is prime?

(b) How do we find the prime factorisation of a given integer n?

The first simple observation that restricts the amount of work required to test whether a given integer n is prime is provided by the following theorem. This says firstly that we only need to test whether n has *prime* factors and secondly that we only need to consider factors up to $\sqrt{n}$ in size.

Theorem 9.10

An integer $n > 1$ is composite if and only if it is divisible by a prime number $p \leqslant \sqrt{n}$.

Proof

If n is divisible by a prime $p \leqslant \sqrt{n}$ then clearly n is composite.

Conversely, suppose that n is composite. Then $n = ab$ where $1 < a < n$ and $1 < b < n$. At least one of a and b is less than or equal to $\sqrt{n}$. (How would you prove this?) Suppose $a \leqslant \sqrt{n}$. If a itself is prime we are finished. If a is not prime then it is divisible by some prime $p \leqslant a \leqslant \sqrt{n}$ which also divides n. □

Example 9.10

We can verify that 101 is prime by showing that it is not divisible by any of the primes $p \leqslant \sqrt{101}$. There are only four such primes, 2, 3, 5, and 7 and it is easy to check that each of these does not divide 101.

For very large integers, this method is inefficient. However there are other much more sophisticated ways of determining whether an integer is prime (which are based on some very advanced number theory). These techniques are beyond the scope of this text.

The second problem mentioned above—finding the prime factors of a large integer n—is much harder than primality testing. For very large integers it is practically impossible to factorise them in a reasonable length of time even using the most powerful computers. We will return to this later.

We conclude this section with a method of finding all the primes p in a particular range $1 < p \leqslant N$, known as the 'Sieve of Eratosthenes'.

Sieve of Eratosthenes

First list the integers $2, 3, 4, \ldots, N$. Then

- circle 2 and strike out all multiples of 2.
- circle 3 (the first number not stuck out) and strike out all remaining multiples of 3.
- circle 5 (the first number not stuck out) and strike out all remaining multiples of 5.

Continue until every integer in the list is either circled or struck out. The primes are the circled numbers.

Example 9.11

We use the Sieve of Eratosthenes to find all primes up to 100.

First list all the integers $2, 3, 4, \ldots, 100$.

	2	3	4	5	6	7	8	9	10
11	12	13	14	15	16	17	18	19	20
21	22	23	24	25	26	27	28	29	30
31	32	33	34	35	36	37	38	39	40
41	42	43	44	45	46	47	48	49	50
51	52	53	54	55	56	57	58	59	60
61	62	63	64	65	66	67	68	69	70
71	72	73	74	75	76	77	78	79	80
81	82	83	84	85	86	87	88	89	90
91	92	93	94	95	96	97	98	99	100

First circle 2 and then strike out all multiples of 2.

	②	3	~~4~~	5	~~6~~	7	~~8~~	9	~~10~~
11	~~12~~	13	~~14~~	15	~~16~~	17	~~18~~	19	~~20~~
21	~~22~~	23	~~24~~	25	~~26~~	27	~~28~~	29	~~30~~
31	~~32~~	33	~~34~~	35	~~36~~	37	~~38~~	39	~~40~~
41	~~42~~	43	~~44~~	45	~~46~~	47	~~48~~	49	~~50~~
51	~~52~~	53	~~54~~	55	~~56~~	57	~~58~~	59	~~60~~
61	~~62~~	63	~~64~~	65	~~66~~	67	~~68~~	69	~~70~~
71	~~72~~	73	~~74~~	75	~~76~~	77	~~78~~	79	~~80~~
81	~~82~~	83	~~84~~	85	~~86~~	87	~~88~~	89	~~90~~
91	~~92~~	93	~~94~~	95	~~96~~	97	~~98~~	99	~~100~~

Next circle 3 and then strike out all multiples of 3 that have not already been struck out.

	②	③	~~4~~	5	~~6~~	7	~~8~~	~~9~~	~~10~~
11	~~12~~	13	~~14~~	~~15~~	~~16~~	17	~~18~~	19	~~20~~
~~21~~	~~22~~	23	~~24~~	25	~~26~~	~~27~~	~~28~~	29	~~30~~
31	~~32~~	~~33~~	~~34~~	35	~~36~~	37	~~38~~	~~39~~	~~40~~
41	~~42~~	43	~~44~~	~~45~~	~~46~~	47	~~48~~	49	~~50~~
~~51~~	~~52~~	53	~~54~~	55	~~56~~	~~57~~	~~58~~	59	~~60~~
61	~~62~~	~~63~~	~~64~~	65	~~66~~	67	~~68~~	~~69~~	~~70~~
71	~~72~~	73	~~74~~	~~75~~	~~76~~	77	~~78~~	79	~~80~~
~~81~~	~~82~~	83	~~84~~	85	~~86~~	~~87~~	~~88~~	89	~~90~~
91	~~92~~	~~93~~	~~94~~	95	~~96~~	97	~~98~~	~~99~~	~~100~~

Then circle 5 and then strike out all multiples of 5 that have not already been struck out.

	②	③	~~4~~	⑤	~~6~~	7	~~8~~	~~9~~	~~10~~
11	~~12~~	13	~~14~~	~~15~~	~~16~~	17	~~18~~	19	~~20~~
~~21~~	~~22~~	23	~~24~~	~~25~~	~~26~~	~~27~~	~~28~~	29	~~30~~
31	~~32~~	~~33~~	~~34~~	~~35~~	~~36~~	37	~~38~~	~~39~~	~~40~~
41	~~42~~	43	~~44~~	~~45~~	~~46~~	47	~~48~~	49	~~50~~
~~51~~	~~52~~	53	~~54~~	~~55~~	~~56~~	~~57~~	~~58~~	59	~~60~~
61	~~62~~	~~63~~	~~64~~	~~65~~	~~66~~	67	~~68~~	~~69~~	~~70~~
71	~~72~~	73	~~74~~	~~75~~	~~76~~	77	~~78~~	79	~~80~~
~~81~~	~~82~~	83	~~84~~	~~85~~	~~86~~	~~87~~	~~88~~	89	~~90~~
91	~~92~~	~~93~~	~~94~~	~~95~~	~~96~~	97	~~98~~	~~99~~	~~100~~

Repeating for 7 gives the following table.

	②	③	~~4~~	⑤	~~6~~	⑦	~~8~~	~~9~~	~~10~~
11	~~12~~	13	~~14~~	~~15~~	~~16~~	17	~~18~~	19	~~20~~
~~21~~	~~22~~	23	~~24~~	~~25~~	~~26~~	~~27~~	~~28~~	29	~~30~~
31	~~32~~	~~33~~	~~34~~	~~35~~	~~36~~	37	~~38~~	~~39~~	~~40~~
41	~~42~~	43	~~44~~	~~45~~	~~46~~	47	~~48~~	~~49~~	~~50~~
~~51~~	~~52~~	53	~~54~~	~~55~~	~~56~~	~~57~~	~~58~~	59	~~60~~
61	~~62~~	~~63~~	~~64~~	~~65~~	~~66~~	67	~~68~~	~~69~~	~~70~~
71	~~72~~	73	~~74~~	~~75~~	~~76~~	~~77~~	~~78~~	79	~~80~~
~~81~~	~~82~~	83	~~84~~	~~85~~	~~86~~	~~87~~	~~88~~	89	~~90~~
~~91~~	~~92~~	~~93~~	~~94~~	~~95~~	~~96~~	97	~~98~~	~~99~~	~~100~~

Note that, at this stage, only three integers were struck out: $49 = 7 \times 7$, $77 = 7 \times 11$ and $91 = 7 \times 13$. In fact, the smallest number struck out when the prime p is circled is p^2 since all smaller integers that are not prime have a factor smaller than p and so have been struck out at an earlier stage.

Therefore, since the next number to be circled is 11 and $11^2 = 121$ is greater than 100, no further numbers will be struck out. Hence all the remaining numbers may be circled which gives the following list of all the prime numbers less than 100.

	(2)	(3)	~~4~~	(5)	~~6~~	(7)	~~8~~	~~9~~	~~10~~
(11)	~~12~~	(13)	~~14~~	~~15~~	~~16~~	(17)	~~18~~	(19)	~~20~~
~~21~~	~~22~~	(23)	~~24~~	~~25~~	~~26~~	~~27~~	~~28~~	(29)	~~30~~
(31)	~~32~~	~~33~~	~~34~~	~~35~~	~~36~~	(37)	~~38~~	~~39~~	~~40~~
(41)	~~42~~	(43)	~~44~~	~~45~~	~~46~~	(47)	~~48~~	~~49~~	~~50~~
~~51~~	~~52~~	(53)	~~54~~	~~55~~	~~56~~	~~57~~	~~58~~	(59)	~~60~~
(61)	~~62~~	~~63~~	~~64~~	~~65~~	~~66~~	(67)	~~68~~	~~69~~	~~70~~
(71)	~~72~~	(73)	~~74~~	~~75~~	~~76~~	~~77~~	~~78~~	(79)	~~80~~
~~81~~	~~82~~	(83)	~~84~~	~~85~~	~~86~~	~~87~~	~~88~~	(89)	~~90~~
~~91~~	~~92~~	~~93~~	~~94~~	~~95~~	~~96~~	(97)	~~98~~	~~99~~	~~100~~

Exercises 9.2

1. (i) Find the prime factorisation of

   (a) 1008

   (b) 792

   (c) 3276

   (d) 22 680.

   (ii) Use the results of part (a) to determine

   (a) $\gcd(1008, 792)$

   (b) $\gcd(1008, 3276)$

   (c) $\gcd(3276, 22\,680)$

   (d) $\gcd(1008, 792, 3276)$.

2. (i) Find the first sequence of three consecutive composite positive integers.

   (ii) Find the first sequence of five consecutive composite positive integers.

   (iii) Prove that, for each integer $n \geqslant 2$, the sequence of $n - 1$ positive integers
   $$n! + 2, n! + 3, n! + 4, \ldots, n! + n$$
   are all composite.

3. Show that the condition that p is prime is a necessary condition for each part of theorem 9.8.

4. Prove the following generalisation of theorem 9.8 (b).

   Let p be prime and let $a_1, a_2, \ldots, a_n \in \mathbb{Z}^+$. If $p|(a_1 a_2 \ldots a_n)$ then $p|a_i$ for some $i = 1, 2, \ldots, n$.

5. For which prime numbers p is $p^2 + 2$ also prime?

6. **Goldbach's conjecture** states that: *every even integer* $n \geqslant 4$ *is the sum of two prime numbers.*

   Verify Goldbach's conjecture for even integers n in the range $4 \leqslant n \leqslant 30$.

   Note. The status of Goldbach's conjecture is unknown. In other words, no proof and no counter-example has been found.

7. Let p be prime and let a and b be positive integers.

   (i) Show that, if $\gcd(a, p^2) = p$ and $\gcd(b, p^2) = p$ then $\gcd(ab, p^4) = p^2$.

   (ii) Show that, if $\gcd(a, p^2) = p$ and $\gcd(b, p^3) = p^2$ then $\gcd(ab, p^4) = p^3$.

   (iii) If $\gcd(a, b) = p$ what are the possible values of $\gcd(a^2, b)$?

   (iv) If $\gcd(a, b) = p$ what are the possible values of $\gcd(a^3, b)$?

8. In example 2.4.1 we proved that $\sqrt{2}$ is irrational. Use the Fundamental Theorem of Arithmetic to prove that, if n is a positive integer that is not a perfect square then $\sqrt{n}$ is irrational.

   Hint. Prove the contrapositive: if $\sqrt{n}$ is rational then n is a perfect square.

9.3 Linear Congruences

We introduced the relation of 'congruence module n' in chapter 4. For convenience, we repeat below the definition given on page 183.

Definition 9.5

Let n be a positive integer.
For integers a and b, we say that a is **congruent to** b **modulo** n, denoted $a \equiv b \mod n$, if their difference $a - b$ is a multiple of n:

$$a \equiv b \mod n \text{ if and only if } a - b = kn \text{ for some } k \in \mathbb{Z}.$$

Other notations for $a \equiv b \mod n$ are $a \equiv b \pmod n$ or $a \equiv_n b$.

Note that another way of phrasing the definition is to say that a and b are congruent modulo n is that n divides their difference:

$$a \equiv b \mod n \text{ if and only if } n|(a - b).$$

In this chapter, we will prefer the notation $a \equiv b \mod n$ rather than $a \equiv_n b$ which we used in chapter 4. The notation $a \equiv b \mod n$ is more standard in number theory where, for example, we may be interested in considering various kinds of 'equations' modulo n. In chapter 4, we were interested in congruence from as a kind of relation so the notation $a \equiv_n b$ was more in tune with the general relation notation $a \mathsf{R} b$.

One reason for considering modulo arithmetic – that is, working modulo n for some n – is that some problems that involve the use of very large integers may be simplified in this way. Actually, we do this all the time in 'everyday life'. For example, what day of the week will it be in 70772 days time? Since $70772 \equiv 2 \mod 7$, the answer is that it will be 2 days later than today; thus, if today is Tuesday then it will be Thursday in 70772 days time. However, none of us will be alive to see it since 70772 days is over 193 years!

Theorem 9.11

Let n be a fixed positive integer. Then, for all $a, b \in \mathbb{Z}$, $a \equiv b \mod n$ if and only if a and b have the same remainder after division by n.

Proof

From the Division Algorithm (or, more properly, the generalization that allows negative integers—see example 9.1.2) there exist integers q, q', r, r' such that

$$a = qn + r \text{ where } 0 \leqslant r < n,$$

$$b = q'n + r' \text{ where } 0 \leqslant r' < n.$$

Therefore

$$a - b = (q - q')n + (r - r') \text{ where } -n < r - r' < n.$$

Suppose $a \equiv b \mod n$. Then n divides $a - b$ so n also divides $(a - b) - (q - q')n = r - r'$. But $-n < r - r' < n$ and the only integer in this range that is a multiple of n is 0. Therefore $r - r' = 0$ so $r = r'$. Hence a and b have the same remainder after division by n.

Conversely, suppose $r = r'$ above; that is, a and b have the same remainder after division by n. Then $a - b = (q - q')n$ where $q - q' \in \mathbb{Z}$. Hence n divides $a - b$, $n|(a - b)$, so $a \equiv b \mod n$. □

Theorem 9.12

Let $n \in \mathbb{Z}$. If n is a perfect square then n is congruent to 0 or 1 modulo 4:

$$n = m^2 \text{ for some } m \in \mathbb{N} \quad \Rightarrow \quad n \equiv 0 \mod 4 \text{ or } n \equiv 1 \mod 4.$$

Proof

For any integer m, we have $m \equiv 0, 1, 2 \text{ or } 3 \mod 4$.

If $m \equiv 0 \mod 4$ then $m^2 \equiv 0 \mod 4$.
If $m \equiv 1 \mod 4$ then $m^2 \equiv 1 \mod 4$.
If $m \equiv 2 \mod 4$ then $m^2 \equiv 0 \mod 4$.
If $m \equiv 3 \mod 4$ then $m^2 \equiv 1 \mod 4$.

Therefore, if $n = m^2$, for some integer m then

$$n \equiv 0 \mod 4 \text{ or } n \equiv 1 \mod 4.$$

□

Example 9.12

636802 is not a perfect square since $636802 = 159200 \times 4 + 2$ so $636802 \equiv 2 \mod 4$.

Note that we are using the contrapositive of theorem 9.12 here: if $n \not\equiv 0$ and $n \not\equiv 1 \mod 4$ then n is not a perfect square.

Recall from chapter 4 that congruence modulo n is an equivalence relation and the equivalence class of $a \in \mathbb{Z}$ is

$$\begin{aligned}[a] &= \{b \in \mathbb{Z} : b \equiv a \mod n\} \\ &= \{\ldots, a-2n, a-n, a, a+n, a+2n, a+3n, \ldots\}.\end{aligned}$$

The equivalence class $[a]$ is often called the **congruence class** of a. For a general n, there are n distinct congruence classes:

$$\begin{aligned}[0] &= \{\ldots, -2n, -n, 0, n, 2n, \ldots\} \\ [1] &= \{\ldots, -2n+1, -n+1, 1, n+1, 2n+1, \ldots\} \\ [2] &= \{\ldots, -2n+2, -n+2, 2, n+2, 2n+2, \ldots\} \\ &\vdots \\ [n-1] &= \{\ldots, -n-1, -1, n-1, 2n+1, 3n+1, \ldots\}.\end{aligned}$$

The set of congruence classes modulo n is denoted

$$\mathbb{Z}_n = \{[0], [1], [2], \ldots, [n-1]\}.$$

We also defined the arithmetic operations of addition and multiplication on the elements of $\mathbb{Z}_n$. Subtraction is similarly defined. If $[a]$ and $[b]$ are congruence classes modulo n then

$$\begin{aligned}[a] +_n [b] &= [a+b] \\ [a] -_n [b] &= [a-b] \\ [a] \times_n [b] &= [ab].\end{aligned}$$

The following theorem shows that these definitions are valid; in other words, the definitions do not depend on the labels of the equivalence classes. See pages 184 and 185 for a more detailed discussion of this point. The proof of parts (i) and (iii) of theorem 9.13 were left as an exercise in chapter 4—see exercise 4.4.11—and the proof of part (ii) is similar.

Theorem 9.13

Let $n \in \mathbb{Z}$. If $a \equiv a' \mod n$ and $b \equiv b' \mod n$ then

(i) $a + b \equiv a' + b' \mod n,$

(ii) $a - b \equiv a' - b' \mod n,$

(iii) $ab \equiv a'b' \mod n.$

Examples 9.13

With a little thought, it is possible to evaluate apparently difficult looking expressions modulo n without resorting to calculator or computer. The following examples illustrate this.

1. Evaluate $3^{10} \mod 14$.

   Solution

   Rather than trying to evaluate 3^{10} (which is $59\,049$), then dividing it by 14 and identifying the remainder, instead we build up to 3^{10} by first evaluating smaller powers.

   First note that $3^3 = 27 \equiv 13 \mod 14$. Now $13 \equiv -1 \mod 14$ so we have $3^3 \equiv -1 \mod 14$.

   From this it follows that $3^9 \equiv (3^3)^3 \equiv (-1)^3 \equiv -1 \mod 14$.

   Hence $3^{10} \equiv 3^9 \times 3 \equiv -1 \times 3 \equiv -3 \equiv 11 \mod 14$.

2. Evaluate $27 \times 13 \times 9 \mod 31$.

   Solution

   Again we begin by evaluating the simpler expression $27 \times 13 \mod 31$ but here, too, a little thought can save effort.

Note that $27 \equiv -4 \mod 31$ so, as in the previous example, we can replace a larger positive value with a smaller negative one which will simplify the subsequent calculation. Hence we have $27 \times 13 \equiv -4 \times 13 = -52 \equiv 10 \mod 31$.

Therefore $27 \times 13 \times 9 \equiv 10 \times 9 = 90 \equiv -3 \equiv 28 \mod 31$.

In this last step we have again switched between equivalent positive and negative values where it is convenient to do so. For example, since 90 is three less than a multiple of 31, we can immediately see that $90 \equiv -3 \mod 31$ and then adding 31 gives $-3 \equiv 28 \mod 31$.

3. Evaluate $5^{302} \mod 31$.

Solution

The expression $5^{302} \mod 31$ looks daunting but something rather nice occurs that makes the evaluation very easy.

First note that $5^2 = 25 \equiv -6 \mod 31$ so that $5^3 \equiv (-6) \times 5 = -30 \equiv 1 \mod 31$.

(Note that we could have obtained this directly since $5^3 = 125 = (4 \times 31) + 1 \equiv 1 \mod 31$.)

Since $5^3 \equiv 1 \mod 31$, it follows that any power of 5^3 is also congruent to 1 modulo 31. In particular $5^{300} = (5^3)^{100} \equiv 1^{100} \equiv 1 \mod 31$. Finally we have

$$5^{302} = 5^2 \times 5^{300} \equiv 25 \times 1 \equiv 25 \mod 31.$$

Equations in Modular Arithmetic

Let $n \in \mathbb{Z}$. We shall consider 'linear congruence equations' or, simply, 'linear congruences'

$$ax \equiv b \mod n$$

where $0 < a < n$ and $0 \leqslant b < n$ and we seek solutions x in the range $0 \leqslant x < n$. We begin by considering some simple examples where we can find the solutions by trial and error.

Examples 9.14

1. Consider the congruence $2x \equiv 4 \mod 5$.

   Working modulo 5 we have $2 \times 0 \equiv 0,\ 2 \times 1 \equiv 2,\ 2 \times \mathbf{2} \equiv 4,\ 2 \times 3 \equiv 1,\ 2 \times 4 \equiv 3$. Hence the congruence has a unique solution $x \equiv 2 \mod 5$.

2. Consider the congruence $2x \equiv 4 \mod 6$.

   Since $2 \times 0 \equiv 0,\ 2 \times 1 \equiv 2,\ 2 \times \mathbf{2} \equiv 4,\ 2 \times 3 \equiv 0,\ 2 \times 4 \equiv 2,\ 2 \times \mathbf{5} \equiv 4$, the congruence has two solutions $x \equiv 2 \mod 6$ and $x \equiv 5 \mod 6$.

3. Consider the congruence $2x \equiv 5 \mod 6$.

   From the calculations above, the congruence has no solution.

4. Consider the congruence $6x \equiv 12 \mod 30$.

   Clearly $x \equiv 2 \mod 30$ is a solution. A little trial and error shows that $x \equiv 7, 12, 17, 22, 27 \mod 30$ are also solutions.

The previous examples show that a linear congruence $ax \equiv b \mod n$ may have no solution, a unique solution or several solutions. Compare this situation with 'ordinary' arithmetic where the equation $ax = b$ has a unique solution (provided $a \neq 0$). The following theorem gives the general situation.

Theorem 9.14 (Solving linear congruences)

Let a, b and n be integers such that $0 < a < n$ and $0 \leqslant b < n$ and let $d = \gcd(a, n)$. Then the congruence

$$ax \equiv b \mod n$$

has a solution if and only if d divides b.
If d divides b and x_0 is a solution of the congruence in the range $0 \leqslant x_0 < n/d$, then there are exactly d solutions in the range $0 \leqslant x_0 < n$, namely

$$x_0,\ x_0 + \frac{n}{d},\ x_0 + 2\frac{n}{d},\ \ldots,\ x_0 + (d-1)\frac{n}{d}.$$

Proof

Let $d = \gcd(a, n)$. By theorem 9.6,

$$\begin{aligned} d|b &\Leftrightarrow b = sa + tn \text{ for some } s, t \in \mathbb{Z} \\ &\Leftrightarrow sa \equiv b \mod n \text{ for some } s \in \mathbb{Z} \\ &\Leftrightarrow ax \equiv b \mod n \text{ has a solution.} \end{aligned}$$

For the second part, suppose that d divides b. Suppose also that x_0 is a solution of $ax \equiv b \mod n$. Then $ax_0 - b = kn$ for some $k \in \mathbb{Z}$. Therefore, for any $s \in \mathbb{Z}$,

$$\begin{aligned} x = x_0 + s\frac{n}{d} &\Rightarrow ax - b = ax_0 + as\frac{n}{d} - b \\ &\Rightarrow ax - b = kn + as\frac{n}{d} \\ &\Rightarrow ax - b = n\left(k + s\frac{a}{d}\right) \\ &\Rightarrow ax \equiv b \mod n \quad \text{since } \frac{a}{d} \in \mathbb{Z}. \end{aligned}$$

Therefore

$$x_0,\ x_0 + \frac{n}{d},\ x_0 + 2\frac{n}{d},\ \ldots,\ x_0 + (d-1)\frac{n}{d}$$

are all solutions of $ax \equiv b \mod n$ and these are all different mod n. □

Examples 9.15

We consider how the solutions to examples 9.14 above fit into the framework of this theorem.

1. $2x \equiv 4 \mod 5$.

   Since $\gcd(2, 5) = 1$ and $1|4$ there is a solution which is unique since $d = \gcd(2, 5) = 1$.

2. $2x \equiv 4 \mod 6$.

   Since $\gcd(2, 6) = 2$ and $2|4$ there is a solution. By the second part of theorem 9.14, there are two solutions since $\gcd(2, 6) = 2$.

   $x_0 = 2$ is a solution in $0 \leqslant x_0 < 6/2 = 3$. The other solution is $2 + \frac{6}{2} = 5$.

3. $2x \equiv 5 \mod 6.$

Since $\gcd(2, 6) = 2$ and 2 does not divide 5, there is no solution.

An important special case of theorem 9.14 is when a and n are coprime. Then $d = 1$, so that $d|b$ for all $b \in \mathbb{Z}^+$ and the congruence has a unique solution. This is captured in the following theorem.

Theorem 9.15

If a and n are coprime and $0 < a < n$ then the congruence $ax \equiv b \mod n$ has a unique solution in the range $0 \leqslant x < n$.

Note that, if n is prime then theorem 9.15 says that any congruence $ax \equiv b \mod n$, where $0 < a < n$, has a unique solution in the range $0 \leqslant x < n$.

Suppose the congruence $ax \equiv b \mod n$ has one or more solutions. How do we *find* the solution(s)? For small values of n, this can be done by testing each of the values x in the range $0 \leqslant x < n/d$ where $d = \gcd(a, n)$. For larger values of n, however, this is tedious.

The following theorem allows us to replace a congruence with one involving smaller numbers.

Theorem 9.16

(a) Suppose m divides each of a, b and n and let $a' = a/m$, $b' = b/m$ and $n' = n/m$. Then

$$ax \equiv b \mod n \text{ if and only if } a'x \equiv b' \mod n'.$$

(b) Suppose a and n are coprime and m divides both a and b. Let $a' = a/m$ and $b' = b/m$. Then

$$ax \equiv b \mod n \text{ if and only if } a'x \equiv b' \mod n.$$

Proof

(a) Suppose m divides a, b and n. Recall that $ax \equiv b \mod n$ if and only if the difference $ax - b$ is a multiple of n:

$$ax \equiv b \mod n \Leftrightarrow ax - b = qn \text{ for some } q \in \mathbb{Z}.$$

In this case, dividing by m gives

$$\frac{a}{m}x - \frac{b}{m} = q\frac{n}{m}$$

which is precisely $a'x - b' = qn'$. This last equation says that the difference $a'x - b'$ is a multiple of n' which, by definition, means $a'x \equiv b' \mod n'$.

We could summarise this argument more symbolically as follows:

$$\begin{aligned} ax \equiv b \mod n &\Leftrightarrow ax - b = qn \text{ for some } q \in \mathbb{Z} \\ &\Leftrightarrow \frac{a}{m}x - \frac{b}{m} = q\frac{n}{m} \text{ where } q \in \mathbb{Z} \\ &\Leftrightarrow a'x - b' = qn' \text{ where } q \in \mathbb{Z} \\ &\Leftrightarrow a'x \equiv b' \mod n'. \end{aligned}$$

(b) Suppose that a and n are coprime and m divides both a and b. As in part (a) we have,

$$\begin{aligned} ax \equiv b \mod n \quad &\Rightarrow \quad ax - b = qn && \text{for some } q \in \mathbb{Z} \\ &\Rightarrow \quad \frac{a}{m}x - \frac{b}{m} = \frac{qn}{m} && \text{where } q \in \mathbb{Z} \\ &\Rightarrow \quad a'x - b' = \frac{qn}{m} && \text{where } q \in \mathbb{Z} \\ &\Rightarrow \quad m|(nq) && \text{since } a', b' \in \mathbb{Z}. \end{aligned}$$

Since m divides a and a and n are coprime so m and n are also coprime. (This is because any common factor of m and n would also be a common factor of a and n.) Therefore, by theorem 9.7, there exist integers r and s such that $1 = rm + sn$. Multiplying by q gives $q = rmq + snq$. We know from the reasoning above that $m|(qn)$ and clearly we also have $m|rmq$. Therefore $m|q$.

Let $q/m = q' \in \mathbb{Z}$. Then the equation $a'x - b' = qn/m$ above becomes $a'x - b' = m'n$. Hence $a'x \equiv b' \mod n$.

The converse is straightforward. If $a'x \equiv b' \mod n$ then $a'x - b' = qn$ for some $q \in \mathbb{Z}$. Multiplying by m gives $ax - b = qmn$, so $ax \equiv b \mod n$. □

Examples 9.16

We use theorems 9.14, 9.15 and 9.16 to help us solve a number of linear congruences.

1. $20x \equiv 12 \mod 14$.

   Solution

   Since $\gcd(20, 14) = 2$ and 2 divides 12, the congruence has two solutions by theorem 9.14.

   Now $2|20$, $2|12$ and $2|14$ so, by part (a) of theorem 9.16, we may 'divide through by 2' and replace the original congruence with

$$10x \equiv 6 \mod 7.$$

   In this new congruence, 10 and 7 are coprime and 2 divides both 10 and 6. By part (b) of theorem 9.16, we may replace it with

$$5x \equiv 3 \mod 7.$$

   This congruence has a unique solution in the range $0 \leqslant x < 7$ by theorem 9.15 since 7 is prime. This solution is easy to find by trial and error:

$$5 \times 0 = 0, \quad 5 \times 1 = 1, \quad 5 \times \mathbf{2} = 10 \equiv 3,$$

   so $x = 2$ is the unique solution of $5x \equiv 3 \mod 7$.

   Therefore, $x = 2$ is also a solution of the original congruence $20x \equiv 12 \mod 14$ in the range

$$0 \leqslant x < \frac{14}{\gcd(20, 14)} = \frac{14}{2} = 7.$$

   By theorem 9.14, the (only) other solution is $x = 2 + 7 = 9$.

2. $18x \equiv 36 \mod 42$.

   Solution

   Since $\gcd(18, 42) = 6$ and $6|36$ the congruence has six solutions by theorem 9.14.

Now $6|18$, $6|36$ and $6|42$ so, by theorem 9.16 (a), we may replace the original congruence with

$$3x \equiv 6 \mod 7.$$

In this new congruence, 3 and 7 are coprime and $3|3$, $3|6$. Hence by theorem 9.16 (b) we may replace it with

$$x \equiv 2 \mod 7.$$

This clearly has the solution $x = 2$ which is the unique solution in the range $0 \leqslant x < 7$.

Therefore, $x = 2$ is also a solution of $18x \equiv 36 \mod 42$ in the range $0 \leqslant x < \frac{42}{6} = 7$.

By theorem 9.14, the remaining solutions in the range $0 \leqslant x < 42$ are obtained by adding multiples of $42/6 = 7$. The full list of solutions is therefore

$$x = 2,\ 2+7 = 9,\ 2+2\times 7 = 16,\ 2+3\times 7 = 23,$$
$$2+4\times 7 = 30,\ 2+5\times 7 = 37.$$

3. $14x = 149 \mod 201.$

Solution

Since $\gcd(14, 201) = 1$ so that 14 and 201 are coprime, there is a unique solution in the range $0 \leqslant x < 201$ by theorem 9.15.

Now $\gcd(14, 149) = 1$ so no further simplification is possible.

How do we solve $14x = 149 \mod 201$ without resorting to trial and error? Answer: use a little cunning!

The idea is we do two things in an attempt to find a solution. Firstly, find an 'approximate' solution; that is, find a multiple of 14 that is close to 149. Secondly, in order to be able to 'adjust' the approximate solution to obtain an exact solution, we find a multiple of 14 that is close to zero modulo 201. In other words, we find a multiple of 14 that is close to the modulus 201.

For the first part, note that $14\times 11 = 154$ which gets 'close' to the desired value of 149.

Also $14 \times 14 = 196 \equiv -5 \mod 201$, which is close to zero. On this occasion, the desired right-hand side 149 is simply the sum of the two values we have obtained: $149 = 154 + (-5)$. Hence the solution is easily obtained:

$$\begin{aligned} & 14 \times 11 + 14 \times 14 \equiv 154 - 5 = 149 && \mod 201 \\ \Rightarrow\ & 14 \times (11 + 14) \equiv 149 && \mod 201 \\ \Rightarrow\ & 14 \times 25 \equiv 149 && \mod 201. \end{aligned}$$

Therefore the solution is $x = 25$.

4. $14x = 156 \mod 201$.

Solution

This is similar to the previous example but the 'adjustment' of the approximate solution to become an actual solution requires a little more work. From the previous example, the approximate solution is provided by

$$14 \times 11 = 154.$$

We now need to find a way of increasing the approximate solution by 2 to obtain the exact solution.

We also know that

$$14 \times 14 \equiv -5 \mod 201.$$

This is a small negative value; taking the 'next' multiple gives a small positive value,

$$14 \times 15 \equiv 9 \mod 201.$$

The idea now is to combine these two 'small' values, -5 and 9, to obtain the difference 2 between the approximate solution $14 \times 11 = 154$ and the actual solution $14x = 156 \mod 201$. The first step is to write the required value 2 as a linear combination of -5 and 9:

$$2 = 3 \times 9 + 5 \times (-5).$$

Working modulo 201 we can replace 9 with 14×15 and we can replace -5 with 14×14. This gives

$$\begin{aligned} & 2 \equiv 3 \times (14 \times 15) + 5 \times (14 \times 14) && \mod 201 \\ \Rightarrow\ & 2 \equiv 14 \times (3 \times 15) + 14 \times (5 \times 14) && \mod 201 \\ \Rightarrow\ & 2 \equiv 14 \times 45 + 14 \times 70 && \mod 201 \\ \Rightarrow\ & 2 \equiv 14 \times 115 && \mod 201. \end{aligned}$$

This is exactly the adjustment we need to make to our approximate solution $14 \times 11 = 154$. Therefore

$$\begin{aligned} 156 &\equiv 154 + 2 &&\equiv 14 \times 11 + 14 \times 115 \\ &\equiv 14 \times (11 + 115) &&\equiv 14 \times 126 \mod 201, \end{aligned}$$

so the required solution is $x = 126$.

Exercises 9.3

1. Show that the last decimal digit of a perfect square cannot be $2, 3, 7$ or 8.

   Is 3 190 493 a perfect square?

2. (i) Determine the congruence classes modulo 6.

   (ii) Draw up tables for addition, subtraction and multiplication for the set $\mathbb{Z}_6$ of congruence classes modulo 6.

3. Show that, in $\mathbb{Z}_3 = \{[0], [1], [2]\}$, $[1] = [4]$ but $[2^1] \neq [2^4]$.

   This example shows that we *cannot* define exponentiation of congruence classes by $[a]^{[b]} = [a^b]$.

4. Using similar techniques to those given in examples 9.13, evaluate each of the following.

   (i) $11 \times 19 \mod 23$,

   (ii) $3^9 \mod 23$,

   (iii) $3^3 \times 17 \mod 23$,

   (iv) $5^{12} \mod 23$.

5. Find the last decimal digit of each of the following.

   (i) 23459^3

   (ii) 29147^5,

   (iii) $1! + 2! + 3! + \cdots + 10!$

6. Suppose $a \equiv b \mod n$ and $k \in \mathbb{Z}$. Does it follow that:

(i) $a^k \equiv b^k \mod n$?

(ii) $k^a \equiv k^b \mod n$?

In each case either give a proof or give a counter-example.

7. Find all the solutions of each of the following congruences by trial and error; that is, by an exhaustive search through the values $0, 1, \ldots, n-1$ (where n is the modulus). Compare this with what happens when we solve these equations in 'ordinary arithmetic'; that is, when we solve them over the real numbers $\mathbb{R}$.

(i) $3x \equiv 4 \mod 6$

(ii) $3x \equiv 4 \mod 7$

(iii) $x^2 \equiv 2 \mod 5$

(iv) $x^2 + 2 \equiv 0 \mod 6$

(v) $x^2 + 2 \equiv 0 \mod 7$

(vi) $x^2 \equiv x \mod 6.$

8. For each of the following linear congruences, determine whether a solution exists. If a solution exists, use the techniques illustrated in examples 9.16 to determine all the solutions in the range $0 \leqslant x < n$ (where n is the modulus).

(i) $12x \equiv 15 \mod 22$

(ii) $5x \equiv 1 \mod 11$

(iii) $19x \equiv 42 \mod 50$

(iv) $18x \equiv 42 \mod 50$

(v) $65x \equiv 27 \mod 169$

(vi) $65x \equiv 39 \mod 169$

(vii) $16x \equiv 301 \mod 595$

(viii) $20x \equiv 101 \mod 637.$

9.4 Groups in Modular Arithmetic

In this section we consider briefly the construction and properties of groups with operation addition modulo n or multiplication modulo n. This builds on some of the examples and exercises in chapter 8.

Addition Modulo n

In example 8.5.5, we saw that the set $\mathbb{Z}_5 = \{[0],[1],[2],[3],[4]\}$ of congruence classes modulo 5 is a group under the operation of addition modulo 5, $+_5$. The Cayley table for this group is given on page 376.

Clearly the congruence class $[1]$ is a generator for $\mathbb{Z}_5$ since each $[n] \in \mathbb{Z}_5$ can be expressed as

$$[n] = n.[1] = \underbrace{[1] + [1] + \cdots + [1]}_{n \text{ terms}}.$$

Hence $(\mathbb{Z}_5, +_5)$ is a cyclic group.

In this case, every non-zero congruence class is a generator. For example, $[2]$ is a generator since

$$\begin{aligned}
1.[2] &= [2] \\
2.[2] &= [2] +_5 [2] = [4] \\
3.[2] &= [4] +_5 [2] = [1] \\
4.[2] &= [1] +_3 [2] = [3] \\
5.[2] &= [3] +_5 [2] = [0].
\end{aligned}$$

Hence every element of $\mathbb{Z}_5$ can be written as $n.[2]$ for some integer n.

Example 9.17

We now consider addition modulo 6. The set of congruence classes is

$$\mathbb{Z}_6 = \{[0],[1],[2],[3],[4],[5]\}$$

and the Cayley table for $+_6$, addition modulo 6, is the following.

$+_5$	[0]	[1]	[2]	[3]	[4]	[5]
[0]	[0]	[1]	[2]	[3]	[4]	[5]
[1]	[1]	[2]	[3]	[4]	[5]	[0]
[2]	[2]	[3]	[4]	[5]	[0]	[1]
[3]	[3]	[4]	[5]	[0]	[1]	[2]
[4]	[4]	[5]	[0]	[1]	[2]	[3]
[5]	[5]	[0]	[1]	[2]	[3]	[4]

It is clear that $(\mathbb{Z}_6, +_6)$ is a group:

- $+_6$ is a binary operation on $\mathbb{Z}_6$ since only elements of $\mathbb{Z}_6$ appear in the table.
- $+_6$ is associative—this follows from the fact that ordinary addition of integers is associative (see example 8.5.5).
- $[0]$ is the identity element since, for all $[n] \in \mathbb{Z}_6$,

$$[0] +_6 [n] = [n] = [n] +_6 [0].$$

- Each $[n] \in \mathbb{Z}_6$ has inverse $[n]^{-1} = [6 - n] \in \mathbb{Z}_6$ since

$$[n] +_6 [6 - n] = [6] = [0].$$

For example, $[4]^{-1} = [2]$ since $[4] +_6 [2] = [0]$.

As with $\mathbb{Z}_5$, it is clear that $[1]$ is a generator for $\mathbb{Z}_6$ so $(\mathbb{Z}_6, +_6)$ is a cyclic group.

In this case, however, the congruence class $[2]$ is not a generator. To see this note that:

$$\begin{aligned}1.[2] &= [2]\\ 2.[2] &= [2] +_6 [2] = [4]\\ 3.[2] &= [4] +_6 [2] = [0].\end{aligned}$$

Hence $[2]$ has order 3 and generates the subgroup $\{[0], [2], [4]\}$ of $\mathbb{Z}_6$.

The general situation for addition modulo n is captured by the following theorem.

Theorem 9.17

Let $n > 1$ be a fixed integer. Then the set of congruence classes modulo n,

$$\mathbb{Z}_n = \{[0], [1], [2], \ldots, [n - 1]\},$$

forms a cyclic group under addition modulo n, $+_n$, with generator $[1]$.

Multiplication Modulo n

Suppose we wish to 'build' a group S whose elements are congruence classes using multiplication modulo n (for some positive integer n). Clearly $[1]$ is the identity element under multiplication modulo n since, for any congruence class $[k]$,

$$[1] \times_n [k] = [1.k] = [k].$$

It is also clear that S cannot contain the congruence class $[0]$. This is because, for all congruence classes $[k]$, we have

$$[0] \times_n [k] = [0]$$

so $[0]$ does not have an inverse under multiplication modulo n. (Any inverse under multiplication would need to satisfy $[k] \times_n [0] = [1]$.) Thus we will always exclude $[0]$ from S and only consider the set of non-zero congruence classes, denoted

$$\mathbb{Z}_n^* = \{[1], [2], \ldots, [n-1]\}.$$

We begin by exploring some examples.

Examples 9.18

1. The Cayley table for $\mathbb{Z}_5^* = \{[1], [2], [3], [4]\}$ under multiplication modulo 5 is the given below.

$\times_5$	$[1]$	$[2]$	$[3]$	$[4]$
$[1]$	$[1]$	$[2]$	$[3]$	$[4]$
$[2]$	$[2]$	$[4]$	$[1]$	$[3]$
$[3]$	$[3]$	$[1]$	$[4]$	$[2]$
$[4]$	$[4]$	$[3]$	$[2]$	$[1]$

It is clear that $\mathbb{Z}_5^* = \{[1], [2], [3], [4]\}$ is a group:

- $\times_5$ is a binary operation on $\mathbb{Z}_5^*$ since only elements of $\mathbb{Z}_5^*$ appear in the table.
- $\times_5$ is associative since ordinary multiplication of integers is associative.
- $[1]$ is the identity element (as noted above).
- Each $[n] \in \mathbb{Z}_5^*$ has an inverse:

$$\begin{aligned}
[1]^{-1} &= [1] \quad \text{since} \quad [1] \times_6 [1] = [1] \\
[2]^{-1} &= [3] \quad \text{since} \quad [2] \times_6 [3] = [1] \\
[3]^{-1} &= [2] \quad \text{since} \quad [3] \times_6 [2] = [1] \\
[4]^{-1} &= [4] \quad \text{since} \quad [4] \times_6 [4] = [1].
\end{aligned}$$

2. We now consider $\mathbb{Z}_6^* = \{[1], [2], [3], [4], [5]\}$ under multiplication modulo 6. The Cayley table is given below.

$\times_5$	[1]	[2]	[3]	[4]	[5]
[1]	[1]	[2]	[3]	[4]	[5]
[2]	[2]	[4]	[0]	[2]	[4]
[3]	[3]	[0]	[3]	[0]	[3]
[4]	[4]	[2]	[0]	[4]	[2]
[5]	[5]	[4]	[3]	[2]	[1]

In this case, the operation $\times_6$ is *not* a binary operation on the set $\mathbb{Z}_6^*$. For example, $[2], [3] \in \mathbb{Z}_6^*$ but $[2] \times_6 [3] = [0]$ which is not an element of $\mathbb{Z}_6^*$. Similarly, $[3], [4] \in \mathbb{Z}_6^*$ but $[3] \times_6 [4] = [0] \notin \mathbb{Z}_6^*$. There are three 'problem elements': $[2], [3]$ and $[4]$. Suppose we remove these three elements from $\mathbb{Z}_6^*$. This gives the set which we denote

$$U(6) = \mathbb{Z}_6^* - \{[2], [3], [4]\} = \{[0], [5]\}.$$

The Cayley table for $(U(6), \times_6)$ is the following.

$\times_5$	[1]	[5]
[1]	[1]	[5]
[5]	[5]	[1]

Although this may appear somewhat trivial, at least multiplication modulo 6 is a binary operation on $U(6)$ since only elements of $U(6)$ appear in the table. Furthermore $(U(6), \times_6)$ is a group as can be readily verified.

Note that the 'problem elements' that were removed—$[2], [3]$ and $[4]$ in this case—are those congruence classes $[k]$ where $\gcd(k, 6) > 1$. For example, $\gcd(4, 6) = 2 > 1$. Equivalently, the elements that are retained in $U(6)$ are the classes $[k]$ where k and the modulus 6 are coprime:

$$U(6) = \{[k] \in \mathbb{Z}_6^* : \gcd(k, 6) = 1\}.$$

3. To explore whether the patterns identified in the previous two examples hold more generally, we now consider multiplication modulo 12. The

Cayley table for $\mathbb{Z}_{12}^* = \{[1], [2], , 3], \ldots, [11]\}$ is given below.

$\times_{12}$	[1]	[2]	[3]	[4]	[5]	[6]	[7]	[8]	[9]	[10]	[11]
[1]	[1]	[2]	[3]	[4]	[5]	[6]	[7]	[8]	[9]	[10]	[11]
[2]	[2]	[4]	[6]	[8]	[10]	[0]	[2]	[4]	[6]	[8]	[10]
[3]	[3]	[6]	[9]	[0]	[3]	[6]	[9]	[0]	[3]	[6]	[9]
[4]	[4]	[8]	[0]	[4]	[8]	[0]	[4]	[8]	[0]	[4]	[8]
[5]	[5]	[10]	[3]	[8]	[1]	[6]	[11]	[4]	[9]	[2]	[7]
[6]	[6]	[0]	[6]	[0]	[6]	[0]	[6]	[0]	[6]	[0]	[6]
[7]	[7]	[2]	[9]	[4]	[11]	[6]	[1]	[8]	[3]	[10]	[5]
[8]	[8]	[4]	[0]	[8]	[4]	[0]	[8]	[4]	[0]	[8]	[4]
[9]	[9]	[6]	[3]	[0]	[9]	[6]	[3]	[0]	[9]	[6]	[3]
[10]	[10]	[8]	[6]	[4]	[2]	[0]	[10]	[8]	[6]	[4]	[2]
[11]	[11]	[10]	[9]	[8]	[7]	[6]	[5]	[4]	[3]	[2]	[1]

The 'problem elements' are again those where $[0]$ appears in the table as this indicates where $\times_{12}$ fails to be a binary operation. For example, $[9], [4] \in \mathbb{Z}_{12}^*$ but $[9] \times_{12} [4] = [0] \notin \mathbb{Z}_{12}^*$. The problem elements are $[2], [3], [4], [6], [8], [9]$ and $[10]$ which are those congruence classes $[k]$ where $\gcd(k, 12) > 1$.

Let $U(12)$ be the set of congruence classes $[k]$ where k is coprime with the modulus 12,

$$U(12) = \{[k] \in \mathbb{Z}_{12}^* : \gcd(k, 12) = 1\} = \{[1], [5], [7], [11]\}.$$

The Cayley table for $(U(12), \times_{12})$ is given below

$\times_{12}$	[1]	[5]	[7]	[11]
[1]	[1]	[5]	[7]	[11]
[5]	[5]	[1]	[11]	[7]
[7]	[7]	[11]	[1]	[5]
[11]	[11]	[7]	[5]	[1]

Clearly $(U(12), \times_{12})$ is a group in which each element is self-inverse.

The following theorem summarises the situation for multiplication modulo n. This theorem gives a solution to the questions posed in exercise 8.2.4 and exercise 8.3.13.

Theorem 9.18

Let $n > 1$ be a fixed integer. Let

$$U(n) = \{[k] \in \mathbb{Z}_n^* : \gcd(k, n) = 1\}$$

be the set of non-zero congruence classes $[k]$ modulo n where k and n are coprime. Then $(U(n), \times_n)$ is a group.
In particular, if p is prime, then

$$\mathbb{Z}_p^* = \{[1], [2], \ldots, [p-1]\}$$

is a group under multiplication modulo p.

9.5 Public Key Cryptography

Cryptography, or **encryption**, is the process of converting a message, or **plaintext**, into an unintelligible sequence of characters, or **ciphertext**, which would be meaningless if intercepted by a third party. The first use of cryptography dates back over two millennia and, over the centuries, many methods, simple and complex, mechanical and mathematical, have been used to devise and to attempt to break encryption schemes. In this chapter, we shall describe one of the more recent encryption schemes, based on simple number theory, that is now in widespread use.

Although the terms such as 'code making' and 'code breaking' are in common everyday use for converting messages into ciphertext and *vice versa*, encryption has quite a different purpose from the codes introduced in the previous chapter. There the emphasis was on introducing redundancy into the encoded words so that errors could be detected and, if possible, corrected. In the present case the emphasis is ensuring that the message words are sufficiently scrambled so that their meaning is hidden from an 'enemy' who may intercept the message.

Suppose Alice wants to send a message M to Bob which she wishes to keep secret from Eve who is eavesdropping†. Alice encrypts the message to give an encrypted message M' which she transmits. When Bob receives M', he decrypts it to obtain the original message M. The general scheme is represented in figure 9.2.

† It is quite common that the protagonists in this scenario are called Alice, Bob and Eve respectively.

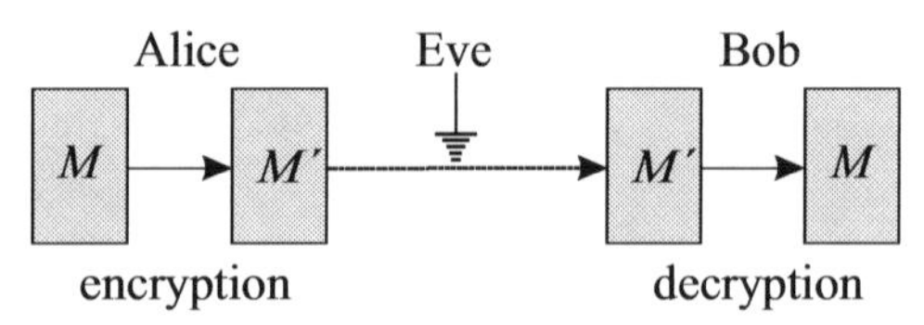

Figure 9.2 Encrypting and decrypting a message

Example 9.19

Suppose Alice encrypts her message by shifting each letter three places along the alphabet. This encryption method is often known as a 'Caesar cypher' after Julius Caesar who is reported to have used it to send messages which had military significance.

Thus, for example, the plaintext message `'send help'` encrypts to the ciphertext `'vhqg khos'`. To decrypt the message Bob needs to shift each letter back three places (or, equivalently, shift along 23 places). So Bob can decrypt the encrypted message `'vhqg khos'` to obtain the original message `'send help'`.

Although very straightforward, the previous example illustrates some general points. To encrypt a message, Alice uses an *encryption algorithm* (here 'shift along n places') and an *encryption key* (here $n = 3$) which is a parameter that specifies how the algorithm operates. To decrypt the message that he receives, Bob also uses a *decryption algorithm* (say, 'shift along m places') and a *decryption key* (here $m = 23$).

The disadvantage with this cypher (apart from being very easy to 'crack'!) is that Bob needs to know Alice's encryption key $n = 3$ to obtain his decryption key $m = 26 - n = 23$. Furthermore, knowing the encryption key allows the decryption key to be calculated. Thus if Eve obtains the encryption key, she can decrypt the encrypted message M'. This means that Alice and Bob have to agree in advance what key they are going to use and then ensure that this information is kept secret.

Until the advent of public key encryption systems, this was a key difficulty (excuse the pun) of cryptography: the sender and receiver needed to agree, in advance, on the encryption key and this information needed to be kept secret. Agreeing or communicating the key could provide a significant security challenge in its own right before the secure communication of any messages could be attempted.

The idea behind public key cryptography is that the *encryption* key can be made public but the *decryption* key is kept secret. For this to be effective, of course, it must be the case that knowing the encryption key does *not* allow the decryption key to be calculated. In this scheme, if Alice wishes to send a message to Bob, she encrypts it using Bob's public key that he publishes to the world. Only Bob, who knows the secret decryption key, can decrypt the message. This solves the problem of Alice and Bob having to agree a key in advance. This scheme is illustrated in figure 9.3.

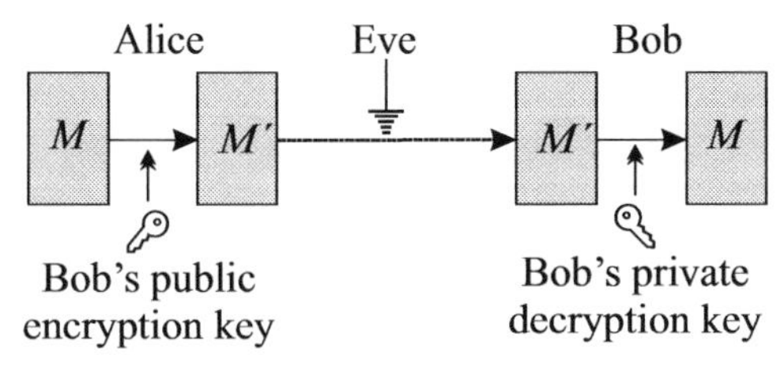

Figure 9.3 Encrypting and decrypting a message using public and private keys

We shall describe a public key encryption system based on number theory, but first we need a little theory.

Euler's Phi Function

Definition 9.6

Euler's phi function $\phi : \mathbb{Z}^+ \to \mathbb{Z}^+$ is defined by

$$\begin{aligned} \phi(n) &= \text{number of integers } a \text{ where } 1 \leqslant a \leqslant n \text{ which are coprime to } n. \\ &= \left|\{a \in \mathbb{Z}^+ : 1 \leqslant a \leqslant n \text{ and } \gcd(a, n) = 1\}\right|. \end{aligned}$$

Example 9.20

$\phi(6) = 2$ since 6 is coprime only with 1 and 5 in the range $1 \leqslant a \leqslant 6$.

$\phi(7) = 6$ since 7 is coprime with every integer in the range $1 \leqslant a < 7$.

$\phi(8) = 4$ since 8 is coprime only with $1, 3, 5$ and 7 in the range $1 \leqslant a \leqslant 8$.

Generalising the result $\phi(7) = 6$ above, if p is prime then every integer in the range $1 \leqslant a < p$ is coprime with p. This gives the following theorem.

Theorem 9.19

If p is prime then $\phi(p) = p - 1$.

In theorem 9.18 in the previous section, we described the group $U(n)$ under multiplication modulo n. Since $U(n)$ is the set of non-zero congruence classes $[k]$ modulo n where k and n are coprime, we have $|U(n)| = \phi(n)$. Using the properties of groups we can establish the following theorem.

Theorem 9.20: Euler's Theorem

If a and n are coprime then $a^{\phi(n)} \equiv 1 \mod n$.

Proof

From exercise 8.4.14 we know that if G is a group of order N, $|G| = N$, and $g \in G$ then $g^N = e$.

We apply this result to $U(n)$ where $|U(n)| = \phi(n)$. If a is coprime with n then the congruence class $[a]$ belongs to $U(n)$. Hence $[a]^{\phi(n)} = [1]$, the identity element of $U(n)$, by the result above. In other words

$$a^{\phi(n)} \equiv 1 \mod n,$$

as required. □

An immediate consequence of theorems 9.19 and 9.20 is the following.

Theorem 9.21: Fermat's little theorem

If p is prime then $a^{p-1} \equiv 1 \mod p$ for all $1 \leqslant a \leqslant p - 1$.

The last theoretical result we need to describe our encryption–decryption system is the following.

Theorem 9.22

If m and n are coprime then $\phi(mn) = \phi(m)\phi(n)$.

Proof

If $m = 1$ or $n = 1$ the result is trivial since $\phi(1) = 1$, so suppose $m > 1$ and $n > 1$.

First arrange the mn integers $1, 2, \ldots, mn$ into an array with n rows and m columns as follows.

$$\begin{array}{ccccc} 1 & 2 & 3 & \ldots & m \\ m+1 & m+2 & m+3 & \ldots & 2m \\ \ldots & \ldots & \ldots & & \ldots \\ (n-1)m+1 & (n-1)m+2 & (n-1)m+3 & \ldots & nm \end{array}$$

Now $\phi(mn)$ is the number of integers a in this array such that $\gcd(a, mn) = 1$. Since m and n are coprime, $\gcd(a, mn) = 1$ if and only if a and m are coprime and a and n are coprime. Thus we need to count the number of integers in the array that are coprime with *both* m and n.

In any column, all of the integers are congruent modulo m. Therefore $\phi(m)$ of the columns contain only integers coprime with m and the remaining $m - \phi(m)$ columns contain only integers a with $\gcd(a, m) > 1$.

Any column of integers coprime with m comprises integers of the form

$$c,\ m + c,\ 2m + c,\ \ldots,\ (n-1)m + c.$$

Since m and n are coprime, these are all different $\mod n$. Therefore such a column contains $\phi(n)$ integers that are coprime with n. Therefore there are $\phi(m)\phi(n)$ integers in the array that are coprime with *both* m and n, so $\phi(mn) = \phi(m)\phi(n)$. □

Example 9.21

Calculate $\phi(56)$.

Note that $56 = 7 \times 8$ where 7 and 8 are coprime. Also $\phi(7) = 6$ (since 7 is prime) and $\phi(8) = 4$ (example 9.20 above). Therefore $\phi(56) = \phi(7)\phi(8) = 6 \times 4 = 24$.

This is clearly simpler than calculating $\phi(56)$ directly. In fact, the 24 integers that are coprime with 56 are:

$1, 3, 5, 9, 11, 13, 15, 17, 19, 23, 25, 27, 29, 31, 33, 37, 39, 41, 43, 45, 47, 51, 53, 55.$

RSA Encryption

RSA† encryption is based on the fact that factorising very large numbers is so time consuming that, for sufficiently large integers, it is practically impossible even using the fastest supercomputers.

The process is as follows. Bob chooses two very large prime numbers p and q (say, 80 digits or more) but keeps these private. Let $n = pq$. Then n is a very large number; it will be made public but, given the practical impossibility in factorising very large numbers, its factors p and q remain private to Bob.

Now

$$\phi(n) = (p-1)(q-1)$$

by theorems 9.19 and 9.22. Bob can use this result to evaluate $\phi(n)$ easily since he knows the factorisation $n = pq$. However, without knowing this factorisation, it will not be possible for anyone (Eve, for example) to evaluate $\phi(n)$.

To illustrate the point but with relatively small numbers, suppose we are required to evaluate $\phi(640\,817)$. Without any further information, this seems very daunting—we need to find how many integers a in the range $1 \leqslant a \leqslant 640\,816$ are coprime with $640\,817$. However, if we are Bob, we constructed $n = 640\,817$ in the first place so we will know that n factorises as

$$n = 773 \times 829$$

which are both prime. Hence we are able to evaluate $\phi(640\,817)$ easily:

$$\phi(640\,817) = 772 \times 828 = 639\,216.$$

Bob then chooses e such that e and $\phi(n)$ are coprime. Bob also calculates d such that

$$de \equiv 1 \mod \phi(n)$$

† This is named after the three people credited with discovering the technique in 1977, Ron **R**ivest, Adi **S**hamir, Leonard **A**dleman. However it is now believed that this scheme was known earlier to cryptographers at the UK Government Communications Headquarters, GCHQ.

but keeps d private. It will not be possible for anyone else to calculate d since, as we have seen, it is not practically possible to find $\phi(n)$ without knowing the factorisation $n = pq$ which only Bob knows.

Bob's public encryption key is the pair (n, e).

Encoding

Suppose Alice has a message M which is an integer in the range $0 \leqslant M < n$. In practice letters are represented as numerical values, for example using their ASCII code. Hence any text message may be split into a sequence of integers so it is sufficient to be able to encrypt integer values. Then M is encoded as M' where

$$M' \equiv M^e \mod n.$$

Note that the encoding only involves the *public* key (e, n). The message M' is transmitted.

Decoding

Bob receives M'. Note that, since $de \equiv 1 \mod \phi(n)$, we have $de = k\phi(n) + 1$ for some $k \in \mathbb{Z}$. Now

$$\begin{aligned}
(M')^d &= (M)^{ed} \\
&= (M)^{k\phi(n)+1} \\
&= ((M)^{\phi(n)})^k \times M \\
&\equiv 1^k \times M \mod n \qquad \text{(by theorem 9.20)} \\
&\equiv M \mod n.
\end{aligned}$$

Therefore, Bob calculates $(M')^d \mod N$ to recover the message M. Bob's private decryption key is the integer d.

Summary

Encoding rule:	$M' \equiv M^e \mod n$ where e and $\phi(n)$ are coprime; n and e are made public.
Decoding rule:	$M \equiv (M')^d \mod n$ where d satisfies $de \equiv 1 \mod \phi(n)$; d and $\phi(n)$ are kept private.

Examples 9.22

1. Our first example is a 'toy example' with small primes just to illustrate the process of encryption and decryption. Let $p = 7$ and $q = 13$ so that $n = pq = 91$ and $\phi(n) = 6 \times 12 = 72$.

Firstly we choose an encryption power $e = 5$, which is coprime with $\phi(n)$, so that the encryption rule is

$$M' \equiv M^5 \mod 143.$$

Since $\phi(n) = 72$, the decryption power d satisfies

$$5d \equiv 1 \mod 72.$$

Since

$$5 \times 14 = 70 \equiv -2 \mod 72$$

and

$$5 \times 15 = 75 \equiv 3 \mod 72,$$

adding the two equations gives

$$5 \times 29 \equiv 1 \mod 72.$$

Hence the decryption power is $d = 29$ and the decryption rule is

$$M = (M')^{29} \mod 91.$$

Let $M = 31$. We will encrypt M to obtain M' and then decrypt M' to obtain the original message M. Encryption is straightforward:

$$M' = 31^5 = 28\,629\,151 \equiv 5 \mod 91.$$

To decrypt the encrypted message $M' = 5$ we need to evaluate 5^{29} mod 91. We will do this in stages, as follows.

$$\begin{aligned}
& (M')^3 &=& \ 125 \\
& &\equiv& \ 34 \mod 91 \\
\Rightarrow\ & (M')^6 &=& \ 34^2 = 1156 \\
& &\equiv& \ 64 \mod 91 \\
\Rightarrow\ & (M')^{12} &=& \ 64^2 = 4096 \\
& &\equiv& \ 1 \mod 91 \\
\Rightarrow\ & (M')^{24} &\equiv& \ 1 \mod 91 \\
\Rightarrow\ & (M')^{29} &\equiv& \ 5^5 = 5^3 \times 5^2 \\
& &=& \ 34 \times 25 = 850 \\
& &\equiv& \ 31 \mod 31.
\end{aligned}$$

Hence we have retrieved the original message $M = 31$.

Suppose instead we had tried the encryption rule $M' \equiv M^{15} \mod 143$. Then the decryption power d satisfies

$$15d \equiv 1 \mod 72.$$

Since $\gcd(15, 72) = 3$ and 3 does not divide 1, this congruence has no solution by theorem 9.14. This illustrates the necessity of selecting the encryption power e that is coprime with $\phi(n)$.

2. Let $p = 89$ and $q = 97$. These are still not very large primes but will illustrate the process in perhaps a slightly more realistic manner. Then $n = pq = 89 \times 97 = 8633$.

Now $\phi(n) = (p-1)(q-1) = 88 \times 96 = 8448$. We need an integer e coprime to 8448. Choose $e = 7$, say. The public encryption key is $(n, e) = (8633, 7)$.

To find the private decryption key d, we need to solve $de \equiv 1 \mod \phi(n)$; in other words $7d \equiv 1 \mod 8448$. Now $7 \times 1207 = 8449 \equiv 1 \mod 8448$, so $d = 1207$.

Therefore we have the following (public) encryption and (private) decryption scheme.

$$\text{Encryption: } M' \equiv M^7 \mod 8633.$$

$$\text{Decryption: } M \equiv (M')^{1207} \mod 8633.$$

For example, suppose a message is numerically represented as $M = 359$. This encodes to

$$359^7 = 768\,530\,557\,342\,240\,919 \equiv 1030 \mod 8633.$$

To decode we need to calculate $1030^{1207} \mod 8633$. This may seem difficult but this can be calculated in stages. For example, one way of doing this is the following.

$$\begin{array}{rrll}
 & 1030^4 & \equiv 4583 & \mod 8633 \\
\Rightarrow & 1030^{12} & = (1030^4)^3 \equiv 4583^3 \equiv 7131 & \mod 8633 \\
\Rightarrow & 1030^{60} & = (1030^{12})^5 \equiv 7131^5 \equiv 5479 & \mod 8633 \\
\Rightarrow & 1030^{300} & = (1030^{60})^5 \equiv 5479^5 \equiv 5870 & \mod 8633 \\
\Rightarrow & 1030^{1200} & = (1030^{300})^4 \equiv 5870^4 \equiv 2036 & \mod 8633 \\
\Rightarrow & 1030^{1207} & = 1030^{1200} \times 1030^7 & \\
 & & \equiv 2036 \times 3564 \equiv 359 & \mod 8633.
\end{array}$$

Hence $1030^{1207} \equiv 359 \mod 8633$ which decodes the message sent.

3. This time we will be a little more ambitious and chose the two primes to be $p = 199$ and $q = 211$. Hence $n = pq = 199 \times 211 = 41\,989$. Then

$$\phi(n) = (p-1)(q-1) = 198 \times 210 = 41\,580.$$

Finally we choose the encryption power to be $e = 13$ so that the encryption rule is

$$M' = M^{13} \mod 41\,989.$$

Firstly, we will encrypt the message $M = 2171$. We do this, firstly, by repeated squaring as follows.

$$\begin{aligned} M^2 &= 2171^2 = 4\,713\,241 \\ &\equiv 10\,473 \mod 41\,989 \\ \Rightarrow \qquad M^4 &= 10\,473^2 = 109\,683\,729 \\ &\equiv 8461 \mod 41\,989 \\ \Rightarrow \qquad M^8 &= 8461^2 = 71\,588\,521 \\ &\equiv 39\,265 \mod 41\,989. \end{aligned}$$

Therefore

$$\begin{aligned} M^{13} &= M^8 \times M^4 \times M \\ &= 39\,265 \times 8461 \times 2171 \\ &= 721\,252\,149\,215 \\ &\equiv 74 \mod 41\,989. \end{aligned}$$

Therefore $M = 2171$ encrypts to $M' = 74$.

Assuming that we are the receiver of the encrypted message, we shall now decrypt M' to retrieve the original message M. First we need to find the decryption power d that satisfies $ed \equiv 1 \mod \phi(n)$. In our case this is

$$13d \equiv 1 \mod 41\,580.$$

Now

$$3198 \times 13 = 41\,574 \equiv -6 \mod 41\,580$$

so

$$3199 \times 13 = 41\,574 \equiv 7 \mod 41\,580.$$

Hence

$$(3198 + 3199) \times 13 = 6397 \times 13 \equiv= 1 \mod 41\,580.$$

Therefore $d = 6397$ so the decryption rule is

$$M = (M')^{6397} \mod 41\,989.$$

We therefore need to evaluate

$$74^{6397} \mod 41\,989$$

which seems extremely daunting but, with persistence, can be achieved by repeatedly squaring as follows.

$$\begin{aligned}
& (M')^2 &=& \ 74^2 = 5476 \\
\Rightarrow\ & (M')^4 &=& \ 5476^2 = 29\,986\,576 \\
& &\equiv& \ 6430 \quad \text{mod } 41\,989 \\
\Rightarrow\ & (M')^8 &=& \ 6430^2 = 41\,344\,900 \\
& &\equiv& \ 27\,724 \quad \text{mod } 41\,989 \\
\Rightarrow\ & (M')^{16} &=& \ 27\,724^2 = 768\,620\,176 \\
& &\equiv& \ 11\,531 \quad \text{mod } 41\,989 \\
\Rightarrow\ & (M')^{32} &=& \ 11\,531^2 = 132\,963\,961 \\
& &\equiv& \ 26\,787 \quad \text{mod } 41\,989 \\
\Rightarrow\ & (M')^{64} &=& \ 26\,787^2 = 717\,543\,369 \\
& &\equiv& \ 35\,337 \quad \text{mod } 41\,989 \\
\Rightarrow\ & (M')^{128} &=& \ 35\,337^2 = 1248\,703\,569 \\
& &\equiv& \ 34\,687 \quad \text{mod } 41\,989 \\
\Rightarrow\ & (M')^{256} &=& \ 34\,687^2 = 1203\,187\,969 \\
& &\equiv& \ 35\,163 \quad \text{mod } 41\,989 \\
\Rightarrow\ & (M')^{512} &=& \ 35\,163^2 = 1236\,436\,569 \\
& &\equiv& \ 28\,475 \quad \text{mod } 41\,989 \\
\Rightarrow\ & (M')^{1024} &=& \ 28\,475^2 = 810\,825\,625 \\
& &\equiv& \ 18\,035 \quad \text{mod } 41\,989 \\
\Rightarrow\ & (M')^{2048} &=& \ 18\,035^2 = 325\,261\,225 \\
& &\equiv& \ 14\,431 \quad \text{mod } 41\,989 \\
\Rightarrow\ & (M')^{4096} &=& \ 14\,431^2 = 208\,253\,761 \\
& &\equiv& \ 30\,310 \quad \text{mod } 41\,989.
\end{aligned}$$

Now we need to write the decryption power 6397 as a sum of powers of 2; essentially, we need to write 6397 as a binary numeral. We have

$$6397 = 4096 + 2048 + 128 + 64 + 32 + 16 + 8 + 4 + 1$$

so that

$$\begin{aligned}
(M')^{6397} = &\ (M')^{4096} \times (M')^{2048} \times (M')^{128} \times (M')^{64} \\
&\times (M')^{32} \times (M')^{16} \times (M')^{8} \times (M')^{4} \times M'.
\end{aligned}$$

Again we need to evaluate this in stages, as follows.

$$\begin{aligned}
(M')^{6144} &= (M')^{4096} \times (M')^{2048} \\
&= 30\,310 \times 14\,431 = 43\,7403\,610 \\
&\equiv 4197 \mod 41\,989 \\
\Rightarrow \quad (M')^{6272} &= (M')^{6144} \times (M')^{128} \\
&= 4197 \times 34\,687 = 145\,581\,339 \\
&\equiv 5476 \mod 41\,989 \\
\Rightarrow \quad (M')^{6336} &= (M')^{6272} \times (M')^{64} \\
&= 5476 \times 35\,337 = 193\,505\,412 \\
&\equiv 20\,100 \mod 41\,989 \\
\Rightarrow \quad (M')^{6368} &= (M')^{6336} \times (M')^{32} \\
&= 20\,100 \times 26\,787 = 538\,418\,700 \\
&\equiv 35\,742 \mod 41\,989 \\
\Rightarrow \quad (M')^{6384} &= (M')^{6368} \times (M')^{16} \\
&= 35\,742 \times 11\,531 = 412\,141\,002 \\
&\equiv 18\,967 \mod 41\,989 \\
\Rightarrow \quad (M')^{6392} &= (M')^{6384} \times (M')^{8} \\
&= 18\,967 \times 27\,724 = 525\,841\,108 \\
&\equiv 12\,861 \mod 41\,989 \\
\Rightarrow \quad (M')^{6397} &= (M')^{6392} \times (M')^{4} \times M' \\
&= 12\,861 \times 6430 \times 74 = 6119\,521\,020 \\
&\equiv 2171 \mod 41\,989.
\end{aligned}$$

Hence we have successfully decrypted $M' = 94$ to obtain the original message $M = 2171$.

Exercises 9.4

1. Evaluate each of the following, where ϕ is Euler's phi function.

(i) $\phi(24)$

(ii) $\phi(30)$

(iii) $\phi(31)$

(iv) $\phi(32)$

(v) $\phi(170)$

(vi) $\phi(195)$

(vii) $\phi(323)$

(viii) $\phi(385)$

2. (i) Evaluate $\phi(9)$, $\phi(25)$ and $\phi(49)$.

   (ii) Conjecture the value of $\phi(p^2)$ where p is prime. Prove your conjecture.

3. An RSA code is designed with $p = 11$ and $q = 19$ (so that $n = pq = 209$).

   The encryption rule for a message M, is $M' \equiv M^7 \mod 209$, where $1 \leqslant M \leqslant 209$.

   (i) Encrypt each of the following 'messages'.

   (a) $M = 7$

   (b) $M = 14$

   (c) $M = 195$

   (ii) Find the integer d (in the range $1 \leqslant d \leqslant \phi(n)$) such that the decryption rule is

$$M \equiv (M')^d \mod 209.$$

   Hence, or otherwise, decrypt each of the following 'messages'.

   (a) $M' = 3$

   (b) $M' = 174$

   (c) $M' = 20$

4. For each of the following RSA encryption schemes, $M \longrightarrow M'$, find the decryption scheme; that is, find d so that $M = (M')^d \mod n$.

   (i) $M' = M^{13} \mod 55$

   (ii) $M' = M^{17} \mod 57$

   (iii) $M' = M^7 \mod 143$

   (iv) $M' = M^{11} \mod 247$

Chapter 10

Boolean Algebra

10.1 Introduction

In chapter 3 we noted the strong similarity between the algebra of sets and that of propositions. In particular, each of the laws listed in §3.5 has a counterpart in §1.5 to which it bears more than a passing resemblance. For example, De Morgan's laws for the propositions p and q are given by $\overline{p \vee q} \equiv \bar{p} \wedge \bar{q}$ and $\overline{p \wedge q} \equiv \bar{p} \vee \bar{q}$. For the sets A and B these laws take the form $(\overline{A \cup B}) = \bar{A} \cap \bar{B}$ and $(\overline{A \cap B}) = \bar{A} \cup \bar{B}$. In this chapter we shall see that the laws common to these two systems are attributable to their relationship to an algebraic structure known as a 'Boolean algebra' and that the properties which they share are those which are common to all Boolean algebras.

The idea of a Boolean algebra was first developed by George Boole† in the middle of the nineteenth century. Boole was chiefly concerned with the algebra of propositions but, in recent years, the subject has been extended and is now a significant component of abstract algebra. An important application of Boolean algebra is in the analysis of electronic circuits and hence in the design of a range of digital devices such as computers, telephone systems and electronic control systems.

† Boole's book *The Laws of Thought* published in 1854 was an attempt to formalize the process of logical thinking.

Definition 10.1

A **Boolean algebra** consists of a set B together with three operations defined on that set. These are:

(a) a binary operation denoted by $\oplus$ referred to as the **sum** (or **join**);
(b) a binary operation denoted by $*$ referred to as the **product** (or **meet**);
(c) an operation which acts on a single element of B, denoted by $^{-}$, where, for any element $b \in B$, the element $\bar{b} \in B$ is called the **complement** of b. (An operation which acts on a single member of a set S and which results in a member of S is called a **unary operation**.)

TWO DISTINCT IDENTITIES

The following axioms apply to the set B together with the operations $\oplus$, $*$ and $^{-}$.

B1. Distinct identity elements belonging to B exist for each of the binary operations $\oplus$ and $*$ and we denote these by $\mathbf{0}$ and $\mathbf{1}$ respectively. Thus we have

$$b \oplus \mathbf{0} = \mathbf{0} \oplus b = b$$
$$b * \mathbf{1} = \mathbf{1} * b = b$$

for all $b \in B$.

Definition 10.1 (continued)

B2. The operations $\oplus$ and $*$ are associative, that is

$$(a \oplus b) \oplus c = a \oplus (b \oplus c)$$
$$(a * b) * c = a * (b * c)$$

for all $a, b, c \in B$.

B3. The operations $\oplus$ and $*$ are commutative, that is

$$a \oplus b = b \oplus a$$
$$a * b = b * a$$

for all $a, b \in B$.

B4. The operation $\oplus$ is distributive over $*$ and the operation $*$ is distributive over $\oplus$, that is

$$a \oplus (b * c) = (a \oplus b) * (a \oplus c)$$
$$a * (b \oplus c) = (a * b) \oplus (a * c)$$

for all $a, b, c \in B$.

B5. For all $b \in B$, $b \oplus \bar{b} = \mathbf{1}$ and $b * \bar{b} = \mathbf{0}$.

A Boolean algebra with underlying set B, binary operations $\oplus$ and $*$, complement operation $^{-}$, and identity elements $\mathbf{0}$ and $\mathbf{1}$ is denoted by $(B, \oplus, *, ^{-}, \mathbf{0}, \mathbf{1})$.

There are a number of alternative notations for the sum and product operations. Some authors use $\vee$ and $\wedge$, others use $+$ and $\times$. However, all of these symbols tend to have connotations associated with their use for specific operations and we therefore prefer to use the more neutral symbols $\oplus$ and $*$ for general binary operations.

A casual glance at axiom B5 may lead you to conclude that $\bar{b}$ is the inverse of b. This is not so. Remember that, if b^{-1} is the inverse of b, then $b * b^{-1}$ gives the identity *with respect to the operation* $*$. However, $b \oplus \bar{b}$ gives the identity *with respect to* $*$ and $b * \bar{b}$ gives the identity *with respect to* $\oplus$, so that $\bar{b}$ is not the inverse of b with respect to either operation.

One final word of caution: note that **0** and **1** are used here as symbols for the two identity elements and not for the numbers which they conventionally symbolize. We must therefore be careful not to make assumptions which are true for the integers 0 and 1 but not necessarily so for identities in general.

Examples 10.1

1. The simplest Boolean algebra (and also the one of most interest to computer scientists, as we shall see later) consists of the set $B = \{0, 1\}$ together with the binary operations $\oplus$ and $*$ and complement operation $^{-}$ defined by the following tables.

$\oplus$	0	1
0	0	1
1	1	1

$*$	0	1
0	0	0
1	0	1

b	$\bar{b}$
0	1
1	0

We leave it as an exercise to verify that the axioms B1–B5 hold.

2. Let S be a non-empty set and consider $\mathscr{P}(S)$, the power set of S, together with the binary operations of union and intersection and the operation of complementation, where, for all $A \in \mathscr{P}(S)$, $\bar{A} = S - A$.

We established the following results in sections 3.5 and 3.6:

(a) the operations $\cup$ and $\cap$ are associative;
(b) the operations $\cup$ and $\cap$ are commutative;
(c) the operation $\cup$ is distributive over $\cap$ and $\cap$ is distributive over $\cup$;
(d) the sets $\varnothing$ and S belong to $\mathscr{P}(S)$ and

$$A \cup \varnothing = \varnothing \cup A = A$$
$$A \cap S = S \cap A = A$$

for all $A \in \mathscr{P}(S)$. Thus $\varnothing$ and S are the identities for $\cup$ and $\cap$ respectively;

(e) for any $A \in \mathscr{P}(S)$, $\bar{A} \in \mathscr{P}(S)$ and $A \cup \bar{A} = S$ and $A \cap \bar{A} = \varnothing$.

Since these are precisely the axioms B1–B5 we can conclude that $(\mathscr{P}(S), \cup, \cap, ^{-}, \varnothing, S)$ is a Boolean algebra. The sum and product operations are union and intersection respectively, and we can write $\mathbf{0} = \varnothing$ and $\mathbf{1} = S$ for the two identities.

In fact we can replace $\mathscr{P}(S)$ by any non-empty family of sets which is closed under the operations of union, intersection and complementation. The resulting structure is also a Boolean algebra.

3. Let B be a set of propositions which is closed under the operations of conjunction, disjunction and negation and where equality of propositions is interpreted as their logical equivalence.

In chapter 1 we showed that the operations $\vee$ and $\wedge$ are associative, commutative and that each is distributive over the other. If we denote a contradiction (a proposition which is always false) by f and a tautology (a proposition which is always true) by t then f and t must belong to B. This is so because, for any proposition p belonging to B, $p \wedge \bar{p} \equiv f$ and $p \vee \bar{p} \equiv t$ belong to B by the closure properties. Further, for any $p \in B$, we have

$$p \vee f \equiv f \vee p \equiv p$$
$$p \wedge t \equiv t \wedge p \equiv p$$

so that f and t are the identities for the binary operations $\vee$ and $\wedge$ respectively. All contradictions are logically equivalent as are all tautologies so that t and f are unique elements of B.

The structure $(B, \vee, \wedge, \bar{}, f, t)$ satisfies the axioms B1–B5 and is therefore a Boolean algebra. The operations $\vee$ and $\wedge$ correspond to $\oplus$ and $*$ respectively and for the identity elements we have $\mathbf{0} = f$ and $\mathbf{1} = t$.

10.2 Properties of Boolean Algebras

In chapters 1 and 3 we considered the duality principle as it applied to the algebras of propositions and of sets. We shall now see that this principle applies to all Boolean algebras.

Given any proposition about a Boolean algebra, we define its **dual** to be the proposition obtained by substituting $\oplus$ for $*$, $*$ for $\oplus$, $\mathbf{0}$ for $\mathbf{1}$ and $\mathbf{1}$ for $\mathbf{0}$.

For example, given the elements a, b of a Boolean algebra, the dual of

$$(a \oplus b) * a * \bar{b} = \mathbf{0}$$

is

$$(a * b) \oplus a \oplus \bar{b} = \mathbf{1}.$$

Each of the Boolean algebra axioms B1–B5 is actually a pair of axioms. Within a pair, each axiom is the dual of the other. Now suppose that, using the axioms, we can prove some theorem about a Boolean algebra. It follows that the dual of that theorem can also be proved by using, in the same sequence, the duals of the axioms used to prove the first. This is the essence of the duality principle. Every time we prove a theorem about a Boolean algebra, we can, by appealing to this principle, simply state that the theorem which is the dual also holds. In Boolean algebra duality gives us 'two theorems for the price of one'.

Duality Principle

For any theorem about a Boolean algebra, the dual is also a theorem.

We will now use the axioms and the duality principle to prove some theorems about the general Boolean algebra $(B, \oplus, *, \bar{\ }, \mathbf{0}, \mathbf{1})$. The first result is simply a restatement of theorem 8.1. This theorem says that, for any binary operation, if an identity element exists, then it is unique.

Theorem 10.1

The identity elements $\mathbf{0}$ and $\mathbf{1}$ are unique.

We now show that the complement of any element of a Boolean algebra (as defined in axiom B5) is also unique.

Theorem 10.2

Given an element $b \in B$, there is only one element $\bar{b} \in B$ such that $b \oplus \bar{b} = \mathbf{1}$ and $b * \bar{b} = \mathbf{0}$ (i.e. which satisfies axiom B5).

Proof

Suppose that $\bar{b}_1$ and $\bar{b}_2$ are both complements of an element b of a Boolean algebra $(B, \oplus, *, \bar{\ }, \mathbf{0}, \mathbf{1})$. This means that

$$b \oplus \bar{b}_1 = \bar{b}_1 \oplus b = \mathbf{1} \qquad b \oplus \bar{b}_2 = \bar{b}_2 \oplus b = \mathbf{1}$$
$$b * \bar{b}_1 = \bar{b}_1 * b = \mathbf{0} \qquad b * \bar{b}_2 = \bar{b}_2 * b = \mathbf{0}.$$

Thus we have

$$\begin{aligned}
\bar{b}_1 &= \bar{b}_1 * \mathbf{1} && \text{(axiom B1)} \\
&= \bar{b}_1 * (b \oplus \bar{b}_2) \\
&= (\bar{b}_1 * b) \oplus (\bar{b}_1 * \bar{b}_2) && \text{(axiom B4)} \\
&= \mathbf{0} \oplus (\bar{b}_1 * \bar{b}_2) \\
&= \mathbf{0} \oplus (\bar{b}_2 * \bar{b}_1) && \text{(axiom B3)} \\
&= (\bar{b}_2 * b) \oplus (\bar{b}_2 * \bar{b}_1) \\
&= \bar{b}_2 * (b \oplus \bar{b}_1) && \text{(axiom B4)} \\
&= \bar{b}_2 * \mathbf{1} \\
&= \bar{b}_2 && \text{(axiom B1).}
\end{aligned}$$

We have shown that $\bar{b}_1 = \bar{b}_2$ and so we can conclude that the complement is unique. □

As we have already noted, the laws common to the algebra of sets and propositions are examples of general results which apply to all Boolean algebras. In fact some authors prefer to include some or all of these in the list of Boolean algebra axioms. This is not necessary since, as we shall now see, all can be proved using only the axioms B1–B5.

Theorem 10.3 Idempotent laws

For all $b \in B$

$$b \oplus b = b \quad \text{and} \quad b * b = b.$$

Proof

For all $b \in B$ we have

$$b = b \oplus \mathbf{0} \qquad \text{(axiom B1)}$$

$$
\begin{aligned}
&= b \oplus (b * \bar{b}) && \text{(axiom B5)} \\
&= (b \oplus b) * (b \oplus \bar{b}) && \text{(axiom B4)} \\
&= (b \oplus b) * \mathbf{1} && \text{(axiom B5)} \\
&= b \oplus b && \text{(axiom B1).}
\end{aligned}
$$

We have proved that $b \oplus b = b$. The result $b * b = b$ follows by the duality principle. □

Theorem 10.4 Identity laws

For all $b \in B$

$$\mathbf{1} \oplus b = b \oplus \mathbf{1} = \mathbf{1} \quad \text{and} \quad \mathbf{0} * b = b * \mathbf{0} = \mathbf{0}.$$

Proof

The proof of this theorem is left as an exercise (see exercise 10.1.5). □

Theorem 10.5 Absorption laws

For all $b_1, b_2 \in B$

$$b_1 \oplus (b_1 * b_2) = b_1 \quad \text{and} \quad b_1 * (b_1 \oplus b_2) = b_1.$$

Proof

For all $b_1, b_2 \in B$

$$
\begin{aligned}
b_1 \oplus (b_1 * b_2) &= (b_1 * \mathbf{1}) \oplus (b_1 * b_2) && \text{(axiom B1)} \\
&= b_1 * (\mathbf{1} \oplus b_2) && \text{(axiom B4)} \\
&= b_1 * \mathbf{1} && \text{(theorem 10.4)} \\
&= b_1 && \text{(axiom B1).}
\end{aligned}
$$

By the duality principle, we have

$$b_1 * (b_1 \oplus b_2) = b_1.$$ □

Theorem 10.6 Involution law

For all $b \in B, \bar{\bar{b}} = b$.

Proof

Since $b \oplus \bar{b} = \bar{b} \oplus b = \mathbf{1}$ and $b * \bar{b} = \bar{b} * b = \mathbf{0}$, it follows that b is the complement of $\bar{b}$. We have already proved (theorem 10.2) that the complement of any element is unique so that $\bar{\bar{b}} = b$. □

Theorem 10.7 De Morgan's laws

For all $b_1, b_2 \in B$

$$(\overline{b_1 \oplus b_2}) = \bar{b}_1 * \bar{b}_2 \quad \text{and} \quad (\overline{b_1 * b_2}) = \bar{b}_1 \oplus \bar{b}_2.$$

Proof

$$\begin{aligned}
(b_1 \oplus b_2) \oplus (\bar{b}_1 * \bar{b}_2) &= [(b_1 \oplus b_2) \oplus \bar{b}_1] * [(b_1 \oplus b_2) \oplus \bar{b}_2] && \text{(axiom B4)} \\
&= [\bar{b}_1 \oplus (b_1 \oplus b_2)] * [(b_1 \oplus b_2) \oplus \bar{b}_2] && \text{(axiom B3)} \\
&= [(\bar{b}_1 \oplus b_1) \oplus b_2] * [b_1 \oplus (b_2 \oplus \bar{b}_2)] && \text{(axiom B2)} \\
&= (\mathbf{1} \oplus b_2) * (b_1 \oplus \mathbf{1}) && \text{(axiom B5)} \\
&= \mathbf{1} * \mathbf{1} && \text{(theorem 10.4)} \\
&= \mathbf{1} && \text{(axiom B1).}
\end{aligned}$$

We have proved that $(b_1 \oplus b_2) \oplus (\bar{b}_1 * \bar{b}_2) = \mathbf{1}$ so that $\bar{b}_1 * \bar{b}_2$ is the complement of $b_1 \oplus b_2$, i.e. $(\overline{b_1 \oplus b_2}) = \bar{b}_1 * \bar{b}_2$.

That $(\overline{b_1 * b_2}) = \bar{b}_1 \oplus \bar{b}_2$ follows from the duality principle. □

Theorem 10.8

$$\bar{\mathbf{0}} = \mathbf{1} \quad \text{and} \quad \bar{\mathbf{1}} = \mathbf{0}.$$

Proof

The proof is left as an exercise (see exercise 10.1.6). □

Exercises 10.1

1. Evaluate the following for the Boolean algebra $(\{0,1\}, \oplus, *, \bar{}, 0, 1)$ as defined in example 10.1.1.

   (i) $(0 \oplus 1) * 0$
   (ii) $0 * \bar{1}$
   (iii) $(1 * 1) \oplus (0 * \bar{0})$
   (iv) $\bar{1} \oplus [(0 * 1) * 1]$
   (v) $[(\overline{0 * 1}) * 1) * (\bar{1} \oplus 1)] \oplus 1$
   (vi) $[1 \oplus (\bar{1} * 1)] * (\bar{0} \oplus 0)$
   (vii) $[(1 * 1) \oplus \bar{0}] * \overline{[(1 \oplus 0) * 1]}$.

2. Consider the set of real numbers $\mathbb{R}$, together with the binary operations of addition and multiplication. Which of the Boolean algebra axioms B1, B2, B3, B4 are not satisfied? Is it possible to define a unary operation on $\mathbb{R}$ so that axiom B5 holds?

3. Let $B = \{1, 2, 3, 5, 6, 10, 15, 30\}$, i.e. the set of divisors of 30. Binary operations denoted by $\oplus$ and $*$ and a unary operation denoted by $\bar{}$ are defined as follows: for all $b_1, b_2 \in B$

$$b_1 \oplus b_2 = \text{the least common multiple of } b_1 \text{ and } b_2$$
$$b_1 * b_2 = \text{the highest common factor of } b_1 \text{ and } b_2$$
$$\bar{b}_1 = 30/b_1.$$

   What are the identity elements with respect to $\oplus$ and $*$? Show that B together with the three operations is a Boolean algebra.

4. Let B be the set of divisors of 24, so that $B = \{1, 2, 3, 4, 6, 8, 12, 24\}$. For all $b_1, b_2 \in B$ define

$$b_1 \oplus b_2 = \text{the least common multiple of } b_1 \text{ and } b_2$$
$$b_1 * b_2 = \text{the highest common factor of } b_1 \text{ and } b_2$$
$$\bar{b}_1 = 24/b_1.$$

   Is B together with these three operations a Boolean algebra?

Suppose B is the set of divisors of 42 with the operations $\oplus$, $*$ and $^{-}$ defined appropriately. Is B together with these operations a Boolean algebra? What about the set of divisors of 45?

5. Prove the identity laws (theorem 10.4), i.e. for all $b \in B$,

$$\mathbf{1} \oplus b = b \oplus \mathbf{1} = \mathbf{1} \quad \text{and} \quad \mathbf{0} * b = b * \mathbf{0} = \mathbf{0}.$$

6. Prove theorem 10.8, i.e. that $\bar{\mathbf{0}} = \mathbf{1}$ and $\bar{\mathbf{1}} = \mathbf{0}$.

7. Although the associativity properties are usually given as two of the axioms of a Boolean algebra, they can be proved from the other axioms. Show this. You may wish to use the absorption laws (theorem 10.5) but this is legitimate since they can be proved without utilizing axiom B2.

   (Hint: Show that, for any $b_1, b_2, b_3 \in B$ $[(b_1 \oplus b_2) \oplus b_3] * [b_1 \oplus (b_2 \oplus b_3)] = (b_1 \oplus b_2) \oplus b_3$ and also that $[(b_1 \oplus b_2) \oplus b_3] * [b_1 \oplus (b_2 \oplus b_3)] = b_1 \oplus (b_2 \oplus b_3)$.)

8. Let $(B, \oplus, *, ^{-}, \mathbf{0}, \mathbf{1})$ be a Boolean algebra. Prove the following results hold for all $b_1, b_2, b_3 \in B$ using the Boolean algebra axioms and any theorem given in §10.2. State the dual of each result:

   (i) $(b_1 \oplus b_2) * \bar{b}_1 * \bar{b}_2 = \mathbf{0}$
   (ii) $b_1 \oplus [\overline{(\bar{b}_2 \oplus b_1) * b_2}] = \mathbf{1}$
   (iii) $(b_1 \oplus b_2) * (\bar{b}_1 \oplus \bar{b}_2) = (b_1 * \bar{b}_2) \oplus (\bar{b}_1 * b_2)$
   (iv) $b_1 * (\bar{b}_1 \oplus b_2) = b_1 * b_2$
   (v) $(b_1 \oplus b_2 \oplus b_3) * (b_1 \oplus b_2) = b_1 \oplus b_2$
   (vi) $(b_1 \oplus b_2) * (b_1 \oplus b_2) = b_1 \oplus b_2$
   (vii) $b_1 \oplus [b_1 * (b_2 \oplus \mathbf{1})] = b_1$.

9. Prove that, in any Boolean algebra $(B, \oplus, *, ^{-}, \mathbf{0}, \mathbf{1})$, $b_1 * \bar{b}_2 = \mathbf{0}$ if and only if $b_1 * b_2 = b_1$.

10. Prove the following **cancellation law**.

    Let $(B, \oplus, *, ^{-}, \mathbf{0}, \mathbf{1})$ be a Boolean algebra and let $b_1, b_2, b_3 \in B$. If $b_1 * b_2 = b_1 * b_3$ and $\bar{b}_1 * b_2 = \bar{b}_1 * b_3$ then $b_2 = b_3$.

    Why are both conditions necessary rather than just one of them?

11. A relation R is defined on the underlying set B of a Boolean algebra $(B, \oplus, *, ^{-}, \mathbf{0}, \mathbf{1})$ as follows:

$$b_1 \,\mathsf{R}\, b_2 \quad \text{if and only if} \quad b_1 * b_2 = b_1.$$

(i) Show that R is a partial order on the set B. (Partial orders are defined in §4.5.)

(ii) Show that $b_1 * b_2 = b_1$ if and only if $b_1 \oplus b_2 = b_2$.

10.3 Boolean Functions

We are already familiar with the idea of a real variable, that is one whose range of possible values is the set of real numbers or some subset of it. The idea of a 'Boolean variable' is similar. It is a variable whose range of possible 'values' is the underlying set B of a Boolean algebra $(B, \oplus, *, ^{-}, \mathbf{0}, \mathbf{1})$.

Definitions 10.2

(a) Given a Boolean algebra $(B, \oplus, *, ^{-}, \mathbf{0}, \mathbf{1})$, a **Boolean variable** is a variable to which can be assigned elements of the set B.

(b) Given a Boolean variable x, the **complement** of x denoted by $\bar{x}$, is a variable which is such that $\bar{x} = \bar{b}$ whenever $x = b$ for any $b \in B$.

(c) A **literal** is a Boolean variable x or its complement $\bar{x}$.

A useful notation for distinguishing literals is to write x^1 for the variable x and x^0 for $\bar{x}$, the complement of x. The two literals associated with the variable x can then be defined by

$$x^e = \begin{cases} \bar{x} & \text{if } e = 0 \\ x & \text{if } e = 1. \end{cases}$$

Just as real variables can be combined to form algebraic expressions using such operations as addition, subtraction, multiplication, etc, so Boolean variables can be combined to form Boolean expressions. The operations appearing in Boolean expressions are of course the Boolean operations of sum, product and complement. A Boolean expression is defined recursively as follows.

Definition 10.3

Given a Boolean algebra $(B, \oplus, *, \bar{}, \mathbf{0}, \mathbf{1})$, the following are **Boolean expressions** in the n Boolean variables $x_1, x_2, \ldots, x_n$:

(a) the identity elements $\mathbf{0}$ and $\mathbf{1}$;
(b) the Boolean variables $x_1, x_2, \ldots, x_n$;
(c) $(X \oplus Y)$, $(X * Y)$ and $\bar{X}$, where X and Y are Boolean expressions.

The following are examples of Boolean expressions: $x_1 \oplus \bar{x}_2$, $(\overline{x_1 * x_2}) \oplus (x_2 * x_3)$, $(x_1 \oplus x_2) * \bar{x}_1$, $\mathbf{1} * x_1 * (\bar{x}_2 \oplus \mathbf{0})$.

Note that a Boolean expression in the n variables $x_1, x_2, \ldots, x_n$ may not necessarily contain all n of the variables.

From this point on we will adopt the common practice of omitting the symbol $*$ in Boolean expressions, although we shall continue to include the symbol $\oplus$. We shall write $x_1 x_2$ for $x_1 * x_2$, $x_1(x_2 \oplus x_3)$ for $x_1 * (x_2 \oplus x_3)$, ${x_1}^2$ for $x_1 * x_1$, etc. This is analogous to the convention we adopted in chapter 8 or to that of dropping the multiplication sign in algebraic expressions so that xy is interpreted as 'x multiplied by y'. As we shall see, Boolean expressions can be lengthy and omitting this symbol is a notational convenience which makes them easier to write and to read.

We shall also follow the convention of evaluating products before sums thereby rendering the use of certain brackets unnecessary. For example $x_1 x_2 \oplus x_3$ is taken to mean $(x_1 * x_2) \oplus x_3$; similarly, $x_1 x_2 \oplus x_3 x_1$ is interpreted as $(x_1 * x_2) \oplus (x_3 * x_1)$, and so on. Again there is the obvious analogy with evaluating algebraic expressions where the rule is that multiplication/division is performed before addition/subtraction. As always, terms which are enclosed in brackets are evaluated before any others.

In the notation which we shall now use the examples of Boolean expressions given above are written: $x_1 \oplus \bar{x}_2$, $\overline{x_1 x_2} \oplus x_2 x_3$, $(x_1 \oplus x_2)\bar{x}_1$, $\mathbf{1} x_1(\bar{x}_2 \oplus \mathbf{0})$.

We have already seen in §10.2 and in exercise 10.1.8 that often the same Boolean expression can be expressed in a number of different forms. This should come as no great surprise since it is a familiar feature of the algebra of real variables. If x and y are real variables $x^2 + 2xy$ and $(x + y)^2 - y^2$ represent equivalent expressions in that one can be derived from the other using the rules of elementary algebra. The situation is the same for Boolean expressions. If one expression can be derived from another using the 'rules' of Boolean algebra then the two expressions are equivalent. Equivalent algebraic expressions are such that their value is the same for any set of values of the variables. So it is with equivalent Boolean expressions. Substitution of the same elements for the variables gives the same result.

Definition 10.4

Two Boolean expressions are said to be **equivalent** (or **equal**) if one can be obtained from the other by a finite sequence of applications of the Boolean algebra axioms.

For instance, $x_1(\bar{x}_1 \oplus x_2)$ and $x_1 x_2$ are equivalent Boolean expressions (see exercise 10.1.8(iv)) and we can write $x_1(\bar{x}_1 \oplus x_2) = x_1 x_2$.

Given a Boolean algebra $(B, \oplus, *, ^{-}, \mathbf{0}, \mathbf{1})$, a Boolean expression can be used to define a function. As with functions of a real variable, the expression provides a 'rule' for evaluating the function for any element of its domain. For example, consider the Boolean expression $(x_1 \oplus x_2)\bar{x}_1$. This defines a function, f, of the two variables x_1 and x_2 as follows:

$$f(x_1, x_2) = (x_1 \oplus x_2)\bar{x}_1.$$

The domain of the function is $B \times B$ and its codomain is B.

The following is the formal definition of a Boolean function.

Definition 10.5

Given a Boolean algebra $(B, \oplus, *, ^{-}, \mathbf{0}, \mathbf{1})$, a **Boolean function** of the n variables $x_1, x_2, \ldots, x_n$ is a function $f : B^n \to B$ such that $f(x_1, x_2, \ldots, x_n)$ is a Boolean expression.

It follows from the definition that equivalent Boolean expressions define the same function. For example, consider the two functions

$$f : B^2 \to B \qquad f(x_1, x_2) = x_1(\bar{x}_1 \oplus x_2)$$

$$g : B^2 \to B \qquad g(x_1, x_2) = x_1x_2.$$

We have $x_1(\bar{x}_1 \oplus x_2) = x_1x_2$ (see exercise 10.1.8), and so the functions f and g are equal.

Since a particular Boolean expression may have a number of equivalent forms, there arises the question of how we can decide whether or not two Boolean expressions are equivalent and hence whether or not two Boolean functions are equal. Of course we could attempt to derive one from the other using the axioms and any theorems proved from them. However, this can be time consuming and, where we are unable to derive one expression from the other, we cannot be sure whether this is because we have not applied the correct sequence of steps or because the two expressions are not equivalent. If the latter is the case then all our efforts are bound to fail. Fortunately there is an alternative method of establishing the equivalence of two Boolean expressions but, before we can consider this, we need some more definitions.

Definition 10.6

A **minterm** (or **complete product**) in the n variables $x_1, x_2, \ldots, x_n$ is a Boolean expression which has the form of the product of each Boolean variable or its complement. Thus a minterm consists of the product of n literals, one corresponding to each variable.

For example, there are eight possible minterms in the three variables x_1, x_2, x_3. These are:

$$\begin{array}{cccc} x_1x_2x_3 & x_1x_2\bar{x}_3 & x_1\bar{x}_2x_3 & x_1\bar{x}_2\bar{x}_3 \\ \bar{x}_1x_2x_3 & \bar{x}_1x_2\bar{x}_3 & \bar{x}_1\bar{x}_2x_3 & \bar{x}_1\bar{x}_2\bar{x}_3. \end{array}$$

Using the definition of x^e given above, we denote a minterm in the n variables $x_1, x_2, \ldots, x_n$ by $m_{e_1e_2\ldots e_n}$ where

$$m_{e_1e_2\ldots e_n} = {x_1}^{e_1}{x_2}^{e_2}\ldots{x_n}^{e_n}.$$

For example, we have

$$\begin{aligned} m_{10110} &= {x_1}^1{x_2}^0{x_3}^1{x_4}^1{x_5}^0 \\ &= x_1\bar{x}_2x_3x_4\bar{x}_5 \end{aligned}$$

and

$$m_{0111} = {x_1}^0{x_2}^1{x_3}^1{x_4}^1$$

$$= \bar{x}_1 x_2 x_3 x_4.$$

For the n variables $x_1, x_2, \ldots, x_n$ there are 2^n possible minterms since each of the n literals in the minterm can take one of two forms, the respective variable or its complement. Of these 2^n minterms, no two are equivalent. This can be verified by appropriate substitution of the values **0** or **1** for each variable. Given two minterms, it is always possible to assign the value **0** or **1** to each variable so that evaluating each minterm gives a different result. For example, consider

$$m_{010} = \bar{x}_1 x_1 \bar{x}_3 \quad \text{and} \quad m_{111} = x_1 x_2 x_3.$$

Substituting $x_1 = \mathbf{0}$, $x_2 = \mathbf{1}$, $x_3 = \mathbf{0}$, we have $m_{010} = \mathbf{1}$ and $m_{111} = \mathbf{0}$. Thus m_{010} and m_{111} are not equivalent Boolean expressions.

Theorem 10.9

Of the 2^n minterms in the variables $x_1, x_2, \ldots, x_n$, no two are equivalent Boolean expressions.

Proof

We first note that $\mathbf{0}^0 = \bar{\mathbf{0}} = \mathbf{1}$ and $\mathbf{1}^1 = \mathbf{1}$ so that, if $x_i = e_i$, ${x_i}^{e_i} = \mathbf{1}$. This means that, given a minterm

$$m = m_{e_1 e_2 \ldots e_n} = {x_1}^{e_1} {x_2}^{e_2} \ldots {x_n}^{e_n}$$

substituting $x_i = e_i$ for $i = 1, 2, \ldots, n$ gives the product of n terms all of which are equal to **1** and so the minterm is equal to **1**.

Now any other minterm contains at least one literal which is the complement of a literal contained in m and so substitution of the values $x_i = e_i$ $(i = 1, 2, \ldots, n)$ as above results in a product which contains at least one zero. Hence, by theorem 10.4, the product is zero.

We have shown that for any two distinct minterms there is at least one set of values of the variables for which the minterms have different values. We can therefore conclude that no two distinct minterms are equivalent. □

Definition 10.7

A **maxterm** (or **complete sum**) in the n variables $x_1, x_2, \ldots, x_n$ is a Boolean expression which has the form of the sum of each Boolean variable or its complement. Thus a maxterm consists of the sum of n literals.

We denote a maxterm in the n variables $x_1, x_2, \ldots, x_n$ by

$$M_{e_1 e_2 \ldots e_n} = {x_1}^{e_1} \oplus {x_2}^{e_2} \oplus \cdots \oplus {x_n}^{e_n}$$

where ${x_i}^{e_i}$ is defined as before.

Thus

$$M_{11010} = x_1 \oplus x_2 \oplus \bar{x}_3 \oplus x_4 \oplus \bar{x}_5$$

and

$$M_{0011} = \bar{x}_1 \oplus \bar{x}_2 \oplus x_3 \oplus x_4.$$

As with minterms, there are 2^n possible maxterms in n variables and (by the duality principle) no two of these are equivalent Boolean expressions.

We now come to an important theorem which will enable us to decide whether two Boolean functions are equal without having to go to the trouble of showing that the Boolean expression defining one of the functions can be transformed to the Boolean expression representing the other using the axioms. What the theorem shows is that any Boolean expression in n variables can be written uniquely as the sum of some or all of the 2^n minterms in these n variables. A function which is defined by a Boolean expression in this form is said to be in **disjunctive normal form**. Since it consists of a sum of minterms, it is sometimes called the **minterm form**. It is also referred to as the **canonical** (or **complete**) **sum-of-products form**.

Before we consider the theorem we will look at an example where we derive the disjunctive normal form of a Boolean expression by applying the axioms and theorems of §10.2. However, although this is a perfectly acceptable method of obtaining the disjunctive normal form, it is not the easiest method and therefore not the one which we shall ultimately adopt.

Example 10.2

Write the following Boolean expression in the three variables x_1, x_2, x_3 in disjunctive normal form: $x_1 x_2 (x_1 \oplus x_3)$.

Solution

$$\begin{aligned}
x_1x_2(x_1 \oplus x_3) &= x_1x_2x_1 \oplus x_1x_2x_3 && \text{(axiom B4)}\\
&= x_1x_1x_2 \oplus x_1x_2x_3 && \text{(axiom B3)}\\
&= x_1x_2 \oplus x_1x_2x_3 && \text{(idempotent laws)}\\
&= x_1x_2\mathbf{1} \oplus x_1x_2x_3 && \text{(axiom B1)}\\
&= x_1x_2(x_3 \oplus \bar{x}_3) \oplus x_1x_2x_3 && \text{(axiom B5)}\\
&= x_1x_2x_3 \oplus x_1x_2\bar{x}_3 \oplus x_1x_2x_3 && \text{(axiom B4)}\\
&= x_1x_2x_3 \oplus x_1x_2\bar{x}_3 && \text{(axiom B3 and idempotent laws).}
\end{aligned}$$

This is the disjunctive normal form since it consists of the sum of the two minterms $x_1x_2x_3$ and $x_1x_2\bar{x}_3$.

We now consider the theorem which guarantees the existence of a disjunctive normal form for any non-zero Boolean function.

Theorem 10.10

Every Boolean function $f(x_1, x_2, \ldots, x_n)$, which is not identically zero, can be written as the sum of all possible Boolean expressions of the form

$$f(e_1, e_2, \ldots, e_n){x_1}^{e_1}{x_2}^{e_2}, \ldots {x_n}^{e_n}$$

where ${x_i}^{e_i}$ has the usual interpretation. Thus we can write

$$\begin{aligned}
f(x_1, x_2, \ldots, x_n) &= \bigoplus_{(e)} f(e_1, e_2, \ldots, e_n){x_1}^{e_1}{x_2}^{e_2}, \ldots {x_n}^{e_n}\\
&= \bigoplus_{(e)} f(e_1, e_2, \ldots, e_n)m_{e_1e_2\ldots e_n}
\end{aligned}$$

where (e) denotes all possible n-tuples $(e_1, e_2, \ldots, e_n)$ where $e_i = \mathbf{0}$ or $\mathbf{1}$ $(i = 1, 2, \ldots, n)$. There are 2^n of these.

Proof

The proof of this theorem, although not difficult, is rather long. We therefore give only an outline and leave the interested reader to fill in the details.

We prove first that the theorem holds for a function of one variable $f(x)$, i.e. that

$$f(x) = f(\mathbf{0})\bar{x} \oplus f(\mathbf{1})x.$$

Now, because it is defined by a Boolean expression in one variable, a function of one variable must take one of the following forms:

(a) $f(x) = \mathbf{0}$ or $f(x) = \mathbf{1}$
(b) $f(x) = x$
(c) $f(x) = \bar{x}$
(d) $f(x)$ consists of the sums and products of terms which are themselves sums or products of x, $\bar{x}$ and the identity elements of B.

It can be shown using the axioms that

$$\mathbf{0}\bar{x} \oplus \mathbf{0}x = \mathbf{0}$$

so that, if $f(x) = \mathbf{0}$,

$$\begin{aligned} f(x) &= \mathbf{0}\bar{x} \oplus \mathbf{0}x \\ &= f(\mathbf{0})\bar{x} \oplus f(\mathbf{1})x. \end{aligned}$$

Similarly, if $f(x) = \mathbf{1}$,

$$\begin{aligned} f(x) &= \mathbf{1}\bar{x} \oplus \mathbf{1}x \\ &= f(\mathbf{0})\bar{x} \oplus f(\mathbf{1})x. \end{aligned}$$

Also, if $f(x) = x$, we have

$$\begin{aligned} f(x) &= \mathbf{0}\bar{x} \oplus \mathbf{1}x \\ &= f(\mathbf{0})\bar{x} \oplus f(\mathbf{1})x \end{aligned}$$

and, if $f(x) = \bar{x}$,

$$\begin{aligned} f(x) &= \mathbf{1}\bar{x} \oplus \mathbf{0}x \\ &= f(\mathbf{0})\bar{x} \oplus f(\mathbf{1})x. \end{aligned}$$

Thus the theorem holds for $f(x)$ in each of the cases (a), (b) and (c) above. That it also holds for case (d) can be established by taking $f_1(x)$ and $f_2(x)$ to be any two functions for which the theorem holds and showing that the theorem also holds for $f_1(x)f_2(x)$ and for $f_1(x) \oplus f_2(x)$. This proves that the theorem applies to any Boolean function of one variable.

Now consider a function of n variables $f(x_1, x_2, \ldots, x_n)$. If we regard this as a function of the single variable x_1 and apply the theorem, we have

$$f(x_1, x_2, \ldots, x_n) = [f(\mathbf{0}, x_2, \ldots, x_n)\bar{x}_1] \oplus [f(\mathbf{1}, x_2, \ldots, x_n)x_1].$$

We now regard $f(\mathbf{0}, x_2, \ldots, x_n)$ and $f(\mathbf{1}, x_2, \ldots, x_n)$ as functions of the single variable x_2 so that applying the theorem again gives

$$f(\mathbf{0}, x_2, \ldots, x_n) = [f(\mathbf{0}, \mathbf{0}, x_3, \ldots, x_n)\bar{x}_2] \oplus [f(\mathbf{0}, \mathbf{1}, x_3, \ldots, x_n)x_2]$$

and

$$f(\mathbf{1}, x_2, \ldots, x_n) = [f(\mathbf{1}, \mathbf{0}, x_3, \ldots, x_n)\bar{x}_2] \oplus [f(\mathbf{1}, \mathbf{1}, x_3, \ldots, x_n)x_2].$$

Thus we have

$$\begin{aligned} f(x_1, x_2, \ldots, x_n) = {} & [f(\mathbf{0}, \mathbf{0}, x_3, \ldots, x_n)\bar{x}_1\bar{x}_2] \oplus [f(\mathbf{0}, \mathbf{1}, x_3, \ldots, x_n)\bar{x}_1 x_2] \\ & \oplus [f(\mathbf{1}, \mathbf{0}, x_3, \ldots, x_n)x_1\bar{x}_2] \oplus [f(\mathbf{1}, \mathbf{1}, x_3, \ldots, x_n)x_1 x_2]. \end{aligned}$$

Continuing in this way, dealing with each variable in turn, gives the result

$$f(x_1, x_2, \ldots, x_n) = \bigoplus_{(e)} f(e_1, e_2, \ldots, e_n){x_1}^{e_1}{x_2}^{e_2}\ldots{x_n}^{e_n}. \qquad \square$$

This result looks a little formidable but it is really quite simple although the notation may make it seem less so. The value of $f(e_1, e_2, \ldots, e_n)$ is obtained by substituting $x_i = e_i$ in the Boolean expression defining the function. Since the e_i values are either zero or one, $f(e_1, e_2, \ldots, e_n)$ will be either zero or one. The theorem shows that a Boolean function can be written as a sum of terms, each of which is the product of $f(e_1, e_2, \ldots, e_n)$ and the corresponding minterm for every set of values of the e_i. If $f(e_1, e_2, \ldots, e_n) = \mathbf{1}$, the product is simply the minterm and if $f(e_1, e_2, \ldots, e_n) = \mathbf{0}$, then the product is zero. Hence the minterms which appear in the disjunctive normal form are those of the form

$${x_1}^{e_1}{x_2}^{e_2}\ldots{x_n}^{e_n}$$

for which $f(e_1, e_2, \ldots, e_n) = \mathbf{1}$.

Some examples may help to clarify this.

Examples 10.3

1. Write the Boolean function $f(x_1, x_2) = x_1 \oplus x_2$ in disjunctive normal form.

Solution

Theorem 10.10 states that $f(x_1, x_2) = x_1 \oplus x_2$ can be written as follows:

$$f(x_1, x_2) = f(\mathbf{0}, \mathbf{0})\bar{x}_1\bar{x}_2 \oplus f(\mathbf{0}, \mathbf{1})\bar{x}_1 x_2 \oplus f(\mathbf{1}, \mathbf{0})x_1\bar{x}_2 \oplus f(\mathbf{1}, \mathbf{1})x_1 x_2.$$

We use a table (with an obvious analogy to a truth table) to calculate $f(e_1, e_2)$ for the various possible assignments of $\mathbf{0}$ and $\mathbf{1}$ to the variables e_1 and e_2.

e_1	e_2	$f(e_1, e_2)$
0	0	0
0	1	1
1	0	1
1	1	1

So we have

$$\begin{aligned} f(x_1, x_2) &= \mathbf{0}\bar{x}_1\bar{x}_2 \oplus \mathbf{1}\bar{x}_1 x_2 \oplus \mathbf{1}x_1\bar{x}_2 \oplus \mathbf{1}x_1 x_2 \\ &= \bar{x}_1 x_2 \oplus x_1\bar{x}_2 \oplus x_1 x_2. \end{aligned}$$

From this example we can see that expressing a Boolean function in disjunctive normal form consists of finding the values of $e_1, e_2, \ldots, e_n$ for which $f(e_1, e_2, \ldots, e_n) = \mathbf{1}$ and writing down the sum of the corresponding minterms.

2. Write $f(x_1, x_2, x_3) = x_2 x_3 \oplus x_3 x_1$ in disjunctive normal form.

Solution

Using a table to calculate $f(e_1, e_2, e_3)$ for all possible values of e_1, e_2, e_3 we have the following.

e_1	e_2	e_3	e_2e_3	e_3e_1	$e_2e_3 \oplus e_3e_1$
0	0	0	0	0	0
0	0	1	0	0	0
0	1	0	0	0	0
0	1	1	1	0	1
1	0	0	0	0	0
1	0	1	0	1	1
1	1	0	0	0	0
1	1	1	1	1	1

Now $f(e_1, e_2, e_3) = \mathbf{1}$ when

(a) $e_1 = \mathbf{0}$, $e_2 = \mathbf{1}$, $e_3 = \mathbf{1}$
(b) $e_1 = \mathbf{1}$, $e_2 = \mathbf{0}$, $e_3 = \mathbf{1}$
(c) $e_1 = \mathbf{1}$, $e_2 = \mathbf{1}$, $e_3 = \mathbf{1}$.

The minterms $x_1{}^{e_1}x_2{}^{e_2}x_3{}^{e_3}$ corresponding to each of these are

(a) $\bar{x}_1 x_2 x_3$
(b) $x_1 \bar{x}_2 x_3$
(c) $x_1 x_2 x_3$.

The disjunctive normal form is therefore given by

$$f(x_1, x_2, x_3) = \bar{x}_1 x_2 x_3 \oplus x_1 \bar{x}_2 x_3 \oplus x_1 x_2 x_3.$$

We now show that there is only one disjunctive normal form for any given Boolean function.

Theorem 10.11

The disjunctive normal form of a given Boolean function is unique (up to reordering of the minterms in the expression).

Proof

The method of proof is by contradiction.

Suppose that the function $f(x_1, x_2, \ldots, x_n)$ can be written in disjunctive normal form in two ways, so that

$$\begin{aligned} f(x_1, x_2, \ldots, x_n) &= P_1 \oplus P_2 \oplus \cdots \oplus P_r \\ &= Q_1 \oplus Q_2 \oplus \cdots \oplus Q_s \end{aligned}$$

where P_i and Q_j ($i = 1, 2, \ldots, r$ and $j = 1, 2, \ldots, s$) are terms of the form

$$x_1{}^{e_1} x_2{}^{e_2} \ldots x_n{}^{e_n}.$$

We will assume, without loss of generality, that $r \geqslant s$.

Now, if the two disjunctive normal forms are not equal then at least one of the P_i must be different from every Q_j. Let us suppose that P_m has this property.

Since P_m is different from all the Q_j, P_m and Q_j must be such that one contains x_k while the other contains $\bar{x}_k$ for some value of k. The expression $P_m Q_j$ then contains the product $x_k \bar{x}_k$ and hence

$$P_m Q_j = \mathbf{0}.$$

Now this is true for each $j = 1, 2, \ldots, s$ so that

$$\begin{aligned} & P_m Q_1 \oplus P_m Q_2 \oplus \cdots \oplus P_m Q_s = \mathbf{0} \\ \Rightarrow \quad & P_m(Q_1 \oplus Q_2 \oplus \cdots \oplus Q_s) = \mathbf{0} \quad \text{(axiom B4)} \\ \Rightarrow \quad & P_m f(x_1, x_2, \ldots, x_n) = \mathbf{0}. \end{aligned}$$

But

$$\begin{aligned} P_m f(x_1, x_2, \ldots, x_n) &= P_m(P_1 \oplus P_2 \oplus \cdots \oplus P_r) \\ &= P_m P_1 \oplus P_m P_2 \oplus \cdots \oplus P_m P_r \quad \text{(axiom B4)} \\ &= P_m P_m \quad \text{(since all the } P_i \text{ are different)} \\ &= P_m \quad \text{(idempotent laws).} \end{aligned}$$

Thus if there are two disjunctive normal forms for $f(x_1, x_2, \ldots, x_n)$ we have

$$\begin{aligned} P_m f(x_1, x_2, \ldots, x_n) &= \mathbf{0} \\ P_m f(x_1, x_2, \ldots, x_n) &= P_m. \end{aligned}$$

This contradiction shows that the disjunctive normal form must be unique. □

It is this theorem which gives us a useful method for establishing the equality or otherwise of two Boolean functions. We simply write the expressions defining the two functions in disjunctive normal form and compare the results. The disjunctive normal form is the 'fingerprint' of a Boolean function. Two or more Boolean functions are equal if and only if they have the same disjunctive normal form.

Example 10.4

Show that $f(x_1, x_2) = x_1 \oplus x_2$ and $g(x_1, x_2) = \bar{x}_1 x_2 \oplus x_1$ are equal functions.

Solution

We shall write each function in disjunctive normal form and, if these are the same, theorem 10.11 allows us to conclude that the two expressions are equivalent and hence that the functions are equal.

Consider first $f(x_1, x_2) = x_1 \oplus x_2$. In example 10.3.1 we derived the disjunctive formal form for this function:

$$f(x_1, x_2) = \bar{x}_1 x_2 \oplus x_1 \bar{x}_2 \oplus x_1 x_2.$$

We now write $g(x_1, x_2) = \bar{x}_1 x_2 \oplus x_1$ in disjunctive normal form. We have the following table.

e_1	e_2	$\bar{e}_1$	$\bar{e}_1 e_2$	$\bar{e}_1 e_2 \oplus e_1$
0	**0**	1	**0**	**0**
0	1	1	1	1
1	**0**	**0**	**0**	1
1	1	**0**	**0**	1

So

$$\begin{aligned} g(x_1, x_2) &= \bar{x}_1 x_2 \oplus x_1 \bar{x}_2 \oplus x_1 x_2 \\ &= f(x_1, x_2). \end{aligned}$$

Hence the two functions are equal.

Note that, to prove that two functions $f(x_1, x_2, \ldots, x_n)$ and $g(x_1, x_2, \ldots, x_n)$ are equal, it is sufficient to prove that $f(e_1, e_2, \ldots, e_n) = g(e_1, e_2, \ldots, e_n)$ for all possible n-tuples $(e_1, e_2, \ldots, e_n)$. Thus a Boolean function is completely determined by the values that it takes for the 2^n combinations of zeros and ones that can be substituted for $e_1, e_2, \ldots, e_n$.

Example 10.5

A Boolean function $f(x_1, x_2)$ is such that $f(\mathbf{0}, \mathbf{0}) = \mathbf{1}$, $f(\mathbf{0}, \mathbf{1}) = \mathbf{0}$, $f(\mathbf{1}, \mathbf{0}) = \mathbf{1}$, $f(\mathbf{1}, \mathbf{1}) = \mathbf{0}$. Find a Boolean expression for this function.

Solution

We are given values of $f(e_1, e_2)$ for all possible choices of e_1 and e_2. The minterms which appear in the disjunctive normal form are those of the form ${x_1}^{e_1} {x_2}^{e_2}$ for which $f(e_1, e_2) = \mathbf{1}$, i.e. $\bar{x}_1 \bar{x}_2$ and $x_1 \bar{x}_2$. So a Boolean expression for the function is

$$f(x_1, x_2) = \bar{x}_1 \bar{x}_2 \oplus x_1 \bar{x}_2.$$

Theorem 10.10 shows that it is possible to write a Boolean function as the sum of minterms. By applying the duality principle to this theorem and to theorem 10.11, we see that it is also possible to write such a function uniquely as the product of maxterms.

Theorem 10.12 (Dual of theorem 10.10)

Every Boolean function $f(x_1, x_2, \ldots, x_n)$, which is not identically one, can be written as the product of all possible Boolean expressions of the form

$$f(\bar{e}_1, \bar{e}_2, \ldots, \bar{e}_n) \oplus {x_1}^{e_1} \oplus {x_2}^{e_2} \oplus \cdots \oplus {x_n}^{e_n}$$

or equivalently, as the product of all Boolean expressions of the form

$$f(e_1, e_2, \ldots, e_n) \oplus {x_1}^{\bar{e}_1} \oplus {x_2}^{\bar{e}_2} \oplus \cdots \oplus {x_n}^{\bar{e}_n}.$$

Therefore

$$\begin{aligned} f(x_1, x_2, \ldots, x_n) &= \underset{(e)}{*}\, f(e_1, e_2, \ldots, e_n) \oplus {x_1}^{\bar{e}_1} \oplus {x_2}^{\bar{e}_2} \oplus \cdots \oplus {x_n}^{\bar{e}_n} \\ &= \underset{(e)}{*}\, f(e_1, e_2, \ldots, e_n) \oplus M_{\bar{e}_1 \bar{e}_2 \ldots \bar{e}_n} \end{aligned}$$

where (e) denotes all possible n-tuples $(e_1, e_2, \ldots, e_n)$ where $e_i = 0$ or 1 $(i = 1, 2, \ldots, n)$.

A function which is written in this form is said to be in **conjunctive normal form**. It is also referred to as the **maxterm form** or the **canonical** (or **complete**) **product-of-sums form**.

Theorem 10.13 (Dual of theorem 10.11)

The conjunctive normal form of a given Boolean function is unique up to a reordering of maxterms.

Example 10.6

Express $f(x_1, x_2) = x_1(x_1 \oplus x_2)$ in conjunctive normal form.

Solution

As with the examples on disjunctive normal forms, we first evaluate $f(e_1, e_2)$ for all values of e_1, e_2.

e_1	e_2	$e_1 \oplus e_2$	$e_1(e_1 \oplus e_2)$
0	0	0	0
0	1	1	0
1	0	1	1
1	1	1	1

Thus

$$f(x_1, x_2) = (\mathbf{0} \oplus x_1 \oplus x_2)(\mathbf{0} \oplus x_1 \oplus \bar{x}_2)(\mathbf{1} \oplus \bar{x}_1 \oplus x_2)(\mathbf{1} \oplus \bar{x}_1 \oplus \bar{x}_2).$$

Now $\mathbf{0} \oplus b = b$ and $\mathbf{1} \oplus b = \mathbf{1}$ for any $b \in B$, where B is the underlying set of a Boolean algebra, so that

$$\begin{aligned} f(x_1, x_2) &= (x_1 \oplus x_2)(x_1 \oplus \bar{x}_2) * \mathbf{1} * \mathbf{1} \\ &= (x_1 \oplus x_2)(x_1 \oplus \bar{x}_2). \end{aligned}$$

This is the conjunctive normal form of $f(x_1, x_2)$.

When expressing a function $f(x_1, x_2, \ldots, x_n)$ in disjunctive normal form the minterms $m_{e_1 e_2 \ldots e_n}$ which are present are those for which $f(e_1, e_2, \ldots, e_n) = \mathbf{1}$. From the example above we can see that the maxterms $M_{\bar{e}_1 \bar{e}_2 \ldots \bar{e}_n}$ which appear in the conjunctive normal form are those for which $f(e_1, e_2, \ldots, e_n) = \mathbf{0}$.

Example 10.7

Express $f(x_1, x_2, x_3) = (\bar{x}_1 \oplus x_2)(\bar{x}_1 \oplus \bar{x}_3)$ in conjunctive normal form.

Solution

e_1	e_2	e_3	$\bar{e}_1$	$\bar{e}_3$	$\bar{e}_1 \oplus e_2$	$\bar{e}_1 \oplus \bar{e}_3$	$(\bar{e}_1 \oplus e_2)(\bar{e}_1 \oplus \bar{e}_3)$
0	0	0	1	1	1	1	1
0	0	1	1	0	1	1	1
0	1	0	1	1	1	1	1
0	1	1	1	0	1	1	1
1	0	0	0	1	0	1	0
1	0	1	0	0	0	0	0
1	1	0	0	1	1	1	1
1	1	1	0	0	1	0	0

The following values of e_1, e_2, e_3 give $f(e_1, e_2, e_3) = \mathbf{0}$:

(a) $e_1 = \mathbf{1}$, $e_2 = \mathbf{0}$, $e_3 = \mathbf{0}$
(b) $e_1 = \mathbf{1}$, $e_2 = \mathbf{0}$, $e_3 = \mathbf{1}$
(c) $e_1 = \mathbf{1}$, $e_2 = \mathbf{1}$, $e_3 = \mathbf{1}$.

The corresponding maxterms $M_{\bar{e}_1\bar{e}_2\bar{e}_3}$ are

(a) $\bar{x}_1 \oplus x_2 \oplus x_3$
(b) $\bar{x}_1 \oplus x_2 \oplus \bar{x}_3$
(c) $\bar{x}_1 \oplus \bar{x}_2 \oplus \bar{x}_3$.

Hence, in conjunctive normal form,

$$f(x_1, x_2, x_3) = (\bar{x}_1 \oplus x_2 \oplus x_3)(\bar{x}_1 \oplus x_2 \oplus \bar{x}_3)(\bar{x}_1 \oplus \bar{x}_2 \oplus \bar{x}_3).$$

Conjunctive normal forms can be used in exactly the same way as disjunctive normal forms for proving whether or not Boolean functions are equal. However, disjunctive normal forms tend to be preferred because of their significance in the application of Boolean algebra to the design of electronic circuits.

Remember that, given the n Boolean variables $x_1, x_2, \ldots, x_n$, there are 2^n possible minterms in these n variables. Now for any Boolean function (not identically zero) $f(x_1, x_2, \ldots, x_n)$ can be expressed uniquely as the sum of some or all of these minterms. A set containing 2^n elements has 2^{2^n} possible subsets including the empty set and the set itself (see theorem 3.5). From this we can deduce that there are 2^{2^n} possible selections of minterms and hence there are just 2^{2^n} distinct Boolean functions of n variables. These include the function which is identically zero and cannot be expressed as the sum of minterms and which can therefore be thought of as the function in which all minterms are absent. The function in which all 2^n minterms are present is $f(x_1, x_2, \ldots, x_n) = \mathbf{1}$.

Exercises 10.2

1. Express each of the following Boolean functions in disjunctive normal form and hence state which of the functions are equal:

   (i) $f(x_1, x_2) = \bar{x}_1 x_2 \oplus x_1 \bar{x}_2$
   (ii) $f(x_1, x_2) = x_1$
   (iii) $f(x_1, x_2) = x_1(\bar{x}_1 \oplus x_2)$
   (iv) $f(x_1, x_2) = x_1 x_2$
   (v) $f(x_1, x_2) = (x_1 \oplus x_2)(x_1 \oplus \bar{x}_2)$
   (vi) $f(x_1, x_2, x_3) = x_2(x_1 x_3 \oplus \bar{x}_1)$
   (vii) $f(x_1, x_2, x_3) = x_1 \oplus \bar{x}_2 \oplus x_3$
   (viii) $f(x_1, x_2, x_3) = x_2(\bar{x}_1 \oplus x_3)$
   (ix) $f(x_1, x_2, x_3) = x_3 \oplus \bar{x}_1 x_2$
   (x) $f(x_1, x_2, x_3) = (x_1 \oplus x_2 \oplus x_3)(\bar{x}_1 \oplus x_3)$.

2. Express each of the following Boolean functions in both disjunctive and conjunctive normal forms:

   (i) $f(x_1, x_2, x_3) = x_1 \oplus x_2 \oplus \bar{x}_3$
   (ii) $f(x_1, x_2, x_3) = x_1 x_2 \oplus \bar{x}_3 \oplus x_1$
   (iii) $f(x_1, x_2, x_3) = \mathbf{1}(x_2 \oplus x_3)x_1$
   (iv) $f(x_1, x_2, x_3) = \bar{x}_1 x_2 \oplus x_1 x_3$.

3. Let F be the set of all Boolean functions in the n variables $x_1, x_2, \ldots, x_n$. (We showed that there are 2^{2^n} such functions.) Suppose that for each $f_i \in F$ $(i = 1, \ldots, 2^{2^n})$ a Boolean expression defining f_i is E_i. We define the following operations on the set F:

   (a) $\bar{f}_i = \bar{E}_i$ for all $f_i \in F$
   (b) $f_i f_j = E_i E_j$ for all $f_i, f_j \in F$
   (c) $f_i \oplus f_j = E_i \oplus E_j$ for all $f_i, f_j \in F$.

   Show that there exist identity elements $f_0, f_1 \in F$ such that

$$f_0 \oplus f_i = f_i \oplus f_0 = f_1$$
$$f_i f_1 = f_1 f_i = f_i$$

for all $f_i \in F$.

Show that $(F, \oplus, *, ^{-}, f_0, f_1)$ is a Boolean algebra.

4. Let $(B, \oplus, *, ^{-}, \mathbf{0}, \mathbf{1})$ be a Boolean algebra and let the binary operation $\circ$ be defined on B by

$$b_1 \circ b_2 = b_1\bar{b}_2 \oplus \bar{b}_1 b_2$$

for all $b_1, b_2 \in B$. Show that $(B, \circ)$ is an abelian group with identity $\mathbf{0}$. What is the inverse of an arbitrary element $b \in B$?

5. Given the Boolean algebra $(B, \oplus, *, ^{-}, \mathbf{0}, \mathbf{1})$, an **atom** is a non-zero element $a \in B$ such that, for all $b \in B$, either $ba = a$ or $ba = \mathbf{0}$.

   (i) Let $S = \{j, k, l, m\}$. What are the atoms of the Boolean algebra $(\mathscr{P}(S), \cup, \cap, ^{-}, \varnothing, S)$?

   (ii) Prove that, if a_1 and a_2 are atoms of the Boolean algebra $(B, \oplus, *, ^{-}, \mathbf{0}, \mathbf{1})$ and $a_1 a_2 \neq \mathbf{0}$, then $a_1 = a_2$.

10.4 Switching Circuits

Many electronic devices such as computers, telephone systems, traffic and train control systems employ as part of their circuitry items known as **switches**. A switch may be viewed as a connection within the circuit such that, when the switch is closed, electric current may pass through it but, when it is open, no current can pass through that point of the circuit. A switch is an example of a **two-state device**, the two states being 'on' and 'off'. A circuit which incorporates one or more switches is known as a **switching circuit**. Diagrammatically, we shall show a switch as follows:

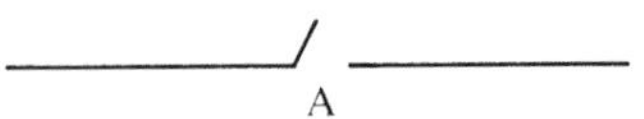

Now imagine that we have a circuit (assumed to include a suitable power source) which contains a switch A. We denote the state of the switch by the variable x where $x = 0$ if A is open and $x = 1$ if A is closed.

Consider now a circuit which contains two switches A_1 and A_2 connected as shown in the diagram below.

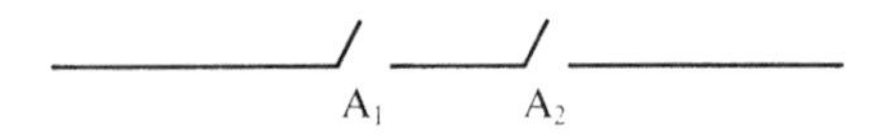

Switches connected to each other in this way are said to be in **series**. It is clear that current will flow across this section of a circuit only if both switches A_1 and A_2 are closed. Let x_1 and x_2 be variables denoting the states of switches A_1 and A_2 respectively. (In each case 0 denotes open and 1 denotes closed.) Let $f(x_1, x_2)$ be a function which has the value 1 for values of x_1 and x_2 which allow current to flow and 0 otherwise. Thus $f : \{0,1\}^2 \to \{0,1\}$ and the value of $f(x_1, x_2)$ for all possible values of x_1 and x_2 is given in the table below.

x_1	x_2	$f(x_1, x_2)$
0	0	0
0	1	0
1	0	0
1	1	1

We can now see that f is the familiar function $f(x_1, x_2) = x_1x_2$ where x_1 and x_2 are variables whose domain is the two-element Boolean algebra $(\{0,1\}, \oplus, *, ^-, 0, 1)$.

Two switches may alternatively be connected in **parallel**, the arrangement shown in the diagram below.

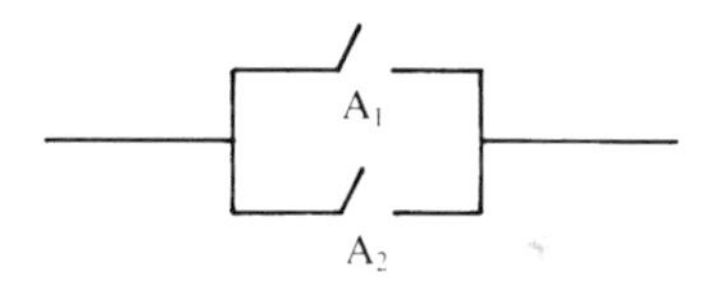

For current to flow around a circuit containing a power source and only two switches connected in this way, it is necessary that one or both of the switches are closed. Defining x_1 and x_2 as before and $g(x_1, x_2)$ exactly as we defined $f(x_1, x_2)$ for switches in series, we have the following table.

x_1	x_2	$g(x_1, x_2)$
0	0	0
0	1	1
1	0	1
1	1	1

Thus $g(x_1, x_2) = x_1 \oplus x_2$ defined on the same Boolean algebra $(\{0,1\}, \oplus, *, ^-, 0, 1)$.

Functions such as the two we have considered describe the behaviour of a circuit according to the states of the switches which are incorporated into that circuit.

Such functions are called **switching functions**. Given n switches $A_1, A_2, \ldots, A_n$ whose states are defined by the n variables $x_1, x_2, \ldots, x_n$ ($x_i = 0$ or 1, $i = 1, 2, \ldots, n$), a switching function $f : \{0,1\}^n \to \{0,1\}$ describes the behaviour of the circuit for all the 2^n possible states of the switches. As we have seen in the examples above, f can be represented by a Boolean expression and hence is a Boolean function.

In the following examples we look at switching functions for more complicated switching circuits.

Examples 10.8

1. Define the switching function f for the circuit incorporating the following arrangement of switches.

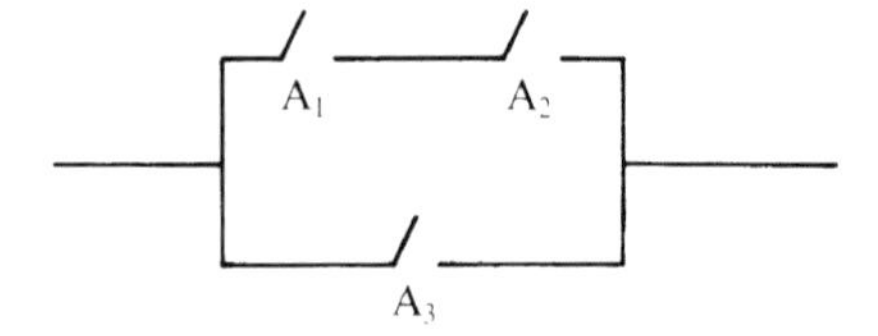

Solution

Let x_1, x_2, x_3 denote the states of the switches A_1, A_2 and A_3 respectively. Let $f_1(x_1, x_2)$ denote the behaviour of the part of the circuit containing the switches A_1 and A_2. Since these are connected in series $f_1(x_1, x_2) = x_1 x_2$. If $f_2(x_3)$ denotes the behaviour of the portion of the circuit containing switch A_3, then clearly $f_2(x_3) = x_3$.

Now the two switches A_1 and A_2 are connected in parallel to the switch A_3 and so, if $f(x_1, x_2, x_3)$ denotes the behaviour of the circuit containing this system of switches, we have

$$\begin{aligned} f(x_1, x_2, x_3) &= f_1(x_1, x_2) \oplus f_2(x_3) \\ &= x_1 x_2 \oplus x_3. \end{aligned}$$

2. Define the switching function f for the circuit incorporating the following system of switches.

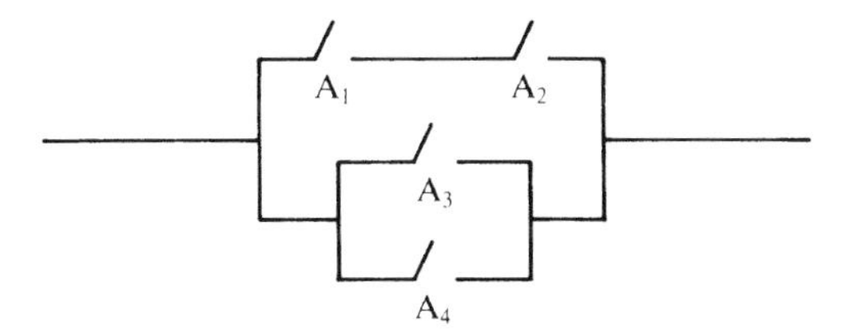

Solution

Let x_1, x_2, x_3, x_4 be the variables denoting the states of the switches A_1, A_2, A_3 and A_4 respectively. Then $f(x_1, x_2, x_3, x_4)$ is the switching function for the circuit.

Switches A_1 and A_2 are connected in series so that $f_1(x_1, x_2) = x_1x_2$ is the switching function for these two switches. Switches A_3 and A_4 are connected in parallel, therefore $f_2(x_3, x_4) = x_3 \oplus x_4$ is the appropriate switching function.

The section of the circuit containing A_1 and A_2 is connected in parallel to the section containing A_3 and A_4 so that

$$\begin{aligned} f(x_1, x_2, x_3, x_4) &= f_1(x_1, x_2) \oplus f_2(x_3, x_4) \\ &= x_1x_2 \oplus x_3 \oplus x_4. \end{aligned}$$

3. Consider the circuit employing the same arrangement of switches as in example 2. However, suppose that the switches A_1 and A_3 are such that they go on or off together. In this case their states are always identical and we can use the single variable x_1 to denote the state of each of them.

   Suppose also that switches A_2 and A_4 are such that when one is on the other is off and vice versa. If x_2 describes the state of A_2, then we can use $\bar{x}_2$ to describe the state of A_4.

   On the diagram we will use the same letter for switches which are always in the same state so that the same variable can be used. If S denotes a switch we shall use $\bar{S}$ to label a switch which is always in the opposite state to S. So for this example the circuit diagram is as follows.

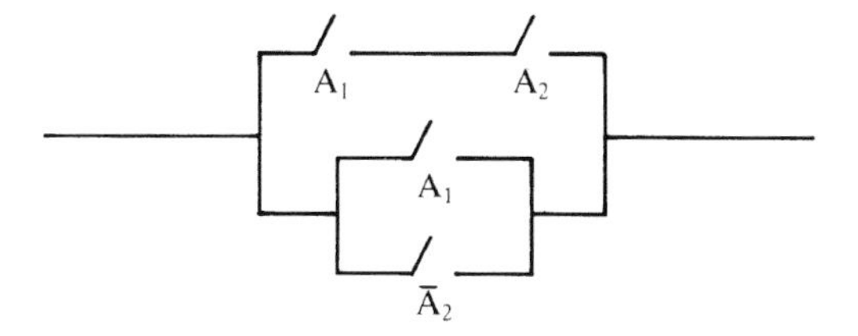

The switching function for this circuit is

$$f(x_1, x_2) = x_1 x_2 \oplus x_1 \oplus \bar{x}_2.$$

We have already seen that a given Boolean expression can be written in a number of equivalent forms so that the functions corresponding to each of these are equal. Applied to switching functions this means that two different arrangements of switches may have equal switching functions. This implies that the behaviour of the two circuits (in terms of whether or not current flows around them) is identical given that the state of corresponding switches incorporated in them is the same. For example, consider the following two systems of switches.

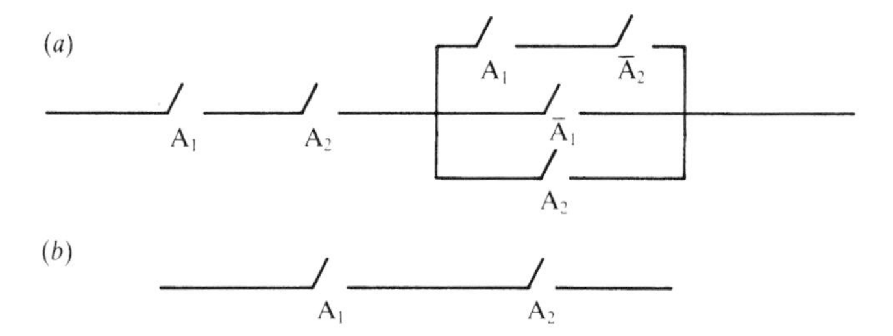

Let x_1 and x_2 describe the states of switches A_1 and A_2 respectively and let $f_1(x_1, x_2)$ be the switching function for the circuit (*a*) and $f_2(x_1, x_2)$ the switching function for (*b*).

Using the same technique as in the previous examples, we have

$$f_1(x_1, x_2) = x_1 x_2(x_1 \bar{x}_2 \oplus \bar{x}_1 \oplus x_2)$$
$$f_2(x_1, x_2) = x_1 x_2.$$

Although it is not immediately obvious, the functions f_1 and f_2 are equal. This can be verified either by rewriting each in disjunctive (or conjunctive) normal form, or by drawing up a table and evaluating each function for the four possible assignments of 0 and 1 to the variables x_1 and x_2.

Since their switching functions are equal, the behaviour of the circuits is identical for any set of states of the switches A_1 and A_2. However, it is clear that the second circuit is very much simpler than the first and also that it is likely to be cheaper to construct and more reliable. In constructing a circuit which is required to behave in a certain way, it is important to be able to recognize whether a particular design is the simplest among all those which are possible. We return to this problem later.

We give one further example to illustrate a familiar practical application of a switching circuit.

Example 10.9

A light bulb located over a flight of stairs is controlled by two wall switches, one at the top of the stairs and the other at the bottom. The switches are such that when the state of either one is reversed, the state of the light is reversed, i.e. it goes on if it was off and off if it was on. Design a circuit which will achieve this.

Solution

First note that the wall switches are not necessarily the switches in the circuit although they will each control one or more circuit switches.

Let us first draw a table showing what is required of the two wall switches S_1 and S_2. We shall arbitrarily suppose that initially both are up and the light is off, i.e. no current flows. When either one or the other is down, current must flow through the circuit, but when both are down there must be no current.

S_1	S_2	Current
Up	Up	No
Up	Down	Yes
Down	Up	Yes
Down	Down	No

Suppose that S_1 and S_2 control switches A_1 and A_2 in the circuit and that when a wall switch is up, the corresponding circuit switch is open. Using the variables x_1 and x_2 to denote the state of the circuit switches in the usual way and $f(x_1, x_2)$ to denote whether or not current flows through the circuit, the table above is equivalent to the following.

x_1	x_2	$f(x_1, x_2)$
0	0	0
0	1	1
1	0	1
1	1	0

From this table we can express $f(x_1, x_2)$ in terms of a Boolean expression in disjunctive normal form:

$$f(x_1, x_2) = \bar{x}_1 x_2 \oplus x_1 \bar{x}_2.$$

The circuit for which this is the switching function incorporates switches $\bar{\mathrm{A}}_1$ and A_2 in series connected in parallel to switches A_1 and $\bar{\mathrm{A}}_2$ in series. The switching system is therefore as shown in the following diagram.

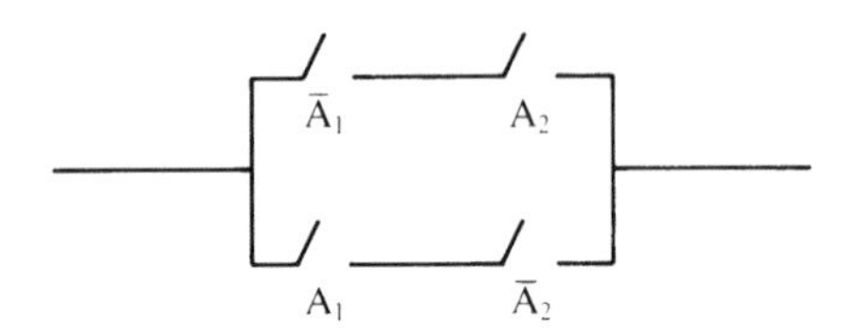

Note that we could equally well have expressed $f(x_1, x_2)$ in conjunctive normal form thus:

$$f(x_1, x_2) = (\bar{x}_1 \oplus \bar{x}_2)(x_1 \oplus x_2).$$

Thus an alternative circuit for achieving the same effect would be as follows.

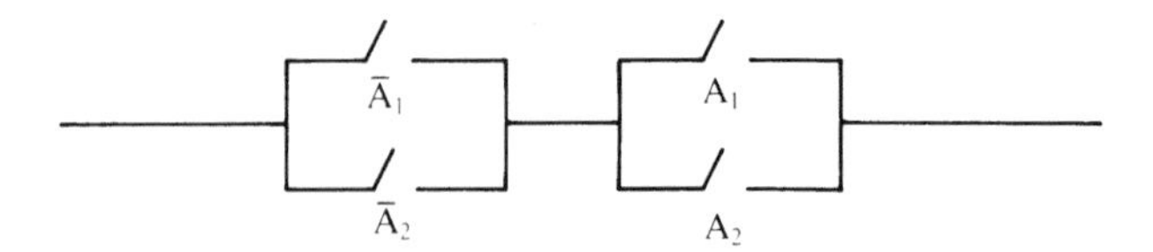

In both these cases some device is necessary so that a single wall switch controls two circuit switches. The exact nature of such a device need not concern us here.

Exercises 10.3

1. Define a switching function for each of the following systems of switches.

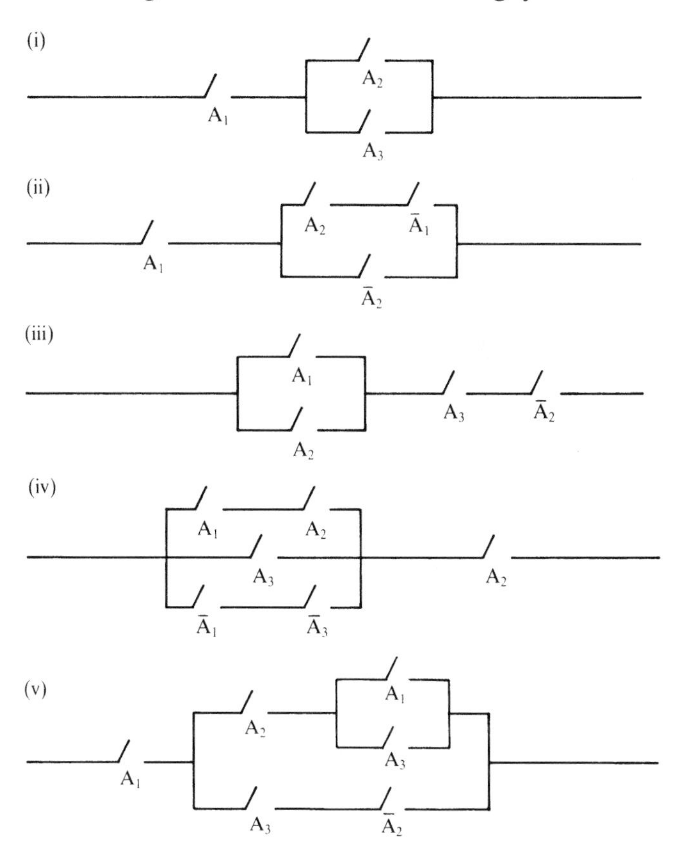

2. For each of the following functions, draw the diagram of a system of switches for which it is the switching function:

(i) $f(x_1, x_2) = (x_1 \oplus x_2)\bar{x}_1$
(ii) $f(x_1, x_2) = x_1 x_2 (x_1 \oplus x_2)$
(iii) $f(x_1, x_2, x_3) = (x_1 \oplus \bar{x}_2 \oplus x_3)\bar{x}_1$
(iv) $f(x_1, x_2, x_3) = (x_1 \oplus x_3)(\bar{x}_2 \oplus \bar{x}_1)$
(v) $f(x_1, x_2, x_3) = \bar{x}_1(x_2 x_3 \oplus \bar{x}_2 x_1)$.

3. Write down the switching function for each of the following systems of switches. By writing the function in disjunctive normal form, design

an equivalent system of switches, that is one having an equal switching function.

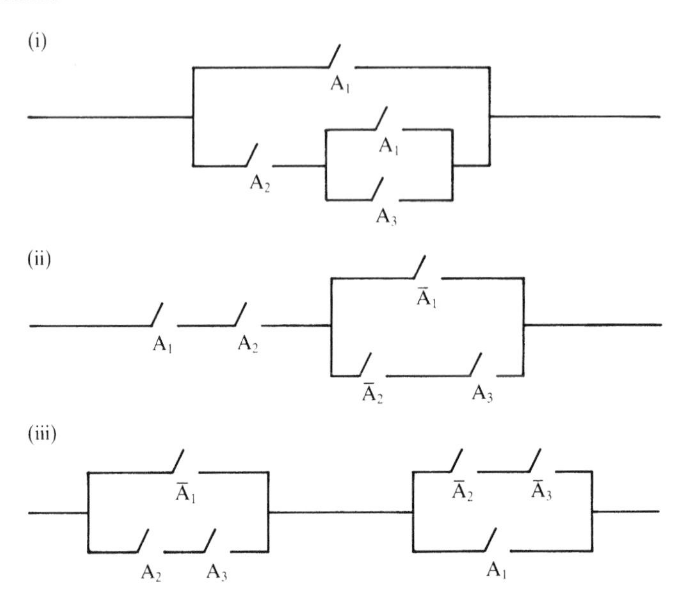

4. Suppose that a light bulb is located as in example 10.9 but that it is controlled by three switches rather than two. Reversing the state of any one of the switches reverses the state of the light. Design a switching system which will achieve this.

5. The central heating system in a small house is to be controlled by three thermostats, one located in each of three rooms. The thermostats are preset at 15°C but in the interests of economy it is desired that the central heating be on only if the temperature in at least two of the rooms falls below 15°C; otherwise the heating must be off.

   Design a switching system, to be operated by the thermostats, which will allow current to flow round a circuit (thereby activating the heating system) only when the temperature in at least two of the rooms falls below 15°C.

   Do you think that your circuit is the simplest one which will achieve the desired effect? If not, try and design a simpler one.

6. A simple burglar alarm system consists of a master switch and two movement sensors. When the master switch is on and either (or both) of the sensors is activated by movement within the room in which it is

located, an alarm bell rings until the master switch is turned off. If the master switch is not turned on, then the bell does not ring, whatever the state of the sensors.

Design a switching circuit incorporating switches activated by the movement sensors and the master switch which will achieve this effect.

Do you think that the circuit you have designed is the simplest one? If not, try to design a simpler one.

10.5 Logic Networks

In this section we deal with 'logic gates'. These are electronic devices which may be viewed as the basic functional components of a digital computer. A **logic gate** is an electronic component, incorporated within a circuit, which operates on one or more inputs to produce one output. Each input and each output can take one of two values (normally low and high voltage) which are denoted by 0 and 1. Because of the 'two-value' nature of the input and output variables, a logic gate is an example of a **binary device**. It also falls in the category of **combinational devices** because the output value depends only on the input values. (This is in contrast to a **sequential device** where the output value is also determined by such factors as the time or the past history of the circuit.)

The three most important types of logic gates are the **AND-gate**, the **OR-gate** and the **NOT-gate** (or **inverter**). These gates are so named because of their association (which will become obvious) with the corresponding logical connective. We use x_i to represent the value of the input(s) to a gate and the variable z to represent its output. The following table summarizes the operation of each of the three gates.

<table>
<tr><th></th><th>AND-gate</th><th>OR-gate</th><th>NOT-gate</th></tr>
<tr><td>Circuit symbol</td><td>x_1, x_2 → z</td><td>x_1, x_2 → z</td><td>x → z</td></tr>
<tr><td>Input/output table</td>
<td><table><tr><th>x_1</th><th>x_2</th><th>z</th></tr><tr><td>0</td><td>0</td><td>0</td></tr><tr><td>0</td><td>1</td><td>0</td></tr><tr><td>1</td><td>0</td><td>0</td></tr><tr><td>1</td><td>1</td><td>1</td></tr></table></td>
<td><table><tr><th>x_1</th><th>x_2</th><th>z</th></tr><tr><td>0</td><td>0</td><td>0</td></tr><tr><td>0</td><td>1</td><td>1</td></tr><tr><td>1</td><td>0</td><td>1</td></tr><tr><td>1</td><td>1</td><td>1</td></tr></table></td>
<td><table><tr><th>x</th><th>z</th></tr><tr><td>0</td><td>1</td></tr><tr><td>1</td><td>0</td></tr></table></td></tr>
<tr><td>Boolean expression</td><td>$z = x_1 x_2$</td><td>$z = x_1 \oplus x_2$</td><td>$z = \bar{x}$</td></tr>
</table>

The AND-gate and the OR-gate each have two inputs and one output. For the AND-gate the output z has the value 1 only when the two input values are 1. The value of z is 0 otherwise. Viewing x_1, x_2 and z as variables whose domain is the underlying set $\{0,1\}$ of the Boolean algebra $(\{0,1\},*,\oplus,^{-},0,1)$ we have $z = x_1x_2$.

The OR-gate has output value 1 only if either or both of the input values are 1 and therefore the Boolean expression for z is given by $z = x_1 \oplus x_2$.

In contrast to the other two gates the NOT-gate has only one input. The gate has the effect of reversing the value of the input variable so that the Boolean expression for z is given by $z = \bar{x}$.

Comparison of the input/output tables with the truth tables for conjunction, disjunction and negation should make it clear why these gates are named as they are.

Within a circuit, a number of these gates may be linked together, the output from one gate acting as the input to one or more others. Such a circuit is termed a **logic network**. We can describe the output of the system of gates by a Boolean expression in terms of the various input variables.

Examples 10.10

1. Give the Boolean expression for the output of the following system of gates.

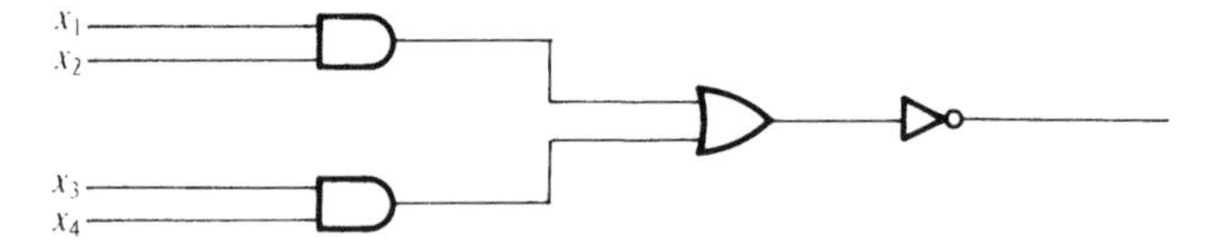

Solution

The output from the AND-gate having x_1 and x_2 as inputs is x_1x_2 and that from the AND-gate having x_3 and x_4 as inputs is x_3x_4. The output from the OR-gate is therefore $x_1x_2 \oplus x_3x_4$ and this is the input to the NOT-gate. The final output is therefore

$$\overline{x_1x_2 \oplus x_3x_4}.$$

We show these stages on the following diagram.

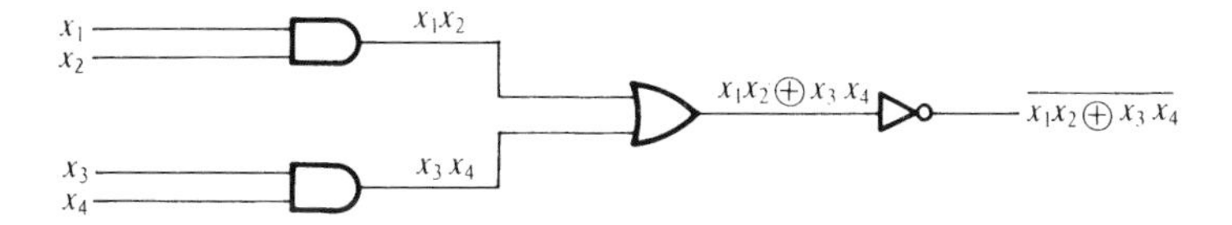

From the Boolean expression for the output, we can determine its value for each of the 16 possible sets of values of the input variables.

2. Design a system of logic gates with input variables x_1, x_2 and x_3 which will produce an output defined by the Boolean expression $\bar{x}_1x_2 \oplus x_1x_3$.

Solution

The final output can be achieved by an OR-gate whose input values are $\bar{x}_1x_2$ and x_1x_3. The first of these expressions is the output of an AND-gate with inputs $\bar{x}_1$ and x_2. The second is the output of an AND-gate with inputs x_1 and x_3.

Thus we have the following diagram.

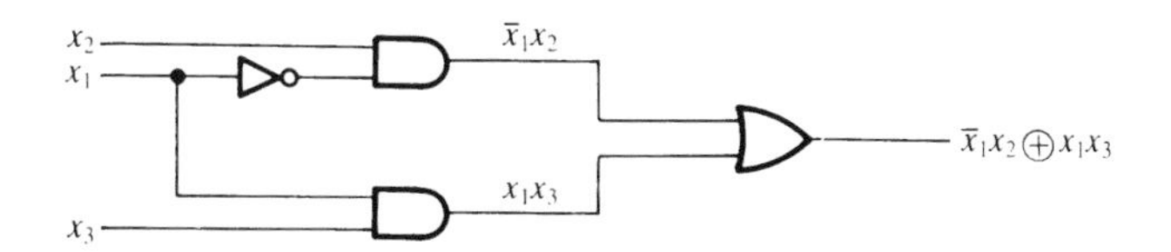

The input variable x_1 branches, one branch passing through the NOT-gate and the other through the AND-gate having the variable x_3 as its other input. The point at which the circuit branches is shown as a filled-in circle on the circuit diagram.

3. Determine the Boolean expression for the output of the following system of gates.

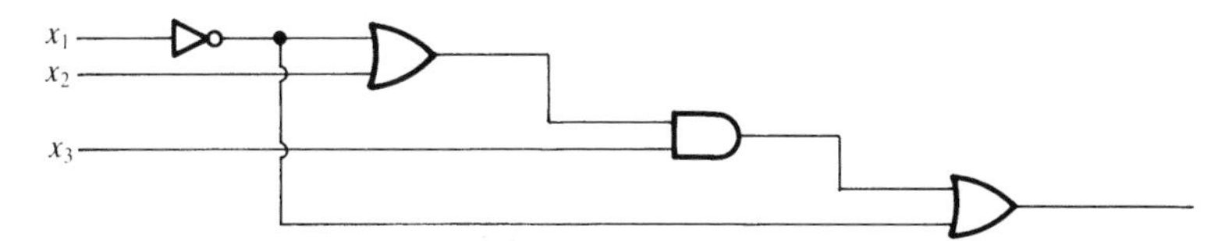

Solution

In this example the output from the inverter branches and one branch is used as one input to the final OR-gate. Proceeding as in the first example we have the following diagram.

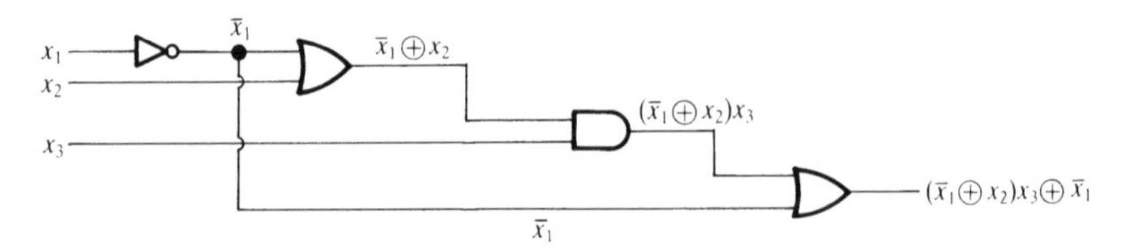

In all these examples the output is represented by a Boolean expression in terms of the input variables. We can, as usual, regard this Boolean expression as defining a function of the variables representing the states of the inputs. As we have seen, a particular function may be defined by a number of equivalent Boolean expressions. It follows therefore that, given any network of logic gates, there may be a number of equivalent networks. By 'equivalent' networks we mean that, given any set of values of the input variables, the output of any of these networks is the same.

For example, it is a simple matter to establish the equivalence of the two Boolean expressions.

$$(x_1 \oplus x_2)(\bar{x}_1 \oplus \bar{x}_2) \quad \text{and} \quad x_1\bar{x}_2 \oplus \bar{x}_1 x_2.$$

(Note that these are the conjunctive and disjunctive normal forms of the expression.) Thus the two logic networks whose outputs can be described by these expressions are equivalent. These are as follows.

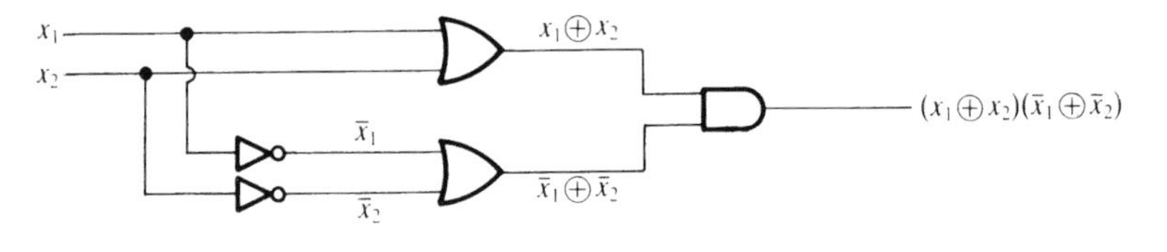

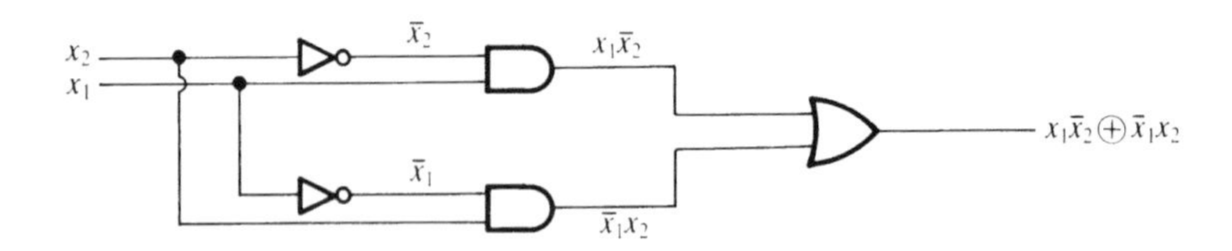

Since the two Boolean expressions are equivalent, the output is the same for any given set of values of the input variables x_1 and x_2.

The fact that a number of different logic networks may be equivalent raises again the question of how we may determine, given a particular Boolean expression describing the output, the simplest network which will do the job. By 'simplest' we mean the one with the fewest logic gates. We turn to this problem in the next section.

Exercises 10.4

1. Give a Boolean expression describing the output of each of the following logic networks.

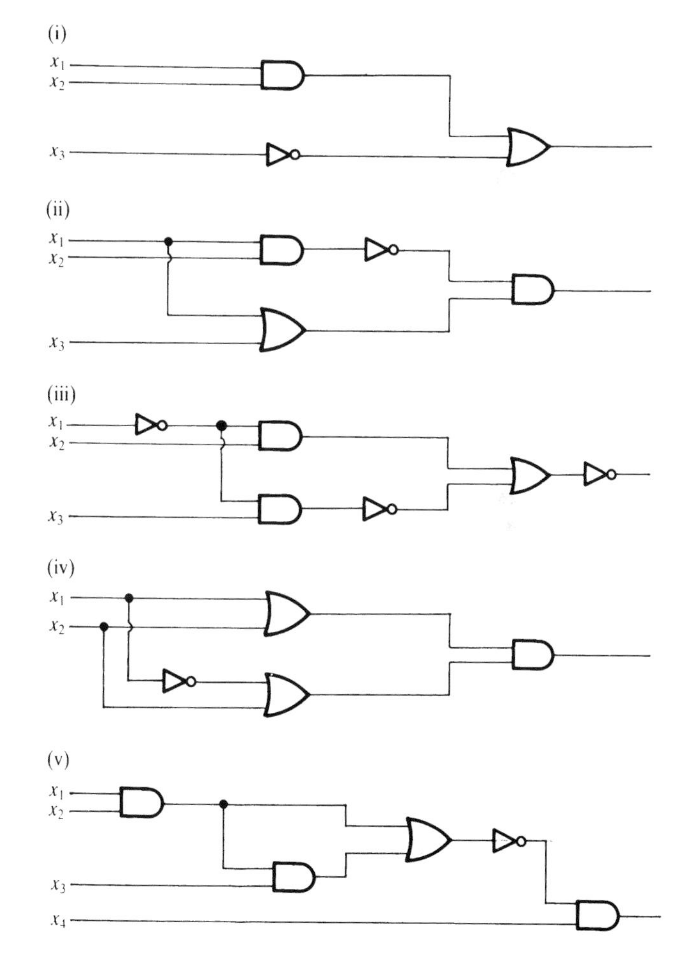

2. Design a logic network for each of the following so that the output is described by the Boolean expression given:

(i) $(x_1 \oplus x_2)(\bar{x}_1 \oplus x_3)$
(ii) $\bar{x}_1 x_2 \oplus x_3 x_2 \oplus x_1$
(iii) $x_1 x_3 \oplus \bar{x}_1 \oplus x_2 \bar{x}_3$
(iv) $x_1 x_2 x_3 \oplus \bar{x}_1 \bar{x}_2 \bar{x}_3$
(v) $(x_1 \oplus \bar{x}_2 \oplus x_3)(\bar{x}_1 \oplus x_3) x_2$.

3. The following circuits have more than one output. Write the Boolean expression for each of the outputs in terms of the input variables.

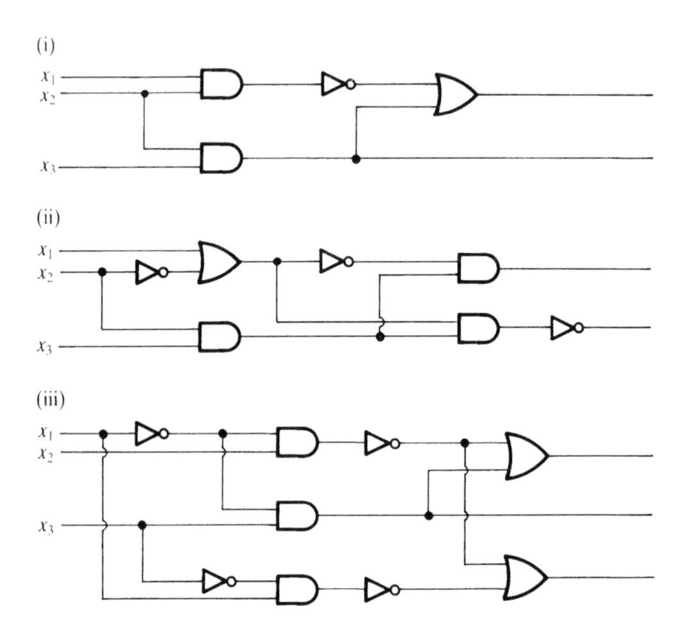

4. The **NAND-gate** is shown symbolically as the following.

The Boolean expression for its output is $\overline{x_1 x_2}$.

For each of the following, design a logic circuit which utilizes only NAND-gates so that the output is defined by the given Boolean expression:

(i) $x_1 \oplus x_2$
(ii) $x_1 x_2$
(iii) $\bar{x}_1$.

Deduce that, for any Boolean expression describing the output, a logic network can be designed using only NAND-gates. (You might find exercise 1.3.10 helpful here.)

5. The **NOR-gate** is shown symbolically as the following.

The Boolean expression for its output is $(\overline{x_1 \oplus x_2})$.

Repeat the last exercise substituting NOR-gate for NAND-gate.

6. (For readers familiar with binary arithmetic.) Suppose that x and y are single-digit binary numerals. The following table gives the sum of x and y as a two-digit binary numeral for all combinations of values of x and y.

x	y	Binary numeral for the sum $x + y$
0	0	00
0	1	01
1	0	01
1	1	10

Design a logic network that has outputs z_1 and z_2 so that the binary number whose first and second digits are z_1 and z_2 (read from left to right) represents the sum of the input variables x and y. (A logic network designed to add two single-digit binary numbers in this way is called a **half-adder**.)

FULL-ADDER HALF-ADDER

7. (Also for readers familiar with binary arithmetic and who have successfully solved the last problem.) A **full-adder** is a logic network which has three inputs x_1, x_2 and x_3 and two outputs z_1 and z_2, the first and second digits respectively of the binary sum of x_1, x_2 and x_3.

(i) Draw a table showing the values of the output variables z_1 and z_2 for each set of values of the input variables x_1, x_2 and x_3.

(ii) Design a logic network which will achieve the output described.

In practice, addition of binary numerals with several digits is achieved by using a half-adder to add the two least significant (rightmost) digits and using the 'carry digit' as input to a full-adder along with the next two digits to be added. The carry-digit from this sum is fed into the next full-adder along with the two digits third from the right in the summands, and so on. The diagram below shows this process applied to the addition of the binary numeral with digits x_1, x_2, x_3 (x_3 being the least significant digit) to another with digits y_1, y_2, y_3. The result is the binary numeral with digits z_1, z_2, z_3, z_4.

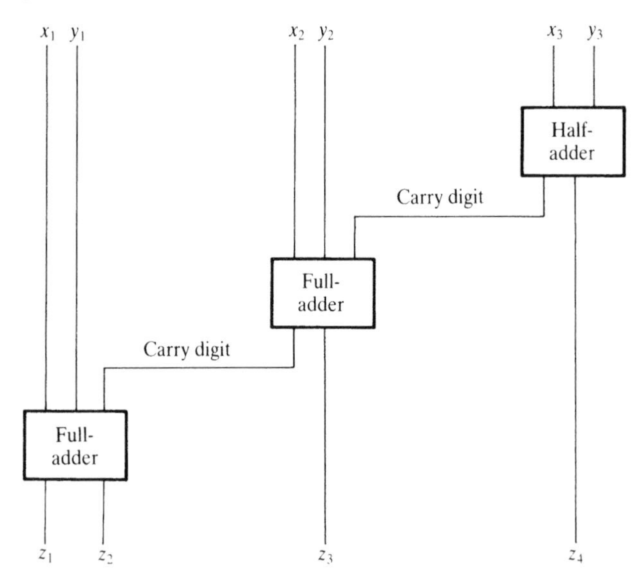

10.6 Minimization of Boolean Expressions

We now consider the following problem—given a Boolean expression (which may represent the output of a logic network or a switching function), what is the 'simplest' equivalent expression? Where the design of switching circuits and logic networks is concerned the question is an important one since it has implications for the cost of producing the circuit and for its efficiency of operation.

Of all the circuits for which a given Boolean expression describes the output, the cheapest to produce is the one having the fewest logic elements (gates or switches) and also the fewest inputs to these elements. An additional advantage for a circuit having as few logic elements as possible is that there is less chance that it will develop a fault.

For our purposes, given a Boolean expression, the 'simplest' Boolean expression equivalent to it will satisfy the following criteria:

(a) it will be expressed as the sum of terms which are themselves the product of literals;
(b) no other equivalent Boolean expression in this form contains fewer terms;
(c) of all the equivalent Boolean expressions in this form which have the same number of terms, none has fewer literals (each literal being counted every time it occurs).

Where these criteria are satisfied by a Boolean expression, we say that it is in **minimal form** (or just **minimal**). There may be several equivalent Boolean expressions which satisfy all three criteria so that the minimal form is not necessarily unique.

The technique which we shall use to obtain the minimal form of a Boolean expression is to start with the disjunctive normal form. This is the sum of terms which are the product of literals and so at least satisfies the first criterion. We shall then aim to reduce the number of terms as far as possible so that the second criterion is satisfied and then to reduce the number of literals. We now consider a systematic method by which this 'pruning' of the disjunctive normal form can be achieved.

Karnaugh Maps

A **Karnaugh map** is a diagrammatic representation of a Boolean expression in disjunctive normal form. It consists of a rectangle divided into subrectangles referred to as **cells** where each cell may be taken to represent a minterm. For a given number of variables the cells within the Karnaugh map represent all the possible minterms which may appear in the disjunctive normal form of a Boolean expression. The minterms are allocated to the cells in such a way that adjacent cells represent minterms in which all the literals are identical except for one which is complemented in one cell but not in an adjacent one. Thus movement around the map from cell to cell (up or down, to left or right, but not diagonally) gives a sequence of minterms where each is different from the last by only one literal.

The minterms represented by the cells in a Karnaugh map for a Boolean expression in the two variables x_1 and x_2 are shown in the diagram below.

	x_2	$\bar{x}_2$
x_1	x_1x_2	$x_1\bar{x}_2$
$\bar{x}_1$	$\bar{x}_1x_2$	$\bar{x}_1\bar{x}_2$

Written in each cell is the corresponding minterm. Notice that the requirement that adjacent cells differ by just one literal also applies at the edges of the map if we view the rightmost column of cells as being adjacent to the left-hand column and also the top and bottom rows as being adjacent. In using the map it is important to realize that the right and left edges are to be regarded as contiguous and so are the top and bottom edges.

The following is a layout for a Karnaugh map for three variables x_1, x_2 and x_3.

	x_2x_3	$\bar{x}_2x_3$	$\bar{x}_2\bar{x}_3$	$x_2\bar{x}_3$
x_1	a	b	c	d
$\bar{x}_1$	e	f	g	h

The cell labelled a represents the minterm $x_1x_2x_3$, cell c represents $x_1\bar{x}_2\bar{x}_3$, f represents $\bar{x}_1\bar{x}_2x_3$, etc. There is more than one way of constructing a Karnaugh map so that the necessary criteria are satisfied. An alternative for three variables is given below.

	x_3	$\bar{x}_3$
x_1x_2	a	d
$x_1\bar{x}_2$	b	c
$\bar{x}_1\bar{x}_2$	f	g
$\bar{x}_1x_2$	e	h

A Boolean expression given as the sum of minterms (i.e. in disjunctive normal form) is represented on the Karnaugh map by placing a one in each cell corresponding to a minterm which is present. For example, the Boolean expression $x_1x_2 \oplus \bar{x}_1x_2$ in the two variables x_1 and x_2 is represented by

	x_2	$\bar{x}_2$
x_1	1	
$\bar{x}_1$	1	

The Boolean expression in the three variables x_1, x_2 and x_3 given by

$$x_1\bar{x}_2x_3 \oplus x_1\bar{x}_2\bar{x}_3 \oplus \bar{x}_1x_2x_3 \oplus \bar{x}_1x_2\bar{x}_3$$

is represented by

	x_2x_3	$\bar{x}_2x_3$	$\bar{x}_2\bar{x}_3$	$x_2\bar{x}_3$
x_1		1	1	
$\bar{x}_1$	1			1

It is important to realize that only Boolean expressions which are the sum of minterms can be represented on a Karnaugh map. However, this need not concern us since we know that any Boolean expression can be written in disjunctive normal form (i.e. as the sum of minterms) and hence we can represent any Boolean expression on a Karnaugh map of appropriate dimensions.

To appreciate how a Karnaugh map will help us to accomplish the process of minimization, note that two adjacent ones (horizontally or vertically) in the map imply that the Boolean expression contains the sum of two minterms which differ only in that one variable is replaced by its complement. Where this is the case, this variable can be eliminated. For example, in the Karnaugh map above, the two adjacent ones in the top row indicate that the Boolean expression represented in the map contains the sum of $x_1\bar{x}_2x_3$ and $x_1\bar{x}_2\bar{x}_3$. Now

$$\begin{aligned} x_1\bar{x}_2x_3 \oplus x_1\bar{x}_2\bar{x}_3 &= x_1\bar{x}_2(x_3 \oplus \bar{x}_3) \quad \text{(axiom B4)} \\ &= x_1\bar{x}_2. \end{aligned}$$

Thus we can replace two terms consisting of a total of six literals by one term consisting of two literals.

We can proceed further: the map has another pair of adjacent ones—those in the bottom row. (Remember that the right and left edges are regarded as coincident.) Applying the same technique we have

$$\begin{aligned} \bar{x}_1x_2x_3 \oplus \bar{x}_1x_2\bar{x}_3 &= \bar{x}_1x_2(x_3 \oplus \bar{x}_3) \\ &= \bar{x}_1x_2. \end{aligned}$$

There are no further adjacent ones and so we have

$$x_1\bar{x}_2x_3 \oplus x_1\bar{x}_2\bar{x}_3 \oplus \bar{x}_1x_2x_3 \oplus \bar{x}_1x_2\bar{x}_3 = x_1\bar{x}_2 \oplus \bar{x}_1x_2.$$

In fact this is a minimal form of the original Boolean expression.

This idea can be extended to larger rectangular groups of ones. Consider, for example, the following Karnaugh map.

	x_2x_3	$\bar{x}_2x_3$	$\bar{x}_2\bar{x}_3$	$x_2\bar{x}_3$
x_1		1	1	
$\bar{x}_1$		1	1	

Here we have

$$\begin{aligned} x_1\bar{x}_2x_3 \oplus x_1\bar{x}_2\bar{x}_3 \oplus \bar{x}_1\bar{x}_2x_3 \oplus \bar{x}_1\bar{x}_2\bar{x}_3 & \\ &= x_1\bar{x}_2(x_3 \oplus \bar{x}_3) \oplus \bar{x}_1\bar{x}_2(x_3 \oplus \bar{x}_3) \\ &= x_1\bar{x}_2 \oplus \bar{x}_1\bar{x}_2 \\ &= \bar{x}_2(x_1 \oplus \bar{x}_1) \\ &= \bar{x}_2. \end{aligned}$$

Similarly, we can show that the rectangular grouping of ones in the Karnaugh map below represents a Boolean expression which is equivalent to the single term x_1.

	x_2x_3	$\bar{x}_2x_3$	$\bar{x}_2\bar{x}_3$	$x_2\bar{x}_3$
x_1	1	1	1	1
$\bar{x}_1$				

Thus a group of four ones arranged in a rectangular block (either 2×2, 1×4 or 4×1) allows replacement of four terms by one and the elimination of two variables. In a similar way we can show that a group of eight ones arranged in any rectangular block indicates that the eight corresponding minterms can be replaced by a single term in which three variables have been eliminated. In all these cases the variable or variables which remain are those which appear unchanged in all cells constituting the block.

It is important to note that only blocks of $2, 4, 8, \ldots$ cells lead to replacement of the appropriate number of minterms by a *single* term, so it is these blocks that we must look for in a Karnaugh map. Further, the larger the rectangular block, the greater the reduction in terms and so we must utilize the larger blocks where we have a choice. (In practice Karnaugh maps become too unwieldy for Boolean expressions in more than about four variables and other techniques for obtaining the minimal form, such as the Quine–McCluskey algorithm, are more appropriate. See, for instance, Gersting (2006).)

Given the criteria for a minimal form of a Boolean expression, our priorities in attempting to simplify an expression given as the sum of minterms are firstly to reduce the number of terms and secondly to reduce the number of literals. Therefore on the Karnaugh map we must aim to group into rectangles all cells containing a one and to use the least possible number of rectangular blocks which include all the marked cells. Since each block results in one term this ensures the minimum number of terms. Furthermore, each cell containing a one must be included in the largest possible block so that the number of literals appearing in the resulting term is as small as possible.

Given a Karnaugh map, the following sequence of steps normally enables identification of a minimal representation of a Boolean expression. Although the method usually gives the minimal form, it is not absolutely foolproof and, having applied it, it is wise to check that there is no other way of grouping the ones, which results in fewer terms or in the same number of terms but fewer literals (see exercise 10.5.5).

(1) Isolate any ones in the map which are not adjacent to any other ones. The terms corresponding to these cells cannot be reduced and will therefore appear unchanged in the minimal form.

(2) Locate any ones that are adjacent to *only one* other cell containing a one and circle the pair. For each of these pairs, the two minterms corresponding to the cells can be represented by a single term consisting of the literals common to both.

(3) Locate any ones which can be allocated to a block of four *in only one way* and circle that block. The corresponding four terms can be represented by one term consisting of the common literals.

(4) Locate any ones which can be allocated to a block of eight and circle that block. The corresponding eight terms can be represented by one term consisting of the common literals.

(5) For any cells containing a one that remain, form the largest possible rectangular groups so that there are as few groups as possible and so that all cells containing a one are enclosed in at least one block.

Note that the process allows for a one to be included in more than one block. This simply means that the term corresponding to that cell is considered as being repeated in the original Boolean expression. Since, for any b belonging to the underlying set B of a Boolean algebra, we have $b \oplus b = b$ (idempotent law), the minterms in a Boolean expression in disjunctive normal form can be repeated any number of times and the resulting expression is equivalent.

We now work through some examples to illustrate the method.

Examples 10.11

1. Find a minimal form of the Boolean expression

$$x_1x_2x_3 \oplus x_1\bar{x}_2x_3 \oplus x_1\bar{x}_2\bar{x}_3 \oplus \bar{x}_1\bar{x}_2\bar{x}_3 \oplus \bar{x}_1x_2\bar{x}_3.$$

Solution

The Karnaugh map for the expression given is as follows.

	x_2x_3	$\bar{x}_2x_3$	$\bar{x}_2\bar{x}_3$	$x_2\bar{x}_3$
x_1	1	1	1	
$\bar{x}_1$			1	1

There are no ones with no 'neighbours' so we proceed to the second step and identify all ones which can be paired with only one other one. These are in the top left and bottom right cells. Thus we have the following.

	x_2x_3	$\bar{x}_2x_3$	$\bar{x}_2\bar{x}_3$	$x_2\bar{x}_3$
x_1	1	1	1	
$\bar{x}_1$			1	1

There are no blocks of four ones so all we have left to do is to allocate the remaining one to a block of two. This can be done in either of two ways:

(a)

	x_2x_3	$\bar{x}_2x_3$	$\bar{x}_2\bar{x}_3$	$x_2\bar{x}_3$
x_1	1	1	1	
$\bar{x}_1$			1	1

(b)

	x_2x_3	$\bar{x}_2x_3$	$\bar{x}_2\bar{x}_3$	$x_2\bar{x}_3$
x_1	1	1	1	
$\bar{x}_1$			1	1

Thus we have two alternative minimal forms of the Boolean expression. They are

$$x_1x_3 \oplus \bar{x}_1\bar{x}_3 \oplus \bar{x}_2\bar{x}_3$$

(corresponding to the grouping in (a)) and

$$x_1x_3 \oplus \bar{x}_1\bar{x}_3 \oplus x_1\bar{x}_2$$

(corresponding to the grouping in (b)).

2. Find a minimal representation of

$$x_1x_2\bar{x}_3\bar{x}_4 \oplus x_1x_2x_3\bar{x}_4 \oplus \bar{x}_1x_2\bar{x}_3\bar{x}_4 \oplus \bar{x}_1\bar{x}_2x_3x_4 \oplus \bar{x}_1\bar{x}_2\bar{x}_3x_4 \oplus x_1\bar{x}_2x_3x_4 \oplus x_1\bar{x}_2\bar{x}_3x_4 \oplus x_1\bar{x}_2x_3\bar{x}_4.$$

Solution

The Karnaugh map is as follows.

	x_3x_4	$\bar{x}_3x_4$	$\bar{x}_3\bar{x}_4$	$x_3\bar{x}_4$
x_1x_2			1	1
$\bar{x}_1x_2$			1	
$\bar{x}_1\bar{x}_2$	1	1		
$x_1\bar{x}_2$	1	1		1

Again there are no isolated ones so we look for those for which there is only one possible 'partner'. There is one of these, in the second row.

	x_3x_4	$\bar{x}_3x_4$	$\bar{x}_3\bar{x}_4$	$x_3\bar{x}_4$
x_1x_2			1	1
$\bar{x}_1x_2$			1	
$\bar{x}_1\bar{x}_2$	1	1		
$x_1\bar{x}_2$	1	1		1

A unique block of four ones exists in the bottom left-hand corner and the remaining ones can be blocked into a pair.

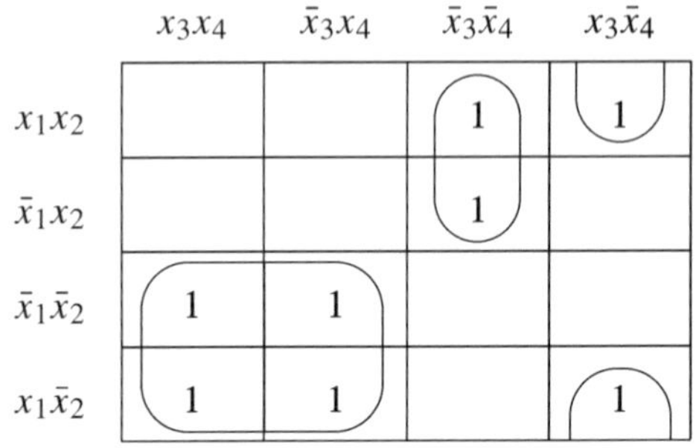

	x_3x_4	$\bar{x}_3x_4$	$\bar{x}_3\bar{x}_4$	$x_3\bar{x}_4$
x_1x_2			1	1
$\bar{x}_1x_2$			1	
$\bar{x}_1\bar{x}_2$	1	1		
$x_1\bar{x}_2$	1	1		1

Thus a minimal representation is

$$\bar{x}_2x_4 \oplus x_2\bar{x}_3\bar{x}_4 \oplus x_1x_3\bar{x}_4.$$

3. Find a minimal representation of the Boolean expression

$$x_1x_2\bar{x}_3x_4 \oplus \bar{x}_1x_2\bar{x}_3x_4 \oplus \bar{x}_1x_2\bar{x}_3\bar{x}_4 \oplus \bar{x}_1x_2x_3\bar{x}_4$$
$$\oplus\, \bar{x}_1\bar{x}_2x_3x_4 \oplus \bar{x}_1\bar{x}_2\bar{x}_3x_4 \oplus \bar{x}_1\bar{x}_2\bar{x}_3\bar{x}_4 \oplus x_1\bar{x}_2\bar{x}_3\bar{x}_4.$$

Solution

Dealing firstly with the pairs we have the following.

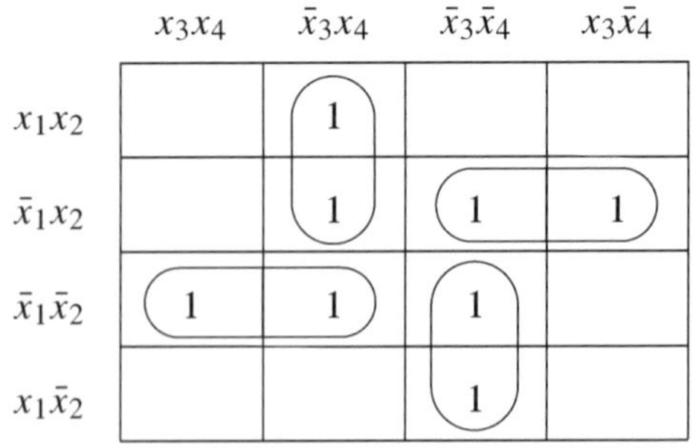

	x_3x_4	$\bar{x}_3x_4$	$\bar{x}_3\bar{x}_4$	$x_3\bar{x}_4$
x_1x_2		1		
$\bar{x}_1x_2$		1	1	1
$\bar{x}_1\bar{x}_2$	1	1	1	
$x_1\bar{x}_2$			1	

All marked cells have been allocated to a block and the minimal representation is

$$\bar{x}_1\bar{x}_2x_4 \oplus x_2\bar{x}_3x_4 \oplus \bar{x}_1x_2\bar{x}_4 \oplus \bar{x}_2\bar{x}_3\bar{x}_4.$$

We might have been tempted in this example to select the block of four ones as follows.

	x_3x_4	$\bar{x}_3x_4$	$\bar{x}_3\bar{x}_4$	$x_3\bar{x}_4$
x_1x_2		1		
$\bar{x}_1x_2$		1	1	1
$\bar{x}_1\bar{x}_2$	1	1	1	
$x_1\bar{x}_2$			1	

However, this would have meant that five terms would have occurred in the reduced expression even if we had 'paired' each of the remaining ones with an adjacent cell. This is not the minimal representation because it has a greater number of terms than our first solution.

4. Find a minimal representation of the Boolean expression

$$x_1x_2x_3x_4 \oplus x_1x_2\bar{x}_3x_4 \oplus \bar{x}_1x_2\bar{x}_3x_4 \oplus \bar{x}_1x_2\bar{x}_3\bar{x}_4 \oplus \bar{x}_1\bar{x}_2\bar{x}_3\bar{x}_4 \oplus \bar{x}_1\bar{x}_2\bar{x}_3x_4 \oplus x_1\bar{x}_2\bar{x}_3x_4 \oplus x_1\bar{x}_2x_3\bar{x}_4.$$

Solution

The Karnaugh map is as follows.

	x_3x_4	$\bar{x}_3x_4$	$\bar{x}_3\bar{x}_4$	$x_3\bar{x}_4$
x_1x_2	1	1		
$\bar{x}_1x_2$		1	1	
$\bar{x}_1\bar{x}_2$		1	1	
$x_1\bar{x}_2$		1		1

Firstly we circle the isolated one in the bottom right-hand corner. Then we pair the ones that have only a single adjacent one. This occurs in the top left-hand corner only.

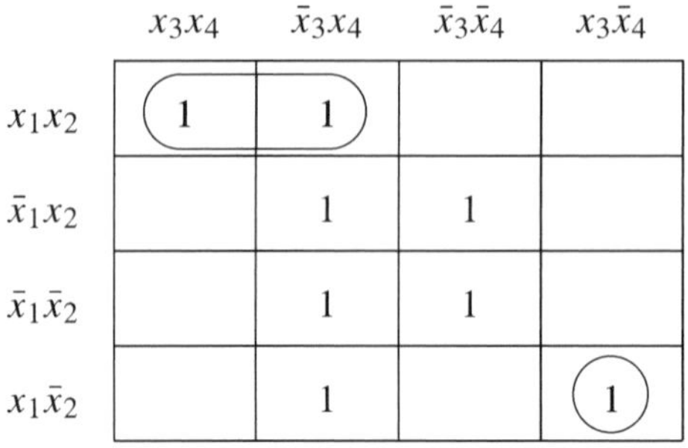

	x_3x_4	$\bar{x}_3x_4$	$\bar{x}_3\bar{x}_4$	$x_3\bar{x}_4$
x_1x_2	1	1		
$\bar{x}_1x_2$		1	1	
$\bar{x}_1\bar{x}_2$		1	1	
$x_1\bar{x}_2$		1		1

For all the other ones there is more than one way of assigning it to a 'pair', so we leave these for the time being.

Now we look for blocks of four—there are two of these.

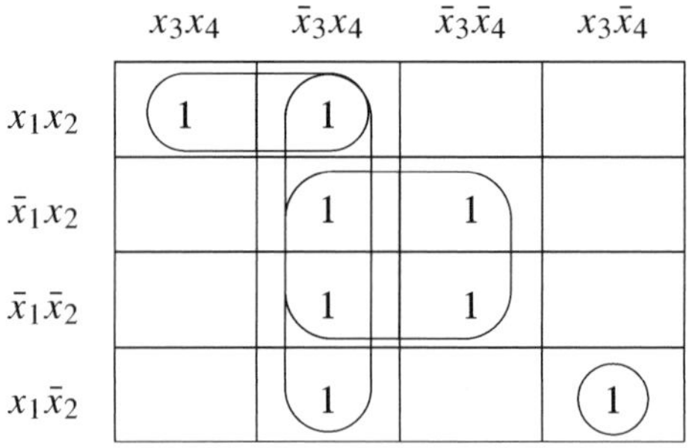

	x_3x_4	$\bar{x}_3x_4$	$\bar{x}_3\bar{x}_4$	$x_3\bar{x}_4$
x_1x_2	1	1		
$\bar{x}_1x_2$		1	1	
$\bar{x}_1\bar{x}_2$		1	1	
$x_1\bar{x}_2$		1		1

All the ones are now covered and so a minimal representation is

$$x_1x_2x_4 \oplus \bar{x}_3x_4 \oplus \bar{x}_1\bar{x}_3 \oplus x_1\bar{x}_2x_3\bar{x}_4.$$

Exercises 10.5

1. Find a minimal representation for each of the following Boolean expressions:

   (i) $x_1x_2x_3 \oplus \bar{x}_1\bar{x}_2x_3 \oplus x_1\bar{x}_2\bar{x}_3 \oplus \bar{x}_1\bar{x}_2\bar{x}_3$

   (ii) $x_1x_2x_3 \oplus x_1\bar{x}_2x_3 \oplus x_1\bar{x}_2\bar{x}_3 \oplus x_1x_2\bar{x}_3 \oplus \bar{x}_1x_2x_3$

   (iii) $x_1\bar{x}_2x_3\bar{x}_4 \oplus x_1\bar{x}_2x_3x_4 \oplus \bar{x}_1x_2x_3x_4 \oplus \bar{x}_1\bar{x}_2\bar{x}_3\bar{x}_4$

   (iv) $\bar{x}_1\bar{x}_2x_3x_4 \oplus \bar{x}_1\bar{x}_2x_3\bar{x}_4 \oplus x_1\bar{x}_2x_3\bar{x}_4 \oplus x_1x_2x_3\bar{x}_4 \oplus x_1x_2\bar{x}_3x_4 \oplus \bar{x}_1x_2x_3\bar{x}_4$

(v) $x_1x_2x_3x_4 \oplus x_1\bar{x}_2x_3x_4 \oplus x_1\bar{x}_2x_3\bar{x}_4 \oplus x_1\bar{x}_2\bar{x}_3\bar{x}_4 \oplus \bar{x}_1x_2x_3\bar{x}_4 \oplus \bar{x}_1\bar{x}_2\bar{x}_3\bar{x}_4$.

2. Find a minimal form of each of the following Boolean expressions by first writing the expression in disjunctive normal form and then using a Karnaugh map:

(i) $x_1(x_2x_3 \oplus \bar{x}_3)$
(ii) $(x_1 \oplus x_2)(\bar{x}_2 \oplus x_3)$
(iii) $(x_1 \oplus x_2 \oplus x_3)(\bar{x}_1 \oplus x_3)$
(iv) $(x_1 \oplus \bar{x}_2 \oplus x_3)(\bar{x}_1 \oplus x_2 \oplus x_3)(x_1 \oplus x_2)$.

3. For each of the Boolean expressions in exercise 10.4.2 obtain a minimal form of the given expression and sketch the corresponding logic network. Compare with the circuits obtained in exercise 10.4.2.

4. For exercises 10.3.5 and 10.3.6, design the 'simplest' switching circuit which will achieve the desired effect where 'simplest' means that the Boolean expression describing the circuit is in minimal form. Compare with the circuits previously obtained.

5. Draw the Karnaugh map for the Boolean expression

$$x_1x_2x_3x_4 \oplus \bar{x}_1x_2x_3x_4 \oplus \bar{x}_1\bar{x}_2x_3x_4 \oplus x_1\bar{x}_2x_3x_4 \oplus x_1x_2\bar{x}_3x_4 \oplus \bar{x}_1\bar{x}_2\bar{x}_3x_4 \oplus x_1\bar{x}_2\bar{x}_3x_4 \oplus x_1x_2\bar{x}_3\bar{x}_4 \oplus \bar{x}_1x_2\bar{x}_3\bar{x}_4 \oplus \bar{x}_1\bar{x}_2\bar{x}_3\bar{x}_4 \oplus x_1x_2x_3\bar{x}_4 \oplus \bar{x}_1x_2x_3\bar{x}_4.$$

Using the method of grouping ones as described in the text, simplify this expression as far as you can. Show that the method does not lead to the minimal form of this Boolean expression, i.e. show that there is a better blocking of the marked cells which gives an expression with fewer terms.

Chapter 11

Graph Theory

11.1 Definitions and Examples

Although generally regarded as one of the more modern branches of mathematics, graph theory actually dates back to 1736. In that year Leonhard Euler† published the first paper on what is now called graph theory. In the paper, Euler developed a theory which solved the so-called Königsberg Bridge problem (see §11.2). Surely few other branches of the subject can be given as precise a 'birthday' as this. However, it must be said that, as a mature subject, graph theory is indeed modern. It came of age, so to speak, exactly 200 years after Euler's paper with the publication in 1936 of the first text in graph theory. (The first 200 years of graph theory are beautifully outlined in Biggs *et al* (1976) which includes extracts from many of the original papers concerned with the development of graph theory.)

Like many of the concepts we have considered, the idea of a graph is very simple. It is probably due to its simplicity that graph theory has found many applications in recent years in fields as diverse as chemistry, computer science, economics, electronics and linguistics.

Before we begin by explaining what a graph is, perhaps we should say what it is *not*. The term 'graph' as used in this chapter and the next does not mean the graph of a function (considered in chapter 5). It is unfortunate that the same term has

† Euler (1707–83) was born in Switzerland and spent most of his long life in Russia (St Petersburg) and Prussia (Berlin). He was the most prolific mathematician of all time, his collected works filling more than 70 volumes. Like many of the very great mathematicians of his era, Euler contributed to almost every branch of pure and applied mathematics. He is also responsible, more than any other person, for much of the mathematical notation in use today.

two quite different meanings, although it is usually clear from the context which meaning is intended.

What, then, is a 'graph'? Intuitively, a graph is simply a collection of points, called 'vertices', and a collection of lines, called 'edges', each of which joins either a pair of points or a single point to itself. A familiar example, which serves as a useful analogy, is a road map which shows towns as vertices and the roads joining them as edges.

For mathematical purposes we require a more precise definition. In order to define a graph, we first need to specify the set of its vertices and the set of its edges. Then we need to say, in precise mathematical terms, which edges join which vertices. An edge is defined as having a vertex at each end, so we need to associate with every edge of the graph its endpoint vertices. The endpoints of an edge are either a pair of vertices (if the edge joins two different vertices) or a single vertex (if the edge joins a vertex to itself). Thus for every edge e of a graph we define a set $\{v_1, v_2\}$ of vertices which specifies that e joins vertices v_1 and v_2, where of course we need to allow the possibility that $v_1 = v_2$. Now this set $\{v_1, v_2\}$, which we denote by $\delta(e)$, is a subset of the set of vertices. Therefore $\delta(e)$ is an *element* of the power set of the vertex set. This leads us to the following formal definition. Its rather technical nature should not be allowed to obscure the essentially simple concept that is being described.

Definition 11.1

An **undirected graph** Γ comprises:

(i) a finite non-empty set V of **vertices**,
(ii) a finite set E of **edges**, and
(iii) a function $\delta : E \to \mathscr{P}(V)$ such that, for every edge e, $\delta(e)$ is a one- or two-element subset of V.

The edge e is said to **join** the element(s) of $\delta(e)$.

Generally we shall use the term 'graph' without qualification to mean an undirected graph. If we need to emphasize a specific graph Γ we will write V_Γ, E_Γ and δ_Γ for the sets V and E and the function $\delta : E \to \mathscr{P}(V)$ respectively.

As we have explained, this function δ is merely a formal way of specifying the ends of edges. In this case where an edge e joins a vertex to itself, the

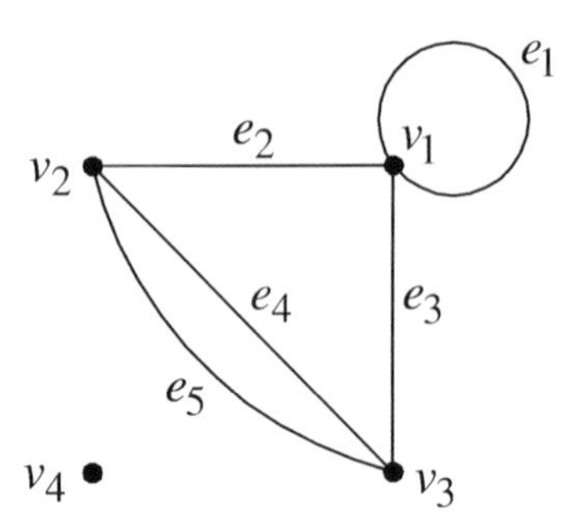

Figure 11.1

set $\delta(e)$ will contain a single element. Consider, for example, the graph Γ represented in figure 11.1. Clearly Γ has vertex set $\{v_1, v_2, v_3, v_4\}$ and edges set $\{e_1, e_2, e_3, e_4, e_5\}$. The function $\delta : E \to \mathscr{P}(v)$ is defined by

$$\begin{aligned}
\delta &: e_1 \mapsto \{v_1\} \\
\delta &: e_2 \mapsto \{v_1, v_2\} \\
\delta &: e_3 \mapsto \{v_1, v_3\} \\
\delta &: e_4 \mapsto \{v_2, v_3\} \\
\delta &: e_5 \mapsto \{v_2, v_3\}.
\end{aligned}$$

This simply indicates that e_1 joins vertex v_1 to itself, e_2 joins vertices v_1 and v_2, etc.

We emphasize that an edge may join a vertex to itself, as in the case of e_1, and a vertex may be connected to no edges at all, as in the case of v_4. Also note that a given pair of vertices may be joined by more than one edge; in this example the edges e_4 and e_5 both connect the vertices v_2 and v_3.

Unfortunately there are many variations on the definition of a graph. Some authors use a definition which excludes the possibility of multiple edges in their graphs; that is, several edges connecting the same pair of vertices. Other definitions exclude the possibility of loops—edges which join a vertex to itself. We shall call a graph which satisfies both these restrictions—that has no loops or multiple edges—a **simple graph**†. The terminology of graph theory is distinctly non-standard. When consulting other texts you are strongly advised to check very carefully the author's definitions and terminology.

There is one restriction which we have placed on a graph which, though common, is not universal; namely, that the sets of vertices and edges are finite. If either (or

† Those authors who define a graph to be what we are calling a simple graph frequently use the term **multigraph** to denote the more general concept (the one which we have called graph).

both) of these are infinite Γ is usually called an **infinite graph**, although we shall not consider these.

We should emphasize at the outset that a graph and a diagram representing it are not the same thing. As we have defined it, a graph consists of two sets together with a function. Figure 11.1 itself is not a graph but a pictorial representation of one. Whilst diagrams are extremely helpful in understanding the properties of graphs, some care needs to be taken in interpreting them. The most significant point to make is that a given graph may be represented by two diagrams which appear very different. For instance, the two diagrams in figure 11.2 represent the same graph Γ as can be observed by writing down the function $\delta : E \to \mathscr{P}(V)$. The diagram in figure 11.2(a) indicates why this graph is sometimes called the **cycle graph** with seven vertices, denoted by C_7. Clearly for every positive integer n there is a corresponding graph C_n, the cycle graph with n vertices and n edges. For each n, the diagram representing C_n can be drawn as a circle with n vertices around its circumference. It should be clear that C_n is simple if and only if $n \geqslant 3$.

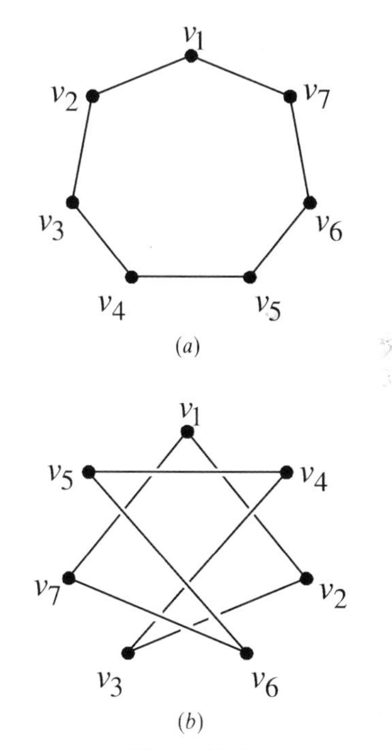

Figure 11.2

When first studying graph theory, one thing soon becomes apparent: there are initially many more definitions than theorems. This is probably because to say, or prove, anything significant about graphs requires reasonably developed terminology. We collect together below a few basic definitions.

Definitions 11.2

(i) A pair of vertices v and w are **adjacent** if there exists an edge e joining them. In this case we say both v and w are **incident** to e and also that e is **incident** to v and to w.

(ii) The edges $e_1, e_2, \ldots, e_n$ are **adjacent** if they have at least one vertex in common.

(iii) The **degree** or **valency**, $deg(v)$, of a vertex v is the number of edges which are incident to v. (Unless stated otherwise, a loop joining v to itself counts *two* towards the degree of v.) A graph in which every vertex has the same degree r is called **regular (with degree** r**)** or simply r**-regular**.

Examples 11.1

1. Let Γ be the graph illustrated in figure 11.1. The vertices v_1 and v_2 are adjacent, because the edge e_2 joins them. Similarly v_1 and v_3 are adjacent, as are v_2 and v_3. The vertex v_4 is adjacent to no other vertex.

   Edges e_1, e_2 and e_3 are adjacent, since they all meet at vertex v_1. Similarly e_2, e_4, e_5 are adjacent, as are e_3, e_4, e_5.

   Note that only *pairs* of vertices may be adjacent, but any number of edges can be adjacent.

   The degrees of the four vertices are given in the following table.

Vertex	Degree
v_1	4
v_2	3
v_3	3
v_4	0

2. A well known 3-regular simple graph is **Petersen's graph**. Two diagrams representing this graph are given in figure 11.3. (We have omitted to label the vertices and edges for the sake of clarity of the diagram.)

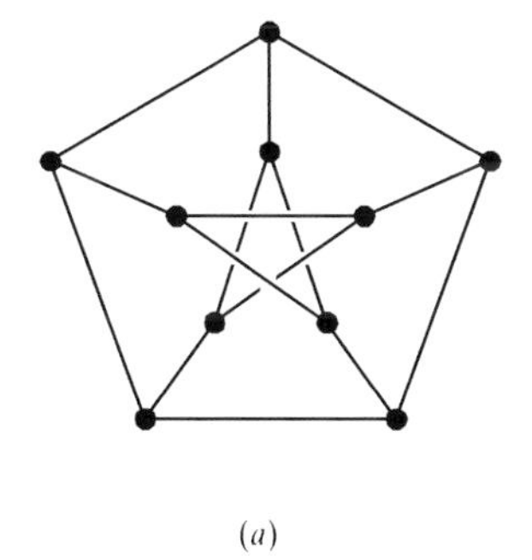
(*a*)

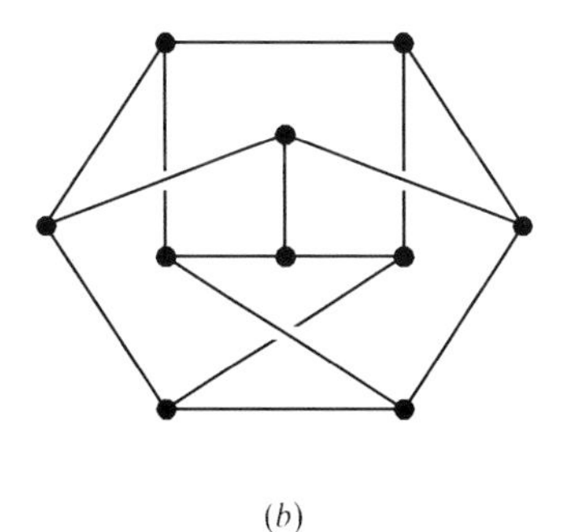
(*b*)

Figure 11.3

In drawing diagrams of graphs we only allow edges to meet at vertices. It is not always possible to draw diagrams *in the plane* satisfying this property (see §11.5), so we may need to indicate that one edge passes underneath another as we have done in figure 11.3.

Definitions 11.3

(i) A **null graph** (or **totally disconnected graph**) is one whose edge set is empty. (Pictorially, a null graph is just a collection of points.)

(ii) A **complete graph** is a simple graph in which every pair of distinct vertices is joined by an edge.

(iii) A **bipartite graph** is a graph where the vertex set has a partition $\{V_1, V_2\}$ such that every edge joins a vertex of V_1 to a vertex of V_2.

(iv) A **complete bipartite graph** is a bipartite graph such that every vertex of V_1 is joined to every vertex of V_2 by a unique edge.

Examples 11.2

1. Since a complete graph is simple there are no loops and each pair of distinct vertices is joined by a *unique* edge. Clearly a complete graph is uniquely specified by the number of its vertices.

The **complete graph K_n with n vertices** can be described as follows. It has vertex set $V = \{v_1, v_2, \ldots, v_n\}$ and edge set $E = \{e_{ij} : 1 \leqslant i < j \leqslant n\}$ with the function δ given by

$$\delta(e_{ij}) = \{v_i, v_j\}.$$

The graph K_n is clearly regular with degree $n - 1$, since every vertex is connected, by a unique edge, to each of the other $n - 1$ vertices.

The complete graphs with three, four and five vertices are illustrated in figure 11.4.

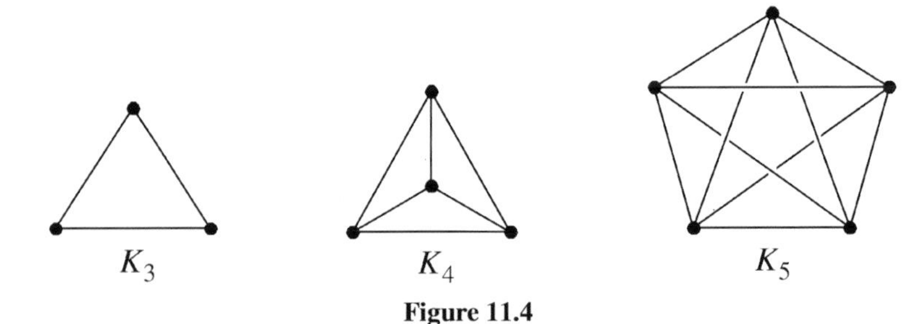

Figure 11.4

2. Let Γ be a bipartite graph where the vertex set V has the partition $\{V_1, V_2\}$. Note that Γ need not be simple. All that is required is that each edge must join a vertex of V_1 to a vertex of V_2. Given $v_1 \in V_1$ and $v_2 \in V_2$, there may be more than one edge joining them or no edge joining them. Clearly, though, there are no loops in Γ.

   A complete bipartite graph is completely specified by $|V_1|$ and $|V_2|$. The **complete bipartite graph on n and m vertices**, denoted $K_{n,m}$, has $|V_1| = n$ and $|V_2| = m$. It is necessarily simple.

   Figure 11.5 shows two bipartite graphs. In each case the vertices of V_1 are indicated by full circles and the vertices of V_2 by crosses. The graph in (*b*) is the complete bipartite graph, $K_{3,3}$.

We have noted that a graph Γ may be represented by diagrams that appear very different. An alternative way of representing a graph, one which is easier for computer representation, for instance, is by its 'adjacency matrix' which we now define.

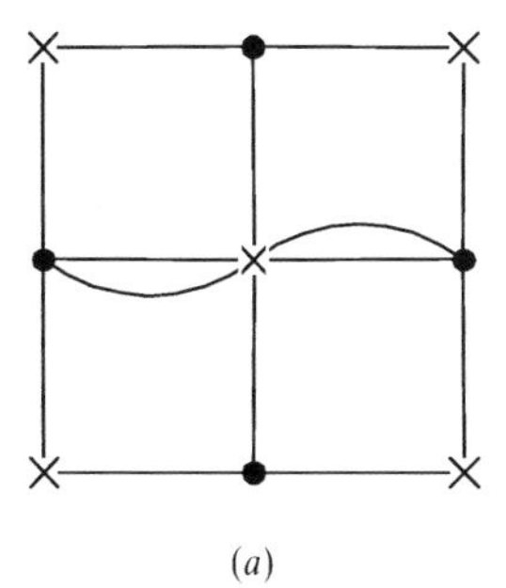
(a)

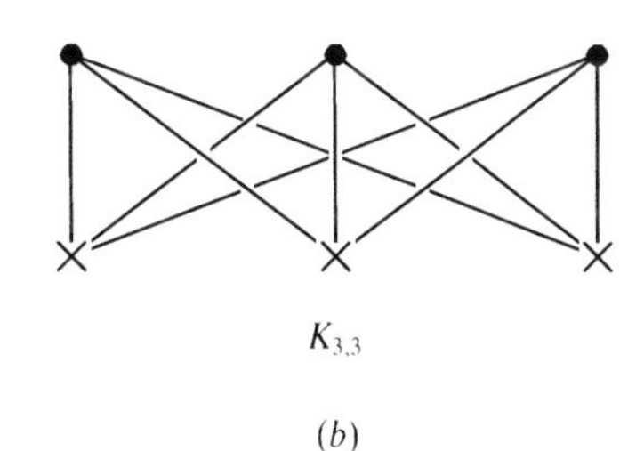

(b)

Figure 11.5

Definition 11.4

Let Γ be a graph with vertex set $\{v_1, v_2, \ldots, v_n\}$. The **adjacency matrix** of Γ is the $n \times n$ matrix $A = A(\Gamma)$ such that a_{ij} is the number† of distinct edges joining v_i and v_j.

The adjacency matrix is necessarily symmetric as the number of edges joining v_i and v_j is the same as the number joining v_j and v_i. The degree of vertex v_i is easily determined from the adjacency matrix. If there are no loops at v_i then its degree is the sum of the entries in the ith column (or ith row) of the matrix. Since every loop counts twice in the degree, when summing the entries in the ith column (or ith row) the diagonal element a_{ii} must be doubled to obtain the degree of v_i.

Examples 11.3

1. The following is the adjacency matrix A of the graph represented in figure 11.1:

$$A = \begin{pmatrix} 1 & 1 & 1 & 0 \\ 1 & 0 & 2 & 0 \\ 1 & 2 & 0 & 0 \\ 0 & 0 & 0 & 0 \end{pmatrix}.$$

† Unfortunately this is another instance where terminology varies. For some authors the adjacency matrix is a binary matrix with $a_{ij} = 0$ if v_i and v_j are not adjacent and $a_{ij} = 1$ if they are adjacent, regardless of the *number* of edges connecting them. For simple graphs, of course, there is no distinction since there can be at most one edge joining any pair of vertices.

Note that $V = \{v_1, v_2, v_3, v_4\}$ and the rows and columns of A refer to the vertices in the order listed. Just as for the binary matrix of a relation (see §4.1), we must always be clear which rows and columns refer to which vertices.

Two properties of the graph are immediately apparent from the matrix. Firstly, by considering the leading diagonal we note that there is only one loop—from v_1 to itself. Secondly, the last row (or column) of zeros indicates that v_4 is an **isolated vertex** connected to no vertices at all (including itself).

The degrees of the vertices are easily calculated from the matrix as follows:

$$\begin{aligned} deg(v_1) &= 2 \times 1 + 1 + 1 = 4 \\ deg(v_2) &= 1 + 2 = 3 \\ deg(v_3) &= 1 + 2 = 3 \\ deg(v_4) &= 0. \end{aligned}$$

2. The null graph with n vertices has the $n \times n$ zero matrix $O_{n \times n}$ as its adjacency matrix, since there are no edges whatsoever.

3. A complete graph has adjacency matrix with zeros along the leading diagonal (since there are no loops) and ones everywhere else (since every vertex is joined to every other by a unique edge).

There is one more piece of terminology which we wish to introduce in this section. The notion of a 'subgraph' of a graph is probably more or less self-evident. The formal definition is the following. (Compare with definition 8.12 of a subgroup, for example.)

Definition 11.5

A graph Σ is a **subgraph** of the graph Γ, denoted $\Sigma \leqslant \Gamma$, if $V_\Sigma \subseteq V_\Gamma$, $E_\Sigma \subseteq E_\Gamma$ and $\delta_\Sigma(e) = \delta_\Gamma(e)$, for every edge e of Σ.

The condition that $\delta_\Sigma(e) = \delta_\Gamma(e)$, for every edge e of Σ, means only that the edges of the subgraph Σ must join the same vertices as they do in Γ. Intuitively,

Σ is a subgraph of Γ if we can obtain a diagram for Σ by erasing some of the vertices and/or edges from a diagram of Γ. Of course, if we erase a vertex we must also erase all edges incident to it.

Example 11.4

Graphs Γ and Σ have vertex sets $V_\Gamma = \{v_1, v_2, v_3, v_4, v_5\}$ and $V_\Sigma = \{v_1, v_2, v_4, v_5\}$ and respective adjacency matrices

$$\begin{pmatrix} 1 & 1 & 0 & 1 & 1 \\ 1 & 0 & 2 & 1 & 0 \\ 0 & 2 & 0 & 0 & 1 \\ 1 & 1 & 0 & 0 & 1 \\ 1 & 0 & 1 & 1 & 0 \end{pmatrix} \quad \text{and} \quad \begin{pmatrix} 1 & 1 & 0 & 1 \\ 1 & 0 & 0 & 0 \\ 0 & 0 & 0 & 1 \\ 1 & 0 & 1 & 0 \end{pmatrix}.$$

Figure 11.6 indicates that we can regard Σ as a subgraph of Γ.

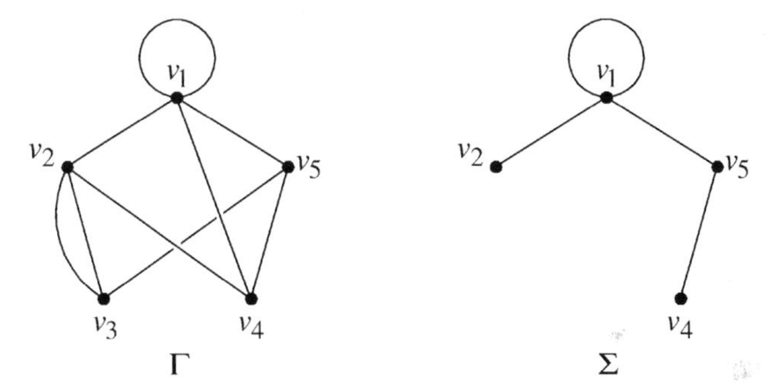

Figure 11.6

Exercise 11.1

1. Draw diagrams to represent the complete graphs K_2 and K_6 and the complete bipartite graphs $K_{2,5}$ and $K_{4,4}$.

2. Draw diagrams to represent each of the graphs whose adjacency matrix is given below. Write down the degree of each vertex, and state whether the graph is (a) simple; (b) regular.

(i) $\begin{pmatrix} 1 & 1 & 1 & 1 \\ 1 & 1 & 1 & 1 \\ 1 & 1 & 1 & 1 \\ 1 & 1 & 1 & 1 \end{pmatrix}$

(ii) $$\begin{pmatrix} 0 & 1 & 0 & 0 & 0 & 1 \\ 1 & 0 & 1 & 0 & 0 & 0 \\ 0 & 1 & 0 & 1 & 0 & 1 \\ 0 & 0 & 1 & 0 & 1 & 0 \\ 0 & 0 & 0 & 1 & 0 & 1 \\ 1 & 0 & 1 & 0 & 1 & 0 \end{pmatrix}$$

(iii) $$\begin{pmatrix} 1 & 2 & 0 & 2 & 1 \\ 2 & 1 & 2 & 0 & 1 \\ 0 & 2 & 1 & 2 & 1 \\ 2 & 0 & 2 & 1 & 1 \\ 1 & 1 & 1 & 1 & 0 \end{pmatrix}.$$

3. Copy figure 11.3(a) of Petersen's graph and label the vertices and edges.

   (i) Write out explicitly the function $\delta : E \to \mathscr{P}(V)$.
   (ii) Write down the adjacency matrix for the graph.

4. Is it possible for a graph to be both null and complete? If so, how is it possible? If not, why not?

5. (i) For each of the graphs in figure 11.5, label the vertices and edges and write down the adjacency matrix of the graph.

   (ii) Let Γ be any bipartite graph whose vertex set is partitioned into the subsets $\{v_1, v_2, \ldots, v_p\}$ and $\{w_1, w_2, \ldots, w_q\}$. What can you say about the adjacency matrix of Γ?

6. (i) Is Petersen's graph bipartite? Justify your answer.
   (ii) For which values of n is the cycle graph C_n bipartite?
   (iii) Devise an algorithm for testing whether a graph is bipartite given a diagram of the graph.

7. (i) Prove the following well known result about graphs.

   The Handshaking Lemma
   In any graph, the sum of the vertex degrees is twice the number of edges,
   $$\sum_{v \in V} deg(v) = 2 \times |E|.$$

   (ii) Deduce that, in any graph, the number of vertices with odd degree is even.

8. The **degree sequence** of a graph is the sequence of its vertex degrees

arranged in non-decreasing order. For example, the degree sequence of the graph shown in figure 11.1 is $(0, 3, 3, 4)$.

(i) Write down the degree sequence of each of the graphs illustrated in figures 11.2–11.6.

(ii) Describe the degree sequence of

(a) a null graph with n vertices;
(b) the complete graph K_n;
(c) an r-regular graph with n vertices;
(d) the complete bipartite graph $K_{n,m}$ where $n \leqslant m$.

(iii) What information about a graph may be deduced from:

(a) the number of entries in its degree sequence?
(b) the sum of the entries of its degree sequence?

9. For each of the following sequences, either draw the diagram of a graph with the given sequence as its degree sequence or explain why no graph has the given sequence as its degree sequence.

(i) $(2, 2, 2, 2, 3, 3, 4)$
(ii) $(1, 2, 2, 2, 3, 3)$
(iii) $(1, 2, 2, 2, 2, 3)$
(iv) $(2, 2, 2, 3, 3, 3, 3)$
(v) $(2, 2, 2, 2, 3, 3, 3)$.

10. For each of the following matrices, draw the diagram of a graph with the given matrix as its adjacency matrix and write down the degree sequence of the graph.

(i) $\begin{pmatrix} 0 & 1 & 0 & 1 & 1 \\ 1 & 0 & 1 & 0 & 0 \\ 0 & 1 & 0 & 1 & 1 \\ 1 & 0 & 1 & 0 & 1 \\ 1 & 0 & 1 & 1 & 0 \end{pmatrix}$ (ii) $\begin{pmatrix} 0 & 1 & 0 & 0 & 1 & 1 \\ 1 & 0 & 1 & 0 & 0 & 1 \\ 0 & 1 & 0 & 1 & 0 & 1 \\ 0 & 0 & 1 & 0 & 1 & 1 \\ 1 & 0 & 0 & 1 & 0 & 1 \\ 1 & 1 & 1 & 1 & 1 & 0 \end{pmatrix}$.

11. For $n \geqslant 2$, the **wheel graph** W_n is the graph obtained from the cycle graph C_n by adding a single new vertex and joining it to each existing vertex of C_n by a unique edge. Diagrams of the graphs of W_5 and W_6 are given below.

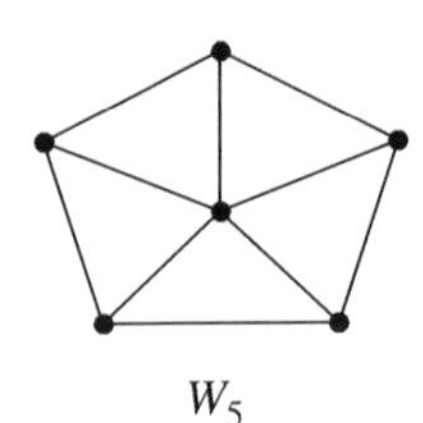

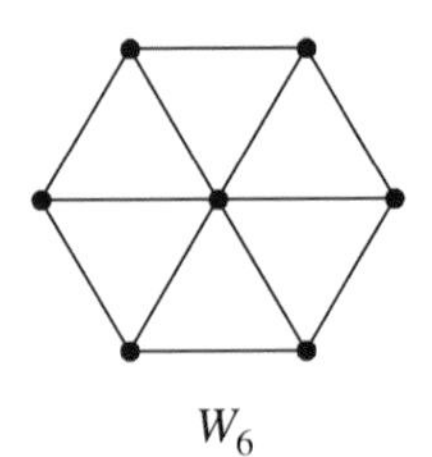

Describe

(i) the degree sequence
(ii) the adjacency matrix

of W_n.

12. Prove that, for $n \geqslant 2$, the complete graph K_n contains K_{n-1} as a subgraph.

13. Show that, if Γ is a simple graph with n vertices, then $|E| \leqslant \frac{1}{2}n(n-1)$. (Hint: think of K_n.)

14. Let Γ and Σ be two graphs with disjoint vertex and edge sets. The **union** of Γ and Σ is the graph denoted $\Gamma \cup \Sigma$ whose vertex and edge sets are respectively the unions of the vertex and edge sets of Γ and Σ with the obvious function δ. The **sum,** $\Gamma + \Sigma$, of Γ and Σ is obtained by taking the union of Γ and Σ and then joining each vertex of Γ to each vertex of Σ by a unique edge.

    What is the sum of (i) two null graphs, (ii) two complete graphs?

15. Explain why any simple graph with n vertices may be regarded as a subgraph of the complete graph K_n.

16. Let Γ be a graph without any loops and with vertex and edge sets $\{v_1, v_2, \ldots, v_n\}$ and $\{e_1, e_2, \ldots, e_m\}$ respectively. The **incidence matrix** of Γ is the $m \times n$ matrix B such that $b_{ij} = 1$ if the edge e_i is incident with the vertex v_j, and $b_{ij} = 0$ otherwise.

    (i) With a suitable choice of labelling if necessary, write down the incidence matrices of the graphs represented in figures 11.1, 11.2, 11.3 and 11.4.

    (ii) What can be said about the sum of the entries in each row?

    (iii) What information is provided by the sum of the entries of each column?

17. Draw a diagram to represent the graph whose incidence matrix is:

$$\text{(i)} \quad \begin{pmatrix} 1 & 1 & 0 & 0 & 0 \\ 1 & 1 & 0 & 0 & 0 \\ 1 & 0 & 0 & 0 & 1 \\ 1 & 0 & 0 & 1 & 0 \\ 0 & 1 & 0 & 0 & 1 \\ 0 & 1 & 0 & 1 & 0 \\ 0 & 0 & 0 & 1 & 1 \\ 0 & 0 & 0 & 1 & 1 \end{pmatrix} \qquad \text{(ii)} \quad \begin{pmatrix} 1 & 1 & 0 & 0 & 0 & 0 \\ 1 & 0 & 0 & 1 & 0 & 0 \\ 0 & 1 & 0 & 1 & 0 & 0 \\ 0 & 1 & 0 & 0 & 1 & 0 \\ 0 & 0 & 1 & 1 & 0 & 0 \\ 0 & 0 & 1 & 0 & 0 & 1 \\ 0 & 0 & 0 & 1 & 1 & 0 \\ 0 & 0 & 0 & 0 & 1 & 1 \end{pmatrix}.$$

18. This question refers to the 'blubs and glugs' axiom system introduced in chapter 2.

    (i) Show that the axiom system may be modelled by a graph Γ with 'blubs' interpreted as vertices and 'glugs' as edges. What special properties must Γ satisfy in order that it is a model for the system?

    (ii) Give an alternative model for the axiom system as a bipartite graph with 'blubs' and 'glugs' interpreted as the elements of the sets of vertices V_1 and V_2 respectively. In this model what is the interpretation of 'lies on'? What special properties must the bipartite graph satisfy in order to serve as a model for the system?

11.2 Paths and Cycles

Using the analogy of a road map, we can consider various types of 'journeys' in a graph. For instance, if the graph actually represents a network of roads connecting various towns, one question we might ask is: is there a journey, beginning and ending at the same town, which visits every other town just once without traversing the same road more than once? As usual, we begin with some definitions.

Definitions 11.6

(i) An **edge sequence of length n** in a graph Γ is a sequence of (not necessarily distinct) edges $e_1, e_2, \ldots, e_n$ such that e_i and e_{i+1} are adjacent for $i = 1, 2, \ldots, n-1$. The edge sequence determines a sequence of vertices (again, not necessarily distinct) $v_0, v_1, v_2, \ldots, v_{n-1}, v_n$ where $\delta(e_i) = \{v_{i-1}, v_i\}$. We say v_0 is the **initial vertex** and v_n the **final vertex** of the edge sequence.

(ii) A **path** is an edge sequence in which all the edges are distinct. If in addition all the vertices are distinct (except possibly $v_0 = v_n$) the path is called **simple**.

(iii) An edge sequence is **closed** if $v_0 = v_n$. A closed simple path containing at least one edge is called a **cycle** or a **circuit**.

CLOSED PATH

An edge sequence is any finite sequence of edges which can be traced on the diagram of the graph without removing pen from paper. It may repeat edges, go round loops several times, etc. Edge sequences are too general to be of very much use which is why we have defined paths. In a path we are not allowed to 'travel along' the same edge more than once. If, in addition, we do not 'visit' the same vertex more than once (which rules out loops), then the path is simple. The edge sequence or path is closed if we begin and end the 'journey' at the same place.

Examples 11.5

1. Let Γ be the graph represented in figure 11.1; examples of edge sequences in Γ are:

   (i) e_1, e_3, e_4, e_5, e_3;
   (ii) e_3, e_3;
   (iii) e_2, e_3, e_4;
   (iv) e_4, e_3;
   (v) e_4, e_5, e_2.

   Sequence (i) is a closed edge sequence beginning and ending at v_1: it determines the vertex sequence $v_1, v_1, v_3, v_2, v_3, v_1$. This edge sequence is not a path because the edge e_3 is traversed twice.

   Sequence (ii) is also closed, but it is ambiguous whether it begins (and ends) at v_1 or v_3. The vertex sequence could be either v_1, v_3, v_1 or v_3, v_1, v_3. This ambiguity will always occur in an edge sequence of the form $e_i, e_i, \ldots, e_i$ where e_1 is not a loop†. Again, it is not a path.

   Sequence (iii) is a cycle: it begins and ends at v_2 and no edge or vertex (except v_2 itself) is repeated.

   Sequence (iv) is a simple path from v_2 to v_1.

   Sequence (v) is a path with initial and final vertices v_2, v_1 respectively. It is not a simple path because vertex v_2 appears twice in the associated vertex sequence.

2. Let Γ be Petersen's graph illustrated in figure 11.3. Beginning at any vertex there is a simple path which passes through every vertex; we leave

† There is another edge sequence whose vertex sequence is ambiguous, namely the empty sequence which has no edges. We regard this has having vertex sequence v_i for any vertex v_i. The empty edge sequence is, in fact, a simple closed path but not a cycle.

it as an easy exercise to find such a simple path. However, there are no cycles which pass through every vertex.

Let Γ be the graph represented in figure 11.1. Its adjacency matrix is

$$A = \begin{pmatrix} 1 & 1 & 1 & 0 \\ 1 & 0 & 2 & 0 \\ 1 & 2 & 0 & 0 \\ 0 & 0 & 0 & 0 \end{pmatrix}.$$

The (i, j)-entry of A is the number of edges joining vertices v_i and v_j. We can think of this as the number of edge sequences of length 1 joining these two vertices. Now the square of the adjacency matrix is

$$A^2 = \begin{pmatrix} 3 & 3 & 3 & 0 \\ 3 & 5 & 1 & 0 \\ 3 & 1 & 5 & 0 \\ 0 & 0 & 0 & 0 \end{pmatrix}.$$

In A^2 the (i, j)-entry represents the number of edge sequences of length 2 joining v_i and v_j. For example, the $(2, 2)$-entry is 5 and there are the following five edge sequences of length 2 joining v_2 to itself: $e_2, e_2; e_4, e_4; e_5, e_5; e_4, e_5; e_5, e_4$.

It is not too difficult to see why this occurs. The (i, j)-entry of A^2 is obtained by 'multiplying' the ith row and the jth column of A. In definition 6.3 we expressed this as

$$\sum_{k=1}^{n} a_{ik}a_{kj}.$$

The rth term in this sum, $a_{ir}a_{rj}$ is the product of the number of edges connecting v_i and v_r with the number connecting v_r and v_j; in other words the number of edge sequences of length 2 joining v_i and v_j via v_r. Summing over all k gives the total number of length 2 edge sequences connecting v_i and v_j.

Similarly the (i, j)-entry of A^3 represents the number of edge sequences of length 3 joining v_i and v_j. For this graph

$$A^3 = \begin{pmatrix} 9 & 9 & 9 & 0 \\ 9 & 5 & 13 & 0 \\ 9 & 13 & 5 & 0 \\ 0 & 0 & 0 & 0 \end{pmatrix}.$$

The nine edge sequences of length 3 joining v_1 and v_2 are: e_1, e_1, e_2; e_2, e_2, e_2; e_1, e_3, e_4; e_1, e_3, e_5; e_3, e_3, e_2; e_2, e_4, e_4; e_2, e_5, e_5; e_2, e_4, e_5; e_2, e_5, e_4.

The following theorem can be proved by induction. The inductive step is similar to the argument used above (exercise 11.2.1).

Theorem 11.1

Let Γ be a graph with vertex set $\{v_1, v_2, \ldots, v_r\}$ and adjacency matrix A. The (i, j)-entry of A^n is the number of edge sequences of length n joining v_i and v_j.

Connectedness

In an intuitively obvious sense, some graphs are 'all in one piece' and others are made up of several pieces. We can use paths to make this idea more precise.

Definition 11.7

A graph is **connected** if, given any pair of distinct vertices, there exists a path connecting them.

An arbitrary graph naturally splits into a number of connected subgraphs, called its **(connected) components**. The components can be defined formally as maximal connected subgraphs. This means that Γ_1 is a component of Γ if it is a connected subgraph of Γ and it is not itself a proper subgraph of any other *connected* subgraph of Γ. This second condition is what we mean by the term maximal; it says that if Σ is a connected subgraph such that $\Gamma_1 \leqslant \Sigma$, then $\Sigma = \Gamma_1$ so there is no *connected* subgraph of Γ which is 'bigger' than Γ_1.

We shall not be too concerned with the formal definition of components as the intuitive idea is clear; the components of a graph are just its connected 'pieces'. In particular, a connected graph has only one component. Decomposing a graph into its components can be very useful. It is usually simpler to prove results about connected graphs and properties of arbitrary graphs can frequently then be deduced by considering each component in turn.

The following is an outline of an alternative way of defining the components of a graph Γ. Define a relation R on V_Γ by

$$v \,\mathsf{R}\, w \quad \text{if and only if } v \text{ and } w \text{ can be joined by a path in } \Gamma.$$

Provided we allow the empty path with no edges, it is easily seen that R is an equivalence relation (exercise 11.2.2). Let $\{V_1, V_2, \ldots, V_p,\}$ be the partition of the vertex set by the equivalence classes of R. We can now form subgraphs Γ_i with vertex V_i and whose edges are those of Γ which join two vertices of V_i. These subgraphs Γ_i are the components of Γ.

Examples 11.6

1. The graph illustrated in figure 11.1 has two components, one of which is the null graph with vertex set $\{v_4\}$.

   All the other graphs which have been considered so far are connected, i.e. have one component.

2. Frequently it is clear from a diagram of Γ how many components it has. Sometimes, however, we need to examine the diagram more closely. For instance, both graphs illustrated in figure 11.7 have two components, although this is not instantly apparent for the graph (*b*).

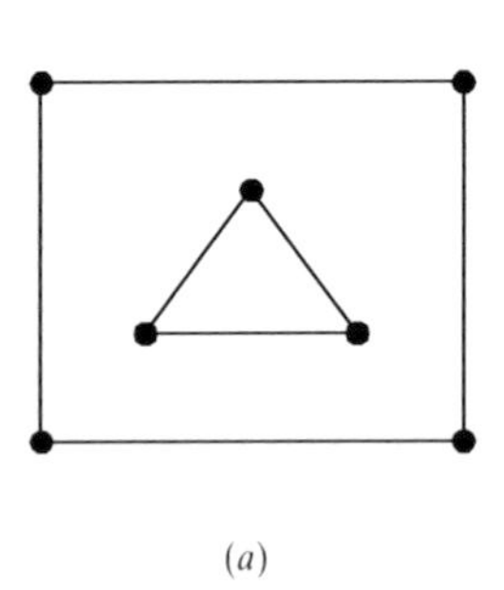

(*a*)

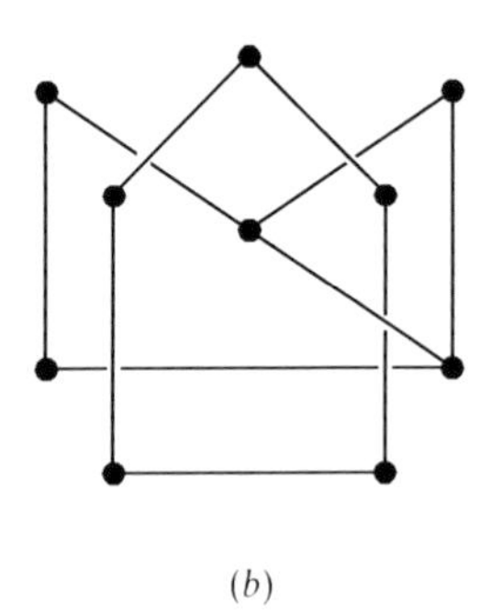

(*b*)

Figure 11.7

Eulerian Paths

We have mentioned Euler's 1736 paper which marked the birth of graph theory. This paper developed a theory which was able to solve the so-called **Königsberg**

Bridge problem, which is the following. The Pregel River flows through the town of Königsberg in Russia. There are two islands in the river, connected to the banks and each other by bridges as shown in figure 11.8(a). The problem for the citizens of Königsberg was whether there was a walk, beginning on one of the banks or islands, which took in every bridge exactly once and finished back at the starting position. They were unable to find such a walk; the problem was either to find such a walk or to show that none existed.

Euler first represented the essential features of Königsberg's geography by a graph, as illustrated in figure 11.8(b). Each of the river banks and islands is represented by a vertex with the edges corresponding to the connecting bridges. In graph-theoretic terms the question is whether there exists a closed path which includes all the edges of the graph.

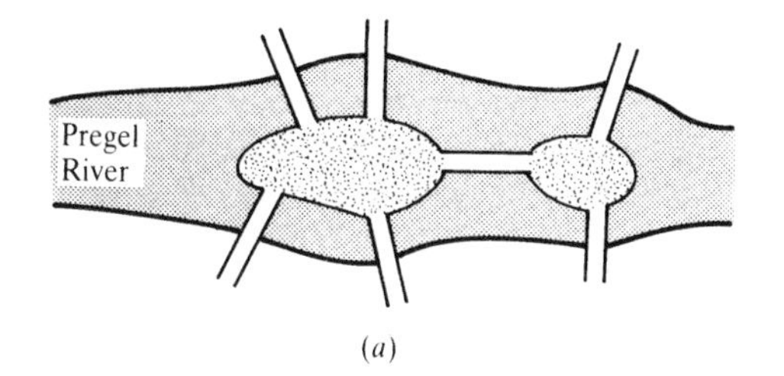

(a)

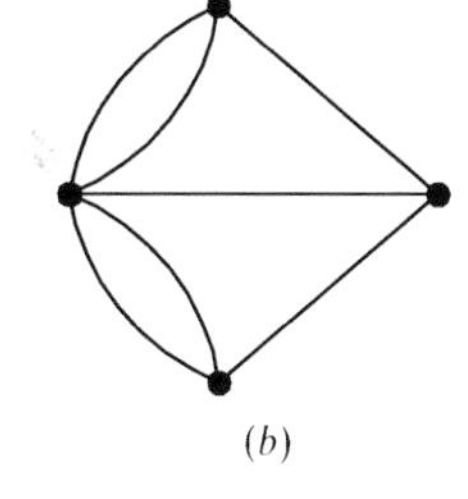

(b)

Figure 11.8

Definition 11.8

An **Eulerian path** in a graph Γ is a closed path which includes every edge of Γ. A graph is said to be **Eulerian** if it has at least one Eulerian path.

Recall that in a path no edge can be traversed more than once. Thus an Eulerian path includes every edge exactly once, although, of course, vertices may be visited more than once.

For a connected graph Γ there is an easily recognized necessary condition for it to have an Eulerian path; namely, every vertex must have even degree. To see this, suppose that Γ is connected and has an Eulerian path. Since Γ is connected, the vertex sequence of the Eulerian path contains every vertex. Whenever the path passes through a vertex it contributes two to its degree (one from the edge 'going in to' and one from the edge 'coming out from' the vertex). Since every edge occurs exactly once in the path, every vertex must have even degree.

The people of Königsberg had not been able to find their Eulerian path for a very good reason—there isn't one. The graph representing the problem, figure 11.8(*b*), is connected but fails the required condition. Every vertex, in fact, has odd degree.

Euler also proved the less obvious fact that, for a connected graph Γ, this necessary condition is also sufficient for it to be Eulerian. A proof of this is indicated in exercise 11.2.15.

Theorem 11.2 (Euler)

A connected graph Γ is Eulerian if and only if every vertex has even degree.

Examples 11.7

1. The complete graph K_n is $(n-1)$-regular—every vertex has degree $n-1$. Since it is connected, K_n is Eulerian if and only if n is odd (so that $n-1$ is even). The graph K_3 has an obvious Eulerian path and we leave it as an exercise to find an Eulerian path in K_5—see figure 11.4. (In fact, K_5 has 264 Eulerian paths.)

2. The complete bipartite graph $K_{4,4}$ is represented in figure 11.9. The vertices have been partitioned into the sets $\{1, 2, 3, 4\}$ and $\{\mathrm{a}, \mathrm{b}, \mathrm{c}, \mathrm{d}\}$. The graph is connected and every vertex has degree 4, so $K_{4,4}$ is Eulerian by theorem 11.2.

   One Eulerian path beginning at the vertex 1 has the following vertex sequence:

$$1, \mathrm{a}, 2, \mathrm{b}, 3, \mathrm{c}, 4, \mathrm{d}, 1, \mathrm{c}, 2, \mathrm{d}, 3, \mathrm{a}, 4, \mathrm{b}, 1.$$

Hamiltonian Cycles

An Eulerian path seeks to travel along every edge of the graph (once) and return to the starting position. An analogous problem is whether we can visit every vertex once, without travelling along any edge more than once, and return to the starting

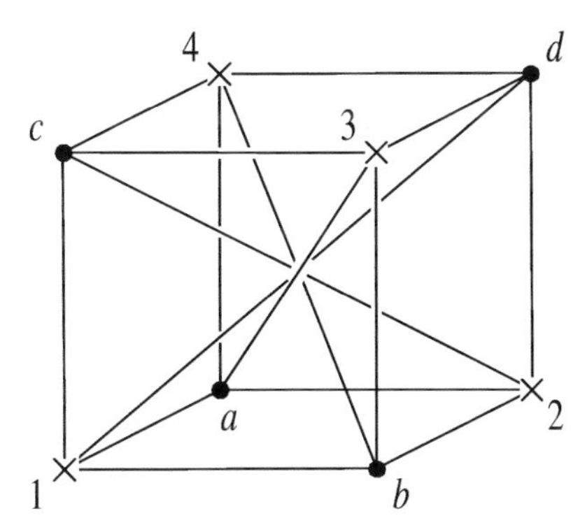

Figure 11.9

position. This problem was considered by Hamilton† (although he was probably not the first to do so in the case of general graphs) and his name is now associated with these paths.

> **Definition 11.9**
>
> A **Hamiltonian cycle** in a graph is a cycle which passes once through every vertex. A graph is **Hamiltonian** if it has a Hamiltonian cycle.

Example 11.8

Figure 11.10 illustrates Hamiltonian cycles in two graphs. The graph (a) is the complete bipartite graph $K_{3,3}$ defined in example 11.2.2, and the graph (b) is called the **dodecahedral graph**‡. In fact, it was cycles in the dodecahedral graph,

† Sir William Rowan Hamilton (1805–65) was Ireland's most gifted mathematician–scientist. As a 22-year old undergraduate he was elected Professor of Astronomy and Astronomer Royal of Ireland. In fact he made little contribution to astronomy; his most significant work was in mathematics and physics. In 1843 he discovered the quaternions—a sort of generalized complex numbers—and he devoted most of the rest of his life to their study. His name is also associated with the Hamiltonian energy operator used in physics, particularly wave mechanics.

‡ The dodecahedron is one of the five regular three-dimensional solids; it has 12 faces, each a regular pentagon, 30 edges and 20 vertices each of degree 3. The dodecahedral graph is a (necessarily distorted) two-dimensional representation of the solid. The other four regular solids can also be represented by graphs—see exercise 11.2.8.

in particular, that interested Hamilton and he invented a board game, The Icosian Game, which explored the cycles in this graph.

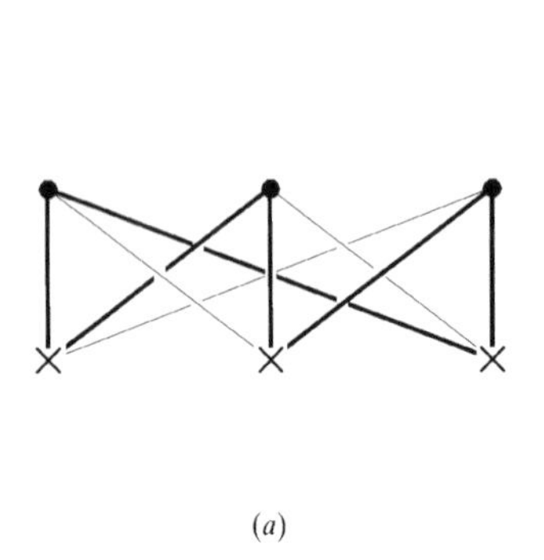

(a)

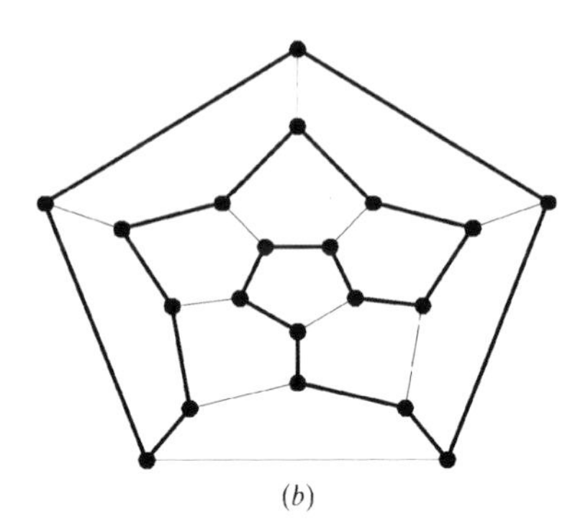

(b)

Figure 11.10

Although Eulerian graphs have a simple characterization, the same is not true of Hamiltonian graphs. Indeed after more than a century of study, no characterization of Hamiltonian graphs is known. (By a 'characterization' of Hamiltonian graphs we mean necessary and sufficient conditions for a graph to be Hamiltonian.) This remains one of the major unsolved problems of graph theory. An obvious necessary condition is that the graph be connected. Various sufficient conditions are also known; most require the graph to have 'enough' edges in some sense. One of the simplest such results is the following.

Theorem 11.3

If Γ is a connected simple graph with n ($\geqslant 3$) vertices and if the degree $deg(v) \geqslant \frac{1}{2}n$ for every vertex v, then Γ is Hamiltonian.

The condition on the degrees, $deg(v) \geqslant \frac{1}{2}n$, is not a necessary condition for Γ to be Hamiltonian, so a graph can be Hamiltonian without satisfying this condition. We can see this by considering the dodecahedral graph—figure 11.10(*b*). The graph has 20 vertices, every vertex has degree 3, but it is still Hamiltonian. In fact the graph of each of the five regular solids has a Hamiltonian cycle—see exercise 11.2.8.

Exercises 11.2

1. Using induction on the number of vertices, prove theorem 11.1.

2. Prove that the relation R on the set V of vertices of graph Γ defined by

$$v \,\mathsf{R}\, w \quad \text{if and only if there exists a path in } \Gamma \text{ joining } v \text{ and } w$$

is an equivalence relation. (You need to allow the empty path with no edges, which can be viewed as joining any vertex to itself.)

3. (i) Is a null graph Eulerian?
 (ii) Is it possible for a non-connected graph to be Eulerian?

4. (i) For which values of n is the complete graph K_n Eulerian?
 (ii) For which values of r and s is the complete bipartite graph $K_{r,s}$ Eulerian?

5. (i) Each of the following matrices is an adjacency matrix of a graph. In each case, determine whether the corresponding graph is Eulerian.

(a) $$\begin{pmatrix} 0 & 1 & 1 & 1 & 1 & 0 \\ 1 & 0 & 0 & 0 & 1 & 0 \\ 1 & 0 & 0 & 1 & 0 & 0 \\ 1 & 0 & 1 & 0 & 1 & 1 \\ 1 & 1 & 0 & 1 & 0 & 1 \\ 0 & 0 & 0 & 1 & 1 & 0 \end{pmatrix}$$

(b) $$\begin{pmatrix} 0 & 1 & 0 & 1 & 1 & 0 \\ 1 & 0 & 0 & 0 & 1 & 0 \\ 0 & 0 & 0 & 1 & 0 & 1 \\ 1 & 0 & 1 & 0 & 1 & 0 \\ 1 & 1 & 0 & 1 & 0 & 1 \\ 0 & 0 & 1 & 0 & 1 & 0 \end{pmatrix}$$

(c) $$\begin{pmatrix} 0 & 0 & 1 & 0 & 1 & 0 \\ 0 & 0 & 0 & 1 & 0 & 1 \\ 1 & 0 & 0 & 0 & 1 & 0 \\ 0 & 1 & 0 & 0 & 0 & 1 \\ 1 & 0 & 1 & 0 & 0 & 0 \\ 0 & 1 & 0 & 1 & 0 & 0 \end{pmatrix}$$

(d) $$\begin{pmatrix} 1 & 1 & 0 & 0 & 0 & 1 \\ 1 & 0 & 1 & 2 & 1 & 1 \\ 0 & 1 & 0 & 1 & 0 & 0 \\ 0 & 2 & 1 & 0 & 1 & 0 \\ 0 & 1 & 0 & 1 & 0 & 0 \\ 1 & 1 & 0 & 0 & 0 & 0 \end{pmatrix}.$$

(ii) Let Γ be a connected graph. How can you determine, from its adjacency matrix, whether or not Γ is Eulerian?

6. Consider the graph Γ whose diagram is given below.

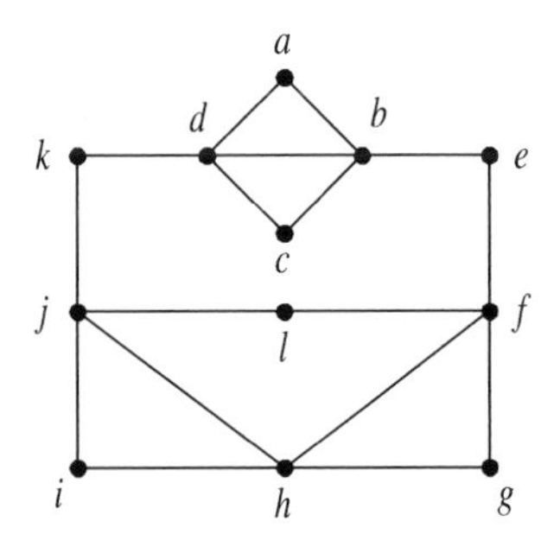

(i) Is Γ Eulerian?
(ii) Is Γ Hamiltonian?

7. A connected graph is called **semi-Eulerian** if it has a (not necessarily closed) path containing all edges. Use theorem 11.2 to prove: Γ is semi-Eulerian, but not Eulerian, if and only if all vertices except two have even degree.

8. Show that each of the graphs of the regular solids shown in figure 11.11 is Hamiltonian.

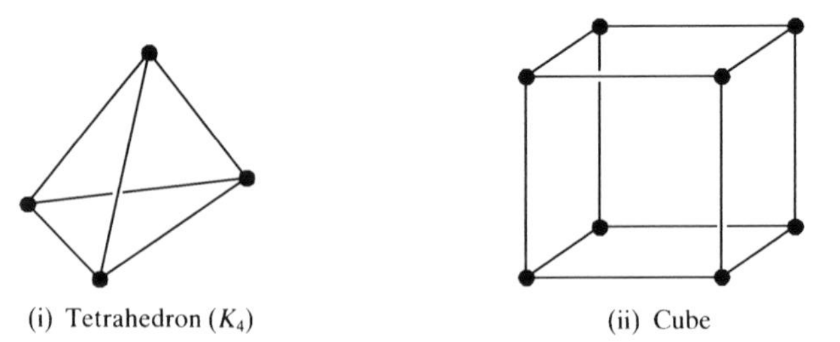

Figure 11.11

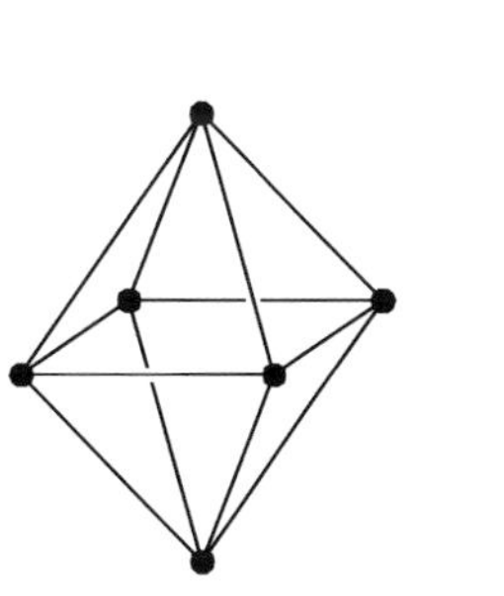

(iii) Octahedron

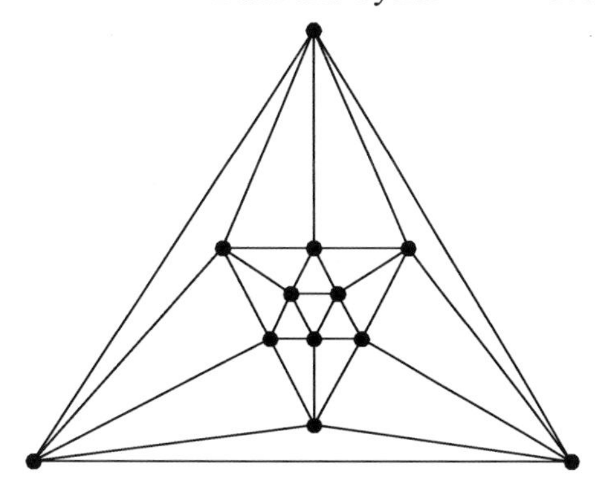

(iv) Icosahedron

Figure 11.11 (Continued)

9. (i) Show that a closed path in a bipartite graph contains an even number of edges.
   (ii) When is $K_{r,s}$ Hamiltonian?

10. A graph is called **semi-Hamiltonian** if there exists a (not necessarily closed) simple path which passes through each vertex.

    Which of the graphs defined in §11.1 are (i) Eulerian, (ii) semi-Eulerian but not Eulerian, (iii) Hamiltonian, (iv) semi-Hamiltonian but not Hamiltonian?

11. (i) What is the minimum number of bridges which would have to have been built in Königsberg so that its graph is (a) semi-Eulerian, (b) Eulerian?
    (ii) Is the graph of the Königsberg bridges Hamiltonian? If not, what is the minimum number of bridges which would have to have been built so that the graph becomes Hamiltonian?

12. For each of the following, determine whether the graph is:

    (a) Eulerian, semi-Eulerian or neither;
    (b) Hamiltonian, semi-Hamiltonian or neither.

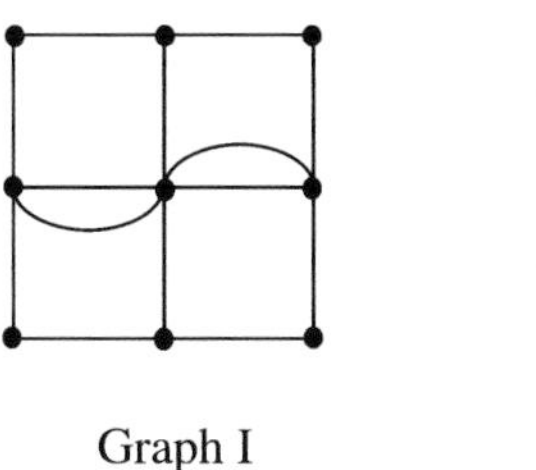

Graph I

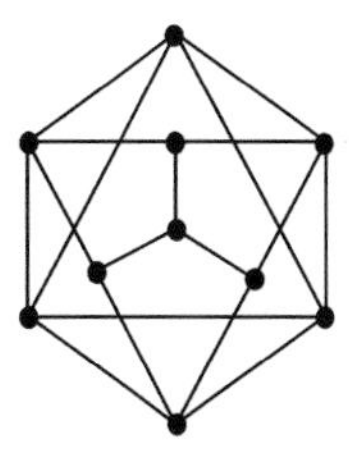

Graph II

13. (i) Prove that a connected graph Γ is Eulerian if and only if it can be split into cycles, no two of which have any edges in common.

(ii) Show that the Eulerian graph below can be split into four cycles, no two of which have any edges in common. How can these cycles be combined to form an Eulerian path?

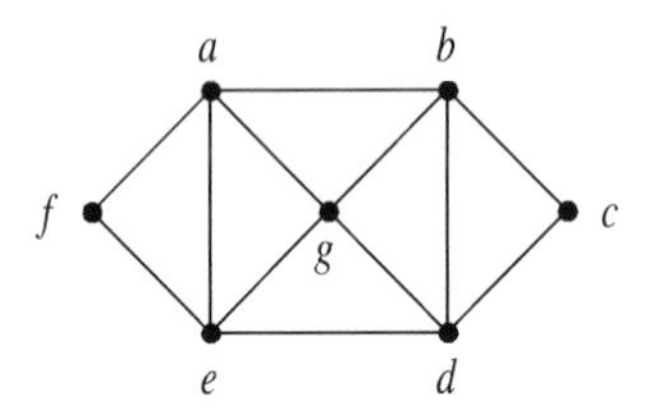

14. **Knight's tour problems**

Can a knight visit each square of a chessboard by a sequence of knight's moves and finish on the same square as it began? A solution to the problem, if it exists, is called a **knight's tour** of the board.

The problem can be modelled by a graph whose vertices represent the squares of the board where two vertices are joined by an edge if and only if there is a knight's move between the corresponding squares on the chessboard.

For example, a 4×4 'chessboard' and its corresponding graph are shown below.

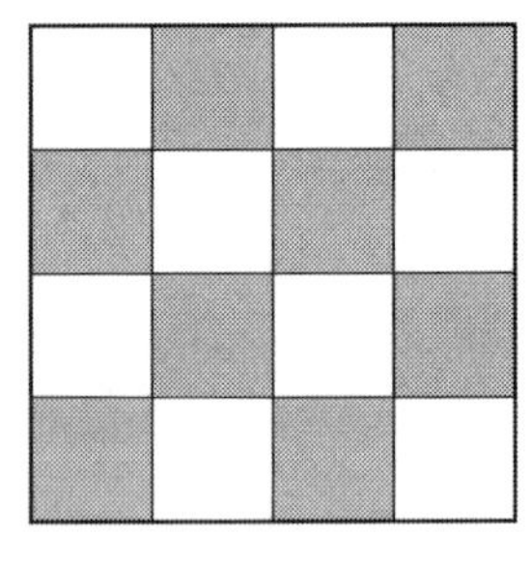

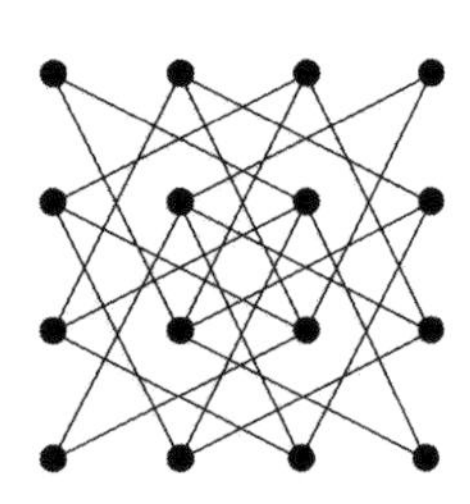

(i) Draw the graph corresponding to a 3×3 chessboard. Deduce that there is no knight's tour on a 3×3 board.

(ii) In fact there is no knight's tour on the 4×4 board—experiment a bit to convince yourself of this.

(iii) Prove that a bipartite graph with an odd number of vertices is not Hamiltonian. Deduce that there is no knight's tour on a 5×5 or a 7×7 board.

(iv) The original knight's tour problem on an 8×8 board does have a solution. See if you can find a knight's tour.

15. Prove that if Γ is connected and every vertex has even degree then Γ is Eulerian. (This completes the proof of theorem 11.2.)

    One method of proof is by induction on $|E|$, the number of edges of Γ. The inductive step is outlined below.

    Firstly, choose any vertex of v of Γ and a closed path P beginning and ending at v.

    If P contains every edge of Γ we are finished; otherwise remove all the edges of P to form a new graph Γ'. This new graph Γ' may be disconnected. Consider each component of Γ' in turn and use the inductive hypothesis to obtain an Eulerian path in each of these components.

    Finally use P and the Eulerian paths in each component of Γ' to piece together an Eulerian path for Γ.

11.3 Isomorphism of Graphs

Consider the two graphs Γ and Σ defined as follows: Γ has vertex set $\{1, 2, 3, 4\}$ and the adjacency matrix A, and Σ has vertex set $\{a, b, c, d\}$ and the adjacency matrix B, where

$$A = \begin{pmatrix} 1 & 2 & 1 & 1 \\ 2 & 0 & 0 & 1 \\ 1 & 0 & 0 & 3 \\ 1 & 1 & 3 & 0 \end{pmatrix} \qquad B = \begin{pmatrix} 0 & 3 & 0 & 1 \\ 3 & 0 & 1 & 1 \\ 0 & 1 & 0 & 2 \\ 1 & 1 & 2 & 1 \end{pmatrix}.$$

Diagrams representing Γ and Σ are given in figure 11.12.

With some thought it should be apparent that the graphs represented in figure 11.12 are essentially the same. If we re-label the vertices a, b, c, d of Σ as 3, 4, 2, 1, in that order, and re-label the edges f_i as e_i for $i = 1, \ldots, 8$, then the two diagrams in figure 11.12 could be regarded as different representations of the same graph. Of course, Γ and Σ are not identical graphs—they have different vertex sets, for instance. However they do have the 'same structure' in some

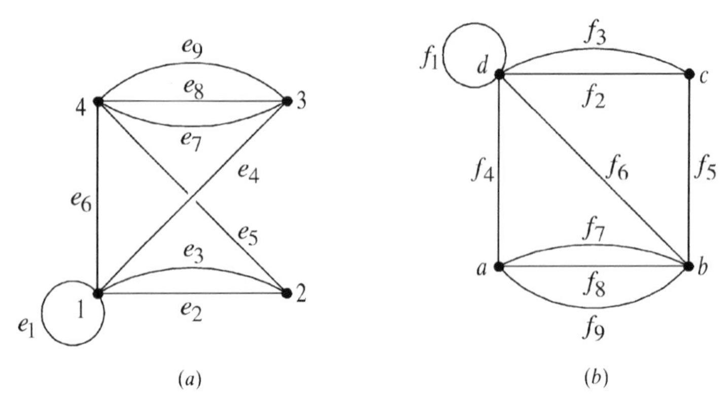

Figure 11.12

sense. We say that Γ and Σ are 'isomorphic' graphs. (The notion of isomorphic graphs is precisely the graph theory equivalent of isomorphic groups considered in chapter 8.)

In relabelling the vertices of Σ we have defined a bijection between the vertex sets of Γ and Σ in such a way that the edge sets also correspond. In other words, if there are n edges joining two vertices in Γ, then there are also n edges joining two vertices in Σ. This correspondence of vertices and edges is called an 'isomorphism' from Γ to Σ. The technical definition is the following.

Definition 11.10

Let Γ and Σ be two graphs. An **isomorphism** from Γ to Σ consists of a pair (θ, ϕ) of bijections

$$\theta : V_\Gamma \to V_\Sigma \quad \text{and} \quad \phi : E_\Gamma \to E_\Sigma$$

such that, for every edge e of Γ, if $\delta_\Gamma(e) = \{v, w\}$ then $\delta_\Sigma(\phi(e)) = \{\theta(v), \theta(w)\}$.

To graphs are said to be **isomorphic**, denoted $\Gamma \cong \Sigma$, if there exists an isomorphism from one graph to another.

The condition that if $\delta_\Gamma(e) = \{v, w\}$ then $\delta_\Sigma(\phi(e)) = \{\theta(v), \theta(w)\}$ is to ensure that the two correspondences between vertices and edges of the two graphs 'match up' in the correct way. In other words, if the edge e of Γ corresponds to the edge

$\phi(e)$ of Σ then their endpoint vertices also correspond (under the vertex bijection θ). This is best illustrated by figure 11.13.

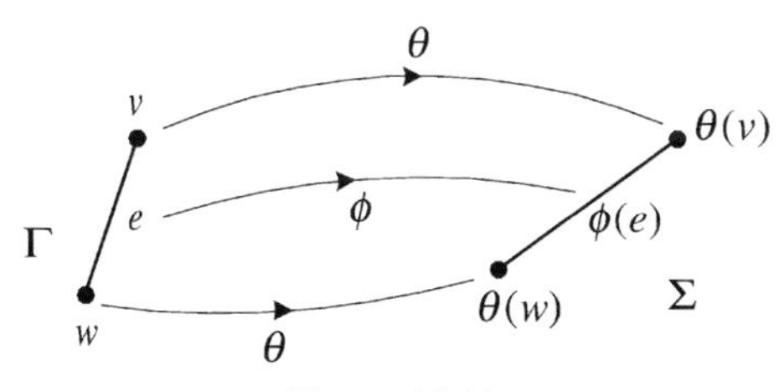

Figure 11.13

If we let $E(v, w)$ denote the set of edges joining the vertices v and w in Γ, then the edge function ϕ defines a bijection $E(v, w) \to E(\theta(v), \theta(w))$. Therefore, for all pairs of edges v, w of Γ, $|E(v, w)| = |E(\theta(v), \theta(w))|$, i.e. the number of edges in Γ joining v and w is the same as the number of edges in Σ joining their corresponding vertices $\theta(v)$ and $\theta(w)$.

For a simple graph Γ, in order to define an isomorphism from Γ to Σ, we need only specify the appropriate vertex bijection $\theta : V_\Gamma \to V_\Sigma$. This is because there is at most one edge joining any pair of vertices, so once θ has been (correctly) defined there is only one function $\phi : E_\Gamma \to E_\Sigma$ with the required properties. (For such an isomorphism to be possible Σ must also be simple—see theorem 11.4(vi) below.)

Since isomorphic graphs have essentially the same structure, any graph-theoretic property which one has, the other must also have. We list, without proof, some of these properties in the next theorem.

Theorem 11.4

Let (θ, ϕ) be an isomorphism from Γ to Σ. Then:

- (i) Γ and Σ have the same number of vertices;
- (ii) Γ and Σ have the same number of edges;
- (iii) Γ and Σ have the same number of components;
- (iv) corresponding vertices have the same degree: for every $v \in V_\Gamma$, $deg(v) = deg(\theta(v))$;
- (v) Γ and Σ have the same degree sequence (see exercise 11.1.8);
- (vi) if Γ is simple, then so too is Σ;
- (vii) if Γ is Eulerian, then so too is Σ;
- (viii) if Γ is Hamiltonian, then so too is Σ.

Proof

See exercise 11.3.10. □

Examples 11.9

1. For the graphs in figure 11.12, an isomorphism is defined by the functions

$$\theta : \{1, 2, 3, 4\} \to \{a, b, c, d\}; \quad 1 \mapsto d, 2 \mapsto c, 3 \mapsto a, 4 \mapsto b$$

and

$$\phi : \{e_1, \ldots, e_8\} \mapsto \{f_1, \ldots, f_8\}; \quad e_i \mapsto f_i \quad \text{for } i = 1, \ldots, 8.$$

2. Consider the graphs K_6 and $K_{3,3}$. Both graphs have six vertices so there certainly exist bijections between their vertex sets. However, none of the possible bijections is an isomorphism, since all the vertices of K_6, for example, have degree 5, but all those of $K_{3,3}$ have degree 3.

3. Determine which of the graphs represented in figure 11.14 are isomorphic. (For convenience we have chosen to label the vertices by upper-case letters rather than $v_1, v_2, \ldots$.)

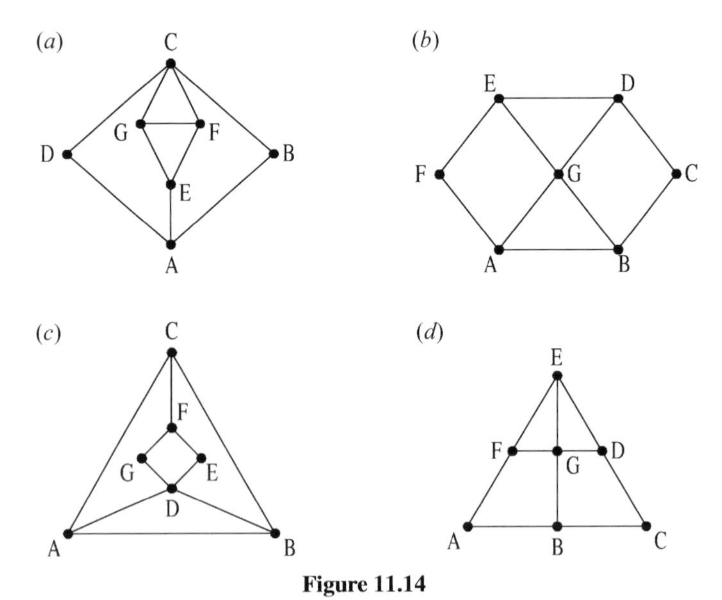

Figure 11.14

Solution

Note first that each graph is simple, connected and has seven vertices and ten edges. Furthermore each has degree sequence $(2, 2, 3, 3, 3, 3, 4)$. Theorem 11.2 shows that each graph is non-Eulerian. Thus the first seven properties listed in theorem 11.4 cannot be used to show that any pair of these graphs is non-isomorphic.

Using the Hamiltonian property, we can show that neither graph (a) nor (c) is isomorphic to graph (b) or (d). It can be proved (by considering the vertices of degree 2) that graphs (a) and (c) are not Hamiltonian; however graphs (b) and (d) are Hamiltonian (with Hamiltonian cycles AGBCDEFA and ABCDGEFA respectively). Since it is difficult in general to show that a graph is not Hamiltonian, we prefer to follow a different approach as follows.

If we look a bit more closely we can see that (a) and (b) are not isomorphic: in graph (a) the vertex of degree 4 is adjacent to two vertices of degree 3, but in graph (b) the vertex of degree 4 is adjacent to four vertices of degree 3. A similar argument also shows that graphs (a) and (d) are not isomorphic.

Graphs (a) and (c) are isomorphic. Labelling their vertex sets V_1 and V_2 respectively, an isomorphism is defined by the following vertex bijection

$$V_1 \to V_2 : \mathrm{A} \mapsto \mathrm{F}, \mathrm{B} \mapsto \mathrm{E}, \mathrm{C} \mapsto \mathrm{D}, \mathrm{D} \mapsto \mathrm{G}, \mathrm{E} \mapsto \mathrm{C}, \mathrm{F} \mapsto \mathrm{A}, \mathrm{G} \mapsto \mathrm{B}.$$

Graphs (b) and (c) are not isomorphic since the latter is isomorphic to graph (a) but the former is not. Similarly graphs (d) and (c) are not isomorphic.

It only remains to determine whether or not graphs (b) and (d) are isomorphic. In fact they are not. We can see this, for instance, by noting that in (d) there is a vertex of degree 3 (vertex E) which is adjacent to two other degree 3 vertices, but in (b) every vertex of degree 3 is adjacent to only one other degree 3 vertex.

In summary, graphs (a) and (c) are isomorphic but no other pairs of graphs are isomorphic.

The examples above illustrate the following general principle. (Compare this with the case of groups.)

Isomorphism principle

To show that two graphs are isomorphic, an isomorphism from one to the other must be found; to show that two graphs are not isomorphic, a graph-theoretic property must be found which one graph has but the other does not.

Exercises 11.3

1. Let A and B be adjacency matrices of two isomorphic graphs. How are A and B related?

2. Are the following two graphs isomorphic? Justify your answer either by finding an isomorphism between them or by showing one has a graph-theoretic property which the other does not have.

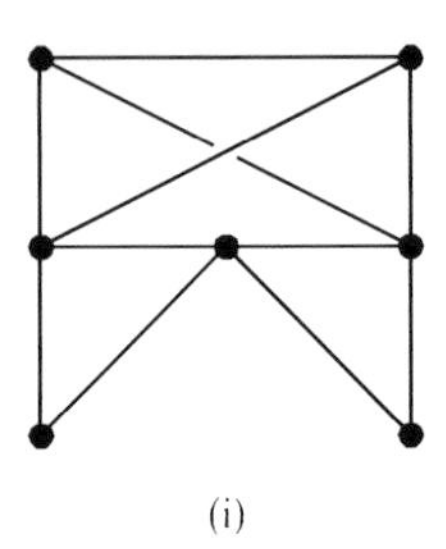

(i)

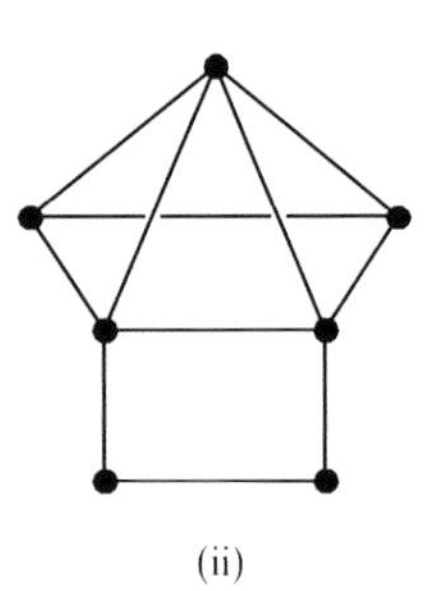

(ii)

3. Show that all of the following graphs are isomorphic by defining explicit isomorphisms between them.

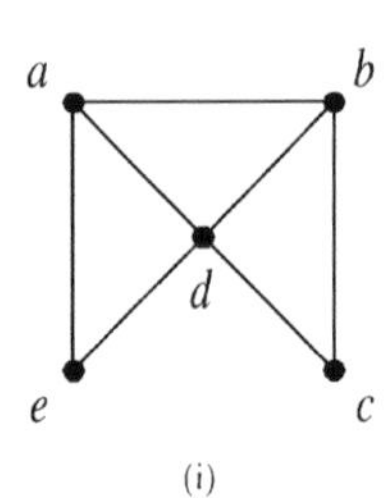

(i)

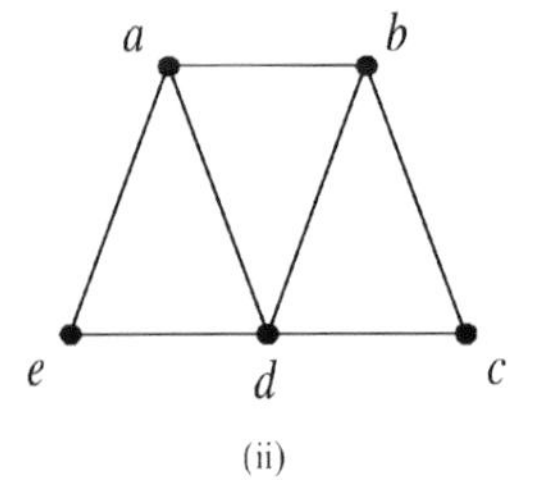

(ii)

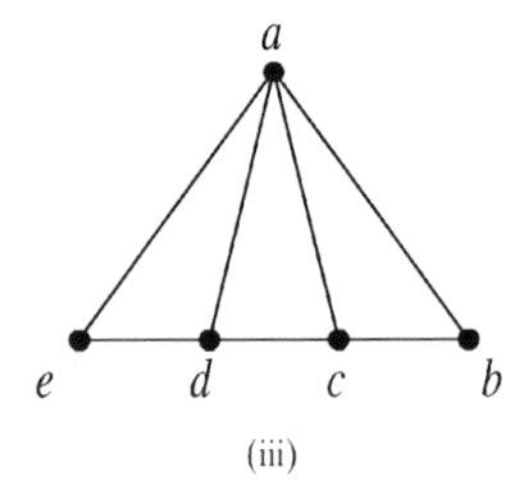

(iii)

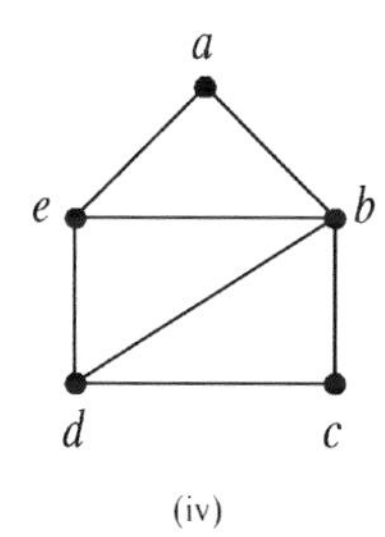

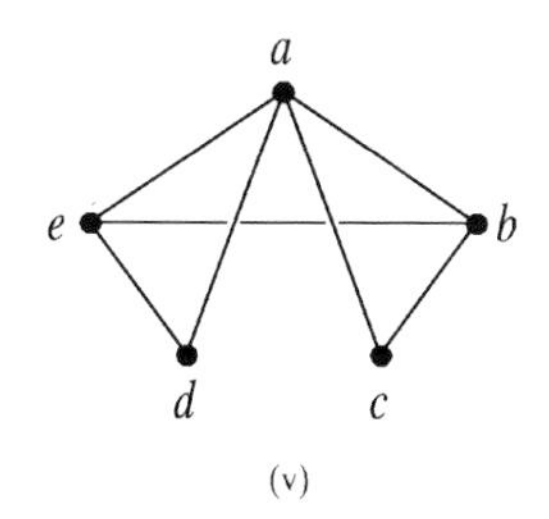

4. (i) Draw the diagrams of two non-isomorphic simple graphs each of which has degree sequence $(2, 2, 3, 3, 4, 4)$. Explain why your two graphs are not isomorphic.

   (ii) Draw the diagrams of two non-isomorphic simple graphs each of which has degree sequence $(2, 2, 2, 4, 4, 5, 5)$. Explain why your two graphs are not isomorphic.

5. Write down the degree sequence of the following disconnected graph Γ.

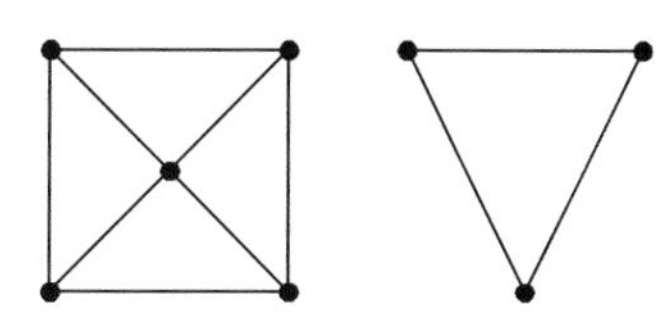

   Draw two non-isomorphic connected graphs with the same degree sequence as Γ and explain why your two graphs are non-isomorphic.

6. Three graphs Γ_1, Γ_2 and Γ_3 have diagrams given below.

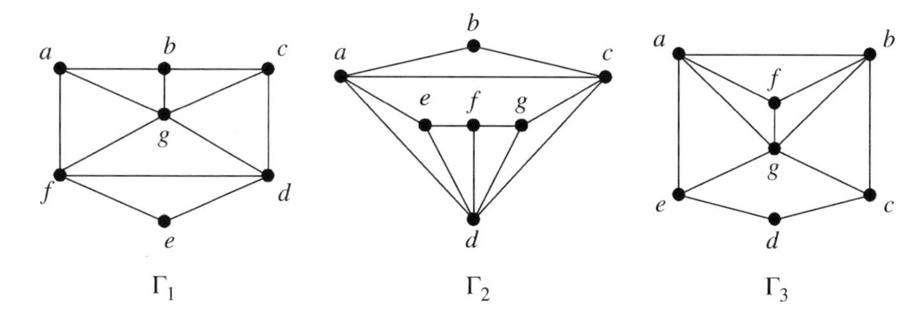

   Show that one and only one pair of the graphs is isomorphic.

7. Which of the following graphs are isomorphic? For those which are isomorphic define isomorphisms and for those which are not isomorphic

give reasons to explain why they are not isomorphic.

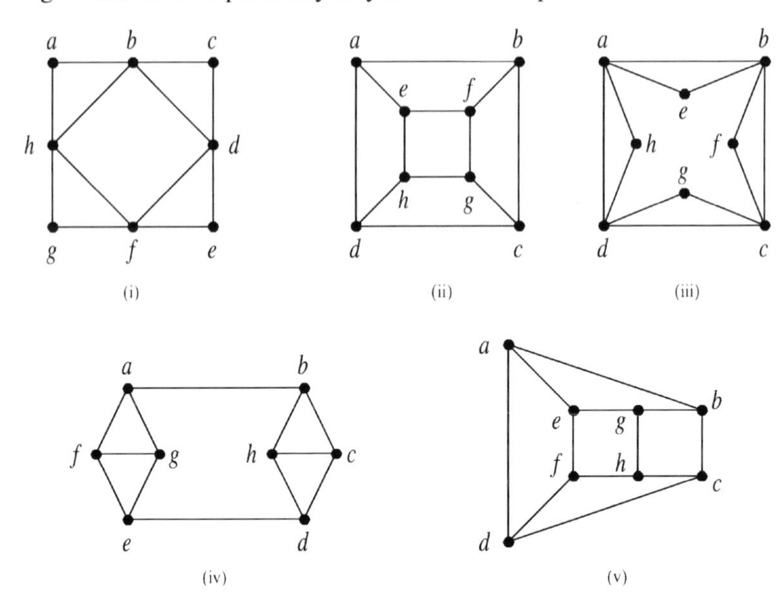

8. (i) There are 11 non-isomorphic simple graphs with four vertices. Draw diagrams to represent these 11 graphs. How many of them are connected?

   (ii) How many non-isomorphic simple graphs are there with five vertices? How many of these are connected?

9. Show that the two graphs illustrated in figure 11.3 are isomorphic.

10. Prove theorem 11.4.

11.4 Trees

In §11.2 we considered paths and cycles in graphs. There is a special class of connected graphs, called 'trees', which have no cycles at all. The first use of trees was by physicist Gustav Kirchhoff in 1847; two years earlier (whilst a student at the University of Königsberg!) Kirchhoff had formulated the laws governing the flow of electricity in a network of wires. The network of wires can be considered as a graph in our sense. The equations which follow from Kirchhoff's laws, as they are now called, are not all independent and Kirchhoff used trees to obtain an independent subset of equations. The term 'tree' was coined by the British mathematician Arthur Cayley† ten years later; Cayley was motivated to study trees by a problem within mathematics itself.

Trees have become important within graph theory for a number of reasons. They also feature in many of the applications of graph theory. Cayley himself provided one of the applications—to the study of isomers in chemistry (see exercises 11.4.13–11.4.15). More recently, computer scientists have found that trees provide a convenient structure for storage and retrieval of certain types of data—using so-called hierarchical databases.

Definition 11.11

A **tree** is a connected graph which contains no cycles.

It is immediately apparent from the definition that a tree has no loops or multiple edges. Any loop is a cycle by itself, and if edges e_i and e_j join the same pair of vertices then the sequence e_i, e_j is also a cycle. Some examples of trees are given in figure 11.15.

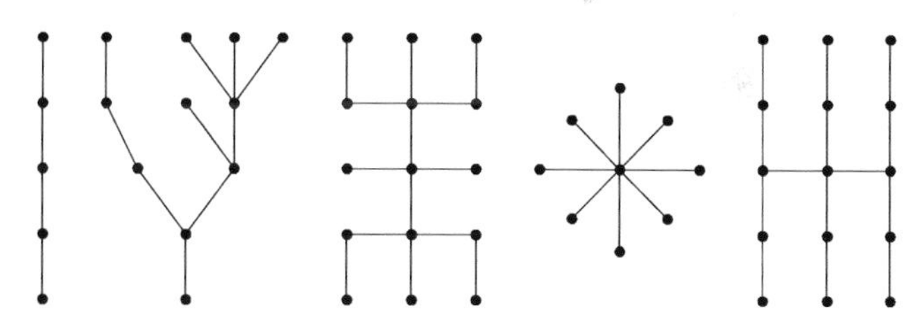

Figure 11.15

One reason for the importance of trees in graph theory itself is that every connected graph contains a tree—called a 'spanning tree'—which connects all its vertices. Amongst other things, a spanning tree provides a convenient set of paths connecting any pair of vertices of the graph.

† Cayley's interest in trees was motivated by some work of his colleague James Joseph Sylvester on operators in differential calculus. It was Sylvester who, in 1877, first used the term 'graph' in the sense we are using here.

Definition 11.12

Let Γ be a connected graph with vertex set V. A **spanning tree** in Γ is a subgraph which is a tree and has vertex set V.

Theorem 11.5

Every connected graph contains a spanning tree.

Proof

Let Γ be a connected graph; if Γ contains no cycle then there is nothing to prove as Γ is its own spanning tree.

Suppose, then, Γ contains a cycle. Removing any edge from the cycle gives a graph which is still connected. If the new graph contains a cycle then again remove one edge of the cycle. Continue this process until the resulting graph T contains no cycles. We have not removed any vertices so T has the same vertex set as Γ, and at each stage of the above process we obtain a connected graph. Therefore T itself is connected; it is a spanning tree for Γ. □

Note that a given connected graph Γ will generally have many different spanning trees. Examples of two spanning trees for the complete graph K_6 are given in figure 11.16.

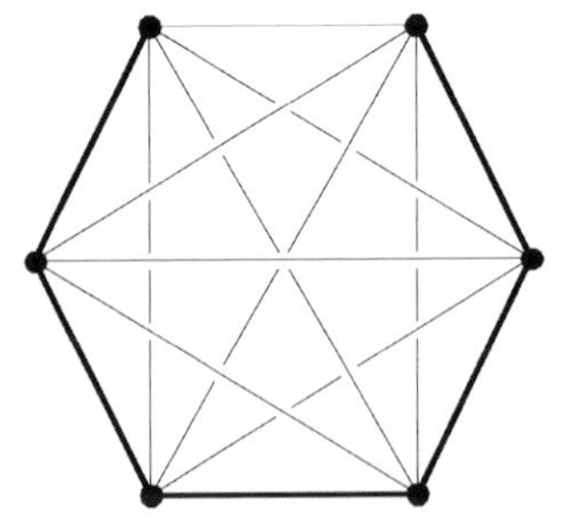

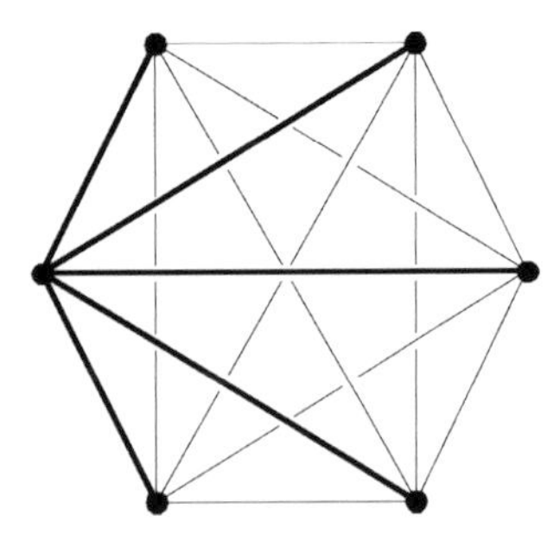

Figure 11.16

The simple structure of trees enables us easily to deduce some elementary facts about them which we give in the following theorem.

Theorem 11.6

Let T be a tree with vertex set V and edge set E. Then:

(i) for every pair of distinct vertices v and w there is a unique path in T connecting them;

(ii) deleting any edge from T produces a graph with two components each of which is a tree†;

(iii) $|E| = |V| - 1$.

Furthermore, a connected graph satisfying any one of these properties is a tree.

Proof

(i) Let v and w be any two disjoint vertices in T; since T is connected, there exists a path $P_1 : e_1, e_2, \ldots, e_n$ joining v to w. Suppose that there is another path $P_2 : f_1, f_2, \ldots, f_m$ also joining v to w. At some point the two paths must diverge; let v^* be the last vertex the two paths have in common before they diverge. Since the two paths both end at w, they must also converge again; let w^* be the first vertex at which P_1 and P_2 converge—see figure 11.17. (We need to take w^* to be the *first* vertex at which they converge because two paths may later diverge once more.) Define a path as follows: take those edges of P_1 joining v^* to w^* followed by those edges of P_2 (in reverse order) joining w^* to v^*. This path joins v^* to itself and repeats no edge; it is a cycle in T. This is a contradiction since T is a tree. Therefore there is a unique path connecting v to w.

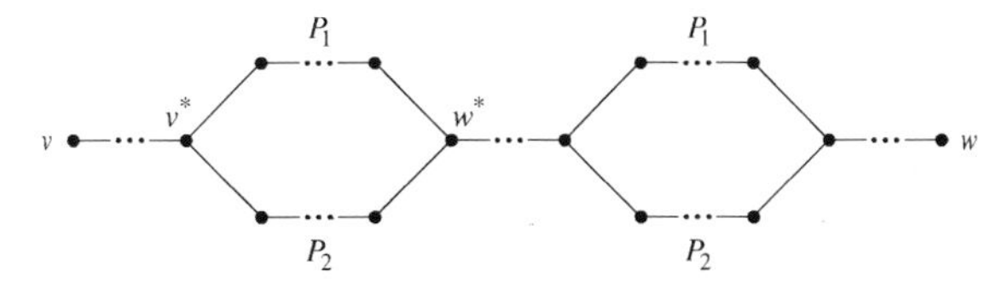

Figure 11.17

† A graph which is the union of non-intersecting trees is, not surprisingly, called a **forest**. A forest could be defined more simply as any graph with no cycles, so that a connected forest is just a tree! Clearly theorem 11.5 generalizes to show that every graph has a spanning forest.

(ii) Let e be any edge in T joining vertices v and w, and let Γ be the graph obtained by removing e from T. Since e is itself the unique path in T joining v to w, there is no path in Γ connecting v and w; thus Γ is not connected.

Let V_1 be the set of vertices of Γ which can be joined by a path (in Γ) to v, and let V_2 be the set of vertices of Γ which can be joined by a path to w. Then $V_1 \cup V_2 = V$ and V_1 and V_2 define two connected subgraphs of Γ. (Exercise: prove this last statement.) Each of these components of Γ must be a tree because any cycle in one of them would also be a cycle in T.

(iii) The proof is by induction on the number of vertices of T and uses part (ii). It is left as an exercise (11.4.8(i)).

For the last part of the theorem, we will again leave property (iii) as an exercise, and prove only that if Γ is a connected graph satisfying either property (i) or property (ii) then Γ is a tree.

Firstly, suppose that Γ is connected and satisfies (i). If there is a cycle in Γ containing a pair of distinct vertices v and w then this cycle provides two distinct paths connecting v and w. Since this contradicts (i), there is no such cycle. There can also be no loops (cycles connecting only one vertex) in Γ. If e is a loop at vertex v, and w is any other vertex, then there are two distinct paths connecting v and w: one path which begins with e and one which does not. Therefore Γ contains no cycles at all and so is a tree.

Finally suppose that Γ is connected and satisfies (ii). If Γ contains a cycle, then we could delete an edge of the cycle without disconnecting Γ, contradicting (ii). Therefore again Γ must contain no cycles at all and so is a tree. □

Exercises 11.4

1. How many non-isomorphic trees are there with:

   (i) three vertices,
   (ii) four vertices,
   (iii) five vertices,
   (iv) six vertices?

   In each case draw diagrams to represent the non-isomorphic trees.

2. Prove that a tree with at least one edge is a bipartite graph.

3. Draw spanning trees for each of the graphs illustrated in figures 11.3, 11.4, 11.5, 11.9, 11.10, 11.11 and 11.12.

4. (i) Draw the diagrams of two graphs with degree sequence $(1, 1, 1, 2, 2, 2, 3)$, one which is a tree and one which is not a tree.

   (ii) Explain why any graph with degree sequence $(1, 1, 2, 2, 2, 3, 3)$ is not a tree.

5. When is $K_{r,s}$ a tree?

6. A **full binary tree** is a tree in which exactly one vertex has degree 2 and all other vertices have degree 1 or 3. The vertex of degree 2 is called the **root** of the tree; vertices of degree 3 are called **decision vertices** and vertices of degree 1 are called **leaf vertices**. Binary trees are frequently used in computer science; we consider some of the applications of binary trees in chapter 11. Examples of binary trees are given in the following diagram.

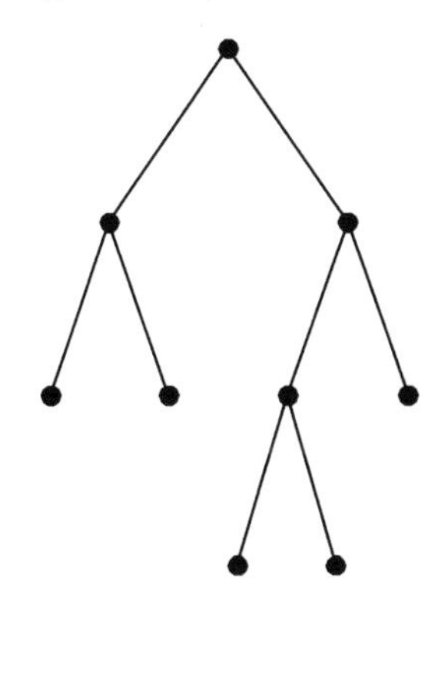

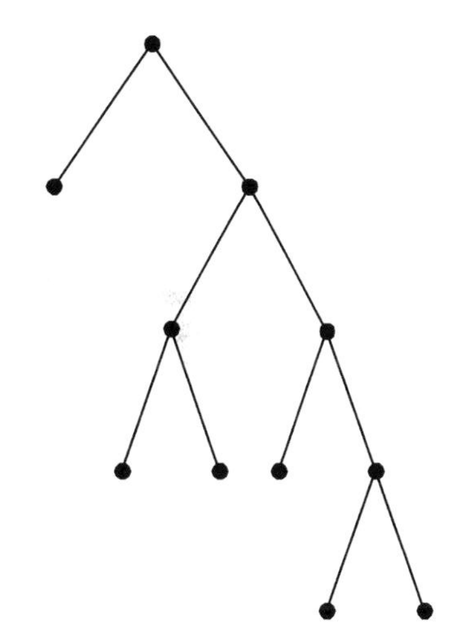

   (i) Show that every full binary tree has an odd number of vertices. (Hint: see exercise 11.1.7(ii).)

   (ii) Let T be a full binary tree with $n \geqslant 3$ vertices. Prove that there are always two more leaf vertices than decision vertices.

   (iii) How many non-isomorphic full binary trees are there with (a) five vertices, (b) seven vertices and (c) nine vertices?

7. Let T be a full binary tree with root r. Let v be any vertex of T. Define the **level** of the vertex v to be the length of the unique path in T joining r and v. Also define the **level** of the binary tree T to be the greatest of the levels of its vertices.

   For example, the following full level 3 binary tree has one level 0 vertex (the root r), two level 1 vertices (a and b), two level 2 vertices (c and d) and two level 3 vertices (e and f).

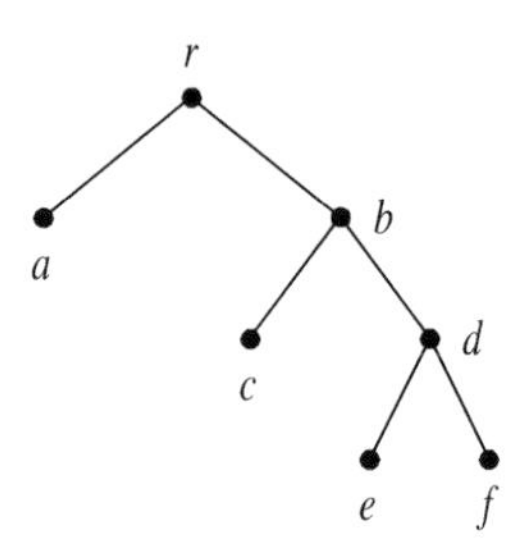

   (i) Draw all the level 1 and level 2 full binary trees.

   (ii) Let a_n denote the number of full level n binary trees. Show that, for $n \geqslant 2$,

$$a_n = 2a_{n-1}(a_0 + a_1 + \cdots + a_{n-2}) + a_{n-1}^2.$$

   (iii) Hence find a_3 and a_4.

8. (i) (Theorem 11.6(iii).) Prove, by induction on the number of edges of T, that if T is a tree then

$$|E| = |V| - 1.$$

   (ii) Use theorems 11.5 and 11.6(iii) to show that if Γ is any connected graph then

$$|E| \geqslant |V| - 1.$$

   (iii) Prove the converse to theorem 11.6(iii) by showing that if Γ is a connected graph which is not a tree then

$$|E| > |V| - 1.$$

9. (i) Let F be a forest with c components (see footnote, page 585. Write and prove an equation connecting $|V|$, $|E|$ and c which generalizes theorem 11.6(iii).

   (ii) Let Γ be any graph. Write and prove an inequality connecting $|V|$, $|E|$ and c which generalizes question 8(ii) above.

10. How many spanning trees are there in $K_{2,n}$? Prove your answer is correct.

11. Let Γ be a connected graph and let $t(\Gamma)$ denote the number of spanning trees in Γ. Let e be an edge of Γ. Then:

   $\Gamma - e$ denotes the graph obtained from Γ by deleting the edge e;

   $\Gamma\backslash e$ denotes the graph obtained from Γ by 'contracting' the edge e; that is, amalgamating the two vertices that are incident with e. This is illustrated in the following diagram.

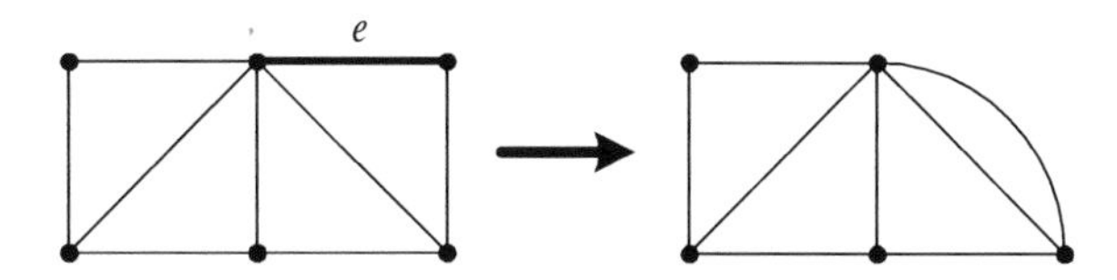

The edge e is a **bridge** if $\Gamma - e$ is disconnected. This is illustrated in the following diagram.

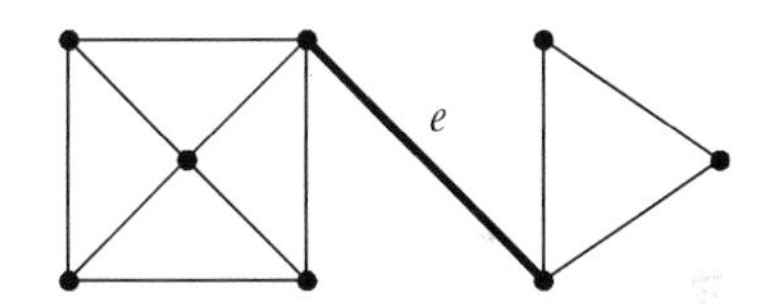

(i) Prove that, if e is an edge of Γ which is not a bridge, then

$$t(\Gamma) = t(\Gamma - e) + t(\Gamma\backslash e).$$

(ii) What is the corresponding result to that in part (i) in the case where e is a bridge? Explain your answer.

(iii) Using the result of part (i), find the number of spanning trees in the following graph.

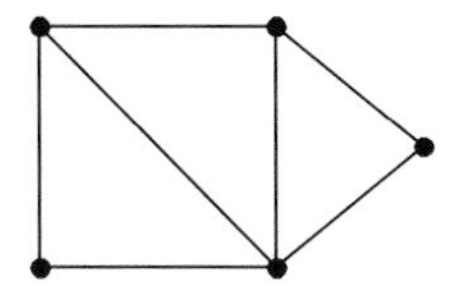

12. (i) How many spanning trees does each of the following graphs have?

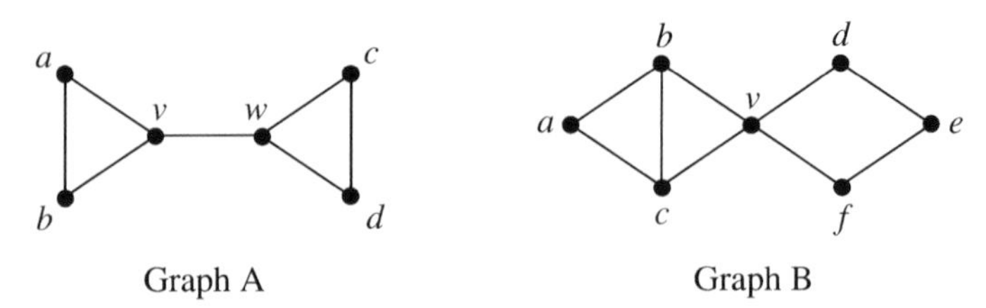

(ii) What general observation can be made about the number of spanning trees in a connected graph which has a bridge or a cut vertex? ('Bridge' is defined in exercise 11.11 above. A 'cut vertex' in a connected graph is a vertex which, if removed from the graph, together with its incident edges, produces a disconnected graph.)

13. In chemistry, graphs are used as symbolic representations of molecules. Each atom is represented as a vertex and each chemical bond is represented as an edge. For example, molecules of ethene (C_2H_4) and ethanol (C_2H_5OH) can be represented by the following graphs.

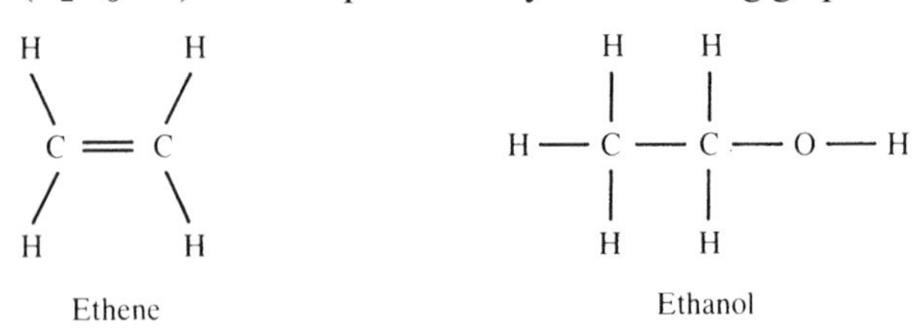

The alkane series of saturated hydrocarbons have chemical formula C_nH_{2n+2}. Each carbon atom (C) always has valency (degree) 4 and each hydrogen atom (H) always has valency (degree) 1.

(i) Draw the graphs to represent methane (CH_4) and ethane (C_2H_6).
(ii) Prove that each alkane is represented by a tree. (Hint: use theorem 11.6.)

A **structural isomer** of a chemical compound is an isomorphism class of its corresponding graph. This means that molecules of different structural isomers of the same compound have non-isomorphic graphs.

(iii) Butane (C_4H_{10}) has two structural isomers; draw their graphs.
(iv) Draw graphs of the different structural isomers of pentane (C_5H_{12}) and hexane (C_6H_{14}).

14. The alkenes and mono-cyclo-alkanes are unsaturated hydrocarbons with chemical formula C_nH_{2n}, where $n \geqslant 2$. (The graph of ethene, the simplest alkene, is given above.)

(i) The compound whose chemical formula is C_3H_6 has two structural isomers; draw their graphs. (One of these is propene and the other is cyclo-propane.)

(ii) Draw the graphs of the different structural isomers of C_4H_8. (Some of these are butenes; others are methyl-cyclo-propane and cyclo-butane.)

15. The hydrocarbon C_5H_8 has many structural isomers. We found 27 (we think!). How many can you find?

 (These isomers delight in such names as pentyne, methyl-butadiene, methyl-cyclo-butene, etc. Several of the isomorphism types do not exist as chemical compounds because the stresses in the molecular structure are too great. Of course, the mathematics cannot tell us that—it is the domain of the chemists.)

11.5 Planar Graphs

Examining the diagrams of graphs earlier in this chapter, it is apparent that some graphs can be represented by diagrams drawn in the plane (without any edges crossing) and some cannot. The object of this section is to examine which graphs can be represented by plane diagrams. Clearly this question is of potential importance in some of the applications of graph theory, notably in the design of electronic circuits which can be printed on boards.

The question of whether a graph can be represented in the plane is illustrated by the so-called **three utilities problem**. Three houses are each to be supplied with three utilities—electricity, gas and water. The problem is whether the three houses can be supplied without any of the utility lines having to cross. The graph which models this situation is the complete bipartite graph $K_{3,3}$. The three vertices of one set represent the three utility outlets and the three of the second set represent the three houses. The edges of the graph represent the utility lines; each utility is connected to each house. In graph-theoretic terms the problem is whether $K_{3,3}$ can be represented in the plane. Our diagram of $K_{3,3}$ (figure 11.5(*b*) on page 555) is not drawn in the plane, but if we were more ingenious perhaps we could have drawn the diagram without the edges crossing.

Intuitively, we say a graph is 'planar' if it can be represented by a diagram in the plane with no edges crossing. The formal definition is the following.

Definition 11.13

A graph whose vertices are points in the plane and whose edges are lines or arcs in the plane which only meet at vertices of the graph is called a **plane graph**. (Thus a plane graph is a certain subset of $\mathbb{R}^2$.)

A graph is a **planar** if it is isomorphic to a plane graph, i.e. if it can be represented by a diagram drawn in the plane with no edges crossing.

A little trial and error should be enough to convince you that $K_{3,3}$ is not planar. In the three utilities problem, the utilities must cross somewhere. Perhaps we would regard a formal proof of this unnecessary, but suppose we were confronted with a significantly more complex graph which we had been unable to represent in the plane; how could we be certain that we hadn't overlooked some configuration of edges and vertices which might show the graph to be planar?

Euler's Formula

Let Γ be a connected planar graph. A diagram of Γ drawn in the plane (technically, a plane graph isomorphic to Γ) divides the plane into regions, usually called **faces**. Referring to figure 11.4, we see that K_3 divides the plane into two regions—one bounded and one unbounded—and K_4 divides the plane into four regions—three bounded and one unbounded. It would also appear from figure 11.4 that K_5 is not planar (although, of course, the diagram does not *prove* this).

It turns out that there is a simple formula connecting the number of vertices, edges and faces of a connected planar graph. To investigate this, the following table provides some evidence which may guide us towards a conjecture.

Graph	Number of vertices	Number of edges	Number of faces
K_3	3	3	2
K_4	4	6	4
Figure 11.2(*a*)	7	7	2
Figure 11.5(*a*)	9	14	7
Figure 11.8(*b*)	4	7	5
Figure 11.12(*b*)	4	9	7
Any tree	n	$n-1$	1

All of these graphs are connected and planar and satisfy the relationship

$$|F| = |E| - |V| + 2$$

where $|F|$, $|E|$ and $|V|$ are the number of faces, edges and vertices respectively. This relationship holds for all connected planar graphs, and is known as **Euler's formula**†.

Theorem 11.7 (Euler's formula)

Let Γ be any connected planar graph with $|V|$ vertices, $|E|$ edges and dividing the plane into $|F|$ faces or regions. Then

$$|F| = |E| - |V| + 2.$$

Proof

The proof is by induction on the number of edges of Γ. If $|E| = 0$ then $|V| = 1$ (Γ is connected, so there cannot be two or more vertices) and there is a single face (consisting of the whole plane except the single vertex), so $|F| = 1$. The theorem therefore holds in this case.

Suppose, now, that the theorem holds for all graphs with fewer than n edges. Let Γ be a connected planar graph with n edges; that is $|E| = n$. If Γ is a tree, then $|V| = n + 1$ (theorem 11.6) and $|F| = 1$, so the theorem holds in this case too. If Γ is not a tree choose any cycle in Γ and remove one of its edges. The resulting graph Γ' is connected, planar and has $n - 1$ edges, $|V|$ vertices and $|F| - 1$ faces. By the inductive hypothesis, Euler's formula holds for Γ':

$$|F| - 1 = (|E| - 1) - |V| + 2$$

so

$$|F| = |E| - |V| + 2$$

as required. □

† The formula has an interesting history, which is related in the book by Imre Lakatos (1976) *Proofs and Refutations*. Lakatos uses an imaginary classroom discussion of Euler's formula to raise philosophical questions about the nature of mathematical discovery. The discussion itself is very lucid and follows, to a large extent, the actual historical development of the formula.

Examples 11.10

1. We can deduce from Euler's formula that $K_{3,3}$ is not planar. Suppose that $K_{3,3}$ is planar. Then it divides the plane into faces, the boundary of each face being a cycle‡. Every edge in a cycle of $K_{3,3}$ forms part of the boundary of *two* faces. Thus the sum of the numbers of edges belonging to the boundaries of all the faces is $2|E|$. In $K_{3,3}$ it is easy to see that every cycle contains at least four edges, so every face must have at least four edges in its boundary. Since every edge belongs to some cycle,

$$2|E| \geqslant 4|F|.$$

Substituting for $|F|$ in Euler's formula gives

$$2|E| \geqslant 4(|E| - |V| + 2).$$

From figure 11.5(b) we see that $|V| = 6$ and $|E| = 9$ so this last inequality becomes

$$18 \geqslant 4 \times (9 - 6 + 2) = 20$$

which is a contradiction. Therefore $K_{3,3}$ is not planar.

2. A similar argument to the above can be used to show that K_5 is not planar. (See figure 11.4.)

Kuratowski's Theorem

The two examples above of non-planar graphs—$K_{3,3}$ and K_5—represent, in a sense, the *only* ways in which a graph can fail to be planar. Clearly if a graph Γ contains a subgraph isomorphic to either $K_{3,3}$ or K_5 then Γ must be non-planar. The converse is almost true, except that we need to replace 'isomorphic' by a slightly weaker relationship.

To understand why we need a slightly different relationship than isomorphism, imagine the graph obtained from K_5, say, by dividing one of its edges in two

‡ For any (finite) planar graph, one of these faces is unbounded. It is perhaps less clear what we mean by the boundary of the unbounded face. An example should clarify this point. The cycle graph in figure 11.2(a) divides the plane into a bounded and an unbounded face which share a common boundary cycle.

by adding a vertex of degree 2 in the middle of the edge. The resulting graph is clearly also non-planar, but it is not isomorphic to K_5 (it has one more vertex and one more edge, for example). We say the resulting graph is 'homeomorphic' to K_5.

Definition 11.14

Two graphs are **homeomorphic** if (an isomorphic copy of) one graph can be obtained from the other by adding and/or deleting vertices of degree 2 into or from its edges.

Example 11.11

All of the graphs shown in figure 11.18 are homeomorphic. To obtain the graph (*b*) from (*a*) we delete two vertices, and to obtain (*c*) from (*b*) delete one vertex and add two. From (*c*) to (*d*) we add one vertex and delete another and from (*d*) to (*e*) we add one vertex. The graphs (*e*) and (*f*) are isomorphic—no vertices need be added or deleted.

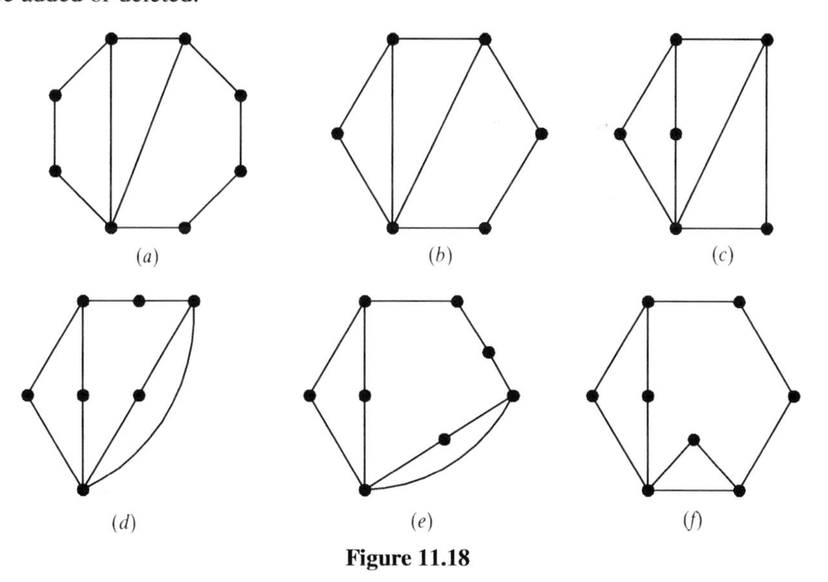

Figure 11.18

There is an alternative, perhaps simpler, way of determining whether two graphs Γ and Σ are homeomorphic. Successively delete vertices of degree 2 from the edges of both graphs until no further deletions can be performed. Let Γ' and Σ' respectively denote the resulting graphs. Then Γ and Σ are homeomorphic if and only if Γ' and Σ' are isomorphic.

Successively deleting vertices of degree 2 from each of the graphs in figure 11.18 produces the graph shown in figure 11.19. This confirms that the graphs in figure 11.18 are all homeomorphic.

Figure 11.19

We can now explain more precisely what we mean by graphs K_5 and $K_{3,3}$ being essentially the only ways a graph can fail to be planar. The following theorem, first proved by the polish mathematician Kuratowski in 1930, provides an elegant characterization of planar graphs. Unfortunately the proof is too long and complicated to include here—see, for example, Gould (1988) or Harary (1969) for a proof.

Theorem 11.8 (Kuratowski's theorem)

A graph is planar if and only if it contains no subgraph homeomorphic to K_5 or $K_{3,3}$.

Exercises 11.5

1. Using an argument similar to that given for $K_{3,3}$ in example 11.10.1, prove that the complete graph K_5 is not planar.

2. (i) Using Kuratowski's theorem, or otherwise, show that K_n is planar for $n \leqslant 4$ and non-planar for $n \geqslant 5$.

   (ii) Show that $K_{1,n}$ and $K_{2,n}$ are planar for all $n \geqslant 1$.

3. Show that one of the two graphs in the following diagram is planar and the other is not.

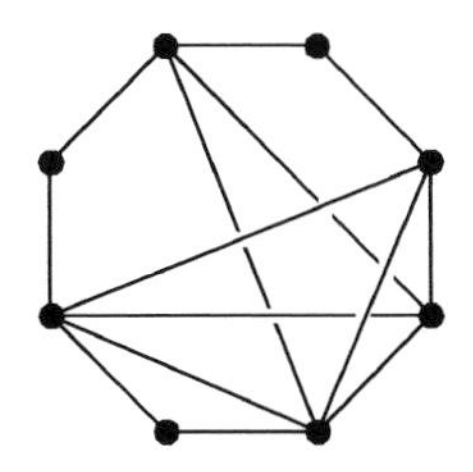
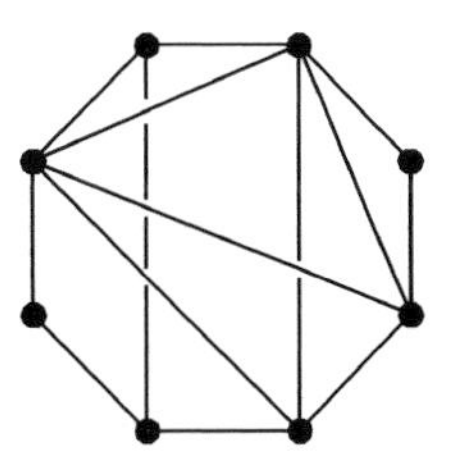

4. (i) Show that the following graph is planar.

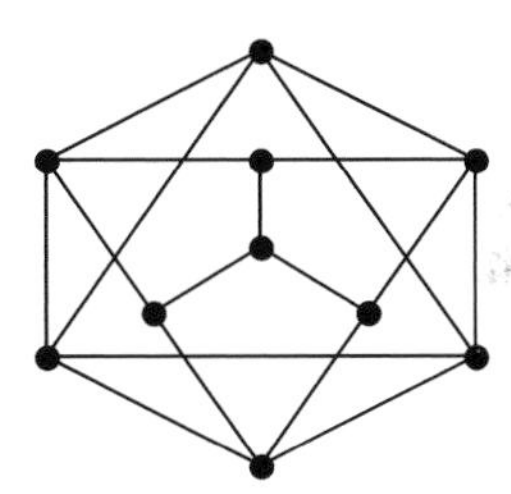

(ii) Determine whether or not each of the following graphs is planar.

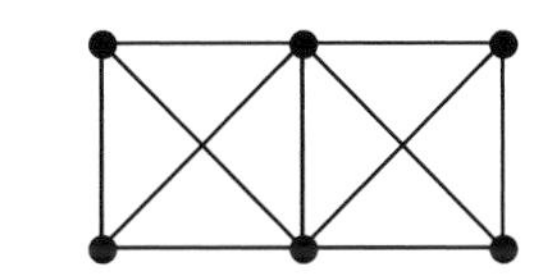

(b)

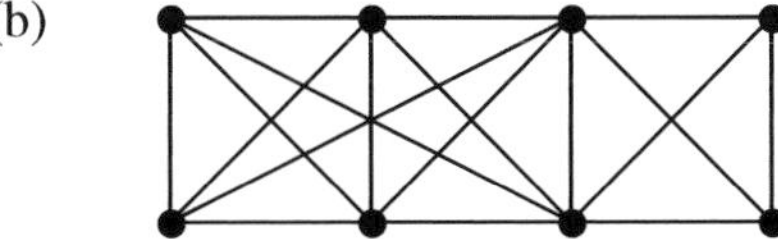

5. Use Kuratowski's theorem to show that each of the following graphs is not planar.

(i)

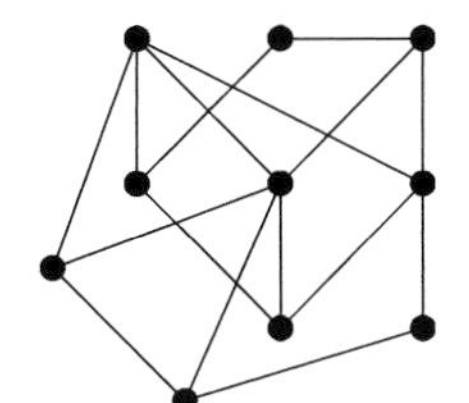

(ii)

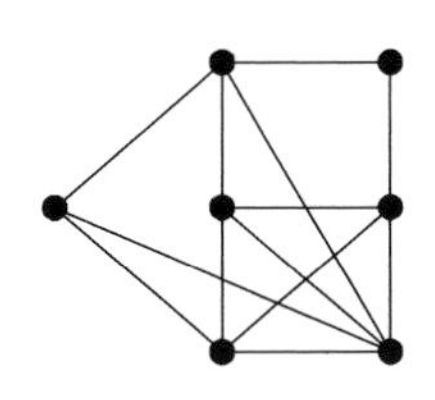

6. Prove that, for $n \geqslant 1$, the cycle graph C_n is homeomorphic to K_3. (Cycle graphs are defined on page 551.) Deduce that any two cycle graphs are homeomorphic.

7. Show that, for each of the pairs of graphs illustrated in the following diagram, the graphs are homeomorphic.

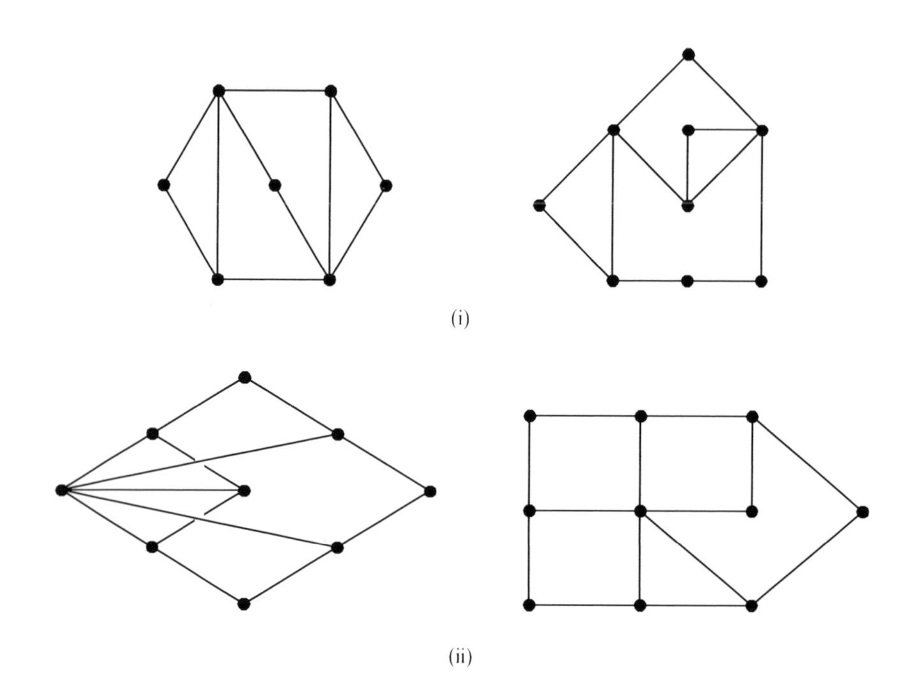

8. (i) Use Euler's formula to show that, for a connected simple planar graph, $3|F| \leqslant 2|E|$. Deduce that $|E| \leqslant 3|V| - 6$. (Hint: use an argument similar to the one in example 11.10.1.)

   (ii) Show that every connected simple planar graph has at least one vertex of degree less than or equal to five. (Hint: use proof by contradiction.)

9. The **dual** Γ^* of a plane graph Γ is also a plane graph obtained firstly by placing a vertex of Γ^* in the interior of each face of Γ. Every edge in Γ separating two faces gives rise to an edge in Γ^* joining the corresponding vertices of Γ^*. This is illustrated in the following diagram.

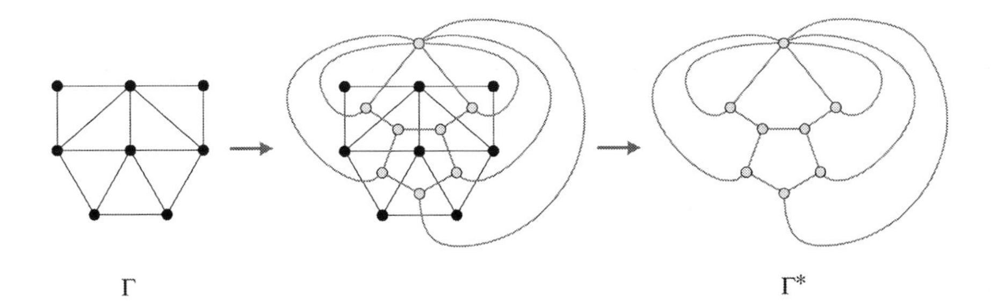

(i) Draw the dual graph of each of the following plane graphs.

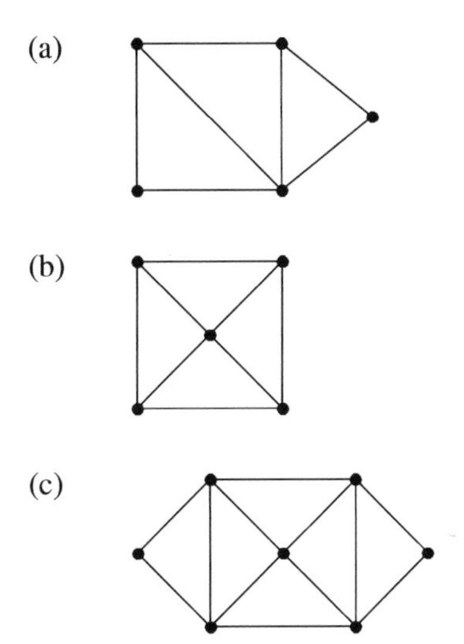

(ii) How do the numbers of vertices, edges and faces of Γ^* relate to the numbers of vertices, edges and faces of Γ?

10. A connected, simple graph Γ has vertex degree sequence $(2, 2, 2, 3, 3, 4, 5, 5)$.

(i) Show that Γ must be planar.
(ii) Find the numbers of edges and faces of Γ.
(iii) Write down the numbers of vertices, edges and faces of Γ^*, the dual of Γ.

11. *Graphs on other surfaces.* Intuitively, a planar graph is one which can be drawn in the plane. We can also consider graphs which can be drawn (without intersecting edges) on other surfaces such as the sphere or torus ('ring doughnut') as in the following diagram.

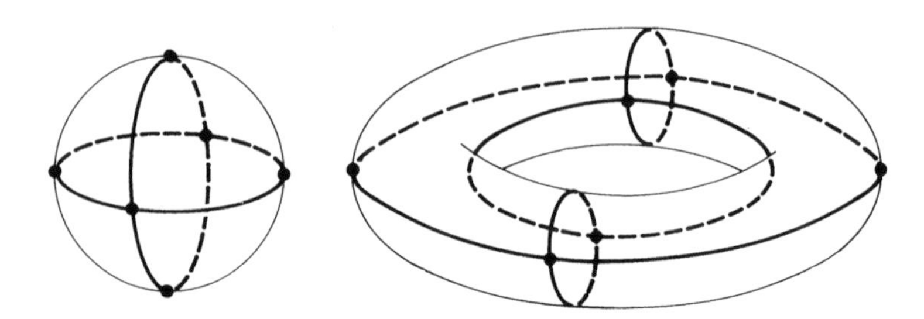

Euler's formula says that $|V| - |E| + |F| = 2$ for any graph drawn in the plane. In fact, the value of $|V| - |E| + |F|$ depends only on the surface on which the graph can be drawn, and not on the graph itself.

(i) Draw several graphs on the surface of a sphere and calculate $|V| - |E| + |F|$ for each.

(ii) Repeat for graphs drawn on the surface of a torus.

12. Show that each of the graphs K_5 and $K_{3,3}$ can be drawn on the surface of the torus.

11.6 Directed Graphs

The graphs we have considered so far have all been undirected graphs, which means that the edges have no preferred direction. Using the analogy of a road map, all our roads have been two-way streets. In some applications, however, we require graphs where the edges do have a specified direction; on a diagram this is indicated by an arrow on each edge. We have already seen examples of the directed graph of a relation (on a finite set) in chapter 4.

There is a simple way of modifying the definition of an undirected graph to give each edge a direction. In definition 11.1 the 'endpoints' of an edge e are defined to be a set of vertices $\delta(e) = \{v, w\}$. A directed edge can be thought of as having an 'initial' and a 'final' vertex, rather than two endpoint vertices of equal status. This can be made precise by defining $\delta(e) = (v, w)$, the *ordered* pair of vertices. The ordering of the vertices then determines the direction of the edge.

Definition 11.15

A **directed graph**, or **digraph**, D consists of a finite non-empty set $V = V_D$ of **vertices**, a finite set $E = E_D$ of **(directed) edges** and a mapping $\delta : E \to V \times V$. If $\delta(e) = (v, w)$ then v is called the **initial vertex** and w the **final vertex** of the edge e.

Given a directed graph D, we can forget about the direction of the edges and obtain an undirected graph Γ, called the **underlying graph** of D. Formally the underlying graph of D has the same vertex and edge sets as D, with mapping δ_Γ defined by $\delta_\Gamma(e) = \{v, w\}$ if $\delta_D(e) = (v, w)$.

A digraph is **simple** if it has no loops (edges e with $\delta(e) = (v, v)$) and multiple edges (edges with the same initial and final vertices). We should point out that the underlying graph of a simple digraph need not be a simple graph. Figure 11.20 shows a simple digraph and its underlying graph. The two edges of the digraph connecting the vertices v_4 and v_5 are not multiple edges because they have different directions—that is, different initial and final vertices. In the underlying graph, however, they are multiple edges.

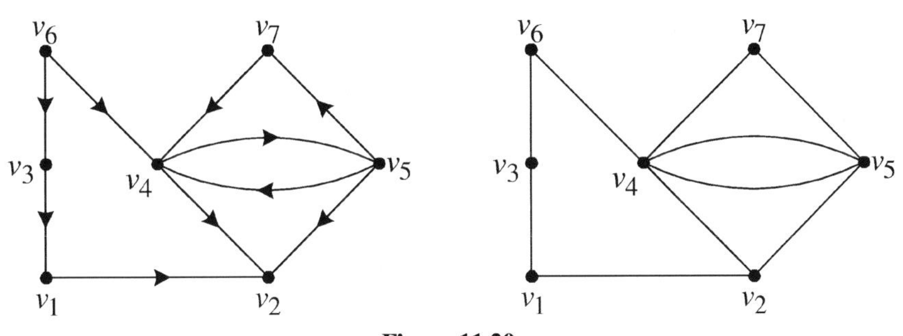

Figure 11.20

Many of the definitions for undirected graphs carry over to directed graphs either unchanged or with obvious modifications to take account of the directions of the edges. In particular, the definitions of **adjacent**, **incident** and **degree** remain unaltered. Thus, for instance, two vertices v and w of a digraph are adjacent if and only if there exists a directed edge either from v to w or from w to v. However, we do modify the definition of the adjacency matrix of a digraph to take account of the directions of the edges. The (i, j)-entry of the adjacency matrix represents the number of edges with initial vertex v_i and final vertex v_j, so the matrix need not be symmetric. The adjacency matrix for the digraph in figure 11.20 is the following, where we have given the vertex set the obvious ordering:

$$\begin{pmatrix} 0 & 1 & 0 & 0 & 0 & 0 & 0 \\ 0 & 0 & 0 & 0 & 0 & 0 & 0 \\ 1 & 0 & 0 & 0 & 0 & 0 & 0 \\ 0 & 1 & 0 & 0 & 1 & 0 & 0 \\ 0 & 1 & 0 & 1 & 0 & 0 & 1 \\ 0 & 0 & 1 & 1 & 0 & 0 & 0 \\ 0 & 0 & 0 & 1 & 0 & 0 & 0 \end{pmatrix}.$$

For a digraph, the degree of a vertex is the sum of the entries in the column *and* the row corresponding to it in the adjacency matrix. The entries in the row corresponding to vertex v represent those edges with v as initial vertex and the entries in the column corresponding to v represent the edges with v as final vertex. (Note that the diagonal elements are automatically counted twice, once when summing a row and once when summing a column. Thus, unlike in the undirected case, we do not need to double the diagonal entries to obtain the correct degree.) The sum of all the entries in the adjacency matrix is, of course, the total number of edges in the digraph.

A **directed edge sequence from v_0 to v_n** in a digraph is a sequence of edges $e_1, e_2, \ldots, e_n$ such that $\delta(e_i) = (v_{i-1}, v_i)$ for $i = 1, \ldots, n$. The definitions of **directed path** and **directed cycle** are the obvious modifications of the undirected definitions. We can define connectivity for a digraph in two different (and non-equivalent) ways.

Definition 11.16

A digraph is **connected** (or **weakly connected**) if its underlying undirected graph is connected. A digraph is **strongly connected** if, for every ordered pair of vertices (v, w), there is a directed path from v to w.

Examples 11.12

1. Figure 11.21 shows two digraphs, both of which are (weakly) connected; they have the same underlying connected graphs. The graph (*a*) is *not* strongly connected because, for example, there is no directed path in (*a*) from v to w: less formally, we cannot travel from vertex v to vertex w along edges *in the direction of the arrows*. The graph (*b*) is strongly

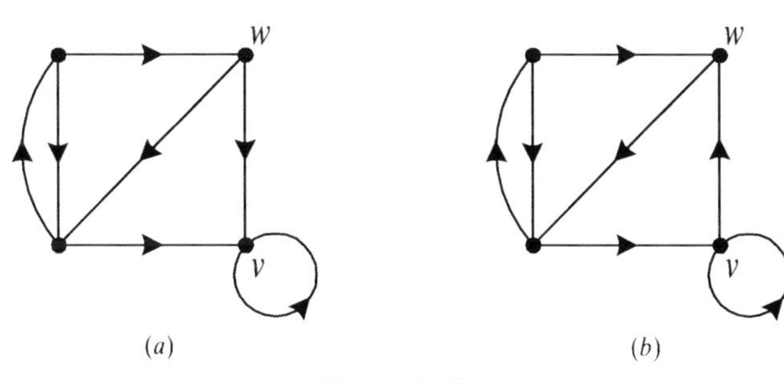

Figure 11.21

connected, however, because we can journey from any vertex to any other by travelling the 'correct' way along edges.

2. A simple situation which can be conveniently modelled using a digraph is a 'round-robin' tournament. This is a tournament in which every competitor plays every other and every match has a winner. It can be modelled by a digraph with vertices representing the players, and with an edge with initial vertex v and final vertex w if and only if player v beats player w. The underlying graph of this digraph is complete because every competitor plays every other. Any digraph whose underlying graph is complete is called a **tournament**.

There are notions of **Eulerian digraph** and **Hamiltonian digraph** defined in the obvious way; an Eulerian digraph has a closed directed path containing every edge and a Hamiltonian digraph has a directed cycle passing through every vertex.

The theorem corresponding to Euler's theorem (theorem 11.2) for digraphs is an easy modification of Euler's theorem for graphs. We define the **in-degree** of a vertex v to be the number of edges with final vertex v, and the **out-degree** of v to be the number of edges with initial vertex v. Thus the in- and out-degrees of a vertex are the number of edges 'going in to' and 'coming out from' the vertex respectively. The proof of the following theorem is a directed version of the proof of Euler's theorem and is left as an exercise (11.6.10).

Theorem 11.9

A connected digraph is Eulerian if and only if the in-degree equals the out-degree for every vertex.

As we might expect from the undirected case, the situation for Hamiltonian digraphs is more complicated. However, for tournaments we can prove the following result.

Theorem 11.10

(i) Every tournament has a directed path containing all the vertices. (In other words every tournament is **semi-Hamiltonian**; see exercise 11.2.10 for the definition of semi-Hamiltonian for undirected graphs.)

(ii) Every strongly connected tournament is Hamiltonian.

Proof

(i) We show that a directed path which does not pass through every vertex can be extended to pass through another vertex. The result then follows because we can begin with any directed path (e.g. containing a single edge) and continue extending it until it passes through every vertex.

Let $P : e_1, e_2, \ldots, e_n$ be any directed path, such that $\delta(e_i) = (v_{i-1}, v_i)$ for $i = 1, 2, \ldots, n$. Suppose P does not pass through every vertex; let v be any vertex through which P does not pass. Since the underlying graph is complete, there is either an edge from v to v_0 or from v_0 to v. In the former case adding this edge at the beginning of P extends it to a path passing through v.

In the latter case let r be the *largest* integer such that there are edges from $v_0, v_1, \ldots, v_r$ to v. If $r < n$ then there are edges from v_r to v and from v to v_{r+1}; we can insert these edges into P between e_r and e_{r+2}. If $r = n$ we can add the edge from v_n to v at the end of P. In either case we have extended P to pass through v.

(ii) We might as well also suppose that $|V| \geqslant 2$—if D has only a single vertex, it must be a null graph, so the empty directed cycle passes through the vertex.

For a strongly connected tournament D, the above argument can be modified to show that a directed cycle not passing through every vertex can be extended to pass through an additional vertex. The result will then follow (as for part (i)) if we can find a directed cycle of any length at all; below we show that D must have a directed cycle of length 3.

Choose any vertex v and let V_1 be the set of all vertices w such that there is an edge e with $\delta(e) = (v, w)$. Similarly let V_2 be the set of all vertices w such that there is an edge e with $\delta(e) = (w, v)$. Since the tournament is strongly connected it follows that V_1 and V_2 are both non-empty. For example, if $V_1 = \varnothing$ then v has out-valency 0, so there is no directed path from v to *any* other vertex of D, which contradicts the assumption that D is strongly connected. A similar argument shows that V_2 is non-empty.

Now there must be at least one directed edge with initial vertex belonging to V_1 and final vertex belonging to V_2, otherwise it would not be possible to join any vertex of V_1 to any vertex of V_2, which again contradicts the strong connectivity of D. Choose an edge e with initial vertex $v_1 \in V_1$ and final vertex $v_2 \in V_2$. By the definitions of V_1 and V_2 there exist edges e_1 from v to v_1 and e_2 from v_2 to v. The edge sequence e_1, e, e_2 is a directed cycle of length 3 in D; the result now follows from the existence of this cycle, as we explained above. □

Exercises 11.6

1. How is the adjacency matrix of a digraph related to the adjacency matrix of its underlying graph?

2. Show that, for the edges of any digraph,

$$\text{sum of in-degrees} = \text{sum of out-degrees} = |E|.$$

3. (i) Explain why a directed graph whose underlying graph is a tree cannot be strongly connected.

   (ii) Is it possible for a strongly connected directed graph to have an underlying graph which is simple? Justify your answer.

4. For each of the following matrices draw a diagram to represent a digraph which has the given matrix as its adjacency matrix:

$$\text{(i)} \quad \begin{pmatrix} 0 & 1 & 1 & 1 \\ 0 & 1 & 0 & 0 \\ 0 & 1 & 0 & 1 \\ 0 & 0 & 1 & 0 \end{pmatrix} \qquad \text{(ii)} \quad \begin{pmatrix} 0 & 1 & 1 & 0 & 0 \\ 0 & 1 & 0 & 0 & 0 \\ 0 & 0 & 2 & 0 & 0 \\ 1 & 0 & 1 & 0 & 2 \\ 1 & 0 & 0 & 0 & 0 \end{pmatrix}.$$

5. For each of the digraphs represented in the following diagram, determine whether the digraph:

   (a) is simple;
   (b) has a simple underlying graph;
   (c) is strongly connected;
   (d) is Eulerian;
   (e) is Hamiltonian.

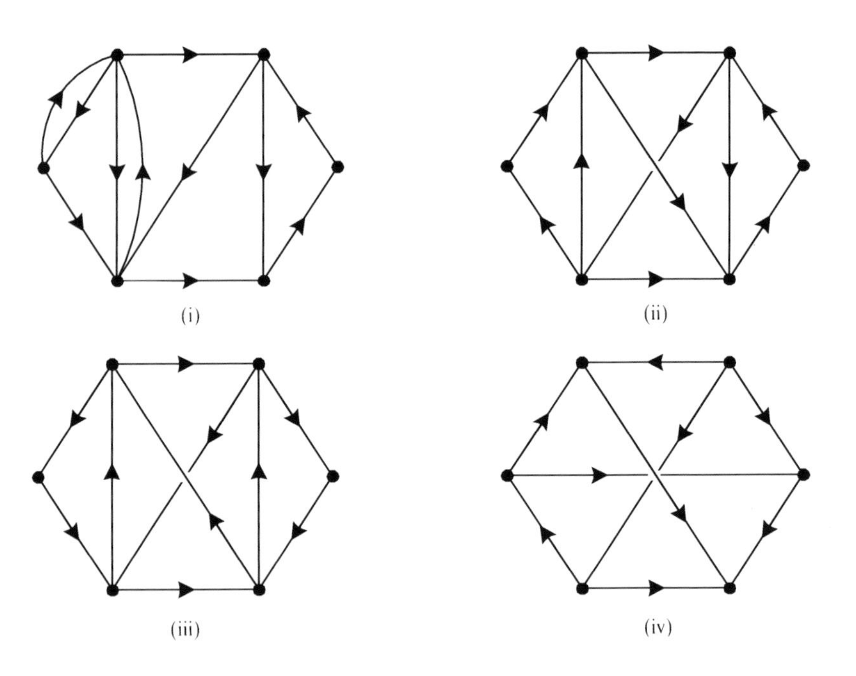

6. Let R be a relation on a set A and let D be the directed graph of R—see chapter 4. Explain why the adjacency matrix of D is equal to the binary matrix of R.

7. A directed graph is said to be **unilaterally connected** if, for every pair of vertices v and w, either there is a directed path from v to w or there is a directed graph from w to v.

(i) Clearly: strongly connected $\Rightarrow$ unilaterally connected $\Rightarrow$ weakly connected. Show that none of these implications reverse.

(ii) Which of the digraphs in question 5 above are unilaterally connected?

8. A graph is **orientable** if it is the underlying graph of some strongly connected digraph. Show that Petersen's graph (example 11.1.2), K_5 and $K_{3,3}$ are all orientable.

9. Let D be a digraph with vertex set $V = \{v_1, v_2, \ldots, v_n\}$ and edge set $E = \{e_1, e_2, \ldots, e_m\}$.

The **incidence matrix** of D is the $m \times n$ matrix $B = (b_{ij})$ where

$$b_{ij} = \begin{cases} 1 & \text{if edge } i \text{ has initial vertex } v_j \\ -1 & \text{if edge } i \text{ has final vertex } v_j \\ 0 & \text{otherwise.} \end{cases}$$

For example, the digraph with diagram

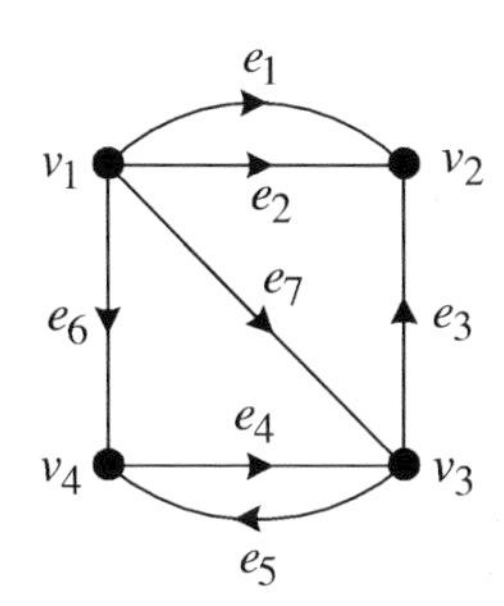

has incidence matrix

$$\begin{pmatrix} 1 & -1 & 0 & 0 \\ 1 & -1 & 0 & 0 \\ 0 & -1 & 1 & 0 \\ 0 & 0 & -1 & 1 \\ 0 & 0 & 1 & -1 \\ 1 & 0 & 0 & -1 \\ 1 & 0 & -1 & 0 \end{pmatrix}.$$

Note: the incidence matrix, as defined here, applies only to digraphs without loops.

(i) Write down the incidence matrix of the following digraph. (You will need to label the edges.)

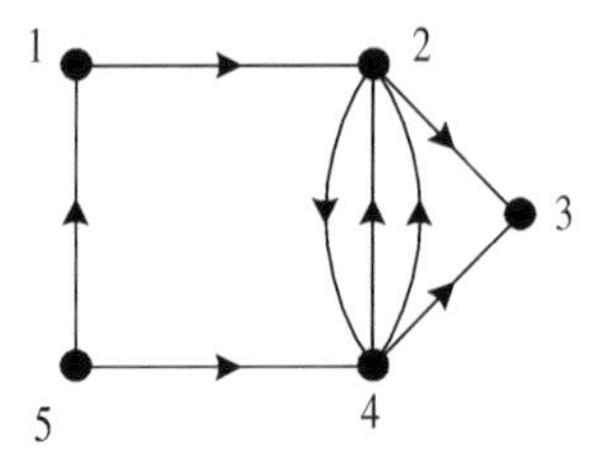

(ii) What information, if any, about a digraph D is provided by the row and column sums of its adjacency matrix?

(iii) What information, if any, about a digraph D is provided by the row and column sums of its incidence matrix?

(iv) Draw a diagram of the digraph which has incidence matrix

$$\begin{pmatrix} 1 & -1 & 0 & 0 & 0 \\ -1 & 1 & 0 & 0 & 0 \\ 1 & 0 & 0 & 0 & -1 \\ -1 & 0 & 0 & 1 & 0 \\ 0 & -1 & 0 & 0 & 1 \\ 0 & 1 & 0 & -1 & 0 \\ 0 & 0 & 0 & -1 & 1 \\ 0 & 0 & 0 & 1 & -1 \end{pmatrix}.$$

(v) How is the adjacency matrix of a digraph related to the adjacency matrix of its underlying graph?

(vi) How is the incidence matrix of a digraph related to the incidence matrix of its underlying graph?

10. Prove theorem 11.9.

11. Define the term **isomorphism** for directed graphs.

If two digraphs are isomorphic, does it necessarily follow that their underlying undirected graphs are isomorphic?

Conversely, if the underlying graphs of two digraphs are isomorphic, does it necessarily follow that the digraphs are themselves isomorphic?

12. The diagrams of three digraphs D_1, D_2 and D_3 are given below.

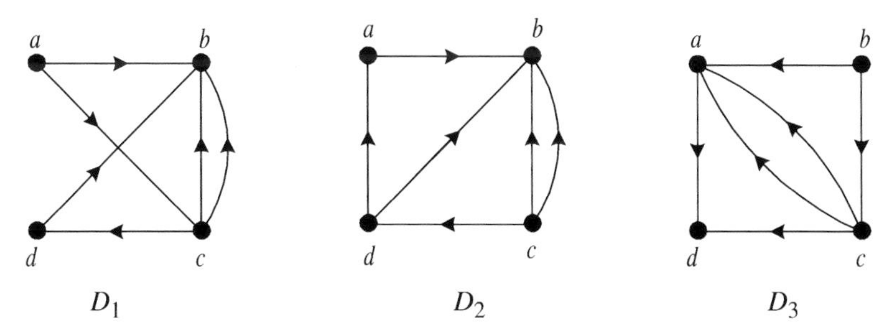

(i) Show that no pair of the digraphs is isomorphic.
(ii) Prove that two of the three digraphs have underlying graphs which are isomorphic.

13. Which of the following digraphs are isomorphic? Justify your answers.

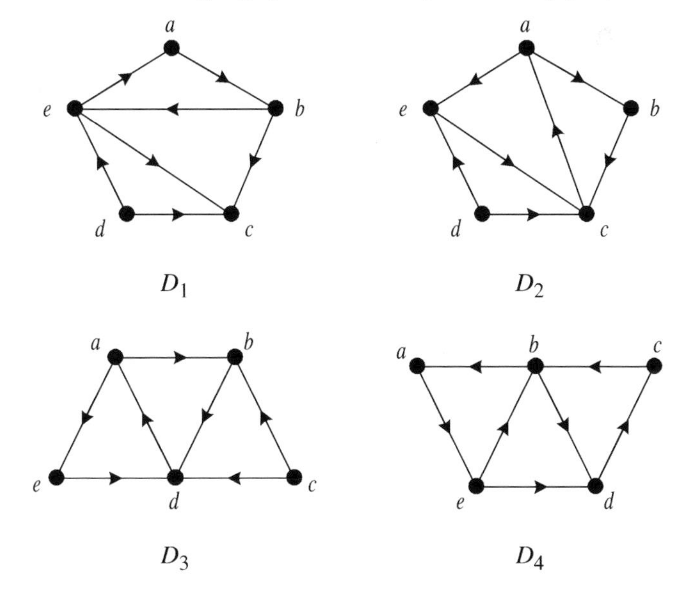

14. The **converse** $\tilde{D}$ of a digraph D is the digraph obtained by reversing the direction of every edge of D.

(i) Give an example of a digraph which is isomorphic to its converse and give an example of a digraph which is not isomorphic to its converse.

(ii) What is the connection between the adjacency matrices of D and $\tilde{D}$?

(iii) What is the connection between the incidence matrices of D and $\tilde{D}$?

15. Let $(G, *)$ be a finite group and S a sub*set* of G. Define a directed graph D of the pair (G, S) as follows. The vertex set of D is G, and there exists a directed edge from g_1 to g_2 if and only if $g_2 = g_1 s$ for some $s \in S$.

Draw the directed graph of (G, S) for each of the following.

(i) $G = \{e, x, x^2, x^3\}$, the cyclic group of order 4.
$S = \{x\}$.

(ii) $G = D_3 = \{r_0, r_1, r_2, m_1, m_2, m_3\}$, the dihedral group of degree 3.
$S = \{r_1, m_1\}$.

Chapter 12

Applications of Graph Theory

12.1 Introduction

In chapter 11 we claimed that graph theory has many applications. Our aim in this chapter is to explain briefly a few of these. Of necessity we can only outline a handful of the many applications of the theory. The interested reader is referred to one of the more specialized texts for more comprehensive treatments and additional uses of graph theory. Broadly, our applications fall into two categories—those in computing and those in a branch of applied mathematics known as combinatorial optimization. However, the distinctions between the two are not entirely clear cut.

Two particular classes of graphs, each of which has some additional structure, figure prominently in this chapter; they are 'rooted trees' and 'weighted graphs'. We shall need to devote some space to the theoretical aspects of these, so this chapter should not be seen as wholly concerned with applying graph theory.

Rooted trees, which occupy the first part of the chapter, are used extensively in computing. Our main applications of rooted trees are in the representation and sorting of data. Weighted graphs feature in the second half of the chapter; they are widely used in certain kinds of optimization problems. As the name suggests, optimization is concerned with finding the optimum or 'best' solution to a problem. The exact meaning of'best' depends on the particular situation under consideration; it may mean longest, shortest, greatest, least, etc. You might be familiar with the use of calculus for finding maxima and minima of functions—this is a branch of optimization. The problems we consider are of a discrete, rather than continuous, nature, so the 'tools' required tend to be different.

Before explaining our applications in more detail, perhaps we should mention briefly what is probably the most famous, if not necessarily the most significant, application of graph theory: the **four-colour conjecture**. In about 1852, Francis Guthrie† pointed out that it appeared that the countries of any map drawn in the plane (or, equivalently, on the surface of a sphere) could be coloured in such a way that countries sharing a common border are coloured differently using only four colours. A map in the plane can be modelled by a (planar) graph whose *vertices* represent the countries and whose edges represent the common borders. At first sight this is somewhat counter-intuitive; the graph corresponding to a particular map looks rather different from the map itself. Some thought should be sufficient to appreciate that this representation captures the essential properties of the original map. Assigning colours to the countries corresponds to colouring the vertices of the graph in such a way that no pair of adjacent vertices are given the same colour.

Over the years, various 'proofs' of the conjecture were given, but each was subsequently found to be flawed. ('Flawed' does not mean 'worthless', however, as many interesting ideas and techniques were developed in several of the unsuccessful attempts at a proof.) Eventually in 1976, Kenneth Appel and Wolfgang Haken completed a proof of the **four-colour theorem** as it then became known. However, their announcement of the proof caused considerable controversy in the mathematical community because, for the first time in mathematics, they had used a computer in an *essential* way to check the many hundreds of possible configurations to which the problem had been reduced. The sheer size of the number of computations required made the use of many hours of mainframe computer time essential. Even today, many mathematicians are reluctant to accept Appel–Haken proof as 'genuine' because it can only be checked by another computer—to work through the details 'by hand' is impossible in practice.

12.2 Rooted Trees

Many of the applications of graph theory, particularly in computing, use a certain kind of tree, called a 'rooted tree'. This is simply a tree where a particular vertex has been distinguished or singled out from the rest. These are the trees used to show the relationships between a person's descendants—the familiar 'family

† At the time Francis Guthrie was a student at the university of London, Francis's brother Frederick drew the problem to the attention of De Morgan who subsequently mentioned it in his lectures and once in print (in a book review). However, the problem did not become so widely known until Cayley introduced it to a meeting of the London Mathematical Society in 1878.

tree'. Figure 12.1 is a typical family tree showing the descendants of great-grandmother Mary who would be the distinguished vertex in this case.

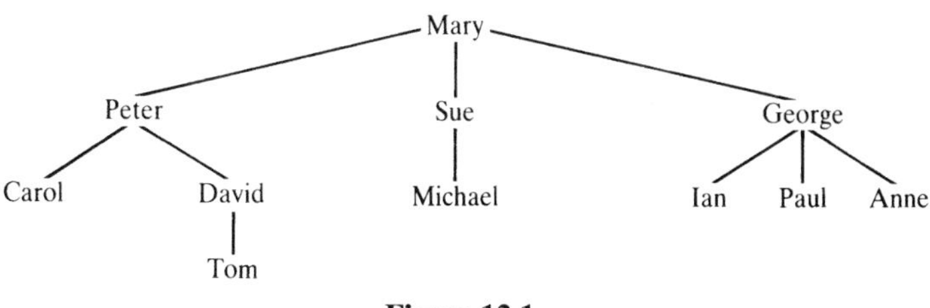

Figure 12.1

Rooted trees are perhaps most familiar in computing as models for the structure of file directories. Figure 12.2 shows part of a typical multi-user file directory, organized as a rooted tree. Directories are organized in this way for two main reasons: so that related files can be grouped conveniently together and, in multi-user systems, to protect the security of the users' files. Each user would usually have a password which would be required to gain access to his or her files.

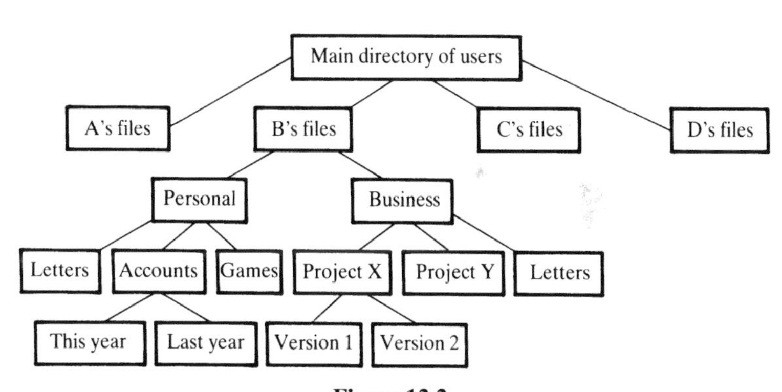

Figure 12.2

Some of the other important uses of rooted trees in computing include the representation of data and the representation of algebraic expressions (see exercises 12.1.10–12.1.14). The rooted tree data structure is particularly appropriate for data where there are hierarchical relationships among the data sets. Data represented as a rooted tree allow various subclasses of data to be accessed readily, often with less processing than some other hierarchical data structures such as linked lists. The detailed study of data structures and tree representations of algebraic expressions is beyond the scope of this book, so the interested reader should consult one of the more specialized texts for further details.

Definitions 12.1

A **rooted tree** is a pair (T, v^*) where T is a tree and $v^* \in V_T$. The distinguished vertex v^* is called the **root** of the tree.

A **leaf** in a rooted tree is a vertex which has degree 1 which is not equal to the root; a **decision vertex** (or **internal vertex**) is a vertex which is neither the root nor a leaf.

The name 'decision vertex' comes from so-called 'decision trees'. These are rooted trees which are used to model multi-stage decision processes where the decision made at one stage affects the possible decisions available at the next stage. The decision vertices represent the points at which decisions need to be made. Sometimes it is convenient to use slightly less formal terminology and refer to a 'rooted tree T with root v^*' rather than having always to use the ordered pair notation (T, v^*) for a rooted tree.

Unlike in nature, it is usual to draw the diagram of a rooted tree so that it 'grows downwards' with the root at the top of the diagram. Figure 12.3(a) shows a tree T with root v^*. (The choice of the root in this example is arbitrary; we could equally well have chosen a different vertex as the root.) The diagram of T is redrawn in figure 12.3(b) growing downwards with the root at the top. The leaf vertices are a, b, d, e, h, j, m, n, q and s, and the decision vertices are c, f, g, k, p and r.

Figure 12.3(b) suggests that we can partition the vertex set V_T into sets of vertices at different 'levels' according to how far they are from the root as follows:

Level 0: $\{v^*\}$
Level 1: $\{g, k, p\}$
Level 2: $\{c, f, h, m, n, r\}$
Level 3: $\{a, b, d, e, j, q, s\}$.

The formal definition of the level of a vertex is the following which relies on the fact (theorem 11.6(i)) that there is a unique path in the tree joining the root to any given vertex.

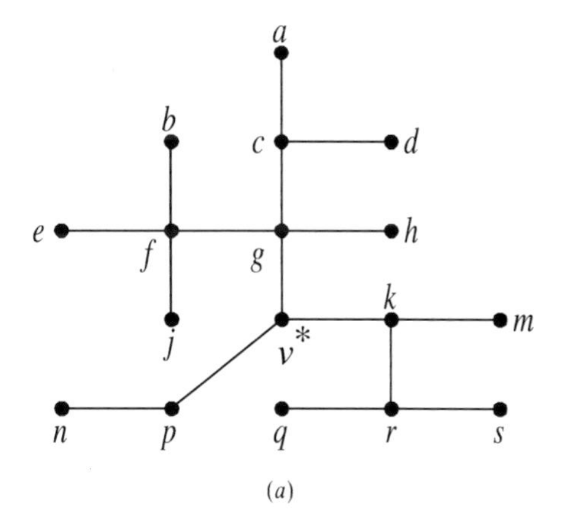

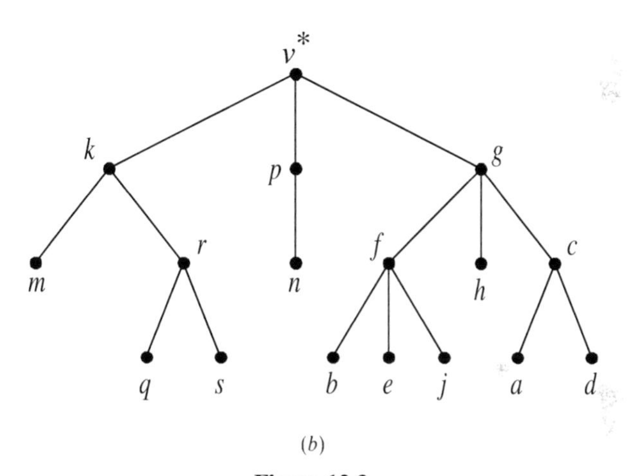

Figure 12.3

> **Definition 12.2**
>
> Let (T, v^*) be a rooted tree. The **level** of a vertex w of T is the length of the (unique) path in T from v^* to w. The **height** of T is the maximum of the levels of its vertices.

Let T be a rooted tree and let p be a vertex of level $k \neq 0$. (This is just another way of saying that $p \neq v^*$.) Since the path in T from v^* to p is unique, it follows that p is adjacent to a unique vertex of level $k - 1$. The terminology of the next definition is motivated by the example of a family tree.

Definition 12.3

Let (T, v^*) be a rooted tree and let p be a vertex of level $k > 0$. The (unique) vertex q of level $k - 1$ which is adjacent to p is called the **parent** of p. Similarly, p is the **child** of q, and any vertex of level k which is also adjacent to q is called a **sibling** of p.

It is clearly possible to define further terms such as **grandparent**, **grandchild**, **ancestor**, **descendant**, etc (see exercise 12.1.4). Also it is sometimes useful to use the term **parent vertex** to refer to a vertex which has children, i.e. a vertex which is either the root or a decision vertex (except in the case where the tree has no edges at all so that the root is childless).

Examples 12.1

1. The rooted tree shown in figure 12.3 has height 3. Also, g is the parent of c, f and h, r is the parent of q and s, and so on. Similarly b, e and j are siblings, a, b, d, e and j are all grandchildren of g, etc.

2. Figure 12.4 represents part of the (line) organizational structure of a large company as a rooted tree whose root is the Managing Director. (In the full organizational structure, of course, each of our leaf vertices would probably have children.)

   The five level 1 directors are siblings; indeed it is clearly always the case that the level 1 vertices of a rooted tree are siblings since they are all children of the root. Of the level 1 directors, the Production Director has the most descendants—three children and three grandchildren. The Research and Personnel Directors are childless and therefore have no descendants. Stretching the family relation definitions still further, we could say that any pair of the level 2 managers are either siblings or 'cousins'. Again this is the case for any rooted tree.

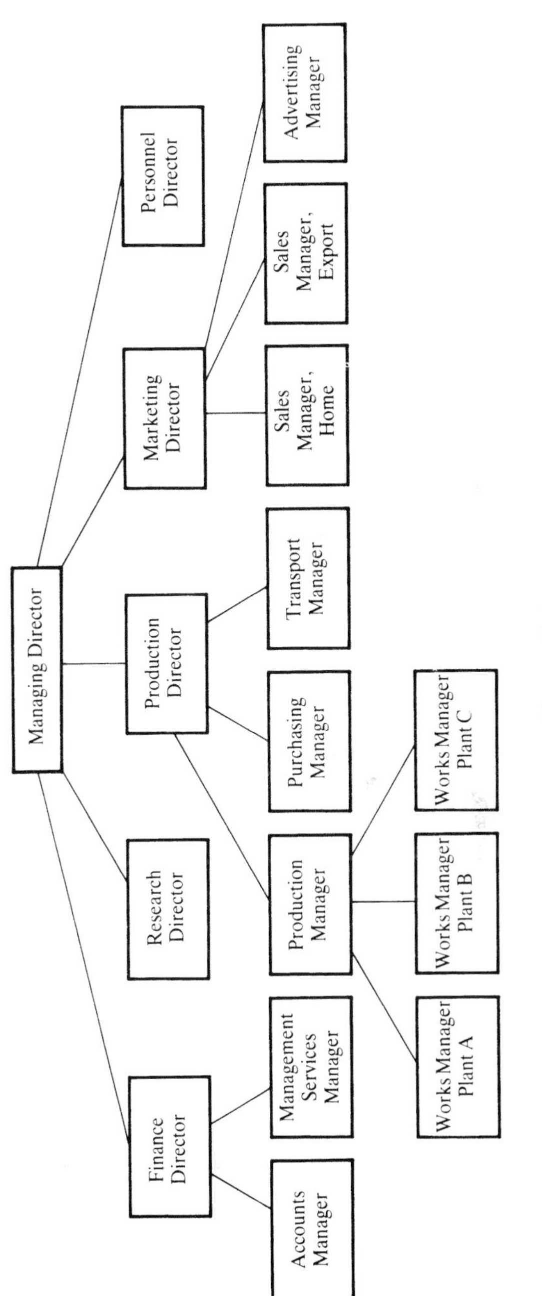
Managing Director
Finance Director
Research Director
Production Director
Marketing Director
Personnel Director
Accounts Manager
Management Services Manager
Production Manager
Purchasing Manager
Transport Manager
Sales Manager, Home
Sales Manager, Export
Advertising Manager
Works Manager Plant A
Works Manager Plant B
Works Manager Plant C

Figure 12.4

Definitions 12.4

(i) An **m-ary tree** is a rooted tree in which every parent vertex has at most m children and some parent vertex has exactly m children. (The terms **binary** and **ternary tree** are used when $m = 2$ and $m = 3$ respectively.)

If every parent vertex of an m-ary tree has exactly m children, we say the tree is **full**.

(ii) A rooted tree of height h is **complete** if all the leaf vertices are at level h.

There are several points to note about these definitions. The first is that every (finite) rooted tree is an m-ary tree for some m—we can simply take m to be the largest number of children of all the vertices of the tree. Some authors define an m-ary tree to be what we called a *full* m-ary tree; that is, one in which every parent has exactly m children. However, for the applications which we consider below, our terminology is more convenient. Note also that a complete rooted tree is not a complete graph (see definition 11.3). Indeed any tree with more than two vertices is not a complete graph. It is a little unfortunate that the word 'complete' is given two different meanings in graph theory, but the context should make it clear which is intended.

Examples 12.2

1. In figure 12.5, diagram (*a*) is a full binary tree and (*b*) is a full ternary tree. Neither tree is complete—(*a*) and (*b*) have leaf vertices at all levels greater than one and zero respectively.

2. Figure 12.6 shows two complete rooted trees—in both cases all the leaf vertices occur at the bottom of the diagram at the highest level of the tree. Clearly this will always be the case in the usual diagram of a complete rooted tree. The tree (*a*) is not full; the tree (*b*) is a complete full binary tree.

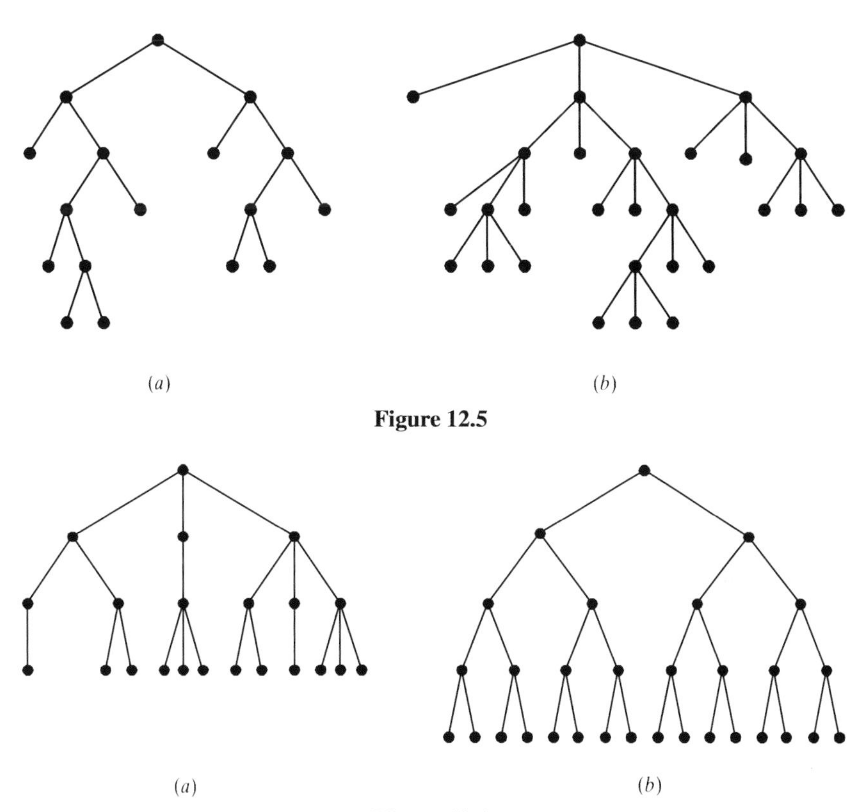

(a) (b)

Figure 12.5

(a) (b)

Figure 12.6

Binary Trees

Binary trees will be used in the sorting procedures outlined in the next section. We therefore need to consider these in a little more detail now. It will be convenient to stretch definition 12.4(i) somewhat and call the tree with no edges a binary tree. Strictly, definition 12.4(i) labels this a 0-ary tree. We could modify the definition to include the tree with no edges as a binary tree, but it would then be somewhat clumsy. We prefer just to adopt the convention that this tree is included in the set of binary trees.

Let T be full binary tree of height greater than zero and with root v^*. Deleting v^* and its two incident edges produces two disjoint binary trees (a binary forest?) whose roots are the two level 1 vertices of T. These are called the **left subtree** and **right subtree** of the root v^*. The roots of these subtrees are called the **left child** and **right child** of v^* and the edges which were deleted are called the **left branch** and **right branch** of v^* respectively. For each of these subtrees, if the subtree has height 1 at least, its left and right subtrees can be defined, and so on throughout the tree. If T is not a *full* binary tree, a left or right subtree may be empty.

There is, of course, a choice to be made in which subtree to call the left subtree and which to call the right subtree at each stage. (In the applications which we consider, the choices will be made in a systematic manner.) Having made these choices we shall always draw the diagram of the tree in the obvious way with the left subtree of each vertex to the left of its corresponding right subtree. The set of choices as to which subtree is 'left' and which is 'right' at each stage actually introduces additional structure to the tree because the choices give the tree a 'left half' and a 'right half' rather than just two halves.

These ideas can be put to use giving the following recursive definition of a binary tree which is rather different from the previous one, but is essentially equivalent to it.

Definitions 12.4′

A **binary tree** comprises a triple of sets (L, S, R) where L and R are binary trees (or are empty) and S is a singleton set. The single element of S is the **root**, and L and R are called, respectively, the **left** and **right subtrees** of the root.

This definition is recursive because it defines a binary tree in terms of the 'components' L, S and R, two of which are themselves binary trees. Thus L and R, if non-empty, are both defined as triples of the form (L', S', R') and so on. This way of defining binary trees is extremely useful for their computer representation. The following example illustrates how to unravel the recursive definition of binary tree to obtain the usual diagram.

Example 12.3

Consider the recursively defined binary tree $(L, \{v^*\}, R)$ where

$$
\begin{array}{lll}
L = (L_1, \{v_1\}, R_1) & \text{and} & R = (L_2, \{v_2\}, R_2) \\
L_1 = (L_3, \{v_3\}, R_3) & \text{and} & R_1 = (\varnothing, \{v_4\}, R_4) \\
L_2 = (L_5, \{v_5\}, R_5) & \text{and} & R_2 = (\varnothing, \{v_6\}, \varnothing) \\
L_3 = (\varnothing, \{v_7\}, \varnothing) & \text{and} & R_3 = (L_8, \{v_8\}, R_8) \\
 & & R_4 = (\varnothing, \{v_9\}, \varnothing)
\end{array}
$$

$$L_5 = (\varnothing, \{v_{10}\}, \varnothing) \quad \text{and} \quad R_5 = (\varnothing, \{v_{11}\}, \varnothing)$$
$$L_8 = (\varnothing, \{v_{12}\}, \varnothing) \quad \text{and} \quad R_8 = (\varnothing, \{v_{13}\}, \varnothing).$$

The triples of sets defining the tree have been grouped into the levels of their roots. Figure 12.7 shows the usual diagram of this binary tree. To draw the diagram, simply work through the triples above systematically. Begin by drawing the root v^*. Its left and right subtrees have roots v_1 and v_2 respectively, so these are the next vertices to add to the diagram. Each of these vertices is then considered in turn.

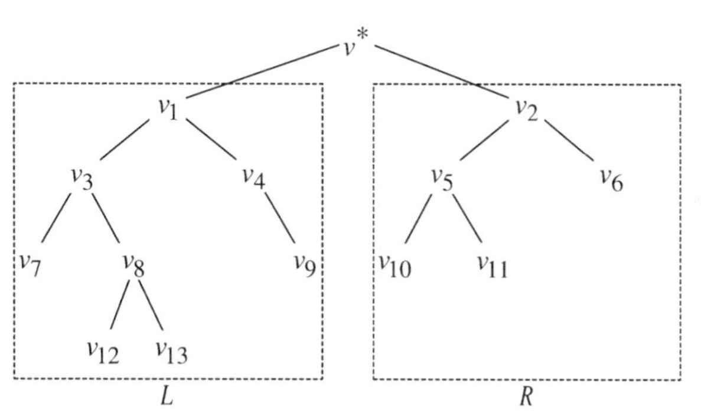

Figure 12.7

The left subtree of v_1 is L_1, with root v_3, and the right subtree is R_1, with root v_4, so these two vertices can then be added to the picture. Continuing in this way produces the figure shown. Note that the binary tree is not full because the vertex v_4 has only one child. Neither is it complete since there are leaf vertices at levels 2 and 3.

Exercises 12.1

1. For each of the following rooted trees, redraw the diagram with the root at the top and the tree growing downwards. Determine:

   (a) the height of the tree;
   (b) whether the tree is complete;
   (c) whether the tree is a full m-ary tree for some m.

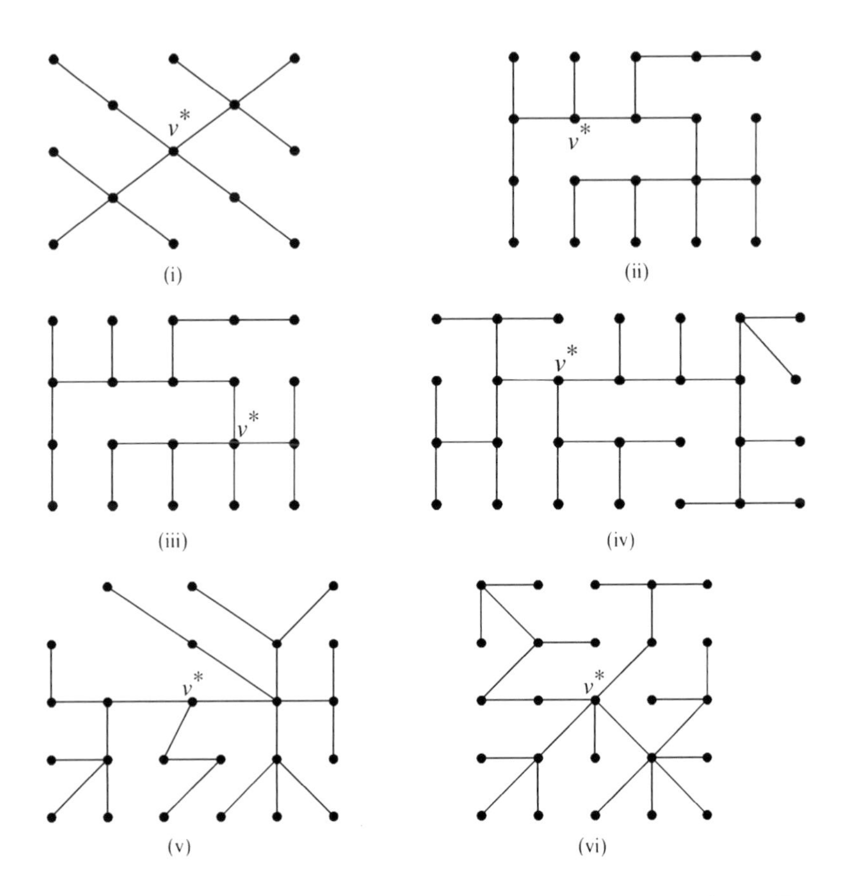

2. Label the vertices of each of the rooted trees in question 1 above. For each tree:

   (i) list the grandchildren of the root;
   (ii) arrange the great-grandchildren of the root into sets of siblings.

3. Let T be a rooted tree with root v^*. Define a relation R on V_T, the set of vertices of T, by:

   $$p \,\mathsf{R}\, q \text{ if and only if the lengths of the paths from } v^* \text{ to } p \text{ and from } v^* \text{ to } q \text{ are equal.}$$

   Show that R is an equivalence relation on V_T and that the equivalence classes are the sets of vertices with equal levels.

4. Let (T, v^*) be a rooted tree. Give formal definitions of the terms **descendant** and **ancestor** applied to vertices. (See definition 12.3.)

Define a relation R on the set V_T of vertices T by:

$$v \,\mathsf{R}\, w \text{ if and only if either } v = w \text{ or } v \text{ is an ancestor of } w.$$

Show that R is a partial order relation on V_T.

5. Let (T, v^*) and (S, w^*) be rooted trees. **A rooted isomorphism** from (T, v^*) to (S, w^*) is an isomorphism of the trees $(\theta, \phi) : T \to S$ such that $\theta(v^*) = w^*$. In other words, a rooted isomorphism of rooted trees is an isomorphism of trees which maps the root of one tree to the root of the other. (See §11.3 for the definition of isomorphism for graphs.)

   (i) There are four non-isomorphic rooted trees with four vertices. Draw a diagram of each of these.

   (ii) There are nine non-isomorphic rooted trees with five vertices. Draw a diagram of each of these.

   (Compare this with the number of unrooted trees with four and five vertices—exercise 11.4.1—and the number of full binary trees with five vertices—exercise 11.4.6.)

6. (i) Give a recursive definition of the binary tree which is represented by the diagram in figure 12.5(*a*).

   (ii) Draw a diagram of the recursively defined binary tree

$$T = (L, \{v^*\}, R)$$

   where:

$$\begin{aligned}
L &= (L_1, \{v_1\}, R_1) & R &= (\varnothing, \{v_2\}, R_2)\\
L_1 &= (L_3, \{v_3\}, \varnothing) & R_1 &= (\varnothing, \{v_4\}, R_4)\\
& & R_2 &= (L_5, \{v_5\}, \varnothing)\\
L_3 &= (\varnothing, \{v_6\}, \varnothing) & &\\
& & R_4 &= (L_7, \{v_7\}, \varnothing)\\
L_5 &= (\varnothing, \{v_8\}, R_8) & &\\
L_7 &= (L_9, \{v_9\}, R_9) & &\\
& & R_8 &= (\varnothing, \{v_{10}\}, \varnothing)\\
L_9 &= (\varnothing, \{v_{11}\}, \varnothing) & R_9 &= (\varnothing, \{v_{12}\}, \varnothing).
\end{aligned}$$

7. Let $T = (L, \{v^*\}, R)$ be a recursively defined binary tree. Give a (recursive) definition of the level of a vertex of T.

8. Let R be a partial order relation on a finite set A. Under what conditions is the Hasse diagram of the poset the standard diagram of a rooted tree (with the root at the top of the tree)?

9. Let T be a full m-ary tree of height h and with vertex set V. Let B denote the set of leaf vertices and let V_k denote the set of vertices at level k.

   Prove each of the following inequalities:

   (i) $m \leqslant |V_k| \leqslant m^k$
   (ii) $hm + 1 \leqslant |V| \leqslant (1 + m + m^2 + \cdots + m^h)$
   (iii) $h(m-1) + 1 \leqslant |B| \leqslant m^h$.

   Find full m-ary tress for which the extremes of the inequalities hold. (These examples show that the inequalities cannot be 'improved'; that is, made more restrictive.)

(Questions 10–14) Rooted Tree Representation of Algebraic Expressions

Let S be a set on which two binary operations, $\oplus$ and $*$, are defined. Without an order of precedence convention an expression such as $x \oplus y * z$ is ambiguous—it could refer either to $(x \oplus y) * z$ or to $x \oplus (y * z)$.

Such expressions can be unambiguously represented using binary trees as follows. An expression can be defined recursively as $X\alpha Y$, where α is (the symbol for) a binary operation and X and Y are either elements of S or expressions. Such an expression can be represented as a binary tree with root α, left child X and right child Y. (Compare this with the definition of a Boolean expression: see definition 9.3.)

For example, the expressions above can be represented by the following binary trees.

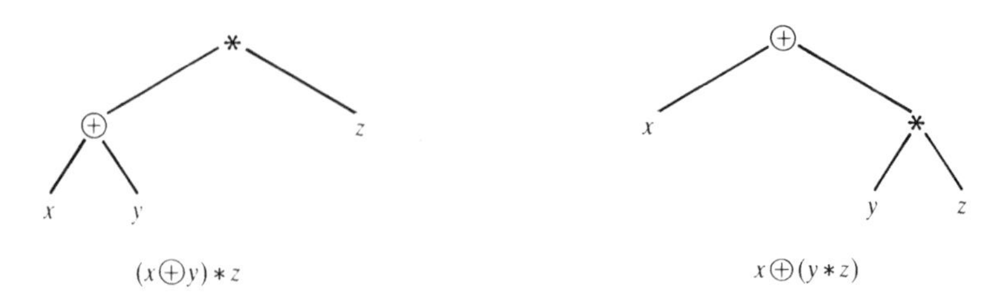

10. (i) Represent each of the following expressions as binary trees:

    (a) $(x \oplus y) * (z \oplus t)$
    (b) $((x \oplus y) * z) \oplus t$
    (c) $(r * s) \oplus ((x \oplus y) * z)$
    (d) $r * (s \oplus ((x \oplus y) * z))$
    (e) $(((r * s) \oplus x) \oplus y) * z.$

    (ii) How many different possible interpretations are there of each of the following expressions?

    (a) $x * y * z$
    (b) $t \oplus x * y * z$
    (c) $t * x \oplus y * z.$

    If $*$ is associative, how many (essentially) different interpretations are there?

11. In the **prefix** (or **Polish**) form of an expression, the symbol for the binary operation is written before the two elements or expressions which it combines. Thus αXY is the prefix form of the expression $X\alpha Y$.

    Parentheses are not necessary when using prefix notation as there is no inherent ambiguity. For example, the expressions $(x\oplus y)*z$ and $x\oplus(y*z)$ have the respective prefix forms $* \oplus xyz$ and $\oplus x * yz$.

    (i) Write each of the expressions in questions 10(i) above in prefix notation.

    (ii) Describe how the binary tree of an expression can be obtained from its prefix form.

12. In the **postfix** (or **reverse Polish**) form of an expression, the symbol for the binary operation is written after the two elements or expressions which it combines. Thus $XY\alpha$ is the postfix form of the expression $X\alpha Y$.

    As for prefix expressions, parentheses are not necessary when using postfix notation. The expressions $(x \oplus y) * z$ and $x \oplus (y * z)$ have the respective postfix forms $xy \oplus z*$ and $xyz * \oplus$.

    (i) Write each of the expressions in question 10(i) above in postfix notation.

    (ii) Describe how the binary tree of an expression can be obtained from its postfix form.

13. Write the usual (or **infix**), the prefix and the postfix forms of the expressions represented by the following binary trees.

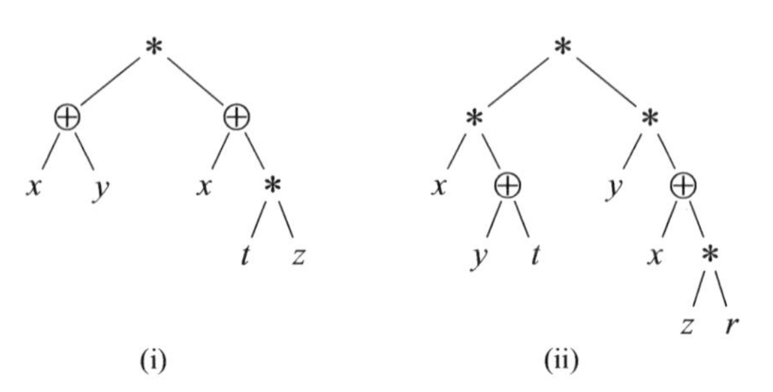

14. Recall from chapter 10 that a Boolean expression may involve two binary operations, $\oplus$ and $*$, as well as the (unary) complement operation $\bar{\ }$.

   (i) In what way may the rooted tree representation of a Boolean expression differ from those considered above?

   (ii) Draw the rooted trees of each of the following Boolean expressions:

   (a) $x_1 \oplus (\overline{x_2 * x_3})$
   (b) $\overline{(\overline{x_1 * x_2}) \oplus (\bar{x}_3 * x_4)}$
   (c) $x_1 \oplus \{x_2 * [x_3 \oplus (x_4 * \bar{x}_5)]\}$.

12.3 Sorting

A common and important operation in data processing is the sorting of data into an appropriate order. The nature of the data and the use to which they are to be put will determine the desired ordering. Commonly this will be numerical or alphabetical order, although the exact nature of the required ordering need not concern us here. However, we shall assume that the order relation is a total order. Recall from §4.5 that a total order R is such that any pair of elements can be 'compared': either $a \mathsf{R} b$ or $b \mathsf{R} a$ (or both). The methods we outline below can be modified to apply to an ordering which is just a partial order, but the results are less meaningful. (See exercise 12.2.10.)

To be more precise, let A be a finite totally ordered set with order relation denoted by $\leqslant$. We write $a \leqslant b$ and 'a is less than or equal to b' rather than $a \mathsf{R} b$ and 'a

is related to b', which we used in chapter 4. Suppose that the elements of A are given as an *unsorted* list $a_1, a_2, \ldots, a_N$ and we wish to obtain a *sorted* list $s_1, s_2, \ldots, s_N$ of the same elements, i.e. a list such that $s_1 \leqslant s_2 \leqslant \cdots \leqslant s_N$. (If A were a partially—but not totally—ordered set, it would not be possible to produce such a list.)

There are many techniques for achieving the desired list. You may be familiar with some of these, for instance 'selection sort', 'bubble sort', 'insertion sort', 'quick sort', etc. The aim of this section is to outline two methods of sorting—'tree sort' and 'heap sort'—both of which use rooted trees as an essential part of the process.

Tree Sort

The tree sort procedure uses a special kind of binary tree, appropriately called a 'sort tree'. This is a binary tree whose vertex set is totally ordered and has the property that, for every vertex v of the tree, the vertices in its left subtree are less than or equal to v, and v is less than or equal to the vertices in its right subtree. The term 'binary search tree' is used by some authors, but we use 'search' in a different context in the next section. The formal definition is the following.

Definition 12.5

A **sort tree** is a binary tree T such that:

(i) the vertex set V is totally ordered (with order relation denoted $\leqslant$), and

(ii) for each $v \in V$, $w_L \leqslant v$ for every vertex w_L in the left subtree of v, and $v \leqslant w_R$ for every vertex w_R in the right subtree of v.

The tree sort procedure occurs in two phases. Suppose we are given the elements of A in an unsorted list $a_1, a_2, \ldots, a_N$. In the first phase, the list is used to construct (or grow!) a sort tree whose vertices are the elements of A and whose root is a_1, the first element of the unsorted list. The second phase obtains the sorted list from the sort tree.

Before describing how the tree sort technique works in general, we illustrate it through an example.

Example 12.4

Suppose the initial list is 6, 2, 9, 4, 15, 1, 12, 7, 20, 10, 3, 11 which is to be sorted (eventually) into increasing order. A sequence of sort trees is produced by adding each element of the list in turn as a leaf vertex. The vertex set of the last sort tree in the sequence contains all the elements of the list.

The first element 6 is defined to be the root. Since $2 \leqslant 6$ we create a left branch from the root with 2 at its end. Next, as $9 \geqslant 6$ we create a right branch from the root and place 9 at its end.

To process the next element 4, first compare it with the root 6. Since $4 \leqslant 6$, we must place 4 in the left subtree of 6. Thus we consider the left subtree whose root is 2. Now $4 \geqslant 2$, so we create a right branch from 2 with 4 as its child.

Inserting the next element 15 as a leaf vertex involves the following comparisons:

$15 \geqslant 6$ so go to the right subtree of 6 whose root is 9;
$15 \geqslant 9$ so create a right branch from 9 with 15 as a level 2 vertex.

At this stage, we have grown the following sort tree.

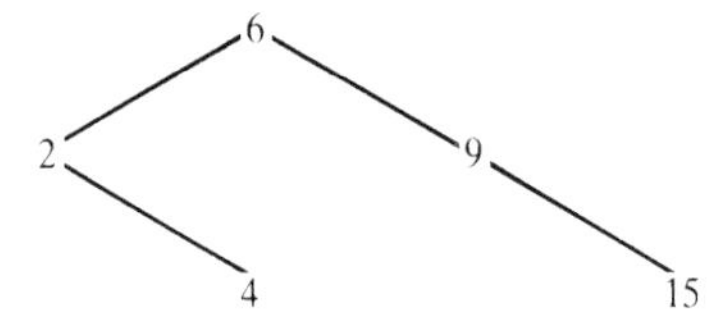

Now continue in the same way adding each element in turn. To insert an element x we first compare it with the root 6. If $x \leqslant 6$ go to the left subtree; otherwise go to the right subtree. Repeat the comparison of x and the root of the new tree. At some stage there is no left or right subtree to 'go to'; then add x as a left or right child as appropriate.

Applying this process to the remaining elements of the unsorted list produces the sequence of sort trees shown in figure 12.8, the last of which is the sort tree of the whole (unsorted) list.

This completes the first phase of the tree sort. Next we use the sort tree to obtain the sorted list. Since every vertex in the left subtree of the root is less than or equal to the root, these need to be listed before the root itself. Similarly all the vertices in the right subtree of the root need to be listed after the root itself. It is important to realize that both the left and right subtrees of the root are themselves

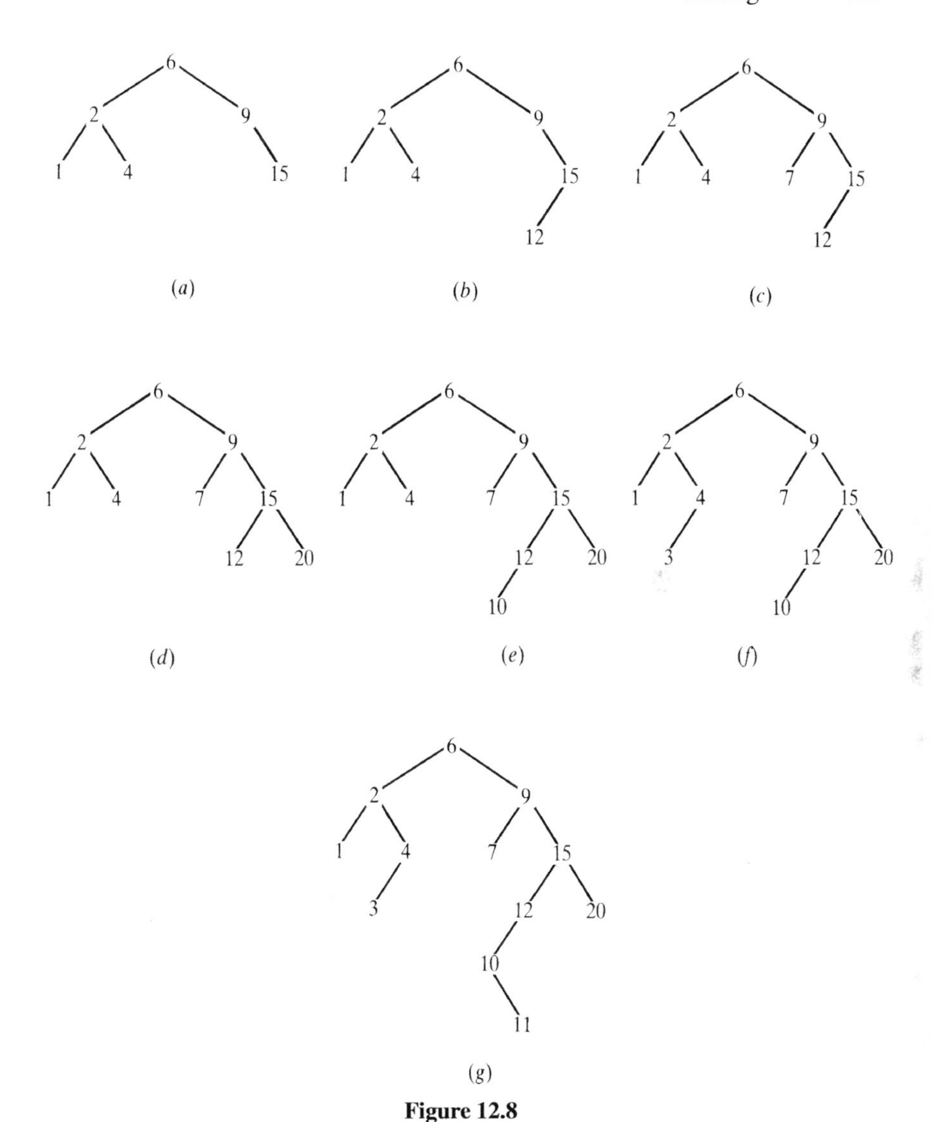

Figure 12.8

sort trees and so *their* left and right subtrees are also sort trees, and so on down through the tree.

The required listing procedure can be described recursively as follows:

(i) list the elements in the left subtree of the root,
(ii) list the root itself, and
(ii) list the elements in the right subtree of the root.

The reason that this is a recursive description of the listing process is that to perform each of the steps (i) and (iii) we need to repeat the *whole process* for the appropriate subtree. (When we do this 'the root' will refer to the root of whichever subtree is being processed, and not the root of the main tree.)

To illustrate how this works, we consider step (i) in a little more detail. For the sort tree in figure 12.8(*g*), the left subtree of the root is the following.

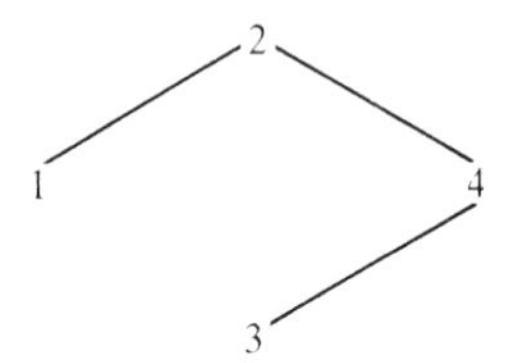

To process this sort tree we list the elements in its left subtree (1), list the root (2) and then list the elements in the right subtree. To list the elements of the right subtree, we apply the same three steps again: list the elements in its left subtree (3), list the root (4) and then list the elements in its right subtree (none). This gives the list 1, 2, 3, 4 and completes step (i) for the main tree.

Step (ii) then lists (6), the root of the main sort tree. Finally we need to list the elements of its right subtree in a similar way. This requires performing our steps several more times, and is left as an exercise. In this way we obtain the sorted list 1, 2, 3, 4, 6, 7, 9, 10, 11, 12, 15, 20.

We shall describe this listing procedure more fully for a general sort tree later.

We now turn to general descriptions of the two phases of tree sort: growing the sort tree from the unsorted list and listing the elements from the sort tree.

Step 1: Growing the Sort Tree

Suppose that $a_1, a_2, \ldots, a_N$ is an unsorted list (with order relation $\leqslant$). Working through the list, the sort tree is grown one branch at a time. Each time an element is processed it is added to the tree as a leaf vertex, although it may subsequently become a parent. To begin, the first element a_1 is defined to be the root vertex v^*. We now have a sort tree T_1 whose root has empty left and right subtrees.

To add a branch with element x at its end proceed as follows. Compare x with the root; if $x \leqslant v^*$ proceed to the root of the left subtree and if $v^* \leqslant x$ proceed to the root of the right subtree. Then we compare x with the root of the appropriate subtree and repeat. Eventually one of the following two possibilities occurs when comparing x with the current root v: either $x \leqslant v$ and v has empty left subtree or

$v \leqslant x$ and v has empty right subtree. In the first case we add x as a left child of v and in the second we add x as a right child of v.

Growing the sort tree from the unsorted list $a_1, a_2, \ldots, a_N$ can be summarized as follows.

Algorithm 12.1

1. Set $n = 1$, a_1 equal to the root and T_1 equal to the sort tree with a_1 as its only vertex. Increase n to $n + 1$.

2. If $n > N$ then end. Otherwise compare a_n with the root.

   If $a_n \leqslant$ root then proceed to the left subtree.
   If the left subtree is empty then add a_n as a left child of the root to form the next sort tree T_n, increase n to $n + 1$ and repeat step 2;
   otherwise repeat step 2 using the left subtree.

   Otherwise proceed to the right subtree.
   If the right subtree is empty then add a_n as a right child of the root to form the next sort tree T_n, increase n to $n + 1$ and repeat step 2;
   otherwise repeat step 2 using the right subtree.

Note the recursive nature of algorithm 12.1. To process step 2 for a particular subtree we may need to process step 2 itself for various smaller subtrees.

Example 12.5

Grow a sort tree from the list *Hawk, Raven, Wren, Falcon, Dove, Eagle, Pelican, Robin, Osprey, Egret, Rook* under alphabetical ordering.

Solution

Firstly, place *Hawk* as the root. Since *Hawk* $\leqslant$ *Raven* a right branch is added with *Raven* as its leaf. Next, since *Hawk* $\leqslant$ *Wren* and *Raven* $\leqslant$ *Wren* we add a right branch from *Raven* with *Wren* as its leaf. *Falcon* $\leqslant$ *Hawk* so it is added as a left child of the root. Similarly, *Dove* $\leqslant$ *Hawk* and *Dove* $\leqslant$ *Falcon* so we add a left branch from *Falcon* with *Dove* as its root. At this stage we have grown the following sort tree.

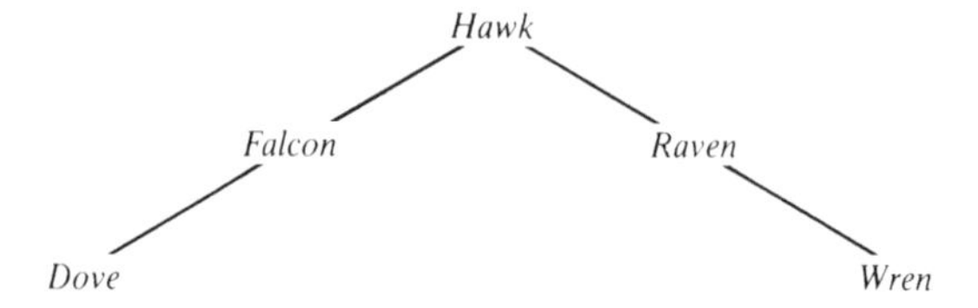

The next element to be added is *Eagle*. Since *Eagle* $\leqslant$ *Hawk* we proceed to the left subtree with root *Falcon*. Now *Eagle* $\leqslant$ *Falcon* as well so we again proceed to the left subtree with root *Dove*. This time *Dove* $\leqslant$ *Eagle* and *Dove* has an empty right subtree so we add *Eagle* as a right branch from *Dove*, and then process *Pelican*, the next element of the unsorted list.

Continuing in this way we eventually obtain the sort tree shown in figure 12.9.

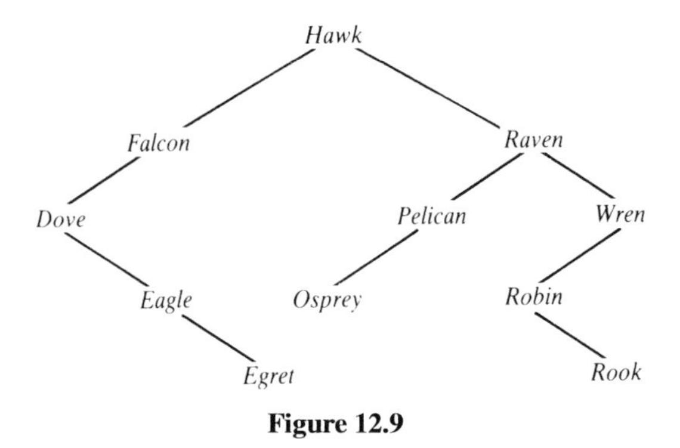

Figure 12.9

Step 2: Listing The Elements of a Sort Tree

Let T be the sort tree grown from the unsorted list $a_1, \ldots, a_N$ according to step 1. To list the vertices of the sort tree in the appropriate order we apply the following three steps.

1. Process left subtree.
2. List the root.
3. Process right subtree.

To process a subtree we repeat each of the three steps with the proviso that an empty tree needs no processing!

Example 12.6

List the elements of the sort tree in figure 12.9.

Solution

Performing the three steps 1, 2 and 3 recursively gives the following. Each level of the process has been indented so that the three steps begin in the same vertical position. Of course, each indentation represents a level of the sort tree, so the maximum number of indentations (four in this case) equals the height of the tree. We have used 'left subtree(X)' to denote the left subtree which has X as its root (and similarly for right subtrees). When no processing is required [Empty] is written after the particular process.

Following through the process, we can see that the elements are listed in the correct alphabetical order: *Dove, Eagle, Egret, Falcon, Hawk, Osprey, Pelican, Raven, Robin, Rook, Wren.*

Step1: Process left subtree(*Hawk*)
Step 1: Process left subtree(*Falcon*)
Step 1: Process left subtree(*Dove*) [Empty]
Step 2: List ***Dove***
Step 3: Process right subtree(*Dove*)
Step 1: Process left subtree(*Eagle*) [Empty]
Step 2: List ***Eagle***
Step 3: Process right subtree(*Eagle*)
Step 1: Process left subtree(*Egret*) [Empty]
Step 2: List ***Egret***
Step 3: Process right subtree(*Egret*) [Empty]
Step 2: List ***Falcon***
Step 3: Process right subtree(*Falcon*) [Empty]

Step 2: List ***Hawk***

Step 3: Process right subtree (*Hawk*)
Step 1: Process left subtree (*Raven*)
Step 1: Process left subtree (*Pelican*)
Step 1: Process left subtree (*Osprey*) [Empty]
Step 2: List ***Osprey***
Step 3: Process right subtree(*Osprey*) [Empty]
Step 2: List ***Pelican***
Step 3: Process right subtree(*Pelican*) [Empty]

Step 2: List ***Raven***
Step 3: Process right subtree(*Raven*)
  Step 1: Process left subtree(*Wren*)
    Step 1: Process left subtree(*Robin*) [Empty]
    Step 2: List ***Robin***
    Step 3: Process right subtree(*Robin*)
      Step 1: Process left subtree(*Rook*) [Empty]
      Step 2: List ***Rook***
      Step 3: Process right subtree(*Rook*) [Empty]
  Step 2: List ***Wren***
  Step 3: Process right subtree(*Wren*) [Empty]

Heap Sort

The heap sort algorithm also uses a special kind of binary tree, called a 'heap', whose vertices are again members of the list to be sorted. The 'shape' of a heap is such that it has the smallest possible height for the number of its vertices. This is achieved by ensuring that, if we ignore its highest level, the remainder of the tree is both complete and full. In addition, the leaf vertices at the highest level are situated as far to the left of the diagram as possible. These two conditions determine the shape of the diagram of a heap (see figure 12.10 below). The last condition in the following definition refers to the relationship between the vertices.

Definition 12.6

A **(descending) heap** is a binary tree of height h with the following properties.

(i) All leaf vertices are at levels $h-1$ or h.
(ii) The leaf vertices at level h are situated as far to the left of the diagram as possible. (This implies that any leaf vertices at level $h-1$ are situated as far to the right as possible. In other words, the subtree formed by deleting the level h vertices—and their incident edges—is a complete full binary tree.)
(iii) The vertex set has a total order, denoted by $\leqslant$.
(iv) If q is a parent of p then $p \leqslant q$, i.e. each child is less than or equal to its parent with respect to the given total order.

In a descending heap the vertices along any path beginning with the root occur in descending order with respect to the given total order. An **ascending heap** can be defined similarly, the only difference being that each parent is less than or equal to each of its children. Of course, in an ascending heap the vertices along any path beginning with the root occur in ascending order.

Example 12.7

Figure 12.10 shows two different (descending) heaps with vertex set $\{1, 2, 3, 4, 5, 6, 7, 8, 9, 10\}$ having the usual ordering $\leqslant$. The heaps have the same 'shape' and differ only in the positions of the elements of the vertex set. In general, conditions (i) and (ii) of the definition ensure that all heaps with a given number of vertices have the same shape.

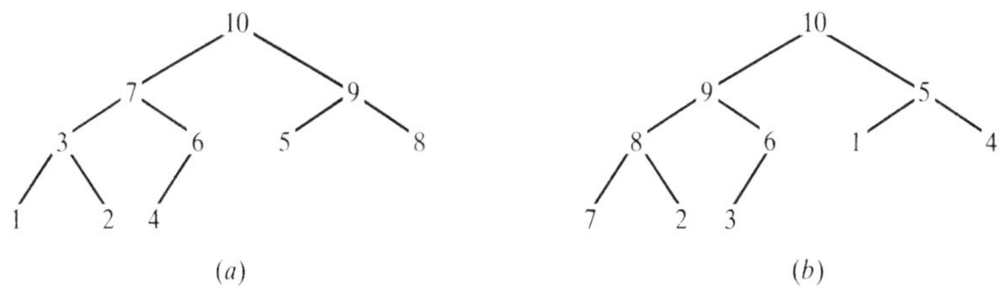

Figure 12.10

The heap sort procedure is similar in overall structure to tree sort. Firstly, a heap is created from the unsorted list and then the sorted list is obtained from the given heap. We outline each phase in the procedure separately.

Step 1: Creating the Heap

As for tree sort, we suppose an unsorted list $a_1, a_2, \ldots, a_N$ is given. Set a_1 as the root, which creates a (rather trivial) heap. Suppose $a_1, a_2, \ldots, a_{k-1}$ have already been formed into a heap and we need to add the next element a_k. There is a unique position to add a_k to satisfy conditions (i) and (ii) of the definition of a heap; this is either immediately to the right of the rightmost vertex at the highest level or, if the highest level is complete, the leftmost vertex at a new higher level.

However, adding a_k in this position does not necessarily produce a heap. For example, suppose $a_k = 17$ is added to the heap in figure 12.11(a) giving the binary tree (b). This new tree is no longer a heap since 17 is greater than its parent 12. In such a case the new vertex is interchanged with its parent, producing the tree (*c*). This has only moved the problem further up the tree as 17 is still greater than its new parent 15. We therefore need to swap 15 and 17 which finally restores the heap—figure 12.11(*d*). In this process of restoring the heap, we say that the newly added vertex is **bubbled up** through the tree.

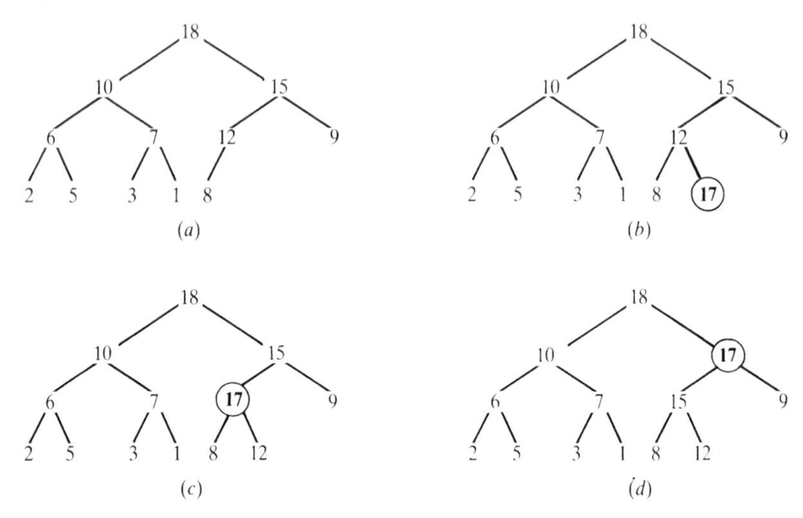

Figure 12.11

The creation of the heap from the (unsorted) list $a_1, a_2, \ldots, a_N$ can be summarized by the following algorithm.

Algorithm 12.2

1. Set a_1 as the root.
2. Add the next element in the (unique) position so that the tree satisfies conditions (i) and (ii) of the definition of a heap.
3. If necessary, restore the heap by bubbling the new vertex up through the tree.
4. Repeat steps 2 and 3 for each element of the list in turn.

Example 12.8

Create a heap from the list *Hawk, Raven, Wren, Falcon, Dove, Eagle, Pelican, Robin, Osprey, Egret, Rook* used in example 12.5.

Solution

(We strongly recommend that you draw diagrams to follow the construction of the heap.)

Firstly, place *Hawk* as the root and then add *Raven* as a left branch. Since *Hawk* $\leqslant$ *Raven* we need to swap the two vertices. Next *Wren* is added as a right child of the root (which is now *Raven*) and then interchanged with *Raven*. Now *Falcon, Dove, Eagle* and *Pelican* can be added in turn at level 2 without any bubbling up through the tree. At this stage we have the following heap.

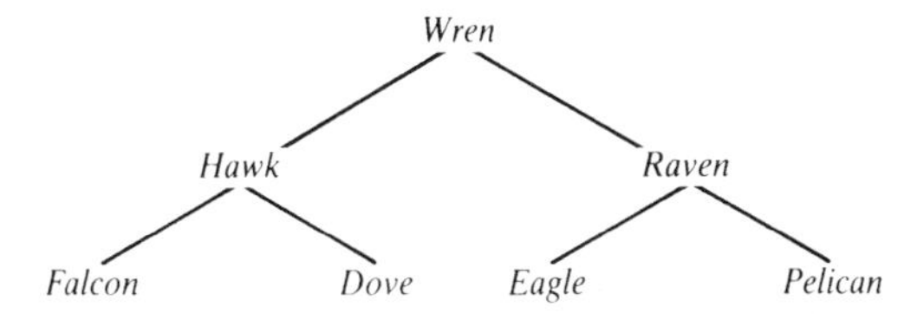

The next element, *Robin*, is added at the leftmost end of the next level. Since *Falcon* $\leqslant$ *Robin* they need to be interchanged; then we have *Hawk* $\leqslant$ *Robin* so these two vertices must also be swapped. The result is that *Robin* has bubbled up to the position previously occupied by *Hawk*, producing the heap in figure 12.12(*a*). Next to be added is *Osprey* which we need to interchange with its parent, *Hawk*, giving figure 12.12(*b*). When *Egret* is added as a left branch from *Dove* and the two vertices are swapped, figure 12.12(*c*) is obtained. Finally, *Rook* is added as a right branch to *Egret* and then is bubbled up to the position previously occupied by *Robin*. This produces the final heap—figure 12.12(*d*).

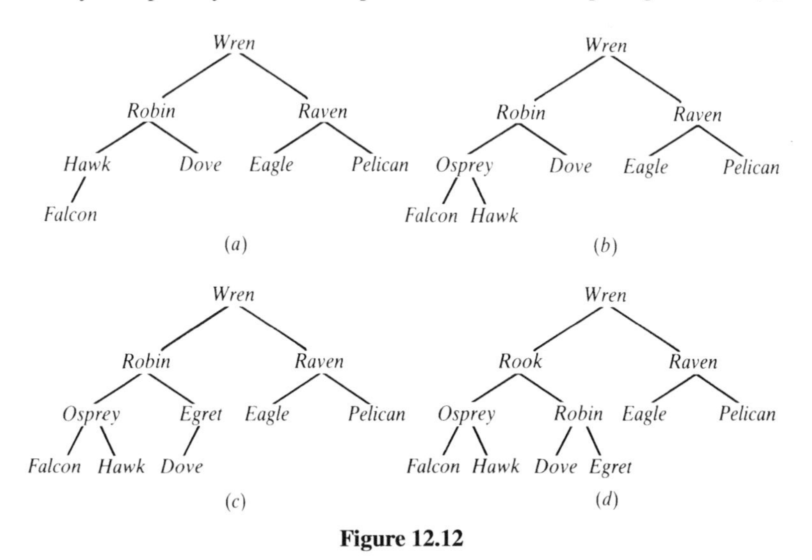

Figure 12.12

Step 2: Obtaining the Sorted List from the Heap

To generate the sorted list from a heap, we first create a binary tree whose vertices, when read from top to bottom and left to right, are in increasing order, i.e. the vertices, read in this way, form the sorted list we are seeking. How this is achieved is illustrated in the next example.

Example 12.9

Suppose the heap in figure 12.13(a) has been created from an unsorted list of numbers with respect to the usual ordering $\leqslant$. We need a process which will create the tree in figure 12.13(b) from this heap. The sorted list 1, 2, 3, 4, 5, 6, 7, 8, 9, 10 can then be read off from the tree (b) by going through the tree from top to bottom and left to right.

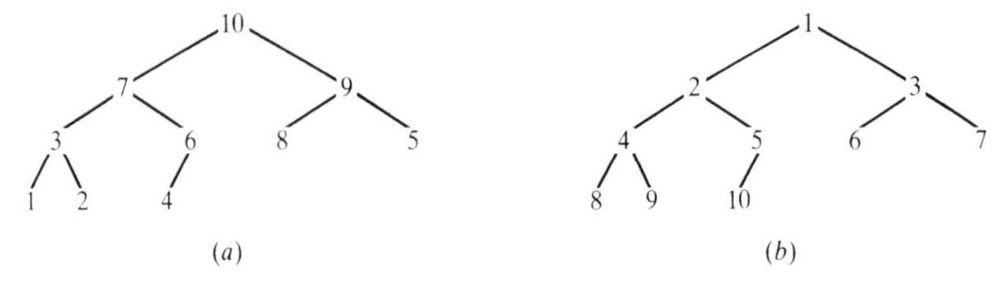

Figure 12.13

Clearly the root of a heap is the maximal element and should therefore be placed at the end of the list. Interchanging it with the last vertex in the heap (i.e. the rightmost vertex at the highest level) places it in the desired position. In figure 12.13(a), we need to interchange 10 and 4. This destroys the heap, of course, because the largest vertex is now a leaf. However, since 10 is now in the desired position of the tree it will be ignored. Even disregarding this vertex, the remainder of the tree is still not a heap because the root 4 is less than each of its children.

We therefore need to restore the heap (remembering to ignore the vertex 10 which is already in its final position). To do this the root 4 needs to be **sunk down** through the tree, by successively interchanging it with the larger of its children. In this instance it is interchanged firstly with its right child 9 and subsequently with its (new) left child 8. This produces the following tree, where the circled vertex is in its final position and the subtree with uncircled vertex set is a heap.

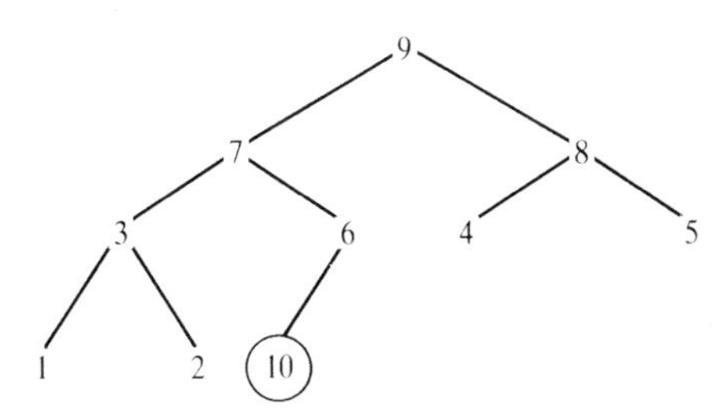

The process is then repeated ignoring the circled vertex 10 which is now in its final position. In general, at each stage we consider only the portion of the tree which is a heap; that is, the portion of the tree whose vertices are not (necessarily) in their final positions. These are the uncircled vertices. Figure 12.14 shows the remaining stages in this process.

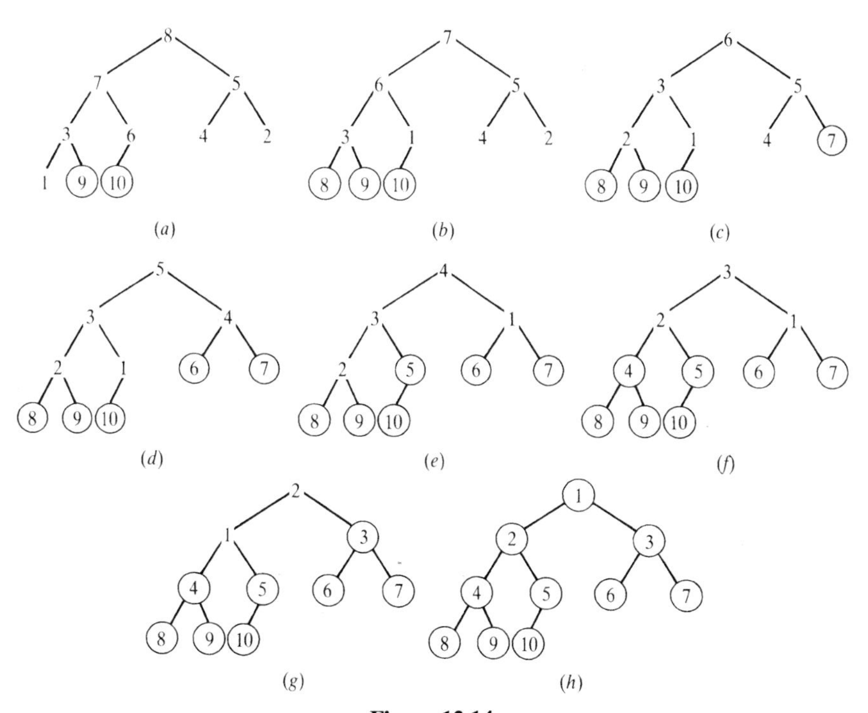

Figure 12.14

The process of converting the heap into the new tree explained in example 12.9 can be summarized as the following algorithm.

Algorithm 12.3

1. Interchange the root with the last vertex, u; that is, the rightmost vertex at the highest level. Fix the position of the (new) last vertex. (By fixing the position of a vertex, we mean that it is to be ignored in all subsequent manipulations of the tree—in the previous example we denoted the fixed vertices by circling them.)

2. Restore the tree (apart from any fixed vertices) to a heap by sinking down the root u as follows.

   (i) Interchange the root u with the largest of its two children.
   (ii) Go to the subtree with u as its root. If this subtree has height 0 (ignoring, of course, any fixed vertices) then go to step 3; otherwise repeat step 2(i) for this subtree.

3. Repeat steps 1 and 2, remembering to ignore any fixed vertices until all vertices are fixed.

We end this section with a few brief comments on the relative performance of various sorting procedures. Unfortunately perhaps, no one sorting method is clearly superior to all the others. Various factors influence the choice of sort process in a given application: size of input data, amount of available memory, whether or not the input data are likely to be 'roughly' sorted, etc. Of the two algorithms we have outlined, tree sort would probably be preferred for small data sets and heap sort for large data sets. From a theoretical point of view heap sort is a 'fast' algorithm. In practice, other algorithms such as quick sort perform better 'on average' than heap sort although on some sets of input data quick sort may be significantly slower than heap sort.

Exercises 12.2

1. Construct a sort tree from each of the following unsorted lists.

(i) 7, 3, 15, 4, 12, 14, 6, 8, 2, 5, 13, 19 ordered by $\leqslant$.

(ii) *when, shall, we, three, meet, again, in, thunder, lightning, or, in, rain* ordered alphabetically. (Note that there is a repeated element in the list. However, the construction of the sort tree described in the text takes care of this possibility.)

(iii) 8, 64, 1, 4, 256, 512, 2, 32, 128, 16 ordered by divisibility.

2. List the elements of each of the following sort trees using the method described in the text. (The ordering in each case is 'less than or equals', $\leqslant$.)

From which of the two sort trees is the sorted list most quickly obtained? Why?

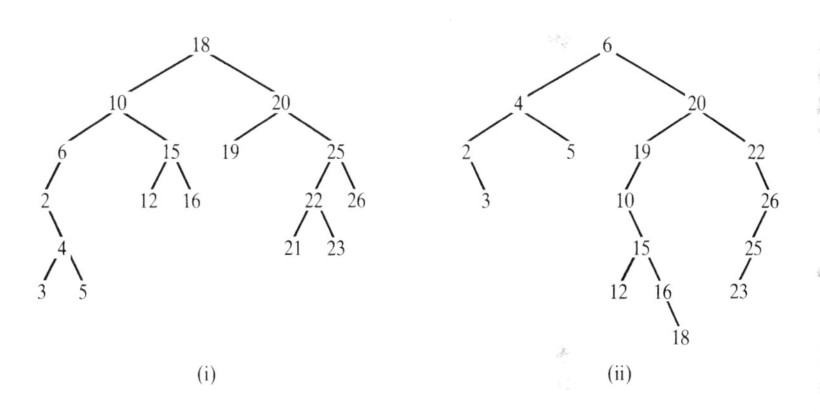

(i) (ii)

3. Work through both steps of tree sort for each of the following unsorted lists.

(i) 17, 31, 5, 23, 2, 7, 19, 29, 11, 3, 13 with order relation $\leqslant$.

(ii) *bca, cbc, acb, bcb, bac, cac, abc, aca, bab, aba, cba, cab* with alphabetical orderings of strings.

4. Construct a heap from each of the unsorted lists in question 1 above.

5. Following the procedure of example 12.9, convert each of the following heaps into a tree whose vertices are in increasing order reading from top

to bottom, left to right.

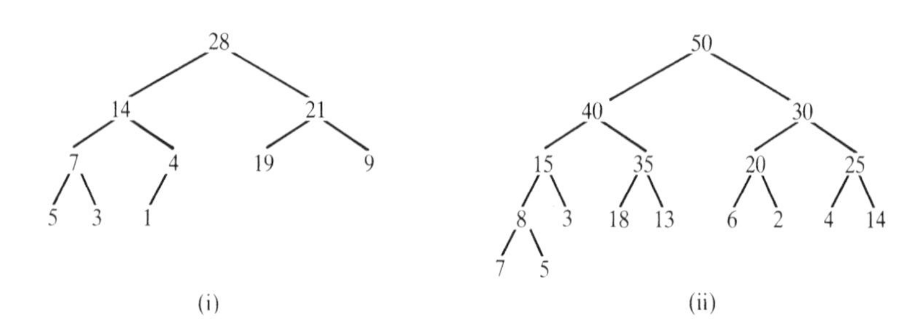

6. Work through both steps of heap sort for each of the unsorted lists in question 3 above.

7. Let A be a set of (English) words. A total order relation $\leqslant$ on A is defined as follows:

   $w_1 \leqslant w_2$ if and only if either $\mathit{length}(w_1) < \mathit{length}(w_2)$ or $\mathit{length}(w_1) = \mathit{length}(w_2)$ and w_1 comes before w_2 in the usual alphabetical ordering of words.

   (Here $\mathit{length}(w)$ is the number of letters in the word w, of course.)

   (i) Work through both steps of tree sort to sort the list *group, morphism, heap, sort, Boolean, algebra, algorithm, tree, discrete* with respect to this order relation.

   (ii) Repeat (i) using heap sort.

8. A heap is defined as a binary tree (with additional properties). Explain why a heap is a fully binary tree if and only if it has an odd number of vertices. When is a heap a complete binary tree?

9. (i) Write a program in any high-level computer language or in pseudocode to grow a sort tree from an unsorted list of numerical data, assuming that the order relation is $\leqslant$.

   (ii) Write a program in any high-level computer language or in pseudocode which inputs the sort tree from your program for part (i) and outputs the sorted list.

   Note: you may wish to use the recursive definition of a binary tree given in §12.2.

10. (Sorting a partially ordered set which is not totally ordered.)

Consider the list 10, 2, 6, 30, 3, 15, 5, 1, 60, 4, 20, 12 (the divisors of 60) ordered by divisibility. The order relation is not a total order since, for example, 5 does not divide 6 and 6 does not divide 5.

Work through both steps of tree sort for this list. (Note that algorithm 12.1 needs to be applied with care: if a_n doesn't divide the root then proceed to the right subtree *even if* the root doesn't divide a_n either.)

Check that your 'sorted' list can be divided into 'blocks', each of which is a chain. (Recall from §4.5 that a chain is a subset which is totally ordered by the given relation.)

12.4 Searching Strategies

A **searching** of the vertices of a connected graph Γ is a systematic procedure for 'visiting' all the vertices of Γ by 'travelling along' its edges. An edge of Γ may be traversed more than once and a vertex may be visited more than once in the search. In other words, a searching of a graph is a method of constructing an edge sequence whose associated vertex sequence includes every vertex of the graph. (Edge sequences and their associated vertex sequences are defined in definition 11.6.)

We can regard a searching of a graph Γ as a systematic procedure for constructing a subgraph which contains all the vertices of Γ. We call such a subgraph a **spanning subgraph** of Γ. (The term 'wide subgraph' is used by some authors.) In fact both of the procedures we consider actually construct a spanning tree in Γ.

The situation is analogous to a game of hide-and-seek. The 'seeker' has to check various potential hiding places (vertices) until all the 'hiders' are found. There are two principal strategies that the seeker may adopt. He or she can move from one potential hiding place to the next, always checking new locations until forced to 'backtrack' in order to continue visiting new hiding places. (One can well imagine an excited child playing the game in this way.) This mode of search is called a 'depth-first search'.

An alternative technique which the seeker may adopt is firstly to check all the potential hiding places in the immediate vicinity of his or her starting position before widening the search area and again checking all the (new) locations in the widened area. The search area is gradually extended until the whole region has been searched. This searching strategy is called a 'breadth-first search'.

We shall consider each of these two methods more precisely beginning with the depth-first approach. We describe this as an algorithm that constructs a sequence of trees, the last of which is the desired spanning tree. In general, Γ will have many spanning trees; the particular tree obtained depends on the choices made in the execution of the algorithm.

Algorithm 12.4 (Depth-first search)

Let Γ be a connected graph and v_0 a vertex of Γ. A **depth-first search** of Γ, with initial vertex v_0, is the following procedure for constructing a spanning tree for Γ.

1. Set $w = v_0$ and $n = 0$; w is called the **current centre** of the search. Let T_0 denote the tree with no edges and vertex v_0; T_0 is the first tree in our sequence.

2. If possible, choose an edge e_n which is incident with w and a new vertex v_{n+1}. (By a new vertex, we mean a vertex which does not appear in T_n, the current tree; in other words, a vertex which has not yet been visited.) Adjoin e_n to the current tree T_n to form the next tree in the sequence, T_{n+1}. Set $w = v_{n+1}$ and increase n to $n + 1$.

3. Repeat step 2 until the current w is not adjacent to any new vertex. If T_n is a spanning tree, the search is complete. If not, backtrack along the last edge to the previous centre. More precisely, set $w = v_{n-1}$ and increase n to $n + 1$. Repeat step 2 (if possible; several backtrackings may be required before a new vertex can be visited).

As with many algorithms, it is only really possible to understand the way that depth-first search constructs the required edge sequence by working through an example.

Example 12.10

We use the depth-first search algorithm to construct a spanning tree for the graph illustrated in figure 12.15. The initial vertex of the search is v_0.

Suppose we begin by choosing the edge e_0 (see figure 12.16). Thus the first tree

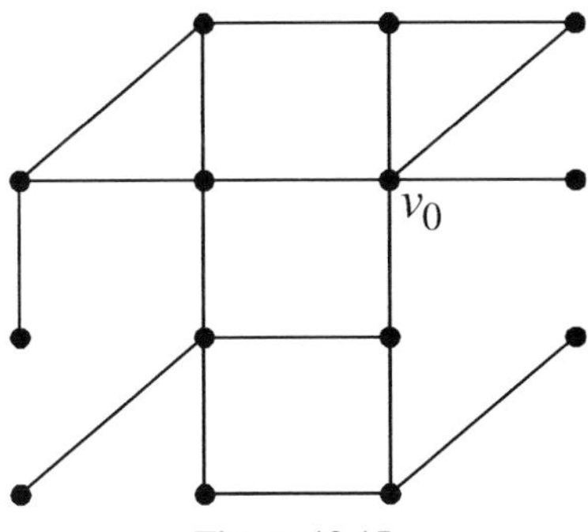

Figure 12.15

is e_0 (together with its vertices) and the vertex v_1 becomes the new centre of the search. This is illustrated by figure 12.16(*a*), where the new centre of the search is circled.

Since v_1 is not adjacent to any new vertex we need to backtrack to v_0. That is, we set $w = v_0$ (again). Now perform step 2 twice choosing to add the edges e_1 and e_2 in turn to the tree. The current centre moves firstly to v_2 and then to v_3. Our status at this stage is reflected in figure 12.16(*b*); again the current centre of the search has been circled.

Step 2 cannot be repeated at this point so backtrack and set $w = v_2$. Applying step 2 four times brings us to the situation represented in figure 12.16(*c*). At this stage the tree has edge set $\{e_0, e_1, e_2, e_3, e_4, e_5, e_6\}$ and the current centre is v_7. To continue, we need to backtrack twice to v_5 before further edges can be added.

The rest of the search is illustrated in figures 12.16(*d*)–(*f*). The diagrams show the search frozen at each stage where backtracking is required, with the current centre circled in each case.

At the end of the search, the last tree contains all the vertices of Γ and so is a spanning tree. The edges have been added to the tree in the order $e_0, e_1, \ldots, e_{12}$, and the vertices have been 'visited' in the order $v_0, v_1, \ldots, v_{13}$. (Of course, had we made different choices we might have obtained a different spanning tree and the vertices would have been visited in a different order.)

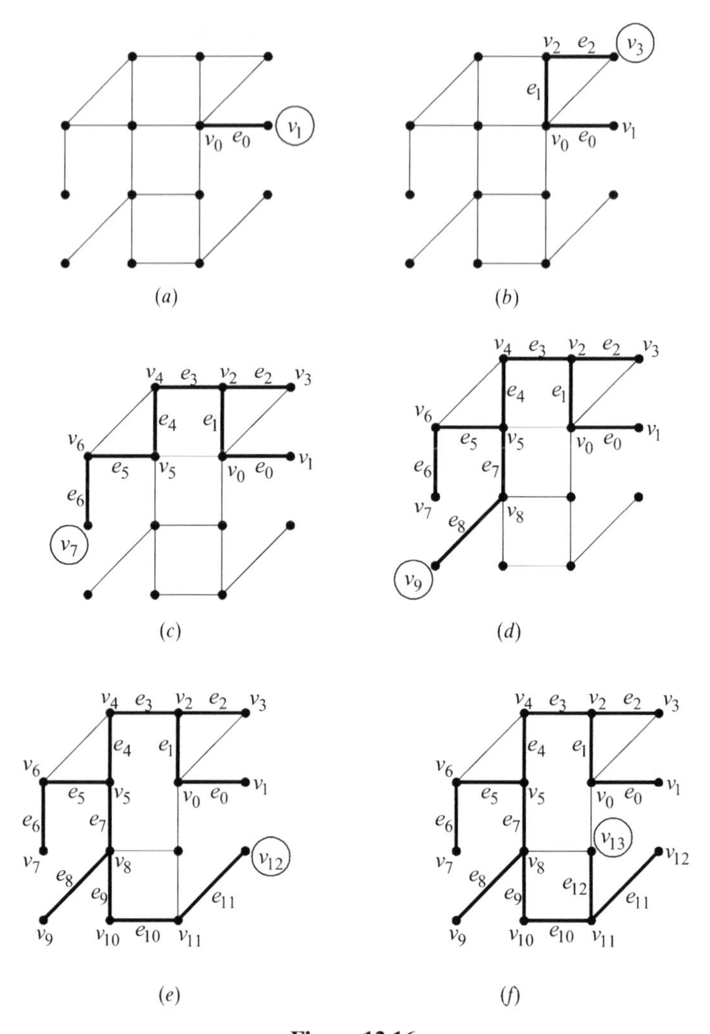

Figure 12.16

We now turn to the breadth-first search, which is again described as an algorithm that constructs a sequence of trees, the last of which is a required spanning tree. The essential difference between breadth-first and depth-first searches concerns the movement of the current centre. In algorithm 12.4 we saw that the current centre moves each time an edge is added to the tree. By contrast, in the following algorithm the current centre moves only when it is 'forced to', i.e. when all possible edges incident with it have already been added to the tree.

Algorithm 12.5 (Breadth-first search)

Let Γ be a connected graph and v_0 a vertex of Γ. A **breadth-first search** of Γ, with initial vertex v_0, is the following procedure for growing a spanning tree.

1. Set $w = v_0$, $n = 0$ and $m = 1$. (As for the depth-first search algorithm, w is called the **current centre** of the search.) Let T_0 denote the tree with no edges and vertex v_0; T_0 is the first tree in the sequence.

2. If possible, choose an edge e_n which is incident with w and a new vertex v_{n+1}. (Recall that a new vertex is one which does not appear in the current tree T_n.) Adjoin e_n to T_n to form the next tree T_{n+1}. Increase n to $n + 1$.

3. Repeat step 2 until there are no further new vertices adjacent to w. If all vertices have now been visited then T_n is a spanning tree and we stop. Otherwise set $w = v_m$, the first of the vertices which has not yet acted as the current centre, and increase m to $m+1$. Repeat step 2.

Example 12.11

To understand how algorithm 12.5 itself works as well as to contrast it with algorithm 12.4, we work through the breadth-first algorithm for the graph in figure 12.15.

Beginning with current centre v_0, we perform step 2 five times, successively adjoining the edges e_0, e_1, e_2, e_3 and e_4. This gives the tree shown in figure 12.7(a), where the convention introduced in example 12.10 of circling the current centre is again employed. (Note that we are using a different labelling of vertices and edges from that used in example 12.10 and figure 12.16.) We cannot adjoin further edges incident to v_0, so the current centre moves to v_1. Clearly there are no edges which can be adjoined with v_1 as the current centre, so the centre moves to v_2 where again step 2 cannot be performed.

Thus the centre then moves on to v_3, where step 2 can be performed once resulting in adjoining the edge e_5. The position at this stage is represented in figure 12.17(b).

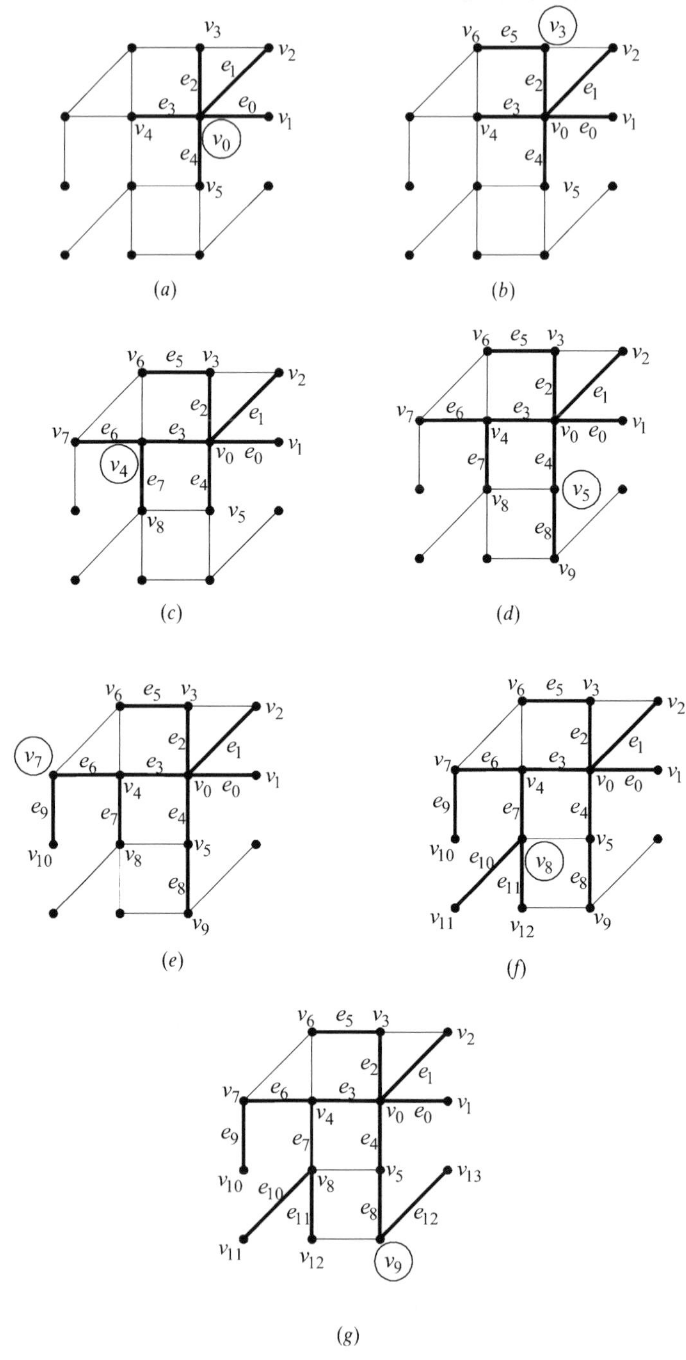

Figure 12.17

The current centre moves to v_4 and the edges e_6 and e_7 are then adjoined to the tree producing figure 12.17(*c*). The remainder of the execution of the algorithm is represented by the graphs in figures 12.17(*d*)–(*f*). The diagrams show the search frozen at each stage where edges have been added to the tree.

We can regard the execution of the breadth-first search algorithm as having several phases. In the first phase we adjoin all possible edges to the initial vertex v_0. This is the situation reached in figure 12.17(*a*). In the second phase each of the vertices added in the first phase becomes the current centre in turn and edges are adjoined where possible. This phase is illustrated by the graphs in figures 12.17(*b*)–(*d*).

In general, during the $(n + 1)$th phase of the execution of the algorithm each vertex added in the nth phase becomes the current centre in turn. If we regard the resulting spanning tree as a rooted tree with root v_0, then the level n vertices are those added during the nth phase of the execution of the algorithm. Our example has been completed in three phases.

Comparing the algorithms for depth-first and breadth-first searches, the significant difference concerns when the current centre moves to a new vertex. In depth-first search, the current centre moves at each step (where possible) to the most recently visited vertex. By contrast, in breadth-first search the centre moves only when all vertices adjacent to the current centre have been visited.

We now have two systematic methods for growing a spanning tree in a connected graph Γ. The obvious question which springs to mind is: 'which is to be preferred?' It can be shown that the depth-first and breadth-first algorithms have the same 'worst-case complexity'. The worst-case complexity of an algorithm is an approximate measure of the maximum number of operations required to perform the algorithm; it is of course, a function of the size of the input data, which would be the graph itself in this case. (The computational complexity of algorithms is explained in a little more detail in §12.6.)

Since algorithms 12.4 and 12.5 have the same worst-case complexity, neither is to be preferred over the other as a general algorithm. For particular graphs, however, one algorithm might produce a spanning tree with much less fuss than the other. To take a simple example, depth-first search would more efficiently on a cycle graph (figure 12.18(*a*)) whilst breadth-first would work more efficiently on the 'Maltese cross' graph (figure 12.18(*b*)). This becomes apparent as soon as you apply the algorithms to each of these graphs, regardless of the choice of initial vertex.

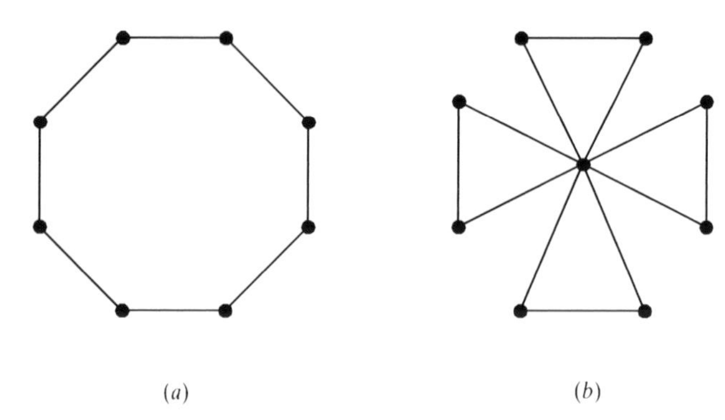

Figure 12.18

Exercises 12.3

1. Find a spanning tree for each of the following graphs using (a) depth-first search and (b) breadth-first search with the indicated initial vertex.

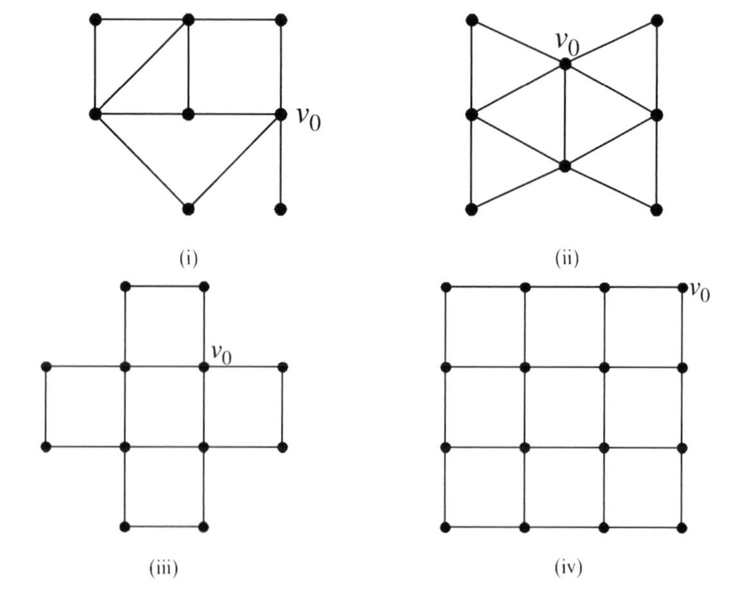

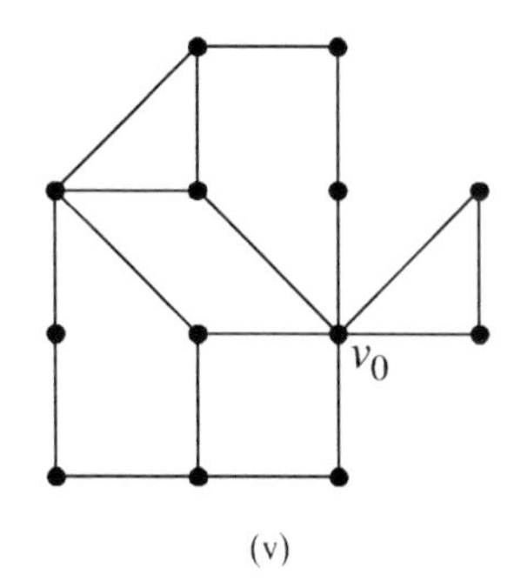

(v)

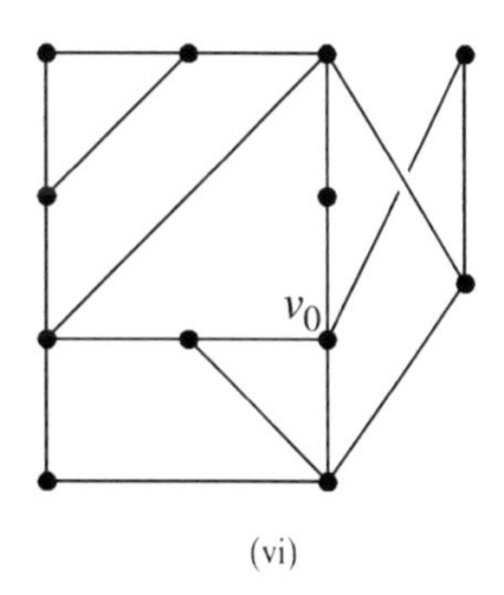

(vi)

2. Show that, if Γ is a connected graph, then the depth-first algorithm 12.4 does indeed produce a spanning tree in Γ.

3. Show that, if Γ is a connected graph, then the breadth-first algorithm 12.5 does indeed produce a spanning tree in Γ.

4. Execute the depth-first and the breadth-first algorithms on the complete graphs K_4 and K_5.

5. The floor plan of a museum is given in the following diagram, where the entrance to the exhibits is from the foyer, labelled E.

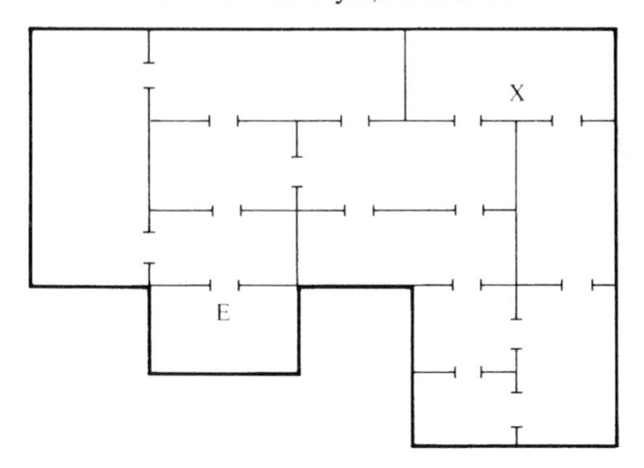

   (i) Draw a graph to represent the geography of the museum by representing each room as a vertex and each doorway as an edge.

   (ii) Perform both a depth-first and a breadth-first search of the museum, with initial vertex E in both cases.

   (iii) A visitor to the museum particularly wishes to see an exhibit in room X. Which of your two searches would you recommend to the visitor?

6. The following is a plan of a maze with entrance A and exit B. Draw a graph to represent the maze as follows. Each letter represents a point in the maze and is represented by a vertex. An edge joins two vertices if and only if there is a path in the maze from one vertex to the other which does not pass another vertex. (For instance, there is an edge in the graph joining J and R, but there is no edge joining J and S since a path in the maze from J to S passes either R or D.)

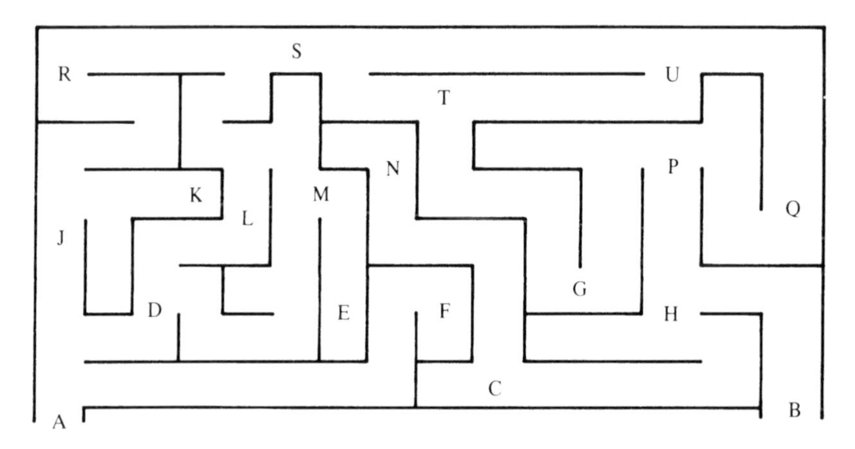

You are lost in the maze at the point E. Perform a depth-first and a breadth-first *partial* search of the maze until you find your way out. (By a partial search, we mean execute the appropriate algorithm only until B, the exit to the maze, is included in the tree.)

Repeat both searches but this time, whenever there is a choice of vertices to visit, choose the one which comes first in the alphabet.

7. Let Γ be a connected graph. Explain why it is possible, with suitable choices (including that of initial vertex), to perform the depth-first search algorithm *without backtracking* if and only if Γ is semi-Hamiltonian. (Semi-Hamiltonian graphs are defined in exercise 11.2.10.)

12.5 Weighted Graphs

In chapter 11 we used the analogy of a road map to illustrate various graph-theoretic concepts. There is one aspect of a road map, however, which is not modelled by a graph—namely, the distances between towns. In many applications of graph theory it is important to be able to attach numbers to the edges of the graph which represent certain physical quantities. For example, if we wish to use a graph to represent an electrical network, it may be important to record the resistances of each of the components represented by the edges of the graph. Similarly, if our graph is a mathematical model of a network of fluid-carrying pipes, we might wish to include information about the capacities of the various different pipes on the graph itself. In probability and decision theory, trees are used where the edges represent possible outcomes of an experiment or possible decisions made. It is often useful to assign probabilities to the outcomes which are represented by the corresponding edges.

The logic networks of §10.5 can be considered as graphs. Boolean expressions, rather than numbers, were attached to the edges of the corresponding graphs. Although such possibilities are not considered here, these graphs could be considered generalizations of those in this section.

To be able to use graphs to represent situations such as these, we need to introduce the concept of a 'weighted graph'. Intuitively a weighted graph is a graph where a non-negative real number $w(e)$, called the 'weight' of e, is 'attached' to each edge e. The formal definition is the following.

Definition 12.7

A **weighted graph** is a graph Γ together with a function

$$w : E_\Gamma \to \mathbb{R}^+ \cup \{0\}.$$

If e is an edge of Γ, then the number $w(e)$ is called the **weight** of e.

Some authors require that the weights of edges should be (non-negative) integers; we will refer to such a weighted graph as an **integer-weighted graph**.

There is an obvious way to represent a weighted graph pictorially—simply write the weight of each edge on the usual diagram of the graph. For example, figure 12.19 is a diagram of an integer-weighted graph.

So far we have only defined the weights of individual edges of a graph. In many, if not most, situations where weighted graphs are applied, it is the sum of the weights of the edges of some subgraph which is important.

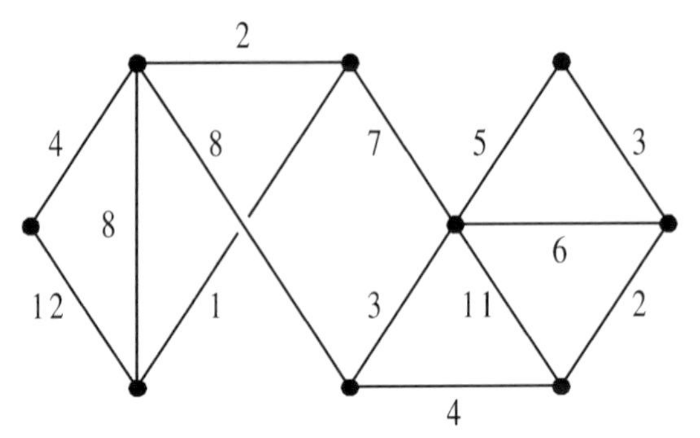

Figure 12.19

Definition 12.8

Let Γ be a weighted graph. For any subgraph Γ' of Γ we define the **weight of Γ'**, $w(\Gamma')$, to be the sum of the weights of its edges. Symbolically,

$$w(\Gamma') = \sum_{e \in E_{\Gamma'}} w(e).$$

In §12.4 two techniques (depth-first and breadth-first searches) for growing spanning trees in a connected graph were considered. One of the reasons why spanning trees are important is that a spanning tree provides a 'complete' set of paths in a connected graph: if T is a spanning tree for Γ, then any two vertices in Γ can be joined by a unique path in T (see theorem 11.6(i)). If Γ is a connected weighted graph, it is often desirable to have a spanning tree with smallest weight.

Definition 12.9

Let Γ be a connected weighted graph. A **minimal spanning tree** for Γ is a spanning tree T which has the smallest possible weight in the sense that if T' is any other spanning tree then

$$w(T) \leqslant w(T').$$

The first thing to observe is that every (finite) connected weighted graph Γ has a minimal spanning tree. Since Γ has only finitely many spanning trees one of them must have minimal weight. Note, however, that a given weighted graph may have more than one minimal spanning tree. Figure 12.20 shows an integer-weighted graph with two minimal spanning trees, both of weight 22.

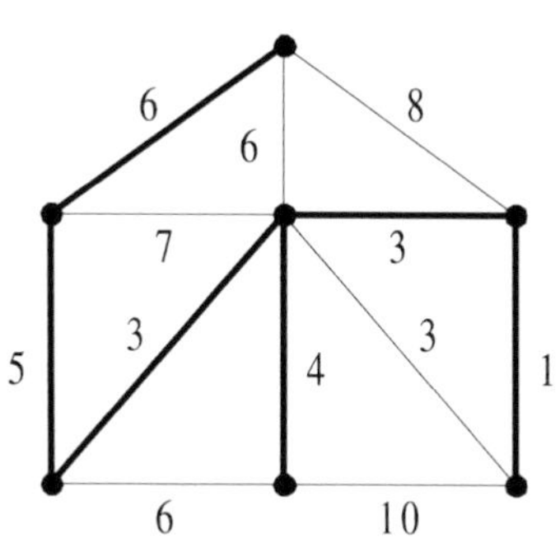

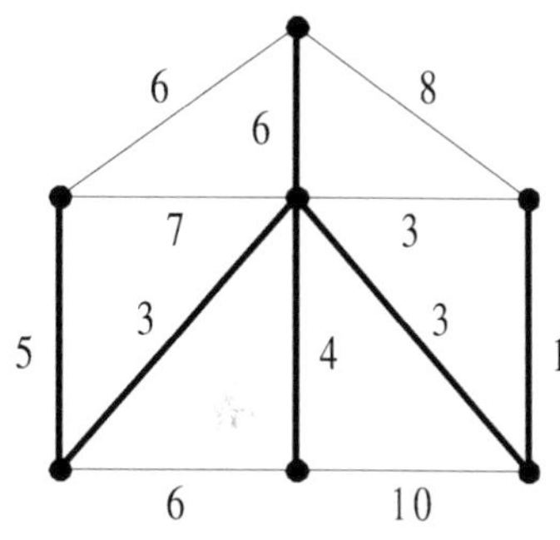

Figure 12.20

In an attempt to define an algorithm which produces a minimal spanning tree, we might try to modify the depth-first and breadth-first algorithms (12.4 and 12.5) so that the edge with the smallest weight amongst those under consideration is added to the tree at each stage. Unfortunately these modifications do not produce the required algorithm because the edge set under consideration at any stage is too restricted. Recall that, in both depth-first and breadth-first searches, only edges incident to the current centre are allowed to be attached to the current tree. (It is an interesting exercise to construct weighted graphs for which these modified depth-first and breadth-first searches do not produce minimal spanning trees. See exercise 12.4.5.)

Widening the possible choice of edges to be added at each stage produces the following simple algorithm for constructing a minimal spanning tree. (Another algorithm is given in exercise 12.4.2.) The algorithm builds up successive subtrees by attaching one edge at a time so that at each stage an edge with the smallest weight is chosen, subject only to the restriction that a sub*tree* results; that is, no cycle is created. The algorithm can be described as follows.

Algorithm 12.6 (Prim's algorithm)

1. First select any vertex. This begins the process with the first subtree T_0 (which has no edges).

2. Consider the set of edges which are incident with one of the vertices of the subtree T_n. Of those edges which do not produce a cycle when added to T_n, select one which has smallest weight. (There may be more than one choice of this edge). Adjoin the chosen edge to T_n to form a new subtree T_{n+1}.

3. Repeat step 2 until it is not possible to adjoin an additional edge to T_n without creating a cycle. The resulting tree is a minimal spanning tree.

Note that it is not immediately apparent that this algorithm actually produces a *minimal* spanning tree. It is conceivable, for instance, that beginning the whole process at a different vertex may produce a tree with a smaller weight. However, working through the algorithm for a few weighted graphs should convince you that it does indeed give the required minimal spanning tree regardless of the choices that are made. For a formal proof see Biggs (1990), for example.

Algorithm 12.6 can, in fact, be used to construct a spanning tree in any connected (unweighted) graph Γ. Firstly, turn Γ into a weighted graph by giving every edge weight 1 and then apply the algorithm. In this situation there will usually be several choices of edges to adjoin at each step in the algorithm as all edges have equal weight.

Example 12.12

Figure 12.21 illustrates the construction of a minimal spanning tree using Prim's algorithm 12.6. We begin by choosing the vertex v as the starting subtree T_0 and add an edge at a time until we produce the minimal spanning tree T_5. Note that in this example there is only one choice of edge at each stage, which means that the minimal spanning tree is unique.

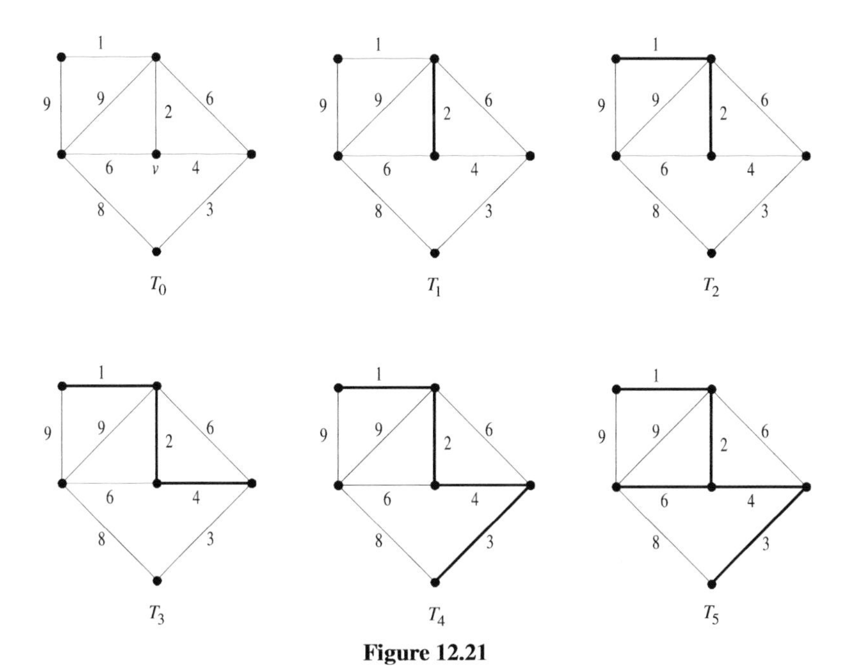

Figure 12.21

Exercises 12.4

1. For each of the following weighted graphs, use Prim's algorithm 12.6, beginning with the indicated vertex v, to find a minimal spanning tree, and give its weight.

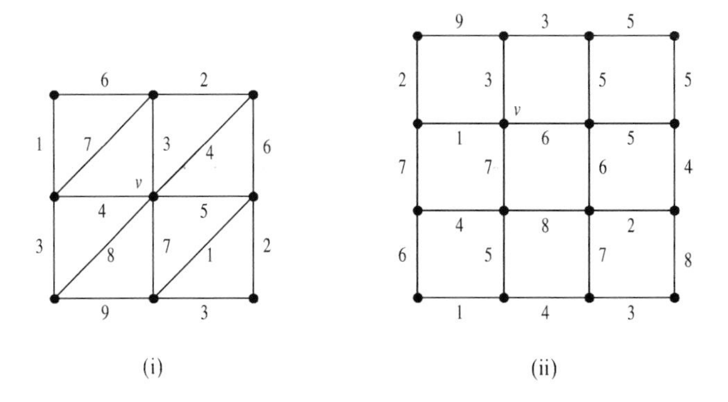

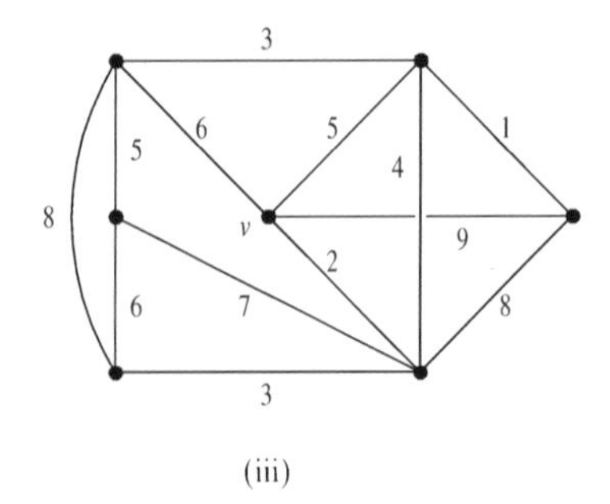

(iii)

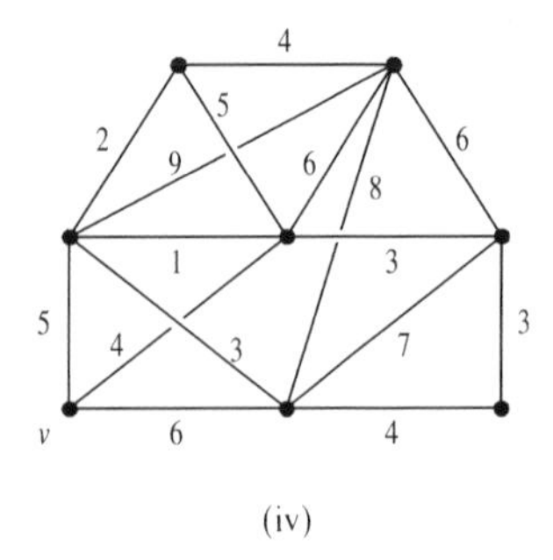

(iv)

2. An alternative algorithm for constructing a minimal spanning tree is **Kruskal's algorithm**†, which can be described as follows.

Beginning with the empty subgraph, form a sequence of (not necessarily connected) subgraphs by adding at each stage an edge with smallest weight which does not form a cycle with the existing

GREEDY ALGORITHMS ARE SHORT SIGHTED

† The essential difference between Prim's and Kruskal's algorithms is that Prim's algorithm constructs a sequence of *connected* subgraphs without cycles (i.e. trees) whereas Kruskal's algorithm does not require the subgraphs to be connected.

Both Kruskal's and Prim's algorithms are examples of **greedy** algorithms. A greedy algorithm is one which always chooses the best option available at each stage, without looking further ahead. Greedy algorithms are 'short sighted' and do not always produce the optimum result because a 'greedy' choice early on may lead to a reduced set of options later. For Prim's and Kruskal's algorithms, however, choosing the best (i.e. smallest weight) edge available at each stage does produce the best (i.e. minimal) spanning tree.

subgraph. When no further edge can be added, the resulting graph is a minimal spanning tree.

Perform Kruskal's algorithm on each of the weighted graphs in question 1 above.

3. Let T be a minimal spanning tree for a weighted graph Γ. Determine whether the following statements are necessarily true.

   (i) The weight of every edge belonging to T is less than or equal to the weight of every edge not belonging to T.
   (ii) If no two edges of Γ have the same weight, then T is unique.

4. Let T be a minimal spanning tree for a weighted graph Γ and let e be an edge not belonging to T which joins distinct vertices v and w. Show that $w(e) \geqslant w(e')$ for every edge e' belonging to the unique path in T joining v and w.

   (Hint: argue by contradiction. Show that if $w(e) < w(e')$ for some e' then a spanning tree T' can be obtained with smaller weight than that of T.)

5. Construct a connected weighted graph and specify a choice of initial vertex for which the application of the depth-first algorithm (modified so that an edge of least weight is added to the tree at each stage) does not produce a minimal spanning tree.

6. (i) Let Γ be a connected weighted graph with v vertices and e edges. The weights of the edges of Γ are the integers $1, 2, \ldots, e$. Determine a lower bound for the weight of a minimal spanning tree.

   (ii) A connected weighted graph Γ has degree sequence $(1, 1, 2, 2, 3, 3, \ldots, n, n)$. The weights of the edges of Γ are the integers $1, 2, \ldots, e$, where e is the number of edges of Γ. Determine a lower bound for the weight of a minimal spanning tree.
   Show, by example, that this lower bound cannot be attained for some weighted graphs Γ of this type.

7. The weighted graph below represents a fire prevention sprinkler system. The vertices represent sprinklers and the edges represent the water pipe connections between them. The connections between sprinklers need checking periodically but some are more difficult and take longer than others due to the location of the sprinklers and the length of the connecting pipe. The weights denote the maintenance costs of checking the corresponding connections.

The sprinkler system will remain effective provided the corresponding graph is connected. The company operating the system wishes to save maintenance costs by removing some of the connecting pipes. What is the maximum saving the company can make?

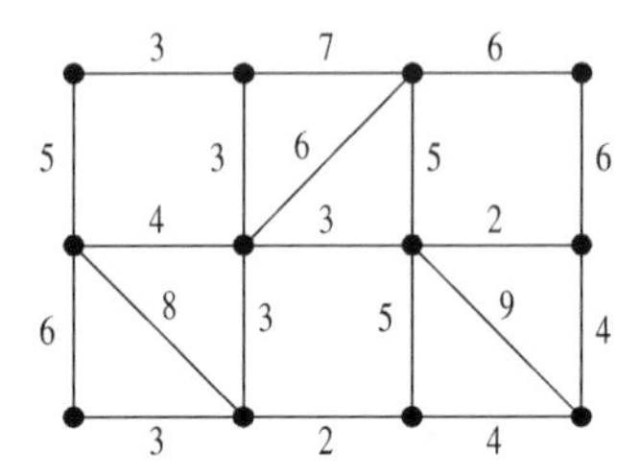

12.6 The Shortest Path and Travelling Salesman Problems

Let Γ be a connected weighted graph and let v, v' be two vertices of Γ. The **shortest path problem** is to find a path joining v and v' with the smallest weight. Of course, since a path is a subgraph of Γ, its weight is defined as the sum of the weights of its edges. As we are dealing with finite graphs, it should be obvious that a shortest path exists, although there may be more than one shortest path joining a given pair of vertices.

There are various methods for finding a shortest path between two given vertices. We describe an algorithm which, like the minimal spanning tree algorithm of the previous section, constructs the path one edge at a time.

The idea is to begin at the vertex v and move through the graph assigning a number $L(u)$ to each vertex u in turn which represents the length of the shortest path *yet discovered* from v to u. These 'length numbers' $L(u)$ are initially considered temporary and may subsequently be changed if we discover a path from v to u which has length less than the currently assigned value $L(u)$. The algorithm is detailed more precisely below. It actually constructs a subtree of the graph containing the vertices v and v'; a shortest path between the two vertices is then the unique path in this tree joining them. (Note that the subtree constructed by algorithm 12.7 need not be a spanning subtree of the graph—the algorithm stops as soon as a shortest path joining v and v' has been found.)

Algorithm 12.7 (Dijkstra's algorithm)

1. First assign $L(v) = 0$ to the starting vertex v. We say that the vertex v has been **labelled** with the value 0. Furthermore, this label is **permanent** as we will not subsequently change its value. Since we are constructing a sequence of trees, we also begin with the tree consisting of vertex v only and no edges.

2. Let u be the vertex which has most recently been given a *permanent* label. (Initially $v = u$ as this is the only vertex with a permanent label.) Consider each vertex u' adjacent to u in turn, and give it a **temporary label** as follows. (Only those vertices u' without a permanent label are considered.)

   If u' is unlabelled, then set $L(u')$ equal to $L(u) + w(e)$ where e is the edge joining u to u'. (If there is more than one such edge e, choose the one with the smallest weight.)

   If u' is already labelled, then again calculate $L(u) + w(e)$ as above and if this is less than the current value of $L(u')$ then change $L(u')$ to $L(u)+w(e)$; otherwise leave $L(u')$ unchanged.

3. Choose a vertex w with the smallest temporary label and make the label permanent. (There may be a choice to be made here as several temporarily labelled vertices may have equal smallest labels. It is also important to realize that w need not be adjacent to u, the most recently labelled vertex.) At the same time adjoin to the tree so far formed the edge which gives rise to the value $L(w)$.

4. Repeat steps 2 and 3 until the final vertex v' has been given a *permanent* label. The path of shortest length from v to v' is then the unique path in the tree thus formed joining v and v'. Its length is the permanent value of $L(v')$.

Example 12.13

We illustrate this algorithm by constructing a shortest path from A to H in the weighted graph illustrated in figure 12.22(a).

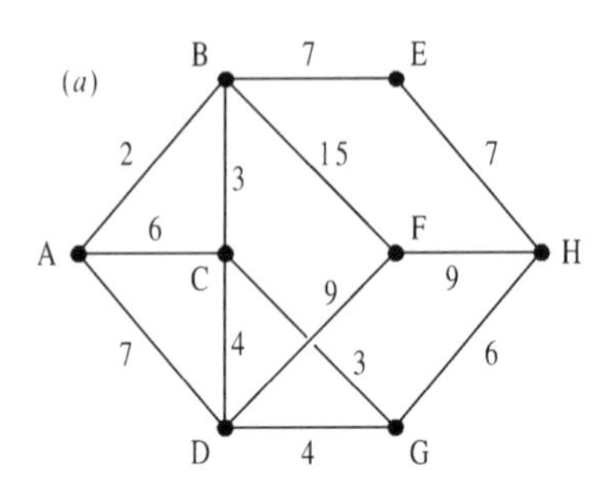
(a)
B
7
E
2
15
7
3
6
F
A
H
C
9
9
7
4
3
6
D
4
G

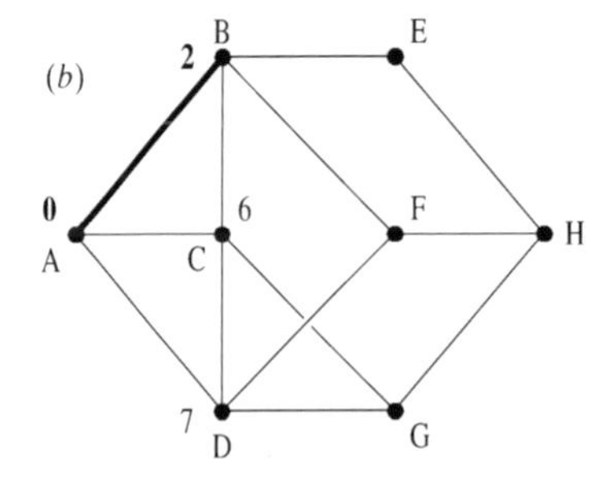
(b)
2
B
E
0
6
F
A
C
H
7
D
G

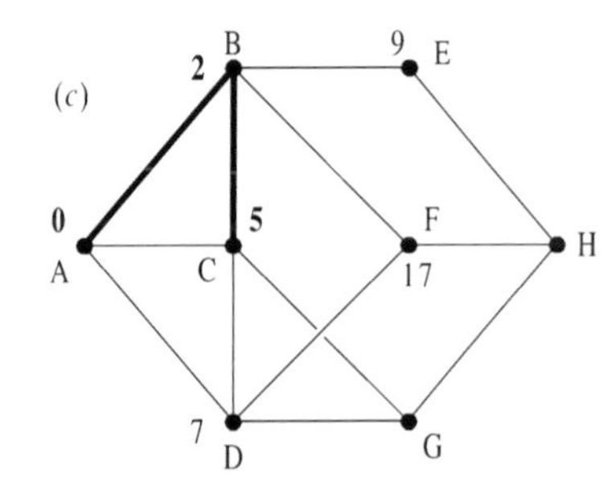
(c)
2
B
9
E
0
5
F
A
C
17
H
7
D
G

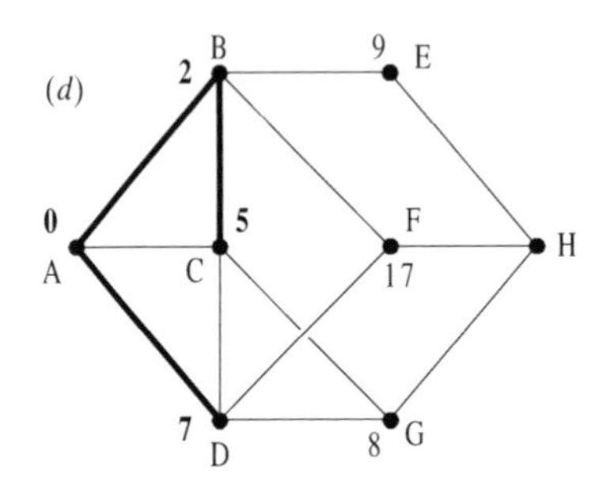
(d)
2
B
9
E
0
5
F
A
C
17
H
7
D
8
G

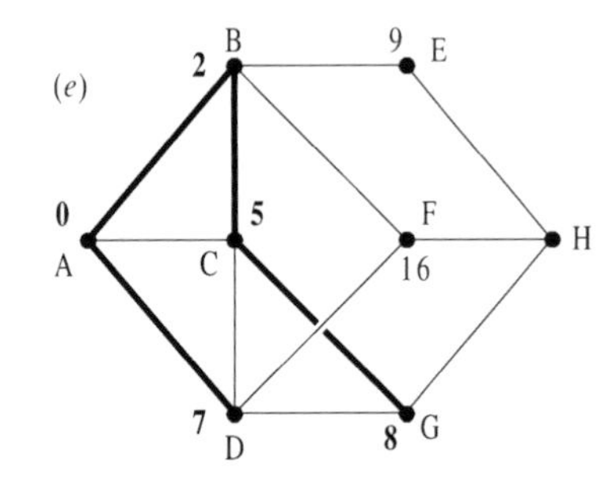
(e)
2
B
9
E
0
5
F
A
C
16
H
7
D
8
G

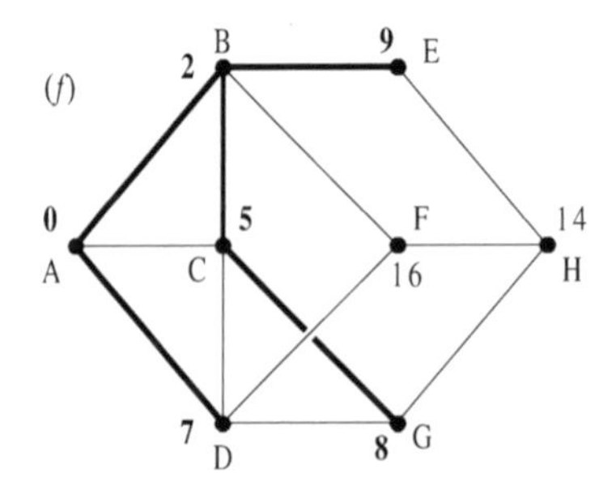
(f)
2
B
9
E
0
5
F
14
A
C
16
H
7
D
8
G

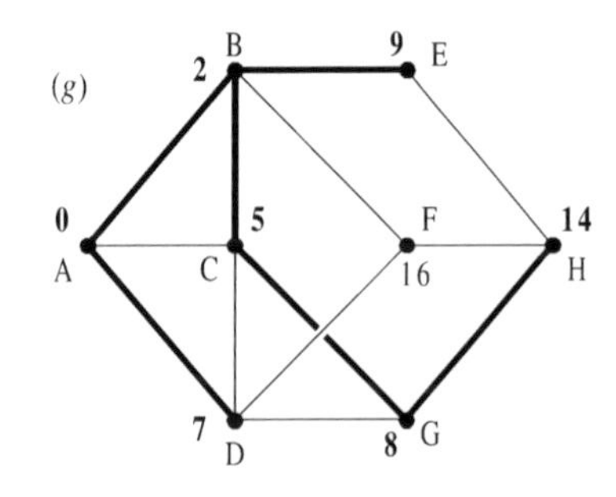
(g)
2
B
9
E
0
5
F
14
A
C
16
H
7
D
8
G

Figure 12.22

We begin by giving a vertex A permanent label 0. Then we give vertices B, C and D temporary labels 2, 6 and 7 respectively. Since 2 is the smallest, we make B's label permanent, which is indicated by printing the label in bold type. At the same time we add the edge joining A to B to the tree we began with (which of course had no edges). This is the stage reached in figure 12.22(*b*).

Next we consider the vertices adjacent to B, the most recently *permanently* labelled vertex. Since E and F are unlabelled, we label them with the result of adding $L(\text{B}) = 2$ to the weight of the edge joining B to each vertex; that is, $2 + 7 = 9$ and $2 + 15 = 17$ respectively. Vertex C is already labelled. However, the current (temporary) label $L(\text{C})$ is greater than $L(\text{B})$ plus the weight of the edge joining B to C, so we need to change $L(\text{C})$ to the smaller value 5. Now C has the smallest temporary label, so we make it permanent and add the edge joining B and C to the tree. We have now obtained figure 12.22(*c*).

Repeating this process for vertex C, we give G the temporary label 8. (Note that the label of vertex D does not change at this stage since the current label, 7, is less than $L(\text{C})$ plus 4, the weight of the edge joining C and D.) Now D has the smallest temporary label, so it is made permanent and the edge joining D to the tree is added to the tree. At this stage—figure 12.22(*d*)—it might appear that, in adjoining the first two edges to the tree, we had made a 'false start'. It is quite possible for this to happen although, as we shall see, in this example these edges will eventually form part of the shortest length path we are seeking.

Repeating the process, we need to consider those vertices (which do not have permanent labels) adjacent to D, the vertex which has most recently received its permanent label. The label on vertex F needs to be reduced to 16 which is the sum of $L(\text{D})$ and the weight of the edge joining D and F. Despite this reduction, it is G which has the smallest temporary label. This label is made permanent and the edge joining C to G is added to the tree producing figure 12.22(*e*).

A further repetition of the process produces figure 12.22(*f*). This does not complete the execution of the algorithm as the label on H is still temporary at this stage. However, one further run through of steps 2 and 3 of the algorithm does complete the construction of the required tree because H receives a permanent label this time—see figure 12.22(*g*).

Now that the tree is complete we can find the shortest path from A to H. It is the unique path in the tree joining the two vertices, passing through vertices B, C and G and with total weight 14 (the permanent label for H).

Using Dijkstra's algorithm 12.7 for a relatively simple graph like the one in the previous example may seem tedious; you could almost certainly find the shortest path by trial and error far more quickly. However, it has the advantage of being a mechanical procedure which will produce the desired path in a more complicated graph for which trial and error would be a considerable trial of endurance and extremely prone to error.

The Travelling Salesman Problem

Suppose a travelling salesman needs to visit each of several towns and return to his starting position. Given the network of roads connecting the various towns on his itinerary, the travelling salesman's problem is to find a route which minimizes his total distance travelled. (Such a route may visit some towns more than once.)

The network of roads can be represented by a weighted graph as follows. Each town is represented by a vertex and each road connecting two towns by an edge joining the corresponding vertices whose weight is the length of the given road. A journey which visits each town and ends back at the starting position is represented by a closed edge sequence in the graph whose associated vertex sequence contains every vertex (see definition 11.6). In graph-theoretic terms, the travelling salesman problem is to find such a closed edge sequence of minimum total weight. We shall assume that the graph is connected so that it is indeed possible for the salesman to visit every town using the roads of the given network.

It is usual to restate the problem in a slightly different graph-theoretic form. The graph described above is replaced by a complete graph with one vertex for every town. (Recall that a complete graph is one in which there is a unique edge joining every pair of distinct vertices.) An edge is given weight equal to the shortest distance between the corresponding towns using the roads of the network. These shortest distances can be found by applying algorithm 12.7 to the original graph.

Figure 12.23 gives an example of the weighted graph representing a network of roads and the corresponding complete weighted graph. Note that the weight of an edge in the complete graph can be less than the weight of an edge joining the same vertices in the original graph. This is because there may be a shorter route than the single-edge one connecting the two vertices. In figure 12.23(a) the weight of the edge joining B to E is 15 which is greater than the weight of the path via C and D joining these vertices. In the complete graph the weight of the edge joining B and E is 13, which is the weight of this path via C and D.

From now on we shall work with the complete graph. In order that we can later recover information about the original, not necessarily complete, graph, we shall need to keep a note of which paths in the original graph gave rise to the various edges of the complete graph.

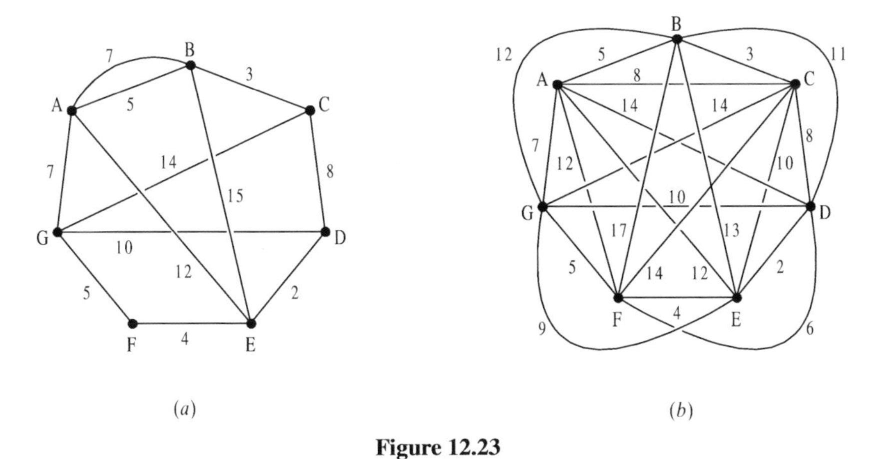

Figure 12.23

The closed edge sequence visiting every vertex of the graph representing the road network corresponds to a Hamiltonian cycle in the complete graph. (Hamiltonian cycles are defined in definition 11.9.) Thus our formal statement of the problem is the following.

> **Travelling Salesman Problem**
>
> Given a connected, weighted, complete graph, construct a Hamiltonian cycle of minimum weight; that is, a **minimal** Hamiltonian cycle.

Recall from chapter 11 that not every graph has a Hamiltonian cycle; however, every complete graph with at least three vertices does have one. This follows from theorem 11.3—it is also easy to see directly. Since our graphs are finite, there can be only a finite number of Hamiltonian cycles which means that there must be (at least) one which is minimal.

Since the weights of the edges in the complete graph are the shortest distances between vertices of the original road-network graph, the complete graph must satisfy the following **triangle inequality**.

For every triple of distinct vertices (v_1, v_2, v_3),

$$w(v_1, v_2) + w(v_2, v_3) \geqslant w(v_1, v_3)$$

where $w(v_i, v_j)$ denotes the weight of the (unique) edge joining v_i and v_j†.

The travelling salesman problem has received considerable attention from graph theorists and others and various algorithms have been given to construct a minimal Hamiltonian cycle. One reason for the interest in the problem is related to the fact that all of the known algorithms which solve the problem are 'computationally inefficient'. To explain precisely what this means is beyond the scope of this book; however, we can give an intuitive explanation. To apply any computational algorithm involves a 'cost', which may be measured, for example, in terms of the number of operations required or the time taken for a computer to execute the algorithm. The cost will of course depend on the 'size' (and nature) of the actual problem being solved, which is frequently measured by the number n of pieces of information required to define the problem.

In 1965 J Edmonds and A Cobham introduced a broad classification of algorithms into those which run in polynomial time and those which run in exponential time. Roughly speaking, an algorithm is **computationally efficient** if its 'cost' is no bigger than some power of n, say n^k for some integer k, where n is the 'size' of the problem. These are the so-called **polynomial time algorithms**. Alternatively, the algorithm is **computationally inefficient** if its cost is an exponential function of n; that is, the cost depends on a^n for some real number $a > 1$. These, of course, are the **exponential time algorithms**. The classification of algorithms as efficient or inefficient corresponds roughly to the practical experience of programmers. A computationally efficient algorithm is likely to be performed in a reasonable time even for fairly large values of the size n of the input data. On the other hand an inefficient algorithm will most likely take too long to be practical for large problems.

Table 12.1 vividly illustrates how a computationally inefficient algorithm can become unmanageable, even for problems of relatively small size. Assuming a processing speed of 10 million operations per second, the approximate computation times of an efficient algorithm (cost $= n^2$) and an inefficient algorithm (cost $= 2^n$) are compared for various sizes of problem. (The situation is not quite as clear cut as the table might suggest—see exercise 12.5.2.)

Returning to the travelling salesman problem, all known algorithms are computationally inefficient with respect to the number of vertices of the graph; that is, all known algorithms are exponential time algorithms. In fact there

† In fact the weight function, considered as a function

$$w : V \times V \to \mathbb{R}^+ \cup \{0\} \qquad (v_1, v_2) \mapsto w(v_1, v_2)$$

is a metric on the set V of vertices of the complete graph. (See §8.7 for the definition of a metric.) Of course, V is the set of towns on the travelling salesman's itinerary.

Table 12.1 Processing times for efficient and inefficient algorithms at a rate of 10^7 operations per second.

Cost (c) of algorithm	Size (n) of problem $n = 10$	$n = 40$	$n = 70$
n^2	0.000 01 s	0.000 16 s	0.000 49 s
2^n	0.000 1 s	30.5 h	37 436 centuries

are many problems of a similar nature in mathematics, for which all known algorithms are inefficient. On the theoretical side, much work has been done on the question of whether the travelling salesman problem is inherently too complex for an efficient algorithm to exist†. On the practical side, however, various efficient algorithms are known which give an approximate solution. In other words they provide a Hamiltonian cycle whose weight is 'close to' the smallest possible, but may not actually be the best possible.

There is a fairly obvious simple 'approximate' algorithm, known as the 'nearest neighbour algorithm'. It is a modification of the depth-first search algorithm given in §12.4. The algorithm starts any vertex and 'travels along' an edge of smallest weight incident with it to visit a new vertex. At each step in the algorithm we proceed from the most recently visited vertex to a new vertex by travelling along an edge of smallest possible weight. (There may be several choices of a 'minimal edge' at each stage.) When all vertices have been visited, we return to the starting position along the unique edge of the complete graph from the last vertex back to the first. As we noted previously, in order to recover information about the original (not necessarily complete) road-network graph, we need to have recorded the paths in the original graph which gave rise to the edges of the complete graph. This is clearly important if we wish to advise a real-life travelling salesman.

Since it is relatively straightforward, we leave the formal description of this algorithm as an exercise (12.5.6). The nearest neighbour algorithm is a greedy algorithm in the sense that it is a nearest vertex which is the one visited at each stage. (See the footnote to exercise 12.4.2 for an explanation of the term 'greedy algorithm'.) It turns out that this algorithm is extremely poor in the following sense. Although it will sometimes produce a minimal Hamiltonian cycle, in

† This is still unknown, so we don't know whether the search for an efficient algorithm is bound to be fruitless. The problem is one of the famous unsolved problems of mathematics. Briefly, the travelling salesman problem belongs to a class of problems known as the 'NP-complete problems'. The existence of an efficient algorithm for *any one* of these problems would imply the existence of efficient algorithms for *all* other NP problems. Thus, proving that the travelling salesman problem has an efficient algorithm would be truly significant both theoretically and practically. Finding an actual efficient algorithm would be even better. We ought to point out, however, that most experts believe that it is very unlikely that any of the NP-complete problems has an efficient algorithm.

general the cycle produced by the algorithm can have weight considerably greater than the minimum possible. In fact the performance of the algorithm is about as bad as one could imagine. Given any positive integer k (no matter how large) there exist graphs for which the weight of the Hamiltonian cycle produced by the nearest neighbour algorithm is greater than k times the weight of the minimal cycle.

We turn instead to another 'approximate' algorithm—the 'nearest insertion algorithm'. Although more complicated to describe, this algorithm guarantees to find a Hamiltonian cycle with total weight no more than twice the minimum—a considerable improvement on the nearest neighbour algorithm. Usually, however, the nearest insertion algorithm will produce a Hamiltonian cycle whose weight is significantly less than twice the minimum.

Algorithm 12.8 (Nearest insertion algorithm)

1. First choose any vertex. Select an edge e with smallest weight incident with the initial vertex and let C be the edge sequence: e, e. We regard C as our starting 'cycle', although strictly speaking it is not a cycle as it repeats an edge.

2. Select an edge with the smallest weight which joins a vertex in C to one not in C and let v be the vertex not in C incident with this edge. (There may be several possible choices for the edge.)

3. The next step is to enlarge the cycle to include the chosen vertex v. To decide how to insert v, consider all pairs u_1, u_2 of adjacent vertices in C and select a pair for which the expression

$$I = w(u_1, v) + w(v, u_2) - w(u_1, u_2)$$

is a minimum, where $w(u, v)$ denotes the weight of the edge joining u and v. This expression I represents the increase in the total weight of C when it is enlarged to include the vertex v. We enlarge C to include v by adjoining the edges connecting u_1 and v and connecting v and u_2, and deleting the edge joining u_1 and u_2.

4. Repeat steps 2 and 3 until the cycle includes all the vertices of the graph.

The basic step in the nearest insertion algorithm is to take a cycle in the graph and enlarge it to include a vertex which is 'closest' to the given cycle. This step is then repeated until all vertices are included in the cycle. The algorithm can be applied to any complete weighted graph satisfying the triangle inequality.

Example 12.14

We illustrate the nearest insertion algorithm for the weighted graph in figure 12.24 (which satisfies the triangle inequality).

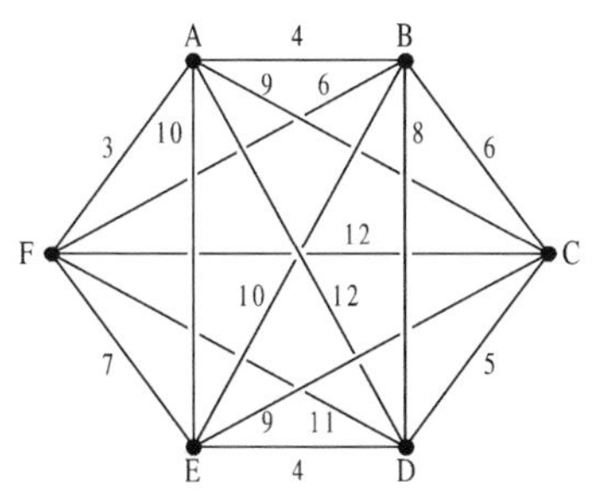

Figure 12.24

We shall adopt the convention that the (unique) edge joining vertices X and Y is denoted e_{XY} or e_{YX}. To perform step 1, we first choose the vertex labelled A in the diagram. (We could, of course, have chosen any of the vertices.) The edge e_{AF} is the edge incident with A which has smallest weight, so the first 'cycle' is $e_{\mathrm{AF}}, e_{\mathrm{FA}}$ from A to F and back to A.

The edge with smallest weight which is incident with either A or F is e_{AB}, so B is the first vertex to be inserted. The shortest way in which B can be inserted using a cycle is to go from A to B and back via F. This gives the cycle $e_{\mathrm{AB}}, e_{\mathrm{BF}}, e_{\mathrm{FA}}$ shown in figure 12.25(*a*).

The vertex which is nearest to a vertex of this cycle is C, so we need to find the best way to insert it. The values of I for the three edges of the current cycle are

$$I(e_{\mathrm{AB}}) = 6 + 9 - 4 = 11$$
$$I(e_{\mathrm{BF}}) = 6 + 12 - 6 = 12$$
$$I(e_{\mathrm{FA}}) = 12 + 9 - 3 = 18.$$

We therefore enlarge the cycle by removing the edge e_{AB} and inserting in its place the edges $e_{\mathrm{AC}}, e_{\mathrm{CB}}$. This gives the cycle shown in figure 12.25(*b*).

Repeating this process twice more, we enlarge the cycle firstly to include D and then to include E as shown in figures 12.25(*c*)–(*d*). The final cycle,

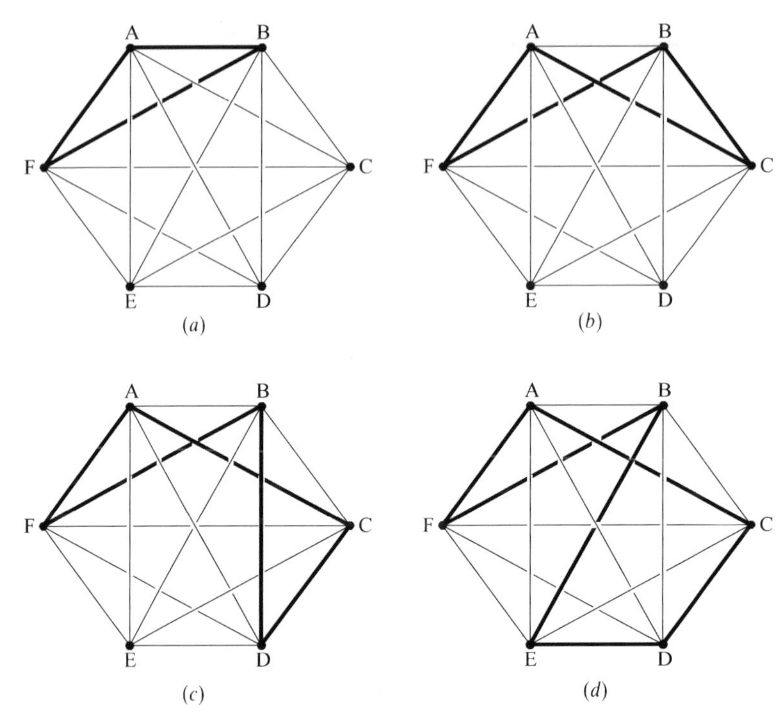

Figure 12.25

$e_{\text{AC}}, e_{\text{CD}}, e_{\text{DE}}, e_{\text{EB}}, e_{\text{BF}}, e_{\text{FA}}$, is the required Hamiltonian cycle with total weight $9 + 5 + 4 + 10 + 6 + 3 = 37$.

This example illustrates the fact that the nearest insertion algorithm may not produce a minimal Hamiltonian cycle. The graph in fact has a unique minimal Hamiltonian cycle—$e_{\text{AB}}, e_{\text{BC}}, e_{\text{CD}}, e_{\text{DE}}, e_{\text{EF}}, e_{\text{FA}}$—with total weight 29.

Exercises 12.5

1. Apply Dijkstra's algorithm 12.7 to obtain the shortest path from v to w in each of the following weighted graphs.

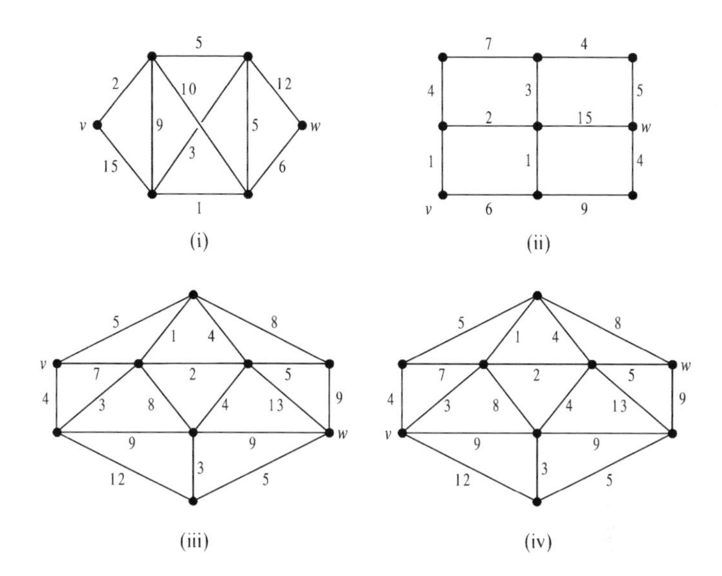

2. Assuming an operation speed of 10 million operations per second (10^7 operations s^{-1}), find the processing times of two algorithms, one with cost $= 10^5 n^6$ and the other with cost $= 10^{-9} 2^n$, for $n = 10$, 40 and 70.

   (These calculations show that calling polynomial time algorithms 'efficient' and exponential time algorithms 'inefficient' is only a rough guide to their expected performance.)

3. (a) Modify the shortest path algorithm 12.7 to produce the shortest directed path connecting two given vertices in a weighted digraph. What will be the result of applying your algorithm if there is no directed path in the graph connecting the two specified vertices?

   (b) Apply your algorithm to find the shortest path from v to w in each of the following weighted digraphs.

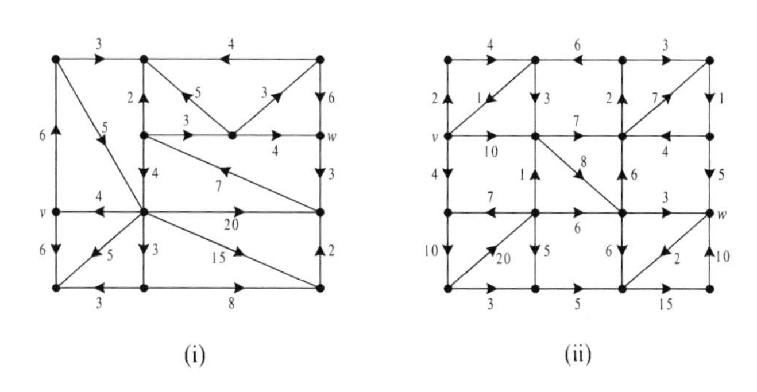

4. A weighted graph has vertex set $\{v_1, v_2, \ldots, v_{10}\}$, edge set $\{e_1, e_2, \ldots, e_{16}\}$ and incidence matrix

$$\begin{pmatrix}
1 & 1 & 0 & 0 & 0 & 0 & 0 & 0 & 0 & 0 \\
1 & 0 & 0 & 0 & 0 & 1 & 0 & 0 & 0 & 0 \\
1 & 0 & 0 & 0 & 0 & 0 & 1 & 0 & 0 & 0 \\
0 & 1 & 0 & 0 & 0 & 1 & 0 & 0 & 0 & 0 \\
0 & 0 & 0 & 0 & 0 & 1 & 1 & 0 & 0 & 0 \\
0 & 0 & 1 & 0 & 0 & 1 & 0 & 0 & 0 & 0 \\
0 & 0 & 0 & 0 & 1 & 1 & 0 & 0 & 0 & 0 \\
0 & 0 & 1 & 0 & 1 & 0 & 0 & 0 & 0 & 0 \\
0 & 0 & 1 & 1 & 0 & 0 & 0 & 0 & 0 & 0 \\
0 & 0 & 0 & 1 & 1 & 0 & 0 & 0 & 0 & 0 \\
0 & 0 & 0 & 1 & 0 & 0 & 0 & 1 & 0 & 0 \\
0 & 0 & 0 & 0 & 0 & 1 & 0 & 1 & 0 & 0 \\
0 & 0 & 0 & 0 & 0 & 0 & 0 & 1 & 1 & 0 \\
0 & 0 & 0 & 0 & 0 & 1 & 0 & 0 & 1 & 0 \\
0 & 0 & 0 & 0 & 0 & 0 & 0 & 0 & 1 & 1 \\
0 & 0 & 0 & 0 & 0 & 0 & 1 & 0 & 0 & 1
\end{pmatrix}.$$

(The incidence matrix of a graph is defined in exercise 11.1.16.)

The weights of the edges are as follows.

Edge:	e_1	e_2	e_3	e_4	e_5	e_6	e_7	e_8	e_9	e_{10}	e_{11}	e_{12}	e_{13}	e_{14}	e_{15}	e_{16}
Weight:	4	8	2	3	7	1	2	3	2	6	5	14	9	7	3	15

Draw a diagram of the graph and find the shortest path (i) from v_1 to v_8, and (ii) from v_4 to v_{10}.

5. Apply the nearest insertion algorithm 12.8 to each of the following complete weighted graphs, with v_0 as the choice of initial vertex. State the weight of the resulting Hamiltonian cycle. You may assume that the triangle inequality holds in each case.

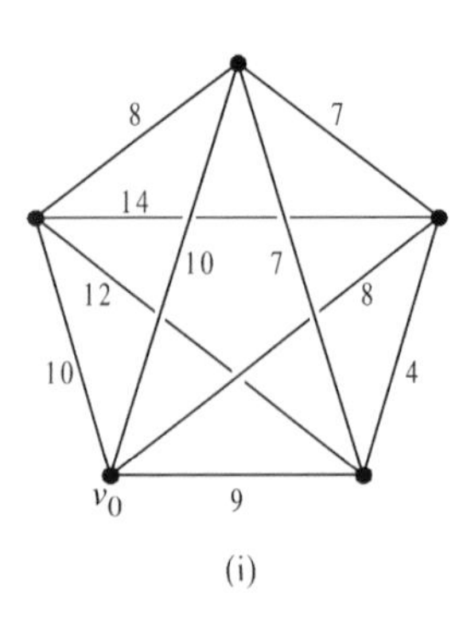

(i)

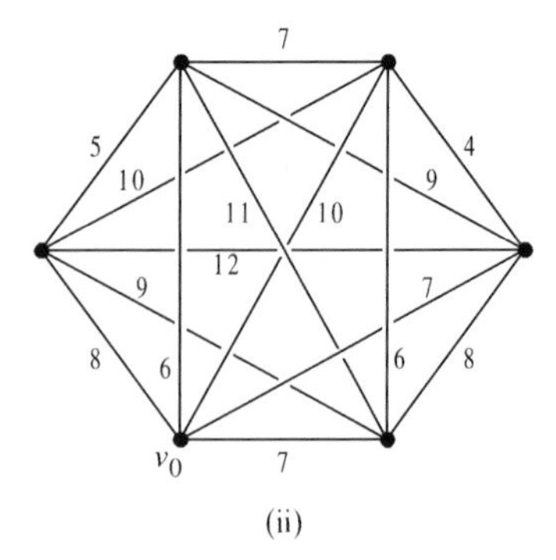

(ii)

Repeat the algorithm with other choices of initial vertex. Do you always obtain the same cycle?

6. Write down in detail the nearest neighbour algorithm (outlined on page 667) for finding a 'near' minimal Hamiltonian cycle.

   Apply the nearest neighbour algorithm to the two graphs in the previous question and compare the cycles obtained with those obtained using the nearest insertion algorithm.

7. Each of the following weighted graphs represents a network of roads. For each graph, construct the complete weighted graph which shows the shortest distances between the various vertices and perform the nearest insertion algorithm on the complete graph to find a 'near' minimal Hamiltonian cycle

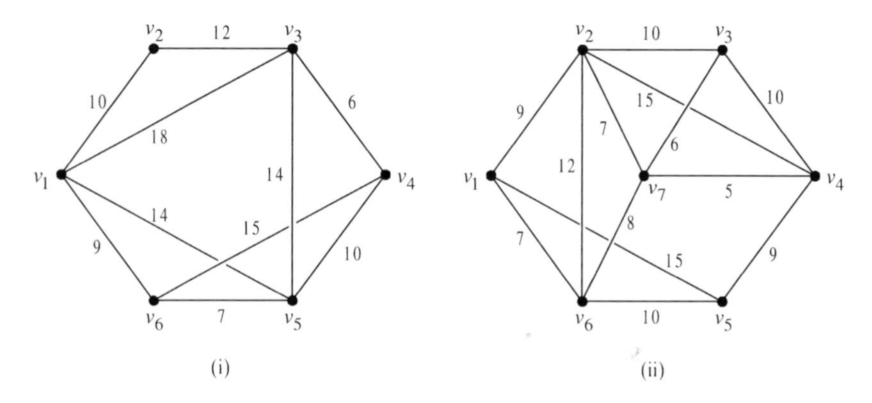

12.7 Networks and Flows

In this section we outline two applications of weighted digraphs. The first of these is to scheduling problems. Suppose a project (e.g. a construction project of some kind) involves the completion of several tasks. Each task takes a certain length of time and some tasks need to be completed before others can begin. The problem is to schedule the various activities to minimize the total time taken to complete the project.

The situation can be modelled using a weighed digraph as follows. Each edge of the digraph represents an activity and its weight represents the time taken to complete the activity. The vertices represent phases of the project, each phase being the completion of one or more activity. (There is an alternative way of modelling the situation using a weighted digraph in which the vertices represent activities. An edge is drawn from vertex v to vertex w if activity v must precede activity w and the edge is given a weight corresponding to the duration of activity v. Although we shall not consider this type of 'activity network', it requires much the same analysis that we will develop below—earliest times, latest times, critical path, and so on.)

Example 12.15

Figure 12.26 represents a project which has 10 activities, $A_1, A_2, \ldots, A_{10}$, whose completion times (in days, say) are indicated as the weights of the edges. The direction of each edge is from start to finish of the relevant task. The vertices S and F represent the start and finish respectively of the project.

The digraph indicates that A_1 must be completed before A_4 or A_5 can begin, both A_2 and A_5 must be completed before either A_7 or A_9 can commence, etc.

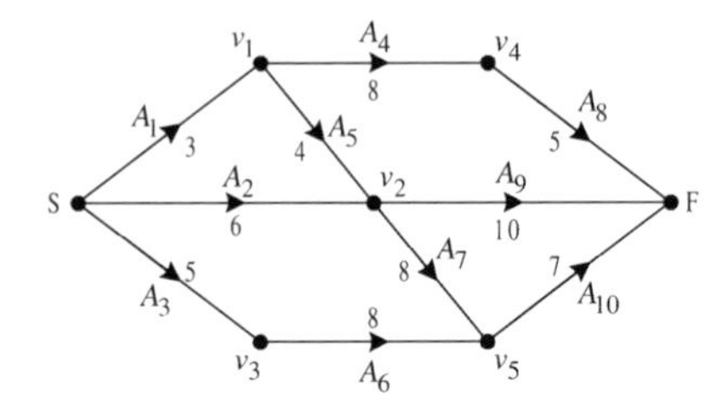

Figure 12.26

Figure 12.26 is called the **scheduling network** or **activity network** for the project. Notice that there is only one vertex (the start S) which has in-degree 0, and only one vertex (the finish F) which has out-degree 0.

Definitions 12.10

(i) Let D be a weighted digraph. A **source** is a vertex with in-degree 0 and a **sink** is a vertex with out-degree 0.

(ii) A **network** is a connected, weighted digraph (with no loops) which has a single source and a single sink.

To avoid repetition we shall assume that all digraphs in this section are connected and contain no loops, although multiple edges are allowed. The restriction that a network should have only one source and one sink is not a substantial one. Suppose D is a weighted digraph which has several sources and/or sinks. We can define a network N by adding two vertices v and w to D, joining v to each source with a directed edge and joining each sink to w with a directed edge. This is illustrated in figure 12.27; N has a single source v and a single sink w. The weights which are appropriate to assign to these additional edges depend on what the network represents. In scheduling problems the additional edges would be given weight 0, so that the completion time for the whole project is not artificially increased.

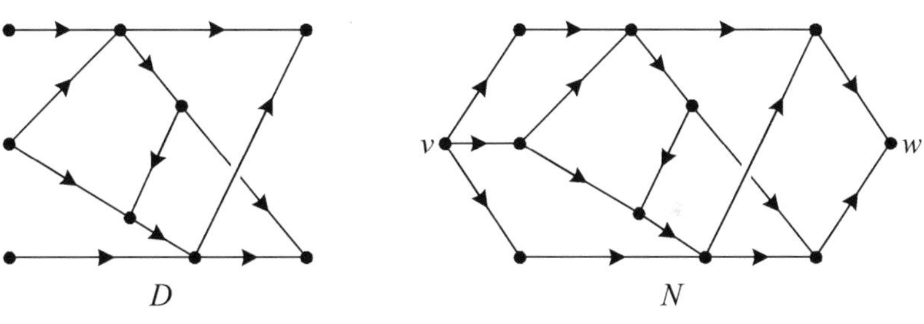

Figure 12.27

Without going into too many generalities, we now consider the scheduling network shown in figure 12.26 and indicate the kind of analysis which can be performed. The first task will be to find the minimum completion time for the project.

To do this we work through the network, labelling each vertex v with the number $E(v)$ which represents the earliest time by which an activity starting at v can commence. Firstly give the source the label $E(\mathrm{S}) = 0$. Next consider the set of those vertices v such that *every* edge ending in v begins at an edge which has already been labelled. For each such vertex, set $E(v)$ equal to the maximum of the values $E(u) + w(e)$, where e is an edge from u to v. Repeat this step until all vertices have been labelled. We need to repeat this step four times in order to label all the vertices in figure 12.26. The vertices labelled at each stage are shown in figure 12.28.

The value $E(\mathrm{F})$, the label on the sink, is the earliest time at which an activity beginning at F could commence. Of course, there are no activities which begin at F, so $E(\mathrm{F})$ is the minimum completion time for the whole project: 22 days in this example.

The next stage in the analysis looks at *how* to schedule the various tasks in order to achieve this minimum completion time. In a similar manner to the assignment

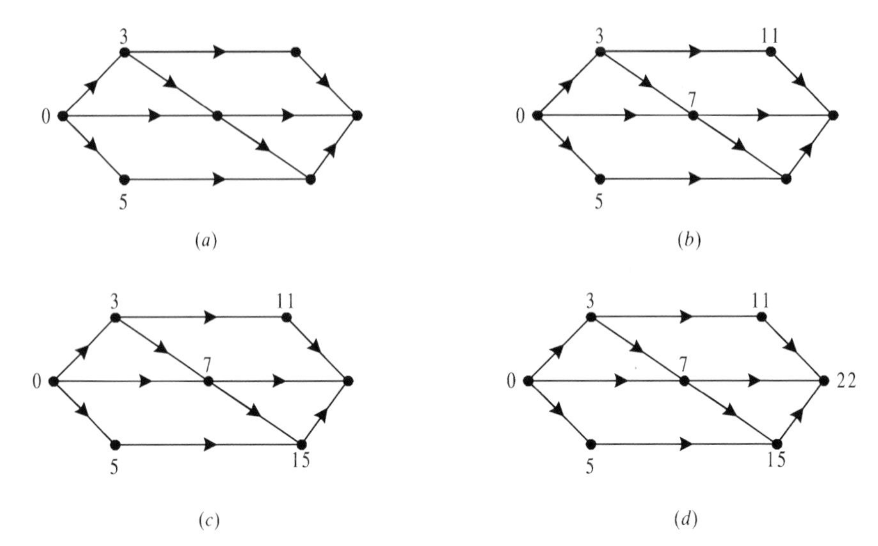

Figure 12.28

of the labels $E(v)$ to the vertices, we next determine for each vertex the latest time at which the activities beginning at the vertex must be started in order to achieve the minimum completion time.

To do this we work 'backwards' through the network from the sink assigning labels $L(v)$ to each vertex in turn. The number $L(v)$ represents the latest time by which the activities beginning at v must be started. Firstly, label the sink with the value $E(\mathrm{F})$, the minimum completion time—22 in our example. Next consider the set of all vertices v such that each edge beginning at v ends at a vertex which is already labelled. Set $L(v)$ equal to the minimum of the values $L(u) - w(e)$ for all the edges e from v to u. Repeat this step until all vertices have been labelled. Again, in our example, four steps are required to label all the vertices: the complete set of labels is shown in figure 12.29(a).

From $E(v)$ and $L(v)$ we now calculate the **float time**

$$\mathit{float}(v) = L(v) - E(v)$$

for each vertex, which represents the maximum possible delay which can occur at the point without increasing the overall completion time. These float times are given in figure 12.29(b). Notice that there is a directed path from source to sink whose weight equals the minimum completion time and which is the longest path from source to sink—see figure 12.29(c). This is called a **critical path** for the network. All activities represented by edges of a critical path must commence without any delay if the minimum completion time is to be achieved. Although it is not entirely obvious, every network will have at least one critical path, and may

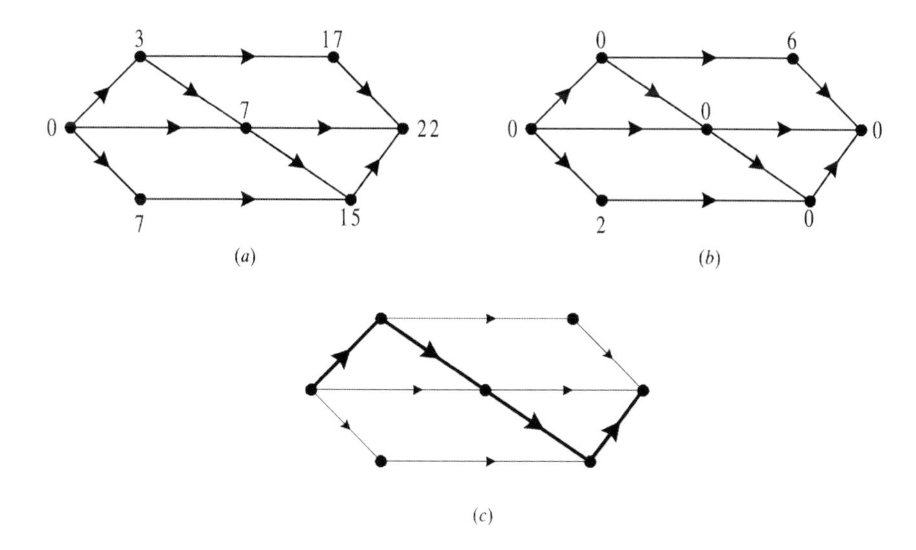

Figure 12.29 (*a*) Latest start times; (*b*) float times; (*c*) critical path.

have several. All of the vertices on any critical path must have zero float time. The converse is not true—there may be paths passing only through vertices with zero float time which are not the longest. However, if there is a path from source to sink which passes through *all* vertices with zero float time, it must be critical.

Flows and Cuts

For our last application we consider networks which represent the 'flow' of a commodity through a series of 'pipelines'. The commodity could actually be a fluid flowing through a network of pipes, but it need not be. The edges just represent parts of a transportation network for the particular commodity and the weights represent their maximum capacities. These could be rail, road or air links for example, or even electrical or fibre-optic cables if the 'commodity' being transported were digital signals.

If the digraph is not a network because there is more than one source and/or sink, we first create a network as illustrated in figure 12.27. In order not to alter the total rate of flow of the commodity through the system, each edge from the new source v to a (previous) source would be given a weight equal to the sum of the weights of the edges beginning at that previous source. Similarly an edge from a (previous) sink to the new sink w would be given a weight equal to the sum of the weights of the edges ending at the previous sink.

The problem is, given a network, to determine the maximum possible flow of the commodity from the source to the sink. To make some headway with this problem we first need some definitions.

Definitions 12.11

Let N be a network.

(i) A **flow in** N is a function $f : E_N \to \mathbb{R}^+ \cup \{0\}$. For each directed edge $e \in E_N$, the non-negative real number $f(e)$ is called the **flow in** e.

(ii) A flow is **conservative** if, for every vertex v except the source and sink, the flow into v equals the flow out from v. More formally,

$$\sum_{\delta(e)=(-,v)} f(e) = \sum_{\delta(e)=(v,-)} f(e)$$

where the summation on the left is over all edges with final vertex v and the summation on the right is over all edges with initial vertex v.

(iii) A flow is **feasible** if, for each edge $e \in E_N$, $f(e) \leqslant w(e)$. In other words, the flow in e is no greater than the weight (capacity) of e.

(iv) The **value** of a flow is the total flow into the sink. If w is the sink, the value of a flow is the sum of the flows in the edges with final vertex w:

$$\sum_{\delta(e)=(-,w)} f(e).$$

We shall henceforth assume that all flows are both conservative and feasible. 'Conservative' means that there is no 'leakage' in the network; apart from the source and the sink, all the material flowing into any vertex must flow out of it. The feasibility of a flow ensures that the capacities (weights) of the edges are not exceeded. The value of a (conservative) flow could equally well be defined as the sum of the flows in the edges with initial vertex equal to the source; whatever flows out of the source must flow into the sink as nothing is lost in the network.

With the notation of definition 12.11, the **maximum flow problem** can be stated as follows.

Given a network, find a (conservative, feasible) flow which has maximum possible value. Such a flow is called **maximal**.

MAXIMUM FLOW PROBLEM

It is easy to see that there may be more than one maximal flow. For example, the maximum value of any flow through the network in figure 12.30(*a*) is 3. There are various ways in which this can be achieved: the flow in both e_1 and e_4 must be 3, but the flows in e_2 and e_3 can be chosen in a variety of ways so that their sum is 3. In other words, the *maximum* value of all possible flows is 3, but there are various different *maximal* flows which achieve this maximum. Two different maximal flows are shown in figures 12.30(*b*) and (*c*). We have adopted a convention in representing flows, which is to show the flow in each edge as a number printed in bold type. (This is to avoid confusing the flow in an edge with the capacity of that edge, which is printed in the usual typeface. When attempting flow problems you are strongly advised to adopt a convention which distinguishes the weight of an edge from the flow in that edge. A convenient way of doing this is to circle the numbers which represent flows in an edge and leave the weights uncircled.)

The maximum value of a flow in a network is closely linked with the idea of a 'cut' of a network, which we now describe. Intuitively, a cut can be thought of as a set of edges which, if 'blocked', would completely stop the flow from source to sink, but if any one edge were unblocked the flow could get through again. The network in figure 12.30 has three different cuts: $\{e_1\}$, $\{e_2, e_3\}$ and $\{e_4\}$.

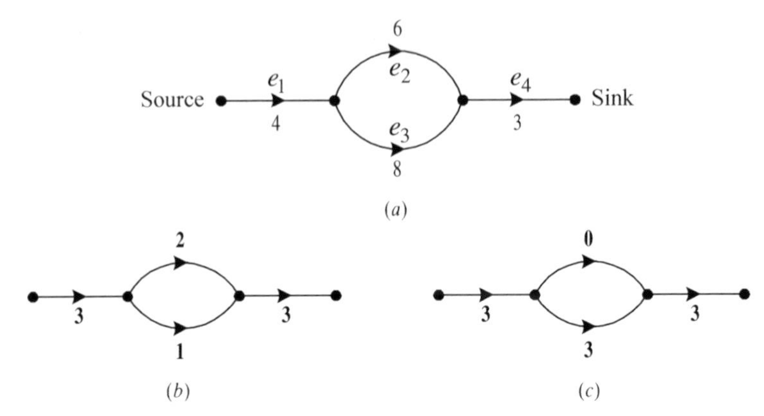

Figure 12.30

Definitions 12.12

Let N be a network.

(i) A **cut of N** is a set of edges which, when removed from N, produces a digraph with two components, one containing the source and the other containing the sink.

(ii) The **capacity** of a cut is the sum of the weights of those of its edges which are directed from the component of N containing the source to the component containing the sink. (Those edges directed from the component containing the sink to that containing the source are ignored when calculating the capacity.)

(iii) A cut is **minimal** if its capacity is less than or equal to the capacity of any other cut.

Figure 12.31 shows a network with source S and sink T. The set of edges $\{AD, DB, BE, EC, CH\}$ is a cut, because removing them from the network separates it into two components, one containing the source and one the sink. Of the edges in the cut, AD, BE, CH are directed from the component containing the source to that containing the sink; the edges DB and EC are directed from the component containing the sink to that containing the source. Therefore the capacity of the cut $\{AD, DB, BE, EC, CH\}$ is $12 + 10 + 9 = 31$.

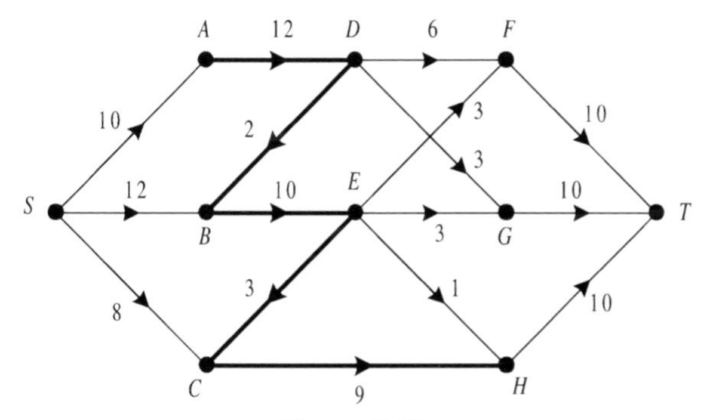

Figure 12.31

The capacities of the three cuts of the network in figure 12.30 are as follows

Cut:	$\{e_1\}$	$\{e_2, e_3\}$	$\{e_4\}$
Capacity:	4	14	3

There is a unique minimal cut in this example —$\{e_4\}$ with capacity 3—although in general there may be several minimal cuts of a given network. Note that the capacity of the minimal cut equals the value of a maximal flow through the network. In fact this is true for any network. This is known as the 'max-flow min-cut theorem'; it was proved in 1955 by L R Ford and D R Fulkerson. As our aim here is to give only a brief outline of the theory of flows, we shall omit the proof.

Theorem 12.1 (The max-flow min-cut theorem)

In a network, the value of any maximal flow is equal to the capacity of any minimal cut.

It should be reasonably clear that the value of any flow cannot be greater than the capacity of any cut. The essential part of the proof of theorem 12.1 involves showing that there is a flow whose value equals the capacity of a minimal cut.

We complete this section by outlining a method for finding a maximal flow. We shall consider the network shown in figure 12.31. The basic idea is to begin with some flow and, if it is not already maximal, improve it. It is relatively straightforward to find *some* flow. The network is shown again in figure 12.32(a). Suppose we begin with the flow shown in figure 12.32(b) which has value 15 and is fairly clearly not maximal.

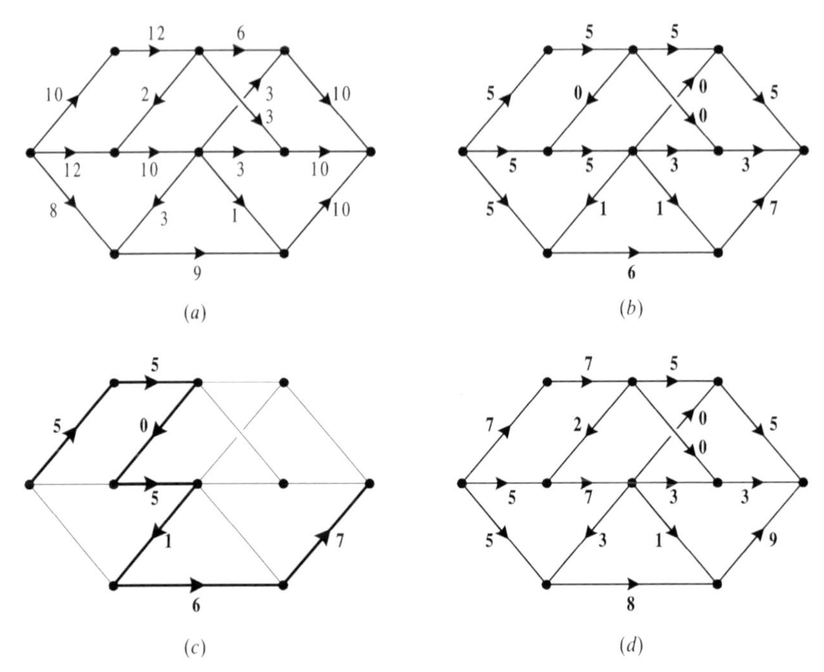

Figure 12.32

To improve the flow, we really mean replace it with a new flow which has greater value. To do this, first look for a directed path from the source to the sink with the property that the flow in every edge of the path is strictly less than its weight. An example of such a path is indicated in figure 12.32(c). (There are other choices of such a path.) For the given path, calculate the minimum value of $w(e) - f(e)$ for its edges. It is possible to increase the flow in every edge of the path by this amount giving a new flow with a larger value than the original. The minimum value of $w(e) - f(e)$ for the edges of the path in figure 12.32(c) is 2. When the flow in each edge of this path is increased by 2, we obtain the flow shown in figure 12.32(d), which has value 17.

The result of several repeats of this process produces the flow, with value 24, shown in figure 12.33(a). For this flow there is no directed path from source to sink with the property that the flow in each edge is strictly less than the capacity of the edge. We may, therefore, be tempted to conclude that this flow is maximal. Unfortunately it is not, but to improve on the flow we need to be slightly more devious.

Consider the path shown in figure 12.33(b). This is not a *directed* path from source to sink; strictly speaking we need to regard it as a path in the underlying undirected graph. The flow in each of the three 'forward' edges is strictly less

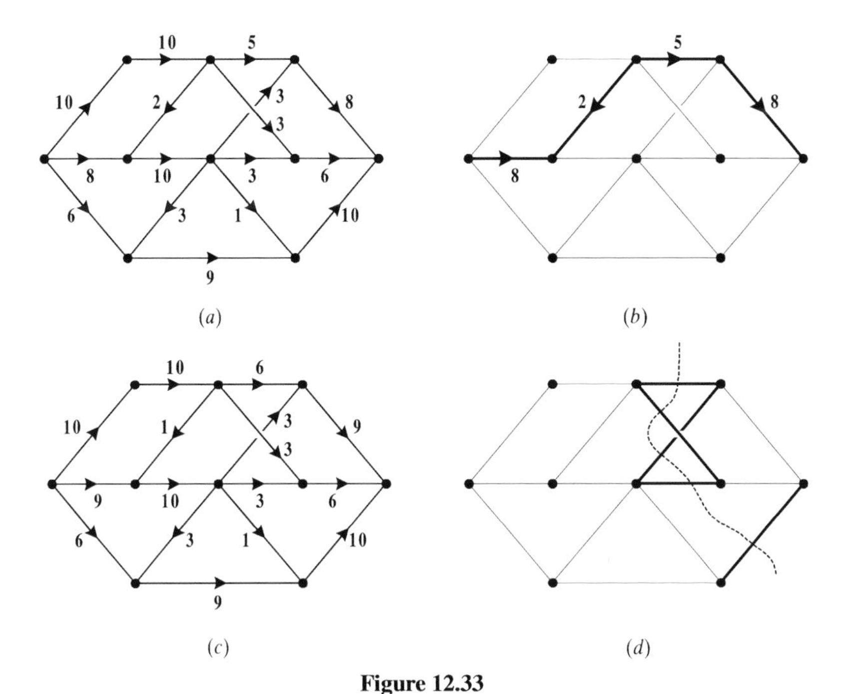

Figure 12.33

than their capacity, and the flow in the 'backward' edge is positive. If we increase the flow in each forward edge by 1 (the minimum value of $w(e) - f(e)$ for its forward edges) and reduce the flow in the backward edge by 1, the resulting flow is still conservative and the value has been increased by 1. The resulting flow, with value 25, is shown in figure 12.33(c).

There are no further (undirected) paths from source to sink of this type where the flow in each forward edge can be increased and not exceed its capacity, and the flow in each backward edge can be reduced and not become negative. This means that the flow shown in figure 12.33(c) is indeed maximal.

In this example it is fairly easy to see that there are no other paths from source to sink which allow the flow to be increased. For larger and more complicated networks, however, it may be difficult to determine with confidence that there are no other such paths. This is where the max-flow min-cut theorem is useful. Since the edges of a cut are those which, if blocked, would completely stop the flow, the value of any flow cannot exceed the capacity of any cut. Therefore the value of a flow can equal the capacity of a cut *only when* the flow is maximal and the cut is minimal.

If we can find a cut whose capacity equals 25, the value of the flow, then we can deduce that our flow is indeed maximal (and, of course, that the cut is minimal). Such a cut is shown in figure 12.33(*d*). The max-flow min-cut theorem then guarantees that the flow in figure 12.33(*c*) is maximal. It is not unique, however. We leave it as an exercise to find another flow which also has value 25.

Exercises 12.6

1. For each of the scheduling networks shown below, determine the minimum completion time, the float times of each vertex and a critical path.

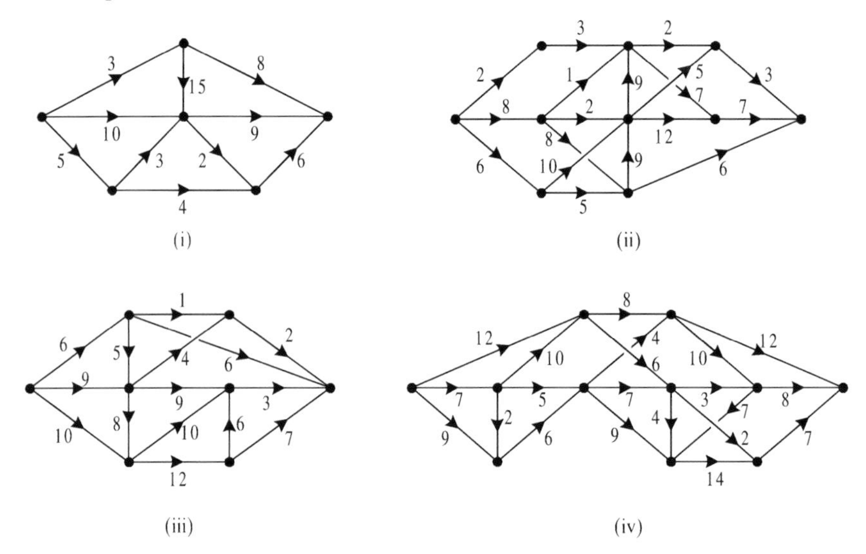

2. For each of the networks below, find a maximal flow, and prove that your flow is maximal by finding a (minimal) cut whose capacity equals the value of your flow.

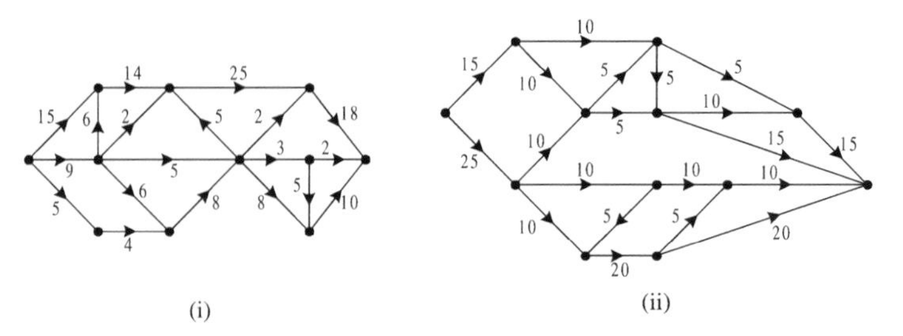

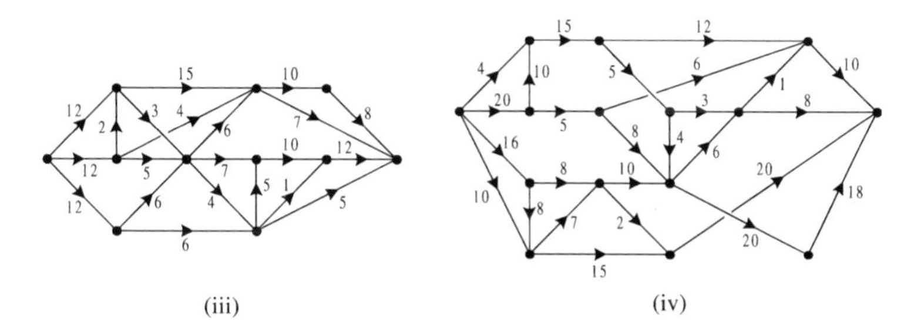

(iii) (iv)

3. For each of the following weighted digraphs, add a unique source and sink as illustrated in figure 12.27.

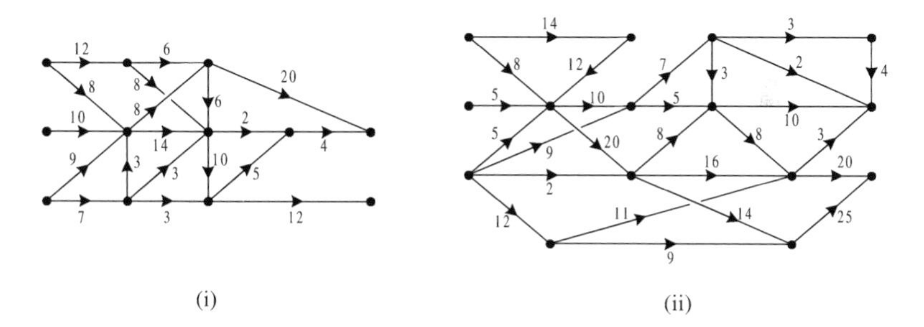

(i) (ii)

(a) If the digraphs represent scheduling problems, give the new edges appropriate weights, and find the minimum completion time and a critical path for the resulting network.

(b) If the digraphs represent flow problems, give the new edges appropriate weights, and find a maximal flow and a minimal cut in the resulting network.

4. Show that, in any network, the set of edges incident with the sink is a cut, and similarly that the set of edges incident with the source is also a cut.

5. Explain why a scheduling network cannot have any *directed* cycle.

6. A certain project requires the completion of 16 activities: $A_1, A_2, \ldots,$ A_{16}. The following table gives the time (in days) required and the prerequisites for each activity.

Activity	Time required	Prerequisite activities
A_1	3	None
A_2	6	None
A_3	5	None
A_4	6	None
A_5	4	A_1, A_2
A_6	8	A_1, A_2
A_7	2	A_3
A_8	5	A_6, A_7
A_9	10	A_5, A_8
A_{10}	7	A_6, A_7
A_{11}	3	A_6, A_7
A_{12}	3	A_3
A_{13}	4	A_4, A_{12}
A_{14}	6	A_4, A_{12}
A_{15}	5	A_{11}, A_{13}
A_{16}	8	A_{14}

(i) Draw a scheduling network for the project.

(ii) Find the minimum completion time and a critical path for the network.

References and Further Reading

The following is a list of books and papers referred to in the text as well as sources of further reading. We have grouped the books into subject areas which relate approximately to the chapters of the book. Those under the heading 'General Discrete Mathematics' cover material found in several of our chapters at a roughly comparable level, but frequently with a different emphasis.

General Discrete Mathematics

Albertson M O and Hutchinson J P 1988 *Discrete Mathematics with Algorithms* (New York: Wiley)

Gersting J L 2006 *Mathematical Structures for Computer Science* 5th edn (New York: Freeman)

Grimaldi R P 2004 *Discrete and Combinatorial Mathematics: an Applied Introduction* 5th edn (Boston: Addison-Wesley)

Grossman P 2002 *Discrete Mathematics for Computing* 2nd edn (New York: Macmillan)

Johnsonbugh R 2004 *Discrete Mathematics* 6th edn (Englewood Cliffs, NJ: Prentice-Hall)

Kolman B, Busby R C and Ross S C 2004 *Discrete Mathematical Structures* 5th edn (Englewood Cliffs, NJ: Prentice-Hall)

Penner R C 1999 *Discrete Mathematics: Proof Techniques and Mathematical Structures* (Singapore: World Scientific)

Piff M 1991 *Discrete Mathematics: an Introduction for Software Engineers* (Cambridge: Cambridge University Press)

Ross K A and Wright C R B 2003 *Discrete Mathematics* 5th edn (Englewood Cliffs, NJ: Prentice-Hall)

Logic and Proof

Ben-Ari M 2001 *Mathematical Logic for Computer Science* 2nd edn (Berlin: Springer)

Franklin J and Daoud M 1996 *Proof in Mathematics: an Introduction* (Quaker Hill Press)

Hamilton A G 1988 *Logic For Mathematicians* revised edn (Cambridge: Cambridge University Press)

Solow D 2004 *How to Read and Do Proofs: an Introduction to Mathematical Thought Processes* 4th edn (New York: Wiley)

Stirling D S G 1997 *Mathematical Analysis and Proof* (Chichester: Horwood Publishing)

Vellerman D J 2006 *How to Prove it: a Structured Approach* 2nd edn (Cambridge: Cambridge University Press)

Wolf R S 1998 *Proof, Logic and Conjecture: The Mathematician's Toolbox* (NY: W H Freeman)

Sets, Relations and Functions

Blyth T S and Robert E F 1984a *Algebra through Practice: Book 1: Sets, Relations and Mappings* (Cambridge: Cambridge University Press)

Ciesielski K 1997 *Set Theory for the Working Mathematician* (Cambridge: Cambridge University Press)

Devlin K J 2003 *Sets, Functions and Logic: an Introduction to Abstract Mathematics* 3rd edn (London: Chapman and Hall)

Fejer P A and Simonvici D A 1991 *Mathematical Foundations of Computer Science: Volume 1: Sets, Relations and Induction* (Berlin: Springer)

Hamilton A G 1982 *Numbers, Sets and Axioms: the Apparatus of Mathematics* (Cambridge: Cambridge University Press)

Nissanke N 1999 *Introductory Logic and Sets for Computer Scientists* (Reading, MA: Addison-Wesley-Longman)

Stoll R R 1979 *Set Theory and Logic* (New York: Dover)

Database Theory

Codd E F 1970 A relation model for large shared data banks *Commun. ACM* **13** 377–87

Connolly T and Begg C 2004 *Database Systems: a Practical Approach to Design, Implementation and Management* 4th edn (Reading, MA: Addison-Wesley)

Date C J 2006 *An Introduction to Database Systems* 8th edn (Reading, MA: Addison-Wesley)
Eaglestone B 1991 *Relational Databases* (Cheltenham: Stanley Thornes)
Elmasri R and Navathe S 2004 *Fundamentals of Database Systems* 4th edn (Reading, MA: Addison-Wesley)
Levene M and Loizou G 1999 *A Guided Tour of Relational Databases and Beyond* (Berlin: Springer)

Matrices and Linear Equations

Anton H 2006 *Elementary Linear Algebra* 9th edn abridged (New York: Wiley)
Blyth T S and Robertson E F 1984b *Algebra through Practice: Book 2: Matrices and Vector Spaces* (Cambridge: Cambridge University Press)
Johnson L W, Dean Reiss R and Arnold J T 2001 *Introduction to Linear Algebra* 5th edn (Reading, MA: Addison-Wesley)
Kaye R and Wilson R 1998 *Linear Algebra* (Oxford: Oxford University Press)
Penney R C 2008 *Linear Algebra: Ideas and Applications* 3rd edn (New York: Wiley)

Algebraic Structures

Asche D 1989 *Introduction to Groups* (Bristol: Institute of Physics Publishing)
Deskins W E 1996 *Abstract Algebra* (New York: Dover)
Foldes S 1994 *Fundamental Structures of Algebra and Discrete Mathematics* (New York: Wiley)
Fraleigh J B 2002 *A First Course in Abstract Algebra* 7th edn (Reading, MA: Addison-Wesley)
Gallian J A 2002 *Contemporary Abstract Algebra* 5th edn (Boston, MA: Houghton Mifflin)
Green J A 1988 *Sets and Groups: a First Course in Algebra* (Dordrecht: Kluwer Academic)
Ledermann W and Weir A J 1996 *Introduction to Group Theory* 2nd edn (Reading, MA: Addison-Wesley)
Rotman J J 1995 *An Introduction to the Theory of Groups* 4th edn (Berlin: Springer)

Coding Theory

Berlekamp E R 1984 *Algebraic Coding Theory* revised edn (Laguna Hills, CA: Aegean Park Press)

Kabatiansky G, Krouk E and Semenov S 2005 *Error Correcting Coding and Security for Data Networks: Analysis of the Superchannel Concept* (Chichester: Wiley)

Lin S and Costello D J Jr 2004 *Error Control Coding : Fundamentals and Applications* 2nd edn (Englewood Cliffs NJ: Prentice Hall)

MacWilliams F J 1993 *The Theory of Error Correcting Codes* (Amsterdam: North-Holland)

Pless V 1998 *Introduction to the Theory of Error Correcting Codes* 3rd edn (New York: Wiley)

Sweeney P 2002 *Error Control Coding: an Introduction* (Englewood Cliffs, NJ: Prentice-Hall)

Number Theory

Burn R P 1997 *A Pathway into Number Theory* 2nd edn (Cambridge: Cambridge University Press)

Berton D M 2007 *Elementary Number Theory* 6th edn (McGraw Hill)

Jones G A and Jones J M 1998 *Elementary Number Theory* (Berlin: Springer)

Shapiro H N 2008 *Introduction to the Theory of Numbers* (New York: Dover)

Rosen K H 1992 *Elementary Number Theory and its Applications* 3rd edn (Reading, MA : Addison-Wesley)

Boolean Algebra, Logic and Switching Circuits

Gregg J 1998 *Ones and Zeros: Understanding Boolean Algebra, Digital Circuits, and the Logic of Sets* (New York: IEEE)

Maxfield C and Waddell P 2003 *Bebop to the Boolean Boogie* 2nd edn (Burlington MA: Newnes)

Rafiquzzaman M 2005 *Fundamentals of Digital Logic and Microcomputer Design* 5th edn (Hoboken NJ: Wiley)

Whitesitt J E 1995 *Boolean Algebra and its Applications* (New York: Dover)

Graph Theory and Applications

Appel H and Haken W 1986 The four-colour proof suffices *Math. Intell.* **8** 10–20
Biggs N L 2002 *Discrete Mathematics* 2nd edn (Oxford: Clarendon)
Biggs N L, Lloyd E K and Wilson R J 1986 *Graph Theory 1736–1936* revised edn (Oxford: Clarendon)
Gould R 1988 *Graph Theory* (Menlo Park, CA: Cummings)
Harary F 1969 *Graph Theory* (Reading, MA: Addison-Wesley)
Lawler E L, Lenstra J K, Rinnooy Kan A H G and Shmoy D B (eds) 1985 *The Travelling Salesman Problem. A Guided Tour of Combinatorial Optimisation* (New York: Wiley)
Saaty T L and Kainen P G 1986 *The Four Colour Problem: Assaults and Conquest* (New York: Dover)
Trudeau R J 1993 *Introduction to Graph Theory* (New York: Dover)
Wilson R J and Watkins J J 1990 *Graphs: an Introductory Approach—a First Course in Discrete Mathematics* (New York: Wiley)
Wilson R J 1996 *Introduction to Graph Theory* 4th edn (Reading, MA: Addison-Wesley-Longman)

Miscellaneous

Hallett M 1984 *Cantorian Set Theory and the Limitation of Size* (Oxford: Oxford University Press)
Hardy G H 1992 *A Mathematician's Apology* New edition (Cambridge: Cambridge University Press)
Lakatos I 1976 *Proof and Refutations: the Logic Of Mathematical Discovery* (Cambridge: Cambridge University Press)

Hints and Solutions to Selected Exercises

Chapter 1

Exercises 1.1

1. (i) Max is not sulking and today is my birthday.
   (iii) If Max is not sulking then today is my birthday.

2. (ii) If and only if Jo shouts and Sally cries, then Mary laughs.
   (iv) Mary laughs or Sally doesn't cry or Jo doesn't shout.

3. (i) $p \to \bar{q}$.
   (iii) $\bar{r} \wedge (p \to \bar{q})$.

5.

		(i)	(ii)	(iv)	(vi)
p	q	$\bar{p} \to q$	$\bar{q} \wedge p$	$(p \to q) \veebar \bar{q}$	$(\bar{p} \wedge q) \veebar (p \vee \bar{q})$
T	T	T	F	T	T
T	F	T	T	T	T
F	T	T	F	T	T
F	F	F	F	F	T

7.

			(ii)	(iv)
p	q	r	$(p \veebar r) \wedge \bar{q}$	$(p \rightarrow (\bar{q} \vee \bar{r})$
T	T	T	F	F
T	T	F	F	T
T	F	T	F	T
T	F	F	T	T
F	T	T	F	T
F	T	F	F	T
F	F	T	T	T
F	F	F	F	T

Exercises 1.2

1. Tautology
2. Neither
3. Tautology
4. Tautology
5. Contradiction
6. Neither
7. Contradiction
8. Tautology
9. Tautology
10. Neither.

Exercises 1.3

2.

p	q	$p \wedge q$	$\overline{p \rightarrow \bar{q}}$
T	T	T	T
T	F	F	F
F	T	F	F
F	F	F	F

Since the last two columns of the truth table are identical, we can conclude that $(p \wedge q) \equiv (\overline{p \rightarrow \bar{q}})$.

5. The truth table for $\bar{q} \rightarrow \bar{p}$ and for $p \rightarrow q$ is as follows.

p	q	$\bar{q} \rightarrow \bar{p}$	$p \rightarrow q$
T	T	T	T
T	F	F	F
F	T	T	T
F	F	T	T

Whenever $\bar{q} \to \bar{p}$ is true (rows 1, 3 and 4), $p \to q$ is true so that $(\bar{q} \to \bar{p}) \vdash (p \to q)$. (In fact $(\bar{q} \to \bar{p}) \equiv (p \to q)$.)

6. (iv) The truth table is as follows.

p	q	r	$(p \to q) \wedge (p \vee r)$	$q \vee r$
T	T	T	T	T
T	T	F	T	T
T	F	T	F	T
T	F	F	F	F
F	T	T	T	T
F	T	F	F	T
F	F	T	T	T
F	F	F	F	F

Whenever $(p \to q) \wedge (p \vee r)$ is true (rows 1, 2, 5 and 7), $q \vee r$ is true. Hence $(p \to q) \wedge (p \vee r) \vdash (q \vee r)$.

(vi)

p	q	$\bar{q}$	$p \vee q$	$(p \vee q) \wedge \bar{q}$
T	T	F	T	F
T	F	T	T	T
F	T	F	T	F
F	F	T	F	F

In each of the cases where $(p \vee q) \wedge \bar{q}$ is true, p is also true. Hence $[(p \vee q) \wedge \bar{q}] \vdash p$.

Exercises 1.4

1. (i) $(p \wedge p) \vee (\bar{p} \vee \bar{p}) \equiv p \vee \bar{p}$ (Idem)

$\equiv t$ (Comp).

(iii) $p \to q \equiv \bar{p} \vee q$ (Imp)

$\equiv \bar{p} \vee \bar{\bar{q}}$ (Inv)

$\equiv \overline{p \wedge \bar{q}}$ (De M).

(v) This can be proved using (in this order): De Morgan's law, an involution law, a commutative law and, finally, an idempotent law.

Exercises 1.5

1. Define the following:

p : You gamble.
q : You're stupid.

The premises are then $p \to q, \bar{q}$ and the conclusion is $\bar{p}$. A truth table shows that the compound proposition $[(p \to q) \wedge \bar{q}] \to \bar{p}$ is a tautology. Hence the argument is valid.

2. Define the following:

p : I leave college.
q : I get a job in a bank.

The compound proposition $[(p \to q) \wedge \bar{p}] \to \bar{q}$ is not a tautology and therefore the argument is not valid.

3. Valid (regardless of whether 'either ... or ...' is interpreted as an inclusive or exclusive disjunction).

4. Not valid.

5. Valid.

6. Not valid.

7. Valid.

8. Valid.

9. Valid.

10. Not valid.

Exercises 1.6

1. (i)

1.	$(p \wedge q) \to (r \wedge s)$	(premise)
2.	p	(premise)
3.	q	(premise)
4.	$p \wedge q$	(3, 4 Conj)
5.	$r \wedge s$	(1, 4. MP)
6.	$s \wedge r$	(5. Comm)
7.	s	(6. Simp)

(iii) A formal proof can be obtained using (in this order): simplification, modus ponens and disjunctive syllogism.

(v) A formal proof can be obtained using (in this order): a commutative law, simplification twice and modus ponens.

(vii)

1.	$(p \vee q) \wedge (q \vee r)$	(premise)
2.	$\bar{q}$	(premise)
3.	$p \vee q$	(1. Simp)
4.	$q \vee p$	(3. Comm)
5.	p	(2, 4. DS)
6.	$(q \vee r) \wedge (p \vee q)$	(1. Comm)
7.	$q \vee r$	(6. Simp)
8.	r	(2, 7. DS)
9.	$p \wedge r$	(5, 8 Conj)

2. (i) Symbolise the premises as follows.

p: The murder was committed by A.
q: The murder was committed by B.
r: The murder was committed by C.

A formal proof of the validity of the argument is then given as follows.

1.	$p \vee (q \wedge r)$	(premise)
2.	$(p \vee q) \wedge (p \vee r)$	(1. Dist)
3.	$(p \vee r) \wedge (p \vee q)$	(2. Comm)
4.	$p \vee r$	(3. Simp)

(iii) A formal proof of validity uses the involution law, the material implication law and finally hypothetical syllogism.

(v) We symbolise the premises:

p: People are happy.
q: People are truthful.

A formal proof of the validity of the argument is then given as follows.

1.	$p \leftrightarrow q$	(premise)
2.	$\overline{p \wedge q}$	(premise)
3.	$(p \wedge q) \vee (\bar{p} \wedge \bar{q})$	(1. Equiv)
4.	$(\bar{p} \wedge \bar{q}) \vee (p \wedge q)$	(3. Comm)
5.	$\bar{p} \wedge \bar{q}$	(4. Simp)

Exercises 1.7

1. (ii) $B(m) \rightarrow C(s)$
   (iv) $\forall x[C(x) \rightarrow F(x)]$
   (v) $\forall x[\{C(x) \wedge \neg B(x)\} \rightarrow F(x)]$.

2. (i) The proposition is symbolized by $\forall x[B(x) \rightarrow C(x)]$ where the predicates are $B(x)$: x is a baby and $C(x)$: x cries a lot. If we define the universe of discourse to be 'babies', the proposition may be shortened to $\forall x C(x)$.

   (iii) We define $S(x)$: x is a student and $G(x)$: x can write a good essay. The proposition is then

   $$\exists x[S(x) \wedge \neg G(x)].$$

   Alternatively, if we define the universe of discourse to be 'students', the proposition is

   $$\exists x[\neg G(x)].$$

   (v) Universe of discourse: people.
   $U(x)$: x has had a university education.
   $P(x)$: x lives in poverty.

   $$\exists x[U(x) \wedge P(x)].$$

   (vii) Universe of discourse: people.
   $F(x)$: x is my friend.
   $N(x)$: x believes in nuclear disarmament.

   $$\forall x[F(x) \rightarrow N(x)].$$

   (ix) Universe of discourse: people in the building.
   $F(x)$: x set off the fire alarm.
   $B(x)$: x left the building.

   $$\exists x F(x) \wedge \forall x B(x).$$

3. (i) Universe of discourse for each variable: people.
   $L(x, y)$: x loves y.
   $$\forall x \exists y L(x, y).$$

   (ii) Universe of discourse and predicate as in (i).
   $$\exists x \forall y L(x, y).$$

4. (i) Somebody doesn't like strawberry jam.
   (ii) All birds can fly.

5. (i) True (iii) True (iv) False
   (v) True (vii) False

Exercises 1.8

1. This argument has the same structure as example 1.15.2.

3. This argument has the same structure as example 1.15.1.

4. Universe of discourse: people.
$G(x)$: x is a gambler.
$R(x)$: x is bound for ruin.
$H(x)$: x is happy.

Premises: $\forall x[G(x) \to R(x)]$
$\forall x[R(x) \to \neg H(x)]$.
Conclusion: $\forall x[G(x) \to \neg H(x)]$.

A summary of the argument is as follows. The propositions $G(a) \to R(a)$ and $R(a) \to \neg H(a)$ both follow by universal specification from the premises and so are true for every a in the universe of discourse. The fact that $(p \to q) \wedge (q \to r)$ logically implies $p \to r$ allows us to deduce $G(a) \to \neg H(a)$ for every a in the universe of discourse. Universal generalization leads to the conclusion.

7. Universe of discourse: alligators

Define: $F(x)$: x is friendly
$S(x)$: x is sociable
$Z(x)$: x lives in the zoo.

Premises: $\exists x[F(x) \wedge S(x)]$
$\forall x[F(x) \to Z(x)]$.
Conclusion: $\exists x[Z(x) \wedge S(x)]$.

1. $\exists x[F(x) \wedge S(x)]$ (premise)
2. $\forall x[F(x) \to Z(x)]$ (premise)
3. $F(a) \wedge S(a)$ (existential specification)
4. $F(a) \to Z(a)$ (universal specification)
5. $F(a)$ (from 3 using simplification)
6. $Z(a)$ (from 5 and 4 using modus ponens)
7. $F(a) \wedge Z(a)$ (from 5 and 6)
8. $\exists x[F(x) \wedge Z(x)]$ (existential generalization)

9. Define the following on the universe of animals:

$S(x)$: x has scales

$D(x)$: x is a dragon
$C(x)$: x has sharp claws.

1.	$\forall x[S(x) \rightarrow D(x)]$	(premise)
2.	$\exists x[\neg D(x) \wedge C(x)]$	(premise)
3.	$\neg D(a) \wedge C(a)$	(existential specification)
4.	$S(a) \rightarrow D(a)$	(universal specification)
5.	$\neg D(a)$	(from 3 using simplification)
6.	$\neg S(a)$	(from 5 and 4, modus tollens)
7.	$C(a)$	(from 3, simplification)
8.	$\neg S(a) \wedge C(a)$	(from 6 and 7)
9.	$\exists x[\neg S(x) \wedge C(x)]$	(existential generalization)

Chapter 2

Exercises 2.2

1. Suppose x and y are consecutive integers with $x < y$. Then

$$\begin{aligned} & y = x + 1 \\ \Rightarrow \quad & x + y = x + x + 1 \\ & = 2x + 1 \\ \Rightarrow \quad & x + y \text{ is odd.} \end{aligned}$$

2. To prove that, if n^2 is odd, then n is odd, we prove the contrapositive, i.e. if n is even, then n^2 is even. This is proved in example 2.2.1.

To prove the converse, suppose that n is odd.

Then

$$\begin{aligned} & n = 2m + 1 \qquad \text{where } m \text{ is an integer} \\ \Rightarrow \quad & n^2 = (2m+1)^2 \\ & = 4m^2 + 4m + 1 \\ & = 2(2m^2 + 2m) + 1 \\ \Rightarrow \quad & n^2 \text{ is odd.} \end{aligned}$$

3. The proof that the product of two consecutive integers is even rests on the fact that, if m is an integer, either m or $m + 1$ is even.

To prove the second result, suppose that the roots of $x^2 + ax + b = 0$ are m and $m + 1$ for some integer m. Then the equation can be written as

$$
\begin{aligned}
& & (x-m)(x-m-1) &= 0 \\
&\Rightarrow & x^2 - (2m+1)x + m^2 + m &= 0 \\
&\Rightarrow & a &= -(2m+1) \\
&\Rightarrow & a &\text{ is odd} \\
&\text{and} & b &= m^2 + m \\
& & &= m(m+1) \\
&\Rightarrow & b &\text{ is even by the first result.}
\end{aligned}
$$

5. m is a factor of $n \Rightarrow m = k_1 n$ where k_1 is a positive integer.
n is a factor of $m \Rightarrow n = k_2 m$ where k_2 is a positive integer.

Therefore

$$
\begin{aligned}
& & n &= k_2 m \\
& & &= k_1 k_2 n \\
&\Rightarrow & k_1 k_2 &= 1 & &\text{(since } n \neq 0\text{)} \\
&\Rightarrow & k_1 = k_2 &= 1 & &\text{(since } k_1 \text{ and } k_2 \text{ are positive integers)} \\
&\Rightarrow & m &= n.
\end{aligned}
$$

6. The contrapositive is 'If n is divisible by 5 then n^2 is divisible by 5'. If n is divisible by 5 then

$$
\begin{aligned}
& & n &= 5k & &\text{where } k \text{ is an integer} \\
&\Rightarrow & n^2 &= 25k^2 \\
& & &= 5(5k^2) & &\text{where } 5k^2 \text{ is an integer} \\
&\Rightarrow & n^2 &\text{ is divisible by 5.}
\end{aligned}
$$

9. If $n - 2$ is divisible by 4 then

$$
\begin{aligned}
& & n - 2 &= 4k & &\text{where } k \text{ is an integer} \\
&\Rightarrow & n + 2 &= 4k + 4 \\
& & &= 4(k+1).
\end{aligned}
$$

Then

$$n^2 - 4 = (n-2)(n+2)$$

$$= 4k \times 4(k+1)$$
$$= 16k(k+1)$$
$$\Rightarrow \qquad n^2 - 4 \text{ is divisible by 16.}$$

10. Assume that an integer n has a smallest factor greater than 1, which is *not* prime, and show that this leads to a contradiction.

Exercises 2.3

2. If $n = 1$, $2^n = 2 > 1$, so that the proposition holds for $n = 1$.

   Assume the proposition holds for $n = k \geqslant 1$, i.e. $2^k > k$.

   Then

$$\begin{aligned} 2^{k+1} &= 2 \times 2^k \\ &> 2k \qquad \text{(by the induction hypothesis)} \\ &= k + k \\ &\geqslant k + 1 \qquad \text{(since } k \geqslant 1\text{)} \end{aligned}$$

   so the proposition holds for $n = k + 1$.

   Hence, by mathematical induction, the proposition holds for all positive integers n.

4. If $n = 0$,

$$\frac{x^{n+1} - 1}{x - 1} = \frac{x - 1}{x - 1} = 1$$

   so that the proposition holds for $n = 0$.

   Suppose that the proposition holds for $n = k \geqslant 0$, i.e.

$$1 + x + \cdots + x^k = \frac{x^{k+1} - 1}{x - 1}.$$

   Then

$$\begin{aligned} 1 + x + \cdots + x^k + x^{k+1} &= \frac{x^{k+1} - 1}{x - 1} + x^{k+1} \\ &= \frac{x^{k+1} - 1 + x^{k+1}(x - 1)}{x - 1} \end{aligned}$$

$$= \frac{x^{k+1} - 1 + x^{k+2} - x^{k+1}}{x - 1}$$
$$= \frac{x^{k+2} - 1}{x - 1}$$

so that the proposition holds for $n = k + 1$.

By mathematical induction, the proposition holds for all integers $n \geqslant 0$.

5. If $n = 1$,

$$\frac{n(n+1)(2n+1)}{6} = \frac{1 \times 2 \times 3}{6} = 1 = 1^2$$

so that the proposition holds for $n = 1$.

Suppose that the proposition holds for $n = k \geqslant 1$, i.e.

$$1^2 + 2^2 + \cdots + k^2 = \frac{k(k+1)(2k+1)}{6}.$$

Then

$$\begin{aligned} 1^2 + 2^2 + \cdots + k^2 + (k+1)^2 &= \frac{k(k+1)(2k+1)}{6} + (k+1)^2 \\ &= \frac{k(k+1)(2k+1) + 6(k+1)^2}{6} \\ &= \frac{(k+1)(k+2)(2k+3)}{6} \\ &= \frac{(k+1)(k+2)(2[k+1]+1)}{6} \end{aligned}$$

so that the proposition holds for $n = k + 1$.

The result follows by mathematical induction.

7. If $n = 1$, $A_1 = 3 \times 1$ so the proposition holds for $n = 1$.

Assume that

$$A_k = 3k \text{ for } k \geqslant 1.$$

Then

$$\begin{aligned} A_{k+1} &= A_k + 3 \\ &= 3k + 3 \\ &= 3(k+1) \end{aligned}$$

so that, if the proposition holds for $n = k$, then it also holds for $n = k+1$.

The result follows by mathematical induction.

10. The proof follows the same lines as example 2.7.2.

11. The result clearly holds for $n = 1$ and $n = 2$.

Assume that it holds for all integers $r \leqslant k$, i.e. $A_r = 5 \times 2^{r-1} + 1$.

For $k \geqslant 2$

$$\begin{aligned} A_{k+1} &= 3A_k - 2A_{k-1} \\ &= 3(5 \times 2^{k-1} + 1) - 2(5 \times 2^{k-2} + 1) \\ &= \tfrac{15}{2} \times 2^k + 3 - \tfrac{5}{2} \times 2^k - 2 \\ &= 5 \times 2^k + 1. \end{aligned}$$

This completes the inductive step and the result follows.

Chapter 3

Exercises 3.1

1. (ii) $\{3, 6, 9, 12, \ldots\}$
   (iv) $\{1/3, -2\}$
   (vi) $\{-2\}$
   (viii) $\{1/2, 1, 3/2, 2, 5/2, 3, \ldots\}$.

2. (ii) $\{0, 1, 2, 3, 4\}$
   (iv) $\{-2, -1, 0, 1, 2\}$
   (vi) $\{-1, 0, 1\}$.

3. (ii) ∞ (iv) 4 (vi) 1 (viii) 3.

4. (ii) $\{x : x$ is an integer multiple of 3 and $3 \leqslant x \leqslant 30\}$
   (iv) $\{x : x$ is a prime number$\}$
   (vi) $\{x : x = n^2 + m^2$ for some integers n and $m\}$
   (viii) $\{x : x = 13n$ for some integer $n\}$
   (x) $\{x : x$ is a play by William Shakespeare$\}$.

Exercises 3.2

1. (i) True (ii) False (iii) False (iv) True (v) True (vi) True (vii) False (viii) True.

3. (i) $x \subseteq A$ (ii) $x \in A$ (iii) $x \subseteq A$
(iv) Both (v) Neither (vi) Neither.

4. (i) $\{\{1,2\},\{1,3\},\{1,4\},\{2,3\},\{2,4\},\{3,4\}\}$
(iv) $\{\{1\},\{1,2\},\{1,3\},\{1,4\},\{1,2,3\},\{1,2,4\},\{1,3,4\},\{1,2,3,4\}\}$.

5. (ii) Both (i.e. $A = B$)
(iv) $B \subseteq A$
(vi) Neither.

7. (i) If $b \in B$ then $b \in A \wedge P(b)$ is true; in particular $b \in A$. Hence $B \subseteq A$, as required.

   If $B \subset A$ then there exists an $a \in A$ such that $P(a)$ is false.

   If $A = B$ then for all $a \in A$, $P(a)$ is true.

8. Suppose that $A \subseteq B$ and $C = \{x : x \in A \wedge x \in B\}$.

   If $x \in A$ then $x \in B$ (since $A \subseteq B$). Therefore $x \in A \wedge x \in B$ is true which means that $x \in C$. Therefore $A \subseteq C$.

   Conversely suppose $x \in C$. Then $x \in A \wedge x \in B$ is true, by definition, so $x \in A$. Therefore $C \subseteq A$.

   Since $A \subseteq C$ and $C \subseteq A$ we conclude that $A = C$.

10. (i) If $A \subseteq B$ and $B \subseteq C$ then $x \in A \Rightarrow x \in B \Rightarrow x \in C$, so $A \subseteq C$.

11. (i) 6 (see exercise 3.2.4(i))
(ii) 8 (see exercise 3.2.4(iv))
(iii) 4
(iv) 3.

12. $A = \{1\}$. Since $\{1\} \notin A$, $A \in R$.

$$B = \{\varnothing, \{\varnothing\}, \{\varnothing, \{\varnothing\}\}, \{\varnothing, \{\varnothing\}, \{\varnothing, \{\varnothing\}\}\},$$
$$\{\varnothing, \{\varnothing\}, \{\varnothing, \{\varnothing\}\}, \{\varnothing, \{\varnothing\}, \{\varnothing, \{\varnothing\}\}\}\}, \ldots\}.$$

$B \notin R$.

If $R \in R$ then by definition $R \notin R$; conversely, if $R \notin R$ then again by definition $R \in R$. R is not a set because we cannot specify either $R \in R$ or $R \notin R$ without obtaining a contradiction.

Exercises 3.3

2. (ii) $\{2, 3, 4, 5, 6, 7, 8\}$
   (iv) $\varnothing$
   (vii) $\mathscr{U} = \{0, 1, 2, \ldots, 8, 9\}$
   (viii) $\varnothing$
   (x) $\{6, 8\}$.

3. (ii) $X \cap Y = \varnothing$
   (iv) $X \subseteq Y$
   (vi) $X \subseteq Y$
   (viii) $X \cap Y = \varnothing$
   (x) $Y \subseteq X$.

4. (i) $\{2, 3, 4, 5, 6, 8, 10\}$
   (iii) $\{1, 2\}$
   (v) $\{3, 4, 5, 6\}$
   (vii) $\{1, 2\}$
   (ix) $\{1, 2, 3, 5, 8, 10\}$
   (xi) $\{2, 3, 5, 8, 10\}$.

5. (i) (a) $\{x : x \text{ is a prime divisor of } 12\}$
   (c) $\{x : x \text{ is an even prime number}\}$, i.e. the singleton set $\{2\}$.

   (ii) (a) $\{1, 2, 3, 4, 5, 6, 7, 11, 12\}$
   (c) $\{8, 10\}$
   (e) $\{1, 4, 5, 6, 7, 8, 9, 10, 11, 12\}$.

6. (ii) $\{x : \neg[P(x) \vee Q(x)]\}$
   (iv) $\{x : \neg P(x) \vee Q(x)\}$
   (vi) $\{x : \neg P(x) \wedge Q(x)\}$.

7. (ii)
$$\begin{aligned} x \in [A \cap (B - C)] &\Leftrightarrow x \in A \text{ and } x \in B - C \\ &\Leftrightarrow x \in A \text{ and } x \in B \text{ and } x \notin C \\ &\Leftrightarrow x \in A \cap B \text{ and } x \notin C \\ &\Leftrightarrow x \in (A \cap B) - C. \end{aligned}$$

(v) $$\begin{aligned} x \in [(A - B) - C] &\Leftrightarrow x \in A - B \text{ and } x \notin C \\ &\Leftrightarrow x \in A \text{ and } x \notin B \text{ and } x \notin C \\ &\Leftrightarrow x \in A \text{ and } x \notin \overline{B \cup C} \\ &\Leftrightarrow x \in A - (B \cup C). \end{aligned}$$

Exercises 3.4

2. $\overline{A - B} = B \cup \bar{A}$, $A - (A \cap B) = A \cap \bar{B}$, $(A - B) \cup (B - A) = (A \cup B) - (A \cap B)$.

3. $(A \cap B) \cup (A \cap C) = A \cap (B \cup C)$, $(A - B) \cap C = (A \cap C) - B$.

4. (ii) $$\begin{aligned} A \cap (B - C) &= A \cap (B \cap \bar{C}) && \text{(definition of difference)} \\ &= (A \cap B) \cap \bar{C} && \text{(associativity of } \cap\text{)} \\ &= (A \cap B) - C && \text{(definition of difference).} \end{aligned}$$

(v) $$\begin{aligned} (A - B) - C &= (A - B) \cap \bar{C} && \text{(definition of difference)} \\ &= (A \cap \bar{B}) \cap \bar{C} && \text{(definition of difference)} \\ &= A \cap (\bar{B} \cap \bar{C}) && \text{(associativity of } \cap\text{)} \\ &= A \cap (\overline{B \cup C}) && \text{(De Morgan's law)} \\ &= A - (B \cup C) && \text{(definition of difference).} \end{aligned}$$

5. (i) $$\begin{aligned} A * \varnothing &= (A - \varnothing) \cup (\varnothing - A) \\ &= A \cup \varnothing \\ &= A \end{aligned}$$
$$\begin{aligned} A * A &= (A - A) \cup (A - A) \\ &= \varnothing \cup \varnothing \\ &= \varnothing. \end{aligned}$$

(iii) There are any number of possible examples of sets with the required properties. For example, if $A = \{1, 2, 3\}$, $B = \{2, 3, 4\}$, $C = \{1, 3, 5\}$ then $A \cup (B * C) = \{1, 2, 3, 4, 5\}$ but $(A \cup B) * (A \cup C) = \{4, 5\}$.

6. (i) $A = \{1\}$, $B = \{1, \{1\}\}$
(ii) $A = \varnothing$, $B = \{\varnothing\}$, $C = \{\varnothing, \{\varnothing\}\}$.

7. (i) $\bar{A} \cup \bar{B} = (\overline{A \cap B})$
(ii) $A \cup B = \mathscr{U}$.

The statement $A \cap B = \varnothing$ is not true for *all* sets A and B so the duality principle does not apply to the statement.

9. (i) 35 (ii) 56 (iii) 7.

   If $|\mathscr{U}| = 150$ then $|\overline{(A \cup B \cup C)}| = 37$.

10. (i) 50 (ii) 165 (iii) 145 (iv) 95.

11. (i) 4 (ii) 35 (iii) 28.

Exercises 3.5

1. (ii) $\{\varnothing, \{\{1\}\}, \{\{1,2\}\}, A\}$

   (iv) $\{\varnothing, \{\varnothing\}, \{\{1\}\}, \{\{2\}\}, \{\{1,2\}\}, \{\varnothing, \{1\}\}, \{\varnothing, \{2\}\},$
   $\{\varnothing, \{1,2\}\}, \{\{1\}, \{2\}\}, \{\{1\}, \{1,2\}\}, \{\{2\}, \{1,2\}\},$
   $\{\varnothing, \{1\}, \{2\}\}, \{\varnothing, \{1\}, \{1,2\}\}, \{\varnothing, \{2\}, \{1,2\}\},$
   $\{\{1\}, \{2\}, \{1,2\}\}, \{\varnothing, \{1\}, \{2\}, \{1,2\}\}\}.$

2. (i) Not a partition since 1 and 2 are not subsets of A.

   (iii) Not a partition since 6 does not belong to any of the subsets (so the first condition of definition 3.4 fails).

   (v) Not a partition since $8 \in \{2,8,10\} \cap \{7,8,9\}$ (so the second condition of definition 3.4 fails).

3. (ii), (iv) and (vi) only are partitions.

4. (i) 15 (ii) None.

5. No; neither condition is satisfied.

6. (i) Not a partition since $\{1,2\}$ and $\{2,3\}$ are both sets in the family but $\{1,2\} \cap \{2,3\} \neq \varnothing$.

   (iv) This is a partition of $\mathbb{Z}$ into the sets of even and odd integers respectively.

7. Only (ii) is a partition.

8. $X_1 = \{\varnothing\}$, $X_2 = \{\varnothing, \{\varnothing\}\}$, $X_3 = \{\varnothing, \{\varnothing\}, \{\varnothing, \{\varnothing\}\}\}$. $|X_n| = n$.
   $X = \{x : x = \varnothing \text{ or } x = y \cup \{y\} \text{ where } y \in X\}$.

9. If $A = \{1, 2\}$ and $B = \{2, 3\}$ then $\mathscr{P}(A) \cup \mathscr{P}(B) \subset \mathscr{P}(A \cup B)$.

Exercises 3.6

1. First some notation: if $(x, y) = \{\{x\}, \{x, y\}\}$ let $\bigcap(x, y) = \{x\} \cap \{x, y\}$ and $\bigcup(x, y) = \{x\} \cup \{x, y\}$. Thus $\bigcap(x, y) = \{x\}$ and $\bigcup(x, y) = \{x, y\}$.

   Now

$$
\begin{aligned}
(x, y) = (a, b) &\Rightarrow \bigcap(x, y) = \bigcap(a, b) \text{ and } \bigcup(x, y) = \bigcup(a, b) \\
&\Rightarrow \{x\} = \{a\} \text{ and } \{x, y\} = \{a, b\} \\
&\Rightarrow x = a \text{ and } \{x, y\} = \{a, b\} \\
&\Rightarrow x = a \text{ and } y = b.
\end{aligned}
$$

   (Note: this argument avoids having to consider the cases $x = y$ and $x \neq y$ separately.)

   The converse is easy.

2. (iii) $\{((1, 2), a), ((1, 2), b), ((1, 2), c), ((1, 2), d), ((1, 2), e)\}$.

3. (ii) $(A \times X) \cap (B \times Y) = \{(3, b), (4, b)\}$.
   (iv) $(A \cap X) \times Y = \varnothing$ since $A \cap X = \varnothing$.
   (vi) $(A \times X) \cup (B \times Y) = \{(1, a), (2, a), (3, a), (4, a), (1, b), (2, b), (3, b), (4, b), (5, b), (3, c), (4, c), (5, c), (3, d), (4, d), (5, d)\}$.

4. (i) Not every possible quadruple in $T \times A \times \mathbb{R}^+ \times \mathbb{Z}$ corresponds to a book in the library's collection. For example, there is no book corresponding to a quadruple of the form $(t, a, x, 3000)$ since no book (yet) has a publication date of the year 3000.

   (iii) S represents the books in the library's collection written by Shakespeare. (More precisely, S represents all ordered quadruples corresponding to those books in the library's collection authored by Shakespeare.)

   (v) It tells us that the library has no books with class number 514.3.

5. No. If X is empty then $X \times Y = \varnothing = X \times Z$ for all sets Y and Z.

6. (i) Square including all edges.
   (ii) Square excluding all edges.
   (iii) Square excluding the bottom and right-hand edges.

(iv) Square excluding the top and bottom edges.

8\. (iii) There are many possibilities. A simple example is $\{(a,1),(b,2)\}$.

9\. (ii)
$$\begin{aligned}(x,y) \in [A \times (X \cup Y)] &\Leftrightarrow x \in A \text{ and } y \in (X \cup Y)\\ &\Leftrightarrow x \in A \text{ and } (y \in X \text{ or } y \in Y)\\ &\Leftrightarrow (x \in A \text{ and } y \in X)\\ &\qquad \text{or } (x \in A \text{ and } y \in Y)\\ &\Leftrightarrow (x,y) \in (A \times X)\\ &\qquad \text{or } (x,y) \in (A \times Y)\\ &\Leftrightarrow (x,y) \in [(A \times X) \cup (A \times Y)].\end{aligned}$$

10\. (i)
$$\begin{aligned}(A \cap B) \times (X \cap Y) &= [A \times (X \cap Y)] \cap [B \times (X \cap Y)]\\ &= (A \times X) \cap (A \times Y)\\ &\qquad \cap (B \times X) \cap (B \times Y).\end{aligned}$$

11\. (ii) Any sets such that $A \not\subseteq B$, $B \not\subseteq A$, $X \not\subseteq Y$, and $Y \not\subseteq X$ will work. For example, $A = \{1\}$, $B = \{2\}$, $X = \{a\}$, $Y = \{b\}$.

13\. First suppose that $(A \times B) \subseteq (X \times Y)$.

If $a \in A$ then (since $B \neq \varnothing$) $(a,b) \in (A \times B)$ for some $b \in B$. By hypothesis this implies that $(a,b) \in (X \times Y)$; in particular $a \in X$. Therefore $A \subseteq X$.

The proof that $B \subseteq Y$ is similar.

Conversely suppose $A \subseteq X$ and $B \subseteq Y$. Then

$$\begin{aligned}(a,b) \in (A \times B) &\Rightarrow a \in A \text{ and } b \in B\\ &\Rightarrow a \in X \text{ and } b \in Y \quad (\text{since } A \subseteq X \text{ and } B \subseteq Y)\\ &\Rightarrow (a,b) \in (X \times Y).\end{aligned}$$

Therefore $(A \times B) \subseteq (X \times Y)$.

Exercises 3.7

1\. (a) (i) *Integer*
(iii) *Boolean*
(v) *Boolean*.

(b) (i) Type checks
(iii) Does not type check
(v) Type checks.

2.

$$\begin{aligned}
Height(_) &: \mathcal{Person} \rightarrow \mathcal{Real} \\
DateOfBirth(_) &: \mathcal{Person} \rightarrow \mathcal{Date} \\
YearOfBirth(_) &: \mathcal{Person} \rightarrow Integer \\
Age(_) &: \mathcal{Person} \rightarrow Integer \\
Mother(_) &: \mathcal{Person} \rightarrow \mathcal{Person} \\
IsOlderThan &: \mathcal{Person}, \mathcal{Person} \rightarrow \mathcal{Boolean} \\
CitizenOf &: \mathcal{Person} \rightarrow \mathcal{Nation} \text{ or} \\
&\quad \mathcal{Person} \rightarrow Set[\mathcal{Nation}] \\
&\qquad \text{if multiple nationality is allowed} \\
Children(_) &: \mathcal{Person} \rightarrow Set[\mathcal{Person}] \\
IsTallerThan &: \mathcal{Person}, \mathcal{Person} \rightarrow \mathcal{Boolean} \\
Qualifications(_) &: \mathcal{Person} \rightarrow Set[\mathcal{Qualification}] \\
&\quad \text{assuming that } \mathcal{Qualification} \\
&\qquad \text{is the type of qualifications} \\
Siblings(_) &: \mathcal{Person} \rightarrow Set[\mathcal{Person}].
\end{aligned}$$

3. (i) False
(iii) True
(v) True
(vii) True
(ix) True.

4. (i) True
(iii) False
(v) True
(vii) False (no-one is 250 years old, for example)
(ix) True (e.g. $n = -2$, $m = -1$)
(xi) True provided we allow *Integer* to be a subtype of $\mathcal{Real}$.

5. (i) $\_ - \_ : n : Integer, m : Integer \rightarrow p : Integer$
postcondition $n = p + m$.

(ii) $-\_ : n : Integer \rightarrow p : Integer$
postcondition $p = 0 - n$
Alternatively $p + n = 0$.

(iii) $\_ > \_ : n : Integer, m : Integer \rightarrow \mathcal{Boolean}$
postcondition $n > m \leftrightarrow IsPositive(n - m)$.

(iv) $IsNegative(\_) : n : Integer \rightarrow \mathcal{Boolean}$
postcondition $IsNegative(n) \leftrightarrow (0 > n)$
Alternatively $IsNegative(n) \leftrightarrow \neg IsPositive(n) \wedge \neg(n = 0)$.

(v) $\_ < \_ : n : \mathit{Integer}, m : \mathit{Integer} \to \mathcal{Boolean}$
postcondition $n < m \leftrightarrow \mathit{IsPositive}(m - n)$
Alternatively $n < m \leftrightarrow m > n$.

(vi) $1/\_ : n : \mathit{Integer} \to r : \mathcal{Real}$
precondition $n \neq 0$
postcondition $r \times n = 1$.

(vii) $\_ \geqslant \_ : n : \mathit{Integer}, m : \mathit{Integer} \to \mathcal{Boolean}$
postcondition $n \geqslant m \leftrightarrow (n > m) \vee (n = m)$
Alternatively $n \geqslant m \leftrightarrow \neg(n < m)$.

(viii) $\_ \leqslant \_ : n : \mathit{Integer}, m : \mathit{Integer} \to \mathcal{Boolean}$
postcondition $n \leqslant m \leftrightarrow (n < m) \vee (n = m)$
Alternatively $n \leqslant m \leftrightarrow \neg(n > m)$.

(ix) $\mathit{IsEven}(\_) : n : \mathit{Integer} \to \mathcal{Boolean}$
postcondition $\mathit{IsEven}(n) \leftrightarrow (\exists m : \mathit{Integer}, n = 2m)$.

(x) $\mathit{IsOdd}(\_) : n : \mathit{Integer} \to \mathcal{Boolean}$
postcondition $\mathit{IsOdd}(n) \leftrightarrow \neg\mathit{IsEven}(n)$
Alternatively $\mathit{IsOdd}(n) \leftrightarrow (\exists m : \mathit{Integer}, n = 2m + 1)$.

(xi) $\_\mathrm{mod}\_ : n : \mathit{Integer}, k : \mathit{Integer} \to p : \mathit{Integer}$
precondition $k \neq 0$
postcondition $(0 \leqslant p < k) \wedge (\exists q : \mathit{Integer}, n = q \times k + p)$.

(xii) $\_|\_ : n : \mathit{Integer}, m : \mathit{Integer} \to \mathcal{Boolean}$
precondition $n \neq 0 \wedge m \neq 0$
postcondition $n|m \leftrightarrow (\exists k : \mathit{Integer}, m = kn)$
Alternatively $n|m \leftrightarrow (m \bmod n = 0)$.

6. (i)
$\mathit{IsMarried}(\_) : \mathcal{Person} \to \mathcal{Boolean}$
$\mathit{IsFemale}(\_) : \mathcal{Person} \to \mathcal{Boolean}$
$\_\mathit{IsChildOf}\_ : \mathcal{Person}, \mathcal{Person} \to \mathcal{Boolean}$
$\_\mathit{IsMarriedTo}\_ : \mathcal{Person}, \mathcal{Person} \to \mathcal{Boolean}$.

(iii) **signature** $\mathit{Sons}(\_) : p : \mathcal{Person} \to A : \mathcal{Set}[\mathcal{Person}]$
Informal
postcondition $A = \mathit{Sons}(p)$ is the set of all male children of p
Formal
postcondition $A = \{q : \mathcal{Person} | \neg\mathit{IsFemale}(q) \wedge q\ \mathit{IsChildOf}\ p\}$.

(v) **signature** $\mathit{FatherInLaw} : p : \mathcal{Person} \to q : \mathcal{Person}$
precondition $\mathit{IsMarried}(p)$

postcondition $\neg IsFemale(q) \wedge (\exists r : \mathcal{P}erson,$
$p\ IsMarriedTo\ r \wedge r\ IsChildOf\ q).$

Chapter 4

Exercises 4.1

1. (iii) (a)

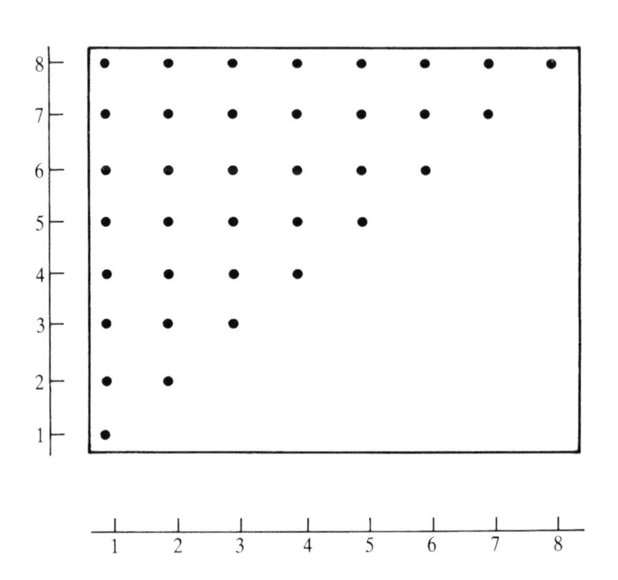

(b)

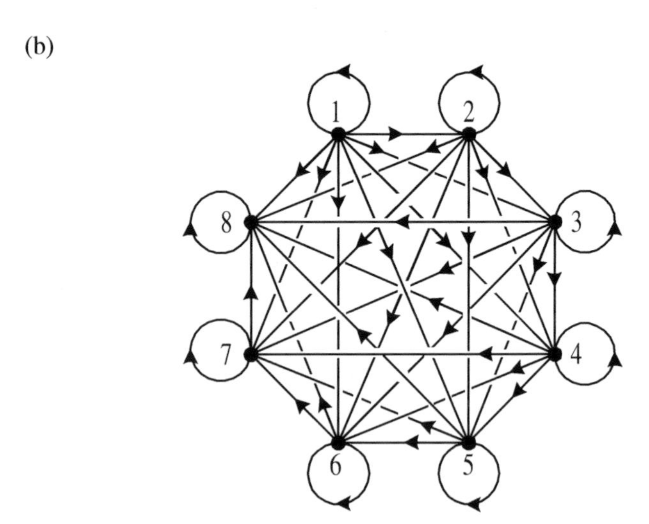

(c)

$$\begin{pmatrix} 1 & 1 & 1 & 1 & 1 & 1 & 1 & 1 \\ 0 & 1 & 1 & 1 & 1 & 1 & 1 & 1 \\ 0 & 0 & 1 & 1 & 1 & 1 & 1 & 1 \\ 0 & 0 & 0 & 1 & 1 & 1 & 1 & 1 \\ 0 & 0 & 0 & 0 & 1 & 1 & 1 & 1 \\ 0 & 0 & 0 & 0 & 0 & 1 & 1 & 1 \\ 0 & 0 & 0 & 0 & 0 & 0 & 1 & 1 \\ 0 & 0 & 0 & 0 & 0 & 0 & 0 & 1 \end{pmatrix}$$

(vi) (a)

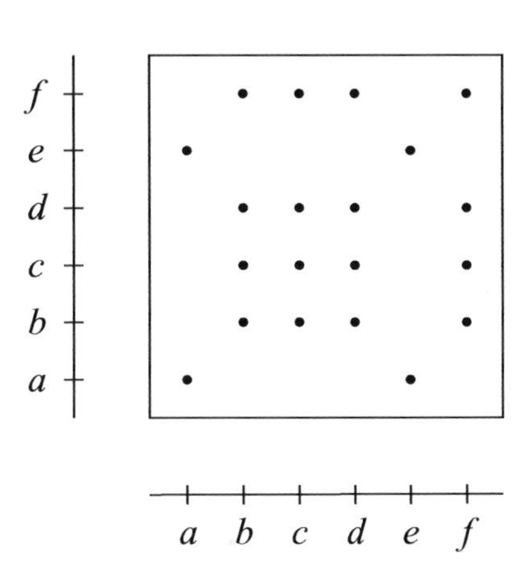

(b)

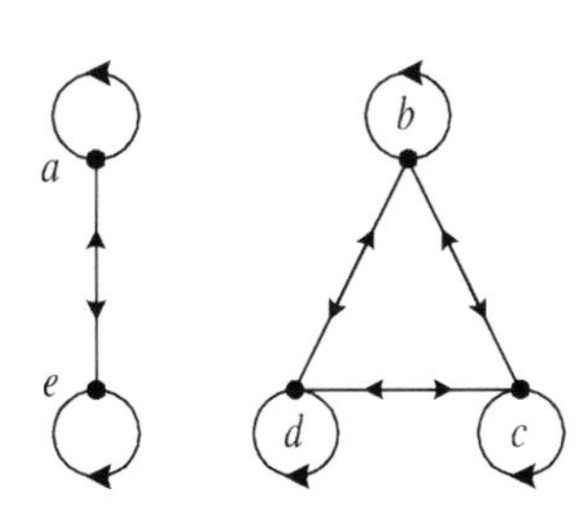

(c) $$\begin{pmatrix} 1 & 0 & 0 & 0 & 1 \\ 0 & 1 & 1 & 1 & 0 \\ 0 & 1 & 1 & 1 & 0 \\ 0 & 1 & 1 & 1 & 0 \\ 1 & 0 & 0 & 0 & 1 \end{pmatrix}.$$

(x) (a)

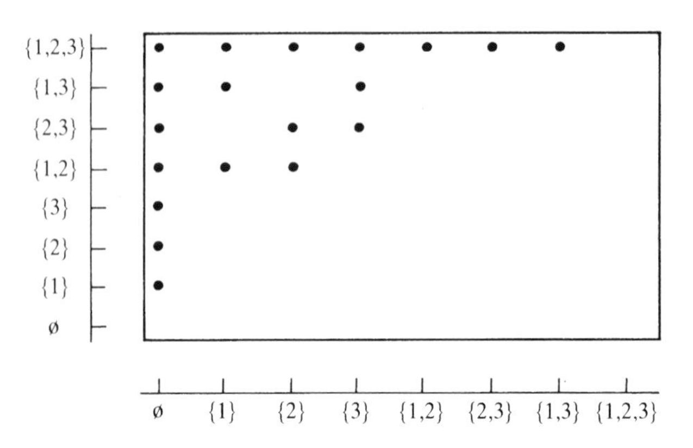

(b)

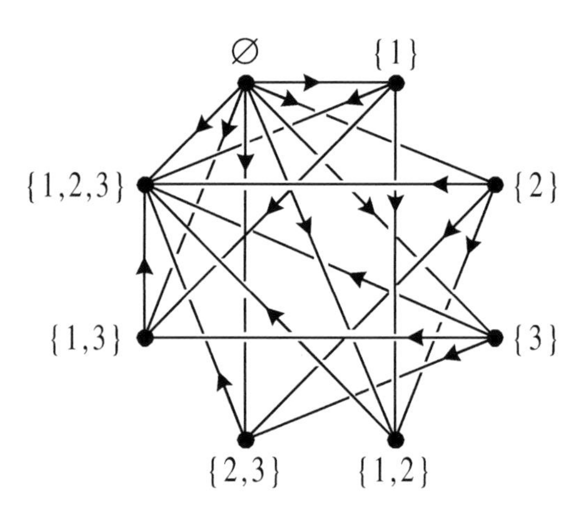

(c) With the rows and columns referring to the elements of A in the order $\varnothing$, $\{1\}$, $\{2\}$, $\{3\}$, $\{1,2\}$, $\{2,3\}$, $\{1,3\}$, $\{1,2,3\}$, the binary matrix is the following.

$$\begin{pmatrix} 0 & 1 & 1 & 1 & 1 & 1 & 1 & 1 \\ 0 & 0 & 0 & 0 & 1 & 0 & 1 & 1 \\ 0 & 0 & 0 & 0 & 1 & 1 & 0 & 1 \\ 0 & 0 & 0 & 0 & 0 & 1 & 1 & 1 \\ 0 & 0 & 0 & 0 & 0 & 0 & 0 & 1 \\ 0 & 0 & 0 & 0 & 0 & 0 & 0 & 1 \\ 0 & 0 & 0 & 0 & 0 & 0 & 0 & 1 \\ 0 & 0 & 0 & 0 & 0 & 0 & 0 & 0 \end{pmatrix}$$

3. (i) $\mathsf{R} = \{(a,c),(a,d),(a,e),(b,e),(c,a),(c,b),(d,c),(d,e),$
$(e,a),(e,b)\}$

(ii)
$$M = \begin{pmatrix} 0 & 0 & 1 & 1 & 1 \\ 0 & 0 & 0 & 0 & 1 \\ 1 & 1 & 0 & 0 & 0 \\ 0 & 0 & 1 & 0 & 1 \\ 1 & 1 & 0 & 0 & 0 \end{pmatrix}.$$

4. $\mathsf{R} = \{(1,\{1\}),(1,\{1,2\}),(1,\{1,3\}),(1,\{1,2,3\}),(2,\{2\}),$
$(2,\{1,2\}),(2,\{2,3\}),(2,\{1,2,3\}),(3,\{3\}),(3,\{2,3\}),$
$(3,\{1,3\}),(3,\{1,2,3\})\}.$

$a \,\mathsf{R}\, b$ if and only if a belongs to b.

5. $\mathsf{R} = \{(\mathrm{A},\mathrm{B}),(\mathrm{A},\mathrm{D}),(\mathrm{B},\mathrm{A}),(\mathrm{B},\mathrm{D}),(\mathrm{C},\mathrm{A}),(\mathrm{C},\mathrm{B})\}$

$$\begin{pmatrix} 0 & 1 & 0 & 1 \\ 1 & 0 & 0 & 1 \\ 1 & 1 & 0 & 0 \\ 0 & 0 & 0 & 0 \end{pmatrix}.$$

7. (i) (a) The set of towns or cities through which the River Thames flows.
(b) The set of rivers which flow through London.

(ii) (a) No river flows through Toronto.
(b) The only river which flows through Washington DC is the Potomac.

(iii) (a) $\{a \in A : a \,\mathsf{R}\, (\text{Paris})\}$.
(b) $\{a \in A : \exists b \in B, a \,\mathsf{R}\, b\}$.

(iv) (a) $\forall a \in A \; \exists b \in B : a \,\mathsf{R}\, b$.
(b) $\forall b \in B \; \exists a \in A : a \,\mathsf{R}\, b$.
(c) $\exists b_1, b_2 \in B : (b_1 \neq b_2) \wedge (\text{Nile} \,\mathsf{R}\, b_1) \wedge (\text{Nile} \,\mathsf{R}\, b_2)$.

8. (i) (a) The graph has a loop from each vertex to itself and no other lines.
(b) The matrix has ones along the diagonal from top left to bottom right and zeros elsewhere. (It is the identity matrix I_n where $n = |A|$; see chapter 6.)

(ii) (a) The graph has a loop from each vertex to itself and every vertex is joined to every other by a bidirectional edge.

(b) Every entry of the matrix is one.

9. (i) 2^{12} (ii) 2^{nm}.

10. (i) The graph of R^{-1} is obtained from that for R by reversing the direction of all the arrows.

(ii) The binary matrix of R^{-1} is obtained from that of R by reflection in the top left to bottom right diagonal. (In other words, the matrix for R^{-1} is the **transpose** of the matrix for R; see chapter 6.)

11. $\{(1,2,2),(1,2,4),(1,2,6),(1,3,3),(1,3,6),(1,4,4),(1,5,5),$
$(1,6,6),(2,3,3),(2,3,6),(2,4,4),(2,5,5),(2,6,6),(3,4,4),$
$(3,5,5),(3,6,6),(4,5,5),(4,6,6),(5,6,6)\}$.

12. Question 1: (i), (ii), (iii), (iv), (vii), (viii) have type $\mathit{Set}[\mathit{Integer} \times \mathit{Integer}]$;
(ix), (x) have type $\mathit{Set}[\mathit{Set}[\mathit{Integer}] \times \mathit{Set}[\mathit{Integer}]]$;
(v), (vi) have type $\mathit{Set}[\mathit{Character} \times \mathit{Character}]$
where $a, b, \ldots, f : \mathit{Character}$.
Question 4: $\mathit{Set}[\mathit{Integer} \times \mathit{Set}[\mathit{Integer}]]$.
Question 5: $\mathit{Set}[\mathit{Team} \times \mathit{Team}]$ where A, B, C, D : Team.
Question 8: If $A : \mathit{Set}[T]$ then $\mathsf{I}_A, \mathsf{U}_A : \mathit{Set}[T \times T]$.

13. (i) $\mathsf{R}^{-1} : \mathit{Set}[T \times S]$.
(ii) $\mathsf{R} : \mathit{Set}[\mathit{River} \times \mathit{Town}]$ where $A : \mathit{Set}[\mathit{River}]$ and $B : \mathit{Set}[\mathit{Town}]$.
R^{-1} can naturally be described as: $b\,\mathsf{R}^{-1}\,a$ if and only if b lies on the banks of a.
$\mathsf{R}^{-1} : \mathit{Set}[\mathit{Town} \times \mathit{River}]$.

Exercises 4.2

2. (i) Anti-symmetric.
(iii) Symmetric.
(v) Anti-symmetric and transitive.
(vii) Reflexive, symmetric and transitive.
(ix) Symmetric (whether the society is monogamous or polygamous).

3. (ii) Reflexive: $(x, x) \in \mathsf{R}$ for all $x \in A$.
Not symmetric: $(a, b) \in \mathsf{R}$ but $(b, a) \notin \mathsf{R}$.
Anti-symmetric: there do not exist $x, y \in A$ where $x \neq y$, $(x, y), (y, x) \in \mathsf{R}$.
Not transitive: $(a, b), (b, c) \in \mathsf{R}$ but $(a, c) \notin \mathsf{R}$.

(iv) Not reflexive: $(a,a) \notin \mathsf{R}$ for example.
Not symmetric: $(a,b) \in \mathsf{R}$ but $(b,a) \notin \mathsf{R}$.
Anti-symmetric: there do not exist $x, y \in A$ where $x \neq y$ and $(x,y), (y,x) \in \mathsf{R}$.
Not transitive: $(a,b), (b,c) \in \mathsf{R}$ but $(a,c) \notin \mathsf{R}$.

4. (ii) Reflexive, anti-symmetric and transitive.
(iv) Reflexive, anti-symmetric and transitive.

5. Symmetric, anti-symmetric and transitive. If $A = \varnothing$ then R is also reflexive.

6. Yes; any subset of the identity relation is symmetric and anti-symmetric.

7. (i) Two—(c,c) and (d,d).
(ii) None—R is anti-symmetric.

8. (i) $\mathsf{R} = \{(a,a), (a,b), (a,c), (b,b), (b,c), (c,c), (d,d)\}$.
(ii) $\mathsf{R} = \{(a,a)\}$.
(iii) $\mathsf{R} = \{(a,b), (b,a), (b,c), (c,b)\}$.

9. (i) Reflexive, symmetric and transitive.
(iii) Symmetric.

10. (i) Reflexive, symmetric and transitive.
(iii) Symmetric.

11. (i) The empty relation on any non-empty set. A more obvious example is $\{(x,x), (x,y), (y,x), (y,y)\}$ on the set $\{x, y, z\}$.

(ii) The problem is essentially in the fourth sentence. Given $a \in A$ there may be no element $b \in A$ such that $a \mathsf{R} b$—the 'proof' implicitly assumes the existence of such an element b. To illustrate this, let $a = z$ in the relation defined in (i).

12. Reflexive: $p \to p$ is true for all propositions.

Transitive: $(p \to q) \wedge (q \to r)$ logically implies $(p \to r)$, so that whenever $(p \to q)$ and $(q \to r)$ are both true, then $(p \to r)$ is also true. Thus, whenever $p \mathsf{R} q$ and $q \mathsf{R} r$, it is also the case that $p \mathsf{R} r$.

Not symmetric: if p has truth value F, and q has truth value T, then $p \mathsf{R} q$ (since $p \to q$ is true), but $q \not\mathsf{R} p$ (since $q \to p$ is false).

Not anti-symmetric (provided the set contains at least three different propositions): let p and q be two different propositions with the same truth value; then $p \mathsf{R} q$ and $q \mathsf{R} p$ (since $p \to q$ and $q \to p$ are both true), but $p \neq q$.

13. R^{-1} inherits each of the properties from R.

Suppose R is symmetric. Then $(a, b) \in \mathsf{R} \Leftrightarrow (b, a) \in \mathsf{R} \Leftrightarrow (a, b) \in \mathsf{R}^{-1}$, so $\mathsf{R} = \mathsf{R}^{-1}$.

Conversely, suppose $\mathsf{R} = \mathsf{R}^{-1}$. Then $(a, b) \in \mathsf{R} \Rightarrow (a, b) \in \mathsf{R}^{-1} \Rightarrow (b, a) \in \mathsf{R}$, so R is symmetric.

Exercises 4.3

2. (i) The directed graphs of $\mathsf{R} \cap \mathsf{S}$ and $\mathsf{R} \cup \mathsf{S}$ both have the same vertex sets as the directed graphs of R and S. The graph of $\mathsf{R} \cap \mathsf{S}$ contains those edges which belong to both the graph of R and the graph of S. The graph of $\mathsf{R} \cup \mathsf{S}$ contains those edges which belong either to the graph of R or to the graph of S (or both).

   (ii) The binary matrices of $\mathsf{R} \cap \mathsf{S}$ and $\mathsf{R} \cup \mathsf{S}$ both have the same dimension as the binary matrices of R and S. The matrix of $\mathsf{R} \cap \mathsf{S}$ has 1s in those positions where there are 1s in both the matrices of R and S (and has 0s elsewhere). The matrix of $\mathsf{R} \cup \mathsf{S}$ has 1s in those positions where there are 1s in either the matrix of R or the matrix of S (and has 0s elsewhere).

3. $\mathsf{R}_1 \subseteq (A_1 \times B_1)$ and $\mathsf{R}_2 \subseteq (A_2 \times B_2)$, so $\mathsf{R}_1 \cap \mathsf{R}_2$ and $\mathsf{R}_1 \cup \mathsf{R}_2$ are both subsets of $(A_1 \times B_1) \cup (A_2 \times B_2)$, which is a subset of $(A_1 \cup A_2) \times (B_1 \cup B_2)$. Therefore $R_1 \cap R_2$ and $R_1 \cup R_2$ are both subsets of $(A_1 \cup A_2) \times (B_1 \cup B_2)$, so both are relations from $A_1 \cup A_2$ to $B_1 \cup B_2$.

4. (i) Suppose R and S are both symmetric. Then $(a, b) \in (\mathsf{R} \cup \mathsf{S})$ implies $(a, b) \in \mathsf{R}$ or $(a, b) \in \mathsf{S}$. If $(a, b) \in \mathsf{R}$ then $(b, a) \in \mathsf{R}$, since R is symmetric, and if $(a, b) \in \mathsf{S}$ then $(b, a) \in \mathsf{S}$, since S is symmetric. Therefore, in either case $(b, a) \in (\mathsf{R} \cup \mathsf{S})$, so $\mathsf{R} \cup \mathsf{S}$ is symmetric.

5. (i) $\mathsf{S} \circ \mathsf{R} = \{(1, 4), (2, 3), (3, 2), (3, 1), (4, 3)\}$
   $\mathsf{R} \circ \mathsf{S} = \{(1, 2), (2, 1), (2, 4), (3, 2), (4, 3)\}$.

   (ii) $\mathsf{R}^{-1} = \{(1, 3), (2, 2), (2, 4), (3, 1), (4, 3)\}$
   $\mathsf{S}^{-1} = \{(1, 4), (2, 1), (3, 2), (4, 3)\}$

$$(\mathsf{S} \circ \mathsf{R})^{-1} = \{(1,3),(2,3),(3,2),(3,4),(4,1)\}$$
$$(\mathsf{R} \circ \mathsf{S})^{-1} = \{(1,2),(2,1),(2,3),(3,4),(4,2)\}.$$

(iii) $\mathsf{R}^{-1} \circ \mathsf{S}^{-1} = \{(1,3),(2,3),(3,2),(3,4),(4,1)\}$
$\mathsf{S}^{-1} \circ \mathsf{R}^{-1} = \{(1,2),(2,1),(2,3),(3,4),(4,2)\}$.

(iv) $\mathsf{R}^{-1} \circ \mathsf{S}^{-1} = (\mathsf{S} \circ \mathsf{R})^{-1}$ and $\mathsf{S}^{-1} \circ \mathsf{R}^{-1} = (\mathsf{R} \circ \mathsf{S})^{-1}$.

6. (i) $\mathsf{R} = \{(a,a),(a,b),(c,g),(d,c),(d,e),(e,d),(e,e),(f,b),$
$(f,c),(g,c),(g,f),(h,a),(h,g)\}$.

(ii) $\mathsf{R} \circ \mathsf{R} = \{(a,a),(a,b),(c,c),(c,f),(d,d),(d,e),(d,g),$
$(e,c),(e,d),(e,e),(f,g),(g,b),(g,c),(g,g),$
$(h,a),(h,b),(h,c),(h,f)\}$.

(iii)

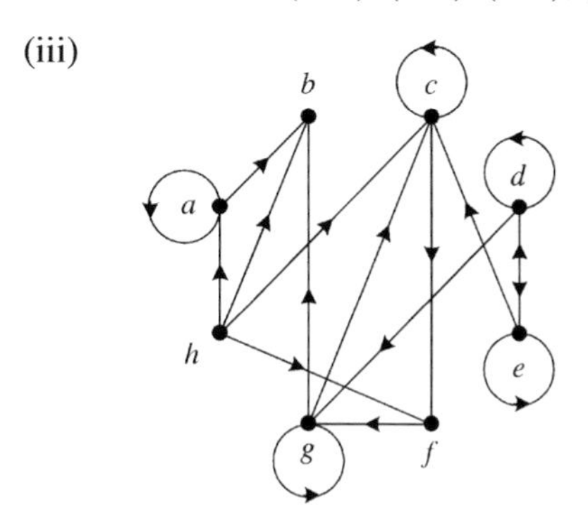

7. (i) $x(\mathsf{S} \circ \mathsf{R})y$ if and only if x is the paternal grandmother of y.
(ii) $x(\mathsf{R} \circ \mathsf{S})y$ if and only if x is the maternal grandfather of y.

8. $n(\mathsf{R}^2)m \Leftrightarrow m = n^4$.

10. $(a,d) \in [(\mathsf{T} \circ \mathsf{S}) \circ \mathsf{R}] \Leftrightarrow (a,b) \in \mathsf{R}$ and $(b,d) \in (\mathsf{T} \circ \mathsf{S})$,
for some $b \in B$.
$\Leftrightarrow (a,b) \in \mathsf{R}$ and $(b,c) \in \mathsf{S}$ and $(c,d) \in \mathsf{T}$,
for some $b \in B, c \in C$.
$\Leftrightarrow (a,c) \in (\mathsf{S} \circ \mathsf{R})$ and $(c,d) \in \mathsf{T}$,
for some $c \in C$.
$\Leftrightarrow (a,d) \in [\mathsf{T} \circ (\mathsf{S} \circ \mathsf{R})]$.

Therefore $(\mathsf{T} \circ \mathsf{S}) \circ \mathsf{R} = \mathsf{T} \circ (\mathsf{S} \circ \mathsf{R})$.

11. If $\mathsf{R} : \mathcal{S}et[\mathcal{T} \times \mathcal{U}]$ and $\mathsf{S} : \mathcal{S}et[\mathcal{U} \times \mathcal{V}]$, then $\mathsf{S} \circ \mathsf{R} : \mathcal{S}et[\mathcal{T} \times \mathcal{V}]$. (This supposes that $A : \mathcal{S}et[\mathcal{T}]$, $B : \mathcal{S}et[\mathcal{U}]$ and $C : \mathcal{S}et[\mathcal{V}]$.)

Exercises 4.4

1. For all $n \in \mathbb{Z}$, $|n| = |n|$; hence $n \mathsf{R} n$ so R is reflexive.

   For all $n, m \in \mathbb{Z}$, $n \mathsf{R} m \Rightarrow |n| = |m| \Rightarrow |m| = |n| \Rightarrow m \mathsf{R} n$ so R is symmetric.

   For all $n, m, p \in \mathbb{Z}$, $n \mathsf{R} m \wedge m \mathsf{R} p \Rightarrow |n| = |m| \wedge |m| = |p| \Rightarrow |n| = |p| \Rightarrow n \mathsf{R} p$ so R is transitive.

   Therefore R is an equivalence relation on $\mathbb{Z}$.

   $[n] = \{m \in \mathbb{Z} : n \mathsf{R} m\} = \{m \in \mathbb{Z} : |n| = |m|\} = \{n, -n\}$. Therefore $[0] = \{0\}$, $[1] = \{1, -1\}$, $[2] = \{2, -2\}, \ldots$.

2. (i) (b) $\left[\frac{1}{4}\right] = \left[0, \frac{1}{2}\right)$, $\left[\frac{1}{2}\right] = \left[\frac{1}{2}, 1\right)$.
   (c) The partition of $\mathbb{R}$ is $\{[k/2, (k+1)/2) : k \in \mathbb{Z}\}$.

3. (i) The equivalence class of a person P contains all those people in A who are the same age as P.

   (ii) The equivalence class of a person P contains all those people in A who were born in the same country as P.

4. For I_A, the equivalence classes are singleton sets, i.e. $[a] = \{a\}$.

   For U_A, the only equivalence class is A itself.

5. (i) $[(a, b)] = \{(x, y) : x = a\}$, the vertical line through (a, b).

   (ii) $[(a, b)] = \{(x, y) : x + y = a + b\}$, the line through (a, b) with gradient -1.

   (iii) $[(a, b)] = \{(x, y) : x^2 + y^2 = a^2 + b^2\}$, the circle centred at the origin with radius $\sqrt{a^2 + b^2}$. If $a = b = 0$ then $[(0, 0)] = \{(0, 0)\}$.

6. For all $(m, n) \in \mathbb{Z}^+ \times \mathbb{Z}^+$, $m + n = n + m$, so $(m, n) \mathsf{R} (m, n)$: R is reflexive.

   $(m, n) \mathsf{R} (p, q) \Rightarrow m + q = n + p \Rightarrow p + n = q + m \Rightarrow (p, q) \mathsf{R} (m, n)$: R is symmetric.

   $(m, n) \mathsf{R} (p, q)$ and $(p, q) \mathsf{R} (r, s) \Rightarrow m + q = n + p$ and $p + s = q + r \Rightarrow m + q + p + s = n + p + q + r \Rightarrow m + s = n + r \Rightarrow (m, n) \mathsf{R} (r, s)$: R is transitive.

   $$[(1, 1)] = \{(m, n) : m = n\} \qquad = \{(1, 1), (2, 2), (3, 3), \ldots\}$$

$$[(2,1)] = \{(m,n) : m = n+1\} = \{(2,1),(3,2),(4,3),\ldots\}$$
$$[(3,1)] = \{(m,n) : m = n+2\} = \{(3,1),(4,2),(5,3),\ldots\}$$
$$[(1,2)] = \{(m,n) : m+1 = n\} = \{(1,2),(2,3),(3,4),\ldots\}.$$

To every integer z there corresponds a unique equivalence class consisting of all pairs (m,n) such that $m - n = z$, and conversely every integer corresponds to an equivalence class. (In the terminology of chapter 5, there is a 'bijection' between $\mathbb{Z}$ and the set of equivalence classes.) Sometimes the integers are defined to be this set of equivalence classes.

7.
$$[2] = \mathbb{Z}$$
$$\left[\tfrac{1}{4}\right] = \left\{n + \tfrac{1}{4} : n \in \mathbb{Z}\right\} = \left\{\ldots, -2\tfrac{3}{4}, -1\tfrac{3}{4}, -\tfrac{3}{4}, \tfrac{1}{4}, 1\tfrac{1}{4}, 2\tfrac{1}{4}, 3\tfrac{1}{4}, \ldots\right\}$$
$$\left[-\tfrac{1}{4}\right] = \left\{n - \tfrac{1}{4} : n \in \mathbb{Z}\right\} = \left\{\ldots, -2\tfrac{1}{4}, -1\tfrac{1}{4}, -\tfrac{1}{4}, \tfrac{3}{4}, 1\tfrac{3}{4}, 2\tfrac{3}{4}, 3\tfrac{3}{4}, \ldots\right\}.$$

8. (i) 5 (ii) 15.

9. (i) $[2] = \{2,4,6,8,10,12,\ldots\} = \{2k : k \in \mathbb{Z}^+\}$
$[3] = \{3,9,15,21,27,\ldots\} = \{3(2k-1) : k \in \mathbb{Z}^+\}$
$[5] = \{5,25,35,55,65,85,95,\ldots\} = \{5(6k+1) : k \in \mathbb{N}\} \cup \{5(6k-1) : k \in \mathbb{Z}^+\}$

(ii) $[2] = \{2,4,8,16,32,\ldots\}$
$[3] = \{3,6,9,12,18,24,27,36,45,48,\ldots\}$
$[5] = \{5,10,15,20,25,30,40,45,50,60,75,\ldots\}$.

12. $n = 3$

$+_3$	[0]	[1]	[2]
[0]	[0]	[1]	[2]
[1]	[1]	[2]	[0]
[2]	[2]	[0]	[1]

$\times_3$	[0]	[1]	[2]
[0]	[0]	[0]	[0]
[1]	[0]	[1]	[2]
[2]	[0]	[2]	[1]

$n = 4$

$+_4$	[0]	[1]	[2]	[3]
[0]	[0]	[1]	[2]	[3]
[1]	[1]	[2]	[3]	[0]
[2]	[2]	[3]	[0]	[1]
[3]	[3]	[0]	[1]	[2]

$\times_4$	[0]	[1]	[2]	[3]
[0]	[0]	[0]	[0]	[0]
[1]	[0]	[1]	[2]	[3]
[2]	[0]	[2]	[0]	[2]
[3]	[0]	[3]	[2]	[1]

For $n = 4$ and 6 there are non-zero elements whose product is zero.

There do not exist non-zero $[a]_n$ and $[b]_n$ such that $[a]_n \times_n [b]_n = [0]_n$ if and only if n is prime.

13. Note that $p \leftrightarrow q$ is true if and only if p and q have the same truth value (i.e. both are true or both are false).

Clearly $p \leftrightarrow p$ for all p so R is reflexive.

If $p \mathrel{\mathsf{R}} q$ then p and q have the same truth value so that $q \mathrel{\mathsf{R}} p$. Hence R is symmetric.

If $p \mathrel{\mathsf{R}} q$ and $q \mathrel{\mathsf{R}} s$ then p, q and s must all have the same truth values so that $p \mathrel{\mathsf{R}} s$. Hence R is transitive.

The equivalence classes are $\{$true propositions in $A\}$ and $\{$false propositions in $A\}$.

14. Let $z = y$ to show that R is symmetric; then show R is transitive.

15. First suppose that $\mathsf{R} \circ \mathsf{S}$ is an equivalence relation.

$$\begin{aligned}(a,b) \in (\mathsf{R} \circ \mathsf{S}) &\Leftrightarrow (b,a) \in (\mathsf{R} \circ \mathsf{S})\\ &\Leftrightarrow (b,x) \in \mathsf{S} \text{ and } (x,a) \in \mathsf{R} \text{ for some } x \in A\\ &\Leftrightarrow (a,x) \in \mathsf{R} \text{ and } (x,b) \in \mathsf{S} \text{ for some } x \in A\\ &\Leftrightarrow (a,b) \in (\mathsf{S} \circ \mathsf{R}).\end{aligned}$$

Therefore $\mathsf{R} \circ \mathsf{S} = \mathsf{S} \circ \mathsf{R}$.

Conversely suppose that $\mathsf{R} \circ \mathsf{S} = \mathsf{S} \circ \mathsf{R}$.

For all $a \in A$, $(a,a) \in \mathsf{S}$ and $(a,a) \in \mathsf{R}$, so $(a,a) \in \mathsf{R} \circ \mathsf{S}$, so $\mathsf{R} \circ \mathsf{S}$ is reflexive.

$$\begin{aligned}(a,b) \in (\mathsf{R} \circ \mathsf{S}) &\Rightarrow (a,b) \in (\mathsf{S} \circ \mathsf{R})\\ &\Rightarrow (a,x) \in \mathsf{R} \text{ and } (x,b) \in \mathsf{S} \text{ for some } x \in A\\ &\Rightarrow (b,x) \in \mathsf{S} \text{ and } (x,a) \in \mathsf{R} \text{ for some } x \in A\\ &\Rightarrow (b,a) \in (\mathsf{R} \circ \mathsf{S}), \text{ so } \mathsf{R} \circ \mathsf{S} \text{ is symmetric.}\end{aligned}$$

$(a,b) \in (\mathsf{R} \circ \mathsf{S})$ and $(b,c) \in (\mathsf{R} \circ \mathsf{S})$

$$\begin{aligned}&\Rightarrow (a,x) \in \mathsf{S} \text{ and } (x,b) \in \mathsf{R} \text{ for some } x \in A\\ &\qquad\qquad \text{and } (b,y) \in \mathsf{S} \text{ and } (y,c) \in \mathsf{R} \text{ for some } y \in A\\ &\Rightarrow (a,x) \in \mathsf{S} \text{ and } (x,y) \in (\mathsf{S} \circ \mathsf{R}) \text{ and } (y,c) \in \mathsf{R} \text{ for some } x,y \in A\end{aligned}$$

$\Rightarrow (a, x) \in \mathsf{S}$ and $(x, y) \in (\mathsf{R} \circ \mathsf{S})$ and $(y, c) \in \mathsf{R}$ for some $x, y \in A$
$\Rightarrow (a, x) \in \mathsf{S}$ and $(x, z) \in \mathsf{S}$ and $(z, y) \in \mathsf{R}$ and $(y, c) \in \mathsf{R}$
for some $x, y, z \in A$
$\Rightarrow (a, z) \in \mathsf{S}$ and $(z, c) \in \mathsf{R}$ for some $z \in A$
$\Rightarrow (a, c) \in (\mathsf{R} \circ \mathsf{S})$,

so $\mathsf{R} \circ \mathsf{S}$ is transitive.

Therefore $\mathsf{R} \circ \mathsf{S}$ is an equivalence relation.

Exercises 4.5

1. For all $n \in \mathbb{Z}^+$, $n|n$, so R is reflexive.

   If $n|m$ and $m|n$ then $m = k_1 n$ and $n = k_2 m$ where $k_1, k_2 \in \mathbb{Z}^+$. Therefore $m = k_1 k_2 m$ so $k_1 k_2 = 1$ and hence $k_1 = k_2 = 1$ (since k_1 and k_2 are positive integers). Therefore $m = n$ so R is anti-symmetric.

   If $n|m$ and $m|r$ then $m = k_1 n$ and $r = k_2 m$ where $k_1, k_2 \in \mathbb{Z}^+$. Therefore $r = k_1 k_2 n$ so $n|r$. Therefore R is transitive.

   The least element is 1 since $1 \,\mathsf{R}\, n$ for all $n \in \mathbb{Z}^+$.

2. (i) R is not reflexive; for example $\{1\} \not\subset \{1\}$.

   (iii) R is not anti-symmetric; for example $1 \,\mathsf{R}\, (-1)$ and $(-1) \,\mathsf{R}\, 1$.

   (v) This depends on the properties of the people in A.

   If there are people P and Q in A such that P is older and shorter than Q then $P \,\mathsf{R}\, Q$ and $Q \,\mathsf{R}\, P$ so R is not anti-symmetric.

   If there exist three people P, Q and S in A such that Q is younger than S is younger than P and S is shorter than P is shorter than Q then $P \,\mathsf{R}\, Q$ and $Q \,\mathsf{R}\, S$ but $P \not\mathsf{R}\, S$; therefore R is not transitive.

3. (i) For all $(x, y) \in \mathbb{R}^2$, $x \leqslant x$ and $y \leqslant y$, so $(x, y) \,\mathsf{R}\, (x, y)$, and R is reflexive.

   If $(x_1, y_1) \,\mathsf{R}\, (x_2, y_2)$ and $(x_2, y_2) \,\mathsf{R}\, (x_1, y_1)$ then $x_1 \leqslant x_2$, $y_1 \leqslant y_2$, $x_2 \leqslant x_1$ and $y_2 \leqslant y_1$. Hence $x_1 = x_2$ and $y_1 = y_2$ so $(x_1, y_1) = (x_2, y_2)$ and R is anti-symmetric.

   If $(x_1, y_1) \,\mathsf{R}\, (x_2, y_2)$ and $(x_2, y_2) \,\mathsf{R}\, (x_3, y_3)$ then $x_1 \leqslant x_2$, $y_1 \leqslant y_2$, $x_2 \leqslant x_3$ and $y_2 \leqslant y_3$. Hence $x_1 \leqslant x_3$ and $y_1 \leqslant y_3$ so $(x_1, y_1) \,\mathsf{R}\, (x_3, y_3)$ and R is transitive.

(ii) The proof is similar to (i).

5. The reflexive and transitive properties are obvious.

Suppose $A \,\mathsf{R}\, B$ and $B \,\mathsf{R}\, A$; then $|A| \leqslant |B|$ and $|B| \leqslant |A|$ so $|A| = |B|$. Since no two of the sets have the same cardinality, this implies $A = B$. Therefore R is anti-symmetric and hence a partial order. Clearly, for any two sets A and B, either $|A| \leqslant |B|$ or $|B| \leqslant |A|$ so R is a total order.

The maximal (minimal) elements are the sets with the largest (smallest) number of elements of all those in $\mathscr{F}$.

6. R is a partial order relation if no two people in A are of the same age. (The proof that, in this case, R is a total order is similar to question 4.5.5 above.)

The greatest and least elements are the oldest and youngest people in A respectively.

7. (i) Minimal elements are the sets in $\mathscr{F}$ with smallest cardinality; maximal elements are the sets in $\mathscr{F}$ with greatest cardinality.

(ii) R is not a total order on $\mathscr{P}(\{1,2,3\})$; for example, $\{1,2\} \not\!\mathsf{R} \{2,3\}$ and $\{2,3\} \not\!\mathsf{R} \{1,2\}$.

11. We are given that $a_1 \mathsf{R} a_2, a_2 \mathsf{R} a_3, \ldots, a_{n-1} \mathsf{R} a_n, a_n \mathsf{R} a_1$. By transitivity (applied several times) $a_1 \mathsf{R} a_n$. Therefore $a_1 \mathsf{R} a_n$ and $a_n \mathsf{R} a_1$ so $a_1 = a_n$. Thus we now have $a_1 \;\mathsf{R}\; a_2, a_2 \;\mathsf{R}\; a_3, \ldots, a_{n-1} \;\mathsf{R}\; a_1$, and repeating the above argument shows that $a_1 = a_{n-1}$.

Continuing in this way we see that $a_1 = a_2 = \cdots = a_n$. (A more formal proof would proceed by mathematical induction.)

13. (i) The proof is by induction on the number of elements of the subsets of A. Let A be a finite totally ordered set. Trivially, every one-element subset has a least element. Suppose that every k-element subset of A $(k < n)$ has a least element and let $B = \{b_1, b_2, \ldots, b_{k+1}\}$ be a $(k+1)$-element subset of A. Then $\{b_1, \ldots, b_k\}$ is a k-element subset and so has a least element, b_i say. Since R is a total order, either $b_i \;\mathsf{R}\; b_{k+1}$ or $b_{k+1} \;\mathsf{R}\; b_i$.

In the first case $b_i \;\mathsf{R}\; b$ for every $b \in B$, so b_i is a least element for B. In the second case, for every $b \in B$, either $b = b_{k+1}$ or $b_i \;\mathsf{R}\; b$. Since $b_i \;\mathsf{R}\; b$, for every $b \neq b_{k+1}$, and $b_{k+1} \;\mathsf{R}\; b_i$ we have, by transitivity, $b_{k+1} \;\mathsf{R}\; b$ for every $b \in B$. Therefore b_{k+1} is a least element.

In either case B has a least element.

Therefore, by induction, every non-empty subset of A has a least element so A is well ordered.

(ii) $\mathbb{Z}^+$ is infinite and well ordered by the usual $<$ relation.

(iii) The set of negative integers is a subset of $\mathbb{Z}$ which has no least element so $\mathbb{Z}$ is not well ordered.

$(0,1) = \{x \in \mathbb{R} : 0 < x < 1\}$ is a subset of $\mathbb{R}^+$ with no least element, so $\mathbb{R}^+$ is not well ordered.

Exercises 4.6

1. (i)

Longest chains: $\{1,2,4,12\}$
$\{1,2,6,12\}$
$\{1,3,6,12\}$

(ii)

Longest chains: $\{1,2,4,20\}$
$\{1,5,10,20\}$
$\{1,2,10,20\}$

(iii)

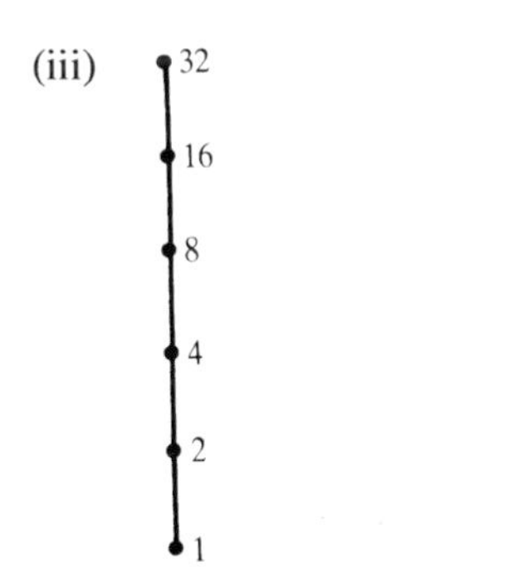

Longest chain: $\{1,2,4,8,16,32\}$

(iv)

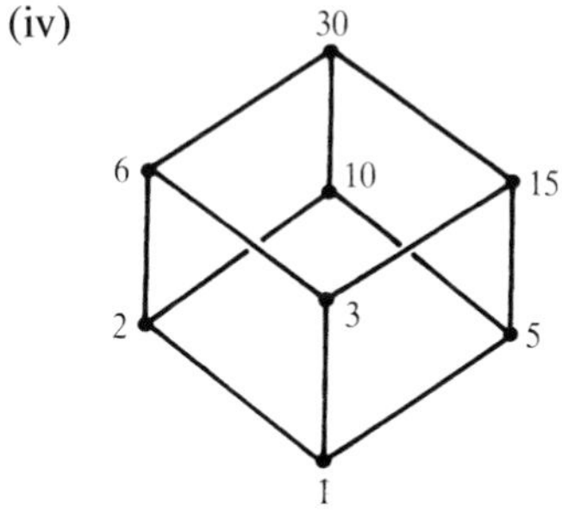

Longest chains: $\{1,2,6,30\}$
$\{1,2,10,30\}$
$\{1,3,6,30\}$
$\{1,3,15,30\}$
$\{1,5,10,30\}$
$\{1,5,15,30\}$.

2. $\{(a,a),(b,b),(c,c),(d,d),(e,e),(f,f),(g,g),(h,h),(i,i),(a,d),$
$(a,e),(a,h),(a,i),(b,d),(b,e),(b,h),(b,i),(c,g),(c,i),(d,e),$
$(d,h),(d,i),(e,h),(e,i),(g,i)\}.$

Maximal elements: f, h and i.
Minimal elements: a, b, c and f.

3. *Three element*: the following are the different possible kinds of Hasse diagrams together with the number of different posets with the given diagram type.

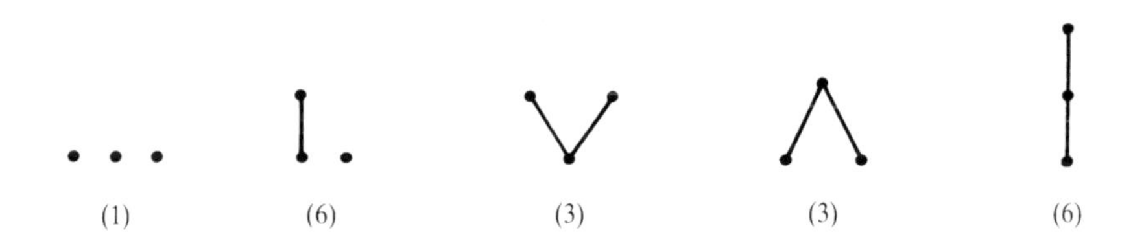

Total number of different order relations is equal to 19.

Four element:

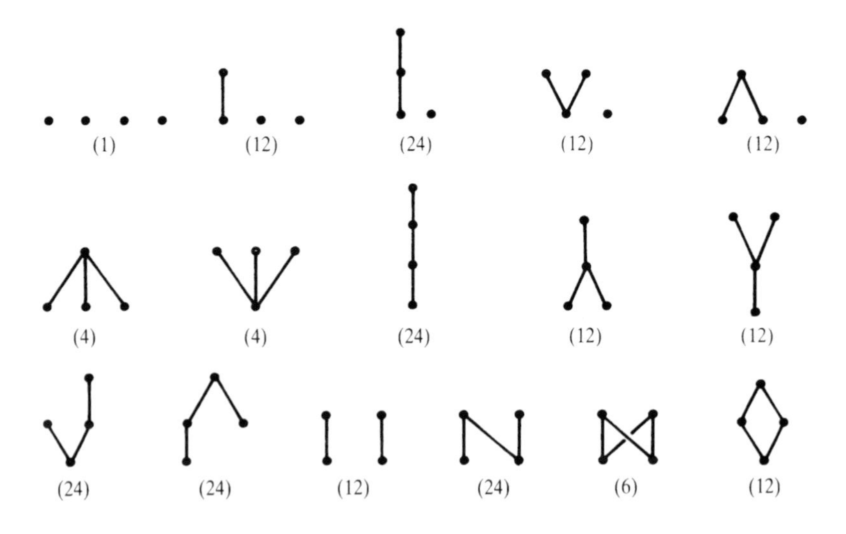

Total number of different order relations is equal to 219.

4.

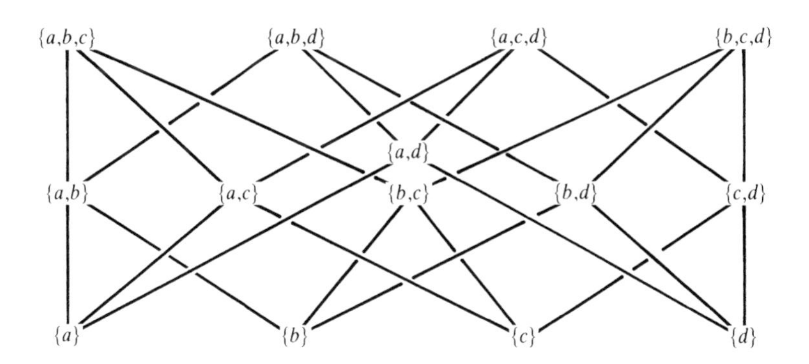

Maximal elements: $\{a, b, c\}, \{a, b, d\}, \{a, c, d\}, \{b, c, d\}$.
Minimal elements: $\{a\}, \{b\}, \{c\}, \{d\}$.
Longest chains: chains of the form $\{x\}, \{x, y\}, \{x, y, z\}$. There are 24 of these.

5. (i)

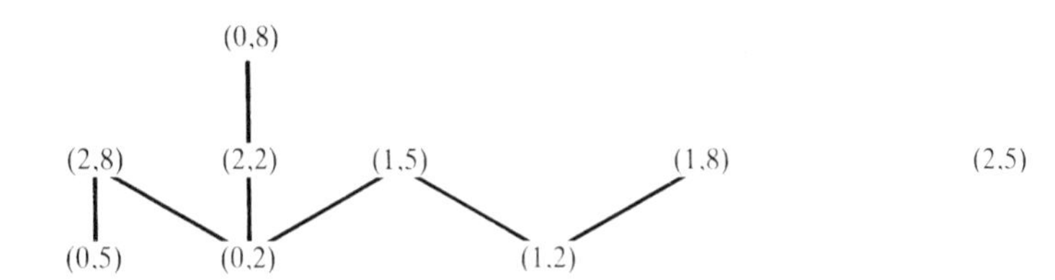

(ii) Maximal elements: $(2, 8), (0, 8), (1, 5), (1, 8)$ and $(2, 5)$.
Minimal elements: $(0, 5), (0, 2), (1, 2)$ and $(2, 5)$.

6. The least element is $\{a\}$ and the greatest element is $\{a, b, c\}$.

7. (i)

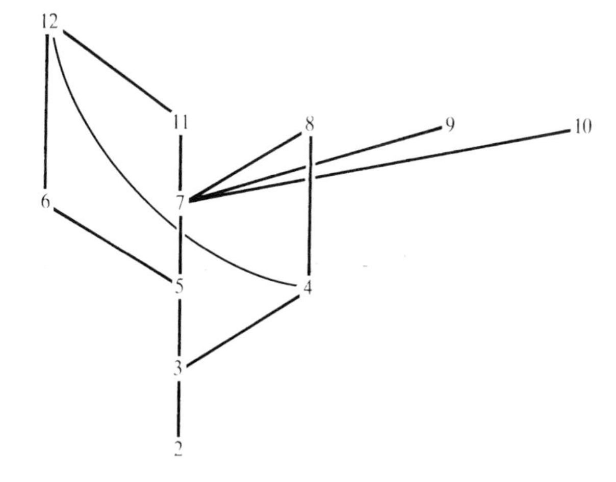

8.

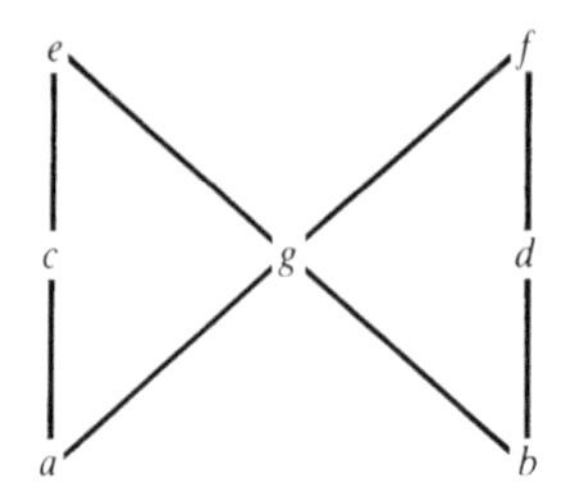

Exercises 4.7

1. (i)

F3286	Johnson, D	15/12/69	1989
M3415	Singer, R	03/10/71	1989
F0278	Williams, L	19/03/70	1989

(iii)

M1452	Adams, K	23/06/71	1990	CompSci
F3286	Johnson, D	15/12/69	1989	Psyc
F5419	Kirby, F	29/07/63	1990	Math/Econ
M3415	Singer, R	03/10/71	1989	Hist
F0278	Williams, L	19/03/70	1989	CompSci/Math

(v)

F3286	Johnson, D
F5419	Kirby, F
F0278	Williams, L

2. F5419 Math100 Math150 Econ110 Econ120.

The selection above does not list students whose B_1, B_2 or B_4 attribute values are 'Econ110'.

To obtain the complete list of Econ110 students, make four selections from CURRENT_COURSE, those record instances whose B_1, B_2, B_3 and B_4 attribute value is 'Econ110' respectively; then take the union of the four resulting tables.

3. Both combinations of selections result in the following table.

M1452	Adams, K	23/06/71	1990	CompSci	Comp100	Math150	Bus_105	Econ110
F3286	Johnson, D	15/12/69	1989	Psyc	Psyc250	Psyc280	Psyc281	Soc_200
F5419	Kirby, F	29/07/63	1990	Math/Econ	Math100	Math150	Econ110	Econ120
M3415	Singer, R	03/10/71	1989	Hist	Hist210	Hist220	Lit_200	Stat120
F0278	Williams, L	19 /03/70	1989	CompSci/Math	Comp210	Comp230	Math205	Math215

4. (i) First take the natural join of PERSONAL and CURRENT_COURSE (or DISCIPLINE and CURRENT_COURSE); then project onto $(A_2, B_1, B_2, B_3, B_4)$.

(iii) First take the natural join of DISCIPLINE and CURRENT_COURSE; then select those record instances whose A_5 attribute value is 'CompSci' or CompSci/*******' or '*******/CompSci'; finally project onto $(A_2, B_1, B_2, B_3, B_4)$.

(Note that the middle 'selection phase' can be accomplished by three separate selections followed by their union in a similar manner to that described in question 4.7.2 above.)

(v) First take the natural join of PERSONAL and CURRENT_COURSE; then select those record instances whose A_3 attribute value is '$**/**/71$'; finally project onto $(A_1, A_4, B_1, B_2, B_3, B_4)$.

Chapter 5

Exercises 5.1

1. (i) 4 (v) -4
   (iii) 3 (vii) $(a+1)^2 - 5 = a^2 + 2a - 4$.

2. (i) $\{3, 4\}$ (v) $\{1, 2\}$
   (iii) $\varnothing$ (vii) A.

3. (i), (iii), (v), (vi) are (vii) are functions.

4. Only (iii) is a function.

5. In general, only (ii) is a function. However if A is a singleton set then (i) and (iii) are also functions.

6. (ii) (a) $f(1) = 5$, $f(2) = 6$, $f(3) = 7$, $f(4) = 8$, $f(5) = 9$, $f(6) = 9$, $f(7) = 9$, $f(8) = 9$, $f(9) = 9$.
   (b) $\{5, 6, 7, 8, 9\}$.
   (c) $\{(1,5), (2,6), (3,7), (4,8), (5,9), (6,9), (7,9), (8,9), (9,9)\}$.

   (v) (a) $f(1) = 7$, $f(2) = 5$, $f(3) = 3$, $f(4) = 1$, $f(5) = 1$, $f(6) = 3$, $f(7) = 5$, $f(8) = 7$, $f(9) = 9$.
   (b) $\{1, 3, 5, 7, 9\}$.
   (c) $\{(1,7), (2,5), (3,3), (4,1), (5,1), (6,3), (7,5), (8,7), (9,9)\}$.

   (viii) (a) $f(1) = 2$, $f(2) = 2$, $f(3) = 3$, $f(4) = 2$, $f(5) = 2$, $f(6) = 2$, $f(7) = 2$, $f(8) = 2$, $f(9) = 2$.

(b) $\{2, 3\}$.
(c) $\{(1,2), (2,2), (3,3), (4,2), (5,2), (6,2), (7,2), (8,2), (9,2)\}$.

7. (i) $\{x \in \mathbb{R} : x \geqslant 2\}$
(iii) $(0, \frac{1}{2}] = \{x \in \mathbb{R} : 0 < x \leqslant \frac{1}{2}\}$
(v) $[-\frac{1}{10}, \frac{1}{2}] = \{x \in \mathbb{R} : -\frac{1}{10} \leqslant x \leqslant \frac{1}{2}\}$.

8. (i) $\{0, 1, 2, 3, 4\}$
(iii) {capital cities of the world}
(v) $\{\{a\}, \{a,b\}, \{a,c\}, \{a,d\}, \{a,b,c\}, \{a,b,d\}, \{a,c,d\}, \{a,b,c,d\}\}$.

9. Question 1 $f : \mathit{Set}[\mathit{Real} \times \mathit{Real}]$
$g : \mathit{Set}[\mathit{Integer} \times \mathit{Real}]$
$h : \mathit{Set}[\mathit{Real} \times \mathit{Integer}]$.
Question 2 $f, g : \mathit{Set}[\mathit{Set}[\mathit{Integer}] \times \mathit{Set}[\mathit{Integer}]]$.
Question 8
(i) $f : \mathit{Set}[\mathit{Set}[\mathit{Character}] \times \mathit{Integer}]$
where $a, b, c, d : \mathit{Character}$.
(ii) $f : \mathit{Set}[\mathit{Integer} \times \mathit{Integer}]$.
(iii) $f : \mathit{Set}[\mathit{Country} \times \mathit{City}]$.
(iv) and (v) $f : \mathit{Set}[\mathit{Set}[\mathit{Character}] \times \mathit{Set}[\mathit{Character}]]$
where $a, b, c, d : \mathit{Character}$.

10. (i) $[0, 9] = \{x \in \mathbb{R} : 0 \leqslant x \leqslant 9\}$
(iii) $\{\frac{1}{2}, 1, 2, 4, 8, 16, 32, 64\}$
(v) $\{1, 2, 7, 11\}$.

11. (i) $[-3, -2] \cup [2, 3]$
(iii) $[-3, 3]$
(v) $\{(x, y) \in \mathbb{R}^+ : x^2 + y^2 \leqslant 1\}$, the 'closed' disc, centred at the origin with radius 1
(vii) A.

12. (i) The domain of f is $\{1, 2, 3, 4, 5, 6\}$.
(iii) The domain of f is $\{1, 2, 3\}$.
(v) The domain of f is $\mathbb{R}^* = \{x \in \mathbb{R} : x \neq 0\}$.
(vii) The domain of f is {integer multiples of 4} $= \{4n : n \in \mathbb{Z}\}$.
(ix) The domain of f is $\{(n, m) \in \mathbb{Z}^+ \times \mathbb{Z}^+ : m \text{ divides } n\}$.

13. (i) Partial function: if n is odd then $f(n)$ is not defined.
(iii) Total function.
(v) Not a function.

14. Let $A = \{a_1, a_2, \ldots, a_n\}$. For each a_i there are m possibilities for $f(a_i)$, namely the m elements of B. Since the image of each element of A can be defined independently, there are $m \times m \times \ldots \times m$ (n times) $= m^n$ possible functions $A \to B$.

15. $[x] = f^{-1}(f\{x\})$, the set of all elements of A which have the same image as x.

Exercises 5.2

1. (i) $(f \circ f)(2) = f(-1) = -4$
   (iii) $(f \circ g)(2) = f(5) = 20$
   (v) $(h \circ g)(3) = h(\frac{15}{7}) = 2$
   (vii) $(f \circ h)(1.5) = f(1) = -4$.

3. (i) $f(2) = 7$
   (iii) $(g \circ f)(2) = g(7) = 50$
   (v) $(f \circ f)(2) = f(7) = 27$
   (iv) $(f \circ g)(2) = f(5) = 19$
   (vii) $(f \circ g \circ f)(3) = f(g(f(3))) = f(g(11)) = f(122) = 487$
   (ix) $(g \circ f)(x) = g(4x - 1) = (4x - 1)^2 + 1 = 16x^2 - 8x + 2$.

4. (i) $(g \circ f)(1) = g(3) = \frac{1}{10}$
   (iii) $(g \circ h)(2) = g(\sqrt{5}) = \frac{1}{6}$
   (v) $(f \circ g)(x) = \dfrac{2}{x^2 + 1} + 1 = \dfrac{x^2 + 3}{x^2 + 1}$
   (vii) $(g \circ h)(x) = \dfrac{1}{x^2 + 2}$
   (ix) $((f \circ g) \circ h)(x) = \dfrac{x^2 + 4}{x^2 + 2}$.

7. *im*$(f) \subseteq C$ and *im*$(g) \subseteq A$.

8. Both expressions are just $h(g(f(x)))$.

9. $f(a) = f(b) = f(c) = a$. (There are other functions with the required property.)

10. (i) If $n \in \mathbb{Z}$ then $\lfloor n \rfloor = n$; since $\lfloor x \rfloor \in \mathbb{Z}$ for all $x \in \mathbb{R}$, it follows that $\lfloor \lfloor x \rfloor \rfloor = \lfloor x \rfloor$.
   (ii) Every real number x can be expressed uniquely as $x = n + \delta$ where $n \in \mathbb{Z}$ and $0 \leqslant \delta < 1$, and then $f(x) = n$. If $k \in \mathbb{Z}$ then

$f(x+k) = f(n+\delta+k) = n+k = f(x)+k$. Conversely, if $f(x+k) = f(x)+k$ then $k = f(x+k) - f(x) \in \mathbb{Z}$, so $k \in \mathbb{Z}$.

(iii) $\bigcup_{n\in\mathbb{Z}} \left[n, n+\frac{1}{2}\right)$.

11. The proof is by induction on n, the case $n = 1$ being trivial. Suppose $f^{[k]}(x) = 2^k x + (2^k - 1)$, for some $k \geqslant 1$. Then

$$\begin{aligned} f^{[k+1]}(x) = f^{[k]}(2x+1) &= 2^k(2x+1) + (2^k - 1) \\ &= 2^{k+1}x + (2^k + 2^k) - 1 = 2^{k+1}x + (2^{k+1} - 1) \end{aligned}$$

which completes the inductive step.

12. (i) For $x \geqslant 3$, $g(x) = x - 2$; but $g(1)$ and $g(2)$ may be chosen arbitrarily.
(ii) Since $1 \notin im(f)$, $f(h(1)) \neq 1$ regardless of the definition of h.
(iii) $f^{[n]}(x) = x + 2n$.

13. (ii)

$$(g \circ f)(x) = \begin{cases} (x+2)/3 & \text{if } x \geqslant 2 \\ |x-1| & \text{if } 1 \leqslant x < 2 \\ (x^3+4)/3 & \text{if } 0 \leqslant x < 1 \\ |x^3+1| & \text{if } x < 0. \end{cases}$$

14. (i) $i_C = \{(c, c) : c \in C\}$.

(ii)
$$\begin{aligned} (x, y) \in f|_C &\Leftrightarrow x \in C \text{ and } y = f(x) \in B \\ &\Leftrightarrow (x, y) \in (C \times B) \text{ and } (x, y) \in f \\ &\Leftrightarrow (x, y) \in (f \cap (C \times B)). \end{aligned}$$

(iii) $f|_C$ and $f \circ i_C$ are both functions $C \to B$ and, for all $x \in C$, $(f \circ i_C)(x) = f(x) = (f|_C)(x)$.

15. (i) f and g are total, $g \circ f$ is partial; domain of $g \circ f$ is $\{1, 2\}$.
(ii) f, g and $g \circ f$ are total.
(iii) f and g are total, $g \circ f$ is partial; domain of $g \circ f$ is $\{0\}$.
(iv) f is total, g is partial, $g \circ f$ is total.
(v) f is partial, g is total, $g \circ f$ is partial; domain of $g \circ f$ is $\{2n : n \in \mathbb{Z}\}$.
(vi) f is total, g is partial, $g \circ f$ is partial; domain of $g \circ f$ is $\{n/2 : n \in \mathbb{Z}\}$.
(vii) f is total, g is partial, $g \circ f$ is total.
(viii) f is total, g is partial, $g \circ f$ is partial; domain of $g \circ f$ is the closed interval $[-1, 1] = \{x \in \mathbb{R} : -1 \leqslant x \leqslant 1\}$.

16.	f	g	$g \circ f$
(i)	*Set* [*Integer* × *Integer*]	*Set* [*Integer* × *Integer*]	*Set* [*Integer* × *Integer*]
(ii)	*Set* [*Integer* × *Integer*]	*Set* [*Integer* × *Integer*]	*Set* [*Integer* × *Integer*]
(iii)	*Set* [*Integer* × *Real*]	*Set* [*Integer* × *Integer*]	*Set* [*Integer* × *Integer*]
(iv)	*Set* [*Real* × *Real*]	*Set* [*Real* × *Real*]	*Set* [*Real* × *Real*]
(v)	*Set* [*Integer* × *Integer*]	*Set* [*Real* × *Real*]	*Set* [*Integer* × *Real*]
(vi)	*Set* [*Real* × *Real*]	*Set* [*Integer* × *Integer*]	*Set* [*Real* × *Integer*]
(vii)	*Set* [*Integer* × *Integer*]	*Set* [*Integer* × *Integer*]	*Set* [*Integer* × *Integer*]
(viii)	*Set* [*Real* × *Real*]	*Set* [*Real* × *Real*]	*Set* [*Real* × *Real*].

Exercises 5.3

1. (i) Not injective: $F(b) = F(e)$; not surjective: for example, $a \notin im(F)$.

   (iii) Injective: all the images are distinct; not surjective: for example, $d \notin im(F)$.

   (v) Not injective: $F(a) = F(d)$; not surjective: for example, $c \notin im(F)$.

2. (i) Injective: $f(n) = f(m) \Rightarrow n - 6 = m - 6 \Rightarrow n = m$.

   Surjective: for each $n \in \mathbb{Z}$ there exists $m = n + 6 \in \mathbb{Z}$ such that $f(m) = m - 6 = (n + 6) - 6 = n$.

   (iii) Not injective: $f(-2) = 4 = f(2)$. Not surjective: for example, $2 \notin im(f)$.

   (v) Not injective: $f(2) = 6 = f(-3)$. Not surjective: for example, $1 \notin im(f)$.

   (vii) Note that $f(n) = n + 1$ if n is even and $f(n) = n - 1$ if n is odd. So n and $f(n)$ always have opposite parity (evenness/oddness).

   Injective: $f(n) = f(m) \Rightarrow n, m$ are both even or n, m are both odd
   $\Rightarrow n + 1 = m + 1$ or $n - 1 = m - 1$
   $\Rightarrow n = m$.

   Surjective: let $n \in \mathbb{Z}$.

   If n is even, let $m = n + 1$. Then m is odd so $f(m) = m - 1 = (n + 1) - 1 = n$.

   If n is odd, let $m = n - 1$. Then m is even so $f(m) = m + 1 = (n - 1) + 1 = n$.

(ix) Not injective: $f(2) = 2 = f(3)$.

Surjective: let $n \in \mathbb{Z}$. Let $m = 2n - 1$; then m is odd so

$$f(m) = \frac{(2n-1)+1}{2} = n.$$

3. (i) Surjective (iii) Neither (v) Injective.

4. (i) Injective: $f(x) = f(y) \Rightarrow \{x\} = \{y\} \Rightarrow x = y$.

Not surjective: for example, $\{1, 2\} \notin im(f)$.

(iii) Not injective: $f(\{1, 2\}) = \{1, 2\} = f(\{1, 2, 3\})$.

Not surjective: for all $X \in B$, $f(X) \subseteq \{1, 2\}$ so $\{1, 2, 3\} \notin im(f)$, for example.

(v) Injective: $f(X) = f(Y) \Rightarrow A - X = A - Y \Rightarrow X = Y$.

Surjective. Given $Y \in B = \mathscr{P}(A)$, let $X = A - Y$. Then $X \subseteq A$, so $X \in B = \mathscr{P}(A)$ and $f(X) = A - X = A - (A - Y) = Y$.

5. (i) Not injective: $f([2]) = [4] = f([3])$.

Not surjective: $im(f) = \{[0], [1], [4]\} \neq \mathbb{Z}_5$.

(iii) $f : [0] \mapsto [3], [1] \mapsto [0], [2] \mapsto [2], [3] \mapsto [4], [4] \mapsto [1]$.

Injective: there are no repeated images.

Surjective: $im(f) = \mathbb{Z}_5$.

(v) Not injective: $f([1]) = [1] = f([5])$.

Not surjective: $im(f) = \{[0], [1], [3], [4]\} \neq \mathbb{Z}_6$.

(vii) Not injective: $f([1]) = [5] = f([4])$.

Not surjective: $im(f) = \{[1], [3], [5]\} \neq \mathbb{Z}_6$.

6. (i) Neither (iii) Injective (v) Surjective
(vii) Injective (ix) Neither.

7. (i) f is injective provided no two people in A are the same age.

f is surjective provided there is at least one person in A of each given age from 0 to 100 inclusive.

(iii) f is injective with no further restrictions on A or B.

f is surjective provided the set B contains only the largest cities (in population terms) of the countries in set A.

(v) f is injective provided it is a well defined function and for this we require $c \leqslant a + 10$ and $d \geqslant b + 10$.

f is surjective provided $c = a + 10$ and $d = b + 10$.

8. (i)
$$\begin{aligned} y \in f(C_1 \cup C_2) &\Leftrightarrow y = f(x) \text{ for some } x \in C_1 \cup C_2 \\ &\Leftrightarrow y = f(x) \text{ for some } x \in C_1 \\ &\qquad \text{or } y = f(x) \text{ for some } x \in C_2 \\ &\Leftrightarrow y \in f(C_1) \text{ or } y \in f(C_2) \\ &\Leftrightarrow y \in f(C_1) \cup f(C_2). \end{aligned}$$

(iii) Suppose that f is injective, and let C_1 and C_2 be subsets of A. By (ii) we only need prove $f(C_1) \cap f(C_2) \subseteq f(C_1 \cap C_2)$, so let $y \in f(C_1) \cap f(C_2)$. Then $y = f(x_1)$ where $x_1 \in C_1$ and $y = f(x_2)$ where $x_2 \in C_2$. Since f is injective, $x_1 = x_2 = x$, say. Therefore $y = f(x)$ where $x \in C_1 \cap C_2$ so $y \in f(C_1 \cap C_2)$. Hence $f(C_1) \cap f(C_2) \subseteq f(C_1 \cap C_2)$, as required.

Conversely, suppose $f(C_1) \cap f(C_2) = f(C_1 \cap C_2)$, for all subsets C_1, C_2 of A. If $f(a) = f(b)$ then $f(\{a\}) = f(\{b\})$ so $\{a\} = \{b\}$ which implies $a = b$, so f is injective.

9. (i) Let $x \in C$. Then $f(x) \in f(C)$, so by definition $x \in f^{-1}(f(C))$. Therefore $C \subseteq f^{-1}(f(C))$.

(ii) Suppose f is injective and let $C \subseteq A$. By (i) we need only show $f^{-1}(f(C)) \subseteq C$. Let $x \in f^{-1}(f(C))$; then by definition $f(x) \in f(C)$, so $f(x) = f(c)$ for some $c \in C$. Since f is injective, $x = c \in C$, so $f^{-1}(f(C)) \subseteq C$ as required.

(iii) Suppose $f^{-1}(f(C)) = C$ for all subsets C of A. If $f(x) = f(y)$ then $f(x) \in f(\{y\})$ so $x \in f^{-1}(f(\{y\}))$. Since $f^{-1}(f(\{y\})) = \{y\}$, this implies $x = y$, so f is injective.

(iv) Suppose f is surjective and let D be a subset of B. Let $y \in f(f^{-1}(D))$. Then $y = f(x)$ for some $x \in f^{-1}(D)$, so $f(x) = y \in D$. Hence $f(f^{-1}(D)) \subseteq D$. Now let $y \in D$. Since f is surjective $y = f(x)$ for some $x \in A$. Since $f(x) \in D$, $x \in f^{-1}(D)$ so $y = f(x) \in f(f^{-1}(D))$. Therefore $D \subseteq f(f^{-1}(D))$, which completes the proof.

11. (i) Suppose f is injective and let X, Y be elements of $\mathscr{P}(A)$ (i.e. subsets of A) such that $\mathscr{P}_f(X) = \mathscr{P}_f(Y)$. In other words, $f(X) = f(Y)$ so that $f^{-1}(f(X)) = f^{-1}(f(Y))$. By question 5.3.9(ii) this implies $X = Y$ so $\mathscr{P}_f$ is injective.

(ii) Suppose f is surjective and let $D \in \mathscr{P}(B)$; that is, $D \subseteq B$. For each $d \in D$ choose $c_d \in A$ such that $f(c_d) = d$. Let $C = \{c_d : d \in D\}$. Now $\mathscr{P}_f(C) = \{f(c_d) : d \in D\} = D$ so $\mathscr{P}_f$ is surjective.

The converse statements are both true.

Exercises 5.4

1. (i) Bijective; $f^{-1} : \mathbb{Z} \to \mathbb{Z}$, $f^{-1}(n) = n + 17$
   (iii) Not bijective
   (iv) Bijective; $f^{-1} = f$.

2. (i) $g \circ f = f \circ g = id_{\mathbb{Z}_5}$; hence $f^{-1} = g$
   (iii) $f([0]) = [0]$, $f([1]) = [1]$, $f([2]) = [3]$, $f([3]) = [2]$ and $f([4]) = [4]$; $f^{-1} = f$.

3. (i) $f^{-1} : \mathbb{R} \to \mathbb{R}$, $f^{-1}(y) = (8y - 3)/5$

   (iii) $f^{-1} : [-2, 2] \to [1, 3]$, $f^{-1}(y) = (y + 4)/2$

   (v) $f^{-1} = f$

   (vii) $f^{-1} : \mathbb{R}^2 \to \mathbb{R}^2$, $f^{-1}(x, y) = ((2x - y)/3, (x - 2y)/3)$

   (ix) $f^{-1} : \mathbb{Z} \to \mathbb{Z}^+$, $f^{-1}(n) = \begin{cases} 2n & \text{if } n \geqslant 1 \\ 1 - 2n & \text{if } n \leqslant 0. \end{cases}$

4. (i) (a) $\delta_C(a) = 0$, $\delta_C(b) = 1$, $\delta_C(c) = 0$, $\delta_C(d) = 1$, $\delta_C(e) = 1$.
   (b) δ_C is injective if $|A| = 2$ and $|C| = 1$ or if $|A| = 1$ (and $|C| = 0$ or 1).
   (c) δ_C is surjective if $|A| \geqslant 2$ and $C \neq \varnothing$, $C \neq A$.

   (ii) The images of the elements of A are: $f(\varnothing) = (0, 0)$, $f(\{a\}) = (1, 0)$, $f(\{b\}) = (0, 1)$, $f(\{a, b\}) = (1, 1)$, so f is clearly a bijection.

   (iii) If $X = \{x_1, x_2, \ldots, x_n\}$ and $C \subseteq X$ then define $f(C) = (\delta_C(x_1), \delta_C(x_2), \ldots, \delta_C(x_n))$.

5. (i) $f(x) = 2x + 1$
   (ii) $f(x) = x/(1 - x)$
   (iii) $f(x) = x - 1/x$
   (iv) $f(x) = (2x - 1)/(x - x^2)$ (the composite of the functions in (ii) and (iii))

(v) $f : \mathbb{Z} \to \mathbb{Z}^+$, $f(n) = \begin{cases} 2n & \text{if } n \geqslant 1 \\ 1 - 2n & \text{if } n \leqslant 0. \end{cases}$

6. (ii) $f(x) = x$, $g(x) = -x$

7. (iii) (a) If $|B| = 1$ there is only one function $A \to B$ which is obviously surjective.

(b) If $|A| = n$ and $B = \{b_1, b_2\}$ there are 2^n functions $A \to B$. The only functions which are not surjective are the function which sends every element to b_1 and the function which sends every element to b_2. Therefore there are $2^n - 2$ surjections $A \to B$.

(c) If $m = n - 1$ then a surjection $A \to B$ is such that two elements a_1, a_2 of A have the same image and all other elements have a unique image. There are $\frac{1}{2}n(n-1)$ ways of choosing the pair $\{a_1, a_2\}$. Then the number of surjections such that this pair has the same image is only the number of bijections from one m-element set to another m-element set. By (i), there are $m!$ such bijections, so the total number of surjections is $\frac{1}{2}n(n-1) \times m!$.

8. Suppose $|A| = |B| = n$ and $f : A \to B$ is injective. Then $|f(A)| = n$, so $f(A)$ is an n-element subset of the n-element set B. Therefore $f(A) = B$, so f is surjective.

To prove the converse we prove its contrapositive: if f is not injective then f is not surjective. So suppose that f is not injective. Then, for some pair of distinct elements a_1, a_2 of A, $f(a_1) = f(a_2)$. Thus $|f(A)| < |A| = n$, so $f(A) \neq B$ and f is not surjective.

11. (i) $f^{-1} : \{1, 2, 3, 4, 5, 6, 7, 8\} \to \{1, 2, 3, 4, 5\}$, $f^{-1} : x \mapsto x - 2$

f^{-1} is partial with domain $\{3, 4, 5, 6, 7\}$.

(iii) $f^{-1} : \mathbb{R} \to \mathbb{R}$, $f^{-1} : x \mapsto (x + 8)/4$; f^{-1} is total.

(v) $f^{-1} : \mathbb{Z}^+ \to \mathbb{Z}$, $f^{-1} = \begin{cases} -\sqrt{m-1} & \text{if } m = n^2 + 1 \\ & \text{for some integer } n \\ \sqrt{m} & \text{if } m = n^2 \\ & \text{for some integer } n. \end{cases}$

f^{-1} is partial with domain $\{n^2 : n \in \mathbb{Z}\} \cup \{n^2 + 1 : n \in \mathbb{Z}\}$.

Exercises 5.5

1. (i) $\aleph_0$. A bijection $\mathbb{Z}^+ \to \{n : n \geqslant 10^6\}$ is $n \mapsto (n + 10^6 - 1)$.

   (ii) c. A bijection $(0,1) \to \mathbb{R}$ is given in exercise 5.4.5(iv).

   (iii) $\aleph_0$. A bijection $\mathbb{Z}^+ \times \{0,1\} \to \mathbb{Z}^+$ is given by

$$f(n,\delta) = \begin{cases} 2n-1 & \text{if } \delta = 0 \\ 2n & \text{if } \delta = 1. \end{cases}$$

   (iv) $\aleph_0$. $|\mathbb{Z}^+ \times \mathbb{Z}^+ \times \mathbb{Z}^+| = |\mathbb{Z}^+ \times \mathbb{Z}^+| \times |\mathbb{Z}^+| = \aleph_0 \times \aleph_0 = \aleph_0$.

2. Suppose $f : A \to \mathscr{P}(A)$ is a bijection and $B = \{x \in A : x \notin A_x\}$ as in the hint. Since f is surjective, $B = f(y)$ for some $y \in A$; that is $B = A_y$. Now if $y \in B$ then $y \notin A_y$ (by definition of B) so $y \notin B$. Conversely, if $y \notin B$ then $y \in A_y$ so $y \in B$ (again by definition of B). This is a contradiction, so there is no bijection $A \to \mathscr{P}(A)$. (Note: this proof actually shows that there is no *surjection* $A \to \mathscr{P}(A)$.)

3. Let $X = \{\text{functions } A \to \{0,1\}\}$; then $|X| = 2^{|A|}$. Define $F : \mathscr{P}(A) \to X$ by $F(B) = f_B$, where f_B is the function $A \to \{0,1\}$ given by $f_B(a) = 0$ if $a \notin B$ and $f_B(a) = 1$ if $a \in B$. It is not too difficult to show that F is a bijection; hence $|\mathscr{P}(A)| = 2^{|A|}$.

4. (i) Let $A = \{n \in \mathbb{Z} : n > k\}$, $B = \{1, 2, \ldots, k\}$. Then $|A| + |B| = |A \cup B| = \aleph_0$, since $A \cap B = \varnothing$ and $A \cup B = \mathbb{Z}^+$. Clearly $|B| = k$ and it is easy to show that $|A| = \aleph_0$; see exercise 5.5.1(i).

   (ii) Let X be the set of all functions $\{0,1\} \to \mathbb{Z}^+$; then $|X| = (\aleph_0)^2$. The function $F : X \to \mathbb{Z}^+ \times \mathbb{Z}^+$ given by $F : f \mapsto (f(0), f(1))$ is a bijection, so $(\aleph_0)^2 = |X| = |\mathbb{Z}^+ \times \mathbb{Z}^+| = \aleph_0$.

5. Let A be a set with cardinality α, and X be the set of all functions $\{0,1\} \to A$. Then $|X| = \alpha^2$ and $|A \times A| = \alpha \times \alpha$. A bijection $F : X \to A \times A$ is given by $F : f \mapsto (f(0), f(1))$.

Exercises 5.6

1. (i) The only functional dependences between single attributes are:

A_1 functionally determines A_3;

A_2 functionally determines A_3.

(ii) $\{A_1, A_2\}$ and $\{A_2, A_4\}$ are the only candidate keys.

(iii) (a) The table is not in second normal form. A_3 is non-prime (since it appears in neither candidate key) and A_2 functionally determines A_3. Thus the non-prime attribute A_3 is functionally dependent on $\{A_2\}$, a proper subset of a candidate key (in fact, a proper subset of both candidate keys).

(b) The table is not in third normal form, since it is not in second normal form.

2. (i) Since there are no record instances with the same values for A_1, A_2 *and* A_4, the set $\{A_1, A_2, A_4\}$ functionally determines A_3 and A_5. We need to check that no proper subset of $\{A_1, A_2, A_4\}$ functionally determines every attribute. $\{A_1, A_2\}$ does not functionally determine A_4 since the record instances in rows 1 and 2 have the same A_1 and A_2 values but different A_4 values. Similarly, $\{A_1, A_4\}$ does not functionally determine A_3 (rows 1 and 3) and $\{A_2, A_4\}$ does not functionally determine A_1 (rows 6 and 7).

(ii) R is in (second and) third normal form. It is easy to check that $\{A_1, A_3, A_4, A_5\}$ is another candidate key for R. Therefore every attribute appears in some candidate key, so there are no non-prime attributes. Hence R satisfies definitions 5.10 and 5.11 albeit rather trivially.

3. Since the key is a single attribute, all three tables are in second normal form.

(i) If no two students have exactly the same name then A_2 = STUDENT_NAME functionally determines A_3, A_4 and A_5. Therefore PERSONAL and DISCIPLINE are not in third normal form. However, CURRENT_COURSE is in third normal form since no B_i attribute $(i = 1, 2, 3, 4)$ functionally determines any other B_i attribute.

(ii) If two students do have exactly the same name then A_2 = STUDENT_NAME does not functionally determine any other attribute and therefore all three tables are in third normal form.

Note: we are assuming here that A_3 = DATE_OF_BIRTH does not functionally determine any other attribute. In other words, we suppose that two different students at the college have the same date of birth, an assumption which is highly probable.

4. (i) The table is in second normal form since the key is a single attribute. It is not in third normal form because DEPARTMENT_# functionally determines WORK_LOCATION. Allowing the possibility that several employees have the same name (and work in the same department with the same job description), there are no other functional dependences.

   The same information can be stored on two tables of attribute types {**EMPLOYEE_#**, EMPLOYEE_NAME, DEPARTMENT_#, JOB DESCRIPTION} and {**DEPARTMENT_#**, WORK_LOCATION} which are both in third normal form.

   (ii) This table is not in second normal form (and hence not in third) because, for example, AIRLINE is functionally dependent on {FLIGHT_#}, which is a proper subset of the primary key. The two tables of types {**PASSENGER_NAME**, **FLIGHT_#**, DATE, CLASS} and {**FLIGHT_#**, AIRLINE, EMBARKATION, DESTINATION} respectively are both in third normal form.

   (iii) PATIENT_HISTORY is not in second normal form since PATIENT_NAME is functionally dependent on the proper subset {PATIENT_#} of the key. PATIENT_CURRENT is in second but not third normal form since both CONSULTANT_NAME and CONSULTANT_PHONE are functionally dependent on {CONSULTANT_#} which does not contain the key as a subset. TREATMENT_CURRENT is similarly in second but not third normal form as DAILY_COST is functionally dependent on {DRUG, QUANTITY}.

   The following tables hold the same information and are in third normal form.

PATIENT:	{**PATIENT_#**, PATIENT_NAME}
CONSULTANT:	{**CONSULTANT_#**, CONSULTANT_NAME, CONSULTANT_PHONE}
HISTORY:	{**PATIENT_#**, **ADMISSION_DATE**, DISCHARGE_DATE, CONDITION}
CURRENT:	{**PATIENT_#**, CONSULTANT_#, CONDITION, WARD_#}
TREATMENT:	{**PATIENT_#**, **DRUG**, **QUANTITY**}
COST:	{**DRUG**, **QUANTITY**, DAILY_COST}.

5. Let R be a table with attributes $(A_1, A_2, \ldots, A_n)$; for each attribute A_i, let X_i denote the set of data items. Let $I = \{i_1, i_2, \ldots, i_k\} \subseteq \{1, 2, \ldots, n\}$ and $J = \{j_1, j_2, \ldots, j_m\} \subseteq \{1, 2, \ldots, n\}$ be disjoint sets of indices. As in the text, for convenience we suppose that $i_1 < i_2 <$

$\cdots < i_k < j_1 < j_2 < \cdots < j_m$. Then $\boldsymbol{A_J} = \{A_{j_1}, A_{j_2}, \ldots, A_{j_m}\}$ is **functionally dependent** on $\boldsymbol{A_I} = \{A_{i_1}, A_{i_2}, \ldots, A_{i_k}\}$ if the projection of R onto $I \cup J$, $p_{I\cup J}(\mathsf{R})$, defines a partial (or total) function $X_{i_1} \times X_{i_2} \times \cdots \times X_{i_k} \to X_{j_1} \times X_{j_2} \times \cdots \times X_{j_m}$.

Chapter 6

Exercises 6.1

1. (i) True (ii) False (iii) False
   (iv) False (v) True.

2.
$$A = \begin{pmatrix} 0 & -1 \\ 1 & 0 \\ 2 & 1 \end{pmatrix} \qquad B = \begin{pmatrix} -1 & -3 \\ 0 & -2 \\ 1 & -1 \end{pmatrix} \qquad C = \begin{pmatrix} 7 & 10 \\ 11 & 14 \\ 15 & 18 \end{pmatrix}.$$

4. Any 1×1 matrix.

6. (i) (a) $\begin{pmatrix} -1 & 2 & 3 \end{pmatrix}$ (c) $\begin{pmatrix} -1 & 0 \\ 6 & 7 \\ 3 & -1 \\ 4 & 3 \end{pmatrix}$.

Exercises 6.2

1. (i) $\begin{pmatrix} 10 & -5 \\ 4 & -3 \end{pmatrix}$ (iii) $\begin{pmatrix} 2 & 1 \\ -2 & 1 \end{pmatrix}$ (v) $\begin{pmatrix} 8 & -11 \\ 4 & -5 \end{pmatrix}$

   (vii) $\begin{pmatrix} 4 & -7 \\ -4 & 5 \end{pmatrix}$ (ix) $\begin{pmatrix} 8 & -2 \\ -14 & 4 \end{pmatrix}$ (xi) $\begin{pmatrix} 2 & -2 \\ 1 & 1 \end{pmatrix}$.

2. (i) $\begin{pmatrix} 9 & 13 \end{pmatrix}$ (iii) Does not exist (v) $\begin{pmatrix} 20 & 10 & -6 \\ 18 & 12 & -6 \\ -2 & -3 & 1 \end{pmatrix}$

   (vii) $\begin{pmatrix} 70 & 40 & -22 \end{pmatrix}$.

3. B is any matrix of the form
$$\begin{pmatrix} a & b \\ 2b & a \end{pmatrix}.$$

6. Let $A = [a_{ij}]$ and $I_n = [b_{ij}]$ where $b_{ii} = 1$ and $b_{ij} = 0$ for $i \neq j$. Then

$$\begin{aligned}(i,j)\text{-entry of } AI_n &= \sum_{k=1}^{n} a_{ik}b_{kj} \\ &= a_{i1}b_{1j} + a_{i2}b_{2j} + \cdots + a_{in}b_{nj} \\ &= a_{ij} \\ &= (i,j)\text{-entry of } A.\end{aligned}$$

Therefore

$$AI_n = A.$$

A similar argument shows that $I_m A = A$.

8. (i) $n = p$, $q = r$. The dimension of ABC is $m \times s$.
   (ii) $s = p$, $q = m$. The dimension of CBA is $r \times n$.
   (iii) $n = q = r$, $m = p$. The dimension of $(A + B)C$ is $m \times s$ (or $p \times s$).

9. Take, for instance,

$$A = \begin{pmatrix} 1 & 1 \\ 1 & 1 \end{pmatrix} \quad B = \begin{pmatrix} 1 & 1 \\ -1 & -1 \end{pmatrix}.$$

12.
$$\begin{aligned}(A + A^{\mathrm{T}})^{\mathrm{T}} &= A^{\mathrm{T}} + (A^{\mathrm{T}})^{\mathrm{T}} && \text{(by 11(ii))} \\ &= A^{\mathrm{T}} + A && \text{(by 11(i))} \\ &= A + A^{\mathrm{T}} && \text{(by commutativity of matrix addition).}\end{aligned}$$

Therefore $A + A^{\mathrm{T}}$ is a symmetric matrix.

Exercises 6.3

1. (i) Yes. Add the elements of row 2 to row 1, or add the elements of column 1 to column 2.
   (iii) Yes. Subtract three times row 3 from row 1 or subtract three times column 1 from column 3.
   (v) No.
   (vii) Yes. Multiply the elements of any row/column by 1.

2. (i) To obtain B from A, interchange the first and third rows. So

$$E_1 = \begin{pmatrix} 0 & 0 & 1 \\ 0 & 1 & 0 \\ 1 & 0 & 0 \end{pmatrix}.$$

(ii) To obtain A from B, interchange the first and third rows. So $E_2 = E_1$ above.

(iii) $E_1E_2 = E_2E_1 = I_3$.

4. (i) To obtain B from A, multiply the second column by -2. So

$$F_1 = \begin{pmatrix} 1 & 0 & 0 \\ 0 & -2 & 0 \\ 0 & 0 & 1 \end{pmatrix}.$$

(ii) $F_2 = \begin{pmatrix} 1 & 0 & 0 \\ 0 & -\frac{1}{2} & 0 \\ 0 & 0 & 1 \end{pmatrix}$.

(iii) $F_1F_2 = F_2F_1 = I_3$.

5.

$$Q = \begin{pmatrix} 0 & \frac{1}{2} & 0 \\ 1 & 0 & 0 \\ 0 & 0 & 1 \end{pmatrix}.$$

7. No, because elementary row operations are not commutative. Consider, for example,

$$E_1 = \begin{pmatrix} 3 & 0 \\ 0 & 1 \end{pmatrix} \quad \text{and} \quad E_2 = \begin{pmatrix} 0 & 1 \\ 1 & 0 \end{pmatrix}.$$

Exercises 6.4

2. (i) $\begin{pmatrix} 2 & 3 \\ 3 & 5 \end{pmatrix}$ (iii) No inverse (v) No inverse

(vii) $\begin{pmatrix} \frac{1}{15} & \frac{4}{15} & -\frac{1}{30} \\ \frac{1}{5} & -\frac{1}{5} & -\frac{1}{10} \\ \frac{1}{3} & -\frac{2}{3} & \frac{1}{3} \end{pmatrix}$.

4. (ii) $A^{-1}B^4A = \begin{pmatrix} 1 & 0 \\ 0 & 16 \end{pmatrix}$.

(iii) The proof is by induction and is outlined below.

The result holds trivially for $n = 1$.

Suppose

$$(A^{-1}BA)^k = A^{-1}B^kA \quad (k \geqslant 1)$$

then

$$\begin{aligned}(A^{-1}BA)^{k+1} &= (A^{-1}BA)^k(A^{-1}BA)\\ &= A^{-1}B^kAA^{-1}BA\\ &= A^{-1}B^kBA\\ &= A^{-1}B^{k+1}A.\end{aligned}$$

6. If $A = [a_{ij}]$ where $a_{ij} = 0$ for $i \neq j$, then $A^{-1} = [b_{ij}]$ where $b_{ij} = 0$ for $i \neq j$ and $b_{ii} = 1/a_{ii}$.

7.

$$\begin{aligned} & (I_n - A)(A_n + A) = O_{n\times n} \\ \Leftrightarrow\quad & (I_n - A)I_n + (I_n - A)A = O_{n\times n} \quad \text{(distributive law)} \\ \Leftrightarrow\quad & I_n - A + A - A^2 = O_{n\times n} \quad \text{(distributive law)} \\ \Leftrightarrow\quad & A^2 = I_n \\ \Leftrightarrow\quad & A \text{ is involutary.} \end{aligned}$$

Chapter 7

Exercises 7.1

1. $x = 4$, $y = 2$.

2. $x = 1$, $y = 2$, $z = -2$.
$x = 0$, $y = 0$, $z = 0$.

3. $x = 0$, $y = 0$, $z = 4$.

Exercises 7.2

1. $x = 1$, $y = 3$, $z = -2$.
$x = y = z = 0$.

2. Inconsistent.

3. $x_1 = t$, $x_2 = 2 + t$, $x_3 = 3t$.

4. $x_1 = -t, x_2 = x_3 = t.$

5. $x_1 = 13 - s - 3t, x_2 = t - 4, x_3 = s, x_4 = t.$

6. $x = z = t, y = 2t.$

7. $x_1 = \frac{53}{49} - \frac{30}{49}t, x_2 = \frac{12}{49} + \frac{8}{49}t, x_3 = \frac{8}{7} + \frac{3}{7}t, x_4 = t.$

Exercises 7.3

1. (i) $x = 2, y = -\frac{3}{2}, z = \frac{5}{2}$
   (ii) $x_1 = -5, x_2 = 0, x_3 = 3$
   (iii) $x = -2, y = 11, z = 4$
   (iv) $x = 1, y = \frac{1}{2}, z = 0$
   (v) $x = y = z = 0.$

2. (i) $x_1 = -6 - 3t, x_2 = 8 - t, x_3 = t$
   (ii) $x = \frac{14}{5}, y = -\frac{12}{5}, z = \frac{2}{5}$
   (iii) $x = 6 - 2t, y = 2 + 3t, z = t$
   (iv) $x = y = z = 0.$

3. (i) $x_1 = 2t, x_2 = 3 + t, x_3 = t$
   (ii) Inconsistent
   (iii) $x = -25, y = -37, z = -14$
   (iv) $x_1 = x_2 = x_3 = 1$
   (v) $x = 0, y = z = t$
   (vi) Inconsistent.

Chapter 8

Exercises 8.1

1. (i) No, because $\mathbb{R}^+$ is not closed under subtraction.
   (iii) Yes. (v) Yes.
   (vi) No, because only matrices having the same dimension can be added.

3. (i) The Cayley table is symmetric about the leading diagonal.
   (ii) The row corresponding to an element x is the same as the column headings and the column corresponding to the same element is the transpose of that row. Then x is the identity.

4. (i) Yes. (ii) A.
(iii) A is the only element with an inverse. (A is self-inverse.)

6. (i) 2^4 (ii) 3^9 (iii) 4^{16} (iv) n^{n^2}.

7. There are six possible Cayley tables, only one of which defines an associative operation.

8. (ii) Yes. (iii) Yes. (iv) Yes: $\varnothing$.
(v) Each element is self-inverse.

Exercises 8.2

1. To show that matrix multiplication is a binary operation on the set of 2×2 non-singular matrices, note that, if A and B are non-singular matrices, then AB is also a non-singular matrix since $(AB)^{-1} = B^{-1}A^{-1}$ (see theorem 6.4). It is a simple matter to show that the group properties hold.

2. Suppose that there are two idempotent elements, the identity (which is clearly idempotent) and another element x. Pre-multiplying the equation $x * e = x * x$ by x^{-1} shows that $x = e$ and hence that e is the only idempotent element.

3. $(\mathbb{Z}_6, \times_6)$ is not a group because each of $[0]$, $[2]$, $[3]$ and $[4]$ has no inverse.

4. $(\mathbb{Z}_n - \{0\}, \times_n)$ is a group if and only if n is prime (or $n = 1$).

5. $(P, *)$ is not a group since $*$ is not associative.

6. $(S, *)$ is a monoid if and only if S is a singleton set.

8. $(\mathbb{N}, *)$ is a monoid but $(\mathbb{N}, \circ)$ is not. The structure $(\mathbb{N}, \circ)$ is a semigroup.

9. (a) Yes. (b) Yes. (c) No.

Exercises 8.3

1. For all $a, b \in G$,

$$\begin{aligned}(ab)(b^{-1}a^{-1}) &= a(bb^{-1})a^{-1} \\ &= aa^{-1}\end{aligned}$$

$$= e.$$

Hence

$$(ab)^{-1} = b^{-1}a^{-1}.$$

The proof that $(a^n)^{-1} = (a^{-1})^n$ is by induction. Note that the first result allows us to write $(a^k a)^{-1} = a^{-1}(a^k)^{-1}$.

2. Most of the Cayley table can be completed using theorem 8.5. However, there are a couple of places where some reasoning is required. For instance:

$$q^2 = (p^2)q = p(pq) = pe = p.$$

4.

$*$	r_0	r_1	m_1	m_2
r_0	r_0	r_1	m_1	m_2
r_1	r_1	r_0	m_2	m_1
m_1	m_1	m_2	r_0	r_1
m_2	m_2	m_1	r_1	r_0

5. The proof utilizes the fact that $\{g^r : r \in \mathbb{Z}\}$ has at most n distinct elements. If $\{g^r : r = 0, 1, \ldots, n\}$ has fewer than n distinct elements, then $g^r = g^s$ for some r and s where $r < s \leqslant n$. This gives $g^{s-r} = e$ and so $m = s - r$. If $\{g^r : r = 0, 1, \ldots, n\}$ has exactly n distinct elements, then $g^{n+1} = g^s$ for some $s \leqslant n$ and $m = n + 1 - s$.

7. For any group, if g is a generator then so is g^{-1}. If $(G, *)$ has only one generator g, then $g = g^{-1}$. This gives $g^2 = e$ and so $G = \{e, g\}$. (The trivial group with the identity as its only element is also cyclic with a single generator.)

8. The generators are [1], [5].

9. The group is cyclic with generators r_1 and r_2.

10. Use the fact that all elements of G are self-inverse and the 'shoes and socks' theorem (exercise 8.3.1).

12. The subsets of $\mathbb{Z}_{10}$ which form groups under $\times_{10}$ are: $\{[1]\}$, $\{[1], [9]\}$ and $\{[1], [3], [7], [9]\}$. All are cyclic.

13. (b) The elements of S are those non-zero members of $\mathbb{Z}_n$ which share no common factors with n.

14. In C_5, each non-identity element generates the group.
In C_6, only g and g^5 generate the group.
In C_9, each of g, g^2, g^4, g^5, g^7, g^8 generates the group.
In C_n, g^r generates the group if and only if r and n share no common factors (other than 1).

Exercises 8.4

1. Let $M' = \{x : x^2 = x,\ \ x \in M\}$. Clearly $M' \subseteq M$ and, since $e^2 = e$, $e \in M$. If $x, y \in M$

$$\begin{aligned} xy &= x^2y^2 \\ &= x(xy)y \\ &= xyxy \qquad \text{(since } (M, *) \text{ is abelian)} \\ &= (xy)^2. \end{aligned}$$

This shows that M' is closed under $*$ and therefore $(M', *)$ is a submonoid of $(M, *)$.

2. (ii) $\{[0], [2], [4], [6]\}$ and $\{[0], [4]\}$.

5. $(\{[0], [3], [6]\}, +_9)$ is a subgroup of $(\mathbb{Z}_9, +_9)$.

7. (i) Let C be the centre of $(G, *)$. $C \neq \varnothing$ since $e \in C$. Suppose $a, b \in C$. Given $g \in G$,

$$\begin{aligned} abg &= agb \quad \text{(since } b \in C) \\ &= gab \quad \text{(since } a \in C) \\ \Rightarrow \qquad ab &\in C. \end{aligned}$$

If $a \in C$, then

$$\begin{aligned} ag &= ga && \text{for all } g \in G \\ \Rightarrow \quad a^{-1}aga^{-1} &= a^{-1}gaa^{-1} && \text{for all } g \in G \\ \Rightarrow \quad ga^{-1} &= a^{-1}g && \text{for all } g \in G. \end{aligned}$$

Therefore

$$a^{-1} \in C.$$

From theorem 8.6 we can conclude that $(C, *)$ is a subgroup of $(G, *)$.

(ii) Centre of $D_3 = \{r_0\}$.

9. To apply theorem 8.6, we must prove that $ab^{-1} \in H$ for all $a, b \in H \Leftrightarrow H$ is closed under $*$ and, for all $a \in H$, $a^{-1} \in H$. To prove that the second proposition implies the first is straightforward. To prove that the first implies the second, begin by establishing that $e \in H$. It then follows that $ea^{-1} \in H$ for all $a \in H$, i.e. that $a^{-1} \in H$ for all $a \in H$. To show that H is closed under $*$: if $a, b \in H$ then $b^{-1} \in H$. Hence $a(b^{-1})^{-1} \in H \Rightarrow ab \in H$.

10. $(H \cup K, *)$ is not necessarily a subgroup of $(G, *)$. Consider, for example, $G = \{r_0, r_1, m_1, m_2\}$, the set of symmetries of a non-square rectangle (see exercise 8.3.4). If $H = \{r_0, r_1\}$ and $K = \{r_0, m_1\}$, then $(H, *)$ and $(K, *)$ are both groups but $(H \cup K, *)$ is not.

11. $(\{[1], [6]\}, \times_7)$ and $(\{[1], [2], [4]\}, \times_7)$.

14. Let G be a group of order n and let $g \in G$.

Suppose that g has order m. Then, by theorem 8.8, g generates a cyclic subgroup of G,

$$H = \{e, g, g^2, \ldots, g^{m-1}\} \text{ where } g^m = e.$$

This subgroup H has order m so we know that m is a factor of n by Lagrange's theorem. Hence $n = km$ for some $k \in \mathbb{Z}^+$ which implies

$$g^n = g^{mk} = (g^m)^k = e^k = e$$

as required.

Exercises 8.5

3. It is easy to show that f is bijective. Also

$$\begin{aligned} f\left[\begin{pmatrix} 1 & n \\ 0 & 1 \end{pmatrix}\begin{pmatrix} 1 & m \\ 0 & 1 \end{pmatrix}\right] &= f\left[\begin{pmatrix} 1 & n+m \\ 0 & 1 \end{pmatrix}\right] \\ &= n + m \\ &= f\left[\begin{pmatrix} 1 & n \\ 0 & 1 \end{pmatrix}\right] + f\left[\begin{pmatrix} 1 & m \\ 0 & 1 \end{pmatrix}\right]. \end{aligned}$$

4. (ii) and (iv) are morphisms. Both are isomorphisms.

5. (i), (iii) and (iv) are morphisms. None are isomorphisms.

6. If e is the identity in $(A, *)$ then

$$\begin{array}{lll} & a * e = e * a = a & \text{for all } a \in A \\ \Rightarrow & f(a * e) = f(e * a) = f(a) & \text{for all } a \in A \\ \Rightarrow & f(a) \circ f(e) = f(e) \circ f(a) = f(a) & \text{for all } f(a) \in f(A) \\ \Rightarrow & f(e) \text{ is the identity in } (f(A), \circ). & \end{array}$$

7. There are two isomorphisms:

$$f : A \mapsto [1], C \mapsto [4], B \mapsto [2], D \mapsto [3]$$
$$g : A \mapsto [1], C \mapsto [4], B \mapsto [3], D \mapsto [2].$$

9. (ii) Suppose that $(G, *)$ is abelian. For all $x, y \in G$,

$$\begin{aligned} f(x * y) &= (xy)^2 \\ &= x(yx)y \\ &= xxyy \\ &= x^2y^2 \\ &= f(x) * f(y) \end{aligned}$$

so that f is a morphism.

Suppose that f is a morphism. For all $x, y \in G$,

$$\begin{array}{lrcl} & f(x * y) &=& f(x) * f(y) \\ \Rightarrow & (xy)^2 &=& x^2y^2 \\ \Rightarrow & xyxy &=& xxyy \\ \Rightarrow & x^{-1}xyxyy^{-1} &=& x^{-1}xxyyy^{-1} \\ \Rightarrow & yx &=& xy \end{array}$$

so that $*$ is commutative.

11. Since every element of a group has a unique inverse, f is a bijective function. If $(G, *)$ is abelian, then

$$f(x * y) = (xy)^{-1} = y^{-1}x^{-1} = x^{-1}y^{-1} = f(x) * f(y).$$

This shows that if $(G, *)$ is abelian then f is a morphism. Now suppose that f is a morphism. For all $x, y \in G$,

$$f(x * y) = f(x) * f(y)$$

$$\Rightarrow \quad (xy)^{-1} = x^{-1}y^{-1}$$
$$\Rightarrow \quad y^{-1}x^{-1} = x^{-1}y^{-1}.$$

So $(G, *)$ is abelian.

13. (a) (i) $\mathit{ker}\, f = \{0\}$;
(ii) $\mathit{ker}\, f = \{0\}$;
(iii) $\mathit{ker} f = \mathbb{R}$.

(b) (i) Suppose f is a monomorphism, i.e. f is injective. If $g \in \mathit{ker}\, f$ then $f(g) = e_2$. But $f(e_1) = e_2$ so that $f(g) = f(e_1)$ and therefore $g = e_1$. This shows that the only member of *ker* f is e_1.
To prove the converse, suppose that *ker* $f = \{e_1\}$. We have

$$\begin{aligned} & f(x) = f(y) \\ \Rightarrow \quad & f(x) \circ [f(y)]^{-1} = e_2 \\ \Rightarrow \quad & f(x * y^{-1}) = e_2 \\ \Rightarrow \quad & x * y^{-1} \in \mathit{ker}\, f \\ \Rightarrow \quad & x * y^{-1} = e_1 \\ \Rightarrow \quad & x = y. \end{aligned}$$

Hence f is injective.

(ii) Since $e_1 \in \mathit{ker}\, f$, $\mathit{ker}\, f \neq \varnothing$. If $g_1, g_2 \in \mathit{ker}\, f$,

$$\begin{aligned} f(g_1 * g_2) &= f(g_1) \circ f(g_2) \\ &= e_2 \circ e_2 \\ &= e_2 \end{aligned}$$

$\Rightarrow$ *ker* f is closed under $*$.

If $g \in \mathit{ker}\, f$, $f(g^{-1}) = [f(g)]^{-1} = e_2{}^{-1} = e_2$, so $g^{-1} \in \mathit{ker}\, f$. Theorem 8.6 allows us to conclude that $(\mathit{ker}\, f, *)$ is a subgroup of $(G, *)$.

(iii)
$$\begin{aligned} f(g^{-1} * x * g) &= f(g^{-1}) \circ f(x) \circ f(g) \\ &= f(g^{-1}) \circ f(g) \quad \text{(since } x \in \mathit{ker}\, f\text{)} \\ &= f(g^{-1} * g) \\ &= f(e_1) \\ &= e_2. \end{aligned}$$

Hence $g^{-1} * x * g \in \mathit{ker}\, f$.

Exercises 8.6

2.

$$G = \begin{pmatrix} 1 & 0 & 0 & 1 & 0 & 0 & 1 & 0 & 0 \\ 0 & 1 & 0 & 0 & 1 & 0 & 0 & 1 & 0 \\ 0 & 0 & 1 & 0 & 0 & 1 & 0 & 0 & 1 \end{pmatrix} = (I_3\, I_3\, I_3).$$

(i) Two (ii) One.

3. (i) Probably correctly transmitted.
(iii) Probably correctly transmitted.
(v) Incorrectly transmitted.

4. The group properties can be established from the Cayley table for the set of codewords under $\oplus$.
(i) Two (ii) One.

5. $m = 4$, $n = 7$.

7. (i) Likely error in the fourth bit.
(ii) Probably correctly transmitted.
(iii) Error in more than one bit and therefore the correct word cannot be determined.
(iv) Likely error in second bit.

8. For the $(1, 3)$ Hamming code,

$$H = \begin{pmatrix} 1 & 1 & 0 \\ 1 & 0 & 1 \end{pmatrix}.$$

For the $(4, 7)$ Hamming code,

$$H = \begin{pmatrix} 1 & 1 & 1 & 0 & 1 & 0 & 0 \\ 1 & 0 & 1 & 1 & 0 & 1 & 0 \\ 1 & 1 & 0 & 1 & 0 & 0 & 1 \end{pmatrix}.$$

(Note that H is not unique.)

Chapter 9

Exercises 9.1

1. (i) (a) gcd(21600, 2970)

$$\begin{aligned} 21600 &= 7 \times 2970 + 810 \\ 2970 &= 3 \times 810 + 540 \\ 810 &= 1 \times 540 + 270 \\ 540 &= 1 \times 270 + 0 \end{aligned}$$

Hence gcd(21600, 2970) = 270.

(c) gcd(2679, 851)

$$\begin{aligned} 2679 &= 3 \times 851 + 126 \\ 851 &= 6 \times 126 + 95 \\ 126 &= 1 \times 95 + 31 \\ 95 &= 3 \times 31 + 2 \\ 31 &= 15 \times 2 + 1 \\ 2 &= 2 \times 1 + 0 \end{aligned}$$

Hence gcd(2679, 851) = 1.

(ii) (a) gcd(21600, 2970) = 270.

$$\begin{aligned} 270 &= 810 - 1 \times 540 \\ &= 810 - 1 \times (2970 - 3 \times 810) \\ &= -2970 + 4 \times 810 \\ &= -2970 + 4 \times (21600 - 7 \times 2970) \\ &= 4 \times 21600 - 29 \times 2970. \end{aligned}$$

(c) gcd(2679, 851) = 1.

$$\begin{aligned} 1 &= 31 - 15 \times 2 \\ &= 31 - 15 \times (95 - 3 \times 31) \\ &= -15 \times 95 + 46 \times 31 \\ &= -15 \times 95 + 46 \times (126 - 1 \times 95) \\ &= 46 \times 126 - 61 \times 95 \\ &= 46 \times 126 - 61 \times (851 - 6 \times 126) \end{aligned}$$

$$
\begin{aligned}
&= -61 \times 851 + 412 \times 126 \\
&= -61 \times 851 + 412 \times (2679 - 3 \times 851) \\
&= 412 \times 2679 - 1297 \times 851.
\end{aligned}
$$

2. (i) If $a|b$ and $b|c$ then $a|c$.

   Proof

   Suppose $a|b$ and $b|c$. Then there exist $m, n \in \mathbb{N}$ such that $b = ma$ and $c = nb$. Therefore $c = (nm)a$ where $nm \in \mathbb{N}$, so $a|c$. □

   (iii) $a|b$ if and only if $ma|mb$.

   Proof

   $(\Rightarrow)$ Suppose $a|b$. Then $b = na$ for some $n \in \mathbb{N}$. Therefore $mb = (mn)a$ where $nm \in \mathbb{N}$, so $ma|mb$.
   $(\Leftarrow)$ Conversely, suppose $ma|mb$. Then $mb = n(ma)$ for some $n \in \mathbb{N}$. Therefore, since $m \neq 0$, $b = na$ so $a|b$. □

3. Theorem 9.2(a)

   Let $a, b, c \in \mathbb{Z}^+$. If $c|a$ and $c|b$ then $c|ma + nb$ for all $m, n \in \mathbb{Z}^+$.

   Proof

   Let $a, b, c \in \mathbb{Z}^+$ and suppose that $c|a$ and $c|b$. Then there exist $p, q \in \mathbb{Z}^+$ such that $a = pc$ and $b = qc$.

   Now, for all $m, n \in \mathbb{Z}^+$, $ma + nb = mpc + nqc = c(mp + nq)$ where $mp + nq \in \mathbb{Z}^+$. Therefore $c|(ma + nb)$. □

7. (i) $3|57$ and $3|45$ so 57 and 45 are *not* coprime.
   In fact $\gcd(57, 45) = 3$ but we don't need to know this; it is sufficient to show that 57 and 45 have a common factor greater than 1.

   (iii) 112 and 117 differ by 5 so 5 is the only possible common prime factor. Since 5 divides neither integer, 112 and 117 are coprime.

8. (i) (b) $\gcd(1092, 1155, 2002)$

$$\gcd(1092, 1155) = 21$$

$$
\begin{aligned}
\Rightarrow \gcd(1092, 1155, 2002) &= \gcd(\gcd(1092, 1155), 2002) \\
&= \gcd(21, 2002) \\
&= 7.
\end{aligned}
$$

9. (ii) Euclid's algorithm to calculate $\gcd(a_n, a_{n-1})$ is as follows.

$$\begin{aligned} a_n &= 1 \times a_{n-1} + a_{n-2} \\ a_{n-1} &= 1 \times a_{n-2} + a_{n-3} \\ a_{n-2} &= 1 \times a_{n-3} + a_{n-4} \\ &\vdots \\ 3 &= 1 \times 2 + 1 \\ 2 &= 1 \times 1 + 1 \\ 1 &= 1 \times 1 + 0. \end{aligned}$$

Hence $\gcd(a_n, a_{n-1}) = 1$.

(iv) In general, for $n \in \mathbb{Z}^+$,

$$1 = (-1)^{n+1} a_n a_{n+2} + (-1)^{n+2} a_{n+1}^2.$$

11. Let a and b be coprime positive integers and let c be a positive integer such that $a|c$ and $b|c$.

Then there exist positive integers k and l such that

$$c = ka \quad \text{and} \quad c = lb.$$

By theorem 9.7 there exist integers n and m such that $1 = na + mb$. Therefore

$$\begin{aligned} c &= c \times 1 \\ &= c(na + mb) \\ &= cna + cmb \\ &= lbna + kamb \quad \text{since } c = lb \text{ and } c = ka \\ &= ab(ln + km) \quad \text{where } ln + km \in \mathbb{Z}^+. \end{aligned}$$

Hence $ab|c$ as required. □

Exercises 9.2

1. (i) (a) $1008 = 4 \times 252 = 4 \times 4 \times 63 = 2^4 \times 9 \times 7 = 2^4 \times 3^2 \times 7$.

(c) $3276 = 6 \times 546 = 2 \times 3 \times 6 \times 91 = 2 \times 3 \times 2 \times 3 \times 7 \times 13 = 2^2 \times 3^2 \times 7 \times 13$.

(ii) (b) $\gcd(1008, 3276) = \gcd(2^4 \times 3^2 \times 7, 2^2 \times 3^2 \times 7 \times 13) = 2^2 \times 3^2 \times 7 = 252$.

2. (ii) The first sequence of 5 consecutive composite positive integers is: $24, 25, 26, 27, 28$.

3. Let $p = 6$.

   (a) Let $a = 8$. Then p does not divide a and p and a are not coprime. So theorem 9.8 (a) fails for $p = 6$.

   (b) Let $a = 3$ and $b = 4$ so $ab = 12$ Then $p|12$ but p does not divide either a or b. So theorem 9.8 (b) fails for $p = 6$.

5. If $p = 3$ then $p^2 + 2 = 11$ is also prime.

   If $p \neq 3$ then $p = 3q + r$ where $r = 1$ or $r = 2$.

   When $r = 1$, $p^2 + 2 = (3q + 1)^2 + 2 = 9q^2 + 6q + 3 = 3(3q^2 + 2q + 1)$ which is divisible by 3.

   When $r = 2$, $p^2 + 2 = (3q + 2)^2 + 2 = 9q^2 + 12q + 6 = 3(3q^2 + 4q + 2)$ which is divisible by 3.

   Therefore, the only prime p for which $p^2 + 2$ also prime is $p = 3$.

7. (i) Let p be prime and a and b be positive integers such that $\gcd(a, p^2) = p$ and $\gcd(b, p^2) = p$.
   Then $a = k_1 p$ where $\gcd(k_1, p) = 1$ and, similarly, $b = k_2 p$ where $\gcd(k_2, p) = 1$. Hence $ab = k_1 k_2 p^2$ where $\gcd(k_1 k_2, p^2) = 1$. Therefore $\gcd(ab, p^4) = p^2$.

   (iii) Let $\gcd(a, b) = p$. Then $a = k_1 p$ and $b = k_2 p$ where $k_1, k_2 \in \mathbb{Z}^+$ are coprime.
   Note that $a^2 = k_1^2 p^2$ so p^2 divides a^2.
   First suppose k_2 is *not* a multiple of p. Then p^2 does not divide $b = k_2 p$ so $\gcd(a^2, b) = p$.
   Now suppose that k_2 *is* a multiple of p. Then p^2 divides $b = k_2 p$ so p^2 divides both a^2 and b. In this case, k_1 is not a multiple of p so p^2 is the highest power of p that divides $a^2 = k_1^2 p^2$. Hence $\gcd(a^2, b) = p^2$ in this case.
   Therefore $\gcd(a^2, b) = p$ or $\gcd(a^2, b) = p^2$.

8. Suppose that $n \in \mathbb{Z}^+$ is such that $\sqrt{n}$ is rational.

   Let $\sqrt{n} = \dfrac{a}{b}$ where a and b are integers.

   By the Fundamental Theorem of Arithmetic, a and b each have (unique) prime factorisations. It will be helpful to write a and b as products of the primes using the *same* primes. This can always be done provided we allow zero powers. For example, we can write $20 = 2^2 \times 5$ and $105 = 3 \times 5 \times 7$ as $20 = 2^2 \times 3^0 \times 5 \times 7^0$ and $105 = 2^0 \times 3 \times 5 \times 7$.

So suppose

$$a = p_1^{e_1} p_2^{e_2} \dots p_r^{e_r} \quad \text{and} \quad b = p_1^{f_1} p_2^{f_2} \dots p_r^{f_r}$$

where $p_1, p_2, \dots, p_r$ are prime and the e_i and f_j are natural numbers. Now

$$\begin{aligned} n = \frac{a^2}{b^2} &= \frac{p_1^{2e_1} p_2^{2e_2} \dots p_r^{2e_r}}{p_1^{2f_1} p_2^{2f_2} \dots p_r^{2f_r}} \\ &= p_1^{2e_1 - 2f_1} p_2^{2e_2 - 2f_2} \dots p_r^{2e_r - 2f_r} \\ &= \left(p_1^{e_1 - f_1} p_2^{e_2 - f_2} \dots p_r^{e_r - f_r} \right)^2 \\ &= m^2 \end{aligned}$$

where $m = p_1^{e_1 - f_1} p_2^{e_2 - f_2} \dots p_r^{e_r - f_r}$.

Therefore n is a perfect square.

Exercises 9.3

1. Any integer n is congruent to one of the numbers $0, 1, 2, \dots, 9$ modulo 10 since these are the possible remainders after division by 10. Now:

$$\begin{aligned} n \equiv 0 \mod 10 &\Rightarrow n^2 \equiv 0 \mod 10 \\ n \equiv 1 \mod 10 &\Rightarrow n^2 \equiv 1 \mod 10 \\ n \equiv 2 \mod 10 &\Rightarrow n^2 \equiv 4 \mod 10 \\ n \equiv 3 \mod 10 &\Rightarrow n^2 \equiv 9 \mod 10 \\ n \equiv 4 \mod 10 &\Rightarrow n^2 \equiv 6 \mod 10 \\ n \equiv 5 \mod 10 &\Rightarrow n^2 \equiv 5 \mod 10 \\ n \equiv 6 \mod 10 &\Rightarrow n^2 \equiv 6 \mod 10 \\ n \equiv 7 \mod 10 &\Rightarrow n^2 \equiv 9 \mod 10 \\ n \equiv 8 \mod 10 &\Rightarrow n^2 \equiv 4 \mod 10 \\ n \equiv 9 \mod 10 &\Rightarrow n^2 \equiv 1 \mod 10 \end{aligned}$$

Therefore n^2 can only be congruent to $0, 1, 4, 5, 6$ or 9 modulo 10. These represent the possible final decimal digits of n^2. Therefore the last decimal digit of n^2 cannot be $2, 3, 7$ or 8.

Hence 3190493 a not a perfect square since its last decimal digit is 3.

4. (ii) $3^3 = 27 \equiv 4 \mod 23$ so $3^9 \equiv 4^3 = 16 \times 4 \equiv -7 \times 4 = -28 \equiv -5 \equiv 18 \mod 23$.

(iv) $5^2 \equiv 2 \mod 23$ so $5^{12} \equiv 2^6 = 2 \times 32 \equiv 2 \times 9 = 18 \mod 23$.

5. (ii) $29147 \equiv 7 \mod 10$, so $29147^2 \equiv 7^2 \equiv 9 \mod 10$. Hence $29147^5 \equiv 9^2 \times 7 \equiv 7 \mod 10$. Therefore the last decimal digit of 23459^5 is 7.

6. (i) This is true.

Proof

Note that $(a-b)|(a^k - b^k)$. This is because

$$a^k - b^k = (a-b)(a^{k-1} + a^{k-2}b + a^{k-3}b^2 + \ldots + ab^{k-2} + b^{k-1}).$$

Therefore

$$\begin{aligned} a \equiv b \mod n \quad &\Rightarrow \quad n|(a-b) \\ &\Rightarrow \quad n|(a^k - b^k) \\ &\Rightarrow \quad a^k \equiv b^k \mod n. \end{aligned}$$

(ii) This is false.

For example, $5 \equiv 1 \mod 4$ but $2^5 = 32 \equiv 0 \mod 4$ so $2^5 \not\equiv 2^1 \mod 4$.

7. (i) $3x \equiv 4 \mod 6$ has no solutions.

(ii) $3x \equiv 4 \mod 7$ has unique solution $x = 6$.

(iii) $x^2 \equiv 2 \mod 5$ has no solutions.

(iv) $x^2 + 2 \equiv 0 \mod 6$ has solutions $x = 2, 4$.

(v) $x^2 + 2 \equiv 0 \mod 7$ has no solutions.

(vi) $x^2 \equiv x \mod 6$ has solutions $x = 0, 1, 3$.

8. (i) $\gcd(12, 22) = 2$ and 2 does not divide 15 so $12x \equiv 15 \mod 22$ has no solutions.

(iii) $\gcd(19, 50) = 1$ so $19x \equiv 42 \mod 50$ has a unique solution. Note that $19 \times 3 \equiv 7 \mod 50$ so $19 \times 18 = (19 \times 3) \times 6 \equiv 7 \times 6 = 42 \mod 50$. Therefore the unique solution is $x = 18$.

(v) $\gcd(65, 169) = 13$ and 13 does not divide 27 so there are no solutions.

(vi) $\gcd(65, 169) = 13$ and $13|39$ so there are 13 solutions. Dividing by 13 gives $5x \equiv 3 \mod 13$ which has unique solution $x = 11$.

Therefore the solutions to $65x \equiv 39 \mod 169$ are of the form $11 + 13q$ where $q = 0, 1, 2, \ldots, 12$. These are:

$$11, 24, 37, 50, 63, 76, 89, 102, 115, 128, 141, 154, 167.$$

(viii) Since $\gcd(20, 637) = 1$ and $\gcd(20, 101) = 1$ there is a unique solution but there is no further simplification possible.
Clearly we can 'get close': $20 \times 5 = 100$ and we need to 'find' another $+1$.
Note that $20 \times 32 = 640 \equiv 3 \mod 637$ so $20 \times 31 \equiv 3 - 20 = -17 \mod 637$.
Therefore $20 \times (5 \times 32 + 31) \equiv 5 \times 3 - 17 = -2 \mod 637$; in other words, $20 \times 191 \equiv -2 \mod 637$.
Hence $20 \times (32 + 191) \equiv 3 - 2 = 1 \mod 637$; ie $20 \times 223 \equiv 1 \mod 637$ so here is the 'other' $+1$ we were seeking.
Finally, $20 \times 228 = 20 \times (5 + 223) = 100 + 1 = 101 \mod 637$, so the solution is $x = 228$.

Exercises 9.4

1. (i) $\phi(24) = \phi(3) \times \phi(8) = 2 \times 4 = 8$.
   (iii) $\phi(31) = 30$ since 31 is prime.
   (v) $\phi(170) = \phi(10) \times \phi(17) = 4 \times 16 = 64$.
   (vii) $\phi(323) = \phi(17) \times \phi(19) = 16 \times 18 = 288$.

2. (i) $\phi(9) = 6$, $\phi(25) = 20$ and $\phi(49) = 42$.
   (ii) *Conjecture*: if p is prime, then $\phi(p^2) = p(p-1)$.

3. (i) (a) We evaluate $M^7 = 7^7 \mod 209$ in stages.

$$\begin{aligned} & 7^3 \equiv 134 \mod 209 \\ \Rightarrow\ & 7^6 \equiv 134^2 \equiv 191 \mod 209 \\ \Rightarrow\ & 7^7 \equiv 7 \times 191 \equiv 83 \mod 209. \end{aligned}$$

   (ii) Since the encryption is $M' = M^7 \mod 209$, we need to solve $7d \equiv 1 \mod \phi(209)$.
   Now $\phi(209) = \phi(11) \times \phi(19) = 10 \times 18 = 180$, so we need to solve

$$7d \equiv 1 \mod 180.$$

   Since $7 \times 25 = 175 \equiv -5 \mod 180$ and $7 \times 26 = 182 \equiv 2 \mod 180$, we have

$$7 \times (3 \times 26 + 25) \equiv 3 \times 2 - 5 = 1 \mod 180.$$

Therefore $d = 3 \times 26 + 25 = 103$ so the decryption scheme is

$$M = (M')^{103} \mod 209.$$

(c) $M' = 20 \Rightarrow M = 20^{103} \mod 209.$

$$\begin{aligned} 20^2 &\equiv 191 \equiv -18 \mod 209 \\ \Rightarrow 20^{10} &\equiv -18^5 \equiv -208 \equiv 1 \mod 209 \\ \Rightarrow 20^{100} &\equiv 1 \mod 209 \\ \Rightarrow 20^{103} &\equiv 20^3 \equiv 58 \mod 209. \end{aligned}$$

Therefore $M = 58$.

4. (i) $M' = M^{13} \mod 55$
Since $n = 55 = 5 \times 11$ we have $\phi(55) = 4 \times 10 = 40$.
Therefore we need to solve the equation $13d \equiv 1 \mod 40$.
Since $13 \times 3 = 39 \equiv -1 \mod 40$ it follows that $d \equiv -3 \equiv 37 \mod 40$.
Therefore the decryption scheme is

$$M \equiv (M')^{37} \mod 55.$$

(iii) $M' = M^7 \mod 143$
Since $n = 143 = 11 \times 13$ we have $\phi(143) = 10 \times 12 = 120$.
Therefore we need to solve the equation $7d \equiv 1 \mod 120$.
Since $7 \times 17 = 119 \equiv -1 \mod 120$ it follows that $d \equiv -17 \equiv 103 \mod 120$.
Therefore the decryption scheme is

$$M \equiv (M')^{103} \mod 143.$$

Chapter 10

Exercises 10.1

1. (ii) 0
(iv) 0
(vi) 1.

4. The identity with respect to $*$ is 24 and the identity with respect to $\oplus$ is 1. There are elements of B for which $b \oplus \bar{b} \neq 24$ so that axiom B5 is not satisfied. For instance, if $b = 4$, $\bar{b} = 6$ and then $b \oplus \bar{b} = 12$.

The set of divisors of 42 together with three operations is a Boolean algebra but the set of divisors of 45 is not.

5.
$$\begin{aligned} \mathbf{1} \oplus b &= (b \oplus \bar{b}) \oplus b && \text{(axiom B5)} \\ &= \bar{b} \oplus (b \oplus b) && \text{(axioms B3, B2)} \\ &= \bar{b} \oplus b && \text{(theorem 9.3)} \\ &= \mathbf{1} && \text{(axioms B3, B5).} \end{aligned}$$

Applying the duality principle gives $\mathbf{0} * b = b * \mathbf{0} = \mathbf{0}$.

8. (i)
$$\begin{aligned} (b_1 \oplus b_2) * \bar{b}_1 * \bar{b}_2 &= (b_1 \oplus b_2) * (\overline{b_1 \oplus b_2}) && \text{(De Morgan's law)} \\ &= \mathbf{0} && \text{(axiom B5).} \end{aligned}$$

Dual: $(b_1 * b_2) \oplus \bar{b}_1 \oplus \bar{b}_2 = \mathbf{1}$.

(iii)
$$\begin{aligned} &(b_1 \oplus b_2) * (\bar{b}_1 \oplus \bar{b}_2) \\ &= [b_1 * (\bar{b}_1 \oplus \bar{b}_2)] \oplus [b_2 * (\bar{b}_1 \oplus \bar{b}_2)] && \text{(axiom B4)} \\ &= (b_1 * \bar{b}_1) \oplus (b_1 * \bar{b}_2) \oplus (b_2 * \bar{b}_1) \oplus (b_2 * \bar{b}_2) && \text{(axiom B4)} \\ &= \mathbf{0} \oplus (b_1 * \bar{b}_2) \oplus (b_2 * \bar{b}_1) \oplus \mathbf{0} && \text{(axiom B5)} \\ &= (b_1 * \bar{b}_2) \oplus (\bar{b}_1 * b_2) && \text{(axioms B1, B3).} \end{aligned}$$

Dual: $(b_1 * b_2) \oplus (\bar{b}_1 * \bar{b}_2) = (b_1 \oplus \bar{b}_2) * (\bar{b}_1 \oplus b_2)$.

(v)
$$\begin{aligned} &(b_1 \oplus b_2 \oplus b_3) * (b_1 \oplus b_2) \\ &\quad = (b_1 \oplus b_2) * (b_1 \oplus b_2 \oplus b_3) && \text{(axiom B3)} \\ &\quad = b_1 \oplus b_2 && \text{(absorption law).} \end{aligned}$$

Dual: $(b_1 * b_2 * b_3) \oplus (b_1 * b_2) = b_1 * b_2$.

10. Both conditions are necessary because $b_1 * b_2 = b_1 * b_3$ holds if $b_1 = \mathbf{0}$ and $b_2 \neq b_3$ and $\bar{b}_1 * b_2 = \bar{b}_1 * b_3$ holds if $b_1 = \mathbf{1}$ and $b_2 \neq b_3$.

11. (ii) If

$$b_1 * b_2 = b_1$$

then

$$\begin{aligned} b_1 \oplus b_2 &= (b_1 * b_2) \oplus b_2 \\ &= b_2 \oplus (b_2 * b_1) && \text{(axiom B3)} \\ &= b_2 && \text{(absorption law).} \end{aligned}$$

Exactly the same line of argument shows that, if $b_1 \oplus b_2 = b_2$, then $b_1 * b_2 = b_1$.

Exercises 10.2

1. (i) $f(x_1, x_2)$ is already in disjunctive normal form.
   (iii) $f(x_1, x_2) = x_1x_2$.
   (v) $f(x_1, x_2) = x_1\bar{x}_2 \oplus x_1x_2$.
   (vi) $\bar{x}_1x_2\bar{x}_3 \oplus \bar{x}_1x_2x_3 \oplus x_1x_2x_3$.
   (ix) $\bar{x}_1\bar{x}_2x_3 \oplus \bar{x}_1x_2\bar{x}_3 \oplus \bar{x}_1x_2x_3 \oplus x_1\bar{x}_2x_3 \oplus x_1x_2x_3$.

   (ii) and (v) are equal and so are (iii) and (iv). Also (vi) and (viii) are equal, so are (ix) and (x).

2. (i) Disjunctive normal form:

$$f(x_1, x_2, x_3) = \bar{x}_1\bar{x}_2\bar{x}_3 \oplus \bar{x}_1x_2\bar{x}_3 \oplus \bar{x}_1x_2x_3 \oplus x_1\bar{x}_2\bar{x}_3 \oplus x_1\bar{x}_2x_3 \oplus x_1x_2\bar{x}_3 \oplus x_1x_2x_3.$$

   Conjunctive normal form:

$$f(x_1, x_2, x_3) = x_1 \oplus x_2 \oplus \bar{x}_3.$$

   (iii) Disjunctive normal form:

$$f(x_1, x_2, x_3) = x_1\bar{x}_2x_3 \oplus x_1x_2\bar{x}_3 \oplus x_1x_2x_3.$$

   Conjunctive normal form:

$$f(x_1, x_2, x_3) = (x_1 \oplus x_2 \oplus x_3)(x_1 \oplus x_2 \oplus \bar{x}_3)(x_1 \oplus \bar{x}_2 \oplus x_3)(x_1 \oplus \bar{x}_2 \oplus \bar{x}_3)(\bar{x}_1 \oplus x_2 \oplus x_3).$$

5. (i) The atoms are $\{j\}, \{k\}, \{l\}, \{m\}$.
   (ii) Since a_1 is an atom and $a_1a_2 \neq \mathbf{0}$, $a_1a_2 = a_1$.
   Since a_2 is an atom and $a_1a_2 \neq \mathbf{0}$, $a_1a_2 = a_2$.
   Therefore $a_1 = a_2$.

Exercises 10.3

1. (i) $f(x_1, x_2, x_3) = x_1(x_2 \oplus x_3)$.
   (iii) $f(x_1, x_2, x_3) = (x_1 \oplus x_2)x_3\bar{x}_2$.
   (v) $f(x_1, x_2, x_3) = x_1[x_2(x_1 \oplus x_3) \oplus x_3\bar{x}_2]$.

2. (ii)

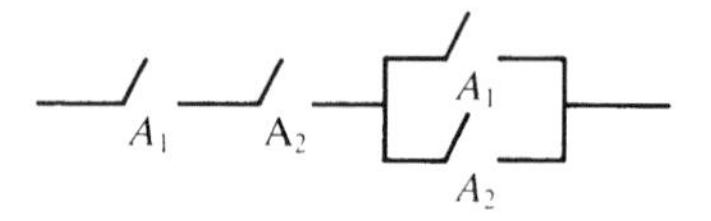

(iv)

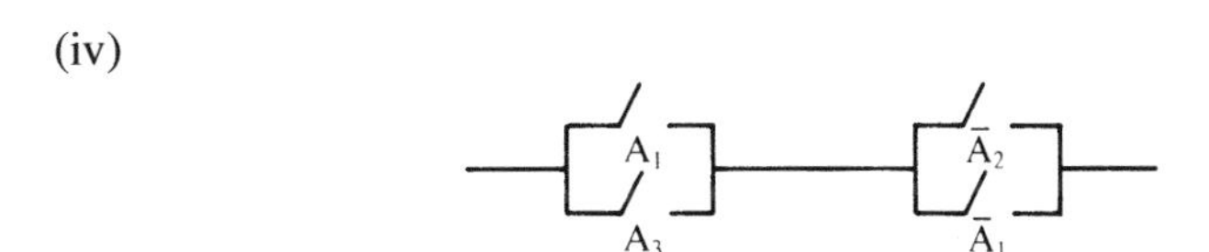

3. (i) $f(x_1, x_2, x_3) = x_1 \oplus x_2(x_1 \oplus x_3)$.

In disjunctive normal form:

$$f(x_1, x_2, x_3) = \bar{x}_1 x_2 x_3 \oplus x_1 \bar{x}_2 \bar{x}_3 \oplus x_1 \bar{x}_2 x_3 \oplus x_1 x_2 \bar{x}_3 \oplus x_1 x_2 x_3.$$

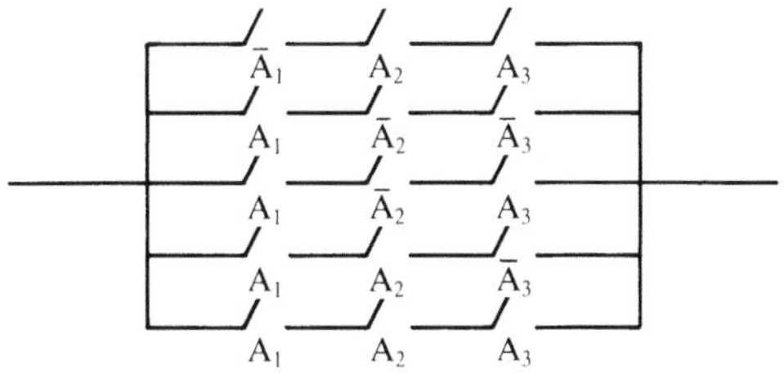

(ii) $f(x_1, x_2, x_3)$ is the zero function. Current never flows through this circuit regardless of the state of the switches.

4. A possible switching system (derived from the disjunctive normal form) is the following.

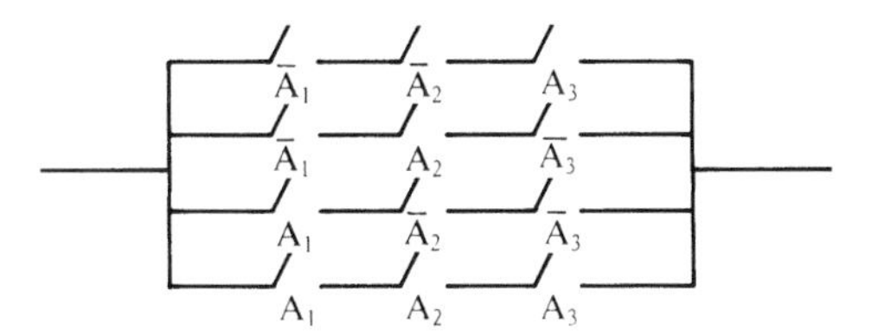

6. A possible switching system (derived from the disjunctive normal form) is the following. Switch A_1 is operated by the master switch and A_2 and A_3 by the sensors.

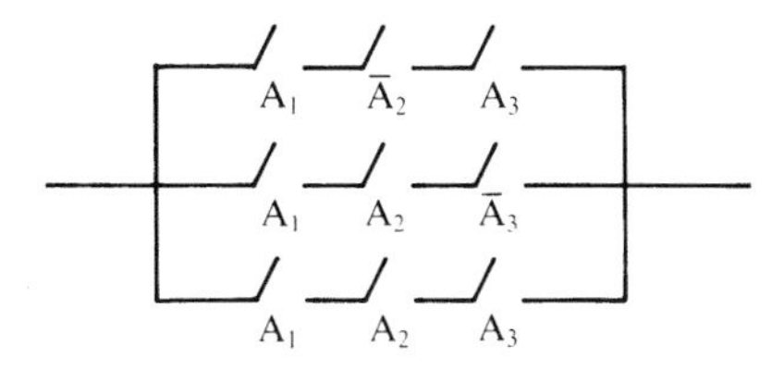

A simpler switching system which will achieve the same effect is as follows.

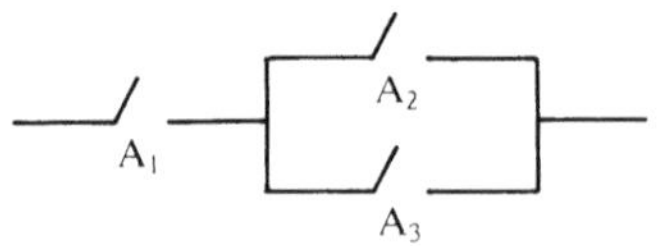

Exercises 10.4

1. (i) $x_1x_2 \oplus \bar{x}_3$

   (iii) $\overline{\bar{x}_1x_2 \oplus \overline{\bar{x}_1x_3}}$

   (v) $(\overline{x_1x_2 \oplus x_1x_2x_3})x_4$.

2. (ii)

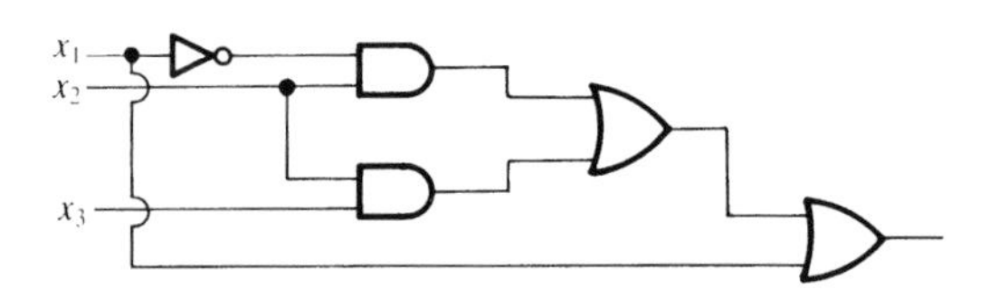

   (iv)

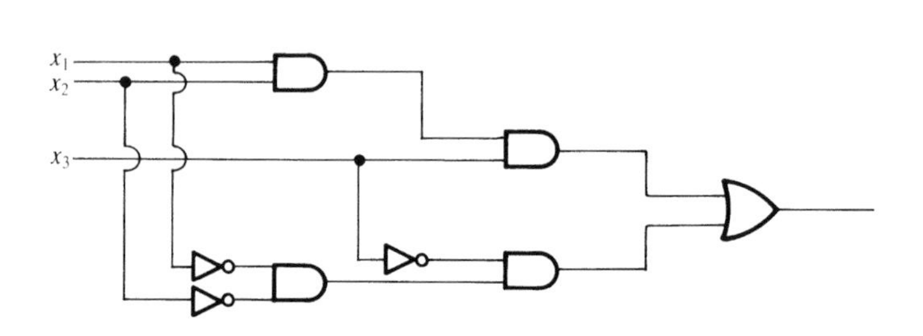

3. (i) $\overline{x_1x_2} \oplus x_2x_3$ and x_2x_3

   (iii) $\overline{\bar{x}_1x_2} \oplus \bar{x}_1x_3$, $\bar{x}_1x_3$ and $\overline{\bar{x}_1x_2} \oplus \overline{\bar{x}_3x_1}$.

4. (i)

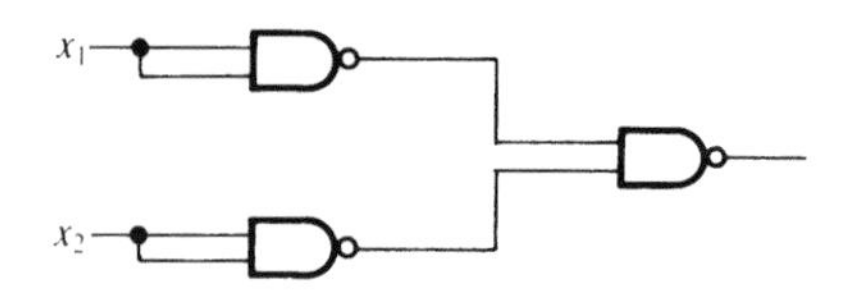

   (ii)

(iii)

6. A simple circuit which achieves the desired output is the following.

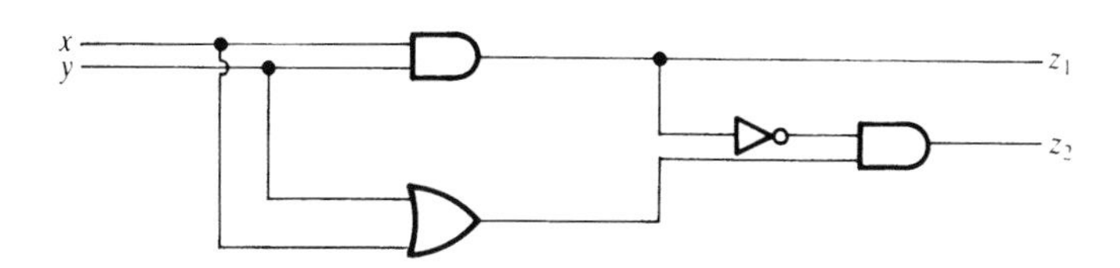

7. (ii) A full-adder consists of two half-adders and an OR-gate arranged as follows.

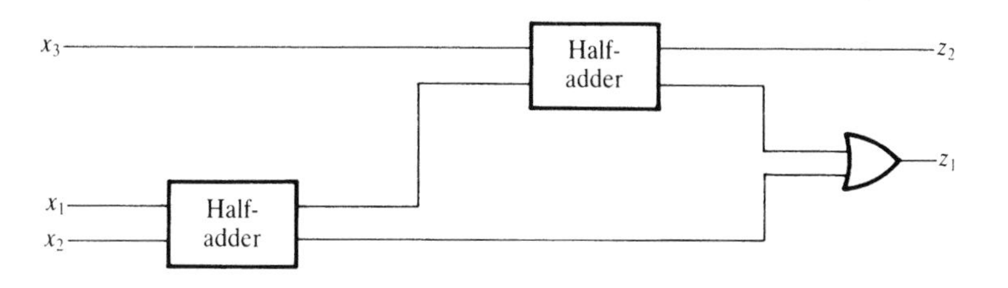

The reader is left to fill in the details of the circuit.

Exercises 10.5

1. (i) $x_1x_2x_3 \oplus \bar{x}_1\bar{x}_2 \oplus \bar{x}_2\bar{x}_3$

(ii)

	x_2x_3	$\bar{x}_2x_3$	$\bar{x}_2\bar{x}_3$	$x_2\bar{x}_3$
x_1	1	1	1	1
$\bar{x}_1$	1			

$x_1 \oplus x_2x_3$

(iii) $\bar{x}_1x_2x_3x_4 \oplus \bar{x}_1\bar{x}_2\bar{x}_3\bar{x}_4 \oplus x_1\bar{x}_2x_3$

(iv)

	x_3x_4	$\bar{x}_3x_4$	$\bar{x}_3\bar{x}_4$	$x_3\bar{x}_4$
x_1x_2		1		1
$\bar{x}_1x_2$				1
$\bar{x}_1\bar{x}_2$	1			1
$x_1\bar{x}_2$				1

$$\bar{x}_1\bar{x}_2x_3 \oplus x_1x_2\bar{x}_3x_4 \oplus x_3\bar{x}_4$$

(v) $x_1x_3x_4 \oplus \bar{x}_1x_2x_3\bar{x}_4 \oplus \bar{x}_2\bar{x}_3\bar{x}_4 \oplus x_1\bar{x}_2\bar{x}_4.$

2. (i) $x_1x_2 \oplus x_1\bar{x}_3$
(ii) $x_2x_3 \oplus x_1\bar{x}_2$
(iii) $x_3 \oplus \bar{x}_1x_2$
(iv) $x_2x_3 \oplus x_1x_3 \oplus x_1x_2.$

3. (i) $x_1x_3 \oplus \bar{x}_1x_2$

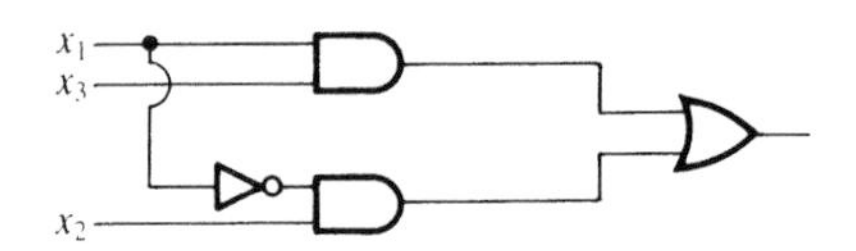

(ii) $x_1 \oplus x_2$

(iv) Circuit as in exercise 9.4.2(iv).

4.

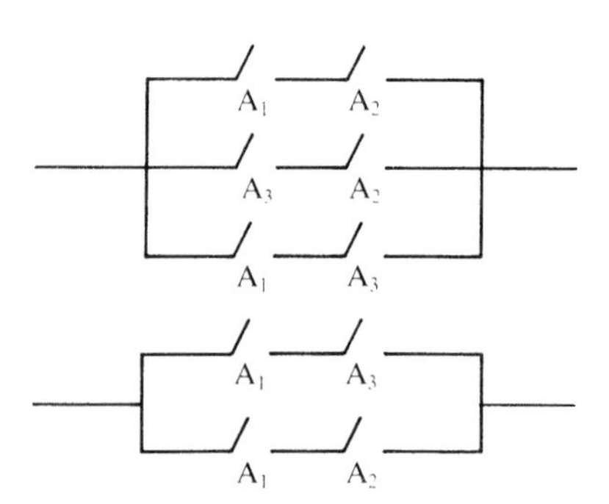

Chapter 11

Exercises 11.1

1.

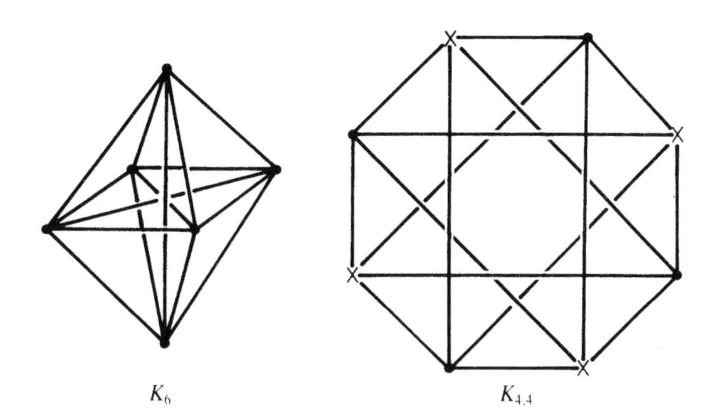

2. (ii) The graph is simple.

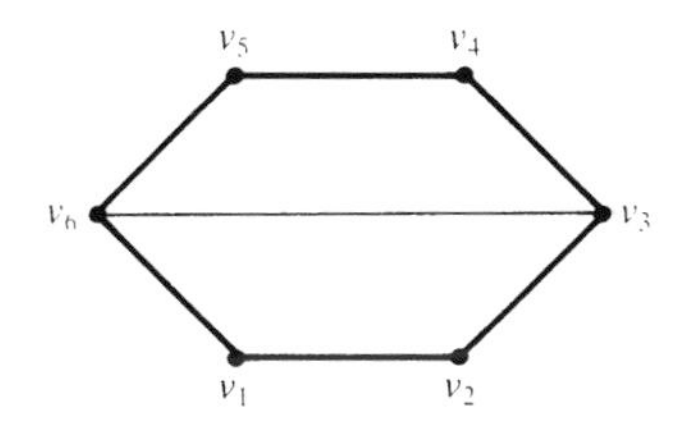

3. (i) With the labelling shown, we can define δ by $\delta(e_{ij}) = \{v_i, v_j\}$.

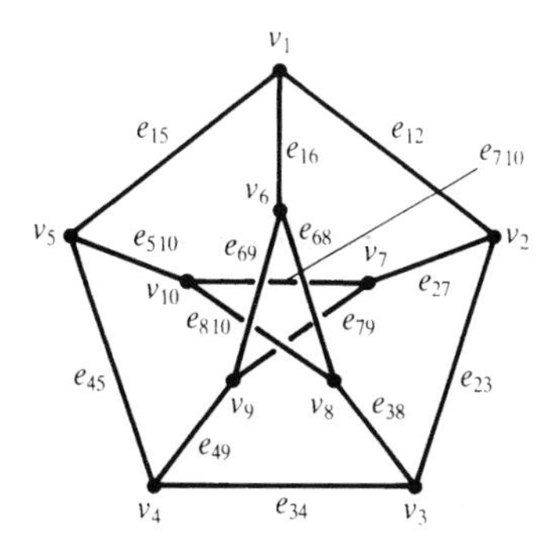

(ii)
$$\begin{pmatrix} 0&1&0&0&1&1&0&0&0&0\\ 1&0&1&0&0&0&1&0&0&0\\ 0&1&0&1&0&0&0&1&0&0\\ 0&0&1&0&1&0&0&0&1&0\\ 1&0&0&1&0&0&0&0&0&1\\ 1&0&0&0&0&0&0&1&1&0\\ 0&1&0&0&0&0&0&0&1&1\\ 0&0&1&0&0&1&0&0&0&1\\ 0&0&0&1&0&1&1&0&0&0\\ 0&0&0&0&1&0&1&1&0&0 \end{pmatrix}$$

5. (ii) If the vertices are written in the order $v_1, v_2, \ldots, v_p, w_1, w_2, \ldots, w_q$ the adjacency matrix is a partitioned matrix with $p \times p$ and $q \times q$ zero matrices as the two submatrices on the diagonal:

$$\begin{pmatrix} O_{p\times p} & A \\ B & O_{q\times q} \end{pmatrix}.$$

6. (i) No—apply the algorithm in (iii) below.
   (ii) C_n is bipartite if and only if n is even.
   (iii)
   1. Choose any vertex and colour it red, and colour all adjacent vertices green.
   2. Consider each green vertex in turn and colour every uncoloured adjacent vertex red.
   3. Consider each red vertex in turn and colour every uncoloured adjacent vertex green.
   4. Repeat steps 2 and 3 until all vertices are coloured.
   5. If any pair of adjacent vertices have the same colour then the graph is *not* bipartite; otherwise it is bipartite.

7. (i) The handshaking lemma follows from the fact that each edge connects two vertices (or a vertex to itself) and so contributes two to the sum of the vertex degrees.

8. (i)

Figure 10.2:		$(2,2,2,2,2,2,2)$
Figure 10.3:		$(3,3,3,3,3,3,3,3,3,3)$
Figure 10.4:	K_3	$(2,2,2)$
	K_4	$(3,3,3,3)$
	K_5	$(4,4,4,4,4)$
Figure 10.5	(a)	$(2,2,2,2,3,3,4,4,6)$
	(b)	$(3,3,3,3,3,3)$
Figure 10.6	Γ	$(3,3,3,4,5)$
	Σ	$(1,1,2,4)$.

(iii) (a) The number of vertices.

(b) Twice the number of edges (by the handshaking lemma—exercise 10.1.7(i)).

9. (i) (iii)

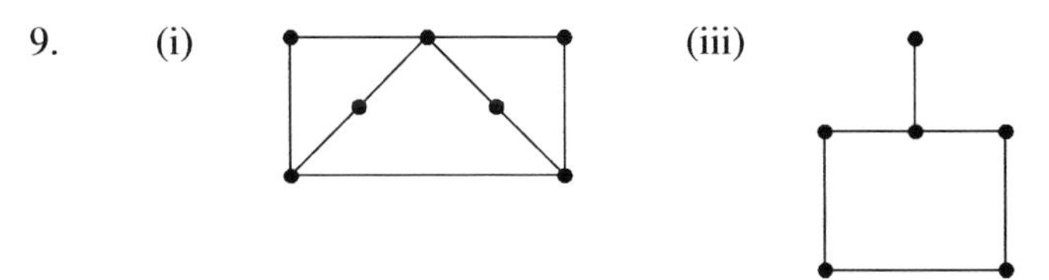

(v) There is no graph with this vertex sequence since, for example, there are three vertices with odd degree (see exercise 10.1.7(ii)).

10. (i) (ii)

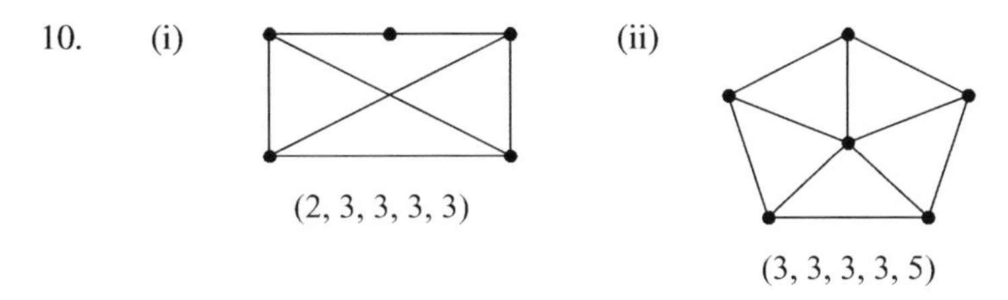

11. (i) $(\underbrace{3, 3, \ldots, 3}_{\longleftarrow n \text{ terms} \longrightarrow}, n)$

(ii) The adjacency matrix is an $(n+1) \times (n+1)$ matrix which can be partitioned as

$$\left(\begin{array}{ccc|c} & & & 1 \\ & \mathbf{A} & & \vdots \\ & & & 1 \\ \hline 1 & \ldots & 1 & 0 \end{array}\right)$$

where $\mathbf{A}$ is the $n \times n$ adjacency matrix of C_n. The matrix $\mathbf{A}$ has 1s immediately above and below its leading diagonal and 0s elsewhere:

$$\mathbf{A} = \begin{pmatrix} 0 & 1 & 0 & & \ldots & & 0 \\ 1 & 0 & 1 & 0 & & & \\ 0 & 1 & 0 & 1 & 0 & & \vdots \\ 0 & \ddots & \ddots & \ddots & \ddots & \ddots & \\ \vdots & & & & & 1 & 0 \\ & & & 0 & 1 & 0 & 1 \\ 0 & & \ldots & & 0 & 1 & 0 \end{pmatrix}$$

13. Show that, for K_n, $|E| = \frac{1}{2}n(n-1)$. Any simple graph with n vertices cannot have more edges than K_n.

14. (i) If N_p denotes a null graph with p vertices, then $N_p + N_q = K_{p,q}$.
    (ii) $K_p + K_q = K_{p+q}$.

16. (ii) The sum of the entries is always two.
    (iii) It is the degree of the vertex.

17. (i) (ii)

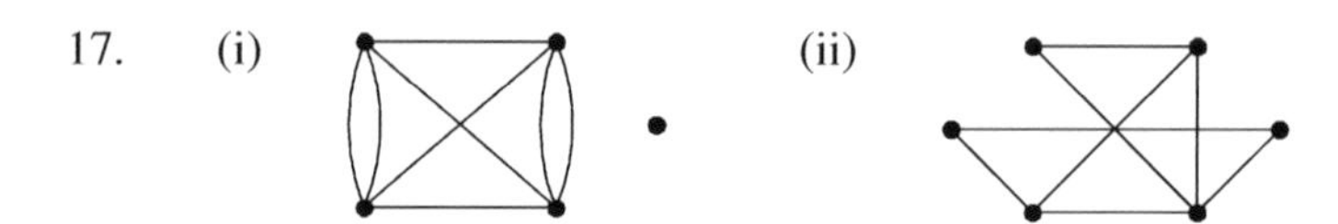

Exercises 11.2

2. The only difficulty is in proving transitivity. If P is a path from u to v and Q is a path from v to w, then the edge sequence 'P followed by Q' is an edge sequence from u to w, but it may not be a path as P and Q may have edges in common. If this is the case the edge sequence needs to be modified by omitting some edges to give the required path from u to w.

3. (i) Yes.
   (ii) Yes, if the graph has a single non-null Eulerian component and its other components are isolated vertices. For example, a null graph with more than one vertex is (trivially) Eulerian since it has no edges.

4. (i) K_n is Eulerian if and only if n is odd.
   (ii) $K_{r,s}$ is Eulerian if and only if both r and s are even.

5. (i) Only (a) and (d) are Eulerian.
   (ii) Define a 'modified row sum' to be the sum of the entries in a row except that the diagonal element is doubled. Then the degrees of the vertices are the modified row sums of the adjacency matrix. Hence, by Euler's theorem (10.2), the (connected) graph is Eulerian if and only if every modified row sum is even.

7. The trick is to notice that adding an appropriate edge to a semi-Eulerian graph produces an Eulerian graph.

8. The following are Hamiltonian cycles in each of the graphs.

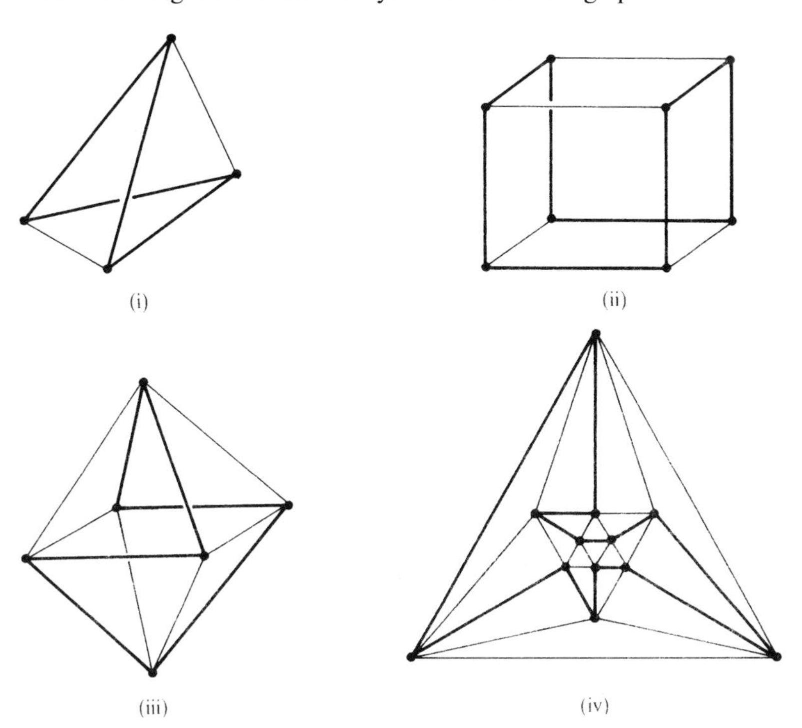

9. (ii) $K_{r,s}$ is Hamiltonian if and only if $r = s \geqslant 2$.

10. (i) Figure 10.2 and K_3 and K_5 (figure 10.4).
    (ii) Figure 10.1, figure 10.5(a) and Σ (figure 10.6).
    (iii) Figures 10.2, 10.4 (all three graphs), 10.5(b) and Γ (figure 10.6).
    (iv) Every graph except figure 10.1 and those in (iii).

11. (i) (a) One (b) Two.
    (ii) Yes, the graph is Hamiltonian.

12. (i) Graph I is semi-Eulerian; graph II is neither Eulerian nor semi-Eulerian.
    (ii) Graph I is semi-Hamiltonian; graph II is Hamiltonian.

13. Let Γ be a connected graph.

    Suppose that Γ is Eulerian. Then every vertex has even degree. We know from the proof of Euler's theorem that we can choose a cycle in Γ. Delete the edges of the cycle to produce a new graph Γ' (which may be disconnected) in which every vertex has even degree. Consider (a component of) Γ' and choose a cycle, which is certainly edge-disjoint from the first chosen cycle. Remove the edges of the cycle, producing a graph Γ'' in which every vertex has even degree. Continue in this way until all the edges of Γ are included in one (and only one) of the chosen cycles.

Conversely suppose that Γ is such that its edges form disjoint cycles. Select one cycle, C_1 say, choose a vertex and start traversing its edges until another a cycle is reached. Traverse the edges of this cycle (as with C_1, breaking off if we meet any other cycle, etc) before resuming traversing the edges of C_1. Continuing in this way we obtain a closed path containing all the edges of Γ.

For the given graph, a possible choice of cycles is: $afea$, $edge$, $bdcb$, $agba$.

Beginning with the cycle $afea$ and starting at a the process described above unfolds as follows.

Traverse af, fe
  (break off to go round $edge$)
  traverse ed
    (break off to go round $bdcb$)
    traverse dc, cb
      (break off to go round $agba$)
      traverse ba, ag, gb
    (resume $bdcb$)
    traverse bd
  (resume $edge$)
  traverse dg, ge
(resume $afea$)
traverse ea.

The resulting Eulerian path is: $afedcbagbdgea$.

Exercises 11.3

1. One matrix can be converted to the other by a reordering of the rows and columns, the same reordering being applied to both rows and columns.

2. No. Each graph has two vertices of degree 2, which are adjacent in (ii) but not in (i).

4. (i) There are various possibilities. For example:

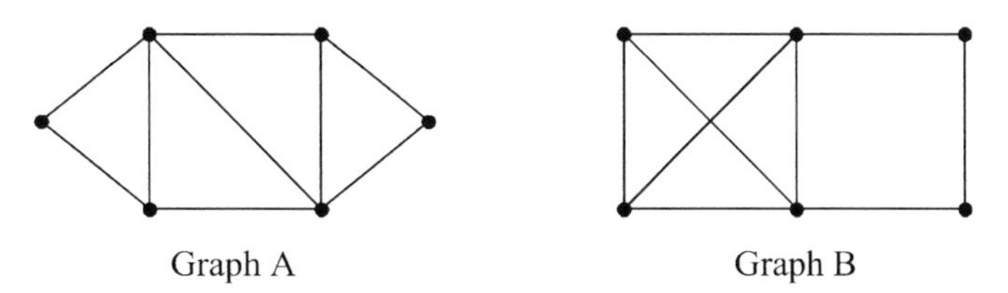

The two vertices of degree 2 are adjacent in graph A but not in graph B; hence the two graphs are not isomorphic.

5. The degree sequence is $(2, 2, 2, 3, 3, 3, 3, 4)$

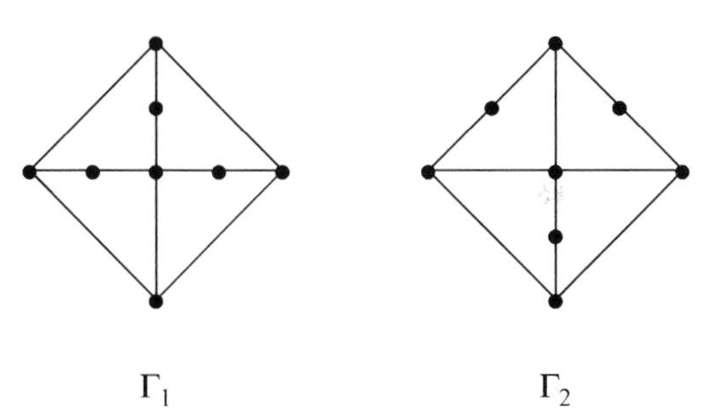

The graphs Γ_1 and Γ_2 are not isomorphic since the vertex of degree 4 in Γ_1 is adjacent to three vertices of degree 2 but the vertex of degree 4 in Γ_2 is adjacent to only one vertex of degree 2.

6. $\Gamma_1 \cong \Gamma_2$.

8. (i)

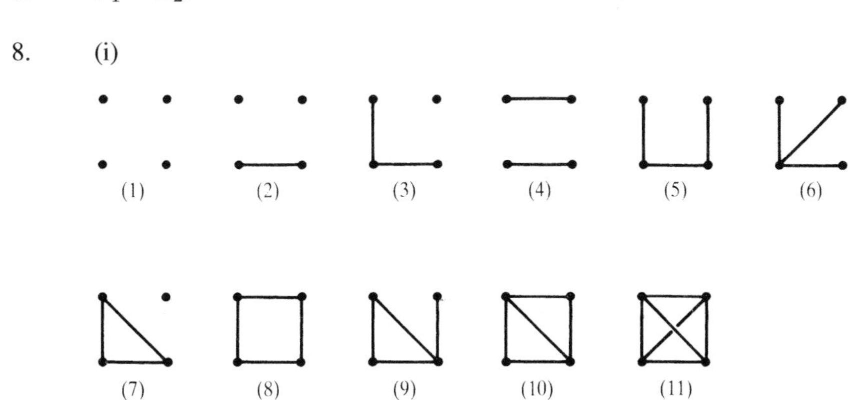

Graphs 5, 6, 8, 9, 10 and 11 are connected.

(ii) There are 34, of which 21 are connected!

9. With the vertex labellings given below, the isomorphism has as its vertex bijection the mapping, $v_i \mapsto w_i$. Since the graphs are both simple, this is sufficient to define the isomorphism.

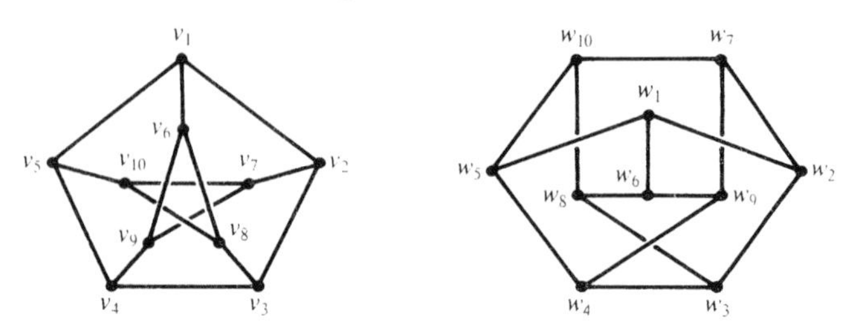

10. (i) and (ii) follow from theorem 5.7(ii) applied to the vertex and edge bijections respectively.

(iii) If $e_1, e_2, \ldots, e_n$ is a path connecting v_1 and v_2 in Γ, then $\phi(e_1), \phi(e_2), \ldots, \phi(e_n)$ is a path connecting $\theta(v_1)$ and $\theta(v_2)$ in Σ. Conversely, if $f_1, f_2, \ldots, f_m$ is a path connecting w_1 and w_2 in Σ, then $\phi^{-1}(f_1), \phi^{-1}(f_2), \ldots, \phi^{-1}(f_m)$ is a path in Γ connecting $\theta^{-1}(w_1)$ and $\theta^{-1}(w_2)$.

Therefore v_1 and v_2 lie in the same component of Γ if and only if $\theta(v_1)$ and $\theta(v_2)$ lie in the same component of Σ. Hence Γ and Σ have the same number of components.

(iv) $deg(v) = \sum_{w \in V_\Gamma} |E(v, w)|$ and $deg(\theta(v)) = \sum_{w' \in V_\Sigma} |E(\theta(v), w')|$.

The result follows since θ is a bijection and the function ϕ defines bijections $E(v, w) \to E(\theta(v), \theta(w))$.

(v) Γ is simple if and only if $|E(v, w)| \leqslant 1$ and $|E(v, v)| = 0$ for all $v, w \in V_\Gamma$. As in (iv) the result follows from the existence of the bijections $E(v, w) \to E(\theta(v), \theta(w))$.

(vi) Follows from (iv) using theorem 10.2.

(vii) If $e_1, e_2, \ldots, e_n$ is a Hamiltonian cycle in Γ then $\phi(e_1), \phi(e_2), \ldots, \phi(e_n)$ is a Hamiltonian cycle in Σ.

Exercises 11.4

1. (i) 1 (ii) 2 (iii) 3 (iv) 6.

2. One method of proof is by induction on the number n of vertices,

beginning with $n = 2$ for which the result is obvious. Suppose every tree with n vertices is bipartite and let T be a tree with $n+1$ vertices. Removing a vertex v of degree 1 and its incident edge from T produces a tree with n vertices which is bipartite, by hypothesis. Let $\{V_1, V_2\}$ be the partition of the vertex set of this tree and suppose that the vertex adjacent to v in T belongs to V_1. Then $\{V_1, V_2 \cup \{v\}\}$ is a partition of the vertex set of T, so T is bipartite.

4. (i)

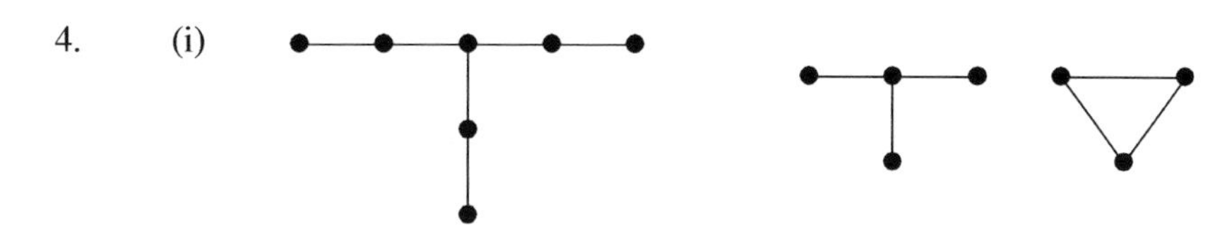

(ii) Such a graph has seven vertices (since there are seven entries in the degree sequence) and $\frac{1}{2}(1+1+2+2+2+3+3) = 7$ edges. Therefore, by theorem 10.6, the graph is not a tree.

5. $K_{r,s}$ is a tree if and only if $r = 1$ or $s = 1$.

6. (iii) (a) 1 (b) 2 (c) 3.

7. (i) Level 1:

Level 2:

(ii) A level n full binary tree looks like the following where $0 \leqslant i$, $j \leqslant n-1$ and at least one of i and j is equal to $n-1$.

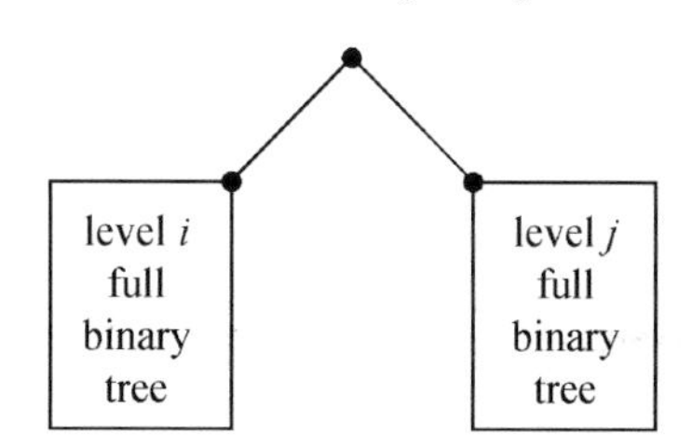

If $j = 0$ then $i = n-1$: there are $a_0 a_{n-1}$ such trees;
if $j = 1$ then $i = n-1$: there are $a_1 a_{n-1}$ such trees;
...

if $j = n - 2$ then $i = n - 1$: there are $a_{n-2}a_{n-1}$ such trees;
if $j = n - 1$ then i can take any value from 0 to $n - 1$ inclusive: there are $a_{n-1}(a_0 + a_1 + \cdots + a_{n-1})$ such trees.

Hence the total number of level n full binary trees is:

$$\begin{aligned} a_n &= a_0 a_{n-1} + a_1 a_{n-1} + \cdots + a_{n-2}a_{n-1} \\ &\quad + a_{n-1}(a_0 + a_1 + \cdots + a_{n-1}) \\ &= 2a_{n-1}(a_0 + a_1 + \cdots + a_{n-2}) + a_{n-1}^2. \end{aligned}$$

(iii) We know $a_0 = 1$ and, from part (i), $a_1 = 1$, $a_2 = 3$. Therefore

$$\begin{aligned} a_3 &= 2a_2(a_0 + a_1) + a_2^2 \\ &= 2 \times 3 \times (1 + 1) + 3^2 = 21, \end{aligned}$$

$$\begin{aligned} a_4 &= 2a_3(a_0 + a_1 + a_2) + a_3^2 \\ &= 2 \times 21 \times (1 + 1 + 3) + 21^2 = 651. \end{aligned}$$

9. (i) $|E| = |V| - c$. (ii) $|E| \geqslant |V| - c$.

10. $K_{2,n}$ has $n \times 2^{n-1}$ spanning trees.

Proof

Suppose the partition of V is $\{\{x, y\}, \{1, 2, 3, \ldots, n\}\}$ so that we may represent $K_{2,n}$ as shown in the following diagram.

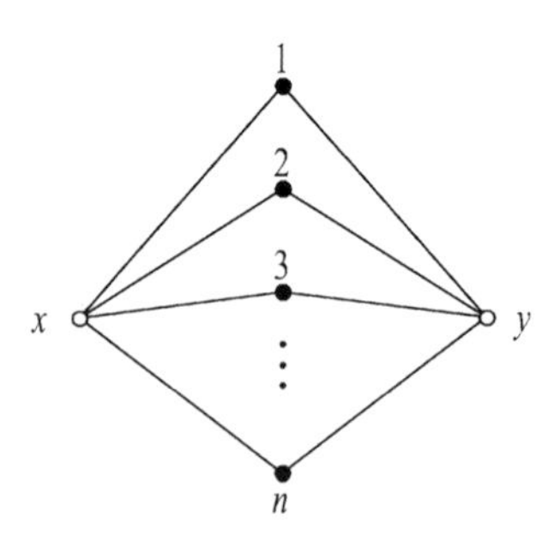

We can define a spanning tree of $K_{2,n}$ as follows. First choose a vertex, k say, in the set $\{1, 2, \ldots, n\}$ that is to be joined to *both* x and y. Then, for each vertex $r \in \{1, 2, \ldots, n\} - \{k\}$, choose either rx or ry to be an edge of the spanning tree. There are n choices for the vertex k and then $2 \times 2 \times \cdots \times 2 = 2^{n-1}$ choices for the remaining edges. One set of choices is illustrated below, where $k = 2$.

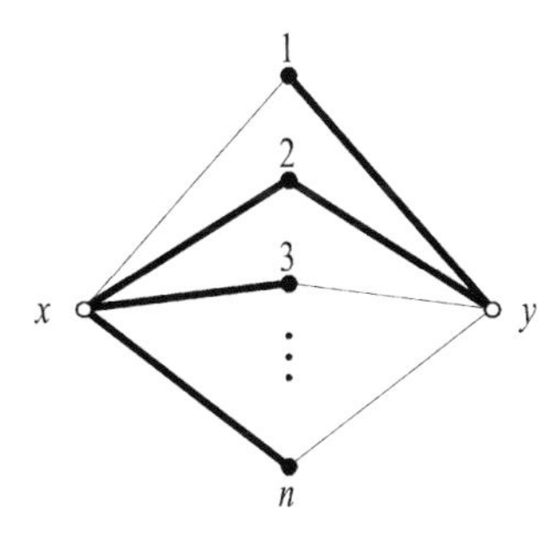

Each of the $n \times 2^{n-1}$ choices gives rise to a unique spanning tree and each spanning tree is obtained in this way; hence there are $n \times 2^{n-1}$ spanning trees.

11. (iii) Γ has 21 spanning trees. There are various ways of using the equation in part (i) to show this. For example:

$t(\;) = t(\;) + t(\;)$

$= 3 \times 3 + t(\;) + t(\;)$

$= 9 + t(\;) + t(\;)$

$+ t(\;) + t(\;)$

$= 9 + 2 + 3 + 3 + 4 = 21.$

12. (i) Graph A has nine spanning trees; graph B has $8 \times 4 = 32$ spanning trees.

(ii) If Γ is a connected graph with a bridge vw (like graph A), then removing vw produces a graph with two components, say Γ_1 and Γ_2. The number of spanning trees in Γ is the product of the numbers of spanning trees in each of these components. Symbolically, if $t(\Gamma)$ denotes the number of spanning trees in Γ (as in exercise 10.4.10), then

$$t(\Gamma) = t(\Gamma_1)t(\Gamma_2).$$

The result is similar for a connected graph Σ with a cut vertex v (such as graph B). However the situation is a little more complicated. Removing v produces a graph with a number of components $\Sigma_1, \Sigma_2, \ldots, \Sigma_m$, but none of these components contains the vertex v itself. We need to 'put v back' into each of these graphs, together with the appropriate edges, before counting the number of spanning trees in each of the resulting subgraphs. The number of spanning trees in Σ is then the product of the numbers of spanning trees in these subgraphs.

13. (i)

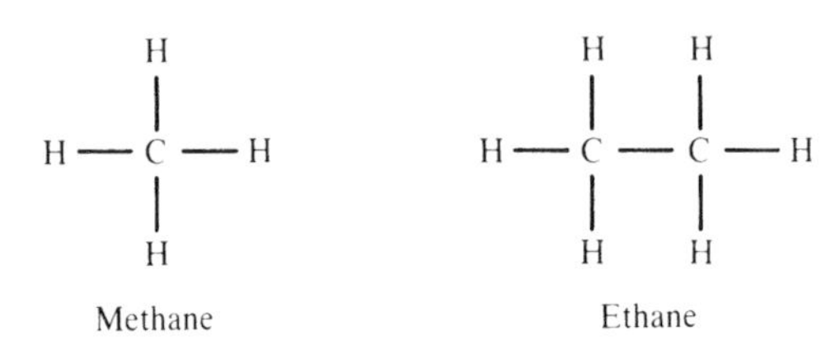

(ii) $|V| = n + (2n+2) = 3n+2$, $|E| = \frac{1}{2}(4n + (2n+2)) = 3n+1$. Therefore $|E| = |V| - 1$, so the graph is a tree, by theorem 10.6(iii).

(iii)

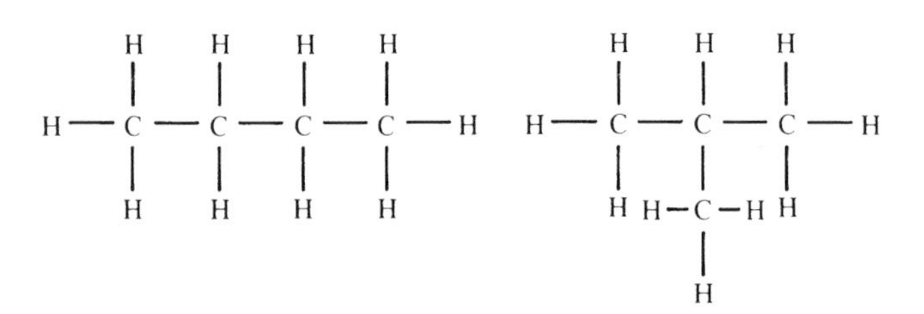

(iv) Pentane has three structural isomers; hexane five.

Exercises 11.5

2. (i) K_2 is a tree and figure 10.4 shows K_3 and K_4 are planar. If $n \geqslant 5$, then K_n contains a subgraph isomorphic (hence homeomorphic) to K_5 so K_n is not planar.

(ii) The following are plane versions of $K_{1,n}$ and $K_{2,n}$.

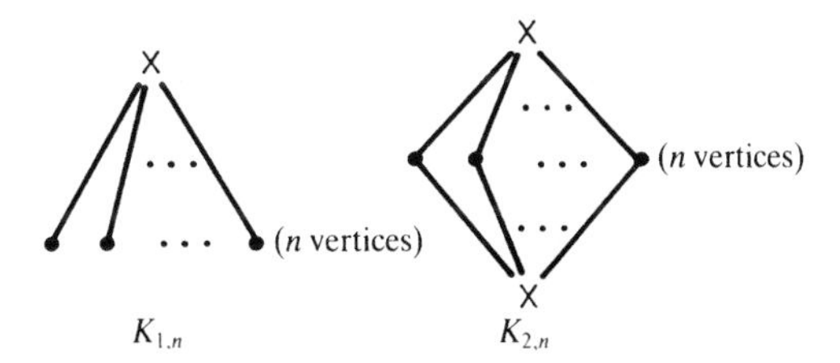

3. The left-hand graph is not planar: deleting the three vertices of degree 2 produces a graph which clearly contains K_5 as a subgraph.

The right-hand graph can be redrawn as follows.

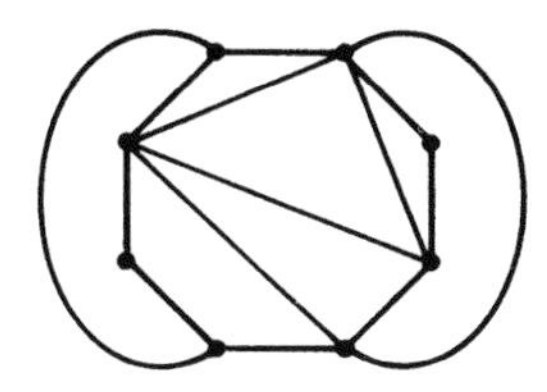

4. (ii) (a) is planar; (b) is non-planar.

5. (i) Since the graph has only three vertices of degree 4 or more, it does not contain a subgraph which is a subdivision of K_5. Therefore we look for a subgraph which is a subdivision of $K_{3,3}$.

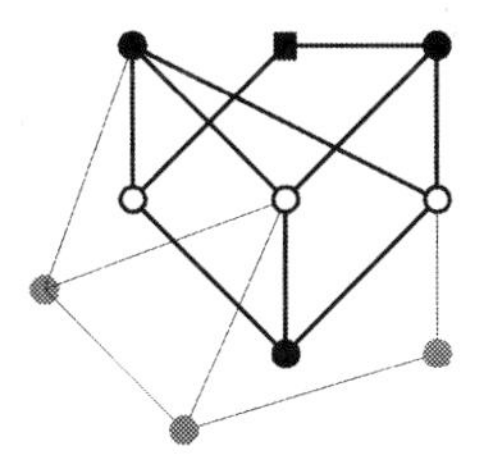

Firstly delete the thin edges and then remove the resulting isolated vertices. Removing the square vertex of degree 2 then gives the graph shown below, which is $K_{3,3}$.

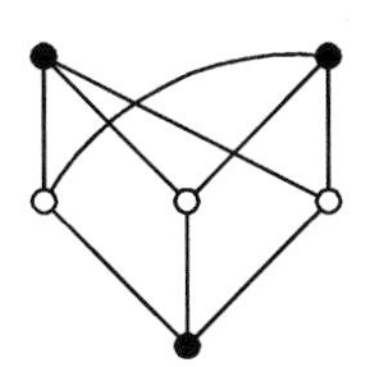

Therefore, the given graph has a subgraph which is a subdivision of $K_{3,3}$, so it is non-planar.

7. (i) Successively deleting vertices of degree 2 produces the following two isomorphic graphs. (The isomorphism has vertex mapping $v_i \mapsto w_i$.)

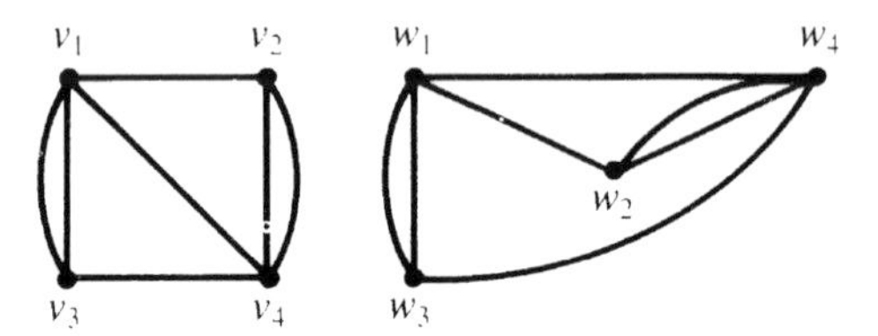

8. (i) Every face is bounded by a cycle which contains at least three edges, and each edge forms the boundary between two faces. Hence twice the number of edges is at least three times the number of faces, $2|E| \geqslant 3|F|$.

   The inequality $|E| \leqslant 3|V| - 6$ comes from substituting $2|E| \geqslant 3|F|$ into Euler's formula.

   (ii) Let Γ be a connected simple planar graph. Suppose that each vertex has degree 6 or more; then $2|E| \geqslant 6|V|$ so $|E| \geqslant 3|V|$. However, from (i) we have $|E| \leqslant 3|V| - 6$, which is a contradiction.

9. (ii) Let v, e, f denote the number of vertices, edges and faces of Γ and similarly let v^*, e^*, f^* denote the number of vertices, edges and faces of Γ^*.

   Then $v^* = f, e^* = e$ and $f^* = v$.

10. (i) Since Γ has only three vertices of degree at least 4, it does not contain a subgraph which is a subdivision of K_5.

   Since Γ has only five vertices of degree at least 3, it does not contain a subgraph which is a subdivision of $K_{3,3}$.

   Therefore, by Kuratowski's theorem 10.8, Γ is planar.

   (ii) Γ has $n = 8$ vertices. The sum of vertex degrees is 26, so Γ has $m = 13$ edges (by the handshaking lemma—exercise 10.1.7(i)).

   By Euler's formula (theorem 10.7), if Γ has f faces, then $8 - 13 + f = 2$ so $f = 7$.

(iii) Using the terminology introduced in solution 9 above, the dual Γ^* has

$$n^* = f = 7 \text{ vertices},$$
$$m^* = m = 13 \text{ edges},$$
and
$$f^* = n = 8 \text{ faces}.$$

11. (i) For all graphs on the sphere, $|V| - |E| + |F| = 2$. (In fact a graph can be drawn on the surface of the sphere if and only if it can be drawn in the plane.)

    (ii) For all graphs on the torus, $|V| - |E| + |F| = 0$.

12. The following are diagrams of K_5 and $K_{3,3}$ drawn on the surface of a torus. (For these graphs we have $(|V|, |E|, |F|) = (5, 10, 5)$ and $(|V|, |E|, |F|) = (6, 9, 3)$ respectively.)

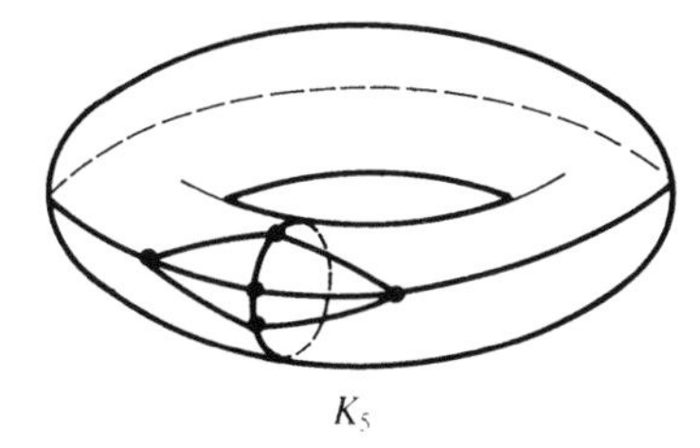

K_5

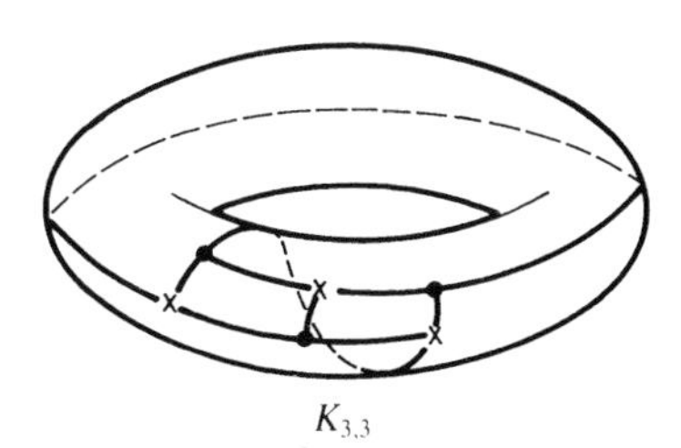

$K_{3,3}$

Exercises 11.6

1. Let A and B be the adjacency matrices of a digraph and its underlying graph respectively. If the digraph has no directed loops, $B = A + A^{\mathrm{T}}$.

   More generally, $B = A + A^{\mathrm{T}} - J$ where J is the diagonal matrix whose diagonal entries are the number of directed loops from the corresponding vertex to itself.

3. (i) In a tree there is a unique path joining any pair of distinct vertices v and w. If, in the digraph, the path is directed from v to w, then there cannot be any directed path from w to v.

   (ii) Yes: a cycle graph with all edges directed clockwise is an example.

4. (i)

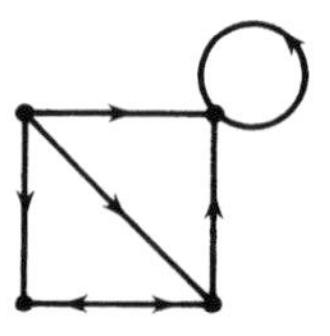

5. (a) (i), (ii), (iii) and (iv) are simple.
   (b) (ii), (iii) and (iv) have simple underlying graphs.
   (c) (i), (ii) and (iii) are strongly connected.
   (d) Only (iii) is Eulerian.
   (e) (ii) only is Hamiltonian.

7. (i) Graph (a) is unilaterally connected but not strongly connected. Graph (b) is weakly connected but not unilaterally connected.

   (ii) Graphs (i), (ii) and (iii) are unilaterally connected—in fact, strongly connected.

8. The following is the diagram of a strongly connected digraph whose underlying graph is Petersen's graph.

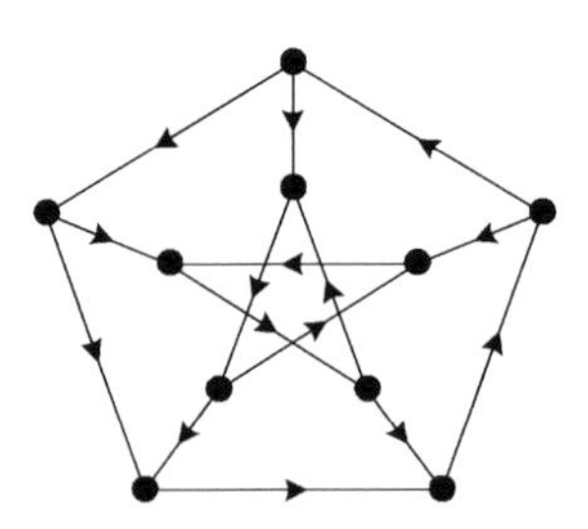

9. (i) Labelling the edges as shown gives the corresponding incidence matrix.

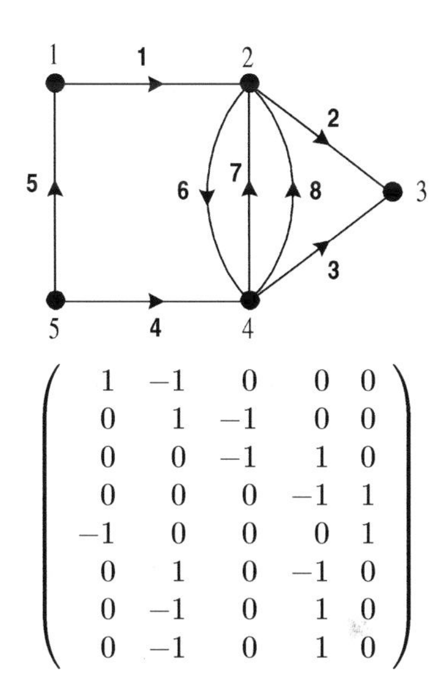

$$\begin{pmatrix} 1 & -1 & 0 & 0 & 0 \\ 0 & 1 & -1 & 0 & 0 \\ 0 & 0 & -1 & 1 & 0 \\ 0 & 0 & 0 & -1 & 1 \\ -1 & 0 & 0 & 0 & 1 \\ 0 & 1 & 0 & -1 & 0 \\ 0 & -1 & 0 & 1 & 0 \\ 0 & -1 & 0 & 1 & 0 \end{pmatrix}$$

(ii) Row sum $=$ out-degree of corresponding vertex.
Column sum $=$ in-degree of corresponding vertex.

(iii) Row sum $= 0$ (so provides no information about the digraph).
Column sum $=$ out-degree $-$ in-degree of corresponding vertex.

(iv)

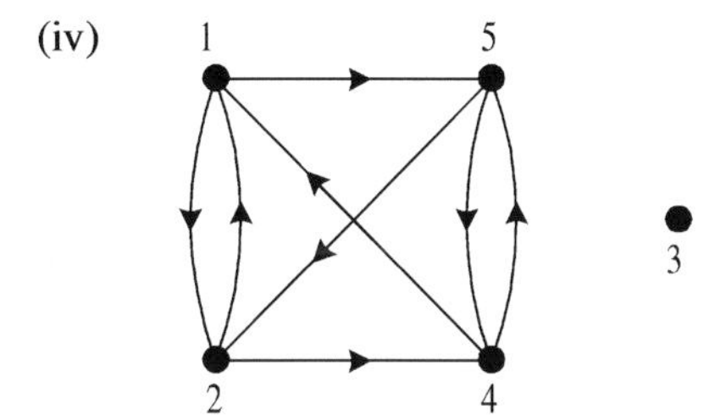

(v) Let A be the adjacency matrix of the digraph and A^{T} be its transpose (see exercise 6.1.6). Let B be the adjacency matrix of the underlying graph.

If the digraph has no loops: $B = A + A^{\mathrm{T}}$.

More generally, $B = A + A^{\mathrm{T}} - J$ where J is a diagonal matrix with diagonal entries equal to the number of loops at each of the vertices.

(vi) Let C be the incidence matrix of the digraph and D be the incidence matrix of the underlying graph.

Then $d_{ij} = |c_{ij}|$.

11. Let D_1 and D_2 be two digraphs. An **isomorphism** $D_1 \to D_2$ is a pair (θ, ϕ) of bijections $\theta : V_1 \to V_2$ and $\phi : E_1 \to E_2$ such that, for every directed edge $e \in E_1$, if $\delta_{D_1}(e) = (v, w)$ then $\delta_{D_2}(\phi(e)) = (\theta(v), \theta(w))$.

    Isomorphic digraphs have isomorphic underlying graphs, but not conversely.

12. (ii) The underlying graphs of D_1 and D_3 are isomorphic.

13. D_2 and D_3 (only) are isomorphic.

14. (i) Let D be the digraph $\bullet \!\longrightarrow\! \bullet$. Then $D \cong \tilde{D}$.

    Let D be the digraph $\bullet \!\longrightarrow\! \bullet \!\longleftarrow\! \bullet$. Then $D \ncong \tilde{D}$.

    (ii) Using $A(D)$ and $A(\tilde{D})$ to denote the adjacency matrices of D and $\tilde{D}$ respectively, $A(D) = A(\tilde{D})^T$.

    (iii) Similarly, using B for the incident matrices, $B(D) = -B(\tilde{D})$.

15. (i) 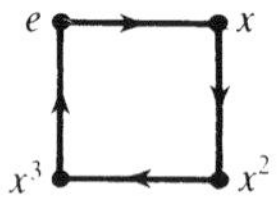 (ii)

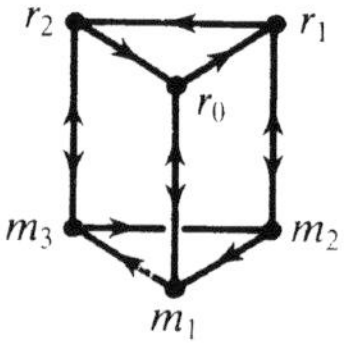

Chapter 12

Exercises 12.1

1. (i) (a) 2 (b) Yes (c) No.
   (iii) (a) 6 (b) No (c) No.
   (v) (a) 3 (b) Yes (c) No.

4. Let v and w be vertices of a rooted tree. Define v to be an **ancestor** of w (and w be a **descendant** of v) if there exist vertices $v_1, v_2, \ldots, v_n$ such that v is the parent of v_1, v_1 is the parent of $v_2, \ldots, v_n$ is the parent of w. (An alternative definition is: v is an **ancestor** of w if the level of v is less than the level of w and the unique path in the tree joining the root and w also contains v.)

   R is (by definition) reflexive. Suppose $v \; \mathsf{R} \; w$ and $w \; \mathsf{R} \; v$. If v is an ancestor of w then w is not an ancestor of v, from which it follows that $v = w$, so R is anti-symmetric. Suppose $u \mathsf{R} v$ and $v \mathsf{R} w$. Then $level(u) \leqslant level(v) \leqslant level(w)$ and the unique path joining the root to v contains u and the unique path from the root to w contains v. Therefore the unique path from the root to w contains u so $u \mathsf{R} w$. Hence R is transitive.

5. (i)

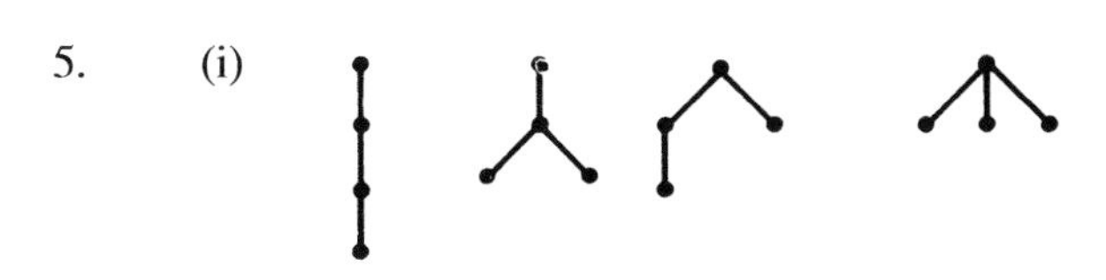

(ii)

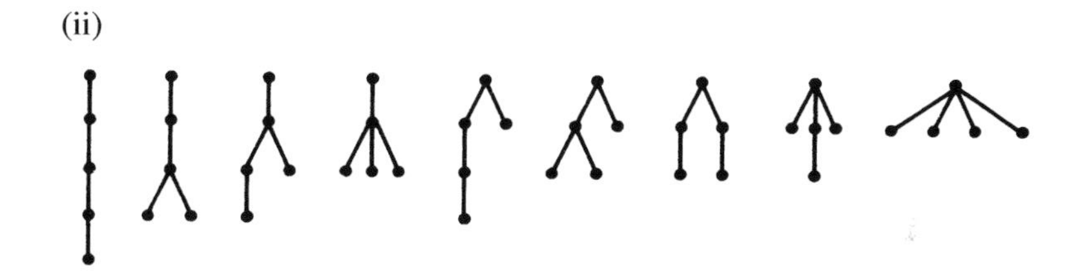

6. (i) $T = (L, \{v^*\}, R)$ where:

$$
\begin{aligned}
L &= (L_1, \{v_1\}, R_1) &\text{and}\quad R &= (L_2, \{v_2\}, R_2)\\
L_1 &= (\varnothing, \{v_3\}, \varnothing) &\text{and}\quad R_1 &= (L_4, \{v_4\}, R_4)\\
L_2 &= (\varnothing, \{v_5\}, \varnothing) &\text{and}\quad R_2 &= (L_6, \{v_6\}, R_6)\\
L_4 &= (L_7, \{v_7\}, R_7) &\text{and}\quad R_4 &= (\varnothing, \{v_8\}, \varnothing)\\
L_6 &= (L_9, \{v_9\}, R_9) &\text{and}\quad R_6 &= (\varnothing, \{v_{10}\}, \varnothing)\\
L_7 &= (\varnothing, \{v_{11}\}, \varnothing) &\text{and}\quad R_7 &= (L_{12}, \{v_{12}\}, R_{12})\\
L_9 &= (\varnothing, \{v_{13}\}, \varnothing) &\text{and}\quad R_9 &= (\varnothing, \{v_{14}\}, \varnothing)\\
L_{12} &= (\varnothing, \{v_{15}\}, \varnothing) &\text{and}\quad R_{12} &= (\varnothing, \{v_{16}\}, \varnothing).
\end{aligned}
$$

(ii)

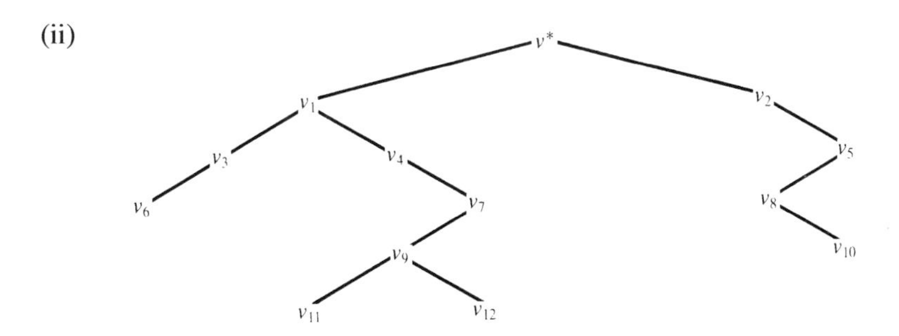

7. First define $level(v^*) = 0$. Given $(L, \{v\}, R)$ where $L = (L', \{v'\}, R')$ and $R = (L'', \{v''\}, R'')$, define $level(v') = level(v'') = level(v) + 1$.

8. A has a greatest element a and for all *minimal* elements $x \in A$ there is exactly one chain containing both x and a.

10. (i) (b)

$\oplus$ $*$ t $\oplus$ z x y

(d)

$*$ r $\oplus$ s $*$ $\oplus$ z x y

(ii) (a) 2 (b) 5 (c) 5.
If $*$ is associative: (a) 1 (b) 3 (c) 4.

11. (i) (a) $* \oplus xy \oplus zt$ (c) $\oplus * rs * \oplus xyz$ (e) $* \oplus \oplus * rsxyz$.

12. (i) (a) $xy \oplus zt \oplus *$ (c) $rs * xy \oplus z * \oplus$ (e) $rs * x \oplus y \oplus z*$.

13. (i) Infix: $(x \oplus y) * (x \oplus (t * z))$
Prefix: $* \oplus xy \oplus x * tz$
Postfix: $xy \oplus xtz * \oplus *$.

14. (i) It is not a full binary tree: any vertex labelled with the complement operation $^-$ has a single child compared with the vertices labelled $\oplus$ or $*$ which have two children (a left child and a right child).

(ii) (b)

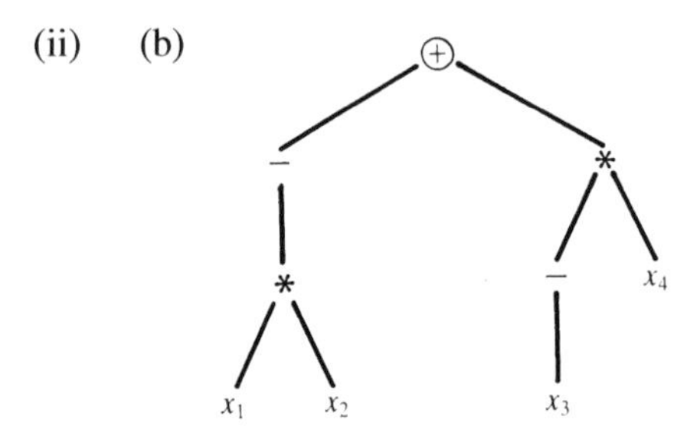

Exercises 12.2

1. (i)

7
3 15
2 4 12 19
6 8 14
5 13

(ii)

when
shall
meet we
again or three
in rain thunder
in lightning

2. The sorted list is more quickly obtained from (i) because the tree has smaller height and is more 'balanced'.

3. (ii) Step 1 produces the following sort tree.

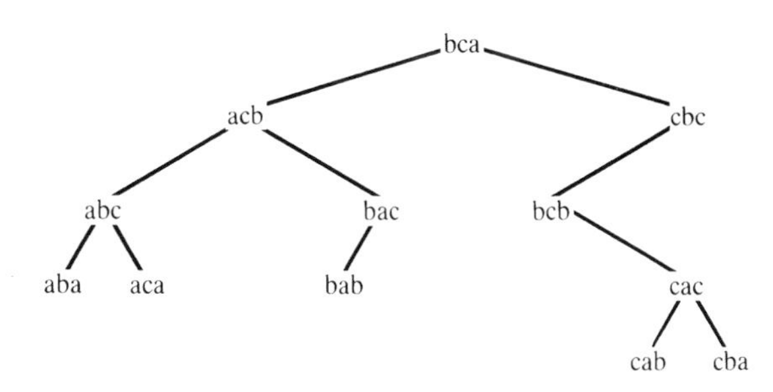

The full listing procedure is the following.

Step 1: Process left-subtree(*bca*)
  Step 1: Process left-subtree(*acb*)
    Step 1: Process left-subtree(*abc*)
      Step 1: Process left-subtree(*aba*) [Empty]
      Step 2: List ***aba***
      Step 3: Process right-subtree(*aba*) [Empty]
    Step 2: List ***abc***
    Step 3: Process right-subtree(*abc*)
      Step 1: Process left-subtree(*aca*) [Empty]
      Step 2: List ***aca***
      Step 3: Process left-subtree(*aca*) [Empty]
  Step 2: List ***acb***
  Step 3: Process right-subtree(*acb*)
      Step 1: Process left-subtree(*bac*)
        Step 1: Process left-subtree(*bab*) [Empty]
        Step 2: List ***bab***
        Step 3: Process right-subtree(*bab*) [Empty]
      Step 2: List ***bac***
      Step 3: Process right-subtree(*bca*) [Empty]
Step 2: List ***bca***
Step 3: Process right-subtree(*bca*)
  Step 1: Process left-subtree(*cbc*)
    Step 1: Process left-subtree(*bcb*) [Empty]
    Step 2: List ***bcb***
    Step 3: Process right-subtree(*bcb*)
      Step 1: Process left-subtree(*cac*)
        Step 1: Process left-subtree(*cab*) [Empty]
        Step 2: List ***cab***

Step 3: Process right-subtree(*cab*) [Empty]
Step 2: List ***cac***
Step 3: Process right-subtree(*cac*)
Step 1: Process left-subtree(*cba*) [Empty]
Step 2: List ***cba***
Step 3: Process right-subtree(*cba*) [Empty]
Step 2: List ***cbc***
Step 3: Process right-subtree(*cbc*) [Empty]

The sorted list is:

$$aba, abc, aca, acb, bab, bac, bca, bcb, cab, cac, cba, cbc.$$

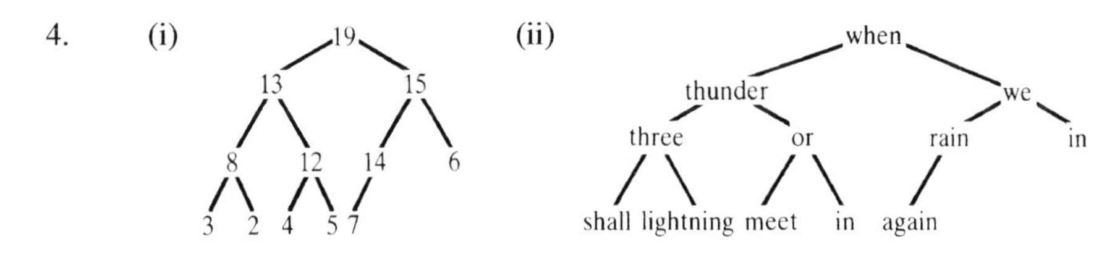

5. (i) The following sequence of rooted trees shows the positions after each new vertex is fixed and the remaining part of the tree restored to a heap.

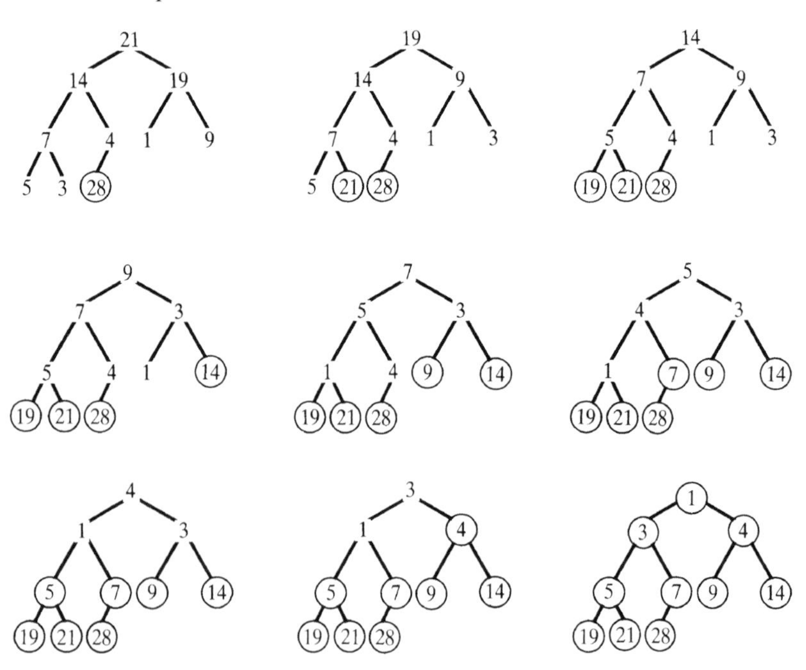

8. In any full binary tree there are an even number of vertices at each level greater than or equal to one. (This can be proved by induction.) Since there is only one root, a full binary tree has an odd number of vertices. The construction of a heap ensures that either it is full or it has a single parent vertex with only one child. The latter cannot occur if there is an odd number of vertices. Hence a heap is a full binary tree if and only if it has an odd number of vertices.

   A complete full binary tree has 2^k vertices at level k. (Again this can be proved by induction.) Therefore if the height of the tree is n, the total number of vertices is $1 + 2 + 2^2 + \cdots + 2^n = 2^{n+1} - 1$. (Once more, prove by induction.) Therefore a heap is a complete full binary tree if and only if it has $2^n - 1$ vertices for some $n \in \mathbb{Z}^+$.

Exercises 12.3

1. (iii) Graph (a) shows a spanning tree produced by a depth-first search where no backtracking is required. Graph (b) shows a spanning tree produced by a breadth-first search completed in three phases. In both cases the vertices are labelled in the order in which they are visited.

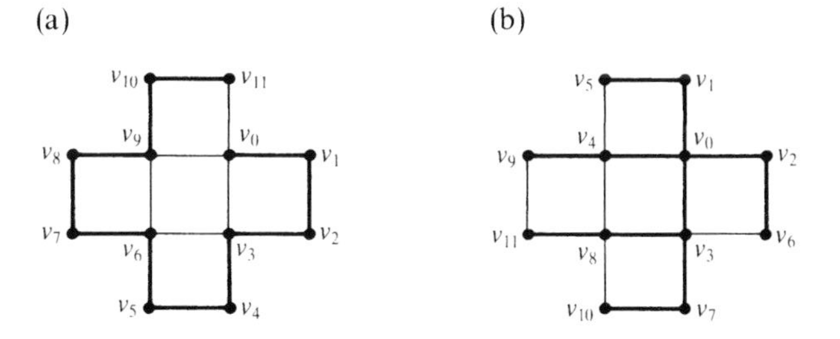

   (vi) Graph (a) shows a spanning tree produced by a depth-first search where two backtrackings were required—these are indicated by the vertices labelled in bold type. (The second of these backtrackings could have been avoided. At least one backtracking appears to be necessary, however.) Graph (b) shows a spanning tree produced by a breadth-first search completed in four phases. Again the vertices are labelled in the order in which they are visited.

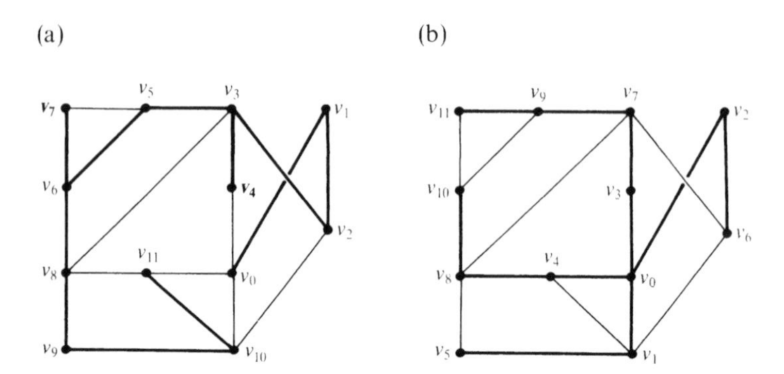

2. It is clear that, if T_n is a tree, then adjoining e_n to form T_{n+1} does not create a cycle, so T_{n+1} is also a tree. Since a single vertex is a tree, the algorithm produces only trees, by induction. Suppose that T_m is the final tree produced by the algorithm and T_m is not a spanning tree. Then there exists a vertex, w say, not belonging to T_m. Let W be the set of vertices in Γ which are adjacent to w. Since Γ is connected, W is non-empty. Then T_m could not have been the final tree because backtracking would have eventually set an element v of W as the current centre and the edge joining v to w could then have been adjoined. This contradiction shows that T_m must be a spanning tree.

3. The proof is similar to that in question 11.3.2 above.

4. The graphs labelled (a) show the result of a depth-first search and those labelled (b) show the result of a breadth-first search. In all cases, the vertices are labelled in the order in which they are visited.

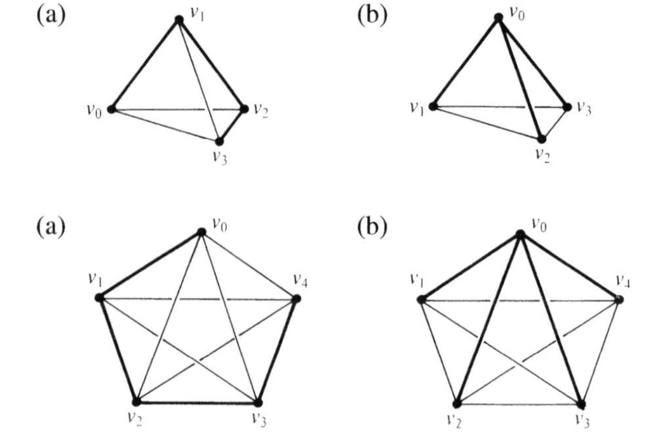

5. (i)

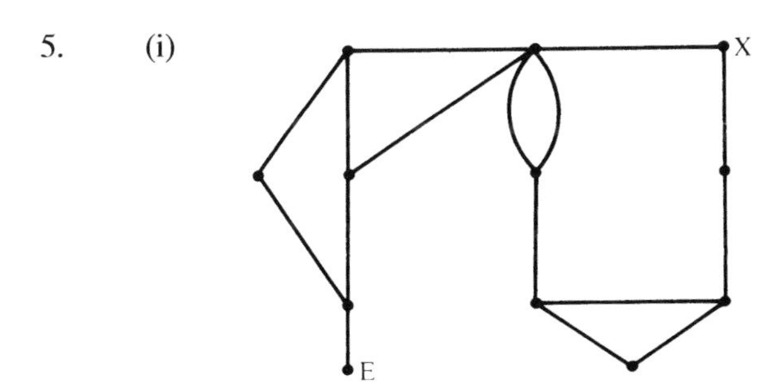

(iii) Which of the two searches will reach X more quickly depends on the choices made. Most probably, though, the depth-first search will be preferred.

6. The following is the graph of a maze. The tree resulting from the partial searches depends on the choices made.

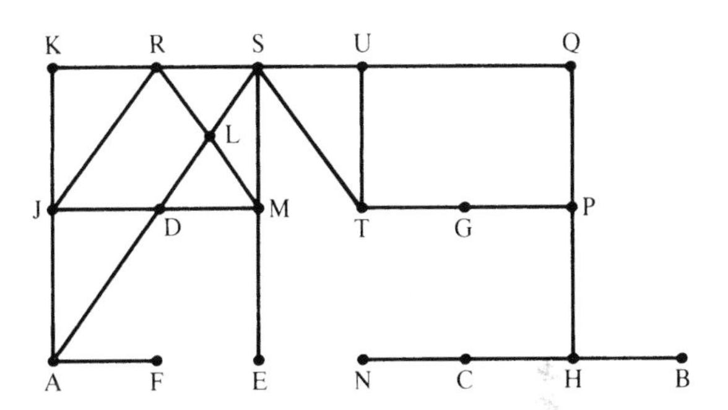

7. If Γ is semi-Hamiltonian there exists a simple path containing all the vertices of Γ. By removing an edge from the path if necessary, we may assume that it is not closed; it is therefore a spanning tree for Γ. The depth-first search algorithm can begin at one end of this path and successively add the next edge in the path, without ever needing to backtrack, until the other end is reached.

Conversely suppose that the depth-first search algorithm can be performed without backtracking to produce a spanning tree T. Since no backtracking takes place, no vertex T can have degree greater than two. Therefore the edges of T, in the order in which they are added, are a simple path which contains every vertex of Γ since T is a spanning tree.

Exercises 12.4

1. (i)

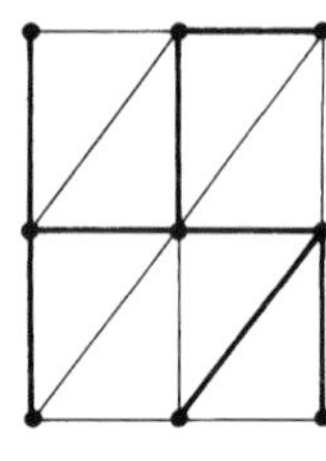

Weight $= 21$

(iii)

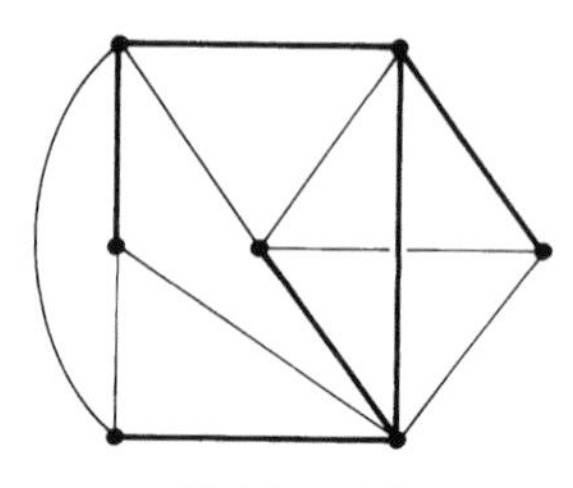

Weight $= 18$

(ii) Weight $= 54$ (iv) Weight $= 20$.

2. For (i), (iii) and (iv) the resulting spanning tree is the same as the one obtained by applying Prim's algorithm. There are choices to be made in (ii) so the spanning tree may be different from the one obtained by applying Prim's algorithm.

3. (i) The statement is not true. The following is a counter-example.

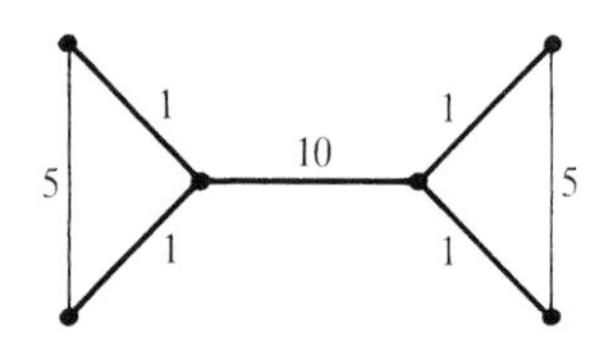

(ii) Yes. Using Kruskal's algorithm (question 2), at each stage there is only one choice of edge to be added and the result is the unique minimal cycle-free subgraph containing the given number of edges.

5.

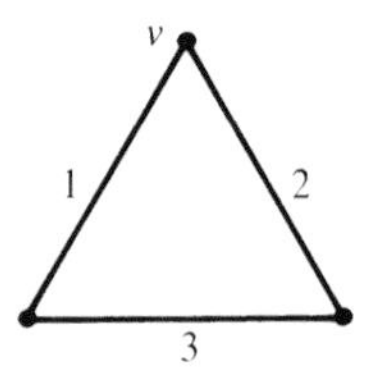

6. (i) A spanning tree T has v vertices and hence $v - 1$ edges (by theorem 10.6). The smallest possible weight for T is $1 + 2 + \cdots + (v - 1) = \frac{1}{2}v(v - 1)$.

(ii) In this case $v = 2n$ so, by part (i), if T is a spanning tree then $w(T) \geqslant n(2n - 1)$. The following graph with $n = 3$ is a suitable example. A minimum spanning tree (shown) has weight 17 but $n(2n-1) = 3 \times 5 = 15$.

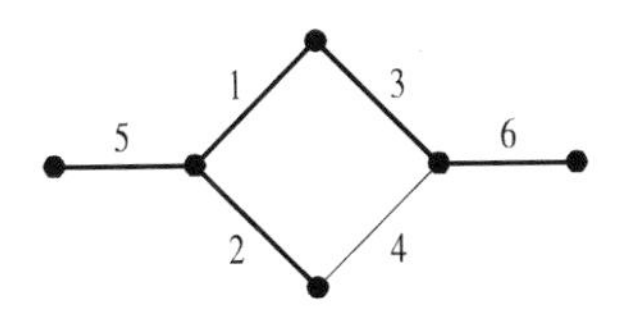

7. We use either Prim's or Kruskal's algorithm to produce the following spanning tree.

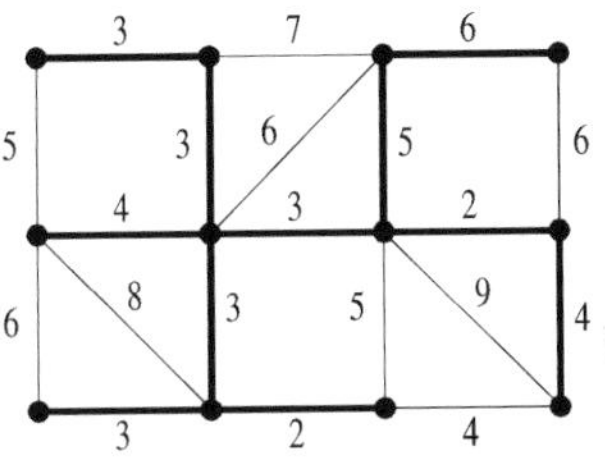

The company can remove the pipes corresponding to edges *not* in the spanning tree, resulting in a total saving of 56 units.

Exercises 12.5

1. (i)

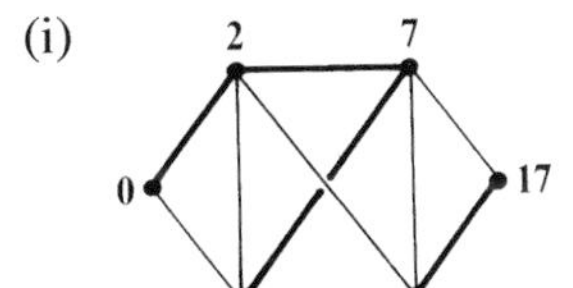

(iii)

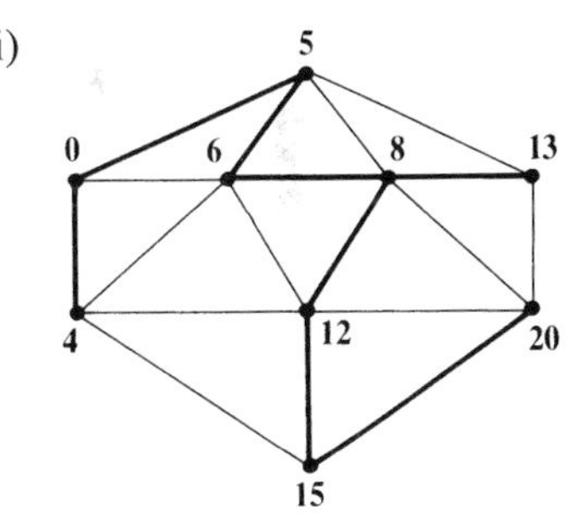

2.

	$n = 10$	$n = 40$	$n = 70$
$10^5 n^6$	2.8 h	474 days	37 years
$10^{-9} 2^n$	10^{-13} s	10^{-4} s	32.8 h

3. (a) When assigning temporary labels (step 2) the only vertices considered are those for which there is an edge directed *away from* the most recently permanently labelled vertex to the given one.

If there is no directed path, a stage will be reached where there are no new edges directed from a permanently labelled vertex to a vertex without permanent label.

(b) (i)

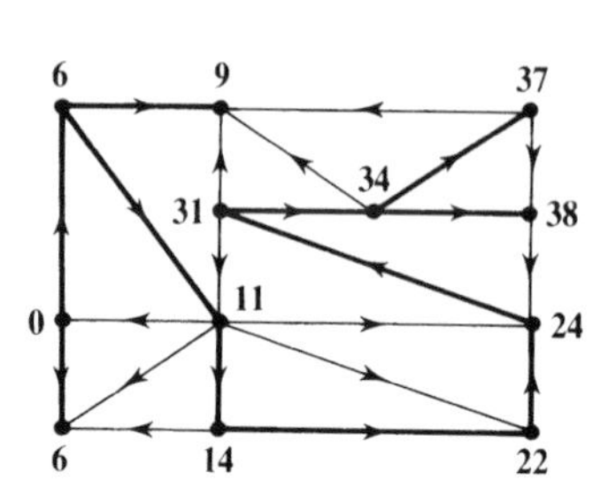

4. (i) The shortest path has length 15.
   (ii) The shortest path has length 13.

Graph (a) below shows a diagram of the graph (without the weights of the edges). Graph (b) shows a tree which contains both the shortest path from v_1 to v_8 and the shortest path from v_4 to v_{10}.

(a)

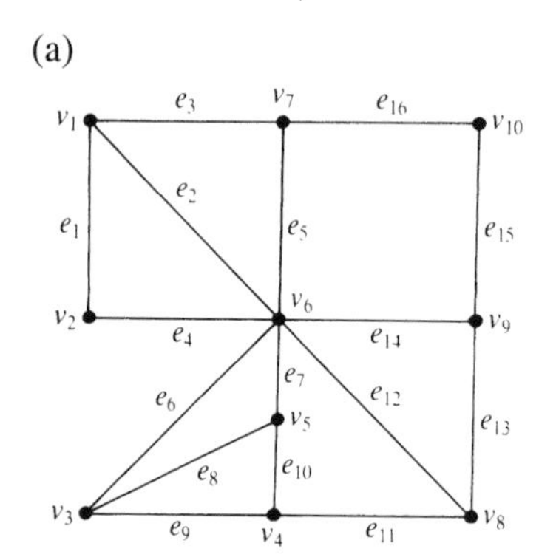

(b)

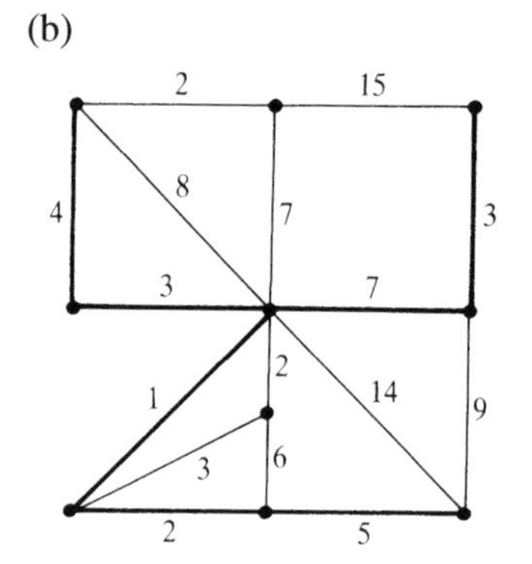

5. (i) (ii)

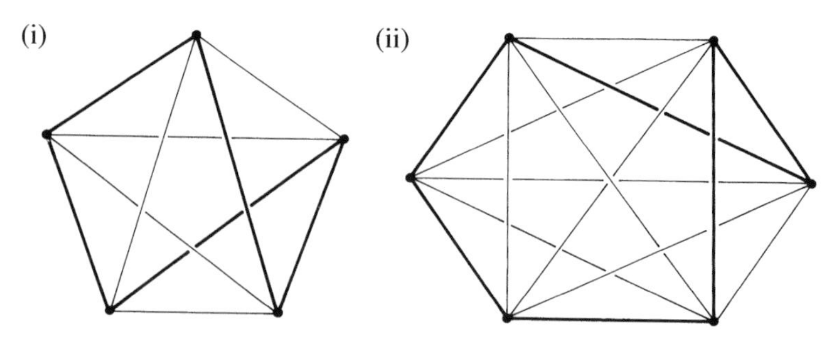

In both cases, starting at a different vertex can result in a different cycle.

7. The following are the 'weight matrices' for the complete graphs: for $i \neq j$ the (i,j)-entry is the weight of the edge joining v_i and v_j.

(i)
$$\begin{pmatrix} 0 & 10 & 18 & 24 & 14 & 9 \\ 10 & 0 & 12 & 18 & 24 & 19 \\ 18 & 12 & 0 & 6 & 14 & 21 \\ 24 & 18 & 6 & 0 & 10 & 15 \\ 14 & 24 & 14 & 10 & 0 & 7 \\ 9 & 19 & 21 & 15 & 7 & 0 \end{pmatrix}$$

(ii)
$$\begin{pmatrix} 0 & 9 & 19 & 20 & 15 & 7 & 15 \\ 9 & 0 & 10 & 12 & 21 & 12 & 7 \\ 19 & 10 & 0 & 10 & 19 & 14 & 6 \\ 20 & 12 & 10 & 0 & 9 & 13 & 5 \\ 15 & 21 & 19 & 9 & 0 & 10 & 14 \\ 7 & 12 & 14 & 13 & 10 & 0 & 8 \\ 15 & 7 & 6 & 5 & 14 & 8 & 0 \end{pmatrix}.$$

Exercises 12.6

1. (i) Minimum completion time $= 27$

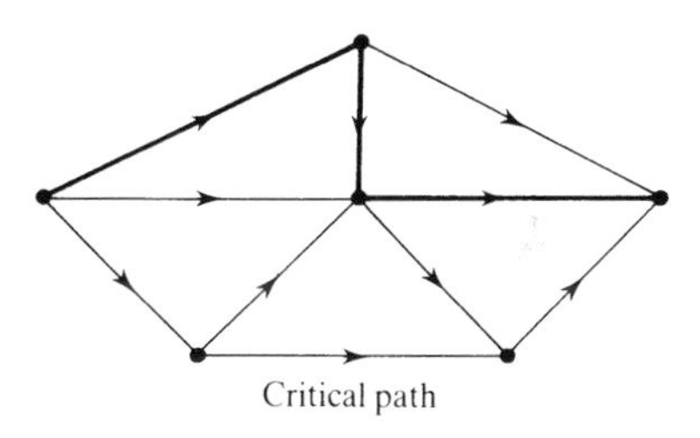

(ii) Minimum completion time $= 48$

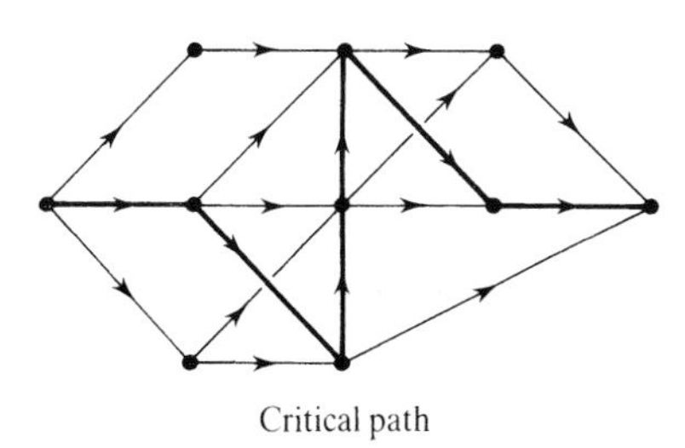

(iii) Minimum completion time $= 40$

(iv) Minimum completion time $= 63$.

2. (i)

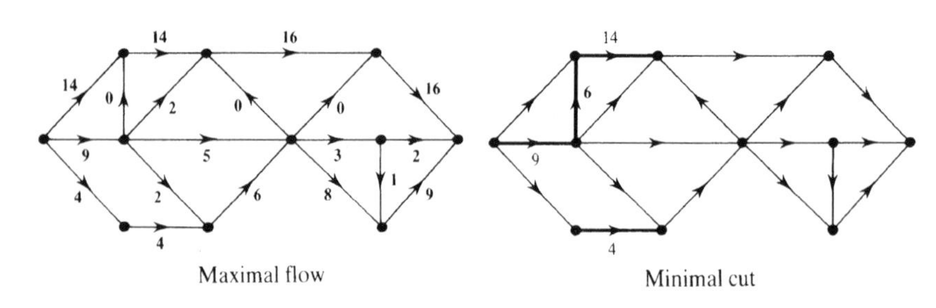

Maximal flow Minimal cut

(ii)

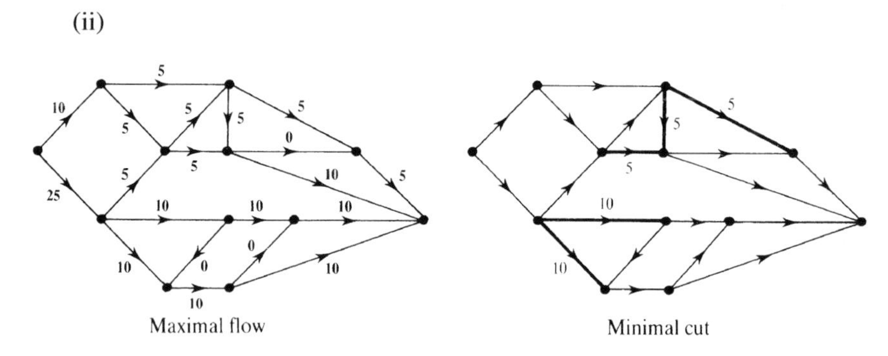

Maximal flow Minimal cut

(iii) Value of a maximal flow/capacity of a minimal cut = 31

(iv) Value of a maximal flow/capacity of a minimal cut = 45.

3. (i) (a) Minimum completion time = 46. The critical path shown is not unique.

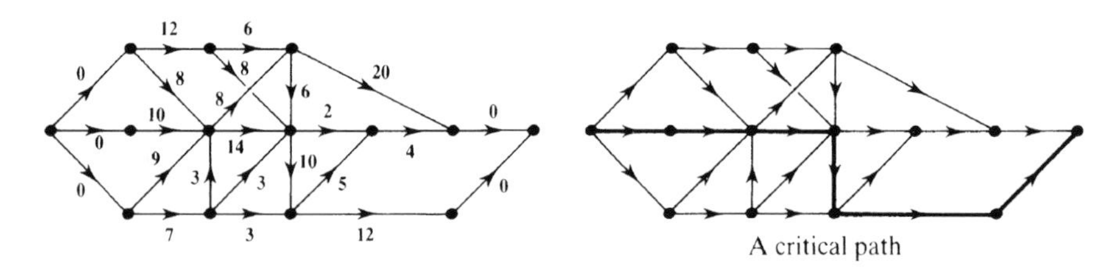

A critical path

(b) Value of maximal flow/capacity of minimal cut = 29.

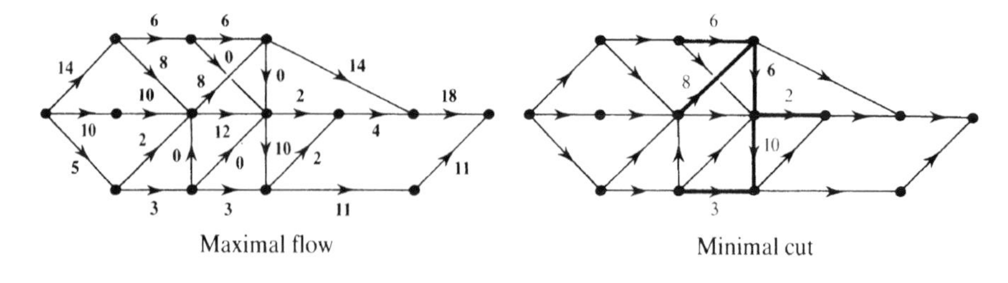

Maximal flow Minimal cut

(ii) (a) Minimum completion time $= 85$
(b) Maximal flow has value 46.

5. Suppose edges $e_1, e_2, \ldots, e_n$ are the edges of a directed cycle representing activities $A_1, A_2, \ldots, A_n$. Then A_1 must be completed before A_2 can begin, A_2 must be completed before A_3 can begin, and so on. In particular, A_1 must be completed before A_n can begin. However, since $e_1, e_2, \ldots, e_n$ are the edges of a directed cycle, A_n must be completed before A_1 can begin, which is not possible.

6. (i) (ii)

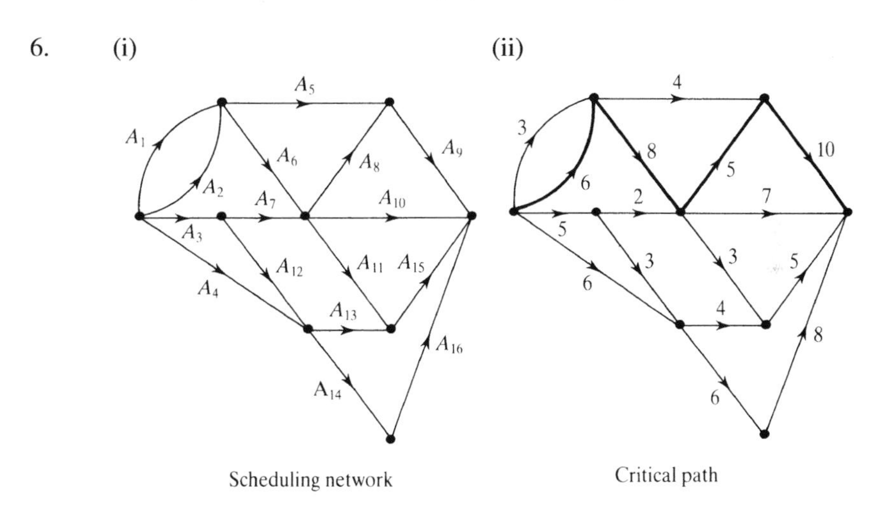

Scheduling network

Critical path

Minimum completion time $= 29$.

Index